# MEMBRANE TECHNOLOGIES *and* APPLICATIONS

# MEMBRANE TECHNOLOGIES *and* APPLICATIONS

*Edited by*

Kaustubha Mohanty
Mihir K. Purkait

CRC Press is an imprint of the
Taylor & Francis Group, an **informa** business

CRC Press
Taylor & Francis Group
6000 Broken Sound Parkway NW, Suite 300
Boca Raton, FL 33487-2742

First issued in paperback 2020



Version Date: 20111026

ISBN-13: 978-0-367-57679-0 (pbk)
ISBN-13: 978-1-4398-0526-8 (hbk)



**Library of Congress Cataloging-in-Publication Data**

Membrane technologies and applications / [edited by] Kaustubha Mohanty and Mihir K. Purkait.
p. cm.
Includes bibliographical references and index.
ISBN 978-1-4398-0526-8 (hardback)
1. Membrane separation. 2. Membranes (Technology) I. Mohanty, Kaustubha. II. Purkait, Mihir K.

TP248.25.M46M467 2011
660'.28424--dc23 2011039816

**Visit the Taylor & Francis Web site at**
**http://www.taylorandfrancis.com**

**and the CRC Press Web site at**
**http://www.crcpress.com**

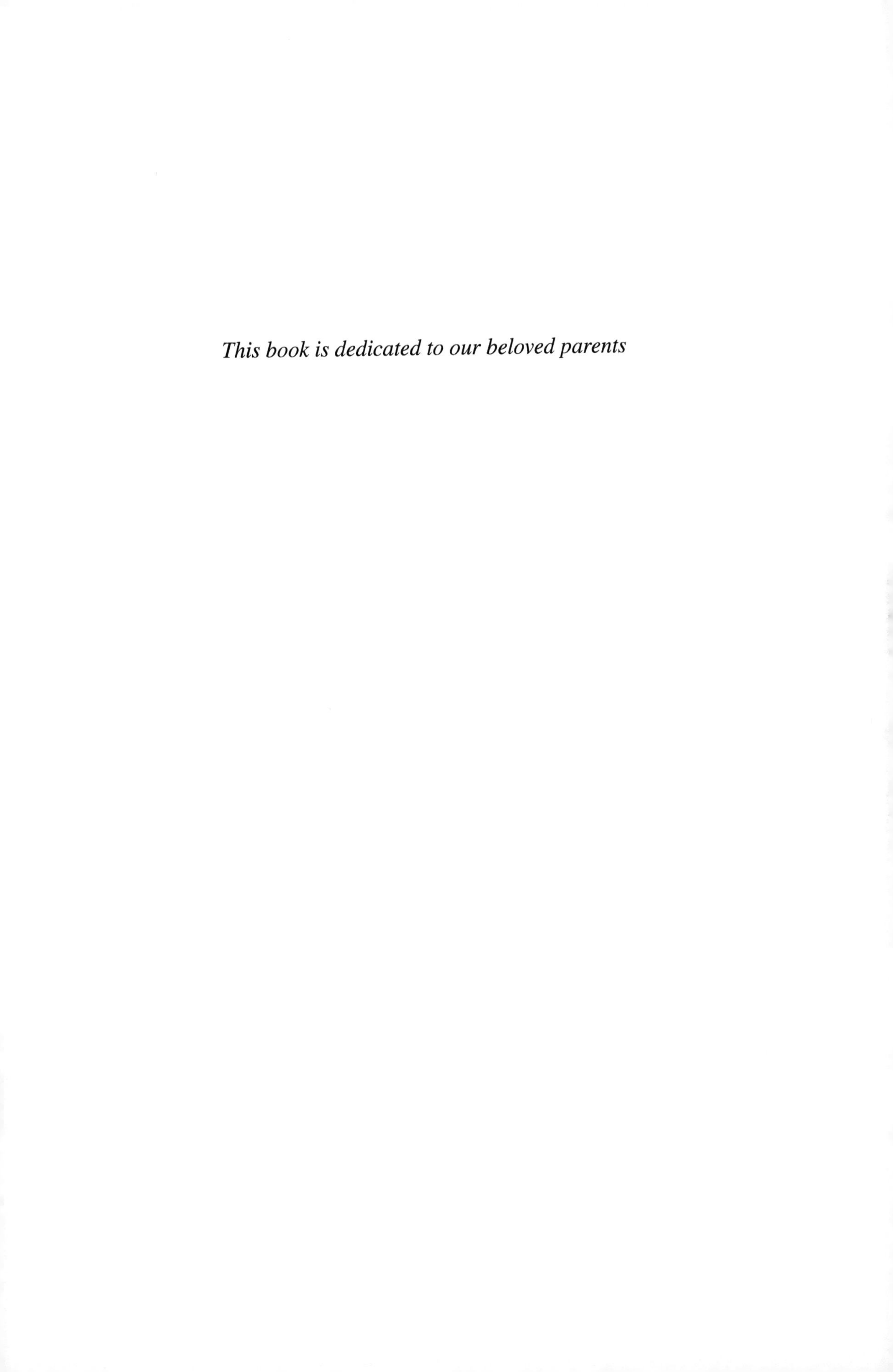

*This book is dedicated to our beloved parents*

# Contents

# Preface

Membranes have shown huge potential as a separating tool since the last few decades. Its application is gradually increasing from traditional fields such as water desalination and purification and food processing to recent applications in industries such as oil and petrochemical, pulp and paper, biopharmaceutical and energy. Chemical industries are the one in which membranes are already being used for various applications; however, there is still enough scope for its use in other applications where it can add new solutions in the near future. Downstream processing in biopharmaceutical industries is a major challenge and recent research findings have shown that proteins and antibodies can be suitably purified by using membrane cascades. Use of membranes is pervasive in the field of alternative energy, such as in fuel cells, as well as to purify biofuels after their synthesis, whereby their contribution to green technology is widely appreciated. Nowadays membranes are always in use in most of industrial processes, and there is huge demand for the supply of commercial membranes. However, optimizing membrane processes and supplying low-cost membranes for various applications is a huge challenge for membrane scientists. Since the research on membrane technology is growing at a faster pace with various developments in the field of membrane preparation as well as applications are reported, there is a need to compile these latest findings in the form of a book.

This book is written to provide in one place the essential data and background materials on various aspects of membrane technology with a major coverage on application. It is intended for the following technologists so they do not need to gather scattered information from the current and past literature: industrial as well as situational researchers, application scientists and engineers with an interest in membrane technologies, and students pursuing advanced separation studies. The next paragraph summarizes the various contributions received for this book.

Chapters 1–3 give detailed procedures for how to prepare different membranes. Examples of production as well as test cases are used to illustrate the niche applications using the unique characteristics of charged organic–inorganic membranes described in Chapter 3. Chapter 4 is a brief review on preparation and applications of zeolite membranes. Composite membranes have recently gained a lot of interest, and in Chapter 5 the authors have listed various technological applications of these membranes. As the emphasis of this book is mainly application oriented, Chapter 6 and onward deals with the various applications of membrane technology, some of them included in the hybrid membrane systems. Treatment of kraft black liquor and refinery wastewater is given in Chapters 6 and 7, respectively. Chapters 8, 9, and 10 deal with the membrane hybrid systems. Advanced oxidation followed by nanofiltration to treat textile mill effluent is elaborately presented in Chapter 8. Similarly, in Chapter 9, a new hybrid membrane system of micellar enhanced ultrafiltration and its various applications is given. Membrane emulsification is one of the latest applications of membrane technology, and the authors have reported the current state-of-the-affairs along with future challenges in Chapter 11. Applications of membrane technology in dairy and food processing industries are well known, and Chapters 12 and 13 present the same. One of the traditional disadvantages of membrane technology is the concentration polarization and subsequent fouling. Chapter 14 details the different types of fouling, their mechanism, and control strategies. Another interesting development is the use of liquids as membrane material. Chapter 15 reports the introduction to liquid membranes and various applications. Supported liquid membranes and their various applications are outlined in Chapter 16. Another recent development is the use of room temperature ionic liquids, one of the greenest solvents in liquid membranes. Both bulk and supported ionic liquid–based membranes are currently being investigated. Chapter 17 gives a brief review of the ionic liquid–based membranes and applications. Membrane processes other than pressure-driven such as electrodialysis and hemodialysis are of great importance for their various applications in chemical as well as biomedical fields. Pervaporation is another non-pressure-driven process that has wider applications especially in separation of liquid mixtures, and Chapter 20 gives the details of the pervaporation process as well as its applications. One of the notable and earlier applications of membrane technology is gas separation. Chapters 21 and 22 deal with the separation of carbon-dioxide and hydrogen sulphide using selective membranes. The challenging field of membrane reactors has emerged as a promising technology when inorganic membranes begin to rise to the horizon of commercial reality. Chapter 23 gives a brief review of membrane reactors along with various applications. Chapter 24 describes the concept of an enzymatic membranes reactor and its application in various industries such as food, biopharmaceutical, and environmental applications. Chapter 25 is a review on membranes for fuel cell application.

It is hoped by the editors that anyone interested in water and wastewater treatment, including problem holders and membrane suppliers as well as students and academics studying in this area, will find this book useful. In particular, the book is meant to provide a practical aid to those readers with an interest in actually selecting, installing, and/or designing systems for recovering and reusing industrial effluent. The idea of this book originated after we offered a short-term course on *Advances in membrane separation technologies* during December 2007. We thank CRC Press for inviting us to write/edit the book, and we hope to stimulate even more the incorporation of membrane technology in chemical processes in the near future. Last but not least we express our thanks to the CRC Press management, which gave us time and resources to edit this book.

Finally, we would like to thank all the contributors of this book, many of whom have submitted their articles in due time. Though there was a delay in submission from many contributors for various reasons, we had a great learning experience in editing those chapters. As with any piece of work, the editors would welcome any comments from readers, critical or otherwise, and our contact details are included in the following section.

**Kaustubha Mohanty**
**Mihir Kumar Purkait**

# Editors

**Dr. Kaustubha Mohanty** is an associate professor of chemical engineering at the Indian Institute of Technology, Guwahati, India. He has obtained his bachelors in chemical engineering from Bangalore University and masters and PhD in chemical engineering from Indian Institute of Technology, Kharagpur. His research interests are primarily in advanced separations (adsorption, membrane, ionic liquids), bioseparation, environmental biotechnology, and biofuels. He has published nearly 35 research papers in international peer-reviewed journals. He is a Life Member of the Indian Institute of Chemical Engineers and a member of Canadian Society for Chemical Engineers and Society of Chemical Industry, London.

**Dr. Mihir Kumar Purkait** is an associate professor of chemical engineering at the Indian Institute of Technology, Guwahati, India. He holds a BSc honors degree in chemistry and BTech degree in chemical engineering from the University of Calcutta. He has obtained his MTech and PhD in chemical engineering from the Indian Institute of Technology, Kharagpur. He is actively involved in research and teaching related to wastewater reclamation and reuse, potable water treatment, water quality issues, and membrane technology. He has authored more than 40 peer-reviewed papers and delivered talks at national and international conferences on aspects of membrane technology.

# Editors

**Dr. Kaustubha Mohanty** is an associate professor of chemical engineering at the Indian Institute of Technology, Guwahati, India. He has obtained his bachelors in chemical engineering from Bangalore University and masters and PhD in chemical engineering from Indian Institute of Technology, Kharagpur. His research interests are primarily in advanced separations (adsorption, membrane, ionic liquids), bioseparation, environmental biotechnology, and biofuels. He has published nearly 35 research papers in international peer-reviewed journals. He is a Life Member of the Indian Institute of Chemical Engineers and a member of Canadian Society for Chemical Engineers and Society of Chemical Industry, London.

**Dr. Mihir Kumar Purkait** is an associate professor of chemical engineering at the Indian Institute of Technology, Guwahati, India. He holds a BSc honors degree in chemistry and BTech degree in chemical engineering from the University of Calcutta. He has obtained his MTech and PhD in chemical engineering from the Indian Institute of Technology, Kharagpur. He is actively involved in research and teaching related to wastewater reclamation and reuse, potable water treatment, water quality issues, and membrane technology. He has authored more than 40 peer-reviewed papers and delivered talks at national and international conferences on aspects of membrane technology.

# Contributors

**Iqbal Ahmed**
Universiti Malaysia Pahang
Pahang, Malaysia

**Muhammad H. Al-Malack**
King Fahd University of Petroleum and Minerals
Dhahran, Saudi Arabia

**Phalguni Banerjee**
Indian Institute of Technology, Kharagpur
Kharagpur, India

**Laurent Bazinet**
Université Laval
Québec, Canada

**M. P. Belleville**
IEM UMR
Montpellier, France

**Mathur S. Bhakhar**
G. H. Patel College of Engineering and Technology
Vallabh Vidyanagar, India

**Chiranjib Bhattacharjee**
Jadavpur University
Kolkata, India

**Remko M. Boom**
Wageningen University
Wageningen, The Netherlands

**M. G. Buonomenna**
University of Calabria
Rende, Italy

**Alvin R. Caparanga**
Chung Yuan Christian University
Taoyuan, Taiwan
Mapua Institute of Technology
Manila, Philippines

**William R. Clark**
Gambro Inc.
Indianapolis, Indiana

**Chandan Das**
Indian Institute of Technology, Guwahati
Guwahati, India

**Sirshendu De**
Indian Institute of Technology, Kharagpur
Kharagpur, India

**Aloke Kumar Ghoshal**
Indian Institute of Technology, Guwahati
Guwahati, India

**A. Gordano**
University of Calabria
Rende, Italy

**W. S. Winston Ho**
Ohio State University
Columbus, Ohio

**Zhongping Huang**
Widener University
Chester, Pennsylvania

**Ani Idris**
Universiti Teknologi Malaysia
Johor Darul Ta'zim, Malaysia

**Somen Jana**
Indian Institute of Technology, Guwahati
Guwahati, India

**Jaya Kandasamy**
University of Technology, Sydney
Broadway, Australia

**A. A. Khan**
Indian Institute of Chemical Technology
Hyderabad, India

**Matthias Kraume**
Technische Universität Berlin
Berlin, Germany

**David P. Lachkar**
Université Laval
Québec, Canada

**Meng-Hui Li**
Chung Yuan Christian University
Taoyuan, Taiwan

**Junsheng Liu**
Hefei University
Hefei, China

**Bishnupada Mandal**
Indian Institute of Technology, Guwahati
Guwahati, India

**Michiaki Matsumoto**
Doshisha University
Kyoto, Japan

**Larisa Melita**
Technical University of Civil Engineering of Bucharest
Bucharest, Romania

**Fangang Meng**
Technische Universität Berlin
Berlin, Germany

**Kaustubha Mohanty**
Indian Institute of Technology, Guwahati
Guwahati, India

**R. Surya Murali**
Indian Institute of Chemical Technology
Hyderabad, India

**Kaushik Nath**
G. H. Patel College of Engineering and Technology
Vallabh Vidyanagar, India

**Nurdiana Mohd Noor**
Universiti Teknologi Malaysia
Johor Darul Ta'zim, Malaysia

**D. Paolucci**
IEM UMR
Montpellier, France

**Maria Popescu**
Technical University of Civil Engineering of Bucharest
Bucharest, Romania

**Yves Pouliot**
Université Laval
Québec, Canada

**Mihir Kumar Purkait**
Indian Institute of Technology, Guwahati
Guwahati, India

**Muhammad Muhitur Rahman**
King Fahd University of Petroleum and Minerals
Dhahran, Saudi Arabia

**P. Rai**
Birsa Agricultural University
Ranchi, India

**Gilbert M. Rios**
IEM UMR
Montpellier, France

**Karin (C.G.P.H.) Schroën**
Wageningen University
Wageningen, The Netherlands

**Dwaipayan Sen**
Jadavpur University
Kolkata, India

**Guoquan Shao**
Hefei University
Hefei, China

**Sumedha Sharma**
Indian Institute of Technology, Guwahati
Guwahati, India

**Hok-Yong Shon**
University of Technology, Sydney
Broadway, Australia

**Allan N. Soriano**
Chung Yuan Christian University
Taoyuan, Taiwan

**S. Sridhar**
Indian Institute of Chemical Technology
Hyderabad, India

**K. Sunitha**
Indian Institute of Chemical Technology
Hyderabad, India

**William H. Van Geertruyden**
EMV Technologies
Bethlehem, Pennsylvania

**Cees J. M. van Rijn**
Wageningen University
Wageningen, The Netherlands

**Saravanamuthu Vigneswaran**
University of Technology, Sydney
Broadway, Australia

**Michael E. Vilt**
Ohio State University
Columbus, Ohio

**Tongwen Xu**
University of Science and Technology of China
Hefei, China

# 1 New Materials, New Devices, New Solutions: How to Prepare a Membrane

*A. Gordano and M. G. Buonomenna*

## CONTENTS



## 1.1 INTRODUCTION

Nowadays, polymeric membranes have been used successfully in a wide range of large-scale industrial applications. The use of membranes in industry is primary the result of two developments: (1) in membrane materials and (2) in membrane structures, both followed by the step of large-scale membrane production. The selection of a suited material and a preparation technique depends on the application in which the membrane is to be used (e.g., gas separation, microfiltration (MF), ultrafiltration (UF), nanofiltration (NF), and reverse osmosis (RO)). In this chapter, a review of up-to-date literature about preparation of polymeric membranes from new materials for different membrane applications by means of phase inversion process is given.

The most important part in any membrane separation process is the membrane itself. A membrane is a thin barrier that allows selective mass transport (Cabasso, 1987). In some applications such as in gas separation and pervaporation, the membrane material as the dense layer plays a crucial role in membrane separation. In other applications such as in liquid separations, MF or UF, wherein the transport is not based on solution-diffusion mechanism but occurs through a porous membrane by means of a sieving process, the membrane material is not quite as important as the membrane structure. In membrane processes such as NF and RO, for which both transport mechanisms (i.e., solution diffusion and sieving) contribute to membrane performance, the choice of membrane material is fundamental. In more recent membrane operations, such as membrane contactors, where the membrane is porous and not selective, the only function of the membrane is to act as a barrier between two nonmixable phases (Drioli et al., 2006). In this case, the main requirement is that the membrane must not be wet, and the suited material is chosen on the basis of surface tension properties.

Different methods of polymeric membrane preparation have been reported in literature (Strathmann, 1981; Lonsdale, 1982; Nunes and Peinemann, 2001). Membranes can be classified according to their bulk structure, morphology, and application (Figure 1.1).

Membranes have either a symmetric (isotropic) or an asymmetric (anisotropic) structure. Symmetric membranes have a uniform structure throughout the entire membrane thickness,

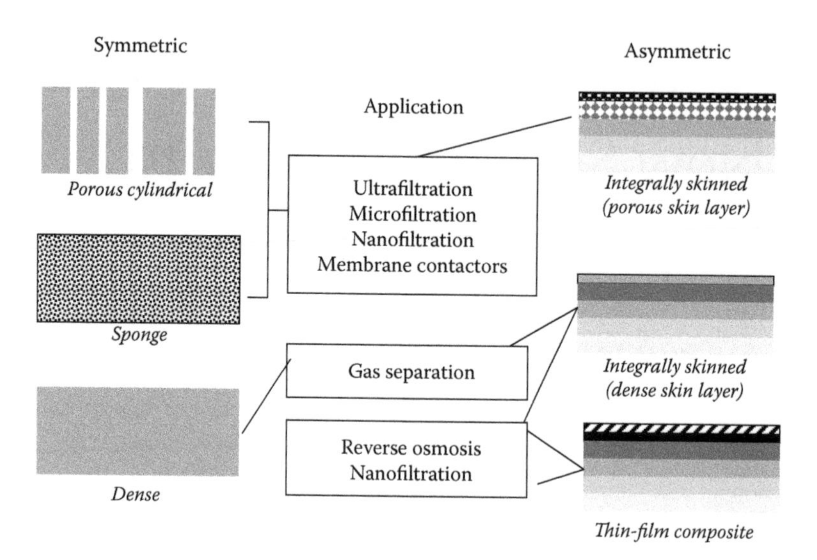


**FIGURE 1.1** Membrane classification according to bulk structure (symmetric, asymmetric), morphology (porous, sponge, dense, integrally skinned, thin-film composite), and application (UF, MF, NF, etc.).

whereas asymmetric membranes have a gradient in structure. This is evident particularly in integrally skinned membranes with a dense top layer. Compared with dense symmetric membranes, for which the separation properties are determined by the overall structure, asymmetric membranes with a dense skin layer on a porous sublayer have the most impact in the membrane market. This fact, in particular for gas separation, derives from Equation 1.1, which relates gas flux through a dense film ($J$) with membrane thickness ($l$), permeability ($P$), and partial pressure difference ($\Delta P$) of penetrant species:

$$J = P/l \, \Delta P. \tag{1.1}$$

In fact, for dense homogeneous membranes, the permeant flow across the membrane is quite low, since a minimal thickness is required to guarantee membrane mechanical stability. Asymmetric membranes combine high permeant flow, which is provided by a very thin selective top layer and a reasonable mechanical stability resulting from the porous sublayer (Nunes and Peinemann, 2001).

The most common method used to prepare polymeric membranes (both symmetric and asymmetric) is the phase inversion process. This method covers a range of different techniques such as solvent evaporation, precipitation by controlled evaporation, thermal precipitation, precipitation from the vapor phase, and immersion precipitation (Mulder, 1984). Particularly in the case of porous membranes as those employed in MF, methods other than phase inversion are applied as sintering of powders, stretching of films, irradiation, and etching of films (Strathmann et al., 2006).

### 1.1.1 Phase Inversion

Polymeric membranes can be prepared by means of phase inversion from virtually all polymers that are soluble at a certain temperature in an appropriate solvent or solvent mixture and can be precipitated as a solid phase by the following (Mulder, 1984):

- Cooling of a polymer solution, which separates at a certain temperature in two phases, that is, a liquid phase forming the membrane pores and a solid phase forming the membrane structure (thermally induced phase inversion).

- Precipitation by solvent evaporation, allowing a dense homogeneous membrane to be obtained.
- Addition of a nonsolvent or a nonsolvent mixture to a homogeneous solution (diffusion-induced phase separation or nonsolvent-induced phase inversion).
- Precipitation from the vapor phase, consisting of a nonsolvent saturated with a solvent: The high solvent concentration in the vapor phase prevents the evaporation of solvent from the cast film. Membrane formation occurs because of diffusion of nonsolvent into the cast film.
- Precipitation by controlled evaporation or dry phase inversion: The polymer is dissolved in a mixture of solvent and nonsolvent. Since the solvent is more volatile than the nonsolvent, the evaporation step leads to polymer precipitation owing to a higher nonsolvent and polymer content.

There are different variations to diffusion-induced phase separation (or nonsolvent-induced phase inversion): an evaporation step to the precipitation is used to change the composition in the surface of the cast film (dry–wet phase inversion) (Pinnau and Koros, 1993; Kawakami et al., 1997; Jansen et al., 2006); an annealing step is applied to change the structure of the precipitated membrane (Mahendran et al., 2004).

## 1.2 NEW MEMBRANES FOR LIQUID SEPARATION

The research of new materials for membrane application in the field of UF, NF, and RO is of great interest, as most recent papers in membrane technology field indicate (Su et al., 2006; Wu et al., 2008a, 2008b; Yang et al., 2006a, 2006b; Rahimpour and Madaeni, 2007; Rahimpour et al., 2008; Tang et al., 2005; Wang et al., 2007).

In particular, two different trends are present: In the first one, design, synthesis, and use of new polymers for membrane preparation constitute the main steps of research efforts. The second one concerns the modification of a membrane formation step by means of additives (Bae et al., 2006; Yang and Wang, 2006; Zhang et al., 2006; Chakrabarty et al., 2008; Qian et al., 2008) or the surface treatment of preformed membranes (Chen and Belfort, 1999; Khayet and Matsuura, 2003; Pieracci et al., 1999; Nunes et al., 1995; Li et al., 2003; Buonomenna et al., 2007).

For both these approaches, the technique used for membrane preparation is phase inversion process, as described above.

Most of the membrane separation system requires an efficient heat exchanger to decrease the temperature of the feed solution in order to avoid damaging the membrane when dealing with a hot stream, such as the effluents from a textile bleaching and dyeing industry, whose temperature can range from 40°C to 95°C (Chen et al., 1997; Yang et al., 2006a). Hence, to develop high-temperature-resistant membranes, the membrane material must have excellent thermal stability.

For this purpose, several new polymers since the 1980s have been synthesized. In Table 1.1, the molecular structure and glass transition temperature (Tg) of polymers characterized by high thermal stability are reported. These polymers are soluble in the most common organic solvents, and therefore, they can be used as membrane materials by means of the easy phase inversion process.

Phenolphthalein poly(ether sulfone) (PES-C), whose Tg is 260°C, was synthesized in the 1980s and exhibited excellent membrane performance when prepared into UF membranes due to the unique structure containing the cardo units (Wang et al., 2007). Poly(phthalazinone ether sulfone ketone) (PPESK) was synthesized by the polycondensation of phthalazinone monomer, bis(4-chlorophenyl)sulfone, and bis(4-fluorophenyl)ketone. It possesses a higher Tg, 285°C (Yang et al., 2006b). PPESK membranes showed interesting permeation performance due to the introduction of the phthalazinone unit.

**TABLE 1.1**
**Some New Aromatic Glassy Polymers Used in the Preparation of Membranes by Nonsolvent-Induced Phase Inversion**

| Polymer | Structure | Tg (°C) |
|---|---|---|
| PES-C | 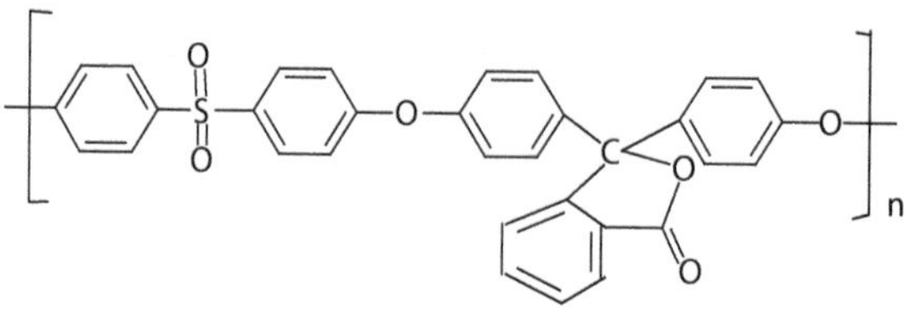 | 260 |
| PPESK | 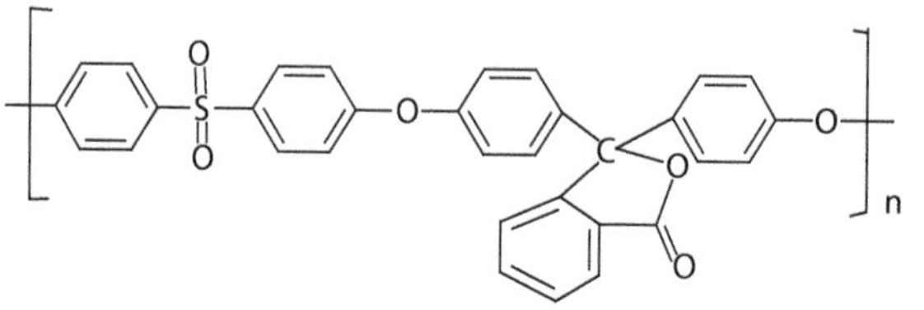 | 285 |
| PPES-B | 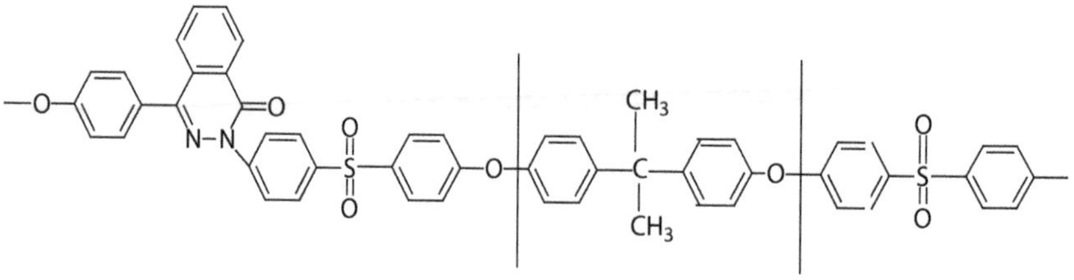 | – |
| PEEKWC | 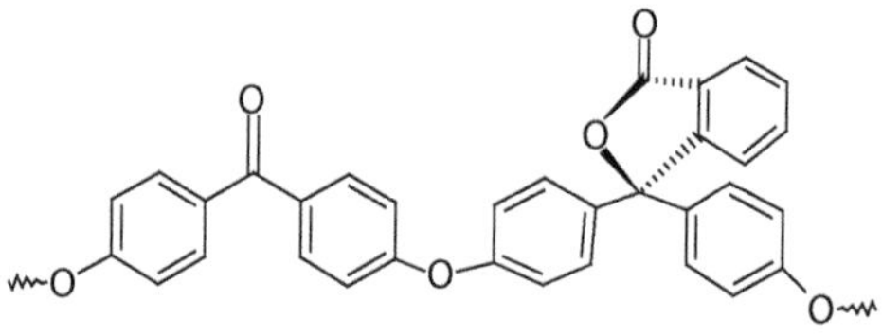 | 228 |

Poly(phthalazinone ether sulfone)-bisphenol (PPES-B) was synthesized by the polycondensation of phthalazinone monomer, bis(4-chlorophenyl)sulfone, and bisphenol-A. It also has a high Tg and perfect mechanical property and chemical stability (Su et al., 2006). Poly(ether ether ketone) modified with a cardo group is a modified poly(ether ether ketone) that has a lactone group attached to the backbone (Gordano et al., 2002; Buonomenna et al., 2004). One of its major advantages is that it preserves the good thermal and mechanical properties of the traditional PEEK but, compared with the latter, has a much higher solubility, which is also found in common organic solvents. Therefore, it is well suited for the preparation of polymeric membranes by phase inversion techniques.

Wang et al. (2007) reported a simple method for the preparation of an environmentally responsive membrane based on PES-C: in situ redox-graft pore-filling polymerization.

Membranes were prepared by means of the classical phase inversion method: PES-C was dissolved in dimethyl sulfoxide (DMSO). After that, methylacrylic acid was grafted successfully onto the membranes. Surface chemical changes and membrane morphology changes before and after graft polymerization were investigated by attenuated total reflectance–Fourier transform infrared spectroscopy and field emission scanning electron microscopy, respectively, to confirm the formation and location of the graft. Permeability and diffusional permeability experiments were carried out by using vitamin $B_{12}$ and KCl as solutes. The grafted membranes exhibited a marked, rapid, and reversible pH response.

Hollow fiber UF membranes were prepared successfully from PPESK using the dry–wet phase inversion technique (Su et al., 2006). For membrane preparation, ethylene glycol methyl ether, diethylene glycol, and methyl ethyl ketone were used as nonsolvent additives, and *N*-methyl-2-pyrrolidone (NMP) was used as a solvent. It was found that the membrane performance is consistent and agrees with the membrane morphology. With the increase of PPESK concentration in the casting solution, the viscosity strongly increases, and it becomes shear-rate dependent. The morphologies of hollow fiber membranes changed from a fingerlike structure to a sponge-shaped structure, and the properties of the UF membrane showed that poly(ethylene glycol) (PEG) 10,000 rejection was above 95% under the operating pressure of 0.1 MPa. PPESK hollow fiber UF membranes prepared were also investigated for their thermal stability at different operating temperatures. When the temperature of the feed solution was raised from 15°C to 100°C, the permeation flux increased more than three times without significant change of rejection.

New PEEKWC-based membranes have been prepared by a dry–wet phase inversion process and have been used in NF operations of organic charged dyes (Buonomenna et al., 2008a). The characteristics and performance of these membranes with NF commercial membranes (N30F and NFTFC50) have been studied. PEEKWC is an amorphous modified poly(ether ether ketone) that has a lactone group attached to the backbone. It was proven that the type of internal nonsolvent (methanol or butanol) and its concentration in the casting solution are an important factor to tune the pore size of membranes and their structure. In Figure 1.2, the morphology of membranes prepared (a) without internal solvent and (b) with methanol (12 phr) and (c) butanol (12 phr) as internal nonsolvent is shown. Cellular structure without pores interconnected as confirmed by gas transport analysis (Buonomenna et al., 2008b) is obtained without the use of an internal solvent after external coagulation in methanol (Figure 1.2). The addition of an internal nonsolvent increased the demixing rate between the solvent and the nonsolvent, and according to the macrovoids theory formation reported by Mulder (1984), structures with fingers have been formed (Figure 1.2b and c) in both cases. The difference in volatility between methanol and butanol determines the difference in terms of morphology and pore size. Separation membrane processes such as UF, NF, and RO derived their names as they have a molecular weight cutoff for uncharged molecules corresponding to pore diameter (Mulder, 1984). In the case of charged molecules, separation properties are influenced by the inherent charge of the membranes (Van der Bruggen et al., 1999).

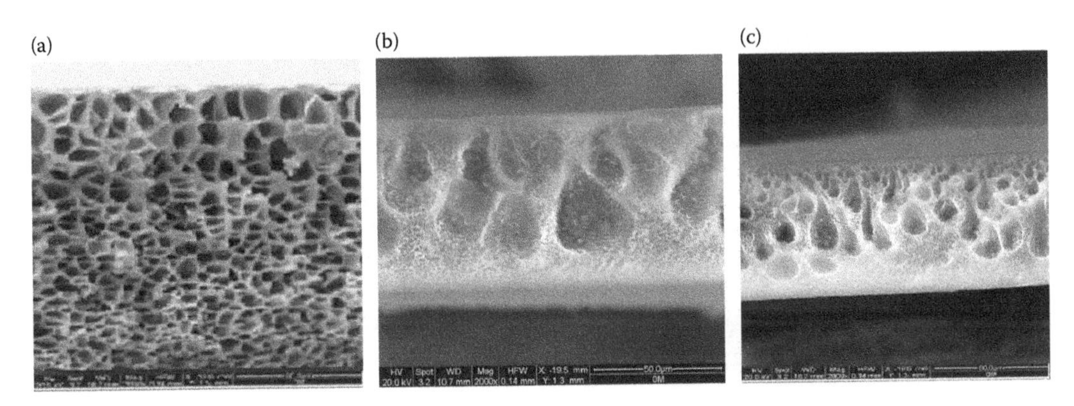


**FIGURE 1.2** Scanning electron microscopy analysis of cross-section of PEEKWC membranes prepared by dry–wet process (a) without internal nonsolvent and (b) with methanol and (c) butanol as internal nonsolvent. (Reprinted from *Microporous and Mesoporous Materials*, 120, Buonomenna, M.G., et al., 147–153, Copyright 2008, with permission from Elsevier.)

Charged membranes can be used to separate charged components with a much smaller size than the pore size according to electrostatic repulsion.

In that case, membranes often have a more open pore structure and a higher permeability. Moreover, the introduction of charged groups can increase the hydrophilicity of a membrane. So, charged membranes have been paid more and more attention from both the scientific and engineering standpoints (Tang et al., 2005).

Membranes based on PEEKWC have been modified via low-temperature plasma treatment to positively introduce ammonia groups to enhance their performance for selective separation of organic dyes (Buonomenna et al., 2008a).

The membranes, prepared by means of dry-wet phase inversion using methanol and 1-butanol as internal nonsolvent, were further plasma-surface-modified (radio frequency 13.56 MHz), introducing amino groups in the membrane and using two different pretreatment gases, Ar and $H_2$. The performances of unmodified and plasma-modified membranes and of a commercial NF membrane (NFTFC50) were tested in treatment of aqueous solution containing two dyes (~320 g/mol), methylene blue and methyl orange, positively and negatively charged, respectively (Figure 1.3).

Low-temperature plasma treatment on poly(vinylidene fluoride) (PVDF)–based membranes has been used as a tool to modify the surface tension of membranes that are able to separate charged organic dyes (Buonomenna et al., 2007). Membranes have been prepared by wet phase inversion process using dimethylformamide (DMF) as the solvent and water as the nonsolvent and using varying exposure times before coagulation to increase humidity absorption and to tune macrovoids formation. The operation conditions used for membrane formation confer to membranes the key characteristics for separation and durability: pore size and morphology, without fingers if possible, as these reduce tensile strength during high-pressure operations. The plasma treatment type is responsible for membrane surface modification, which modifies surface tension and provides membranes with a positive charge, which enables the selective retention of small positively charged compounds.

PVDF is one of the most popular materials for microporous and UF membranes due to its good mechanical properties, thermal and chemical stability, and radiation resistance (Qian et al., 2008; Huang et al., 2007). However, PVDF membranes are easily fouled by proteins, oil, and sewage, and they may need a high driving force for water filtration due to their hydrophobic properties. As a result, modifications to improve the hydrophilicity are required to obtain PVDF membranes with good performance characteristics.

The main modification methods focus on surface modification and blending with hydrophilic components. Surface modification methods include smearing surfactant, surface grafting via ultra violet spectroscopy, and plasma and high-energy particle technologies. Surface modification can deal with only the membrane surface layer. However, in cases where it is necessary to modify the

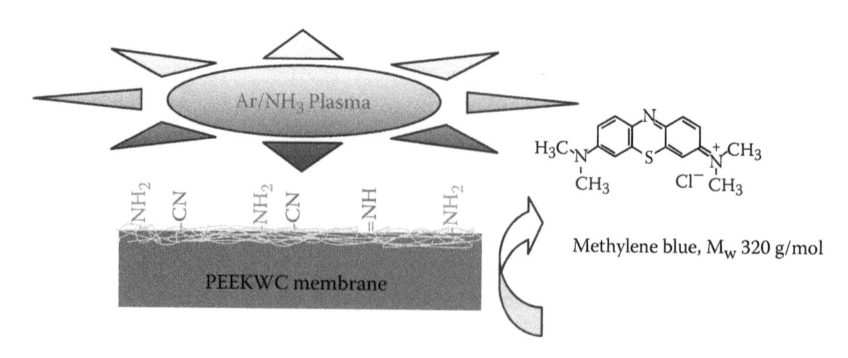


**FIGURE 1.3** Schematization of surface functionalization via plasma treatment of PEEKWC membranes prepared by nonsolvent-induced phase inversion. (Reprinted from *Microporous and Mesoporous Materials*, 120, Buonomenna, M.G., et al., 147–153, Copyright 2008, with permission from Elsevier.)

hydrophobic characteristics not only on the surface but also inside the membranes, the design of polymers containing charged groups (Tang et al., 2005) or blending with hydrophilic additives (polymers or metal oxides) appears as a viable route.

Hydrophilic polymers frequently used as poly(vinyl pyrrolidone) (PVP) or PEG can act as pore-inducing components too (Kim and Lee, 1998, 2004; Chen and Hong, 2002). Recently, comblike amphiphilic copolymers were used for improving the hydrophilicity of membranes (Qian et al., 2008). The successful compatibilization between the PVDF and the copolymer is guaranteed by the strong attraction between hydrophobic segments of the amphiphilic copolymer and the hydrophobic matrix (PVDF); the comblike hydrophilic segments endow the blend membrane with hydrophilicity (Figure 1.4).

The membranes have been prepared by means of classical nonsolvent-induced phase inversion: DMF has been used as solvent for PVDF and copolymer polysiloxane blend, with water as a coagulant.

Another recent example of polymeric blend that involves PVDF is that with poly(methyl methacrylate) (PMMA) (Lin et al., 2006). Porous PMMA membranes are very brittle, and this drawback has prevented it from being applied to separation processes such as MF, UF, and NF. On the other hand, PVDF membranes possess good mechanical strengths. Thus, it is of great interest to combine the advantages offered by PMMA and PVDF to prepare membranes that have both good strength and porous morphology that meet the requirements of real fine separation practice. In addition, these two polymers are known to have excellent compatibility in their amorphous phases, minimizing the possibility of polymer–polymer phase separation during the membrane precipitation process.

Nonsolvent-induced phase process has been used to prepare the membranes: DMSO has been chosen as an optimal solvent for both polymers of blend, and mixtures of DMSO/water as a different coagulant to modify membrane morphology.

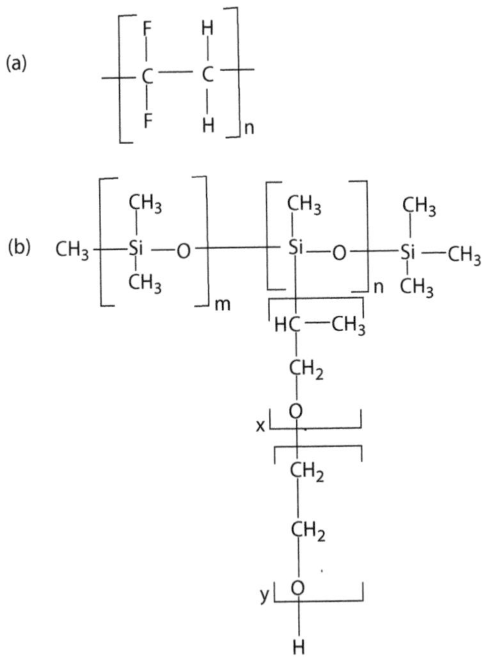

**FIGURE 1.4** Structures of (a) PVDF and of (b) copolymer polysiloxane containing polyether side chains used for the polymeric blend. (With kind permission from Springer Science+Business Media: *Front. Chem. China*, 3(4), 2008, 432, Qian, Y., et al.)

As an alternative to PMMA, polyethersulfone (PES) has been used as a copolymer for PVDF-based polymeric blends (Wu et al., 2006). The effect of PES concentration, content proportion of PVDF/PES on blend membrane shrinkage, pure water flux, and retention to bovine serum albumin were investigated. Four different materials, DMSO, DMF, NMP, and dimethylacetamide (DMAc), were used as the solvents. PVP was used as a pore-forming additive.

The influence of the solvent type and the concentration of PVP on the properties and morphology of PVDF/PES blend membranes is investigated. The shrinkage ratio of blend membranes is much reduced, and the pure water flux increases enormously, which may offer the blend membranes a prospective application in plasma separation and the ascites UF concentration process.

PVDF and its copolymer poly(vinylidene fluoride-*co*-hexafluoropropylene) (PVDF–HFP) have been reported as suitable membrane materials for lithium-ion batteries (Seol et al., 2006; Zhang et al., 2008; Pu et al., 2006).

The membrane separator plays an important role in rechargeable lithium batteries since it can keep the positive and negative electrodes apart to prevent electrical short circuits (Figure 1.6). Many requirements for the high performance of the separator include the following: electronic insulation, minimal electric resistance, sufficient tensile strength to allow easy handling, chemical resistance to degradation by electrolyte, and good wettability with electrolyte. Generally, poly(ethylene) (PE)-based membrane separators have been used in conventional lithium secondary batteries (Figure 1.5). In fact, PE has good mechanical properties, and it can also effectively prevent thermal runaway that results from the short-circuiting caused by rapid overcharging of the battery. However, PE has a drawback, and that is poor compatibility with liquid electrolytes.

Therefore, it is necessary to consider polymers as PVDF, which is suitable as a separator in lithium secondary batteries.

Seol et al. (2006) reported the preparation of PVDF-based membranes by phase inversion method using DMF and water as the solvent and the nonsolvent, respectively. Normal porous PVDF membrane, as a reference, was prepared using the simple phase inversion method. Inserting an evaporation step at 90°C prior to coagulation, to eliminate part of casting solvent, the structure of the membrane can become denser and the crystallinity of the membrane can increase. The solvent pre-evaporation condition alters the ratio between the amorphous and crystal phases of PVDF, which has an important effect on the physical properties of the membrane.

An additional uniaxial stretching process was used: ionic conductivity of the membrane was increased through a high uptake of electrolyte solution in the pores resulting from uniaxial stretching. The tensile strength of the membrane was also increased by the introduction of an additional solvent pre-evaporation step, which induced the formation of a more compact membrane structure.

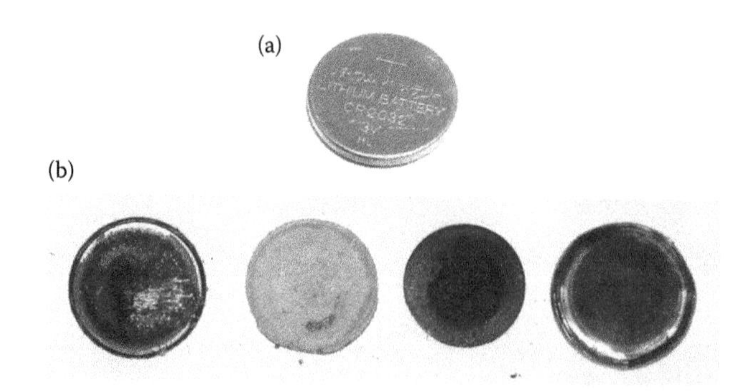


**FIGURE 1.5** (a) Lithium battery. (b) Disassembled battery, from left: first, anode cup, upside down, spent lithium partially scratched off; second, separator, a thin layer of porous membrane soaked with electrolyte—lithium salt in an organic solvent; third, cathode, a tablet of manganese dioxide; fourth, cathode can, with current collector (carbon layer) on its bottom and a gasket around its inner edge. (Courtesy of Wikimedia Commons.)

PVDF–HFP is also used as separator material in the field of rechargeable lithium-ion batteries. Compared with PVDF, PVDF–HFP has relatively lower crystallinity due to the copolymerization effect. Therefore, PVDF–HFP contained more amorphous domains capable of trapping large amounts of liquid electrolytes, whereas the crystalline regions provided mechanical integrity for the processing of free-standing films.

Pu et al. (2006) prepared PVDF–HFP by means of nonsolvent-induced-based membranes suitable as a separator for lithium secondary batteries.

The problem of solvent removing, in this case, has been overcome by using in the casting solution a volatile solvent, with acetone and water as the nonsolvent. In this way, gel microporous membranes can be directly obtained after the evaporation of acetone.

Zhang et al. (2008) prepared gel polymer electrolyte membrane (Li-GPEM) based on a PVDF–HFP matrix and PEG by means of a typical phase inversion process.

PVDF–HFP copolymer and PVP were dissolved in DMAc at 70°C for 8 h, with mechanical stirring to form a homogeneous casting solution. The resulting solution was cast onto a clean glass plate, and the glass plate with the nascent membrane was immersed in an ultrafiltrated water bath for 30 s at 30°C to solidify. The formed membrane was then thoroughly washed with a water–ethanol–hexane sequence for 24 h and was completely dried at 50°C in a vacuum oven for 12 h to obtain the membrane. The obtained porous PVDF–HFP membrane was immersed in liquid electrolyte containing PEG diacrylate and an initiator to absorb the liquid electrolyte and was then thermally cross-linked at 60°C. The Li-GPEM had better overall performance than the liquid and blend gel systems used as conductive media in lithium batteries.

## 1.3 NEW MATERIALS FOR GAS SEPARATION MEMBRANES

Separation of gases using polymeric membranes competes effectively with well-established processes such as cryogenic distillation, absorption, and pressure-swing adsorption (Baker and Lokhandwala, 2008).

One trend that has emerged in commercial gas separation membranes is the use of composite membranes (see Figure 1.1; thin-film composite), in which a thin selective layer is deposited on a highly porous support. In this way, it is possible to use two different materials: one characterized by sufficient mechanical strength and low cost for the support (generally made of either polyetherimide or polyacrylonitrile) and the other one made of a polymer that performs the separation. In this way, it is possible to optimize separately each membrane component.

The general procedure followed by most research to form flat composite membranes is as follows:

(1) Preparation of homogeneous polymer/solvent mixture
(2) Casting the solution on the support
(3) Evaporation of the solvent
(4) Sometimes, annealing at high temperatures to remove the residual solvent

For the choice of the dense separation layer, the key parameters are the permeance (*P*/*l* of Equation 1.1) and membrane selectivity (Freeman and Pinnau, 1999). To introduce the selectivity concept, Equation 1.1 can be rewritten as Equation 1.2:

$$J = DS/l \cdot \Delta P, \tag{1.2}$$

where permeability (*P*) of Equation 1.1 has been expressed as the product of diffusivity (*D*) and solubility (*S*). Diffusivity is an indication of the mobility of the molecules in the membrane material, and solubility is an indication of the concentration of molecules dissolved in the membrane material. *P*, the product of *D* and *S*, is a measure of the membrane's ability to permeate gas.

Membrane selectivity is the measure of a membrane's ability to separate two gases $A$ and $B$, and it can be written as in Equation 1.3:

$$\alpha_{A/B} = \frac{P_A}{P_B} = \left[\frac{D_A}{D_B}\right] \times \left[\frac{S_A}{S_B}\right], \tag{1.3}$$

where $D_A/D_B$ is the diffusivity or mobility selectivity, and $S_A/S_B$ is the solubility selectivity.

The mobility selectivity is proportional to the ratio of the molecular size of the two permeant gases. The solubility of a component increases with the condensability of the component.

So, membrane materials can be classified as (1) materials based on diffusivity selectivity and (2) materials based on solubility selectivity (Freeman and Pinnau, 1999).

Materials of class 1 are used in the separation of small penetrant gases such as $O_2$, $CO_2$, and $H_2$ from air, natural gas, and a variety of chemical and petrochemical processes (Baker and Lokhandwala, 2008). In these cases, membrane materials research focuses on developing materials with high diffusivity and high diffusivity selectivity.

Materials of class 2 are polymers that are more permeable to larger penetrants than to smaller gas components and are used for the selective removal of volatile organic compounds from air and the selective removal of higher hydrocarbons such as propane and butane from natural gas (Baker and Lokhandwala, 2008).

Transport of small gas molecules through polymers occurs by diffusion through transient free-volume elements or cavities formed by random, thermally stimulated motion of the flexible organic chains. Cavity shapes and sizes are not uniform in amorphous polymers (Budd et al., 2005).

For rubbery polymers, with their high free volumes, the mobility selectivity ($D_A/D_B$) is low. Rubbery polymers are commonly used for their ability to separate large organic vapor molecules from smaller molecules, because of their high solubility selectivity.

Conventional, low-free-volume, glassy polymers show a high mobility selectivity, with smaller molecules diffusing more rapidly than larger ones. However, glassy polymers usually exhibit low permeability, and it has been found that modifications to the polymer structure that improve permeability generally lead to a decrease in selectivity and vice versa.

Robeson (1991, 2008) demonstrated an upper bound in the plot of selectivity logarithm versus permeability logarithm for a wide range of polymers: the objective of much research is to generate polymers that perform in the upper right of Robeson's plot. Freeman and Pinnau (1999) have developed a theory to justify an empirical principle: to obtain the best permeability/selectivity properties up to a limit, a polymer structure with a stiff backbone disrupting interchain packing (with consequent generation of high free volume) to improve permeability has been designed. This principle is taken for perfluoropolymers (Merkel et al., 2006), unconventional mixed matrix membranes (MMMs) (Chung et al., 2007), and polymers of intrinsic microporosity (PIMs) (Budd et al., 2005).

The high free volumes of these systems reduce the importance of diffusivity selectivity, so that solubility selectivity becomes dominant for the overall separation process. This approach to tune membrane separation properties through solubility selectivity changes is an alternative to the conventional approach, which is to make changes to the polymer structure to alter diffusion selectivity (Freeman and Pinnau, 1999).

In Table 1.2, the structures and properties of some ultrahigh free volume glassy polymers are reported.

Poly(1-trimethylsilyl-1-propyne) (PTMSP) is known as the most permeable of all polymers to gases. The presence of bulky trimethylsilyl and methyl groups causes the polymer chain to be twisted. Sorption measure on this twisted, inflexible polymer indicates that this polymer is porous, with a void fraction of 0.29; the free volume elements are connected to give a micropore structure (Freeman and Pinnau, 1999).

**TABLE 1.2**
**Structures and Properties of Some Ultrahigh Free Volume Glassy Polymers**

| Polymer | Structure | Tg (°C) | FFV | $P_{O_2}$ (Barrer) | $\alpha_{O_2/N_2}$ |
|---|---|---|---|---|---|
| PTMSP | $-[C(CH_3)=C(Si(CH_3)_3)]_n-$ | >250 | 0.29 | 6100 | 1.8 |
| PMP | $-[C(CH_3)=C(CH(CH_3)_2)]_n-$ | – | – | 2700 | 2.0 |
| Teflon AF[a] | $[\text{perfluoro-2,2-dimethyl-1,3-dioxole } (CF_3, CF_3, O, O, F, F)]_x[CF_2-CF_2]_{1-x}$ | 240 | 0.33 | 1600 | 2.0 |
| Hyflon AD[b] | $[\text{dioxole ring } (F, F, O, O, OCF_3, F)]_x[CF_2-CF_2]_{1-x}$ | 134 | 0.23 | 67 | 2.0 |
| Cytop | $-[CF_2-CF(-(CF_2)_x-)CF-(CF_2)_y-]_n$, ring: $CF-O-(CF_2)-CF_2-CF$ | 108 | 0.21 | 16 | 3.1 |

[a] Teflon AF2400, x = 0.87; another commercial grade offered by DuPont is Teflon AF1600, x = 0.65.
[b] Hyflon AD80, x = 0.80; another commercial grade offered by Solvay Solexis is Hyflon AD60, x = 0.60.

The discovery in the past 20 years of amorphous, solvent-processible perfluoropolymers such as Teflon® AF, Cytop™, and Hyflon® AD (Table 1.2) has created new opportunities for membrane separations.

The relative permeability of the perfluoropolymers is well described by the fractional free volume (FFV), a common measure of the free space in a polymer matrix available for molecular transport. On the basis of data reported in Table 1.2, both gas permeability and FFV exhibit the following order:

Teflon AF > Hyflon AD > Cytop

They can be fabricated into thin, high-flux composite or anisotropic membranes, retaining the excellent stability typical of fluorinated material (Figure 1.6).

Of note is that for these polymers, an unusual tendency to retain the solvent used when preparing membranes by solution casting and solvent evaporation has been observed.

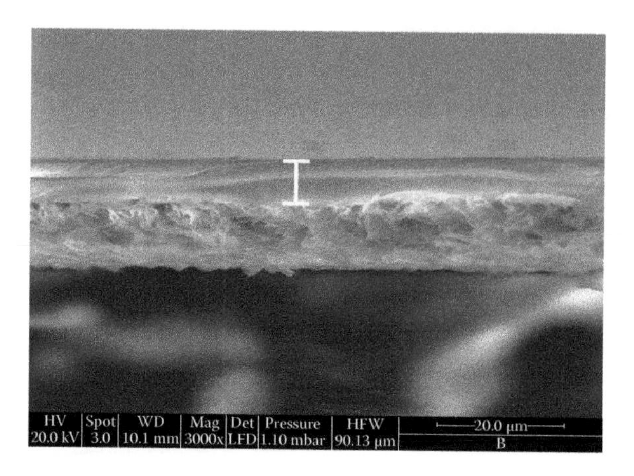


**FIGURE 1.6** Asymmetric Hyflon AD60X–based membrane prepared by nonsolvent-induced phase inversion using Galden HT 55 as solvent and *n*-hexane as nonsolvent.

This phenomenon occurred for Hyflon AD60X, which retained its own solvent, Galden® HT 55, in spite of the low boiling point and high volatility of the solvent (Jansen et al., 2006). The same experimental evidence was obtained by Pinnau and Toy (Robeson, 2008) with Teflon AF2400. The permeabilities of the solution-cast film were between 30% (helium) and 80% (methane) higher than those reported for the melt-pressed film. Thermal analysis reveals that residual solvent plasticizes the polymer and strongly affects its gas transport properties, increasing noticeably the diffusion coefficient of the gases with a high molecular size and reducing the membrane's permselectivity. The solvent retention is kinetically determined rather than thermodynamically; that is, it is a result of the slow diffusion of the solvent molecules through the polymer matrix. The solvent retention in Hyflon might further be related to the absence of strong cohesive forces in the polymer and the consequently high FFV.

Despite their unexceptional $O_2/N_2$ selectivity compared to polyimide and polycarbonate membranes (Robeson, 2008), perfluoropolymer membranes have been considered for oxygen-enrichment applications because of their high permeability combined with their chemical and thermal stability (Baker and Lokhandwala, 2008). In particular, low hydrocarbon vapor solubility in perfluoropolymers renders them more resistant to sorption-induced plasticization than hydrocarbon-based membranes (Merkel et al., 2006). Plasticization of glassy, size-sieving membranes is caused by the sorption of large condensable species and dramatically reduces the diffusion selectivity of a polymer.

Under industrial operating conditions (at high, near-saturation pressures), the use of plasticization-resistant perfluoropolymers represents an alternate strategy for membrane-based natural gas treatment.

It is important to emphasize the concept that if the proportion of free volume is increased to the extent that it is effectively interconnected, the polymer may take on the characteristics of microporous materials (with pores < 2 nm), such as zeolites and carbon molecular sieve materials. Organic polymers generally have enough conformational and rotational freedom to pack space efficiently; as a result, they do not have large internal surface areas.

Contrary to conventional MMMs consisting of porous fillers and polymeric matrices of similar selectivity, nonporous nano-sized particles have been introduced in the polymer matrix of high-free-volume glassy polymers such as poly(4-methyl-2-pentyne) (PMP) and Teflon AF (Chung et al., 2007) to systematically manipulate the molecular packing of the polymer chains (unconventional MMMs). In this way, the average size of the free volume has been increased, and compared with neat polymer membranes, the transport of *n*-butane over methane was enormously enhanced in both

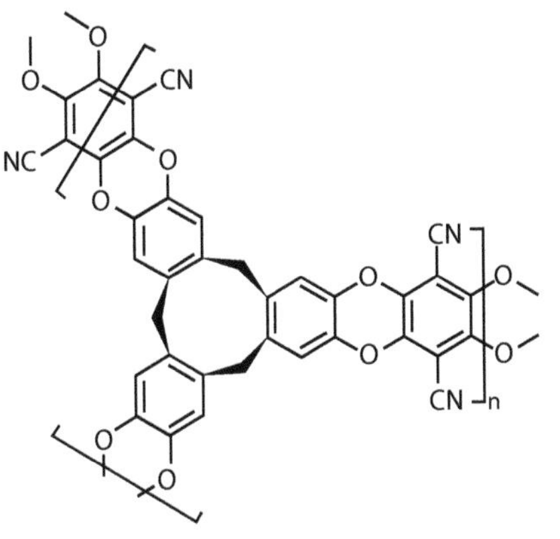

FIGURE 1.7 Polymer incorporates bowl-shaped cyclotricatechylene monomer (PIM). (Budd, P.M., N.B. McKeown, and D. Fritsch. 2005. *J Mater Chem* 15: 1977.)

permeability and selectivity. The enhanced free volume that was created weakened the size selectivity, and solubility selectivity favoring the hydrocarbon transport dominated the separation process.

Budd et al. (2005) reported the synthesis of a novel family of microporous polymers that have exceptionally high internal surface areas. The internal surface areas of these so-called PIMs are greater than 800 $m^2$ per gram of material.

Some PIMs are prepared by using similar dioxane-forming reactions involving readily prepared cyclotricatechylene and chlorinated hexaazatrinaphthylene precursors. Cyclotricatechylene (Figure 1.7) is a bowl-shaped receptor monomer that leads to enhanced ultramicroporosity.

A recent paper reports the generation of an unusual microstructure tailored by thermally driven segment rearrangement, named TR polymers (Park et al., 2007)—polymers with a fixed cavity size, a narrow cavity size distribution, and a shape reminiscent of bottlenecks connecting adjacent chambers. Central to this approach is controlled free volume element formation through spatial rearrangement of the rigid polymer in the glassy phase, PIOFG-1 (Figure 1.8).

The changes in chain conformation and topology create well-connected, narrow-sized distribution cavities for molecular separations. The precursor polymer, PIOFG-1, synthesized from 4,4′-(hexafluoroisopropylidene)-diphthalic anhydride (6FDA) and 2,2′-bis(3-amino-4-hydroxylphenyl) hexafluoropropane (bisAPAF), and corresponding thermally rearranged samples (i.e., TR-1-350, TR-1-400, and TR-1-450, respectively) show that the polymer undergoes microstructural change depending on the extent of rearrangement. The cavity radius of PIOFG-1 polymer (of about 0.28 nm and is very broad) increases to ~0.4 nm, with the distribution of cavity sizes becoming very narrow as the thermal rearrangement temperature increases to 450°C.

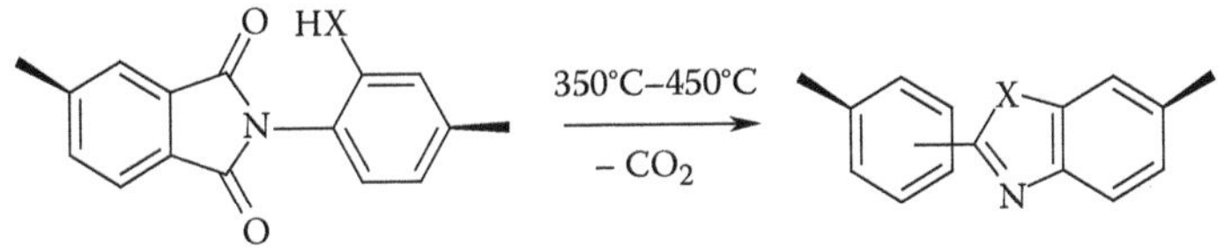

FIGURE 1.8 Structural change during thermal chain rearrangements of polyimides containing an ortho-positioned functional group (X = O or S) to form TR polymers. (Park, H.B., Jung, C.H., Lee, Y.M., Hill, A., Pas, S.J., Mudie, S.T., Van Wagner, E., Freeman, B.D., Cookson, D.J. 2007. *Science* 318: 254.)

## REFERENCES


Bae, T.H., I.C. Kim, and T.M. Tak. 2006. Preparation and characterization of fouling-resistant $TiO_2$ self-assembled nanocomposite membranes. *J Membr Sci* 275: 1.

Baker, R.W., and K. Lokhandwala. 2008. Natural gas processing with membranes: An overview. *Ind Eng Chem Res* 47(7): 2109.

Budd, P.M., N.B. McKeown, and D. Fritsch. 2005. Free volume and intrinsic microporosity in polymers. *J Mater Chem* 15: 1977.

Buonomenna, M.G., A. Figoli, J.C. Jansen, et al. 2004. Preparation of asymmetric PEEKWC flat membranes with different microstructures by wet phase inversion. *J Appl Polym Sci* 92: 576.

Buonomenna, M.G., L.C. Lopez, P. Favia, et al. 2007. New PVDF membranes: The effect of plasma surface modification on retention in nanofiltration of aqueous solution containing organic compounds. *Water Res* 41: 4309.

Buonomenna, M.G., A. Gordano and Drioli, E. 2008a. Characteristics and performance of new nanoporous PEEKWC films. *Eur Polymer J* 44: 2051.

Buonomenna, M.G., L.C. Lopez, M. Davoli, et al. 2008b. Polymeric membranes modified via plasma for nanofiltration of aqueous solution containing organic compounds. *Micropor Mesopor Mat* 120: 147–153.

Cabasso, I. 1987. Liquid crystalline polymers to mining applications. In *Encyclopedia of Polymer Science and Engineering* Vol. 9, ed. H.F. Mark, pp. 509–579. New York: John Wiley & Sons.

Chakrabarty, B., A.K. Ghoshal, and M.K. Purkait. 2008. Preparation, characterization and performance studies of polysulfone membranes using PVP as an additive. *J Membr Sci* 315: 36.

Chen, G., X. Chai, P.L. Yue, et al. 1997. Treatment of textile desizing wastewater by pilot scale nanofiltration membrane separation. *J Membr Sci* 127: 93.

Chen, H., and G. Belfort. 1999. Surface modification of poly(ether sulfone) ultrafiltration membranes by low-temperature plasma induced graft polymerization. *J Appl Polym Sci 72*: 1699.

Chen, N.P., and L. Hong. 2002. Surface phase morphology and composition of the casting films of PVDF–PVP blend. *Polymer* 43: 1429.

Chung, T.S., L.Y. Jiang, Y. Li, et al. 2007. Mixed matrix membranes (MMMs) comprising organic polymers with dispersed inorganic fillers for gas separation. *Prog Polym Sci* 32: 483.

Drioli, E., A. Criscuoli, and E. Curcio. 2006. Mass transfer with chemical reaction. In *Membrane Contactors: Fundamentals, Applications and Potentialities*. Chapter 10. Amsterdam: Elsevier.

Freeman, B.D., and I. Pinnau. 1999. *Polymer Membranes for Gas and Vapour Separation*. Washington: American Chemical Society.

Gordano, A., G. Clarizia, A. Torchia, et al. 2002. New membranes from PEEK-WC and its derivatives. *Desalination* 145: 47.

Huang, S., G. Wu, and S. Chen. 2007. Preparation of microporous poly(vinylidene fluoride) membranes via phase inversion in supercritical $CO_2$. *J Membr Sci* 293: 100.

Jansen, J.C., M.G. Buonomenna, A. Figoli, et al. 2006. Asymmetric membranes of modified poly(ether ether ketone) with an ultra-thin skin for gas and vapour separations. *J Membr Sci* 272: 188.

Kawakami, H., M. Mikawa, and S. Nagaoka. 1997. Formation of surface skin layer of asymmetric polyimide membranes and their gas transport properties. *J Membr Sci* 137: 241.

Khayet, M., and T. Matsuura. 2003. Application of surface modifying macromolecules for the preparation of membranes for membrane distillation. *Desalination* 158: 51.

Kim, I.C., and K.H. Lee. 2004. Effect of poly(ethylene glycol) 200 on the formation of a polyetherimide asymmetric membrane and its performance in aqueous solvent mixture permeation. *J Membr Sci* 230: 183.

Kim, J.H., and K.H. Lee. 1998. Effect of PEG additive on membrane formation by phase inversion. *J Appl Polym Sci* 138: 153.

Li, B., W.G. Chen, X.G. Wang, et al. 2003. PNIPAAm-grafted layers on polypropylene films II. Surface properties and temperature sensitivity study. *Acta Polym Sin* 1: 7 (in Chinese).

Lin, D.J., C.L. Chang, C.K. Lee, et al. 2006. Preparation and characterization of microporous PVDF/PMMA composite membranes by phase inversion in water/DMSO solutions. *Eur Polym J* 42: 2407.

Lonsdale, L.K. 1982. Review—the growth of membrane technology. *J Membr Sci* 10: 81.

Mahendran, R., R. Malaisamy, and D. Mohan. 2004. Preparation, characterization and effect of annealing on performance of cellulose acetate/sulfonated polysulfone and cellulose acetate/epoxy resin blend ultrafiltration membranes. *Eur Polym J* 40: 623.

Merkel, T.C., I. Pinnau, R. Prabhakar, et al. 2006. Gas and vapour transport properties of perfluoropolymers. In *Materials Science of Membranes for Gas and Vapour Separation*. New York: Wiley & Sons.

Mulder, M. 1984. *Basic Principles of Membrane Technology*. Dordrecht: Kluwer.

Nunes, S.P., and K.V. Peinemann (Eds.). 2001. *Membrane Technology in the Chemical Industry*. Weinheim: John Wiley & Sons.

Nunes, S.P., M.L. Sforca, and K.V. Peinemann. 1995. Dense hydrophilic composite membranes for ultrafiltration. *J Membr Sci* 106: 49.

Park, H.B., C.H. Jung, Y.M. Lee, et al. 2007. Polymers with cavities tuned for fast selective transport of small molecules and ions. *Science* 318: 254.

Pieracci, J., J.V. Crivello, and G. Belfort. 1999. Photochemical modification of 10 kDa polyethersulfone ultrafiltration membranes for reduction of biofouling. *J Membr Sci* 156: 223.

Pinnau, I., and W.J. Koros. 1993. A qualitative skin layer formation mechanism for membranes made by dry/wet phase inversion. *J Appl Polym Sci, Polym Phys* 31:419.

Pu, W., X. He, L. Wang, et al. 2006. Preparation of PVDF–HFP microporous membrane for Li-ion batteries by phase inversion. *J Membr Sci* 272: 11.

Qian, Y., J. Wang, B. Zhu, et al. 2008. Modification effects of amphiphilic comb-like polysiloxane containing polyether side chains on the PVDF membranes prepared via phase inversion process. *Front Chem China* 3(4): 432.

Rahimpour, A., S.S. Madaeni, and S. Mehdipour-Ataei. 2008. Synthesis of a novel poly(amide-imide) (PAI) and preparation and characterization of PAI blended polyethersulfone (PES) membranes. *J Membr Sci* 311: 349.

Rahimpour, A., and S.S. Madaeni. 2007. Polyethersulfone (PES)/cellulose acetate phthalate (CAP) blend ultrafiltration membranes: Preparation, morphology, performance and antifouling properties. *J Membr Sci* 305: 299.

Robeson, L.M. 1991. Correlation of separation factor versus permeability for polymeric membranes. *J Membr Sci* 62: 165.

Robeson, L.M. 2008. The upper bound revisited. *J Membr Sci* 320: 390.

Seol, W.H., Y.M. Lee, and J.K. Park 2006. Preparation and characterization of new microporous stretched membrane for lithium rechargeable battery. *J Power Sources* 163: 247.

Strathmann, H. 1981. Review—membrane separation processes. *J Membr Sci* 9: 121.

Strathmann, H., L. Giorno, and E. Drioli. 2006. *An Introduction to Membrane Science and Technology*. Rome: Consiglio Nazionale delle Ricerche.

Su, Y., X. Jian, S. Zhang, et al. 2006. Preparation of novel PPES-B UF membrane with good thermal stability: The effect of additives on membrane performance and cross-section morphology. *J Membr Sci* 271: 205.

Tang, B., T. Xu, M. Gong, et al. 2005 A novel positively charged asymmetry membranes from poly(2,6-dimethyl-1,4-phenylene oxide) by benzyl bromination and in situ amination: membrane preparation and characterization. *J Membr Sci* 248: 119.

Van der Bruggen, B., J. Schaep, D. Wilms, et al. 1999. Influence of molecular size, polarity and charge on the retention of organic molecules by nanofiltration. *J Membr Sci* 156: 29.

Wang, M., O.F. Ana, L.G. Wu, et al. 2007. Preparation of pH-responsive phenolphthalein poly(ether sulfone) membrane by redox-graft pore-filling polymerization technique. *J Membr Sci* 287: 257.

Wu, G., S. Gan, L. Cui, et al. 2008a. Preparation and characterization of PES/$TiO_2$ composite membranes. *Appl Surf Sci* 254: 7080.

Wu, C., S. Zhang, C. Liu, et al. 2008b. Preparation, characterization and performance of thermal stable poly(phthalazinone ether amide) UF membranes. *J Membr Sci* 311: 360.

Wu, L., J. Sun, and Q. Wang. 2006. Poly(vinylidene fluoride)/polyethersulfone blend membranes: Effects of solvent sort, polyethersulfone and polyvinylpyrrolidone concentration on their properties and morphology. *J Membr Sci* 285: 290.

Yang, Y., X. Jian, D. Yang, et al. 2006a. Poly(phthalazinone ether sulfone ketone) (PPESK) hollow fiber asymmetric nanofiltration membranes: Preparation, morphologies and properties. *J Membr Sci* 270: 1.

Yang, Y., and P. Wang. 2006. Preparation and characterizations of a new PS/$TiO_2$ hybrid membranes by sol–gel process. *Polymer* 47: 2683.

Yang, Y., D. Yang, S. Zhang, et al. 2006b. Preparation and characterization of poly(phthalazinone ether sulfone ketone) hollow fiber ultrafiltration membranes with excellent thermal stability. *J Membr Sci* 280: 957.

Zhang, M., A. Zhang, Z. Cui, et al. 2008. Preparation and properties of gel membrane containing porous PVDF–HFP matrix and cross-linked PEG for lithium ion conduction. *Front Chem Eng China* 2(1): 89.

Zhang, Y., H. Li, J. Lin, et al. 2006. Preparation and characterization of zirconium oxide particles filled acrylonitrile–methyl acrylate–sodium sulfonate acrylate copolymer hybrid membranes. *Desalination* 192: 198.

# 2 Asymmetric Polyethersulfone Membranes: Preparation and Application

*Ani Idris, Iqbal Ahmed, and Nurdiana Mohd Noor*

## CONTENTS



## 2.1 INTRODUCTION

### 2.1.1 Overview

In recent years, the membrane separation technique was proven to be a very fast, efficient, practical, and cost-effective process than other conventional techniques. Although the technology started as a laboratory tool, it has since then developed into an industrial process with significant considerable technical and commercial impact. Various types of membrane are available in the market; among them, ultrafiltration (UF) is recognized as having the largest range of applications in diverse industries due to its high efficiency and low energy consumption separation technology (Nunes and Pienemann, 2006).

Since the discovery of UF and reverse osmosis (RO) membranes, extensive studies are carried out as an effort to develop new UF membranes that meet the industrial requirements and thus can be commercialized. Today, most of the commercial membranes are asymmetric membranes made of synthetic polymers, copolymers, or blends formed by the phase inversion method.

Asymmetric membranes usually have long fingerlike pores or voids that reach to the membrane surface, but the pores are reduced in size upon reaching the thin skin layer on the membrane surface (Scott, 1995). The very thin-skinned top layers of asymmetric cellulose acetate (CA), polysulfone (PSf), or polyethersulfone (PES) membranes are usually achieved by the phase inversion process (Kim et al., 1996).

### 2.1.2 Configurations of Commercial Ultrafiltration Membranes

Currently, several membrane configurations are available both commercially and at the laboratory scale. Table 2.1 lists some commercially available membranes for UF. The membranes itself must satisfy a number of mechanical, hydrodynamic, and economic requirements.

Recently, Millipore Corporation has introduced and commercialized their next generation UF membranes under the trade name of Biomax (void-free PES family). These are based on novel

**TABLE 2.1**
**Commercially Available Membranes for UF**

| Manufacturer | Brand Name | Material (Module) | MWCO (kDa) | Flux ($L \cdot m^{-2} \cdot h^{-1}$) | Application |
|---|---|---|---|---|---|
| AMI (DOW/FILMTEC) (USA) | M-U4040 PES | PES (SW) | 10 | 4.5–18.2 at 3.2 bar | Pharmaceutical and food industry |
| AMI (DOW/FILMTEC) (USA) | MU2540 PAN | Hydrophilic PAN (SW) | 20 | 50 at 3.5 bar | Waste water |
| AMC (USA/China) | AC 120 R01 | Modified PES (HF) | 15 | 112 at 4.1 bar | Pharmaceutical and food industry |
| Aquious (UK) | UltraBar | PES (HF) | 100–150 | 27 | Surface-water treatment |
| AsahiKasei (Japan) | UF AP series | Hydrophilic PAN (HF) | 69 | 16 | Pharmaceutical and food industry |
| Dainippon (Japan) | PF004D | PMP (HF) | — | 36 | Surfactant solved water |
| KOCH (USA) | HFK-131 | SMP PES (SW) | 10 | 24–53 | Waste water |
| KOCH | HFK-328 | SMP PES (SW) | 5 | 24–53 | Waste water |
| KOCH | *HF 8H-72-35-PMPW | SMP PSf (HF) | 100 | 32 | PVC separation |
| Luxx Ultra-Tech Inc. (USA) | L" Series | PVDF, PES PS (Tubular) | 5 | 27–45 | Waste water and food industry |
| Millipore (USA) | Ultracel Biomax PB | PES (HF) | 5 | 35-45 at 1 bar | Protein purification |
| Millipore | Amicon | CA (HF) | 10 | 97.2 at 1 bar | Protein purification |
| Membrane elements | M-series | Mod. Hydro. PAN (HF) | 0.03–1 μm | At 9.3 bar | Oil–water separation |
| Nitto Denko (Japan) | Hydracap | Hydro. PES (HF) | 150 | 51–128 | Waste water |
| PALL | BTS | Asym. PES Casseette | 0.5–10 | 187.2 at 3 bar | Pharmaceutical and food industry |
| Polymem | polymem | PSf HF | 6 | 313.2 at 1 bar | Protein purification |
| Sterlitech (USA) | Sterl UF | PES HF | 0.04–23 | 9.2–210 | Pharmaceutical and food industry |
| Synder (Canada) | PES 100 | PES (SW) | 70 | 51 | Gelation separation |
| TriSep (Canada) | UE10 | PES (SW) | 10 | 2.1 | Dairy and food industry |

casting processes that provide void-free, produce high process flux membranes with enhanced product retention, and significantly increased mechanical resistance.

Dow/FILMTEC is currently commercializing UF series M-U4040 PES for pharmaceutical, food beverage, and waste water treatment. American Membrane Corporation (AMC) has launched the Accupor membrane, which is a highly microporous membrane composed of modified hydrophilic PES specifically designed for biological, analytical, electronic, pharmaceutical, beverage, and sterilizing filtration applications. Nitto Denko and Asahi Kasei commercialized Hydracap (hydrophilic PES), for waste water and the pharmaceutical and food industries.

Recently, KOCH has supplied a series of UF HFK-131/138 (PES) spiral wound and hollow fiber membranes for waste water treatment. PALL has introduced BTS highly asymmetric PES membranes with "cut off" layer in the BTS membranes of only about 10 μm thick, versus conventional membranes with cut off layers of approximately 100–125 μm thickness. This difference in thickness results in the BTS highly asymmetric membrane having significantly higher flow rates with much lower pressure drop for pharmaceutical purposes.

## 2.2 MEMBRANE MATERIALS: PAST, PRESENT, AND FUTURE

In membrane research history, the first membrane materials were reported and described by Reid and Breton and also by Loeb and Sourirajan in the late 1950s. Since then, numerous materials have been developed in order to improve the capacity and performances of the existing membrane technology.

Membranes can be classified as biological and synthetic. Synthetic membranes are divided into organic membranes, (e.g., polymeric or liquid), and inorganic membranes (e.g., ceramic or metal membranes). Synthetic membranes can be made from different kinds of materials of which polymers and ceramics are the most important (Baker, 2004; Mulder, 1996).

Synthetic polymeric membranes can be divided into hydrophobic and hydrophilic classifications. Membranes can also be classified by their structure. Structural classification is very important because it is the structure that determines separation mechanisms and membrane application. Membranes can be further classified as symmetric or asymmetric. The symmetric membranes can be porous, cylindrical porous and homogeneous (non porous). The asymmetric membranes can be porous, porous with top layer and composite that is consisting of a porous membrane part and a dense top layer. The thickness of the top layer in asymmetric membranes is 0.1 to 0.5 μm and is supported on a porous sub layer with a thickness of about 50 to 150 μm.

Various polymers and organic materials can be used to yield hydrophobic and hydrophilic surface. Table 2.2 shows the various hydrophilic and hydrophobic polymers for membrane production. Furthermore common commercial hydrophilic copolymers are made of polyethylene oxide (PEO) and crystallizable polyamide, polyurethanes and polyester (PBT). These materials can be used to make a hydrophobic polymer more hydrophilic. Hydrophobic materials such as cellulose esters, polycarbonate (PC), polysulfone (PSf), polyethersulfone (PES), poly vinylidene fluoride (PVDF) polyimide (PI), polyetherimide (PEI), aliphatic polyamide (PA) and polyetherketone (PEEK) are also common polymeric materials (Kesting, 1985; Lloyd, 1985; Mulder, 1996).

Each year, research papers in polymer and membrane science present many new examples of materials that demonstrate semi-permeable qualities at some scale. However, only a very limited number of these potential candidates make it to the commercial environment (Matsuura, 1994; Nunes and Peinemann, 2006). Among all those materials CA, PSf and PES are the most commonly used polymers for UF membranes (Idris and Ahmed, 2007a).

### 2.2.1 POLYETHERSULFONE

Polyethersulfone (PES) is an amorphous polymeric material, which is cost-effective with sufficient selectivity and good process ability, is one type of dominating material in the membrane separation

**TABLE 2.2**
**Hydrophobic and Hydrophilic Membranes**

| | | | |
|---|---|---|---|
| | | **Hydrophilic Polymers** | |
| Poly(vinyl alcohol) | PVAL | As well as cellulose and its derivative | |
| Poly(vinyl chloride) | PVC | Cellulose acetate | CA |
| Polyamide | PA | Cellulose acetate butyrate | CAB |
| Poly(acrylic acid) | PAA | Cellulose acetate propionate | CAP |
| Poly(ethylene oxide) | PEOX | Cellulose nitrate | CN |
| Polyacrylonitrile | PAN | Cellulose propionate | CP |
| Poly(vinyl acetate) | PVAC | Ethyl cellulose | EC |
| Poly(vinyl butyral) | PVB | Carboxymethyl cellulose | CMC |
| Poly (p-hydroxystyrene) | PHS | | |
| | | **Hydrophobic Polymers** | |
| Polysulfone | PSf | Poly tetrafluoro ethylene | PTFE |
| Polyethersulfone | PES ($\theta \leq 90$) | Polyethylene | PE |
| Poly vinylidene fluoride | PVDF | Silicone | |
| Polycarbonate | PC | Polyphenylene oxide | PPO |
| Polypropylene | PP | Polyphenylene sulfide | PPS |
| Poly(methyl methacrylate) | PMMA | Polystyrene | PS |

technology. It offers the greatest promise for both industrial application and fruitful academic research. PES is characterized as an ultra-performance resin because it offers exceptional performance such as high continuous use temperature of 180°C, low water sorption (0.8% at 50% relative humidity), its broad resistance to chemical attack, toughness and excellent hydrolytic stability and good resistance to environmental. It is also easy to fabricate into a wide variety of configurations and modules with a wide range of pore sizes available for UF applications ranging from 10 Å to 0.2 μm. An improved PES UF membrane with less susceptibility to fouling would have strong economic impact on the membrane industry, in addition to being commercially available and relatively inexpensive (Idris and Ahmed, 2007b). The specific gravity of the PES polymer used in these membranes is relatively high at 1.37. PES shows excellent properties in dielectric constant and low dielectric loss (dielectric dissipation factor) of 0.006 at 1 MHz and 0.001 at 50 MHz (Kesting, 1985).

PES is a closely related derivative of polysulfone which is totally devoid of aliphatic hydrocarbon groups and consists of phenylene ring structures connected together as bridge with sulfone groups ($SO_2$) or ether linkages (–O–) in the backbone chain to form a polymer as depicted in Figure 2.1. In "bridging" moieties (Figure 2.2), the sulfone groups tend to make the polymer stiff with a high glass transition temperature (230°C) and, together with the ring structures, tend to make the polymer chemically resistant and relatively hydrophobic (Kesting, 1985).

However, application of PES membranes is often limited because of its hydrophobic property. PES is uncharged and hydrophobic in nature. Hydrophilic or the surface wettability is an important characteristic in membrane studies as it is one of the aspects that can improve the usefulness of the PES membranes. Research on improvement of its flux and retention behavior has started in the early 1980s. One of the methods used to improve the structure and performance of the membrane is to

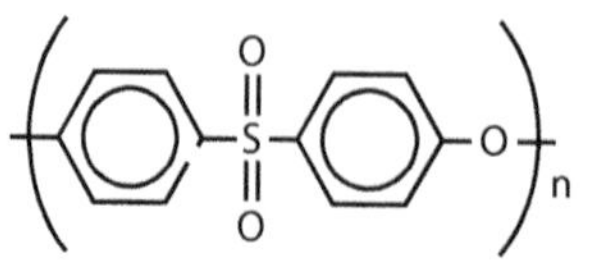

**FIGURE 2.1** Primary structure of polyethersulfone.

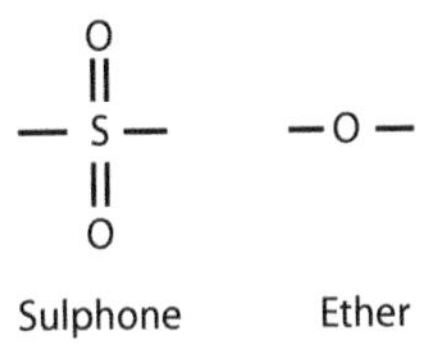

Sulphone Ether

**FIGURE 2.2** Bridging moieties.

introduce low molecular weight organic additives having different functional groups in the polymer solution (Kesting, 1985).

### 2.2.2 Choice of Additives

In order to increase the usefulness of the hydrophobic PES membranes, hydrophilicity or surface wettability of the membrane need to be improved. Based on the fundamental concept that the surface layer of the asymmetric polymeric membrane is strongly influenced by the additives or that of their aggregates which are in the casting solution, there is always an ongoing research in exploring new suitable additives for membrane making. In brief, additives used in the fabrication of PES membranes can be broadly categorized into polymeric additives such as polyvinylpyrrolidone (PVP), polyethylene glycol (PEG) and weak solvents such as glycerol. In fact, the addition of PVP and PEG has become a standard method or approach to obtain "hydrophilized" membranes (Ahmed et al., 2010). Organic acid such as acetic acid, propionic acid causes macrovoid formation (Idris and Ahmed, 2007b; Lee et al., 2002). Other less common additives used are low-molecular-weight inorganic salts such as lithium chloride (LiCl), zinc chloride ($ZnCl_2$), magnesium chloride ($MgCl_2$), calcium chloride ($CaCl_2$), magnesium perchlorate ($Mg(ClO_4)_2$) and calcium perchlorate $Ca(ClO_4)_2$ (Ahmed et al., 2010). The list of additives used for PES membranes is tabulated in Table 2.3.

In addition, additive creates a spongy membrane structure by prevention of macrovoid formation, enhance pore formation, improve pore interconnectivity and/or introduce hydrophilicity (Liu et

**TABLE 2.3**
**List of Additives Used for PES Membranes**

| Polymer | Inorganic Additives | Solvent | Application | Reference |
|---|---|---|---|---|
| PAN, PSf, PVC, PC, PMMA | $ZnCl_2$, $FeCl_3$, LiCl, $LiNO_3$, $(Al)_2(NO_2)_3$, NaCNS, $CuNO_2$ | DMF, DMSO, DMAc, NMP, acetone | UF | Michaels, 1971; Shinde et al., 1999; Lai et al., 1992 |
| SPSf | LiCl. $LiNO_3$, $Mg(C10_4)_2$ | NMP | UF comp. | Ikeda et al., 1989 |
| PSf | $ZnCl_2$ | DMF | UF | Katarzyna, 1989 |
| PVDF, PES | LiCl | DMAc | UF comp | Allegrezza and Burke, 1989 |
| CPSf | LiCl, $LiNO_3$, $MgCl_2$ | DMAc, NMP | RO | Guiver et al., 1990 |
| PES | LiCl | Butanol+NMP | Composite UF | Bellantoni and Loya, 1997 |
| PSf | $ZnCl_2$ | NMP | UF | Kim et al., 1996 |
| PES | LiCl | NMP | Composite UF | Bellantoni and Loya, 1997 |
| SPSf, CPSf | LiCl | NMP | UF | Möckel et al., 1999 |
| PES-PEES | $AgNO_3$ | NMP | UF | Kutowy and Sterlez, 2003 |
| PES | LiCl | DMF | UF | Idris et al., 2007 |
| PES | LiCl + PVP | DMAc | Microcapsules | Wang et al., 2006 |
| SPPO-PES | LiCl | DMF | UF | Yin et al., 2007 |
| PES | LiF, LiBr, LiCl | DMF+Acetone | UF | Idris et al., 2010 Ahmed et al., 2010 |

al., 2002). These membranes possess properties that are greatly different from the pure membrane forming polymer. There are also other frequently used additives reported in membrane making processes such as alcohols, dialcohols, water, formamide, polyethylene oxide (PEO), surfactants, mineral fillers or the mixture of them. There are several mechanisms through which such additives can affect the final membrane properties.

Kim and Lee (1998) investigated the effect of PEG additive as a pore former on the structure formation of PES membranes and their permeation properties. Wechs (1990) also studied on the influence of glycerol and PVP as additives on the performance of integral asymmetric PES membrane for UF and microfiltration (MF) application. Report showed high permeability of PES hollow fiber gas separation membranes could be prepared with nonsolvent additive (Wang et al., 2000a). Studies also showed the presence of formamide as an additive in the solutions helps to produce membranes with good selectivity characteristics (Pertov, 1996).

Higher performance membrane was reported achievable when inorganic salts were added into the casting solutions (Kim et al., 1996; Kraus et al., 1979; Munari et al., 1980). As documented by Hoehn (1985), lyotropic salts are said to be those with lithium, zinc, calcium and magnesium as the cation and chloride, bromide, iodide, nitrate, thiocyanate and perchlorate as the anion. They are greatly known to form complexes with the carbonyl group in polar aprotic solvent via ion–dipole interaction such as with acetone, dimethyl acetate (DMAc), dimethylformamide (DMF), dimethylsulfoxide (DMSO) and N-methyl pyrrolidone (NMP) (Kesting, 1990).

Certain inorganic salt are able to increase the permeation rates of semipermeable membranes of CA and is associated to the capacity of the component ions to swell the cellulosic substrate (Kesting, 1965). Inorganic additives such as zinc chloride ($ZnCl_2$), zinc bromide (ZnBr) can stabilize the membrane for desalination application (Watson et al., 1965). The effect of low molecular weight additives on the properties and structure of aromatic polyamide reverse osmosis membranes is reported by Kraus et al. (1979). He discovered that highly dissociated salts such as $LiClO_4$ or $Mg(ClO_4)_2$ exert a strong influence than the commonly used LiCl. Other studies had also been carried out only to reveal that the "salt effect" in aromatic polyamide membranes is due to a general effect on solvent activity.

### 2.2.3 Selection of Solvent

The choice of solvent is principally important in the nucleophilic aromatic substitution polymerization reaction. There are three major requirements for the solvent to be used in the dope solution preparation. Firstly, it is important that the solvent does not undergo any reaction with the reactants resulting to side products. On the other hand, the solvation power of the solvent must be able to ensure sufficient solubility reaction between the reactants and products. Lastly, the solvent should be able to assist the dissociation of the nucleophilic anion from the metal cation associated with it. Nucleophilic substitution reactions are usually conducted in polar aprotic solvents as opposed to protic solvents. This is because the nucleophile is highly dependent on solvent interactions (Mecham, 1997). Acetone, alcohol and water are examples of protic solvents which highly solvate the nucleophile and consequently reduce he nucleophile ability to react with the activated carbon.

Some frequently used polar and dipolar solvents are N-methylpyrrolidone (NMP), dimethylacetamide (DMAc), dimethylformamide (DMF), sulfolane, diphenyl sulfone, dimethylsulfoxide (DMSO), acetone, γ-butyrolactone and ε-caprolactam acid solvents such as acetic acid and formic acid (Idris and Ahmed, 2007b). Some of the solvents are believed to reduce the solvation of the nucleophiles and enhance the nucleophilic properties. This can be done by strongly solvating the cations associated with the nucleophiles (Mecham, 1997). The use of two solvents for polymeric membrane dope solution has also been reported. Baker (1971) prepared the first high flow polysulfone anisotropic membranes from a mixture of two solvents. It was reported that a mixture of polar, aprotic and volatile solvent such as dioxane and acetone causes rapid evaporation on the surface, leading to the formation of a dense layer on the surface. Ahmed et al. (2010) recently used a mixture of acetone/DMF for the dissolution of PES polymer for UF membranes. The use of solvent–non-solvent mixture changes the solubility parameter of the solvent system, which in turn changes the polymer–solvent interaction in the ternary-phase polymer system. Subsequently, these changes have altered the polymer morphology of the surface layer and sub-layer (Yanagishita et al., 1994).

## 2.3 MEMBRANE PREPARATION PROCESS

Generally polymeric membrane production is a complicated process since it involves many steps namely; material selection, drying process, dope solution preparation, casting or hollow fiber spinning, phase inversion process, and posttreatment. These steps are illustrated in Figure 2.3. Among the various steps, the dope dissolution process for membrane production is expensive and time consuming particularly when membranes are prepared from glassy amorphous polymers such as PVDF, PSf, PES, PI, PA, PP and polyetherketone.

### 2.3.1 Polymer Dissolution

The dissolution of a polymer in a pair of nonsolvent was first observed in the 1920s during research related to cellulose nitrate solution systems. It was then found that solvent and nonsolvent for a polymer, when mixed in some specific compositions, might actually function as a solvent for that polymer (Cheng et al., 2000). Generally the dissolution of polymer solids or powder samples are usually carried out in reaction vessels containing the sample volume of polymer solution, typically at laboratory level of 200 to 1000 ml. Traditionally, the mixture is heated for long periods of time using a hot plate, heating mantle, or oven. Normally as the temperature of a casting solution increases so does the average diameter of the pores in the resulting membrane, all other variables being constant (Wrasidlo, 1986). If the temperature of a casting solution is too high or low, the resulting membrane can have undesirable characteristics.

Besides, the frequently available dope solution methods for membrane fabrication, the equipment, man power and energy required to exercise an effective control and minute observation regarding their much needed properties put the most common methods for making microporous membrane beyond reality and financial reach (Wrasidlo, 1986). These traditional heating techniques are slow and time-consuming, and sometimes can lead to overheating and decomposition of the substrate

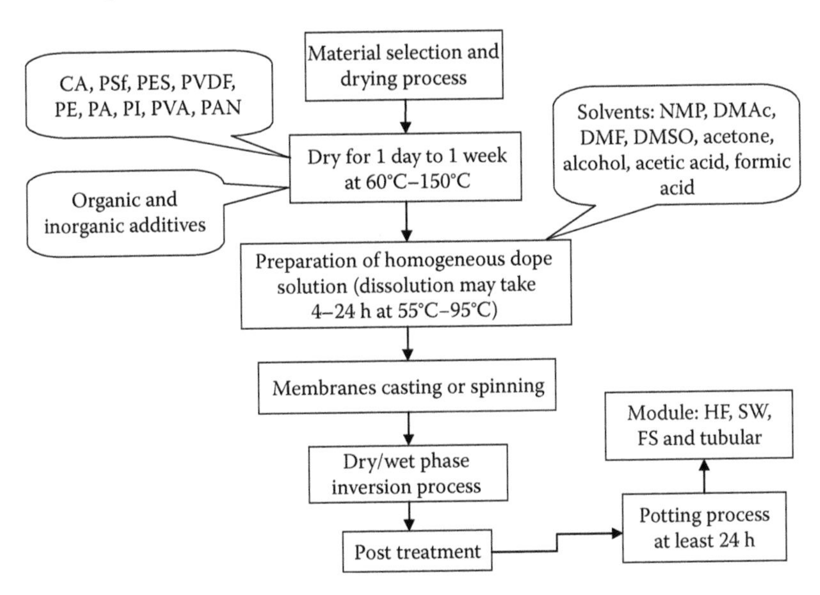


**FIGURE 2.3** Polymeric membrane production steps.

and product. Heating is terminated when the analyst decides that the dissolution of the polymer is sufficiently complete. This type of reaction vessel digestion has many drawbacks, which include the use of large volumes (and multiple additions) of materials, a large potential for contamination of the sample by materials and laboratory environment (Richter, 2003). While the dissolving rate can be increased either by the use of high temperatures or intense agitation, these practices are undesirable. If high shear agitation is employed to enhance the dissolving rate, the shearing forces can rupture or break the polymer chains thus reducing the molecular weight of the polymer in solution (Hadermann et al., 1985).

In cases, when multi solvents and additives are used, the dissolution process becomes even more difficult. The dissolution of amorphous polymers becomes more difficult by physical media at the presence of inorganic salts. These low molecular weight inorganic salt additives in casting solution are considered to change the solvent properties and/or the interaction between the macromolecule chains of polymer. These low molecular weight inorganic salts are particularly interesting as additives for membrane casting solutions because it interacts strongly to form complexes with solvents commonly used for membrane preparation (Bottino et al., 1988; Kraus et al., 1979; Shibata et al., 2000; Tweddle et al., 1983; Wang et al., 2000b). This strong inorganic–solvent interaction would increase the viscosity of the casting solution but reduce the solvency power (Phadke et al., 2005). However, in practice, the addition of inorganic salts to casting solutions was reported to be very effective to prepare membranes with higher performance (Bottino et al., 1988; Kraus et al., 1979; Tweddle et al., 1983). Besides that, the addition of cosolvents induces the change in the solvent quality, which would affect the interaction between polymer and solvent (Wang, 1999). Therefore, the use of multisolvent and additives are motivated by several factors such as to reduce the cost and improve the membranes performance. The use of the microwave technique is recently introduced so as to reduce the amorphous polymer dissolution time and also in certain cases reduce the viscosity of the solution. Consequently, spinning process becomes much easier (Idris and Ahmed, 2008a).

Since the mid-1980s, the applications of microwave in chemical synthesis have been widely investigated. Many inorganic (Komarneni et al., 1992) and organic (Baghurst and Mingos, 1992) reactions could proceed under microwave radiation at a much higher rate than conventional methods. Besides the rapid reaction rate, microwave heating has some other advantages (Mingos, 1994). Molecular sieve membranes consisting of NaA zeolite crystals have been successfully synthesized on symbol $\alpha$-$Al_2O_3$ substrate by means of microwave heating and membranes obtained are stable and dense, and their thickness is well controlled (Han et al., 1999). Bryjak et al. (2002) have produced plasma treatment porous polymer membranes by microwave technique while Boey and Yap (2001) have used microwave technique for curing epoxy-amine system.

Thermoplastic polymers such as PES constitute of long chains with a large number of segments, forming tightly folded coils, which are entangled to each other. Numerous cohesive and attractive both intra and intermolecular forces hold these coils together, such as dispersion, dipole–dipole interaction, induction, and hydrogen bonding. Based on these features, one may expect noticeable differences in the dissolution behavior shown by polymers. Due to their size, coiled shape, and the attractive forces between them, polymer molecules dissolved quite impulsively by microwave irradiation than conventional heating. Billmeyer (1984) pointed out that there are two stages involved in physical media process: (1) the polymer swelling and (2) the dissolution step itself. Moreover, the trend in material development for better solid liquid separation membranes is mainly toward improving the properties of existing polymers, which is attained via chemical and/or physical modification of the polymers to favor the transport properties of the solvents of interest.

## 2.4 PHASE INVERSION

Phase inversion process is considered as one of the most important methods to prepare asymmetric membrane. Besides phase inversion technique, there are also other different techniques to prepare membranes such as sintering, stretching, track-etching, template leaching and also coating.

Most of the polymeric membranes are produced by phase inversion, an exchange process between liquid and solid phase (Scott, 1995). The preparation of membrane structures with controlled pore size involves several techniques, which are tricky but simple principles. Thus, the phase inversion technique is said to produce various types of membrane morphologies.

Phase inversion process can be categorized as dry phase inversion, wet phase inversion and dry/wet phase inversion method.

1. Dry phase inversion method is introduced when a mixture of polymer solution is casted onto a glass piece or a nonwoven fabric and left to vaporize. The solvent and nonsolvent components are both vaporized at the same time during the process.
2. As for the wet phase inversion method the membrane is immediately soaked into a nonsolvent medium after the casting process. During this process, phase separation occurs as the effect of reaction between solvent and nonsolvent components. This process is widely used to produce asymmetric membrane for microfiltration, ultrafiltration, reverse osmosis and gas permeation.
3. In the dry/wet inversion method, after the casting process, the membrane is left to vaporize for a moment before immersion into a nonsolvent medium. The membrane structure is formed by the phase separation between solvent and nonsolvent component and it helps in vaporization process and polymer clotting.

## 2.5 DESCRIPTION OF PES MEMBRANE PREPARATION

### 2.5.1 Dope Preparation

The polymer can be dissolved using two different methods: the conventional electrical heating (CEH) or the recently developed microwave heating (MW) method (Idris and Ahmed, 2008a).

### 2.5.2 Conventional Electrical Heating Method

The polymer dissolution process was carried out in a 1-L, four-necked, round-bottomed vessel with stirrer and condenser placed over the electrical thermal heater. The dope temperature was kept constant at 90°C–95°C by stirring and the dissolution of polymer may take 4 to 7 h depending on the type of formulation used. After the polymer was fully dissolved, the polymer solution was cooled, poured into a storage vessel and degassed to remove any micro bubbles present.

### 2.5.3 Microwave Setup

The microwave technique (MW) which was recently introduced for polymer dissolution process in membrane production involved modifications of the domestic microwave oven set up (Idris et al., 2008b). The microwave oven is modified in order to provide closed heating microwave irradiation system for polymer solution preparation. The same setup is also used for drying and membrane posttreatment purposes.

This retrofitted microwave apparatus gives a simple and inexpensive assembly to prepare polymeric membrane dope solution, which involves the dissolution of polymeric resins and additives in the selected solvent via dielectric heating of microwave irradiation in a very short time.

The dissolution process is performed in the glass vessel setup, which has glass connectors attached to the reflux condenser, thermocouple to control the temperature and a stirrer inside the vessel to ensure homogeneity as described in Figure 2.4. The temperature of the dope solutions was kept at 80°C–85°C, which is digitally recorded using digital thermocouple attached to a computer. After the irradiation heating stops, the solution is left to remain inside the vessel to allow natural heat transfer until it cools to room temperature. Then, the solution is stored in an appropriate bottle

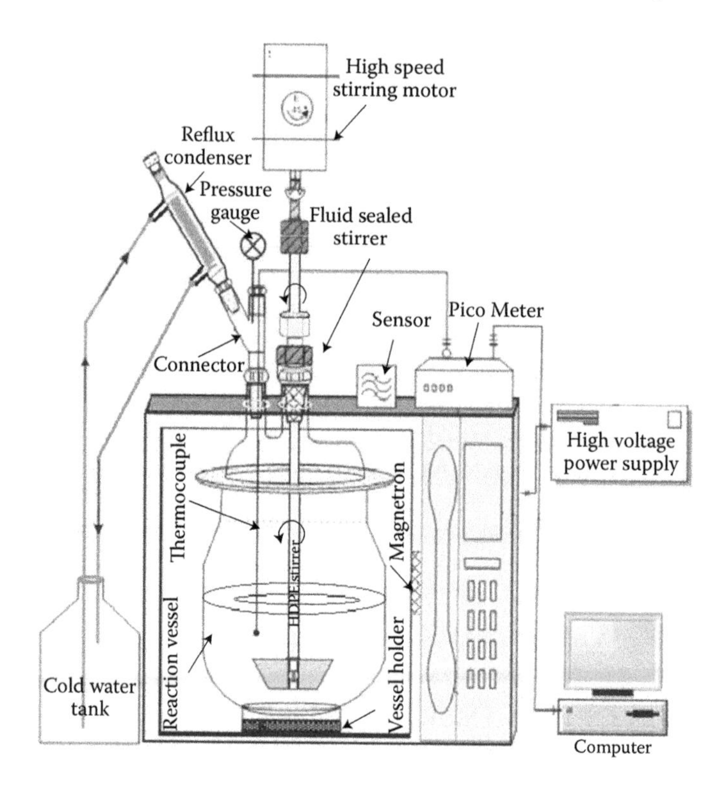


**FIGURE 2.4** Schematic experimental setup for the microwave. (Idris et al., 2008b). Microwave assisted polymer dissolution apparatus for membrane production. PI 20080270.)

for membrane casting. Heating time by microwave was 15 minutes at low to high pulse, and the dissolution time was kept to a maximum at 1 h. The vessel is made of quartz glass with 0.5 cm glass thickness and fluid sealed stirring assemble is made of Teflon and high density polyethylene (HDPE) material which can tolerate high speed agitation (~1200 rpm).

## 2.6 MEMBRANE CASTING/SPINNING

The polymeric solutions can be casted into flat sheet membranes or spun into hollow fibers. The solution is subjected to ultrasonification process for several hours to ensure removal of air bubbles before casting the membrane. This is crucial as it helps to minimize the cavity formation on the surface of membranes during the casting process.

### 2.6.1 Flat Sheet Membranes

Membranes are casted on a glass plate using the casting knife, which is measured and fixed at 200 μm. Immediately after casting, the glass plate with the solution film was immersed into the coagulation bath of distilled water at room temperature. The phase inversion process takes place, and after a few minutes, a thin polymeric film is separated out from the glass. All membranes are then inspected for defects and good areas are chosen for evaluation.

### 2.6.2 Hollow Fiber Membranes

The PES hollow fiber membranes are spun via the dry–wet phase inversion method. The equipment is illustrated in Figure 2.5. The dope solution that is contained in the pressure vessel is

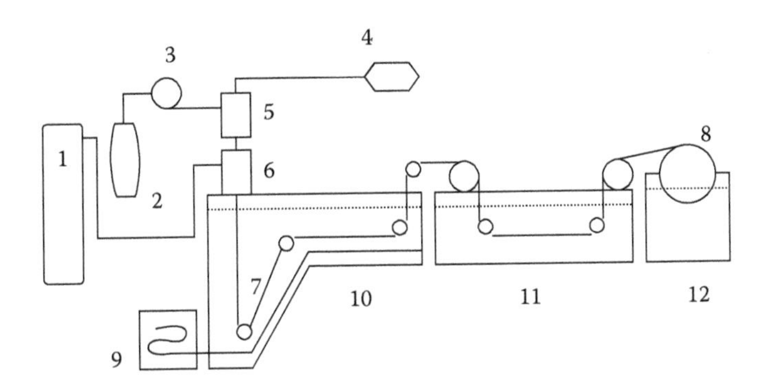


**FIGURE 2.5** Hollow fiber spinning system: (1) nitrogen cylinder; (2) dope vessel; (3) gear pump; (4) syringe pump; (5) spinneret; (6) forced convective tube; (7) roller; (8) wind up drum; (9) refrigeration/heating unit; (10) coagulation bath; (11) washing bath/treatment bath; (12) wind up drum. (Reprinted from *Journal of Membrane Science*, 205, Idris, A., et al., 223–237, Copyright 2002, with permission from Elsevier.)

subsequently pumped at the desired dope extrusion rate to the spinneret by means of a gear pump. The pressure of the dope solution reservoir is kept at 1 atm. The internal coagulant is pumped to the spinneret using a high pressure precision metering pump. The nascent polymeric fiber that emerges from the spinneret is partially solidified by the distilled water used as the internal coagulation fluid. The spinneret is positioned above the coagulation bath such that the outer surface of the fiber is exposed to air. This is to allow partial evaporation of solvent before being immersed in the coagulation bath. Coagulation is allowed to occur on the outer surface of the membrane as a result of solvent–nonsolvent exchange. Through this, asymmetric hollow fibers are obtained.

## 2.7 POSTTREATMENT OF ASYMMETRIC MEMBRANE

The posttreatment step can be performed by immersing the membranes in de-ionized water placed a vessel and heating it in a microwave oven to improve the performance of the membranes. Then, the glass container is placed inside a microwave oven for approximately 10 minutes at medium high pulse with the temperature recorded using Pico data logger. Other posttreatment methods such as chemical posttreatments can also be used.

## 2.8 PERFORMANCE EVALUATION OF MEMBRANES

Ultrafiltration PES membrane performance are characterized in terms of pure water permeation (PWP), permeation rate (PR) and solute separation of polyethylene glycols (PEG) of various molecular weight cut off (MWCO) ranging from 600 to 36,000 Da. The concentration of PEG solution used is kept in the range of 500–1000 ppm. Besides PEG, dextran solutions can also be used. The concentration of PEG can be determined by several methods such as HPLC or analytical method described by Sabde et al. (1997) and Idris et al. (2007). Pure water permeation fluxes (PWP) and solutes water permeation fluxes (PR) of membranes are calculated using the following equation:

$$J = \frac{Q}{\Delta t \times A}$$

where $J$ is the permeation flux of membrane for PEG solution ($L \cdot m^{-2} \cdot h^{-1}$) or pure water and $Q$ is the volumetric flow rate of permeate solution. $\Delta t$ is the permeation time (h) and $A$ is the membrane surface area ($m^2$). The membrane rejection (SR) is defined as

$$SR\ (\%) = 1 - \frac{C_p}{C_f} \times 100$$

where $C_f$ and $C_p$ are the polyethylene glycol concentrations in the feed solution and permeate solution, respectively. The concentration of PEG was determined based on absorbency in a UV-spectrophotometer at a wavelength of 535 nm.

## 2.9 INFLUENCE OF ADDITIVES ON PERFORMANCE OF PES MEMBRANES

Additive has an influence on the performance of membranes and performance is described in terms of pure water permeation (PWP), permeation (PR) and rejection rate of solutes. Studies on the effect of using different molecular weight PVP as additives such as PVP K10, PVP K30, PVP K90, and PVP K360 on performance of PES membranes were performed by several researchers (Xu et al., 1999; Kim and Lee, 1998; Kraus et al., 1979; Munari et al., 1980; Kim et al., 1996; Lafreniere et al., 1987) and it was found that addition of small quantities of PVP of different molecular weights to a phase inversion UF membrane resulted in an increased in permeability without significant changes in selectivity (Kraus et al., 1979). PVP proved to act as a pore forming agent, which is important in attaining appreciable porosity and pore size of the membrane (Hoehn, 1985; Kesting, 1965).

Polyethylene glycol 400 (PEG 400) has also been used as an additive to enhance PES hollow fiber membranes and the results showed that PEG 400 can be used very well as polymeric additive to increase the polymer dope viscosity and enhance the pore interconnectivity when added in appropriate amounts (Cabasso et al., 1976). It was reported that PEG 400 also acts as macrovoids suppressor, producing nice spongy structure and give the membrane its hydrophilic character.

In another study by Idris et al. (2007), the effect of using PEG of different molecular weights, namely, PEG 200, PEG 400, and PEG 600, as additives on PES UF membranes was investigated. Results revealed that the addition of different PEG as additives in casting solution influenced the viscosity and the performance of pure water permeation rate (see Table 2.4 and Figure 2.6). It was found that UF membrane with PEG 200 as additive exhibits the lowest value of MWCO and mean pore size compared to other membranes. The value of molecular weight cut off for PES UF membranes with PEG 200, PEG 400, and PEG 600 are 26,000, 36,000, and 45,000 Da, respectively. The

**TABLE 2.4**
**PES Ultrafiltration Formulations and Their Viscosities**

| | Composition of Casting Solution (wt%) | | | |
|---|---|---|---|---|
| Type of PEG Additive | Polymer | Solvent | PEG | Viscosity (cp) |
| Without additive | 20 | 80 | 0 | 521 |
| PEG 200 | 20 | 65 | 15 | 1190 |
| PEG 400 | 20 | 65 | 15 | 1270 |
| PEG 600 | 20 | 65 | 15 | 1590 |

*Source:* *Desalination*, 207, Idris, A., et al., 324–339, Copyright 2007c, with permission from Elsevier.

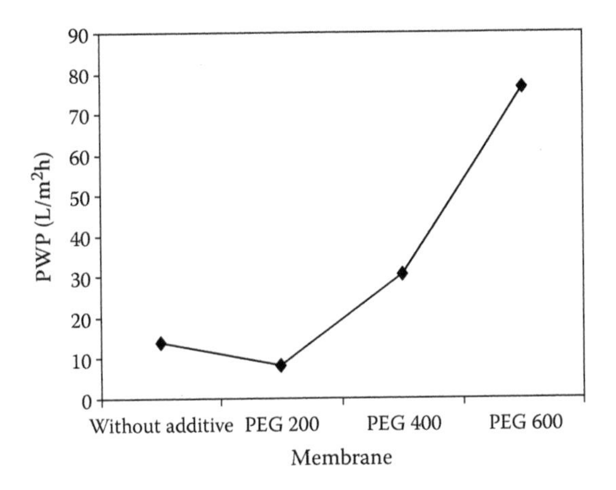


**FIGURE 2.6** Pure water permeation for PES membranes without and with different additives. (Reprinted from *Desalination*, 207, Idris, A., et al., 324–339, Copyright 2007c, with permission from Elsevier.)

findings revealed that as concentration of PEG 400 and PEG 600 in the casting solution is increased solute separation decreased while flux increased (see Figures 2.7 and 2.8). However, PES UF flat sheet membranes with PEG 200 as additive displayed a different behavior, which exhibits the highest solute separation but lowest flux rate when concentration is increased from 5 to 25 wt% in casting solution. In addition to performance, the morphology of membranes also changes with addition of different molecular weight additives in the casting solution.

In a recent study, Idris et al. (2010) studied the influence of low molecular lithium halides (Li-halides) on the performance of membrane and found that these additives have the ability to improve the PES membrane performance. Figures 2.9 and 2.10 show the influence of three different Li-halides additives, i.e., LiBr, LiCl, and LiF, on the PWP and PR of PES membranes. The presence of these Li-halides additives improves both the PWP and PR of the membranes. Increase

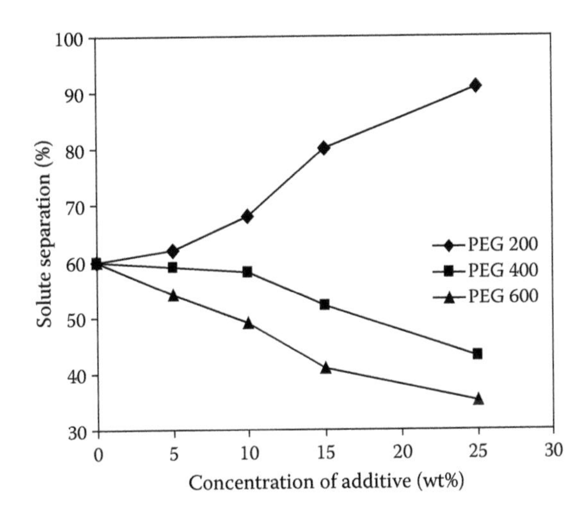


**FIGURE 2.7** The changes of solute separation with the concentration of additives using PEG 10 kDa solution. (Reprinted from *Desalination*, 207, Idris, A., et al., 324–339, Copyright 2007c, with permission from Elsevier.)

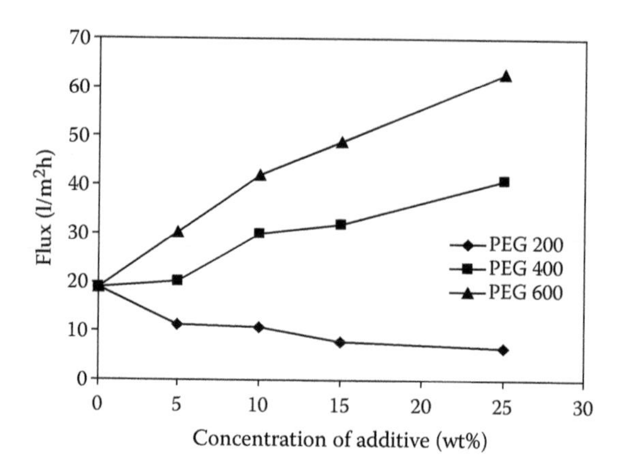


**FIGURE 2.8** Flux rates versus concentration of various additives using PEG 10 kDa solution. (Reprinted from *Desalination*, 207, Idris, A., et al., 324–339, Copyright 2007c, with permission from Elsevier.)

in concentration of the LiBr increases the PWP but only up to a certain concentration, which, in this case, was up to 2 wt% LiBr. Increasing concentration of LiBr beyond this value resulted in a decrease in the PWP additive as depicted in Figure 2.9. Membranes containing LiF and LiCl additives have almost similar PWP rates. As the amount of LiCl and LiF is increased, the PWP also increased.

As illustrated in Figure 2.10, among the three Li-Halides, the membrane with the 2 wt% LiBr exhibits the highest PR. LiBr seemed to have improved the hydrophilic properties of the PES membranes and this is proven by the contact angle measurement results shown in Figure 2.11. LiBr membrane has the lowest contact angle value thus indicating the membrane surface is the most hydrophilic.

The PEG separation graphs for the PES/DMF/LiF, PES/DMF/LiCl, PES/DMF/LiBr membranes are depicted Figure 2.12. The presence of additive has not only improved the permeation rates but also the separation rates. Among these lithium halides, LiBr seems to exhibit very good separation with sharp molecular weight cut off (MWCO) at 90% separation. Separation rates obtained using LiCl additives are better compared to LiF. The use of LiCl has also been reported by other researchers (Kim et al., 1996).

Further investigation on the value of the mean pore size ($\mu_p$), standard deviation ($\sigma_p$) and the MWCO of the PES membranes revealed that PES/DMF/LiBr membrane has the minimum MWCO of 2.82 kDa and smallest pore size of 0.2466 nm with standard deviation of 0.3012 (Idris et al.,

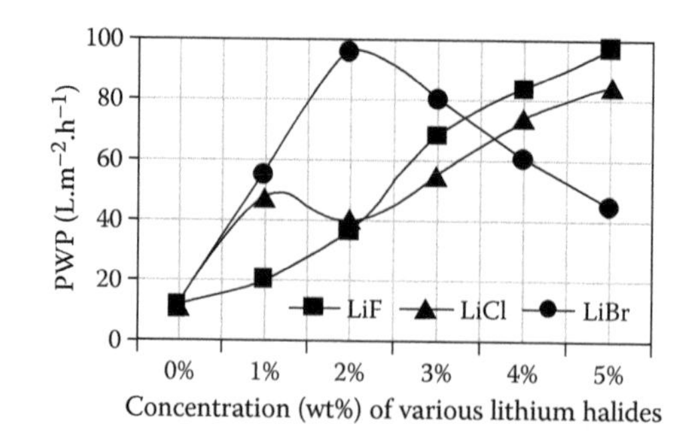


**FIGURE 2.9** Pure water permeation rates versus concentration of lithium halides.

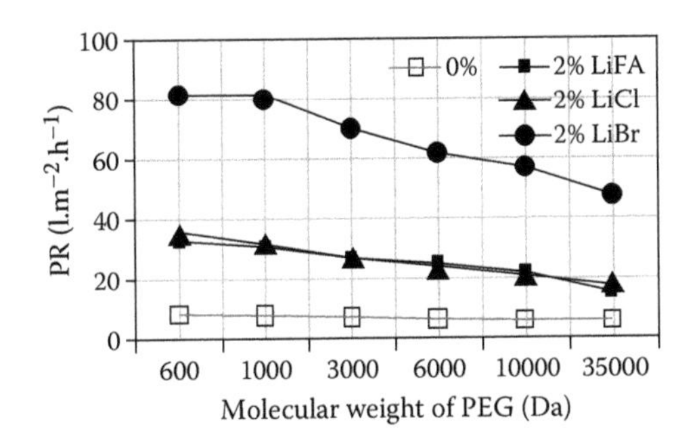


**FIGURE 2.10** Permeation rates of the various PEG solutions for the lithium halides.

2010). The mean pore size is defined as the pore diameter when solute separation is 50% (Idris et al., 2007). This explains for the highest separation rate obtained. Results also revealed that the membrane with LiBr additive at a concentration of 2 wt% not only exhibits good permeation but also excellent separation with high rejection rates. This is followed by LiCl and LiF membranes with MWCO of 9.161 and 10.992 kDa, respectively.

Among the lithium halides—LiBr, LiCl, and LiF—LiBr has the highest molecular weight at 86.845 g/mol, followed by LiCl with 42.39 g/mol, and LiF with 25.94 g/mol. Unlike LiF and LiCl, which leach out of the membrane during the precipitation of the polymer solution and act as a pore former (Huang and Feng, 1995), LiBr seems to behave differently. Previous study has revealed that among the three dope solutions containing LiBr, LiCl and LiF, the dope solution containing LiBr has a moderate viscosity between LiF and LiCl and this is clearly observed in Figure 2.13 (Idris and Ahmed, 2008a). It is believed that the difference in viscosity and the molecular weights can influence the nonsolvent–solvent exchange and the velocity of phase separation.

In addition, the volume concentration of the polymer in the solution at the precipitation point is determined by the rate of flow of solvent out of the cast polymer solution. In this case the moderate viscosity LiBr dope solution and the salt's high molecular weight results in the moderate diffusion of the solvent during the coagulation process and this promotes delayed demixing, thus suppressing the formation of macrovoids, thereby resulting in a change in the porosity of the membrane and at the same time a thin skin layer is formed. Unlike LiCl and LiF, the presence of LiBr affects mainly the structure of the skin layer rather than the structure of the entire membrane, it is also

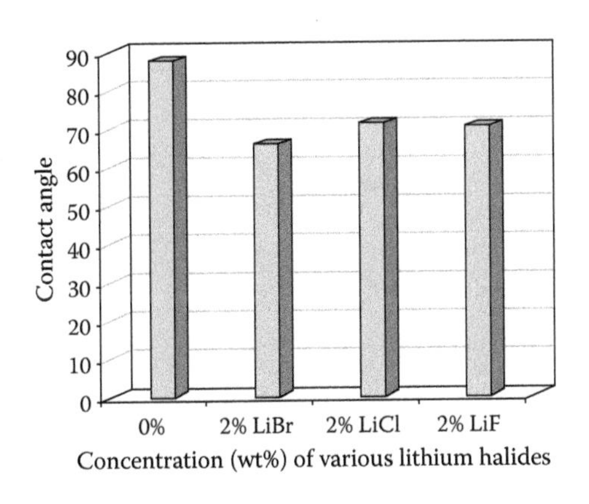


**FIGURE 2.11** Contact angle for the PES/DMF.

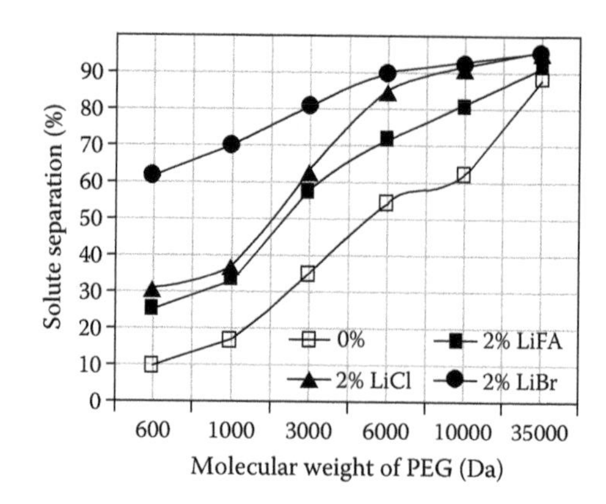


**FIGURE 2.12** Solute separations for the PES/DMF.

proven by the contact angle measurement depicted in Figure 2.11. The low contact angle measurement indicates that LiBr additive has improved the hydrophilic properties of the membranes.

The pore size on the surface of the membrane becomes smaller because the LiBr gives rise to an association between the moieties units of PES, and its nucleophilicity is greater in dipolar aprotic solvents like dimethyl sulfoxide and dimethylformamide than in protic solvents like water or alcohols. For this reason, DMF is often participating as a solvent for carrying out nucleophilic substitutions of $Br^{-1}$ ions thereby decreasing the mobility of the polymer chain. Therefore, the LiBr additive acted as a pore inhibitor rather than a pore former, leading to the simultaneous decrease of pore size and increase in hydrophilicity as depicted in Figure 2.11. However, when the LiBr concentration in the casting solution increased to more than 2 wt%, the permeation rates declined as shown in Figure 2.10. This is probably attributed to the increase in packing density in the polymer matrix at higher concentrations of LiBr. The experimental results indicated that the increasing membrane pore density was the major factor, and decreasing membrane pore radius is the secondary factor. During the phase inversion process, the low molecular weight LiF probably diffused rapidly, thus promoting the macrovoid formation and also acting as a pore former. A similar phenomenon seems to occur for the LiCl additive.

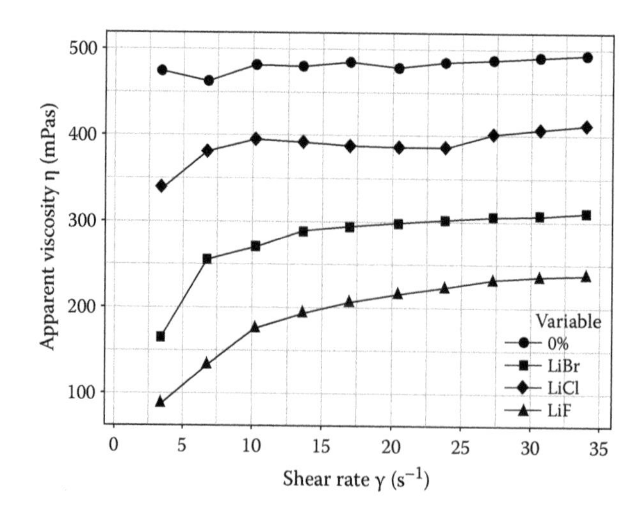


**FIGURE 2.13** Apparent viscosities of the various membranes.

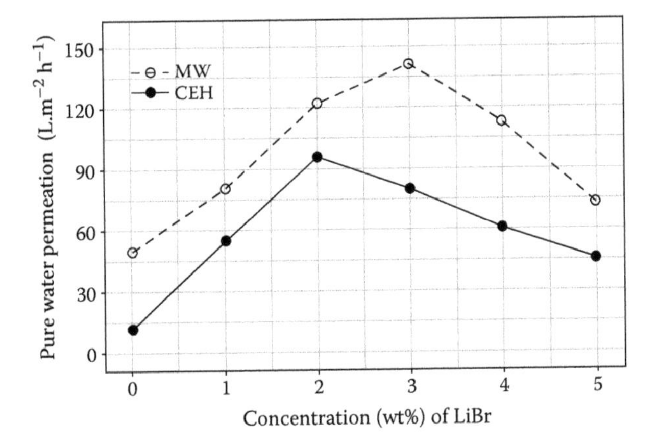


**FIGURE 2.14** Pure water permeation rates versus concentration of LiBr of for PES/DMF/LiBr membranes prepared using MW and CEH.

## 2.10 INFLUENCE OF MICROWAVE TECHNIQUE ON PERFORMANCE OF PES MEMBRANES

In a very recent study (Ahmed, 2009), it was revealed that the microwave (MW)-prepared PES membranes containing various concentrations of LiBr have higher PWP and PR compared to the CEH membranes as illustrated in Figures 2.14 and 2.15. These results apparently seem to indicate that the MW prepared membranes are more hydrophilic. The differences could be due to different solubility parameters as well as salt solvent interaction. Highest PR was obtained when LiBr concentration was at 3 wt% and prepared using MW.

The permeation rates for membranes with LiBr are approximately 83.9% higher than those without LiBr. Permeation rates of 5.26- to 3.8-fold increments are achieved when 3 wt% and 4 wt% LiBr are used for MW and CEH membranes respectively and this means increase in productivity. With the participation of LiBr, which has high swelling properties, the PES becomes more hydrophilized

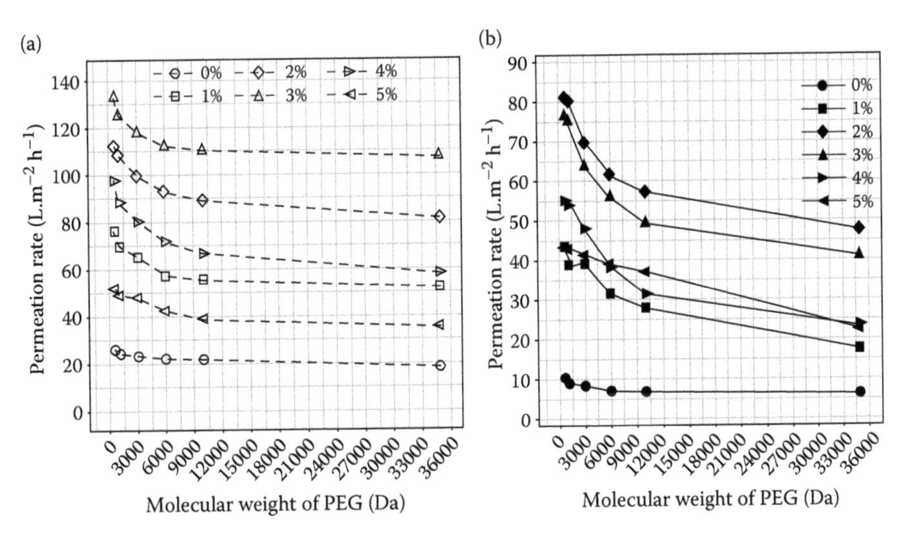


**FIGURE 2.15** Permeation rates of the membranes with various concentrations of LiBr, (a) membranes prepared by MW and (b) membranes prepared by CEH.

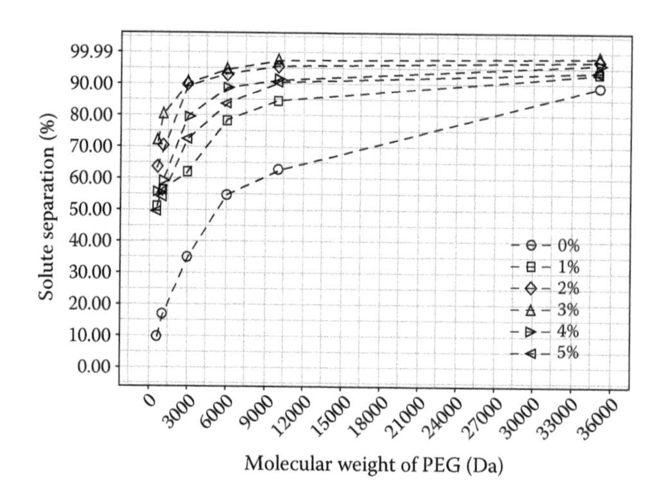


**FIGURE 2.16** Molecular weight cutoff profile of MW assisted PES/DMF membranes with various concentrations of LiBr.

and this hydrophilicity become more pronounced at 3 wt% LiBr in both the MW and CEH prepared membranes. There is the possibility that at this concentration, the balance of the hydrophilic and hydrophobic moieties has prevailed.

The results also showed that MW assisted membranes exhibited higher rejection and permeation performance than CEH assisted membranes. Figure 2.16 shows the rejection performance of MW assisted PES membranes composition with various concentration of LiBr. Similar PEG rejection trends are also observed in Figure 2.17 for CEH assisted membranes respectively. However, a comparison of Figures 2.16 and 2.17 showed that MW assisted membranes have higher PEG rejection rates compared to all CEH assisted membranes. This higher PEG rejection with increasing LiBr content in PES might be due to the good homogeneity or best solubility arising as a result of the LiBr content creating hydrophilicity in the PES membranes due to the nucleophilic substitutions, which is carried by electrophilic dipolar solvents such as DMF.

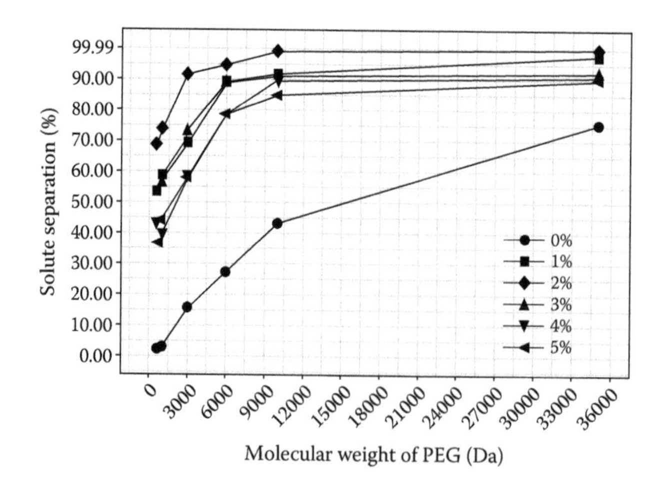


**FIGURE 2.17** Molecular weight cutoff profile of CEH prepared PES/DMF membranes with various concentrations of LiBr. "•" represents PES/DMF membranes without LiBr.

## 2.11 APPLICATION

PES membranes can be used for a wide variety of applications, including dairy, chemical, paper, semiconductor, food, pharmaceutical industries, protein purification, gelation separation, surface water treatment, and waste water treatment. Some of the applications are listed in Table 2.1. As for medical applications, PES membranes are used for hemodialysis treatment and also for bacteria and particulate removal. Due to the superior properties, especially its chemical, mechanical, and thermal resistance, PES based membranes are now increasingly used in industrial demand, especially in the water treatment technology.

The PES/DMF/LiBr membranes fabricated have also been successfully evaluated for palm oil mill effluent (POME) treatment. The results revealed that the membranes are capable of reducing the turbidity, chemical oxygen demand (COD), and biochemical chemical demand (BOD) of the POME feed by of 96.7%, 53.12%, and 76.7% respectively (Idris et al., 2010).

## REFERENCES

Ahmed, I. 2009. High performance ultrafiltration polyethersulfone membrane using microwave assisted technique; PhD Thesis, Universiti Teknologi Malaysia.

Ahmed, I., A. Idris, and N.F. Che Pa. 2010. Novel method of synthesizing polyethersulfone membrane containing two solvents and lithium chloride additive and its performance. *J Appl Polym Sci* 115: 1428–1437.

Allegrezza, A.E., and E.T. Burke. 1989. Composite ultrafiltration membranes. U.S. Patent 4,824,568.

Baghurst, D.R., and D.M.P. Mingos. 1992. Superheating effects associated with microwave dielectric heating. *J Chem Soc, Chem Comm* 9: 674–677.

Baker, R.W. 1971. Process for making high flow anisotropic membranes. U.S. Patent 3,567,810.

Baker, R.W. 2004. *Membrane Technology and Applications*. John Wiley & Sons Ltd., England.

Bellantoni, E.C., and R.S. Loya. 1997. Composite ultrafiltration membrane. U.S. Patent 5,698,281.

Billmeyer, F.W. 1984. In: *Textbook of Polymer Science*, 3rd ed. pp. 212–214. New York: John Wiley & Sons.

Boey, F.Y.C., and B.H. Yap. 2001. Microwave curing of an epoxy-amine system: Effect of curing agent on the glass-transition temperature. *Polym Test* 20(4–5): 837–845.

Bottino, A., G. Capanelli, S. Munari, et al. 1988. High performance ultrafiltration membranes cast from LiCl doped solutions. *Desalination* 68(2–3): 167–177.

Bryjak, M., I. Gancarz, G. Poźniak, et al. 2002. Modification of polysulfone membranes 4. Ammonia plasma treatment. *Eur Polym J.* 38(4): 717–726.

Cabasso, I., E. Klein, and J.K. Smith. 1976. Polysulfone hollow fibers. I: Spinning and properties. *J Appl Polym Sci* 20(1976): 2377–2394.

Cheng, L.P., D.J. Lin, and K.C. Yang. 2000. Formation of mica-intercalated-nylon 6 nanocomposite membranes by phase inversion method. *J Membr Sci* 172(1–2): 157–166.

Ellen, C.B., and R.S. Loya. 1997. Composite ultrafiltration membrane. U.S. Patent 5,698,281.

Guiver, M.D., A.Y. Tremblay, and C.M. Tam. 1990. Method of manufacturing a reverse osmosis membrane and the membrane so produced. U.S. Patent 4,894,159.

Hadermann, A. F., M.D. Ijansville, C.T. Jery, et al. 1985. Dissolving water soluble polymer. U.S. Patent 4,501,828.

Han, Y., H. Ma, S. Qiu, et al. 1999. Preparation of zeolite A membranes by microwave heating. *Micropor Mesopor Mater* 30(2–3): 321–326.

Hoehn, H.H. 1985. Aromatic polyamide membranes. *ACS Symposium Series 269*. pp. 81–97. Washington, DC: American Chemical Society.

Huang, R.Y.M., and X. Feng. 1995. Studies on solvent evaporation and polymer precipitation pertinent to the formation of asymmetric polyetherimide membranes. *J Appl Polym Sci* 57(1–2): 613–621.

Idris, A., and I. Ahmed. 2007a. Performance of cellulose acetate – polyethersulfone blend membrane prepared using microwave heating for palm oil mill effluent treatment. *Wat Sci Technol* 56(8): 169–177.

Idris, A., and I. Ahmed. 2007b. A production of polyethersulfone asymmetric membranes using mixture of two solvents and lithium chloride as additive. *Jurnal Teknologi* 47(F) Dis.: 25–34.

Idris, A., and I. Ahmed. 2008a. Viscosity behavior of microwave-heated and conventionally heated poly(ether sulfone)/dimethylformamide/lithium bromide polymer solutions. *J Polym Sci* 108(1): 302–307.

Idris, A., I. Ahmed, and M.A. Limin. 2010. Influence of lithium chloride, lithium bromide and lithium fluoride additives on performance of polyethersulfone membranes and its application in the treatment of palm oil mill effluent. *Desalination* 250(2): 805–809.

Idris, A., I. Ahmed, and M.Y. Noordin. 2008b. Microwave assisted polymer dissolution apparatus for membrane production. *PI 20080270.*

Idris, A., M.Y. Noordin, A. Ismail, et al. 2002. Optimization of cellulose acetate hollow fiber reverse osmosis membrane production using Taguchi method. *J Membr Sci* 205: 223–237.

Idris, A., N.M. Zain, and M.Y. Noordin. 2007c. Synthesis, characterization and performance of asymmetric polyethersulfone (PES) ultrafiltration membranes with polyethylene glycol of different molecular weights as additives. *Desalination* 207: 324–339.

Ikeda, K., S. Yamamoto, and H. Ito. 1989. Sulfonated polysulfone composite semipermeable membranes and process for producing the same. U.S. Patent 4,818,387.

Katarzyna, M.N. 1989. Synthesis and properties of polysulfone membranes. *Desalination* 71(2): 83–95.

Kesting, R.E. 1965. Semipermeable membranes of cellulose acetate for desalination in the process of reverse osmosis. Part 1. Lytropic swelling of secondary cellulose acetate. *J Appl Polym Sci* 9(2–3): 663–673.

Kesting, R.E. 1990. The four tires of structure in integrally skinned phase inversion membranes and their relevance to the various separation regimes. *J Appl Polym Sci* 41(9): 2739–2752.

Kesting, R.E. 1985. *Synthetic Polymeric Membranes.* A structural perspective, 2nd ed, pp. 237–261. John Wiley & Sons: New York, Mcgraw Hill.

Kim, J., and K. Lee. 1998. Effect of PEG additive on membrane formation by phase inversion. *J Membr Sci.* 138(1–2): 153–165.

Kim, S.R., K.L. Lee, and M.S. Lee. 1996. The effect of $ZnCl_2$ on the formation of polysulfone membrane. *J Membr Sci* 119(1): 59–64.

Komarneni, S., R. Roy, and Q.H. Li. 1992. Microwave-hydrothermal synthesis of ceramic powders. *Mater Res Bull* 27: 1393–1401.

Kraus, M.A., M. Nemas, and M.A. Frommer. 1979. The effect of low molecular weight additives on the properties of aromatic polyamide membranes. *J Appl Polym Sci* 23(1): 445–457.

Kutowy, O., and C. Sterlez. 2003. Intrinsically bacteriostatic membranes and systems for water purification. U.S. Patent 6,652,751.

Lafreniere, L.Y., D.F. Talbot, T. Matsuura, et al. 1987. Effect of polyvinylpyrrolidone additive on the performance of polyethersulfone ultrafiltration membranes. *Ind Eng Chem Res* 26: 2385–2389.

Lai, J.Y., S.J. Huang, and S.H. Chen. 1992. Poly(methyl methacrylate/(DMF-metal salt) complex membrane for gas separation. *J Membr Sci* 74(1): 71–86.

Lee, H.J., J. Won, H. Lee, et al. 2002. Solution properties of poly(amic acid)–NMP containing LiCl and their effects on membrane morphologies. *J Membr Sci* 196(2–3): 267–277.

Liu, Y., G.H. Koops, and H. Strathmann. 2002. Characterization of morphology controlled polyethersulfone hollow fiber membranes by the addition of polyethylene glycol to dope and bore liquid solution. *J Membr Sci* 223(1–2): 187–199.

Lloyd, D.R. 1985. Materials science of synthetic membranes. *ACS Symposium Series 269*. pp. 1–18. Washington, DC: American Chemical Society.

Matsuura, T. 1994. *Synthetic Membranes and Membrane Separation Process.* Chapter 5. CRC Press.

Mecham, S. 1997. Synthesis and characterization of phenylethynyl terminated poly(arylene ether sulfone)s as thermosetting structural adhesives and composite matrices. Dissertation, Doctor of Philosophy in Chemistry. Faculty of the Virginia Polytechnic Institute and State University.

Michaels, A.S. 1971. High flow membrane. U.S. Patent 3,616,024.

Mingos, D.M.P. 1994. The application of microwaves in chemistry. *Res Chem Intermed* 20: 85–91.

Möckel, D., E. Staude, and M.D. Guiver. 1999. Static protein adsorption, ultrafiltration behavior and cleanability of hydrophilized polysulfone membranes. *J Membr Sci* 158(1–2): 63–75.

Mulder, M. 1996. *Basic Principles of Membrane Technology*, 2nd ed., Chapter 1. Dordrecht, The Netherlands: Kluwer Academic.

Munari, S., F. Vigo, G. Capannelli, et al. 1980. Method for the preparation of asymmetric membranes. U.S. Patent 4,188,354.

Nunes, S.P., and K.V. Peinemann. 2006. *Membrane Technology in the Chemical Industry*, 2nd ed. Chapters 1–3. Wiley-VCH Verlag GmbH, Weinheim FRG.

Pertov, S.P. (1996). Conditions for obtaining ultrafiltration membranes from a solution of polyacrylonitrile in dimethylformamide in the presence of formamide. *J Appl Polym Sci* 62(1): 267–277.

Phadke, M.A., D.A. Musale, S.S. Kulkarni, et al. 2005. Poly(acrylonitrile) ultrafiltration membranes. I. Polymer–salt–solvent interactions. *J Polym Sci B, Polym Phys* 43: 2061–2073.

Richter, R. 2003. *Clean Chemistry: Techniques for the Modern Laboratory.* Chapter 6. Monroe, CT: Milestone Press.

Sabde, A.D., M.K. Trivedi, V. Ramachandran, et al. 1997. Casting and characterization of cellulose acetate butyrate based UF membranes. *Desalination* 114(3): 223–232.
Scott, K. 1995. *Handbook of Industrial Membranes*, 1st ed. Oxford: Elsevier Advanced Technology.
Shibata, M., T. Kobayashi, and N. Fujii. 2000. Porous nylon-6 membranes with dimethylamino groups for low pressure desalination. *J Appl Polym Sci* 75(8): 1546–1553.
Shinde, M.H., S.S. Kulkarni1, D.A. Musale, et al. 1999. Improvement of the water purification capability of poly(acrylonitrile) ultrafiltration membranes. *J Membr Sci* 162(1): 9–22.
Tweddle, T.A., O. Kutowy, W.L. Thayer, et al. 1983. Polysulfone ultrafiltration membranes. *Ind Eng Chem Prod Res Dev* 22: 320–326.
Wang, I.F. 1999. Highly asymmetric polyethersulfone filtration membranes. U.S. Patent 5,869,174.
Wang, D., K. Li, and W.K. Teo. 2000a. Highly permeable polyethersulfone hollow fiber gas separation membranes prepared using water as non-solvent additive. *J Membr Sci* 176(2): 147–158.
Wang, D., K. Li, and W.K. Teo. 2000b. Porous PVDF asymmetric hollow fiber membranes prepared with the use of small molecular additives. *J Membr Sci* 178(1): 13–23.
Wang, G.J, L.Y. Chu, M.Y. Zhou, et al. 2006. Effects of preparation conditions on the microstructure of porous microcapsule membranes with straight open pores. *J Membr Sci* 284(1–2): 301–312.
Watson, E.R., H.W. Heidsman, and B. Keilin. 1965. Stabilization of desalination membranes. U.S. Patent 3,250,701.
Wechs, F. 1990. Integral asymmetric polyether-sulfone membrane, process for its production, and use for ultrafiltration and microfiltration. U.S. Patent 4,976,859.
Wrasidlo, W.J. 1986. Asymmetric membranes. U.S. Patent 4,629,563.
Xu, L., T.S. Chung, and Y. Huang. 1999. Effect of polyvinylpyrrolidone molecular weights on morphology, oil/water separation, mechanical and thermal properties of polyetherimide/polyvinylpyrrolidone hollow fiber membranes. *J Appl Polym Sci* 74(9): 2220–2227.
Yanagishita, H., T. Nakane, and H. Yoshitome. 1994. Selection criteria for solvent and gelation medium in the phase inversion process. *J Membr Sci* 89(3): 215–230.
Yin, H.H., H. Huan, T. Shibiao, et al. 2007. Comparison of the free volume of LiCl-added SPPO membrane and SPPO-PES blend membrane by positron annihilation method. *Plasma Sci Technol* 9(5): 575–577.

# 3 Preparations and Applications of Inorganic–Organic Charged Hybrid Membranes

## *A Recent Development and Perspective*

*Junsheng Liu, Tongwen Xu, and Guoquan Shao*

## CONTENTS



## 3.1 INTRODUCTION

Membranes have gained an important place in chemical technology and are expanding their applications to many aspects in industrial processes and environmental area, such as chemical separation, wastewater treatment, microfiltration (MF), ultrafiltration (UF), nanofiltration (NF), reverse osmosis (RO), pervaporation (PV), electrodialysis (ED), and gas separation (Baker, 2004). Among these membranes, charged membranes (usually named ion exchange membranes (IEMs)) fix ionic groups on the molecular backbone and can be used for separation of ionic species from aqueous solutions. Therefore, they have attracted great attention since they were invented in 1950 by Juda and McRae (1950). Presently, they have been widely used in electrodialysis for desalination of brackish water, production of table salt, recovery of valuable metals from the effluents of metal plating industry, and many other purposes (Xu, 2005).

Traditionally, charged membranes can be sorted into various categories (Ohya et al., 1995; Wu et al., 2006a, 2006b; Xu and Yang, 2003; Sforca et al., 1999; Lu et al., 2003; Cornelius and Marand, 2002; Kogure et al., 1997; Zhang et al., 2006; Kusakabe et al., 1998). Nevertheless, if the main ingredients incorporated into the membrane matrix are considered, charged membranes can be further divided into two major categories: organic charged membranes and inorganic/organic charged hybrid membranes. Organic charged membranes are referred to those mainly composed of pure organic polymers. Whereas inorganic/organic charged hybrid membranes, just as its name implies, are referred to those composed of both organic and inorganic moieties in the hybrid matrix. The hybrid effect will supply these charged membranes with combined properties as compared with the pure organic analogue (Xu, 2005; Ohya et al., 1995; Wu et al., 2006a, 2006b; Xu and Yang, 2003). Similar to organic charged membranes, charged hybrid membranes can be further classified as positively charged hybrid membranes (i.e., hybrid anion exchange membranes) (Xu and Yang, 2003; Wu et al., 2003a, 2003b), negatively charged hybrid membranes (i.e., hybrid cation exchange membranes) (Wu et al., 2003a, 2003b; Shahi, 2007), and zwitterionic hybrid membranes, which contain both anion and cation exchange groups and arrange a pendant-side structure on the molecular chains.

Different from the pure organic charged membranes, which possess lower mechanical strength and thermal stability, it is now well accepted that charged hybrid membranes have not only high mechanical and thermal stabilities but also excellent flexibility and film-forming properties. Meanwhile, these charged hybrid membranes also exhibit special electronic and mechanical properties, electro-optic effect, structural flexibility, and high thermal stability (Wu et al., 2006a, 2006b; Xu and Yang, 2003; Sforca et al., 1999). Consequently, they are expected to be applied in some harsh conditions, such as higher temperature and strongly oxidizing circumstances for industrial applications (Lu, 2003; Cornelius and Marand, 2002). Many researchers (Wu et al., 2003a, 2003b; Shahi, 2007; Matsumoto et al., 2002; Yamauchi et al., 1987; Neihof and Sollner, 1950; Miyaki et al. 1984; Linder and Kedem, 2001; Liu et al. 2006a, 2006b, 2006c; Innocenzi et al., 2002; Liu et al., 2005a, 2005b, 2005c; Nagarale et al., 2006) have focused on this field for many years. Figure 3.1 presents the chronology of publications on charged hybrid membranes and confirms an increase in the research of charged hybrid membranes over the last 15 years.

To date, numerous strategies have been proposed to prepare inorganic–organic charged hybrid membranes, such as liquid-phase coupling (Ohya et al., 1995), free radical polymerization, blending, sol-gel, and epoxide ring opening. In general, when both inorganic and organic species have unsaturated bonds, free radical polymerization probably occurs in the presence of the initiator AIBN or BPO; and the inorganic ingredient can thus be grafted onto the molecular chains of organic polymer. The hybrid copolymer prepared through free radical polymerization can be used as the hybrid precursor for charged hybrid membranes. The membranes constructed from such hybrid copolymer

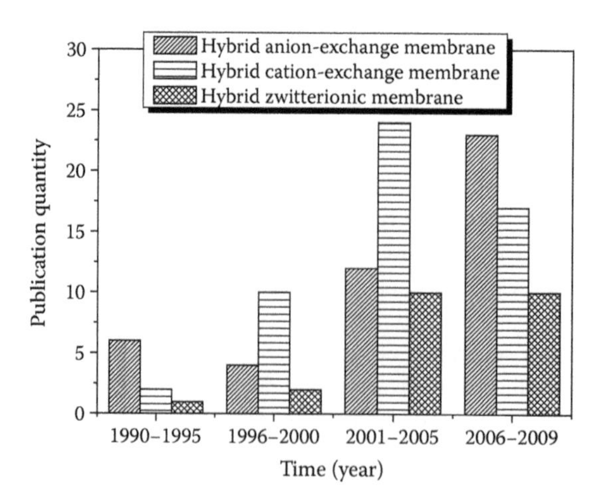


**FIGURE 3.1** Chronology of hybrid ion-exchange membrane publications. Source: Web of Science (http://pcs.isiknowledge.com) [Search settings: TITLE-ABS-KEY (hybrid anion-exchange membrane/hybrid cation-exchange membrane/hybrid zwitterionic membrane). Search date: Mar. 25, 2009].

will obtain high flexibility and thermal stability as compared with the pure organic copolymer. The advantage of this method is that the charged groups can be directly or indirectly introduced into the polymeric backbones, which makes a simple membrane preparation. On the other hand, the major problem with this technique is the selection of organic and inorganic monomers because they must be suitable for preparation of charged hybrid membranes and meanwhile the functional groups can be directly grafted on the molecular chains or conveniently transformed into ionic groups. Since the types of organic and inorganic monomers are relatively deficient, the amount of charged hybrid membranes prepared through free radical polymerization is insufficient so far. For this reason, in most cases, free radical polymerization is used together with another membrane preparation method, such as sol-gel and blending, to achieve the desirable charged hybrid membranes. In contrast, when the organic and inorganic compounds both can be dissolved into one solvent and form a homogeneous solution, blending is considered a suitable choice to introduce inorganic ingredients into the organic phase. Homogeneous hybrid membranes are thus obtained when there is a good compatibility between organic and inorganic moieties. The intractable problem with blending is the incompatibility or precipitation between inorganic and organic phases. For example, when excessive inorganic ingredient is added into the organic phase, the brittleness or crack will occur on the membrane surface; the membrane performances will thus be impacted. Consequently, the major difficulty in blending is the screening of cosolvent for both organic and inorganic compounds, and control over the inorganic/organic ratio. With regard to the sol-gel process, it is usually conducted

**TABLE 3.1**
**Primary Techniques for Preparation of Charged Hybrid Membranes**

| Techniques | Advantages | Disadvantages | Applications | Reference |
|---|---|---|---|---|
| Liquid-phase coupling | Higher ion permselectivity | High electrical resistance | Preparing hybrid anion exchange membranes | Kogure et al. (1997) |
| Sol-gel process | Low operating temperature; eliminating the incompatibility between the inorganic and organic matrix | – | Preparing negatively charged hybrid membranes, positively charged hybrid membranes, and zwitterionic hybrid ones | Zhang et al. (2006); Wu et al. (2003a, 2003b); Wen et al. (1996) |
| Epoxide ring-opening and sol-gel process | Can be easily transformed into ionic groups | Insufficient types of polymerizing species | Preparing negatively and positively charged hybrid ones | Liu et al. (2009a); Wu et al. (2006b) |
| Free radical polymerization and sol-gel process | Combined the merits of free radical polymerization and sol-gel process | Need initiator, and the types of monomers are insufficient | Preparing negatively and positively charged hybrid ones | Liu et al. (2009b) |
| Blending and sol-gel process | Combined the advantages of blending and sol-gel process | The incompatibility or precipitation between the inorganic and organic phase | Preparing negatively and positively charged hybrid ones | Nagarale et al. (2005) |
| Blending and epoxide ring-opening | Combined the advantages of blending and epoxide ring-opening | The incompatibility or precipitation between the inorganic and organic phase | Preparing negatively and positively charged hybrid ones | Zuo et al. (2009) |
| UV/thermal curing | Produce the polymeric network | Insufficient types of polymerizing species | Preparing negatively and positively charged hybrid membranes | Choi et al. (2004); Wu et al. (2008a) |

in the starting material of metal or nonmetal alkoxide precursors $M(OR)_n$ (M= Si, Ti, Zr, etc.) and operated at low temperature. Then, these precursors are subject to hydrolysis and polycondensation in the presence of water or nonaqueous solvent such as organic alcohols. Subsequently, they undergo a gelation and curing process, and the final hybrid material or membrane can be achieved (Wu et al., 2006a, 2006b; Wen and Wilkes, 1996; Hench and West, 1990). Since the sol-gel process is mainly conducted at low temperature, the incompatibility between the inorganic and organic matrixes produced during blending can be highly eliminated during membrane preparation. Consequently, sol-gel process is regarded as one of the most important approaches.

It should be pointed out that the combination of the above-mentioned techniques for preparation of charged hybrid membranes is more common in the newly developed routes because the main purpose is to combine their respective advantages. The prevailing combinations are as follows: free radical polymerization and sol-gel process, blending and sol-gel process, blending and epoxide ring-opening. The combination of various techniques has greatly extended the preparation approaches. The primary techniques for preparation of charged hybrid membranes are listed in Table 3.1. Based on these techniques, a variety of charged hybrid membranes have been prepared. However, no

**TABLE 3.2**
**Main Production Methods of Ionic Groups in the Molecular Backbone of the Investigated Charged Hybrid Membranes**

| Membranes | Main Production Methods of Ionic Groups | Anionic Group | Cationic Group | Reference |
|---|---|---|---|---|
| Negatively charged hybrid membranes | oxidization of –SH | $-SO_3H$ | | Wu et al. (2003a); Wen et al. (1996) |
| | sulfonation of phenyl groups | $-SO_3H$ | | Wu et al. (2005); Wu et al. (2004); Vona et al. (2005) |
| | direct grafting of acidic groups through free radical polymerization | –COOH | | Liu et al. (2009b) |
| | epoxide ring-opening | $-SO_3H$ | | Liu et al. (2009a) |
| | primary amine groups | $-PO_3H_2$ | | Binsu et al. (2005) |
| Positively charged hybrid membranes | epoxide ring-opening; changed into the quaternary amine groups | | $N(CH_3)_3H^+Cl^-$ | Wu et al. (2006b); Wu et al. (2009b) |
| | quarteramination reaction of ternary amine groups | | $-N^+(C_2H_5)_2$; $-N^+(CH_3)_2$ | Wu et al. (2003b) |
| | dispersing anion exchange resin particles in the hybrid matrix | | anion | Nagarale et al. (2005b) |
| | chemical grafting of 4-vinyl pyridine (4-VP) | | positively charged groups | Nagarale et al. (2005a) |
| Zwitterionic hybrid membranes | epoxide ring-opening of 1,4-butyrolactone | –COOH | $-N^+-$ | Liu et al. (2005a); Liu et al. (2006b) |
| | epoxide ring-opening of 1,3-propanesultone | $-SO_3^-$ | $-NH_2^+-$ | Liang et al. (2006) |
| | sulfonation and epoxide ring-opening of 1,4-butyrolactone | –COOH and $-SO_3^-$ | $-N^+-$ | Liu et al. (2006c) |
| | oxidization of –SH group and epoxide ring-opening of 1,4-butyrolactone | –COOH and $-SO_3^-$ | $-N^+-$ | Liu et al. (2008) |

matter what the membrane preparing methods are used, one important issue to be considered is how to produce ionic groups in the membrane matrix, which will be discussed later (cf. Table 3.2).

Presently, the developments of organic charged membranes have been extensively reviewed (Xu, 2005; Nagarale et al., 2006). Unfortunately, few refers to the preparations and applications of inorganic/organic charged hybrid membranes; therefore, in the following section, we will mainly focus on the preparations and potential applications of newly developed charged hybrid membranes, including negatively charged hybrid membranes, positively charged hybrid membranes, and zwitterionic hybrid membranes.

In this chapter, the current developments on the preparation strategies of inorganic–organic charged hybrid membranes, including the newly developed in our group, will be introduced firstly. Then, the potential applications of charged hybrid membranes, especially application in adsorption and fuel cell fields, will be highlighted. In the last part of this chapter, the challenges and perspective are given to facilitate their further investigations.

## 3.2 PREPARATION OF INORGANIC–ORGANIC CHARGED HYBRID MEMBRANES

### 3.2.1 Negatively Charged Hybrid Membranes

The above-mentioned methods can be used to prepare negatively charged hybrid membranes. The key of these routes is how to generate the negatively charged groups (anionic groups) in the membrane matrix. Conventionally, they can be produced by oxidization of mercapto group (–SH) (Wu et al., 2003a, 2003b; Nagarale et al., 2004) or sulfonation of phenyl groups (Wu et al., 2005; Wu et al., 2004; Vona et al., 2005). Besides these methods, many alternative novel routes were recently developed to create negatively charged groups in the molecular backbone. This is equivalent to direct grafting of acidic groups through free radical polymerization (Liu et al., 2009a, 2009b) and the indirect transformation of functional groups, such as the epoxide ring-opening reaction (Liu et al., 2009a, 2009b; Choi et al., 2004) and transformation of primary amine groups (Binsu et al., 2005).

For example, direct grafting of acidic groups through free radical polymerization was reported to produce negatively charged groups (–COOH) in the polymeric chains (Liu et al., 2009a, 2009b), in which the hybrid precursors were mainly created by using γ-methacryloxypropyl trimethoxy silane (MAPTMS) and acrylic acid (AA) monomers. Figure 3.2 presents such synthetic procedure. The thermal stability of these hybrids could reach up to 420°C. The stable molecular structure of these hybrids could be controlled by adjusting the silica and AA contents in the molecular chains.

Different from the direct grafting of acidic groups on the molecular chains mentioned above, indirect transformation via the epoxide ring-opening reaction is also able to generate negatively charged groups (Liu et al., 2009a, 2009b). Figure 3.3 presents such a novel route to generate $-SO_3H$ groups. TGA and DrTGA analyses indicated that their thermal stabilities could reach near 300°C and the optimum molar ratio of GPTMS and TEOS is 1:1. The testing results confirmed that their IECs were related to the content of ionic groups produced in the charged hybrid precursor.

Moreover, primary amine groups can be transformed into the negatively charged groups. For example, Binsu et al. (2005) prepared the hybrid proton exchange membranes by transformation of primary amine groups, in which the $-NH_2$ groups were transformed into $-PO_3H_2$ groups. These hybrid proton exchange membranes obtained favorable hydrophilicity and proton conductivity.

### 3.2.2 Positively Charged Hybrid Membranes

Similar to the preparation of negatively charged hybrid membranes discussed above, the key for preparation of positively charged hybrid membranes is how to generate the positively charged groups in the hybrid matrix. Presently, positively charged groups can be produced through such methods as quarteramination of ternary amine groups (Wu et al., 2003), epoxide ring-opening reaction (Wu

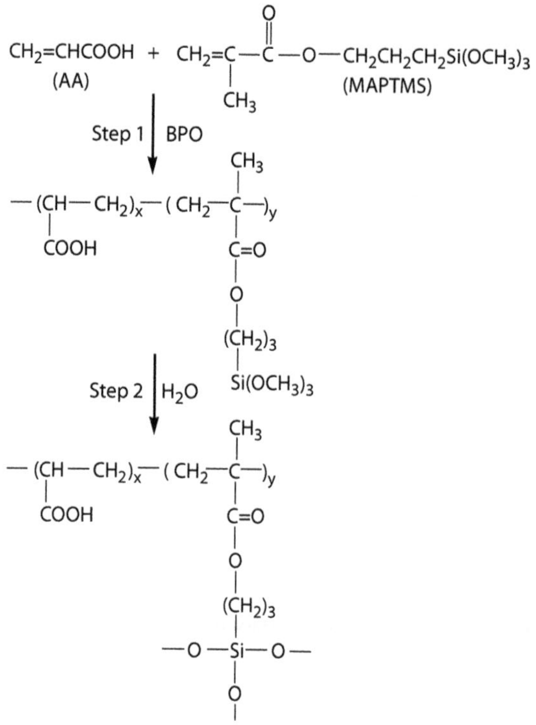

$$CH_2{=}CHCOOH + CH_2{=}C(CH_3){-}C({=}O){-}O{-}CH_2CH_2CH_2Si(OCH_3)_3$$

(AA) (MAPTMS)

Step 1 BPO

$$-(CH(COOH)-CH_2)_x-(CH_2-C(CH_3)(C{=}O{-}O{-}(CH_2)_3{-}Si(OCH_3)_3))_y-$$

Step 2 $H_2O$

$$-(CH(COOH)-CH_2)_x-(CH_2-C(CH_3)(C{=}O{-}O{-}(CH_2)_3{-}Si(-O-)_3))_y-$$

**FIGURE 3.2** Preparing route for negatively charged hybrid copolymers: step 1 is free radical polymerization between AA and MAPTMS monomers; step 2 is hydrolysis and condensation of alkoxysilane via sol-gel reaction. (Reprinted from *Separation and Purification Technology*, 66, Liu, J.S., et al., 135–142, Copyright 2009, with permission from Elsevier.)

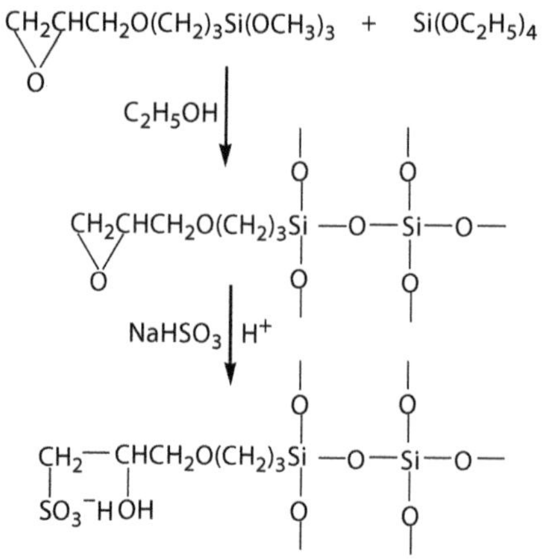

$$\underset{\text{epoxide}}{CH_2CHCH_2O(CH_2)_3Si(OCH_3)_3} + Si(OC_2H_5)_4$$

$C_2H_5OH$

$$CH_2CHCH_2O(CH_2)_3Si(-O-)_3-O-Si(-O-)_3$$

$NaHSO_3$ $H^+$

$$CH_2(SO_3^-H)-CH(OH)CH_2O(CH_2)_3Si(-O-)_3-O-Si(-O-)_3$$

**FIGURE 3.3** The preparing route for negatively charged hybrid materials: step 1 is the sol-gel process, and step 2 is the epoxide ring-opening reaction. (Liu, J.S., et al., *J. Appl. Polym. Sci.*, 2009, 112: 2179–2184. Copyright Wiley-VCH Verlag GmbH & Co. KGaA. Reproduced with permission.)

et al., 2003, 2009), dispersing anion exchange resin particles in the hybrid matrix (Nagarale et al., 2005), chemical grafting, and UV/thermal curing (Nagarale et al., 2005; Wu et al., 2008a, 2008b).

Some researchers have prepared a series of positively charged hybrid membranes by these preparing techniques. For example, to optimize the preparation conditions of the epoxide ring-opening reaction, the influence of different precursors on membrane formation were recently investigated (Wu et al., 2009). It is reported that the hybrid precursor with a lower molecular weight is more suitable to obtain a homogenous and flexible hybrid membrane. The membranes produced in this case possess a relatively strong hydrophobicity and high mechanical strength. SEM images reveal that the membrane surface is compact and homogenous even if the silica content incorporated in the membranes is as high as 27.0–29.8%.

Furthermore, the preparation of positively charged hybrid membranes can also adopt the method of dispersing anion exchange resin particles in the hybrid matrix (Nagarale et al., 2005). Typically, the positively charged groups in the membrane matrix are physically blended in this case. Consequently, incompatibility between the charged components and the membrane matrix might be the major consideration during membrane preparation.

Alternatively, chemical grafting and UV/thermal curing are also newly invented for preparation of positively charged hybrid membranes (Nagarale et al., 2005; Wu et al., 2008a, 2008b). For example, Nagarale et al. (2005) employed an aqueous dispersion polymerization method to prepare hybrid anion exchange PVA–$SiO_2$ membranes, in which the positively charged (anion exchange) groups were introduced by chemical grafting of 4-vinyl pyridine (4-VP). Moreover, another alternative route to prepare anion exchange hybrid membranes is sol-gel reaction and UV/thermal curing, in which the positively charged group was created by using the positively charged alkoxysilane and the alkoxysilane-containing acrylate or epoxy groups as the starting precursors. The crosslinking reaction between the partial epoxy and acrylate groups occurred during UV curing. The experimental results reveal that their thermal degradation temperatures are in the range of 212°C–226°C and the IEC values within 0.9–1.6 mmol $g^{-1}$. Meanwhile, these charged hybrid membranes exhibit high permeability to anions, suggesting their potential application in electrochemical separations (Wu et al., 2008a, 2008b).

### 3.2.3 Zwitterionic Hybrid Membranes

#### 3.2.3.1 Zwitterionic Hybrid Membranes Containing Single Acidic Groups

For preparation of zwitterionic hybrid membranes containing single acidic groups, the ring-opening reaction of a lactone reagent, such as 1,3-propanesultone or 1,4-butyrolactone (we simply named it zwitterionic process), is considered one of the most effective methods and they are commonly used to create both positively and negatively charged groups (i.e., ion pairs). Through the ring-opening reaction of a lactone reagent, various routes to prepare zwitterionic hybrid membranes have been proposed in recent years (Liu et al., 2005a, 2005b, 2005c, 2006a, 2006b, 2006c; Liang et al., 2006). For example, our group proposed a novel approach to prepare zwitterionic hybrid membranes by incorporating silicone into the membrane precursors (Liu et al., 2005a, 2005b, 2005c), and both the positively charged groups ($N^+$) and the negatively charged groups (–COOH) were created by zwitterionization of 1,4-butyrolactone in the hybrid precursors. Based on the sol-gel and zwitterionic process, this route started from a charged hybrid precursor, which was obtained by coupling of 3-glycidoxypropyltrimethoxysilane (GPTMS) with N-[3-(trimethoxysilyl) propyl] ethylene diamine (TMSPEDA), and was finalized with a subsequent reaction with 1,4-butyrolactone to create ion pairs in the polymer chains. The thermal stability of these zwitterionic hybrids could reach up to 350°C. However, due to its lower organic component incorporated into the prepared hybrid precursor, cracks occurred on the membrane surface. Consequently, it is difficult to elevate the flexibility of zwitterionic hybrid precursors prepared from lower molecular weight of silicone.

To improve the flexibility of the zwitterionic hybrid membranes, an alternative approach by using chemical grafting of organic polymer with a high molecular weight on the hybrid precursor was put forward, and both the positively charged groups (–$N^+$–) and the negatively charged groups (–COOH) were also generated by zwitterionization of 1,4-butyrolactone in the hybrid precursors

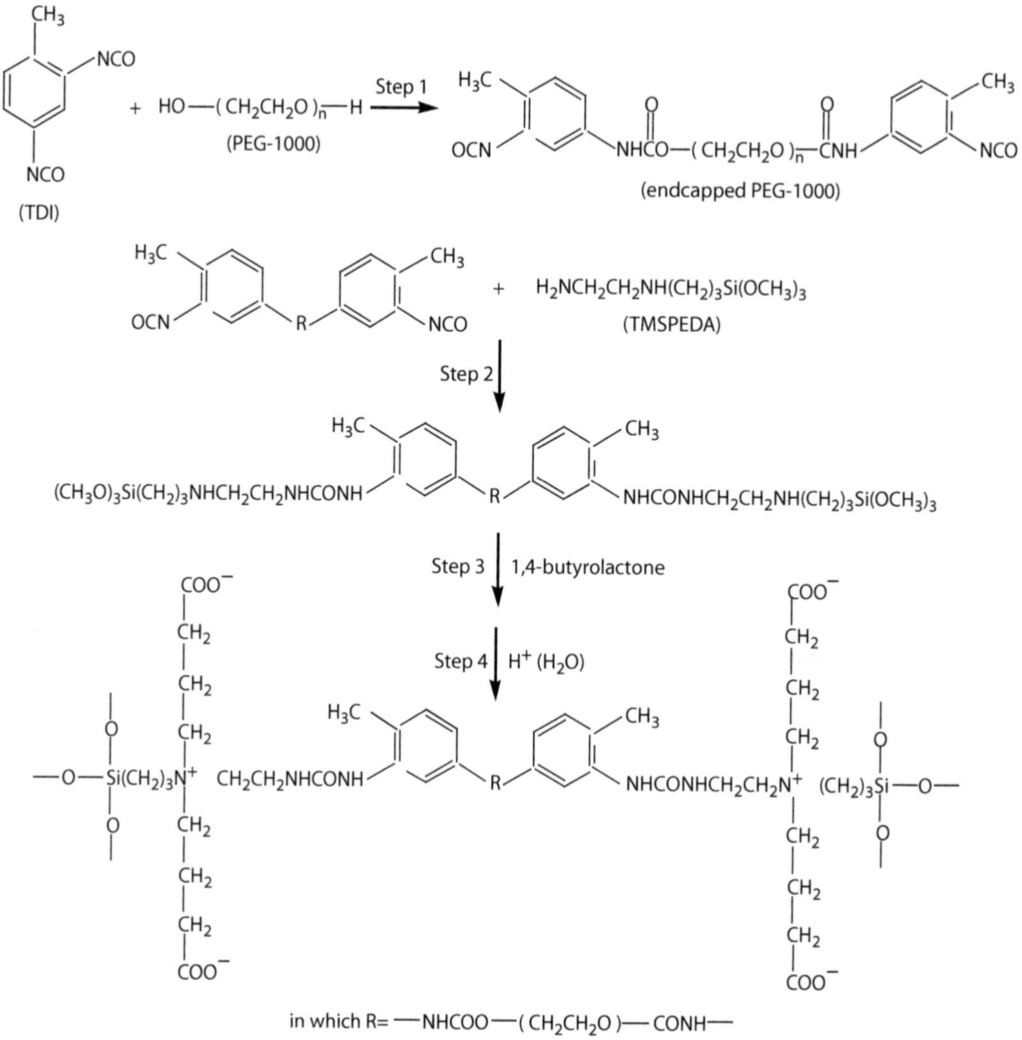

**FIGURE 3.4** The route for preparing zwitterionic hybrid membranes with higher organic ingredients: step 1 is endcapping PEG with TDI; step 2 is coupling endcapped PEG with TMSPEDA; step 3 is zwitterionic process of PEG–[Si(OCH3)3]2 with 1,4-butyrolactone (BL) to create hybrid zwitterionic copolymer, i.e., PEG–[Si(OMe)3]2–COO–; step 4 is hydrolysis and condensation of the resulting product. (Reprinted from *J. Membr. Sci.*, 283, Liu, J.S., et al., 190–200, Copyright 2006, with permission from Elsevier.)

(Liu et al., 2006a, 2006b, 2006c). Figure 3.4 presents the newly developed route to prepare highly–flexible zwitterionic hybrid membranes. The IEC measurements revealed that their cation exchange capacities (CIECs) were in the region of (1.9–2.5)×$10^{-2}$ meq·$cm^{-2}$ when the membranes were coated for one to four times. TGA thermal analysis showed that the thermal stability of these zwitterionic hybrid membranes could reach around 240°C, indicating a higher thermal stability than the organic analogue.

Apart from chemical grafting of organic ingredients on the molecular backbone, a physical process for preparing zwitterionic hybrid membranes was also proposed by Liang et al. (2006). These zwitterionic hybrid membranes were prepared by incorporation of PEG into the zwitterionic siloxane (ZS) and 3-glycidoxypropyltrimethoxysilane (GPTMS) hybrid matrix, in which the 1,3-propanesultone conducted the epoxide ring-opening reaction to produce the $-SO_3^-$ and $-NH_2-$ groups. Their thermal stabilities could reach as high as 250°C.

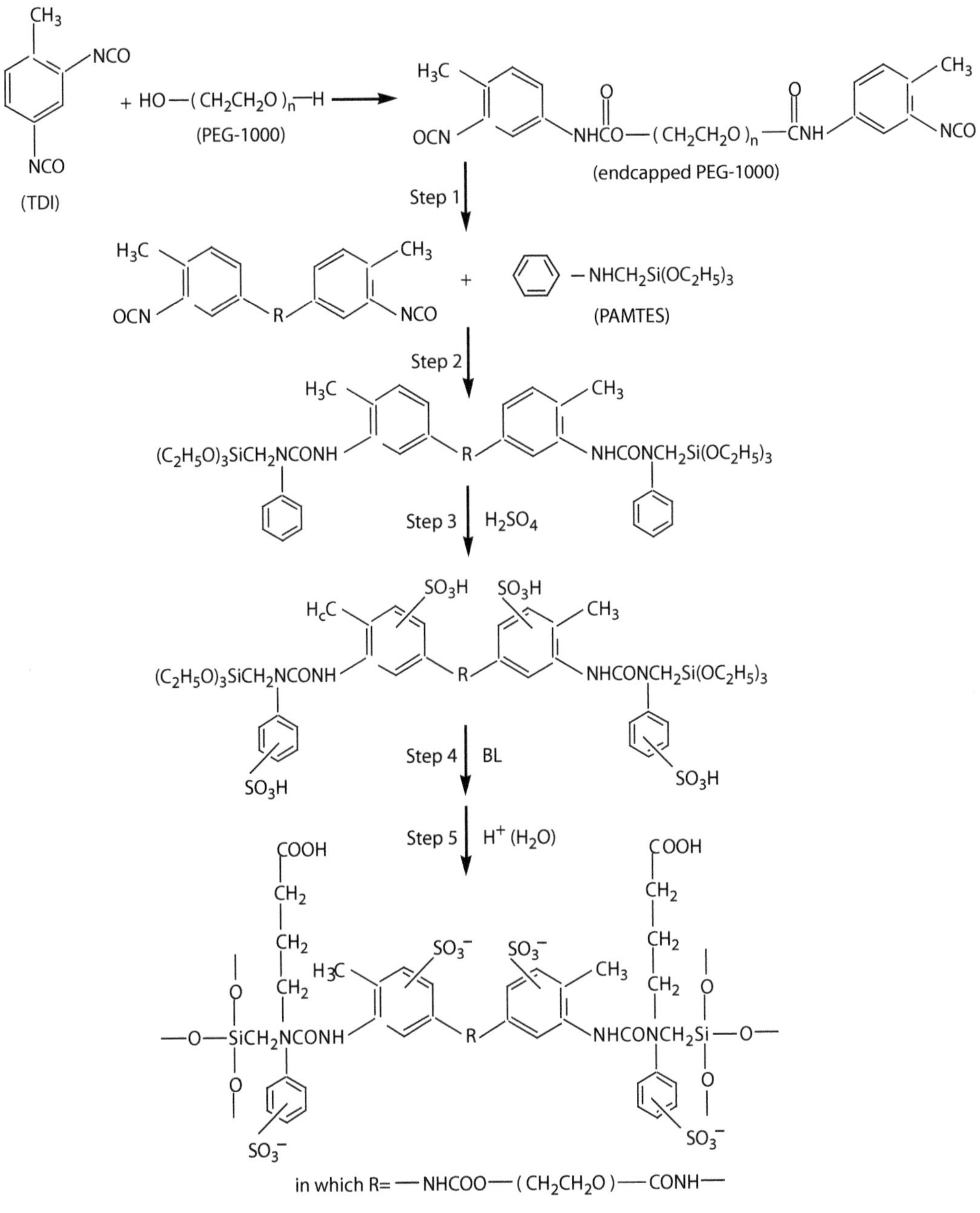

in which R= —NHCOO—( $CH_2CH_2O$ )—CONH—

**FIGURE 3.5** Possible reactions for preparation of zwitterionic hybrid copolymer: step 1 is endcapping PEG with TDI; step 2 is coupling endcapped PEG with PAMTES to produce PEG–$[Si(OEt)_3]_2$; step 3 is sulfonation of PEG–$[Si(OEt)_3]_2$ with fuming sulfuric acid (20% $SO_3$) (completely sulfonation reaction was shown for simplicity); step 4 is zwitterionic process of PEG–$[Si(OEt)_3]_2SO_3H$ with BL to produce zwitterionic hybrid precursor; step 5 is hydrolysis and polycondensation of the hybrid precursor to generate zwitterionic hybrid copolymer. (Reprinted from *European Polymer Journal*, 42, Liu, J.S., et al., 2755–2764, Copyright 2006; with permission from Elsevier.)

### 3.2.3.2 Zwitterionic Hybrid Membranes Containing Different Acidic Groups

The aforementioned techniques for creating the negatively or positively charged groups can also be selectively combined to prepare the zwitterionic hybrid membranes, which contain different acidic groups. These methods include oxidization of –SH group, sulfonation of phenyl groups, the epoxide ring-opening reaction, and quarteramination of amine groups. Typical examples are as follows.

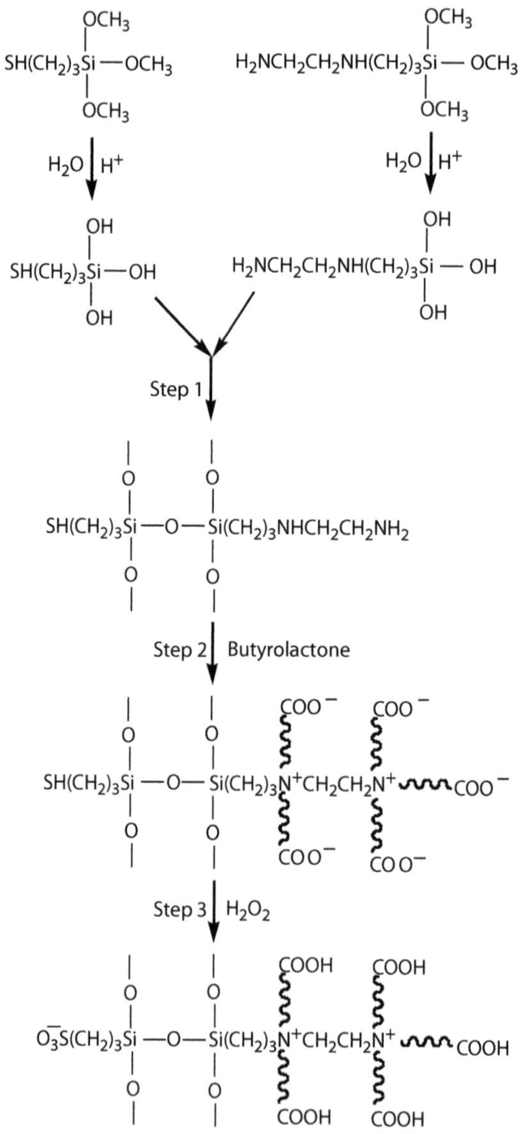

**FIGURE 3.6** Preparation of zwitterionic hybrid membranes containing both sulfonic and carboxylic acid groups: step 1 is hydrolysis and condensation of N-[3-(trimethoxysilyl)propyl] ethylene diamine (TMSPEDA) and 3-(mercaptopropyl) trimethoxysilane (MPS), respectively; step 2 is zwitterionic process of the prepared hybrid precursors to create ion pair grafted on the polymer chain; step 3 is the oxidization of the mercapto (–SH) group to produce the sulfonic acid group in the above-prepared hybrid membranes. (Liu, J.S., et al., *J. Appl. Polym. Sci*, 2008, 107, 3033–3041. Copyright Wiley-VCH Verlag GmbH & Co. KGaA. Reproduced with permission.)

The zwitterionic hybrid membrane precursors containing both sulfonic and carboxylic acid groups were recently prepared by sulfonation of phenyl groups and zwitterionization, and the negatively charged groups (–COOH and $-SO_3^-$ groups) are placed on the pendant-side chain (Liu et al., 2006a, 2006b, 2006c). Figure 3.5 displays such preparing route. Based on the TGA and DrTGA thermal analyses, the thermal stabilities of these zwitterionic hybrid membranes increased with an increase in the zwitterionic extent. DSC curves indicated that both the glass transition temperature ($T_g$) and the melting temperature ($T_m$) increase as the content of ion pairs increases and the relative content of PEG in the copolymers decreases. MALDI-TOF measurements suggested that the molecular structure of the zwitterionic hybrid PEG-$[Si(OEt)_3]_2SO_3H$-COOH was more stable than that of the neutral hybrid PEG-$[Si(OEt)_3]_2$ or that of the negatively charged hybrid PEG-$[Si(OEt)_3]_2SO_3H$.

Another approach to prepare zwitterionic hybrid membranes containing both sulfonic acid and carboxylic acid groups involves the oxidization of –SH group and zwitterionization, and the negatively charged groups ($-SO_3^-$) are placed on the main chain (Liu et al., 2008). The reaction was initiated by a coupling reaction between the N-[3-(trimethoxysilyl)propyl] ethylene diamine (TMSPEDA) and 3-(mercaptopropyl) trimethoxysilane (MPS), and was finalized with a reaction with 1,4-butyrolactone to create ion pairs in the polymer chain. The $-SO_3^-$ groups were produced by oxidization of –SH groups in the membrane matrix. Figure 3.6 illustrates the reaction steps. The results indicated that the anion exchange capacity (AIEC), total cation exchange capacity ($CIEC_{total}$), and the CIEC of the sulfonic acid groups ($CIEC_{sulf}$) of these zwitterionic hybrid membranes were in the range of 0.017–0.12, 0.1–0.53 and 0.029–0.14 mmol $g^{-1}$, respectively, when coated for one to three times.

As discussed above, various innovative techniques were recently developed to prepare charged hybrid membranes. However, no matter what the membrane preparation methods are used, how to generate ionic groups in the membrane matrix will be the main consideration in the preparation of charged hybrid membrane. This will be a new topic in the area of membrane preparation. To sum up, Table 3.2 lists the main methods of generating ionic groups in the backbones of the explored charged hybrid membranes.

## 3.3 APPLICATIONS OF INORGANIC–ORGANIC CHARGED HYBRID MEMBRANES

Because of their unique properties and practical requirements in industrial separation, inorganic–organic charged hybrid membranes have drawn escalating interests since their appearance in the 1980s (Wu et al., 2006a, 2006b). At present, they have found many potential applications, such as adsorption separation (Liu et al., 2009a, 2009b; Lin et al., 2009), fuel cells (Liang et al., 2006; Wu et al., 2008a, 2008b; Umeda et al., 2009; Chen et al., 2008; Mosa et al., 2009; Chen et al., 2007; Wu et al., 2009), pressure-driven membrane separation (Nagarale et al., 2005; Yu et al., 2009; Zuo et al., 2009; Liu et al., 2007), biomedical systems (Chang et al., 2008; Qiu et al., 2005; Shirosaki et al., 2005, 2009; Chiu et al., 2007), and optical and electrical devices (Brusatin et al., 2004; Wang et al., 2004).

### 3.3.1 Application in Adsorption Separation

Heavy metal ions in aqueous solutions can be separated using negatively charged hybrid membranes (Liu et al., 2009a, 2009b). For example, a novel route to prepare the negatively charged hybrid precursors containing –COOH was recently proposed in our group for removal of $Pb^{2+}$, $Cu^{2+}$ from aqueous solutions (Liu et al., 2009a, 2009b) (cf. Figure 3.2). The $Cu^{2+}$ adsorption data revealed that the adsorption behaviors fitted well with the Freundlich isotherm model and Lagergren second-order kinetic model, suggesting that the adsorption mechanism of such negatively charged hybrids for $Cu^{2+}$ ion is the heterogeneous surface adsorption. Moreover, another route to prepare the negatively charged hybrid precursors containing $-SO_3H$. Their thermal stabilities could arrive at near 300°C. Their IEC values could be adjusted by control of silica in the hybrid matrix. The adsorption behavior for $Pb^{2+}$ and $Cu^{2+}$ ions demonstrated that they can be used to absorb and separate heavy metal ions. These findings suggest that negatively charged hybrid membranes can be used for separation and recovery of environmentally hazardous heavy metal ions and have potential applications in environmental protection.

Besides the adsorption separation of heavy metal ions, negatively charged hybrid membranes can also be used to separate cationic dye. For example, Lin et al. (2009) investigated the adsorption of cationic dye (methyl violet) on porous PMMA/$Na^+$–montmorillonite cation exchange membranes. The experimental results demonstrated that 95% of the methyl violet could be absorbed in 2 h, suggesting relatively higher adsorption efficiency.

### 3.3.2 Applications in Fuel Cells

Because of the scarcity of energy sources derived from the diminished mineral resources, the oil price throughout the world has sharply risen to a new height in the past. Accordingly, the searching of new substitutes for natural resources has become urgent. As one of the most promising alternatives to mineral resources, fuel cells have become a focus. For fuel cell applications, the membrane separates the module of fuel cell into two parts and transport protons from one cell to another, and thus is the crucial part. Currently, it is accepted that this type of membrane should bear higher proton conductivity and thermal stability, lower permeability to fuel (methanol) (Wu et al., 2006a, 2006b; Liang et al., 2006). Consequently, it is important to improve the membrane performances so that the synthesized membranes can meet the requirements for application in fuel cells. Among them, charged hybrid membrane is considered one of the most promising ones. Presently, zwitterionic hybrid membranes, both the positively and negatively charged hybrid membranes, and hybrid acid–base polymer membranes are developed for such purpose. Typical examples are as follows.

Liang et al. (2006) prepared a series of zwitterionic hybrid membranes for fuel cell applications based on a zwitterionic siloxane precursor (ZS). The proton conductivity could arrive at $3.5 \times 10^{-2}$ S $cm^{-1}$ at 85°C and relative humidity 95%, demonstrating that these zwitterionic hybrid membranes are the promising solid electrolytes for fuel cell applications.

Furthermore, a series of positively charged PEO–$SiO_2$ hybrid membranes with higher flexibility, good mechanical strength, and high temperature tolerance were prepared based on sol-gel process for fuel cell applications (Wu et al., 2008a, 2008b). The thermal degradation temperature of these positively charged hybrid membranes in air was in the range of 220°C–240°C, suggesting relatively higher thermal stability. Meanwhile, the conductivity of the membranes could approach to near 0.003 S $cm^{-1}$. Therefore, these positively charged hybrid membranes can be potentially applied in fuel cells as alkaline membranes.

Umeda et al. (2009) synthesized a series of negatively charged hybrid membranes containing phosphonic acid groups. Their thermal stabilities could reach 200°C. The conductivities of the HEDPA/GPTMS/PhTES membranes were $1.0 \times 10^{-1}$ and $4.5 \times 10^{-4}$ S $cm^{-1}$ at 130°C when the relative humidities are 0 and 100%, respectively. This suggests that they can be potentially applied as electrolyte membranes for polymer electrolyte fuel cells (PEFCs). In addition, Chen et al. (2008) also prepared a negatively charged hybrid membrane via the $\gamma$-ray preirradiation, hydrolysis condensation, and a subsequent sulfonation. The proton conductivity of them could reach 0.11 S $cm^{-1}$ at room temperature. The methanol permeability of them was almost six times lower than that of Nafion®, suggesting their potential application in fuel cells.

Besides the positively charged and negatively charged hybrid membranes, hybrid acid–base polymer membranes, which contain both the acidic and basic groups in the membrane matrix, can also be applied in fuel cells. For example, a series of hybrid acid–base polymer membranes were prepared through physical blending of sulfonated poly(2,6-dimethyl-1,4-phenylene oxide) (SPPO) with (3-aminopropyl)triethoxysilane via the sol–gel process (Wu et al., 2009). The measurements of fuel cell performances revealed that these hybrid membranes had obtained higher proton conductivity and lower methanol permeability when compared with Nafion 117. Consequently, they are the promising charged hybrid membranes for fuel cell applications.

### 3.3.3 Applications in Pressure-Driven Membrane Separation

Pressure-driven membrane separation using charged hybrid membranes is an effective method to separate mixtures. Currently, positively charged hybrid membranes, zwitterionic hybrid membranes, and negatively charged hybrid have been successfully used to separate inorganic salts from the mixture by pressure-driven (applied pressure) membrane separation (Nagarale et al., 2005; Liu et al., 2005a, 2005b, 2005c; Yu et al., 2009; Zuo et al., 2009). Negatively charged hybrid membranes were also applied to separate organic species from their organic mixtures by pervaporation (Qiu et al., 2005), which is usually operated under vacuum.

Presently, special electrolyte species can also be separated via positively charged hybrid membranes, zwitterionic hybrid membranes, and negatively charged hybrid membranes. For example, Nagarale et al. (205) measured the electroosmotic permeability of a NaCl solution by imposing a known potential difference across different hybrid anion exchange PVA–$SiO_2$ membranes. The electroosmotic flux (Jv) increased with the applied current, suggesting the impact of coulomb electricity on electroosmotic permeability. These findings imply that these positively charged hybrid membranes have potential applications in separation of inorganic electrolyte solution.

Moreover, to separate KCl from an aqueous solution, a piezodialysis using zwitterionic hybrid membranes was carried out (Liu et al., 2005a, 2005b, 2005c). The conductivity of permeate (KCl solution) decreased with the elevating negative charge sites in the prepared zwitterionic hybrid membranes, suggesting the increasing retention of ions with an increase in anionic groups in membrane. Consequently, these types of zwitterionic hybrid membranes can be used to separate and recover electrolytes from the organic and inorganic salts. Furthermore, Yu et al. (2009) prepared a series of negatively poly(vinylidene fluoride) hybrid membranes filled with different weight fractions of $SiO_2$ nanoparticles for application in electro-membrane processes. These negatively charged hybrid membranes exhibited high thermal stability, improved selectivity for monovalent and bivalent ions, and moderate membrane conductivity, which are favorable for electro-driven separation processes.

As described above, the separation of inorganic salts from aqueous solutions by charged hybrid membranes are primarily performed under applied pressure. It is difficult to meet the requirements of organic mixture separation because the organic solvent has a considerable impact on the performances of these prepared membranes. In contrast to the separation of inorganic salts, organic mixture separations by pervaporation are mainly operated at minus pressure (vacuum). Hence, high thermal stability, mechanical strength, and solvent resistance are the crucial properties for charged hybrid membranes. Presently, negatively charged hybrid membranes have been successfully developed to separate organic mixtures by pervaporation. Liu et al. (2007) prepared the chitosan/ polytetrafluoroethylene (PTFE) composite membranes for dehydration of isopropanol. The permeation flux across such chitosan/PTFE composite membrane was 1730 g $m^{-2}$ h and the separation factor was 775°C at 70°C on pervaporation dehydration of a 70 wt% isopropanol aqueous solution. This indicates the excellent separation performances of negatively charged hybrid membranes for organic mixtures.

### 3.3.4 Applications in Biomedical Systems

Currently, numerous efforts have been made to extend the application of charged hybrid membranes in biomedical systems because these membranes can considerably improve the physicochemical properties of proteins. Some successful examples can be found in separation of DNA and RNA species (Chang et al., 2008), platelet adhesion (Qiu et al., 2005), bone regeneration (Shirosaki et al., 2005, 2009), isolation of lysozyme from egg albumen (Chiu et al., 2007).

Chang et al. (2008) used the hybrid positively charged hybrid membranes to separate DNA and RNA species. It was reported that in batch adsorption, the corresponding adsorption capacity for plasmid DNA has the following order: commercial polymeric SB6407 > modified glass fiber > modified alumina membrane. For RNA adsorption, the order became as follows: modified glass fiber ≈ SB6407 > modified alumina membrane. However, in membrane chromatography, the plasmid DNA and RNA could be separated from the feed of plasmid DNA+ RNA mixture. The overall recovery of plasmid DNA by the modified glass fiber membrane was in the range of 98–106%, which is significantly higher than that of SB6407 membrane (91–96%). These findings suggest that hybrid anion exchange membranes can be used in biomedical systems.

In addition, Qiu et al. (2005) applied zwitterionic hybrid films as platelet adhesion. The platelet assay in vitro had excellent nonthrombogenicity with the glass slides bearing sulfobetaine structure on their surfaces, demonstrating that zwitterionic interfacial molecular structures of biomaterials were thermodynamic advantage for maintaining the natural plasma proteins/cells conformation.

Furthermore, Shirosaki et al. (2009) investigated the osteocompatibility and cytocompatibility of chitosan–silicate hybrid membranes for medical applications. The adhesion and proliferation of the MG63 osteoblast cells cultured on the chitosan–GPTMS hybrid surface were improved as compared to those on the chitosan. Meanwhile, human bone marrow osteoblast cells proliferated on the chitosan–GPTMS hybrid surface and formed a fibrillar extracellular matrix with numerous calcium phosphate globular structures.

Chiu et al. (2007) explored the isolation of lysozyme from hen egg albumen via glass fiber-based cation exchange membranes with sulphonic acid groups. The prepared hybrid cation exchange membranes have higher purification efficiency than those of commercial ones and thus have a potential application in protein separation.

### 3.3.5 Applications in Optical and Electrical Devices

Currently, zwitterionic hybrid membranes and negatively charged hybrid membranes can be applied to prepare optical and electrical devices, such as nonlinear optical systems (Innocenzi et al., 2002; Brusatin et al., 2004), and single ion conductors (Choi et al., 2004; Wang et al., 2004).

Innocenzi et al. prepared a series of zwitterionic hybrid films and verify the feasibility of their applications in nonlinear optical system (Innocenzi et al., 2002; Brusatin et al., 2004). In particular, the zwitterionic hybrid films were prepared by incorporation of dihydroxy-functionalized zwitterionic push–pull chromophores into the hybrid organic–inorganic matrixes derived from 3-glycidoxypropyltrimethoxysilane (GPTMS) and *N*-[(3-trimeth-oxysilyl) propyl]-ethylenediamine (TMESPE). The poling measurement indicated that the second harmonic (SH) signal could be observed during the poling process performed at 85°C. This finding suggests that zwitterionic hybrid films can be produced as waveguide devices.

Moreover, negatively charged hybrid membranes can be used as a new functional lithium salt to fabricate single ion conductors. Choi et al. (2004) prepared the new nanosized silica cross-linked single ion conductor by UV irradiation, in which the nanosized fumed silica reacted with 1,3-propane sultone to produce the $-SO_3H$. The introduction of the modified silica as a lithium ion source was able to increase the ion conductivity. Furthermore, Wang et al. (2004) attempted to lithiate and codissolve Nafion® 117 film with copolymer poly-(vinylidene fluoride) hexafluoropropylene (PVDF–HFP) and to prepare the single ion conductor as lithium polymer electrolyte. The fumed silica was used as filler to improve the mechanical strength, ionic conductivity, and interfacial stability of such single ion conductor. It was confirmed that the resulting materials exhibited good film formation, solvent-maintaining capability and dimensional stability.

Just the same as their preparations, the potential applications of charged hybrid membranes have received much attention. However, little information is given on their practical applications so far. Consequently, an attempt needs to be made for their practical applications. As a summary of this chapter, Table 3.3 lists the main functions and advantages of charged hybrid membranes in various fields.

**TABLE 3.3**
**Main Applications and Potential Advantages of Charged Hybrid Membranes**

| Applications | Main Functions | Potential Advantages | Reference |
|---|---|---|---|
| Adsorption separation | Ion exchange; adsorption | high adsorption capacity | Liu et al. (2009a, 2009b); Lin et al. (2009) |
| Fuel cells | Electrolyte membranes | High thermal stability and proton conductivity, low methanol permeability | Liang et al. (2006); Wu et al. (2008b); Wu et al. (2009a) |
| Pressure-driven membrane separation | Pressure-driven permeation | Permselectivity; high permeation flux and separation factor | Yu et al. (2009); Liu et al. (2007) |
| Biomedical systems | Adsorption; adhesion and proliferation | High adsorption capacity; excellent nonthrombogenicity; improving osteocompatibility and cytocompatibility | Chang et al. (2008); Shirosaki et al. (2009); Shirosaki et al. (2005); Chiu et al. (2007) |
| Optical and electrical devices | waveguide device; single ion conductor | Good temporal stability; high mechanical strength, ionic conductivity, and interfacial stability | Brusatin et al. (2004); Wang et al. (2004) |

## 3.4 CHALLENGES AND PERSPECTIVE

As novel and promising candidates, inorganic–organic charged hybrid membranes are becoming more attractive. The corresponding research has been extending the potential applications to various aspects besides the new materials science, chemical engineering, and environmental protection.

When it comes to the preparation of charged hybrid membranes, many achievements have been acquired in recent years. However, the practical applications of charged hybrid membranes in industry are unsatisfactorily so far. There still remain some unknown things to be explored, especially about the performances of charged hybrid membranes. To resolve these questions and bring the application to practice, additional efforts are required. Considering these insufficient investigations, further research will focus on the following aspects.

1. Further enhancement of the performances of charged hybrid membranes.
2. Design and development of novel routes to prepare charged hybrid membranes.
3. Matching between the properties of charged hybrid membranes and the application ranges. It is especially necessary to establish the relationship between the membrane performances obtained from laboratory scale experiments and industrial applications.

With regard to the applications of charged hybrid membranes in current fields, further extension and improvement will be a great challenge to the researchers and engineers in the coming years.

## ACKNOWLEDGMENTS

The authors appreciate the financial support from the Significant and Key Foundations of Educational Committee of Anhui Province (ZD2008002 and KJ2008A072), Anhui Provincial Natural Science Foundation (090415211), the Opening Project of Key Laboratory of Solid Waste Treatment and Resource Recycle (SWUST), Ministry of Education (09zxgk03), Specialized Research Fund for the Doctoral Program of Higher Education (200803580015), the National Basic Research Program of China (973 program, 2009CB623403), and the National Natural Science Foundation of China (21076055). Special thanks will be given to Dr. C. H. Huang for proofreading the English.

## REFERENCES

Baker, R.W. 2004. *Membrane Technology and Applications*, 2nd ed. John Wiley & Sons.

Binsu, V.V., R.K. Nagarale, and V.K. Shahi. 2005, Phosphonic acid functionalized aminopropyl triethoxysilane–PVA composite material: organic–inorganic hybrid proton-exchange membranes in aqueous media, *J. Mater. Chem.* 15: 4823–4831.

Brusatin, G., P. Innocenzi, A. Abbotto, et al. 2004. Hybrid organic–inorganic materials containing poled zwitterionic push–pull chromophores, *Journal of the European Ceramic Society* 24: 1853–1856.

Chang, C.-S., H.-S. Ni, S.-Y. Suen, et al. 2008. Preparation of inorganic–organic anion-exchange membranes and their application in plasmid DNA and RNA separation, *J. Membr. Sci.* 311: 336–348.

Chen, J., M. Asano, Y. Maekawa, and M. Yoshida. 2007. Polymer electrolyte hybrid membranes prepared by radiation grafting of *p*-styryltrimethoxysilane into poly(ethylene-*co*-tetrafluoroethylene) films, *J. Membr. Sci.* 296: 77–82.

Chen, J., M. Asano, Y. Maekawa, and M. Yoshida. 2008. Chemically stable hybrid polymer electrolyte membranes prepared by radiation grafting, sulfonation, and silane-crosslinking techniques, *J. Polym. Sci. Part A: Polym. Chem.* 46: 5559–5567.

Chiu, H.-C., C.-W. Lin, and S.-Y. Suen. 2007. Isolation of lysozyme from hen egg albumen using glass fiber-based cation-exchange membranes, *J. Membr. Sci.* 290: 259–266.

Choi, N.-S., Y.M. Lee, B.H. Lee, J.A. Lee, and J.-K. Park. 2004. Nanocomposite single ion conductor based on organic–inorganic hybrid, *Solid State Ionics* 167: 293–299.

Cornelius, C.J., and E. Marand. 2002. Hybrid silica-polyimide composite membranes: gas transport properties. *J. Membr. Sci.* 202: 97–118.

Hench, L.L., and J.K. West. 1990. The sol-gel process. *Chem. Rev.* 90: 33–72.

Innocenzi, P., E. Miorin, G. Brusatin, et al. 2002. Incorporation of zwitterionic push–pull chromophores into hybrid organic–inorganic matrixes, *Chem. Mater.* 14: 3758–3766.

Juda, M., and W.A. McRac. 1950. Coherent ion-exchange gels and membranes, *J. Am. Chem. Soc.* 72: 1044–1049.

Kogure, M., H. Ohya, R. Paterson, et al. 1997. Properties of new inorganic membranes prepared by metal alkoxide methods. Part II: New inorganic–organic anion-exchange membranes prepared by the modified metal alkoxide methods with silane coupling agent, *J. Membr. Sci.* 126:161–169.

Kusakabe, K., S. Yoneshige, and S. Morooka 1998. Separation of benzene/cyclohexane mixtures using polyurethane-silica hybrid membranes, *J. Membr. Sci.* 149: 29–37.

Liang, W.J., C.P. Wu, C.Y. Hsu, and P.L. Kuo. 2006. Synthesis, characterization, and proton-conducting properties of organic–inorganic hybrid membranes based on polysiloxane zwitterionomer, *J. Polym. Sci., Part A—Polymer Chemistry* 44 (11): 3444–3453.

Lin, R.-Y., B.-S. Chen, G.-L. Chen, J.-Y. Wu, H.-C. Chiu, and S.-Y. Suen. 2009. Preparation of porous PMMA/$Na^+$–montmorillonite cation-exchange membranes for cationic dye adsorption, *J. Membr. Sci.* 326: 117–129.

Linder, C., and O. Kedem. 2001. Asymmetric ion exchange mosaic membranes with unique selectivity, *J. Membr. Sci.* 181: 39–56.

Liu J.S., T.W. Xu, X. Zhu, and Y.X. Fu. 2006a. Membrane potentials across hybrid charged mosaic membrane in organic solutions, *Chinese Journal of Chemical Engineering* 14(3): 330–336.

Liu, J.S., T. Li, K.Y. Hu, and G.Q. Shao. 2009a. Preparation and adsorption performances of novel negatively charged hybrid materials, *J. Appl. Polym. Sci.* 112: 2179–2184.

Liu, J.S., X.H. Wang, T.W. Xu, and G.Q. Shao. 2009b. Novel negatively charged hybrids. 1. copolymers: Preparation and adsorption Properties, *Separation and Purification Technology* 66: 135–142.

Liu, J.S., T.W. Xu, and X.Y. Fu. 2005a. Fundamental studies of novel inorganic–organic charged zwitterionic hybrids. 2. Preparation and characterizations of hybrid charged zwitterionic membranes, *J. Membr. Sci.* 252:165–173.

Liu, J.S., T.W. Xu, and X.Y. Fu. 2005b. Fundamental studies of novel inorganic–organic zwitterionic hybrids. 1. Preparation and characterizations of hybrid zwitterionic polymers, *J. Non-Crystalline Solids* 351:3050–3059.

Liu, J.S., T.W. Xu, and Y.X. Fu. 2008. A novel route for the preparation of hybrid zwitterionic membranes containing both sulfonic and carboxylic acid groups, *J. Appl. Polym. Sci.* 107: 3033–3041.

Liu, J.S., T.W. Xu, M. Gong, and Y.X. Fu. 2005c. Fundamental studies of novel inorganic–organic charged zwitterionic hybrids. 3. New hybrid charged mosaic membranes prepared by modified metal alkoxide and zwitterionic process, *J. Membr. Sci.* 260: 26–36.

Liu, J.S., T.W. Xu, M. Gong, F. Yu, and X.Y. Fu. 2006b. Fundamental studies of novel inorganic–organic charged zwitterionic hybrids 4. New hybrid zwitterionic membranes prepared from polyethylene glycol (PEG) and silane coupling agent, *J. Membr. Sci.* 283: 190–200.

Liu, J.S., T.W. Xu, X.Z. Han, and Y.X. Fu. 2006c. Synthesis and characterizations of a novel zwitterionic hybrid copolymer containing both sulfonic and carboxylic groups via sulfonation and zwitterionic process, *European Polymer Journal* 42(10): 2755–2764.

Liu, Y.-L., C.-H. Yu, K.-R. Lee, and J.-Y. Lai. 2007. Chitosan/poly(tetrafluoroethylene) composite membranes using in pervaporation dehydration processes, *J. Membr. Sci.* 287: 230–236.

Lu, Z.H., G.J. Liu, and S. Duncan. 2003. Poly(2-hydroxyethyl acrylate-co-methyl acrylate)/$SiO_2$/ $TiO_2$ hybrid membranes, *J. Membr. Sci.* 221: 113–122.

Matsumoto, H., Y. Koyama, and A. Tanioka. 2002. Interaction of Organic Molecules with Weak Amphoteric Charged Membrane Surfaces: Effect of Interfacial Charge Structure, *Langmuir* 18: 3698–3703.

Miyaki, Y., H. Nagamatsu, M. Iwata, et al. 1984. Artificial membranes from multiblock copolymers. 3. Preparation and characterization of charge-mosaic membranes, *Macromolecules* 17: 2231–2236.

Mosa, J., A. Durán, and M. Aparicio. 2009. Proton conducting sol–gel sulfonated membranes produced from 2-allylphenol, 3-glycidoxypropyl trimethoxysilane and tetraethyl orthosilicate, *J. Power Sources* doi:10.1016/j.jpowsour.2008.12.126.

Nagarale, R.K., G.S. Gohil, and V.K. Shahi. 2006. Recent developments on ion-exchange membranes and electro-membrane processes, *Advances in Colloid and Interface Science* 119: 97–130.

Nagarale, R.K., G.S. Gohil, V.K. Shahi, and R. Rangarajan. 2005a. Preparation of organic–inorganic composite anion-exchange membranes via aqueous dispersion polymerization and their characterization, *J. Colloid Interface Sci.* 287: 198–206.

Nagarale, R.K., G.S. Gohil, V.K. Shahi, and R. Rangarajan. 2004. Organic–inorganic hybrid membrane: Thermally stable cation-exchange membrane prepared by the sol-gel method, *Macromolecules* 37: 10023–10030.

Nagarale, R.K., V.K. Shahi, and R. Rangarajan. 2005b. Preparation of polyvinyl alcohol–silica hybrid heterogeneous anion-exchange membranes by sol–gel method and their characterization, *J. Membr. Sci.* 248: 37–44.

Neihof, R., and K. Sollner. 1950. A quantitative electrochemical theory of the electrolyte permeability of mosaic membranes composed of selectively anion-permeable and selectively cation-permeable parts, and its experimental verification. I. An outline of the theory and its quantitative test in model systems with auxiliary electrodes, *J. Phys. Chem.* 54: 157–176.

Ohya, H., R. Paterson, T. Nomura, et al. 1995. Properties of new inorganic membranes prepared by metal alkoxide methods. Part I: A new permselective cation exchange membrane based on Si/Ta oxides, *J. Membr. Sci.* 105: 103–112.

Qiu, Y., D. Min, C. Ben, et al. 2005. A novel zwitterionic silane coupling agent for nonthrombogenic biomaterials, *Chinese Journal of Polymer Science* 23: 611–617.

Sforca, M.L., I.V.P. Yoshida, and S.P. Nunes. 1999. Organic–inorganic membranes prepared from polyether diamine and epoxy silane, *J. Membr. Sci.* 159: 197–207.

Shahi, V.K. 2007. Highly charged proton-exchange membrane: Sulfonated poly(ether sulfone)-silica polyelectrolyte composite membranes for fuel cells, *Solid State Ionics* 177: 3395–3404.

Shirosaki, Y., K. Tsuru, S. Hayakawa, et al. 2005. In vitro cytocompatibility of MG63 cells on chitosan–organosiloxane hybrid membranes, *Biomaterials* 26: 485–493.

Shirosaki, Y., K. Tsuru, S. Hayakawa, et al. 2009. Physical, chemical and in vitro biological profile of chitosan hybrid membrane as a function of organosiloxane concentration, *Acta Biomaterialia* 5: 346–355.

Umeda, J., M. Moriya, W. Sakamoto, and T. Yogo. 2009. Synthesis of proton conductive inorganic–organic hybrid membranes from organoalkoxysilane and hydroxyalkylphosphonic acid, *J. Membr. Sci.* 326: 701–707.

Vona, M.L.D., Marani, D., D'Epifanio, A. et al. 2005. A covalent organic/inorganic hybrid proton exchange polymeric membrane: synthesis and characterization, *Polymer* 46: 1754–1758.

Wang, M.K., F. Zhao, and S.J. Dong. 2004. A single ionic conductor based on Nafion and its electrochemical properties used as lithium polymer electrolyte, *J. Phys. Chem. B* 108: 1365–1370.

Wen, J., and G.L. Wilkes. 1996. Organic/inorganic hybrid network materials by the sol-gel approach, *Chem. Mater.* 8: 1667–1681.

Wu, C.M., T.W. Xu, and J.S. Liu. 2006a. Charged hybrid membranes by the sol-gel approach: Present states and future perspectives. In: *Focus on Solid State Chemistry*, edited by Newman, A. M. *Nova Science Publishers Inc., USA*, 1–44.

Wu, C.M., Y.H. Wu, T.W. Xu, and Y.X. Fu. 2008a. Novel anion-exchange organic–inorganic hybrid membranes prepared through sol-gel reaction and UV/thermal curing, *J. Appl. Polym. Sci.* 107: 1865–1871.

Wu, C.M., T.W. Xu, and W.H. Yang. 2003a. A new inorganic–organic negatively charged membrane: membrane preparation and characterizations, *J. Membr. Sci.* 224: 117–125.

Wu, C.M., T.W. Xu, and W.H. Yang. 2003b. Fundamental studies of a new hybrid (inorganic–organic) positively charged membrane: membrane preparation and characterizations, *J. Membr. Sci.* 216: 269–278.

Wu, C.M., T.W. Xu, and W.H. Yang. 2004. Synthesis and characterizations of new negatively charged organic–inorganic hybrid materials: effect of molecular weight of sol-gel precursor, *Journal of Solid State Chemistry* 177: 1660–1666.

Wu, C.M., T.W. Xu, M. Gong, and W.H. Yang. 2005. Synthesis and characterizations of new negatively charged organic–inorganic hybrid materials. Part II. Membrane preparation and characterizations, *J. Membr. Sci.* 247: 111–118.

Wu, D., T.W. Xu, L. Wu, and Y.H. Wu. 2009a. Hybrid acid–base polymer membranes prepared for application in fuel cells, *Journal of Power Sources* 186: 286–292.

Wu, Y.H., C.M. Wu, M. Gong, and T.W. Xu. 2006b. New anion exchanger organic–inorganic hybrid materials and membranes from a copolymer of glycidylmethacrylate and γ-methacryloxypropyl trimethoxy silane, *J. Appl. Polym. Sci.* 102: 3580–3589.

Wu, Y.H., C.M. Wu, T.W. Xu, and Y.X. Fu. 2009b. Novel anion-exchange organic–inorganic hybrid membranes prepared through sol–gel reaction of multi-alkoxy precursors, *J. Membr. Sci.* 329: 236–245.

Wu, Y.H., C.M. Wu, F. Yu, T.W. Xu, and Y.X. Fu. 2008b. Free-standing anion-exchange PEO–$SiO_2$ hybrid membranes, *J. Membr. Sci.* 307: 28–36.

Xu, T.W. 2005. Ion exchange membranes: State of their development and perspective, *J. Membr. Sci.* 263: 1–29.

Xu, T.W., and W.H. Yang. 2003. A novel positively charged composite membranes for nanofiltration prepared from poly(2,6-dimethyl-1,4-phenylene oxide) by in situ amines crosslinking, *J. Membr. Sci.* 215(1–2): 25–32.

Yamauchi, A., Y. Okazaki, R. Kurosaki, Y. Hirata, and H. Kimizuka. 1987. Effect of ionic change on the electrical properties of an amphoteric ion exchange membrane, *J. Membr. Sci.* 32: 281–290.

Yu, S.L., X.T. Zuo, R. L. Bao, X. Xu, J. Wang, and J. Xu. 2009. Effect of $SiO_2$ nanoparticle addition on the characteristics of a new organic–inorganic hybrid membrane, *Polymer* 50: 553–559.

Zhang, S.L., T.W. Xu, and C.M. Wu. 2006. Preparation and characterizations of new anion exchange hybrid membranes based on Poly (2,6-dimethyl-1,4-phenylene oxide) (PPO), *J. Membr. Sci.* 269: 142–151.

Zuo, X.T., S.L. Yu, X. Xu, R.L. Bao, J. Xu, and W.M. Qu. 2009. Preparation of organic–inorganic hybrid cation-exchange membranes via blending method and their electrochemical characterization, *J. Membr. Sci.* 328: 23–30.

# 4 Preparation and Applications of Zeolite Membranes

## *A Review*

*Somen Jana, Mihir Kumar Purkait, and Kaustubha Mohanty*

**CONTENTS**



## 4.1 INTRODUCTION

Attempts are being made to replace all thermally driven separation processes by membrane-based processes, as energy is becoming an important issue day by day. Main advantages of inorganic membranes are that they can be operated at elevated temperature (metal membrane can tolerate

500°C–600°C and some ceramic membrane can withstand more than 1000°C), are chemically inert, and are inert to microbiological degradation. However, this is not the only reason why inorganic membranes are currently of prime interest. Inorganic membranes applications in many new areas like fuel cell, oil and petrochemical industry, pulp and paper manufacturing industry and biopharmaceutical industries are increasing because of their flexibility in pore size and operating conditions. The recent R&D is focused on mainly three different types of inorganic membrane: (1) zeolite, (2) sol-gel based microporous membranes, and (3) dense membranes (Caro et al., 2000). Especially when a small and narrow pore size distribution is the need, then zeolite membrane is supposed to be the best as they have very specific pore size. A critical review of literature says that there is a continuous increase in the number of scientific publications on the zeolite-based membrane, which proves that zeolite membranes are showing huge potential in the separation technology day by day. One of the first commercial uses of zeolites was the LTA zeolite, which was utilized by Mitsui & Co. for drying of ethanol and isopropanol. Another important application of zeolites (functional zeolite) would be in the area of nanotechnology due its small pore size (Caro et al., 2008).

In the last decade, zeolites have been used not only as a separative membrane but also in catalytic membrane reactor, chemical sensor, electrode, open electronic device and insulating materials. Another some of the very recent pioneering application of zeolites include molecular sensor (Vilaseca et al., 2003), and especially in small or micro scale applications (Coronas et al., 2003). Very recently, many zeolite membranes were prepared and progresses were reported. These are LTA (Shah et al., 2003; Yamazaki et al., 2000), FAU (Nikolakis et al., 2001; Lee et al., 2001), MOR (Lin et al., 2000), FER (Li et al., 2004), SAPO-35 (Miachon et al., 2006) and mixed tetrahedral–octahedral oxides.

## 4.2 PREPARATION OF ZEOLITE MEMBRANES

Different methodologies are used for the preparation of composite zeolite membranes. These techniques are listed in Figure 4.1 (Caro et al., 2000).

### 4.2.1 In Situ Crystallization

This is a one-step process for preparing composite zeolite membrane. In a very brief way, the support is made contact with the zeolite sol and then fed for hydrothermal treatment (Miachon et al., 2006). The gel layer either can create a surface layer or can penetrate in to the pores. In the second

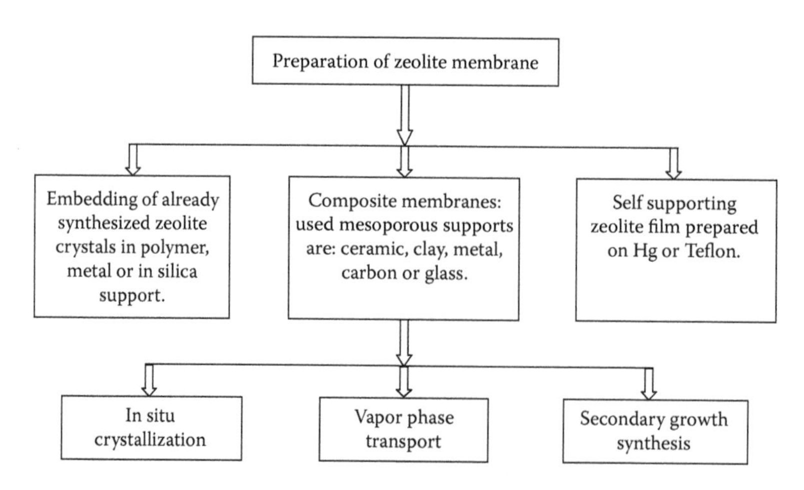


**FIGURE 4.1** Concepts of zeolite membrane preparation. (Reprinted from *Microporous and Mesoporous Materials*, 38, Caro, J., et al., 3–24, Copyright 2000, with permission from Elsevier.)

case, the separation properties are better but the flux rate is less. So there are attempts for preventing this penetration. One of such is reported that ZSM-5 zeolite is layered over $\alpha$-$Al_2O_3$ discs. The discs are adsorbed by furfuryl alcohol and tetraethylorthosilicate followed by a carbonation and partial carbon burning. In this way, these discs develop a diffusion barrier to prevent this penetration (Caro et al., 2000). However, this simple strategy has some drawbacks. The homogeneity of the membrane depend on different factors like the local surface properties and the grown crystals depend on the nuclei, and for obtaining a complete homogenous film, a thick zeolite layer have to develop (Miachon et al., 2006).

### 4.2.2 Vapor Phase Transport

This is a two-step method. In the first step, the zeolite sol is deposited on the support and in the second step; the sol is undergone to hydrothermal treatment by the water coming from the bottom. The template may be present in the water or in the sol itself. The advantage of this process is that, the amount of zeolite deposited can be controlled in a well manner (Miachon et al., 2006).

### 4.2.3 Secondary Growth Synthesis

This seeded supported crystallization process also have two steps viz. a) nucleation step and b) crystal growth. In the first step, a zeolite suspension is prepared (generally of sub µm range) which is used as seed crystal. These crystals are attached with the support. This attachment sometimes done simply by rubbing and sometimes the attachment is enhanced by electrostatic forces or by changing the pH to increase the zeta potential. In the second step, the crystals grow on the seeded crystals and the seed characteristics (size, shape) affect the original zeolite layer. MFI, LTA, zeolite L can be prepared by this procedure. In most of the cases the growth of the crystals are in the perpendicular to the direction of the support (c-orientation). The layer growth can be changed by changing the temperature (Caro et al., 2000, 2008).

Two principle methods are used for nucleation: (1) Dry gel conversion, where the SDA (structure directing agent) is in the vapor phase not in the gel (Matsufuji et al., 2000) and (2) steam assisted crystallization where the SDA remains in the gel (Alfaro et al., 2001). There is a general problem of penetrating sol into the pores of the support and growth of the zeolite crystals into the pores. To overcome this problem sometimes the support is given a sophisticated pretreatment of PMMA wax (Caro et al., 2008). The first research paper published for this method is by K. Horii, K. Tanaky, K. Kita, and K. Okamoto, and first patent is taken by W.F. Lai, H.W. Deckman, J.A. McHenry, and J.P. Verduijn (U.S. Patent 5,871,650). Four methods for the attachment of zeolite crystals to the support are:

1. By controlling the pH, the surface charges are adjusted in such a way so that the seeds are electro statically attached to the support.
2. The surface can be charged by different cationic polymer like poly-DADMAC or redifloc to change the zita potential (Hedlund et al., 1997). This is specially applied for the negatively charged nanoparticles. In general, the MFI zeolite layer is prepared as c-orientation crystals, but by this method, b-orientation crystal layer also can be prepared.
3. Electrophoretic deposition of nanosized seeds on solid support (Berenguer-Murcia et al., 2005).
4. Dipping the support into the seed solution and followed by a heat treatment by which the organic additives are burned out and seeds are fixed via dehydroxilation of the support.

### 4.2.4 Intergrowth Supporting Substances

Permporometric and mixture permeation studies on MFI membrane is telling that the intercrystalline defect is linked to the Al content (Noack et al., 2006). In Figure 4.2, this process is represented in a proper way.

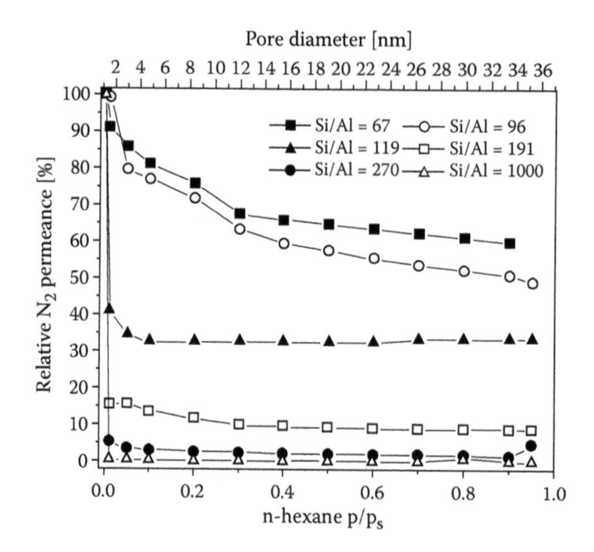


**FIGURE 4.2** Permporometry characterization of MFI membranes of different Si/Al-ratio showing that the residual nitrogen permeance as a measure of the defect concentration increases with the Al content. (Reprinted from *Microporous and Mesoporous Materials*, 115, Caro J., and M. Noack, 215–233, Copyright 2008, with permission from Elsevier.)

The scheme of the permporometry can be explained stepwise: (1) measuring of an inert noncondensable gas like He or $N_2$ through a porous membrane and setting it as 100% and (2) selection of another vapor, which is adsorbed by the zeolite, and, by mixing the vapor with the nonadsorbing gas, sending the mixture through the membrane. Tuning the ratio $p/p_s$ the relative decrease of the gas is measured.

### 4.2.5 Crystallization by Microwave Heating

Currently, microwave heating is getting acceptance as a very efficient tool for preparing zeolite within a short period of time (Cundy et al., 1998). Pioneer papers are reported as successful synthesis of MFI and FAU (Arafat et al., 1993), LTA (Jansen et al., 1992). The advantage of this process is that larger crystals can be developed in a short period of time but up to then there was a mystery about the molecular understanding of the microwave heating zeolite synthesis.

Further, slowly, it was understood that microwave heating brings the sol quickly to the crystal forming temperature and suppresses kinetically the formation of nuclei (Weh et al., 2002).

In very recent years, progress on microwave synthesis is on silicalite-1 crystals on membrane support or direct silicalite-1 crystal. Two main parameters are considered here, viz. heating temperature and heating time. As an example for the maximum permselectivity of n/i-butane the treatment time is 2 h at 160°C was found experimentally. By heating the seeded support in microwave, silicalite-1 membranes can be derived with a controllable thickness and high crystallinity. Extra time heating can damage the membrane. Recent attempts have been done to prepare hydrophilic H-SOD membrane and MFI membrane with titanium used as the additive in the framework of the membrane. Rapid heating rate, differential microwave absorption and nonequilibrium contributions to reactions may limit the formation of undesired phases in comparison with conventional heating (Caro et al., 2008).

Afterward, Yang et al. gives a remarkable progress in the microwave synthesis of zeolite (Chen et al., 2005; Xu et al., 2004; Li et al., 2006b) by in situ aging. In the first stage, the clear synthesis solution is contacted with the support and the gel layer formed by in-situ aged in an air conditioned

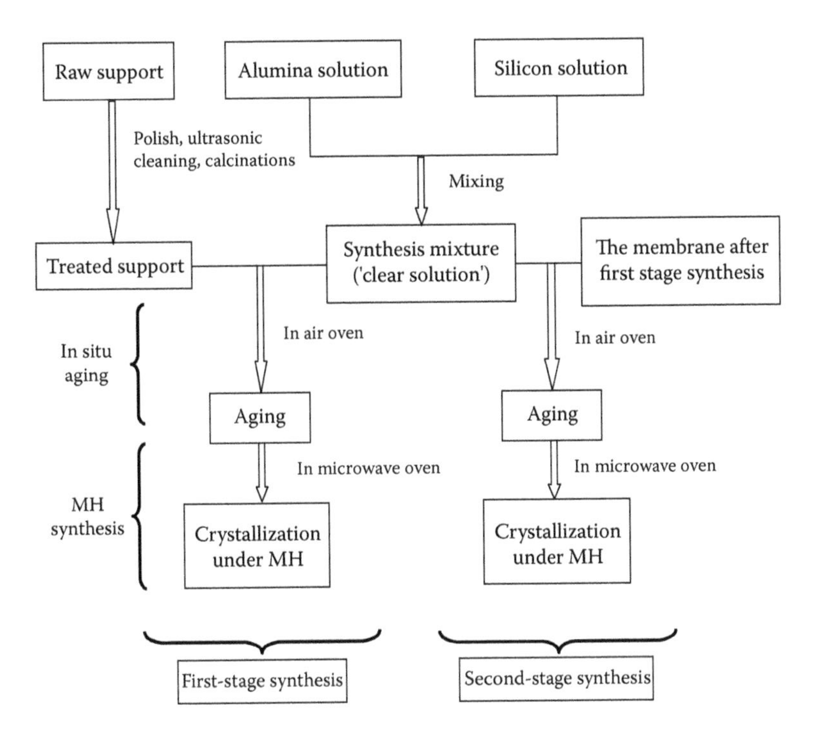


**FIGURE 4.3** Detail of the microwave synthesis method.

oven. Then the membrane is send to microwave oven for zeolite synthesis. The process is repeated in second stage synthesis. This process is explained in Figure 4.3. High quality LTA zeolite membranes can be formed in this way. This procedure takes into account that the support does not absorb microwaves and remains unheated but the microwaves selectively couple with the gel layer because of its higher dielectric loss factor. The gel layer first formed on the support after in situ ageing with plenty of pre nuclei. During the microwave heating they convert in larger crystals. In this way finally same size of zeolite crystals can be formed. However, despite this remarkable progress, the permselectivities of the LTA membranes are only slightly superior to the Knudsen separation factor.

### 4.2.6 Silica Nanoblock

Crystallization mechanism of zeolite is a large research topic from the beginning. TEM gives much more clear idea regarding this matter (Kumar et al., 2007). The structures of these nanoparticles are still controversial. Typical for this new type of hierarchical materials is their bi-porosity containing not only the micropores inside the zeolitic nanoblocks, but at the same time many small mesopores of a few nanometer in between the stacked nanoblocks. But properties of this type of generated materials are different from the original zeolite. In this way stacking of these blocks generates a membrane with potentially thin top layer and extra high flux.

Coating is given on the alumina layer of pore size 100 nm. To prevent the entering of the zeolite nanoparticles there first a titanium sol-gel layer is given. Afterward, the coating is given with the help of the surfactants and by dip coating. Figure 4.4 explains this fact.

However the performance of this membrane is very poor. The cutoff is only 250 Dalton. The challenge lies in the proper stacking of nanoblocks on the surface after removing the surfactant molecules. The recently reported combined UV and thermal treatment of spin-on silicalite-1 films could be implemented for the soft detemplating of the membrane with a reduced tendency of crack formation (Caro et al., 2008).

**FIGURE 4.4** Cross section of a supported silicalite-1 membrane made by coating with nanoblocks.

A very recent and very interesting finding is published in the *4th International Zeolite Membrane Meeting* by E. Biemmi et al., claiming that when LTA zeolite is mixed with HCl, crystals are dissolved and a nanoblock structure is formed. They applied a layer of nanoblock on the vicor glass, applied the separation of hydrogen from different gas mixture, and found that the separation factors are quite good. For $H_2/CH_4 > 1200$, $H_2/N_2 > 1600$, $H_2/CO > 600$ (Nishiyama et al., 2007).

### 4.2.7 Oriented Zeolite Membrane

The orientation of zeolites also changes the properties of membrane. In the science of porous material it is also a challenging issue. There exist different concepts for generation of pore orientation. Experiment shows that c-oriented MFI zeolite is less applicable (Caro et al., 1993) rather than a- or b-oriented MFI crystals because they gives higher fluxes. Recently, even a 2.5-m-thick c-oriented MFI zeolite layer was also obtained by epitaxial growth. For b-oriented crystals tetrapropyl ammonium hydroxide has been used as SDA. For depositing the seed in b-orientation, the stainless steel supports have to be smoothened by silica because the seed monolayer is attached to the silica by covalent bond. Another finding is telling that the b-oriented crystals can be grown on chitosan modified alpha-alumina disc also (Caro et al., 2008).

Both in situ and secondary growth can give different orientation of crystals but for MFI generally c-orientation is obtained perpendicular to the surface. Usually, the nanocrystallites used as seeds do not show developed crystal faces, and, therefore, these crystallites are randomly oriented. If the crystal growth is anisotropic, those crystallites with their fastest growth direction pointing away from the seeded surface will grow more rapidly than crystallites in other orientations. Finally, the crystals with the fastest growth direction perpendicular to the plane of the membrane will dominate. For MFI, c orientation is the longest dimension and so generally the crystals grow as a columnar structure. Recently, Tsapatsis et al. prepared b-oriented MFI membrane in large crystals ($0.5 \times 0.2 \times 0.1\ \mu m^3$). As SDA, they have used di- and trimers of TPAOH. The resulting polycrystalline films are 1 lm thin and consist of large b-oriented single crystals with straight channels running down the membrane thickness. Another disadvantage is that for xylene separation, c-orientation gives a separation factor nearly 1.0 and for b-orientation separation factor is nearly 500. Another interesting point is that in the later case the separation factor is increasing with the temperature. The experimental findings are characterized by the interplay of adsorption and diffusion effects. At low temperatures the zeolite pores are filled and the more mobile *p*-xylene cannot move faster through the pore network than the less mobile *o*-xylene (Lai et al., 2003a). Another situation occurs at low pressure. At this point, *p*-xylene can move more easily at the presence of *o*-xylene. But according to Snyder and Tsaptatsis's recent findings, the best separation performance for xylene isomers is obtained from b-oriented films, while c-oriented hardly exhibit any selectivity for *p*-xylene. This result contrasts with the previous findings and from this conclusion can be drawn that many complex things are occurring within the pores.

### 4.2.8 Bilayered Zeolite Membrane

Different aims are there for preparation of multilayered zeolite membrane, viz. i) improved separation selectivity by repetitive crystallization of same zeolite, ii) novel properties by using layers of different type zeolites and iii) application of zeolite layer with other membrane type.

First, Vroon et al. (1996) proposed that for improving the quality of MFI, repeat the crystallization process. It is observed that two repetitions improve the quality, but afterward, no improvement occurs, and chances of crack formation increase due to increment of thickness. Many works has been done on LTA, FAU and MFI as multi layered zeolite membrane and those works has been shown enough potential to increase the application. Especially when one layer shows catalytic property with other layer as inert, those can be used as membrane reactor (Sterte et al., 2001; Li et al., 2001). Lai et al. (2000 patented the fabrication of multilayered zeolite membranes and demonstrated both seeded and epitaxial growth of ZSM-5 on silicalite-1 layers supported on porous alumina and stainless steel supports. Some researcher has prepared silicalite-1 layer on LTA layer on trumen supports that allows the growth of the second layer without dissolution of the first. Very recently ZSM-5 and silicalite-1 bilayer has been prepared by Li et al. (2002). A second ZSM-5 layer on the first layer of silicalite-1 can be applied. The crystallization of shape-selective silicalite-1 layers on LTA and FAU layers turned out to be even more complicated.

## 4.3 PROBLEMS ASSOCIATED WITH CRYSTAL STRUCTURE

It is a challenge of membrane separation technology to do separation at high temperature. Zeolite membranes are very promising regarding this matter, but they have some problems that occur especially at high temperature due to the crystal structure.

### 4.3.1 Irregular Thermal Expansion

Almost all thermal expansion of zeolite shows a peculiar behavior. It expands steadily with increments of temperature, and during a particular temperature region, it shrinks. Like MFI and fluoride synthesized silicalite shows a positive volume expansion nearly in the order of $+10 \times 10^{-6}$, and a negative volume expansion coefficient in the order of $-10 \times 10^{-6}$. For MFI, the temperature range is 75°C to 120°C, and for fluoride synthesized silicalite, the range is 300°C to 400°C. Another interesting behavior is that the expansion coefficients are different in different directions. For example, for MFI crystals, the expansion coefficients are 20, −22 and $0.5 \times 10^{-6}$ in a, b, and c directions, respectively.

So for a composite membrane, the support expands continuously and the zeolite layer contracts. For overcoming this problem, different researchers have proposed different solutions like to calcine the MFI membrane maximal at 400°C or to use a porous intermediate layer in between the ceramic or metal support.

### 4.3.2 Anisotropy of Mass Transport

The main reasons of anisotropic mass transport through the zeolite layer are the anisotropic pore geometry and the orientation of the microcrystals. For example, MFI zeolite has elliptical pore structure (0.51 × 0.545 nm) extending along y direction and cross linked with almost circular

(0.54 nm) cross section in x direction. Therefore, mass transport in the z direction can proceed only by repeated change in the x and y directions. From the mass transport theory,

$$\frac{c^2}{D_z} = \frac{a^2}{D_x} + \frac{b^2}{D_y}$$

where, $D_x$, $D_y$, and $D_z$ are diffusivities in the x, y, and z directions, and mass transport in the z direction will be at least 4.4 times less than the transport in the x or y direction.

### 4.3.3 Influence of Crystal Boundaries

For migrating to permeate side from feed side a molecule have to overcome several boundaries. Experiment has done on the propagation of n-heptanes for single and twinned crystal system under same condition and it is observed that the permeation flux in case of single crystal is almost 60 times greater than the twinned crystals. This observation attributed to the hindered diffusion through the MFI crystals. From another experiment by Muller et al. for studying sorption and diffusion of toluene within ZSM-5 crystals strongly depend on the degree of crystal intergrowth. The results show that the diffusion coefficient in single crystal is three orders of magnitude higher than the polycrystalline sample. So this can be concluded that the thickness of the zeolite layer should be as low as possible.

## 4.4 USE OF ZEOLITE MEMBRANES

Various applications of zeolite membranes are summarized below.

### 4.4.1 Water Separation

Hydrophilic LTA zeolite membrane is very well applied for separation of water from organic solution by steam permeation and pervaporation. As a blend for gasoline the water content must be reduced less than 2000 ppm. Or for production of bio ethanol the water content must be less than 500 ppm. In those cases LTA is applied successfully. LTA was successfully prepared by Bussan Nanotech Research Institute Inc. and applied by Mitsui & Co., Japan, for dehydration of ethanol. The separation factor of nearly 10,000 with a flux of 7 kg $m^{-2}$ $h^{-1}$ is observed. In alpha-alumina support, two different membranes are produced. The first one is 16 mm in outer diameter and 1 m in length; it is used for dewatering bioethanol. The second one is used for recovering *i*-propanol and is 12 mm in diameter and 0.8 m in length.

Mitsui & Co. prepared a pilot scale plant from April to September 2003, and afterward, they prepared a large scale plant at Daurala Sugar Works (Uttar Pradesh, India) with an LTA membrane area of 30 sq. m. This plant is successfully operating since January 2004.

Another important application is that smaller Zeospec membranes are used for the separation of azeotrope mixture of *i*-propanol and water. For the mixture of 90% *i*-propanol and 10% water the separation factor is nearly 10,000. Therefore, Mitsui-BNRI used the same support as FAU zeolite (Si/Al ratio between 1.5 and 1.6) and tested successfully for the dewatering of IPA solution. The water flux was 7–10 kg $m^{-2}$ $h^{-1}$, and the separation factor was approximately 300. The same membrane is also applied for the separation of alcohol and ether separation: ethanol/ETBE mixture of 5/95 ratio is separated with fluxes of 2.2 and 4.1 kg $m^{-2}$ $h^{-1}$ and selectivity of 2800 and 1600 at 90°C and 110°C, respectively. It is observed that the stability of the hydrophilic FAU membranes could be increased having a Si/Al ≈ 2.2 by using USY seed crystals. Further increase in stability of hydrophilic membranes can be done by increase Si/Al ration to 120.

Despite these technologies, zeolites are also used for the adsorptive separation of water mainly from ethanol. By the so called DELTA-T technology even less than 100 ppm water also can be separated from ethanol using zeolite 3A. For removing low amount of water from the mixture, adsorption is better option because heat management also can be done by this. As the azeotrope *i*-propanol/water contains more water (12.1 wt.%), hydrophilic organic or inorganic membranes can be used with steam permeation and has a better probability for in comparison with the ethanol/water mixture.

With these innovations, there experiments have been done on the development of novel hydrophilic membrane. LTA membranes can be substituted by ZSM-5. PHI and FAU membranes increase the stability but decrease the water fluxes due to a reduced hydrophilicity. One innovative way to produce hydrophilic compound is to input a heteroatom like boron in the framework and then remove it by thermal treatment. As a result, a lattice vacancy forms which is a nest of four hydroxyls that provides the deborated silicalite-1 membrane hydrophilicity. Another remarkable progress is the H-SOD membrane with a thickness of 2 micrometer and on the support on alpha–alumina (Khajavi et al., 2007). The membrane pore size is about 0.265 nm and successfully tested for pervaporation of water and *i*-propanol. The fluxes are 2.25 kg $m^{-2}$ $h^{-1}$ at 150°C. A recently developed PHI membrane shows a mixture separation factor nearly 3.000 with water fluxes > 0.3 kg $m^{-2}$ $h^{-1}$ for pervaporation of a 90 wt.% EtOH/10 wt.% $H_2O$ mixture.

### 4.4.2 Gas Separation

The separation ability of a zeolite membrane is not only depends on the pore size but also with the mixture adsorption and mixture diffusion data. Separation depending on molecular sieving is relatively simple but it requires perfect membrane. In the case of separation of $N_2/SF_6$ through MFI membrane, it appears like the shape-selective separation, but detailed investigation shows that it is due to a sorption–diffusion mechanism.

Another remarkable result is shown for the separation of n-heptanes from other hydrocarbon through $AlPO_4$-5 system. It shows, if both the component is smaller than the pore size of the membrane, then the flux is reduced than that of the pure n-heptanes due to interaction of the two different types of molecules. On the other hand if one component is larger than the pore size then flux rate is not disturbed.

Parallel pores provides much more flux than the three dimensional system. Like in $AlPO_4$ model the flux is almost two magnitudes higher than the ZSM system for a single molecule (Caro et al., 2000).

As maintained before, there is a big advantage of zeolite membrane is that it provides diffusion controlled separation also. If both the components are smaller than the pore but there relative size varies very much then molecular mobility becomes the prime factor. But for this mechanism the pores of the zeolite should be filled by a low or medium pressure. In two different experiments on separation of *p*-/*o*-xylene shows that if the pressure is higher (6120 Pa) then no separation is occurred on the other hand under medium pressure (500 Pa) it shows a separation factor nearly 100. Same type of result observed when n-butane (molecular size = 0.43 nm) and i-butane (molecular size = 0.5 nm) through MFI zeolite membrane (pore diameter = 0.55 nm). But in the case of separation factor, $\alpha\text{-}_{\text{n-butane/i-butane}}$ decreases with the partial pressure of n-butane/i-butane.

Another important factor that plays an important role in the gas separation is adsorption. The strongly adsorbing molecule favors over the weekly adsorbing species. From published data the flux of hydrogen or methane (weekly bonded molecule) can drop a number of magnitudes in the presence of n-butane or $SF_6$. For H2/SF6 a high permselectivity of 136 is determined where as the separation factor is only 13. For the case of CH4/n-butane the permselectivity is 3 whereas separation factor is nearly 0.06 (Caro et al., 2000).

#### 4.4.2.1 Hydrogen Separation

Gas polymer, metal membranes (Pd, Ag, Cu, Nb, V) even microporous membrane like silica, zeolite membranes are used for hydrogen separation. Except carbon molecular sieve MFI zeolites are most common use. As the kinetic diameter of hydrogen is around 0.29 nm, so the pore size should be larger than this value but lower than the other molecule from which hydrogen have to separate. In general, microporous membranes prepared from silica and zirconia prepared by sol-gel method and by dip coating have very low hydrothermal stability, which is unusable for separation of hydrogen from the mixture of steam. However, only the sol–gel chemistry of SiO2 allows forming narrow pores but SiO2 is hydrothermally unstable, ZrO2 and TiO2 are more stable but their sol–gel chemistry does not allow forming narrow pore systems suitable for shape-selective gas separation.

Comparing to these amorphous metal oxide membranes, crystalline zeolite membranes offer much better thermal and hydrothermal stability. Hydrogen separation through silicalite-1 is done by researchers (Aoki et al., 2000; Bernal et al., 2002). In these cases the separation effect is based on the interplay of mixture adsorption and mixture diffusion effects. The separation factor of hydrogen from i-butane is 1.5 at room temperature, and it increases to 70 at 500°C. The explanation is like this: According to kinetic diameter, both the components can pass through the pores of silicalite-1 membrane. At low temperature, i-butane is adsorbed in the pores of the silicalite-1 and blocks the passage of hydrogen, but at high temperature, the adsorption decreases, so hydrogen can pass through (Lin et al., 2007).

In a nutshell, silicalite-1 is the only zeolite that can be compared with other inorganic or organic membranes. However zeolite membranes can be an effective tool for separation of hydrogen. A six-ring structure zeolite membrane with silica is developed with a 0.3 nm pore size. The narrow pore size zeolite structures and compactness viz. density of pores per unit area is less of the suitable zeolite structures for hydrogen sieving require thin membrane layers for sufficient fluxes.

Except MFI, in recent days, DDR membranes are also showing hydrogen separation characteristics (Lin et al., 2007). Single gas permeance for silicalite membrane was reported by template-free secondary growth method and for DDR-type zeolite membrane prepared by NGK Insulators Ltd Japan. The permeation data of He, $H_2$, $CO_2$, CO, $SF_6$ through silicalite-1 were studied in detail. The kinetic diameters of these gases are 2.6, 2.89, 3.3, 3.76 and 5.5 Å respectively. $SF_6$ has shown much lower gas permeability ($2 \times 10^{-9}$ to $3 \times 10^{-9}$ mol $m^{-2}$ $s^{-1}$ $Pa^{-1}$ depending upon the operating pressure). CO and $CO_2$ are shown ($10^{-8}$ to $2 \times 10^{-8}$ mol $m^{-2}$ $s^{-1}$ $Pa^{-1}$), and $H_2$ and He were shown ($2 \times 10^{-8}$ to $3 \times 10^{-8}$ mol $m^{-2}$ $s^{-1}$ $Pa^{-1}$) depending upon the operating pressure. At room temperature, $H_2/SF_6$ permselectivity is 90, which is much higher than the Knudsen selectivity (Caro et al., 2008).

#### 4.4.2.2 Carbon Dioxide Separation

The main drawback for $CO_2$ separation by polymeric membrane is that it suffers from the problem of swelling (Wind et al., 2004). Zeolite membranes do not have this kind of problems. With this, polymer membranes can separate $CO_2/CH_4$ effectively but not $CO_2/N_2$ (the kinetic diameter of $CO_2$, $CH_4$ and $N_2$ are, respectively, 0.33 nm, 0.38 nm and 0.36 nm) (Sebastian et al., 2006). Few zeolite membranes have already shown promising results for separation of $CO_2$ from $CH_4$ and $N_2$ (Hasegawa et al., 2002; Cui et al., 2004). As the separations of these membranes are based on competitive adsorption, the selectivity's are low. For the separation of $CO_2/N_2$, the methodology is the stronger electrostatic quadrupole moment for $CO_2$ than $N_2$. The maximum separation factor of 12–13 is obtained within 6–16 bar. The selectivity depends mainly on two things: (1) the support and (2) the modification of the MFI structure. For different kind of membranes different permeability and selectivity are reported. And for $CO_2/CH_4$ separation, silicalite-1 membranes give better result because $CO_2$ have better favorable adsorption (Zhu et al., 2005). DDR (0.36 nm × 0.44 nm) and SAPO-34 (0.38 nm) have pores that are sililar in size of $CH_4$ (0.38 nm) but larger than $CO_2$. So it is expected that these membranes will show high separation factor. With T type membrane, some researchers (Cui et al., 2004) have found separation factor of 400 and permeance of $4.6 \times 10^{-8}$ mol $m^{-2}$ $s^{-2}$ $Pa^{-1}$ at 35°C. On the other hand, Tomita et al. (2004) obtained a separation factor of 220 with a permeance of $7 \times 10^{-8}$ $m^{-2}$ $s^{-2}$ $Pa^{-1}$ on a DDR membrane.

Another important finding is that it is seen that selectivity drops with the increase of temperature and pressure (Figure 4.5). The selectivity drops with the increase of temperature is the cause of the fact that the diffusivity depend on the total loading (Van den Bergh et al., 2007).

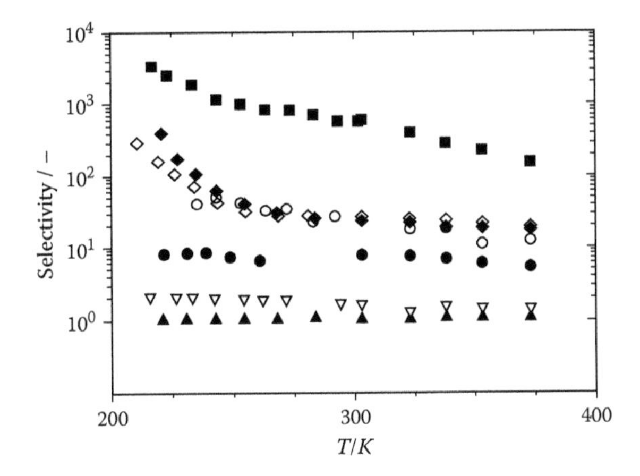


**FIGURE 4.5** Selectivity of equimolar mixtures through the DDR membrane as a function of the temperature constant total feed pressure of 101 kPa, sweep gas He at 101 kPa. Legend: $CO_2/CH_4$ (■), $N_2/CH_4$ (○), $CO_2$/Air (◆), $N_2O$/Air (◇), Air/Kr (●), $O_2/N_2$ (▽), $N_2O/CO_2$ (▲) (Van den Bergh et al., 2007b).

The high selectivity is the effect of molecular sieving and some extent to the preferential adsorption. Recently many mixture gas permeation results for high silica percentage DDR membranes are published (Van den Bergh et al., 2007). Figure 4.5 represents all the data in detail. The selectivity decreases with increasing feed pressure. Also, hydrotalcites, a layered double hydroxide compound, are good candidates for being $CO_2$–selective membranes due to their desirable properties (Yang et al., 2002; Gardner et al., 2002).

SAPO-34 membrane on stainless steel support by in-situ crystallization is synthesized (Li et al., 2006a). For this membrane with a Si/Al ratio of 0.1, the $CO_2/CH_4$ selectivity of 170 with a $CO_2$ permeance of $1.2 \times 10^{-7}$ mol $m^{-2}$ $s^{-2}$ $Pa^{-1}$ was found at 22°C. With decrease the selectivity increase and the separation factor is 560 at –21°C. At 7 MPa, this membrane shows the selectivity of 100 for a 50%–50% mixture over one week. The overall performance of SAPO-34 membranes for separation of $CO_2$ and $CH_4$ is nearly industrial requirement. From these results a general conclusion can be drawn that small pore hydrophobic membranes are appropriate to separate $CO_2$ from humid gas. On the other hand, DDR membranes show high $CO_2$ flux and selectivity but with a negligible water permeance.

Very recently, an ion exchange study for $CO_2$ separation by SAPO-34 membrane has been done (Hong et al., 2006). The membranes are prepared inside SS tubes (Li et al., 2006). $Li^+$, $Na^+$, $K^+$, $NH_4^+$, and $Cu^+$ are used as ions and the selectivity for both $CO_2$ and $CH_4$ are increased by almost 60%. The separation factor is in the range of 60–90.

In some cases, contradictory results have been found for separation of $CO_2$ by zeolite membrane. For example, Clet et al. (2001) found that the separation factor is between 2 to 3 for equimolar mixture of $CO_2/N_2$ by FAU membrane at 30°C, whereas Weh et al. (2002) found the opposite selectivity viz. $N_2/CO_2$ to be between 5 and 8. This contradiction can be defended by the role of moisture. Shin et al. (2005) found that for ZSM-5, selectivity for $CO_2/N_2$ is 54.3 at 25°C and 14.9 at 100°C. They explained this fact by the mechanism that moisture occupies large pores and only through those nitrogen flows.

### 4.4.3 *p*-Xylene Separation

Attempts are being made to replace the processes for separation of xylene isomers like fractional crystallization, adsorption or distillation by membrane process, because these existing processes are energy consuming process. Silica membranes with some organic SDA, can shift the pore size of 0.3 to 0.4 nm to 0.5 to 0.6 nm. The permselectivity of *o*-/*p*-xylene in these membranes are in

**TABLE 4.1**
**Single Component Pervaporation Results of the Xylene Isomers on an MFI Membrane at 25°C**

| Orientation | *p*-Xylene Flux (kg $m^{-2}$ $h^{-1}$) | *o*-Xylene Flux (kg $m^{-2}$ $h^{-1}$) | Ideal *p*-/*o*-Xylene Permselectivity |
|---|---|---|---|
| Random | 27.1 | 1.1 | 21 |
| c-Oriented | 22.9 | 6.3 | 3.6 |
| c-Oriented | 35.2 | 9.9 | 3.6 |
| h, o, h-Oriented | 24 | 9.9 | 2.4 |
| h, o, h-Oriented | 32.6 | 11.5 | 2.8 |

*Source:* O'Brien-Abraham et al. (2007).

the order of 10–20 (Xomeritakis et al., 2003). However, X-ray amorphous silica membranes suffer from hydrothermal instability. A crystalline silicalite-1 membrane with a pore size of about 0.55 nm should be perfectly suited to separate the *p*-xylene (kinetic diameter 0.58 nm) from the bulkier *o*- and m-xylenes (kinetic diameters about 0.68 nm) and to show a higher stability in comparison with silica.

Some literature found that b-oriented silica crystals provide almost two magnitude higher mixture separation factor of p/o xylene. The b-oriented crystal prepared with TPA as a template gives permeation of *p*-xylene of about $2 \times 10^{-7}$ mol $m^{-2}$ $s^{-1}$ $Pa^{-1}$ with a *p*-/*o*-xylene separation factor of up to 500. These experiments are carried out in extremely low partial pressure of xylene. It is known that the separation factor of *p*-xylene decreases with increasing partial pressure (Xomeritakis et al., 2001).

Another study tells that silicalite-1 membrane prepared without template, by two stage dip-coating method. This membrane provides an excellent separation factor of 40 and a *p*-xylene separation flux of 0.137 kg $m^{-2}$ $h^{-1}$, which can be explained by the absence of intercrystalline defects because in the case of the template free synthesis, no thermal burning have to apply (Yuan et al., 2004). Comparisons between different orientation and fluxes are represented in Table 4.1.

Except for silicalite-1, a tubular MFI membrane (O'Brien-Abraham et al., 2007) has shown a separation factor of 7.7 of *p*-xylene from m-xylene and *o*-xylene. The permeation flux of *p*-xylene is $6.8 \times 10^{-6}$ mol $m^{-2}$ $s^{-1}$ at 250°C and atmospheric pressure.

### 4.4.4 Alcohol Separation

It is well known that some hydrophilic membrane like LTA and FAU can separate moisture from alcohol. It is also proved that same zeolite can be used for separation of two alcohols (if one is more polar than the second one) or an alcohol from organics (Kita et al., 2001).

Taking the challenge from opposite side is also done. Fermentation processes stop after 15% production of alcohol. This is separated by MFI type membrane. The flux is 1 kg $m^{-2}$ $h^{-1}$ with a separation factor of 57. For increasing the ethanol flux, many attempts have been done. Very recently TS-1 membrane has shown 5% ethanol/water separation factor of 127 and a total flux of 0.77 kg $m^{-2}$ $h^{-1}$ (Coronas and Santamaria, 2004a).

### 4.4.5 Pervaporation

The first criterion for use in pervaporation applications is that the zeolite should be hydrophilic. Mitsui & Co. from Tokyo has used zeolite A as semi commercial membrane. Mitsui-LTA was reported to shown a water flux of 16 kg water/$m^2$ h for the separation of water/*i*-propanol mixture at 120°C. They have also applied zeolite X, zeolite Y and ZSN-5 for pervaporation on different types of separation like isopropanol, ethanol water mixture, methanol and acetone separation and water dioxane mixture.

Due to the problem of instability of zeolite A attempts have been made to use highly stable zeolites like silicates, but these zeolites are known as hydrophobic membranes, i.e., the organic compounds passes through the zeolite and water will retain in the retentate.

### 4.4.6 As Membrane Reactor

Membrane reactor is a multifunctional apparatus by which a chemical reaction especially catalytic reaction taken place in the presence of a membrane. In case of equilibrium limited reaction the conversion can be improved. But main problem in the application of catalytic membrane reactor is that still now those cannot be used in high temperature applications. Organic polymer membranes can be used as reactor but the applicability is only at low temperature (<150°C). Metallic or ceramic membranes could be used for high temperature (400°C–600°C), but the challenges are to increase the module reliability under extreme temperature conditions (Caro et al., 2000).

Zeolite membrane reactors are used to enhance chemical reactions like dehydrogenation, partial oxidation, isomerization or esterification. Traditionally, zeolite membranes are used for shifting the equilibrium point or for replacing or removing the inhibitors (Coronas and Santamaria, 2004b). Due to the molecular sieving properties, zeolite membrane reactors are also known as "membrane extractor reactor" removing small product molecule like water or hydrogen created by dehydration or dehydrogenation. Zeolite films also can increase the selectivity of reaction, viz. Co/$SiO_2$ catalyst pellet by a HZSM-5 shell enhances the selectivity of middle i-paraffines and suppresses the formation of long chain hydrocarbon by the Fisher–Tropsch reactions (He et al., 2005).

MFI membranes can be used for the catalytic dehydrogenation of i-butane. Hydrogen and i-butane both can pass through MFI membrane (kinetic diameters are 0.29 and 0.50 nm respectively) because the pore size is 0.55 nm. Experimental results for the silicalite-1 assisted dehydrogenation of i-butane were described in some literatures (Casanave et al., 1999; Schäfer et al., 2003). In an experiment on conventional packed bed was done and found that as hydrogen is removed from the silicalite-1 membrane, the i-butane conversion increases by 15%. Figure 4.6 represent these results (Illgen et al., 2001).

Removal of hydrogen leads to hydrogen-depleted conditions, which have two positive effects: (1) conversion of i-butane increased due to of the reverse reaction and (2) increase in the selectivity of i-butane due to suppression of hydrogenolysis. So, at the beginning of the reaction the i-butane

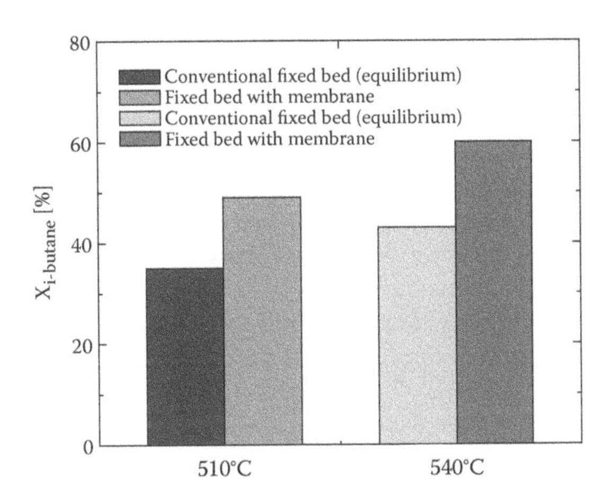


**FIGURE 4.6** Increase of the i-butane conversion above the equilibrium limit if hydrogen is removed through a silicalite-1 membrane. Conditions: WHSV = 1 $h^{-1}$, $Cr_2O_3/Al_2O_3$ catalyst (Süd-Chemie), membrane area per unit mass of catalyst = 20 $cm^2$ $g^{-1}$, data after 20 min time on stream. (Reprinted from *Catalysis Communications,* 2, Illgen, U., et al., 339–345, Copyright 2001, with permission from Elsevier.)

yield is higher by about 1/3 than the conventional packed bed. And during hydrogen removal, cocking is promoted and after 2 h approximate time on stream the olefin yield of the membrane reactor drops below that of the general packed bed (Schäfer et al., 2003). However, after an oxidative regeneration the activity and selectivity completely restored.

During the esterification, yield is increased in many ways. Most frequent technique is to remove the low boiling ester by reactive distillation. Another concept is to keep the water content low by using hydrophilic component like LTA zeolite or using any chemical reaction which consumes water. Another case like low temperature esterification of methanol or ethanol with short chain monovalent carboxylic acid, hydrophilic polymer membrane can be used for removal of water. On the other hand, esterification at higher temperature, inorganic membrane with high stability against acid has to be used. ZSM-5 zeolite membrane is one of them. The advantage of esterification reaction in membrane reactor is that the yield can be increased in a greater way.

The drawback is that this ZSM-5 membrane is stable up to pH = 1 but the hydrophilicity is low due to low Al content. The water flux is 72 g $m^{-2}$ $h^{-1}$ $bar^{-1}$ is too much low with respect to technical application. In contrast, a recently developed PHI membrane has shown a water flux of 0.3 kg $m^{-2}$ $h^{-1}$ $bar^{-1}$ for the mixture of 10 wt% water and 90 wt% ethanol with a separation factor of greater than 3000 and that can be a better choice (Kiyozumi et al., 2007). Another example of very efficient membrane reactor is separation of water during the esterification reaction of lactic acid and ethanol. This reaction reached almost 100% conversions in 8 h (Kita et al., 2001).

Conventionally, hydrophilic membranes are used for dehydration of solvent. With this all – silica DDR membranes are also efficient for separation of water from organic solvent under pervaporation condition. Excellent performance for removing water from ethanol and methanol is done at the flux of 5 kg $m^{-2}$ $h^{-1}$ and selectivity of 9. Even 1.5% concentrated water also can be removed. The water easily passes through DDR pores and organic molecules faces hindrance (Kuhn et al., 2008).

Different studies to remove water from Fischer–Tropshe reaction were reported in literature by the use of hydrophilic membrane (Espinoza et al., 2000), or sol-gel silica based membrane or ceramic supported polymeric membrane (Rohde et al., 2003). Among all, the application of more hydrophilic zeolite membranes such as LTA (Zhu et al., 2005) and H-SOD (Khajavi et al., 2007) were most promising.

### 4.4.7 Micromembrane Reactor

The combination of the concepts of process miniaturization and membrane reactor itself draws the micro membrane reactor. These reactors are expected to be used in micro pharmacies or miniature factories or for the production of high value added products. The addition of zeolite in the micro reactor also includes catalytic properties (Chau et al., 2002).

A very recent fine reaction carried out in micro membrane reactor is the Knoevenagel condensation reaction, where the selective removal of water increases the conversion by 25% (Lai et al., 2003b). Another system used a very intelligent design for the reaction of benzaldehyde and ethyl cyanoacetate to produce ethyl-2-cyano-3-phenylacrylate was catalyzed by a CsNaX zeolite and water was selectively pervaporated through LTA zeolite (Lai et al., 2003c). The performance of the micro membrane reactor is limited mainly by the kinetics (Zhang et al., 2004). A fourfold increase in conversion is observed when CsNaX-$NH_2$ is used instead of CsNaX (Yeung et al., 2005). Placing the separation unit just after the catalytic unit further increases the performance. Other Knoevenagel condensation reactions are also beneficial in the same way when water is removed from the mixture (Lau et al., 2007).

The multichannel micro reactor is used for continuous oxidation of aniline on TS-1 nanoparticles. The higher surface area to volume ratio can be obtained in microreactor (up to 3000 $m^2/m^3$). Azobenzene production from nitrobenzene and aniline is observed in the micro reactor. It is observed that the increasing temperature is beneficial for both selectivity and yield (Wan et al.,

2004). But beyond 67°C, use of micro reactor is inefficient due to bubble formation and $H_2O_2$ separation (Wan et al., 2005).

Syntheses of other advance materials are also prepared in zeolite enclosed micro reactor. Hollow silica nanosphere was prepared successfully within the zeolite enclosed micro channels, fabricated by selective etching of the silicon below the zeolite membrane. The nanospheres were then prepared by using silicon substrate as silica source and the ferrocene as catalyst.

### 4.4.8 Functional Films

With the uses discussed above, zeolite films can be used as functional films in chemical sensors or as electrode or as optoelectronic device. It also can be used as low dielectric constant material or corrosion resistant coating (Mitra et al., 2002) or as proton exchange membrane (Holmberg et al., 2005).

By the method of dry conversion, the precursor gel can be converted to a zeolite layer in the presence of vapor (Matsukata et al., 1999). The vapor can be consist of only steam or steam with SDA like TPAOH. In contrast to the conventional steam-assisted crystallization in which the substrate is coated with all the nutrients and then steam-treated, by a novel steam-assisted method, the oxidized surface layer of a silicon wafer can be transformed into a silicalite-1 zeolite film (Alfaro et al., 2001). First, the silicon wafer is coated with 1 M TPAOH and then zeolite films are coated over it. Silicon wafer serves as the support for the zeolite film and as the Si source for the silicalite-1. For a-orientation or b-orientation silicalite-1 crystals, 100°C to 200°C is best. This procedure is best for preparation of chemical sensor and optoelectronics.

Silica materials are used as insulators in on-chip interconnectors for their high porosity, hydrophobicity, acceptable heat conductivity and low dielectric constant. A remarkable improvement could be achieved by a post treatment of UV on silicalite-1 film to induce hydrophobization (Eslava et al., 2007). During the removal of the organic template the UV radiation decrease the silanol drastically. So the crack formation during the synthesis can be minimized. Pure-silica zeolites have a remarkably higher mechanical strength and hydrophobicity than amorphous porous silica due to their crystalline structure making them a likely dielectric material for enabling smaller feature sizes in future generation of microprocessors (Lew et al., 2007).

For transdermal drug delivery, zeolite membranes are formed in the form of micro needles. The intergrown zeolites have shown excellent mechanical strength and successfully pierced through skin without damage (Wong et al., 2007). Through zeolite permeable wall controlled drug delivery can be achieved. A very innovative study has been done by preparing zeolite membrane hollow sphere by growing a thin zeolite shell on polymer bead, which is used to store and release bioreactive molecule (Wong and Yeung 2007).

Another use of zeolite nanocrystals is to make the surface smooth up to 20 nm surface roughness. The coating shows a broadband antireflection effect with less than 1% average reflection over the visible range. With the proper control of the thickness shifting of the reflection can be done to achieve a neutral color (Chiang et al., 2007).

The high silica zeolites can be used for corrosion protection also (Beving et al., 2007). The as-synthesized SDA-containing MFI films are nonporous and allow the coating of complex shapes and in confined spaces by in-situ crystallization (Cheng et al., 2007).

### 4.4.9 Mixed Matrix Membranes

Just after the successful preparation of the idea of mixed matrix membrane viz. binding the zeolite crystals in the polymer matrix is appeared (Wyllie and Patnode, 1950). The property of the matrix is used as well as the peculiar properties of the inorganic fillers are also used (Koros and Maharajan, 2001; Moore et al., 2003). Currently polymeric membrane researches have reached to a saturation level so attempts are being done to accumulate zeolite nanoparticles in polymeric bed. This is generally done by spinning technology.

Early studies have been focused on small pore zeolite like LTA, but a main drawback of LTA for moist gas separation is that water blocks the pores of the LTA zeolite thus reducing the flux (Moore et al., 2003). Recent studies have focused, therefore, on more hydrophobic zeolites with a high molar $SiO_2/Al_2O_3$ ratio. As an example, zeolite CHA with a $SiO_2/Al_2O_3$ ratio >30 with a particle size < 1 lm can be processed to mixed matrix membrane layers of about 1 μm thickness (Zones et al., 2004).

Low selectivities have been found for zeolite–polymer mixed matrix membrane probably due to poor wetting (Vu et al., 2003). It is observed that the polymer can penetrate into the pores of membrane. As a result the glass temperature and young's modulus both increase for the mixed matrix. In the comparing of the mixed matrix and pure polymer, the permselectivity and the permeability both are increase for the mixed matrix.

Another type mixed matrix is made from zeolite/carbon. A polyimide precursor containing MFI crystals was slip cast onto a stainless steel support and calcined at 580°C in nitrogen. For the medium of oxygen–nitrogen mixture, selectivity of 4–6 with relative high oxygen permeance of about $10^{-8}$ mol $m^{-2}$ $s^{-1}$ $Pa^{-1}$ is obtained.

## 4.5 PROBLEMS IN SUCCESSFUL COMMERCIALIZATION

Application of new organic polymeric membranes for industrial gas separation like $N_2$ or $H_2$ production from purge gas or separation of $CO_2$ from methane are increasing day by day. But the problem with this is that the membrane is poisoned by higher hydrocarbons that are present with methane.

And in the case of zeolite membrane, those get hydrolyzed even under slightly acid and hydrothermal condition. So for hydrophilic/hydrophobic material separation, the membrane should also be slightly hydrophilic/hydrophobic. By incorporating charged zeolite particles into polymeric membranes, the separation specificity can be maintained. Another very popular application of inorganic membranes excluding pervaporation is the separation of hydrogen. Generally this type of separation is induced by pore diffusions rather than molecular exclusions. On a silicalite membrane at 400°C, a diffusion based separation of $\alpha_{H2/CH4} = 15$ has been achieved. This factor is really small in the point of view of commercial applications. On the other hand polymeric membranes have shown better separation factor as well as higher flux for separation of hydrogen in low temperatures.

In the point of view of current position, eight-member ring zeolite like CHA, GIS, or PHI has shown potential for separation of hydrogen by molecular sieve mechanism.

Another problem with inorganic membrane is that the cost of installation and upholds cost should not be much more where as cost of a zeolite module is \$ 3000/$m^2$.

## 4.6 CONCLUSIONS

During the last 25 years, different research has been carried on zeolite membrane and as a conclusion it can be said that zeolite membranes are narrow to their commercial production and application. Only a few applications like LTA zeolite is commercially applied in dewatering of bioethanol and *i*-propanol. The main field of the application will be size/shape-selective operation. Recently for developing membranes with improved selectivity and permeate, researches on molecular dynamics simulation considering basis chemical and physical interactions are being developed.

Seeding supported technique has found to be the best successful and flourishing method for zeolite membrane preparation. MFI and MOR type zeolites can be coated on carbon, $Al_2O_3$, $ZrO_2$ or on mullite.

From this literature review, we have identified the following R&D scopes:

- Preparation of metal or ceramic supported composite zeolite membrane with less than 1 μm thickness in disc shape or hollow fiber shape.
- Try to make the zeolite layer anisotropic from the point of mass transport and thermal stress.

- Formation of zeolite layers which consist of oriented single crystals in the direction of the flux (perpendicular to the support) to avoid mass transport barriers due to crystal imperfections like twinning and crystal intergrowth.
- Focusing on template free synthesis and seeding technique for minimizing the thermal stress.
- Formation of small pore size and high Si zeolite as hydrophilic membrane with high chemical and thermal stability. They also can be applicable for shape selective separation of small molecule like hydrogen.
- The proper uses of microwave prepared zeolites have to be found.
- Studies related to or using permporometry and confocal laser microscopy can be done.
- Reducing the cost of the multilayered asymmetric support is an open challenge.

Thin supported zeolite layer on different carriers such as capillaries, fibers, tubes or monoliths seems promising. With the proper handling of the anisotropic thermal expansion, oriented zeolite membrane can enable high flux membrane. Corma's membrane (using high Si in LTA structure) can be a promising candidate for membrane development. A seeding support secondary growth can be applied for the functional layers. For separation of small molecules like hydrogen or water, six-ring SOD (0.28 nm), eight-ring CHA (0.34 nm) or DDR (0.44 nm) can be applied. The zeolite nanoblock is also showing potential.

## REFERENCES

Alfaro, S., M. Arruebo, J. Coronas, M. Memendez, and J. Santamaria. 2001. Preparation of MFI type tubular membranes by steam-assisted crystallization. *Microporous and Mesoporous Materials* 50: 195–200.

Aoki, K., V.A. Tuan, J.L. Falconer, and R.D. Noble. 2000. Gas permeation properties of ion-exchanged ZSM-5 zeolite membranes. *Microporous and Mesoporous Materials* 39: 485–492.

Arafat, A., J.C. Jansen, A.R. Ebaid, and H. van Bekkum. 1993. Microwave preparation of zeolite Y and ZSM 5. *Zeolites* 13: 162–165.

Berenguer-Murcia, Á., E. Morallón, D. Cazorla-Amorós, and A. Linares-Solano. 2005. Preparation of silicalite-1 layers on Pt-coated carbon materials: a possible electrochemical approach towards membrane reactors. *Microporous and Mesoporous Materials* 78: 159–167.

Bernal, M.P., J. Coronas, M. Menéndez, and J. Santamaria. 2002. Characterization of zeolite membranes by temperature programmed permeation and step desorption. *Journal of Membrane Science* 195: 125–138.

Beving, D., C. O'Neill, and Y.S. Yan. In: R. Xu, Z. Gao, J. Chen, and W. Yan (Eds.). 2007. Zeolites to Porous MOF Materials—the 40th Anniversary of International Zeolite Conference. *Studies in Surface Science and Catalysis.* Elsevier, Amsterdam, Netherlands.

Caro, J., M. Noack, F. Marlow, D. Petersohn, M. Griepentrog, and J. Kornatowski. 1993. Selective sorption uptake kinetics of n-hexane on ZSM5—a new method for measuring anisotropic diffusivities. *Journal of Physical Chemistry* 97: 13685–13690.

Caro, J., M. Noack, P. Kolsch, and R. Schafer. 2000. Zeolite membranes—state of their development and perspective. *Microporous and Mesoporous Materials* 38: 3–24.

Caro J., and M. Noack. 2008. Zeolite Membranes—Recent Development and Progress. *Microporous and Mesoporous Materials* 115: 215–233.

Casanave, D., P. Ciavarella, K. Fiaty, and J.A. Dalmon. 1999. Zeolite membrane reactor for isobutane dehydrogenation: Experimental results and theoretical modeling. *Chemical Engineering Science* 54: 2807–2815.

Chau, J.L.H., Y.S.S. Wan, A. Gavriilidis, and K.L. Yeung. 2002. Incorporating zeolites in microchemical systems. *Chemical Engineering Journal* 88: 187–200.

Chen, X., W. Yang, J. Liu, and L. Lin. 2005. Synthesis of zeolite NaA membranes with high permeance under microwave radiation on mesoporous-layer-modified macroporous substrates for gas separation. *Journal of Membrane Science* 255: 201–211.

Cheng, X.L., Z.B. Wang, and Y.S. Yan. 2001. Corrosion-resistant zeolite coatings by in situ crystallization. *Electrochemical and Solid-State Letters* 4 (2001) B23–B26.

Chiang, A.S.T., L.-J. Wong, S.-Y. Li, S.-L. Cheng, C.-C. Lee, K.-L. Chen, S.-M. Chen, and Y.-J. Lee. In: R. Xu, Z. Gao, J. Chen, and W. Yan (Eds.). 2007. Introduction to Zeolite Molecular Seieves. *Studies in Surface Science and Catalysis.* Elsevier, Amsterdam, Netherlands.

Clet, G., L. Gora, N. Nishiyama, J.C. Jansen, H. van Bekkum, and T. Maschmeyer. 2001. An alternative synthesis method for zeolite Y membranes. *Chemical Communications* 2: 41–45.

Coronas, J., and J. Santamaria. 2004a. The use of zeolite films in small-scale and micro-scale applications. *Chemical Engineering Science* 59: 4879–4885.

Coronas, J., and J. Santamaria. 2004b. State-of-the-Art in Zeolite Membrane Reactors. *Topics in Catalysis* 29: 29–44.

Cui, Y., H. Kita, and K. Okamoto. 2004. Preparation and gas separation performance of zeolite T membrane. *Journal of Materials Chemistry* 14: 924–932.

Cundy, C.S. 1998. Microwave techniques in the synthesis and modification of zeolite catalysts. A review. *Collection of Czechoslovak Chemical Communications* 63: 1699–1723.

Eslava, S., F. Iacopi, M.R. Baklanov, C.E.A. Kirschhock, K. Maex, and J.A. Martens. In: R. Xu, Z. Gao, J. Chen, and W. Yan (Eds.). 2007. *Studies in Surface Science and Catalysis*. Elsevier.

Espinoza, R.L., E. Du Toit, J.M. Santamaria, M.A. Menendez, J. Coronas, and S. Irusta. In: A. Corma, F.V. Melo, S. Mendioroz, and J.L.G. Fierro (Eds.). 12th International Congress on Catalysis. *Studies in Surface Science and Catalysis*. Elsevier, Amsterdam, Netherlands. 130 A (2000) 389.

Gardner, E., K.M. Huntoon, and T.J. Pinnavaia. 2002. Direct synthesis of alkoxide-intercalated derivative of layered double hydroxide suspensions and transparent thin films. *Advance Materials* 13: 1263–1266.

Hasegawa, Y., T. Tanaka, K. Watanabe, K. Kusakabe, and S. Morooka. 2002. The separation of $CO_2$ using Y-type zeolite membranes ion-exchanged with alkali metal cations. *Korean Journal of Chemical Engineering* 19: 309–313.

He, J., Y. Yoneyama, B. Xu, N. Nishiyama, and N. Tsubaki. 2005. Designing a Capsule Catalyst and Its Application for Direct Synthesis of Middle Isoparaffins. *Langmuir* 21: 1699–1702.

Hedlund, J., B.J. Schoeman, and J. Sterte. In: H. Chon, S.K. Ihm, and Y.S. Uh (Eds.). 1997. Synthesis of ultra thin films of molecular sieves by the seed film method. *Surface Science Catalysis* 105: 2203–2210.

Holmberg, B.A., S.J. Hwang, M.E. Davis, and Y.S. Yan. 2005. Synthesis and proton conductivity of sulfonic acid functionalized zeolite BEA nanocrystals. *Microporous and Mesoporous Materials* 80: 347–356.

Hong, M., S. Li, H. Funke, J.L. Falconer, and R.D. Noble. 2007. Ion-exchanged SAPO-34 membranes for light gas separations. *Microporous and Mesoporous Materials* 106: 140–146.

Illgen, U., R. Schäfer, M. Noack, P. Kölsch, A. Kühnle, and J. Caro. 2001. Membrane supported catalytic dehydrogenation of iso-butane using an MFI zeolite membranes. *Catalysis Communications* 2: 339–345.

Jansen, J.C., A. Arafat, A.K. Barakat, and H. Van Bekkum. In: M.L. Occelli, and H. Robson (Eds.). 1992. *Synthesis of Microporous Materials, vol. I,* Van Nostrand Reinhold, New York.

Khajavi, S., J.C. Jansen, and F. Kapteijn. In: R. Xu, Z. Gao, and J. Chen, W. Yan (Eds.). 2007. *Studies in Surface Science and Catalysis*, Elsevier.

Kita, H., K. Fuchida, T. Horida, H. Asamura, and K. Okamoto. 2001. Preparation of Faujasite membranes and their permeation properties *Separation and Purification Technology* 25: 261–268.

Koros, W.J., and R. Mahajan. 2001. Pushing the limits on possibilities for large scale gas separation: which strategies? *Journal of Membrane Science* 175: 181–196.

Kuhn J., K. Yajima, T. Tomita, J. Gross, and F. Kapteijn. 2008. Dehydration performance of a hydrophobic DD3R zeolite membrane. *Journal of Membrane Science* 321: 344–349.

Kumar, S., T.M. Davis, H. Ramanan, R.L. Penn, and M. Tsapatsis. 2007. Aggregative growth of silicalite-1. *Journal of Physical Chemistry B.* 111: 3398–3403.

Lai, Z., G. Bonilla, I. Diaz, J.G. Nery, K. Sujaoti, M.A. Amat, E. Kokkoli, O. Terasaki, R.W. Thompson, M. Tsapatsis, and D.G. Vlachos. 2003a. Microstructural optimization of a zeolite membrane for organic vapor separation. *Science* 100: 456–460.

Lai, S.M., R. Martin-Aranda, and K.L. Yeung. 2003b. Knoevenagel condensation reaction in a membrane microreactor. *Chemical Communications* 2: 218–219.

Lai, S.M., C.P. Ng, R. Martin-Aranda, K.L. Yeung. 2003c. Knoevenagel condensation reaction in zeolite membrane microreactor. *Microporous and Mesoporous Materials* 66: 239–252.

Lai, W., and E. Corcoran 2000. Compositions having two or more zeolite layers. U.S. Patent 6,037,292.

Lau, W.N., K.L. Yeung, X.F. Zhang, and R. Martin-Aranda. 2007. *Studies in Surface Science and Catalysis* Elsevier. 170 B 1460.

Lee, Y., and P.K. Dutta. 2002. Charge transport through a novel zeolite **Y** membrane by a self-exchange process. *Journal of Physical Chemistry B.* 106: 11898–11904.

Lew, C.M., and Y.S. Yan. In: R. Xu, Z. Gao, J. Chen, and W. Yan (Eds.). 2007. *Studies in Surface Science and Catalysis.*

Li, S.G., J.L. Falconer, and R.D. Noble. 2004. SAPO-34 membranes for $CO_2/CH_4$ separation. *Journal of Membrane Science* 241: 121–135.

Li, S., J.L. Falconer, and R.D. Noble. 2006a. SAPO-34 membranes for $CO_2/CH_4$ separations: Effect of Si/Al ratio. *Advanced Materials* 18: 2601–2603.

Li, Q., J. Hedlund, D. Creaser, and J. Sterte. 2001. Zoned MFI films by seeding. 2001. *Chemical Communications* 6: 527–528.
Li, Q. J. Hedlund, J. Sterte, D. Creaser, and A.-J. Bons. 2002. Synthesis and characterization of zoned MFI films by seeded growth. *Microporous and Mesoporous Materials* 56: 291–302.
Li, Y.S., J. Liu, and W.S. Yang. 2006b. Formation mechanism of microwave synthesized LTA zeolite membranes. *Journal of Membrane Science* 281: 646–657.
Lin, X. E. Kikuchi, and M. Matsukata. 2000. Preparation of mordenite membranes on α-alumina tubular supports for pervaporation of water-isopropyl alcohol mixtures. *Chemical Communications* 11: 957–958.
Lin, Y.S., and M. Kanezashi. In: R. Xu, Z. Gao, J. Chen, and W. Yan (Eds.). 2007. *Studies in Surface Science and Catalysis.* Elsevier.
Matsufuji, T., N. Nishiyama, K. Ueyama, and M. Matsukata. 2000. Permeation characteristics of butane isomers through MFI-type zeolitic membranes. *Catalysis Today* 56: 265–273.
Matsukata, M., M. Ogura, T. Osaki, P.R.H. Prasas Rao, M. Nomura, and E. Kikuchi. 1999. Conversion of dry gel to microporous crystals in gas phase. *Topics in Catalysis* 9: 77–92.
Miachon, S., E. Landrivon, M. Aouine, Y. Sun, I. Kumakiri, Y. Li, O. Pachtová Prokopová, N. Guilhaume, A. Giroir-Fendler, H. Mozzanega, and J.A. Dalmon. 2006. Nanocomposite MFI-alumina membranes via pore-plugging synthesis Preparation and morphological characterization. *Journal of Membrane Science* 281: 228–238.
Mitra, A., Z.B. Wang, T.G. Cao, H.T. Wang, L.M. Huang, and Y.S. Yan. 2002. Synthesis and corrosion resistance of high-silica zeolite MTW, BEA, and MFI coatings on steel and aluminum *Journal of Electrochemical Society* 149: B447–B472.
Moore, T.T., T. Vo, R. Mahajan, S. Kulkarni, D. Hasse, and W.J. Koros. 2003. Effect of humidified feeds on oxygen permeability of mixed matrix membranes. *Journal of Applied Polymer Science* 90: 1574–1580.
Nikolakis, V., G. Xomeritakis, A. Abibi, M. Dickson, M. Tsapatsis, and D.G. Vlachos. 2001. Growth of a faujasite-type zeolite membrane and its application in the separation of saturated/unsaturated hydrocarbon mixtures. *Journal of Membrane Science* 184: 209–219.
Nishiyama, N., M. Yamaguchi, T. Katayama, Y. Hirota, M. Miyamoto, Y. Egashira, K. Ueyama, K. Kakanashi, T. Ohta, A. Mizusawa, and T. Satoh. 2007. Hydrogen-permeable membranes composed of zeolite nanoblocks. *Journal of Membrane Science* 306: 349–354.
Noack, M., P. Kölsch, A. Dittmar, M. Stöhr, G. Georgi, R. Eckelt, and J. Caro. 2006. Effect of crystal intergrowth supporting substances (ISS) on the permeation properties of MFI membranes with enhanced Al-content. *Microporous and Mesoporous Materials* 97: 88–96.
O'Brien-Abraham J.L., M. Kanezashi, and Y.S. Lin. In: R. Xu, Z. Gao, J. Chen, and W. Yan (Eds.). 2007. *Studies in Surface Science and Catalysis.* Elsevier. 170: 967.
Rohde, M.P., D. Unruh, and G. Schaub. 2005. Membrane Application in Fischer–Tropsch Synthesis to Enhance $CO_2$ Hydrogenation. *Industrial Engineering Chemical Resources* 44: 9653–9658.
Schäfer, R., M. Noack, P. Kölsch, M. Stöhr, and J. Caro. 2003. Comparison of different catalysts in the membrane-supported dehydrogenationof propane. *Catalysis Today* 82: 15–23.
Sebastian, V., I. Kumakiri, R. Bredesen, and M. Menendez. 2006. Zeolite membrane for $CO_2$ removal: operating at high pressure. *Desalination* 199: 464–465.
Shah, D., K. Kissick, A. Ghorpade, R. Hannah, and D. Bhattacharyya. 2000. Pervaporation of alcohol–water and dimethylformamide–water mixtures using hydrophilic zeolite NaA membranes: mechanisms and experimental results. *Journal of Membrane Science* 179: 185–205.
Shin, D.W., S.H. Hyun, C.H. Cho, and M.H. Han. 2005. Synthesis and $CO_2/N_2$ gas permeation characteristics of ZSM-5 zeolite membranes. *Microporous and Mesoporous Materials* 85: 313–323.
Sterte, J., J. Hedlund, D. Creaser, O. Ohrman, W. Zheng, M. Lassinantti, Q. Li, and F. Jareman. 2001. Application of the seed-film method for the preparation of structured molecular sieve catalysts. *Catalysis Today* 69: 323–329.
Tomita, T., K. Nakayama, and H. Sakai. 2004. Gas separation characteristics of DDR type zeolite membrane. *Microporous and Mesoporous Materials* 68: 71–75.
Van den Bergh, J., W. Zhu, J.C. Groen, F. Kapteijn, J.A. Moulijn, K. Yajima, K. Nakayama, T. Tomita, and S. Yoshida. In: R. Xu, Z. Gao, J. Chen, and W. Yan (Eds.). 2007. *Studies in Surface Science and Catalysis.*
Vilaseca, M., J. Coronas, A. Cirera, A. Cornet, J.R. Morante, and J. Santamaria. 2003. Use of zeolite films to improve the selectivity of reactive gas sensors. *Catalysis. Today.* 82: 179–185.
Vroon, Z.A.E.P., K. Keizer, M.J. Gilde, H. Verweij, and A.J. Burggraaf. 1996. Transport properties of alkanes through ceramic thin zeolite MFI membranes. *Journal of Membrane Science* 180: 127–134.
Vu, D.Q., W.J. Koros, and S.J. Miller. 2003. Mixed matrix membranes using carbon molecular sieves: I. Preparation and experimental results. *Journal of Membrane Science* 211: 311–334.

Wan, Y.S.S., K.L. Yeung, and A. Gavriilidis. 2004. *Studies in Surface Science and Catalysis*. Elsevier. 154A–C: 285–293.

Wan, Y.S.S., K.L. Yeung, and A. Gavriilidis. 2005. TS-1 oxidation of aniline to azoxybenzene in a microstructured reactor. *Applied Catalysis A* 281: 285–293.

Weh, K., M. Noack, I. Sieber, and J. Caro. 2002. Permeation of single gases and gas mixtures through faujasite-type molecular sieve membranes. *Microporous and Mesoporous Materials* 54: 27–36.

Wind, J.D., D.R. Paul, and W.J. Koros. 2004. Natural gas permeation in polyimide membranes. *Journal of Membrane Science* 228: 227–236.

Wong, L.W., W.Q. Sun, N.W. Chan, W.Y. Lai, W.K. Leung, J.C. Tsang, Y.H. Wong, and K.L. Yeung. In: R. Xu, Z. Gao, J. Chen, and W. Yan (Eds.). 2007. Recent Progresses in Mesostructured Materials. *Studies in Surface Science and Catalysis*. Elsevier.

Wong, L.W., and K.L. Yeung. In: R. Xu, Z. Gao, J. Chen, and W. Yan (Eds.). 2007. *Studies in Surface Science and Catalysis*. Elsevier.

Wyllie, M.R.J., and H.W. Patnode. 1950. The development of membranes from artificial cation-exchange materials with particular reference to the determination of sodium-ion activity. *Journal of Physical Chemistry* 54: 204–227.

Xomeritakis, G., Z.P. Lai, and M. Tsapatsis. 2001. Separation of Xylene Isomer Vapors with Oriented MFI Membranes Made by Seeded Growth. *Industrial Engineering Chemistry Research* 40: 544–552.

Xomeritakis, G., S. Naik, C.M. Braunbarth, C.J. Cornelius, R. Parday, and C.J. Brinker. 2003. Organic-templated silica membranes: I. Gas and vapor transport properties. *Journal of Membrane Science* 215: 225–233.

Xu, X., Y. Bao, C. Song, W. Yang, J. Liu, and L. Lin. 2004. Microwave-assisted hydrothermal synthesis of hydroxy-sodalite zeolite membrane. *Microporous and Mesoporous Materials* 75: 173–181.

Yamazaki, S., and K. Tsutsumi. 2000. Synthesis of A-type zeolite membrane using a plate heater and its formation mechanism. *Microporous and Mesoporous Materials* 37: 67–80.

Yang, W., Y. Kim, P.K. Liu, M. Sahami, and T.T. Tsotsis. 2002. A study by in situ techniques of the thermal evolution of the structure of a Mg–Al–$CO_3$ layered double hydroxide. *Chemical Engineering Science* 57: 2945–2953.

Yeung, L., X.F. Zhang, W.N. Lau, and R. Martin-Aranda. 2005. Experiments and modeling of membrane microreactors. *Catalysis Today* 110: 26–37.

Yuan, W., Y.S. Lin, and W. Yang. 2004. Molecular Sieving MFI-Type Zeolite Membranes for Pervaporation Separation of Xylene Isomers. *Journal of the American Chemical Society* 126: 4776–4777.

Zhang, X.F., S.M. Lai, R. Martin-Aranda, and K.L. Yeung. 2004. An investigation of Knoevenagel condensation reaction in microreactors using a new zeolite catalyst. *Applied Catalysis A* 261: 109–118.

Zhu, W., L. Gora, A.W.C. van den Berg, F. Kapteijn, J.C. Jansen, and J.A. Moulijn. 2005. Water vapour separation from permanent gases by a zeolite-4A membrane. *Journal of Membrane Science* 253: 57–66.

Zhu, W., P. Hrabanek, L. Gora, F. Kapteijn, and J.A. Moulijn. 2006. Role of Adsorption in the Permeation of $CH_4$ and $CO_2$ through a Silicalite-1 Membrane. *Industrial & Engineering Chemistry Research 45*(2): 767–776.

Zones, S.I., L. Yuen, and S.J. Miller. 2004. Small crystallite zeolite CHA. U.S. Patent 6,709,644.

# 5 Technological Applications of Composite Membranes

*Maria Popescu and Larisa Melita*

## CONTENTS



## 5.1 INTRODUCTION

In recent years, membrane-based processes have attracted considerable attention as a valuable technology for many industries. This significant gain in momentum is driven in part by spectacular advances in membrane development, wider acceptance of the technology as opposed to conventional separation processes, increased environmental awareness and most of all stricter environmental regulations and legislation. In addition to consuming less energy than conventional processes, membrane systems are compact and modular, enabling easy retrofit to existing industrial processes (Nghiem et al., 2006). For a few applications techniques, membranes may be more desirable than traditional procedures; close boiling or azeotropic mixtures, for example, are difficult to separate with distillation. In addition, membranes can be combined with distillation or other traditional methods into hybrid processes to decrease costs and increase productivity.

### 5.1.1 Composite Membranes: Modules

The most important element in a membrane process is the membrane itself. The membrane may be molecularly homogeneous, uniform in composition and structure, or it may be chemically or physically heterogeneous, for example, containing holes or pores of finite dimensions or consisting of some form of layered structure. The principal types of membrane are shown schematically in Figure 5.1.

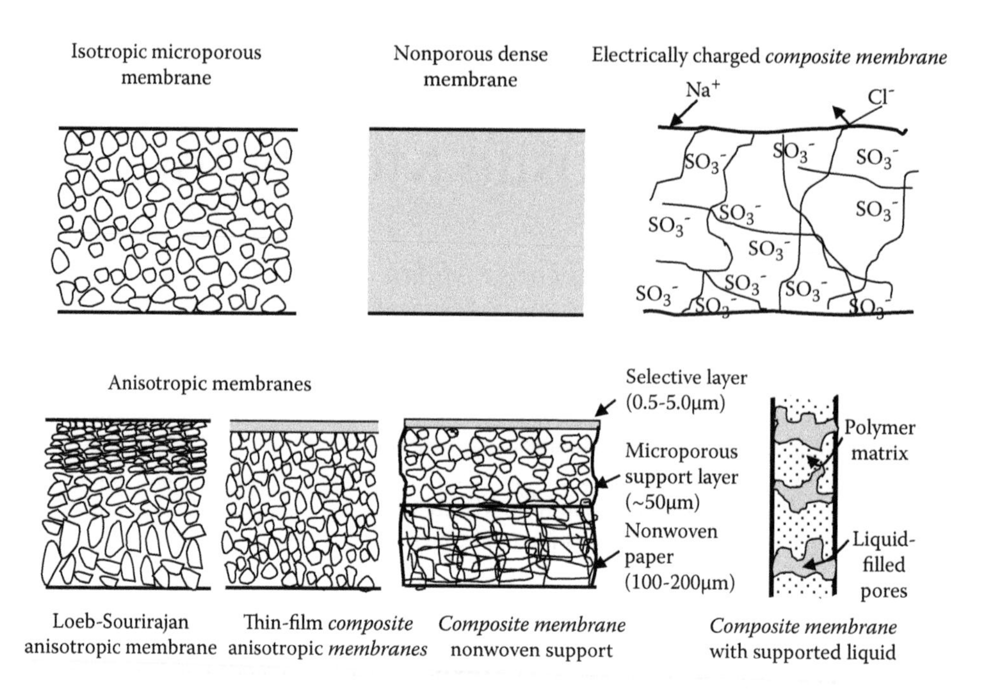


**FIGURE 5.1** Schematic diagrams of principal types of membranes. (Baker, R.W.: *Membrane technology and applications.* 2004. Copyright Wiley-VCH Verlag GmbH & Co. KGaA. Reproduced with permission.)

Anisotropic membranes have a thin, finely microporous or dense permselective layer, supported on a much thicker, porous substructure. The surface layer (dense polymer) and its substructure may be formed in a single operation or separately. In composite membranes, the layers are usually made from different polymers; the separation properties and permeation rates of the membrane are determined exclusively by the dense polymer, and the substructure functions as a mechanical support. The dense polymer of a membrane is a selective barrier layer but has poor mechanical strength or poor film-forming properties. The optimal thickness of these films is, generally, too thin (0.1–1 μm) to be mechanically stable, so the films must be supported.

Composite membranes combine dense polymers and inorganic/organic membrane support into one material with a complex structure.

One class of composite membranes is represented by skinned asymmetric membranes, where a thin dense polymer layer is deposited onto an asymmetric membrane acting as a support. Each part of the composite membrane (i.e., the support and dense polymer) can be optimized independently for greater membrane performance with respect to selectivity, flux, and chemical and thermal stability. In addition, hard/soft segmented composite membranes, which employ a copolymer of a polyimide and either a polycarbonate or a polyester were developed; the hard segment (the support) governs the temperature stability and solvent resistance of the membrane, while the soft segment (a dense polymer) governs selectivity and flux. Composite membranes can also consist of a polymer within the pores/non-woven fibers of a support. Plasma graft-filling polymerization is another technique used to produce composite membranes; these membranes are composed of a porous substrate and another polymer, which fills the pores of the substrate. The composite membranes with liquid support have gained significance in combination with the facilitated transport, which utilizes a selected "carrier," transporting certain components, such as metal-ions selectively, at a high rate across the liquid membrane interphase. It is relatively easy to form a thin fluid film but it is difficult to maintain and control this film and its properties during a mass separation process.

However, most support membranes usually have an asymmetric structure in which the top coating layer is dense.

The choice of substrate material is important; a commercially available, rigid material that would not swell in organic solvents used to perform transport experiments was desired. A general strategy for the selection of polymers to be used in composition membranes utilizes concepts from gas chromatography (GC) and data that are readily available in the literature. Solutes with little affinity for the membrane will stay mostly in the feed phase, while those with higher affinities for the membrane material will be more likely to partition into the membrane and diffuse down their concentration gradient.

In addition, studies need to be completed in order to correlate transport properties with the structure of materials. A five-step mechanism is generally assumed to describe the mass transfer of the penetrants through a membrane by permeation. The penetrants first diffuse from the bulk of the feed to the feed–membrane interface and then are absorbed into the membrane. They further diffuse through the membrane to the downstream and desorb into the vapor at the permeate side. Finally, the penetrants diffuse from the vapor–membrane interface to the bulk of the vapor phase (Kim et al., 2002).

The same is applicable to estimate the effective permeability of gases and liquids through mixed matrix membranes, for example prepared with different polymer matrices, as a continuous phase, and organic or inorganic compounds, such as filler, even at relatively high volume fractions of these compounds. A comparison between estimated values of the effective permeability and experimental data reported in the literature generally shows good agreement although there are a number of observations that are not easily explained with any of models (Gonzo et al., 2006).

Industrial membrane plants often require hundreds to thousands of square meters of membrane to perform the separation required on a useful scale. Before a membrane separation can be performed industrially, therefore, methods of economically and efficiently packaging large areas of membrane are required. These packages are called membrane modules. Besides economic considerations, chemical engineering aspects are of prime importance for the design of membrane modules. The earliest designs were based on a simple filtration technology and consisted of flat sheets of membrane held in a type of filter press; these are called plate-and-frame modules. Membrane feed spacers and product spacers are layered together between two end plates. The feed mixture is forced across the surface of the membrane. The tubular membrane module consists of membrane tubes placed into porous stainless steel or fiber glass reinforced plastic pipes. The pressurized feed solution flows down the tube bore, and the permeant is collected on the outer side of the porous support pipe. The capillary membrane module (hollow fiber) consists of a large number of membrane capillaries with an inner diameter of 0.2 to 3 mm, arranged in parallel as a bundle in a shell tube.

The problems of membranes that remain under the focus of membrane research today are: selectivity, productivity/cost, and operational reliability. The selectivity of a membrane is the ability to make the required separation. The productivity, or the separation performance per unit cost, is an issue in all membrane separation processes. In some processes, membrane and module costs stand for more than 50% of the operating costs. The operational reliability varies from process to process. For example, in typical membrane gas permeation, a critical factor is to develop one defect per square meter of membrane to essentially destroy the efficiency of the process. The solution usually appears to be a combination of a number of factors, such as better membrane materials, better module designs, improved cleaning and antifouling procedures, and better process designs.

### 5.1.2 Membrane Processes

In interaction with a membrane, a high degree of permeability coupled with a large selectivity of a specific gaseous species ensures superior performances in the gas processing industry. In the past 10 years, the membrane gas separation technology has advanced significantly and can now be regarded as a competitive industrial gas separation method.

In membrane separation processes there are two basic forms of mass transport. The simplest form is the "passive transport." Here, the membrane acts as a physical barrier through which all components are transported under the driving force of a gradient in their electrochemical potential. The second form of mass transport through the membrane interphase is the "facilitated transport or active transport." Facilitated transport is coupled with a carrier in the membrane interphase and is found mainly in the membranes of living cells.

Four developing industrial membrane separation processes are: gas permeation (gas separation), pervaporation, ion exchange processes and carrier-facilitated transport or active transport (facilitated transport and coupled transport).

The gas permeation is the more advanced of these techniques. In gas permeation, a gas mixture, at an elevated pressure, is passed across the surface of a membrane; that is selectively permeable to one of the components of the feed mixture. The membrane permeate is enriched in this species. Pervaporation is a relatively new process. In pervaporation, a liquid mixture contacts one side of a membrane and permeate is removed as a vapor from the other side. The driving force for the process is the low vapor pressure on the permeated side of the membrane generated by cooling and condensing the permeated vapor. In ion exchange processes, the charged membranes are used to separate ions from aqueous solutions, under the driving force of an electrical potential difference. A number of other industrial membrane processes are placed in the category of to-be-developed technologies. The most important of these is carrier-facilitated transport, which employs liquid membranes, containing a complexing or carrier agent (Porter, 1990; Baker et al., 1991; Baker, 2004; Le Cloirec, 1998; Popescu et al., 1998, 2006; Paul and Yampl'skii, 1994; Basu et al., 2004).

## 5.2 TECHNOLOGICAL APPLICATIONS OF COMPOSITE MEMBRANES

Polymeric thin-film composite membranes are being increasingly implemented during municipal and industrial water purification for removing hardness, synthetic organic compounds, natural organic matter, disinfection by-product precursor, mono and multivalent ions, etc.

Research in the general areas of gas permeation, pervaporation, ion exchange membrane processes and carrier-facilitated transport or active transport (facilitated transport and coupled transport) was ranked substantially higher than the other technology areas (Porter, 1990; Baker et al., 1991; Baker, 2004; Sharma et al., 2003).

### 5.2.1 Applications of Gas Permeation (Gas Separation) (PG)

Gas permeation research topics are divides into two areas. The first deals with methods of making better, high performance membranes, and the second deals with the development of membrane materials with improved selectivity and permeability.

The benefits of "Membrane Technology" include a higher degree of recovery of the desired gaseous effluent that can be reused for multiple purposes. Various experimental and analytical models related to current and future membrane technologies include regular, liquid and hybrid systems (Basu et al., 2004; Le Cloirec, 1998; Kang et al., 2006, 2008; Kim et al., 2004; Hess et al., 2006; Jiang et al., 2008; Chenar et al., 2008; Hao et al., 2008; Anson et al., 2004; Ismail and David, 2001; Jha and Way, 2008; Ismail et al., 2008; Sadeghi et al., 2008; Li et al., 2008a, 2008b; Molyneux, 2008; Anderson et al., 2008; Muñoz et al., 2008; Iqbal et al., 2008; Kosuri and Koros, 2008; Majundar et al., 2003; Obuskovic et al., 2003; Teplyakov et al., 2003; Daisley et al., 2006; Vu et al., 2003; Liu et al., 2006; Yeom et al., 2002a, 2002b, 2002c).

#### 5.2.1.1 Olefins/Paraffins Gas Separation (Permeation)

Ethylene and propylene are primarily used for the production of polyethylene, polypropylene, styrene, ethylbenzene, ethylene dichloride, acrylonitrile, and isopropanol. Large quantities of ethylene and propylene are used as feedstock in the production of plastics such as polyethylene or polypropylene. The most widely used production process for olefins is the cracking of $C_4$-hydrocarbon fractions, followed by a dehydrogenation reaction. The typical conversion equilibrium of the dehydrogenation reaction is around 50%–60%. An important step in the manufacture of olefins is the large-scale separation of the olefin from the corresponding paraffin. Currently, this separation is carried out by distillation, an extremely energy-intensive process due the difficult separation because of the similar physical properties of the saturated /unsaturated hydrocarbons.

Membrane-based gas separation devices shown in Figure 5.2, with the introduction of membrane units into the conventional separation process will lead, depending on the separation characteristics of the membrane material, to a significant reduction of the steam brought to the splitter. A possible reduction of the splitter column might be of highest interest because the olefin/paraffin separation train represents more than half of the total cost of an olefin plant.

However, four problems have limited the commercial application of facilitated transport membranes: (1) poor mechanical stability, (2) the difficulty in preparing thin, high-flux composite membranes, (3) the requirement of a water-vapor–saturated feed to provide mobility for the olefin-selective carrier, and (4) poor chemical stability due to carrier poisoning.

Generally, the best polymeric membranes exhibit ethylene/ethane selectivities of only 4–5 while the separation factors of propylene/propane mixture through various polymeric membranes, without carriers, are: 1 for silicon rubber, 3.8 for cellulose acetate, 1.1 for PDMS and 1.7 for 1, 2-polybutadiene, and so on.

An improving facilitated transport membrane is based on solid polymer electrolytes made from rubbery, for example, poly(ethylene oxide) (PEO), containing dissolved metal ions ($Cu^+$, $Li^+$, $Ag^+$, $Au^+$); thus, both anions and cations are sufficiently mobile in the rubbery polymer matrix without the need of a solvent or plasticizer to promote conductivity. The mechanism by which facilitated transport takes place in a liquid membrane containing ions $Ag^+$, for example, and is based on the reversibly reaction of an olefins with a metal cation, results from the interaction of the olefin $\pi$-orbital with the $\sigma$- and $\pi$-orbitals of the metal. On the high-pressure side of the membrane, both ethylene and ethane are sorbed into the membrane. However, only ethylene forms a complex with the silver-ion carrier, providing an increase in the solubility of ethylene in the membrane. Desorption of ethylene, at the low pressure side of membranes leads to a dissociation of the ethylene–silver complex. The ethylene–silver ion complex is believed to be sufficiently labile for transport of the ethylene molecules hopping from side to side through the rubbery polymer matrix. A PEO-based electrolyte membrane, containing 80 wt. % $AgBF_4$ had an ethylene/ethane selectivity of 120–240. However, the practical use of this class of membranes is still limited by the poor chemical stability of currently available olefin carriers. For example, $Ag^+$ degrades easily if trace amounts of sulfur compounds, such as $H_2S$, are present in the feed stream. Furthermore, $Ag^+$ can react with acetylene to form an

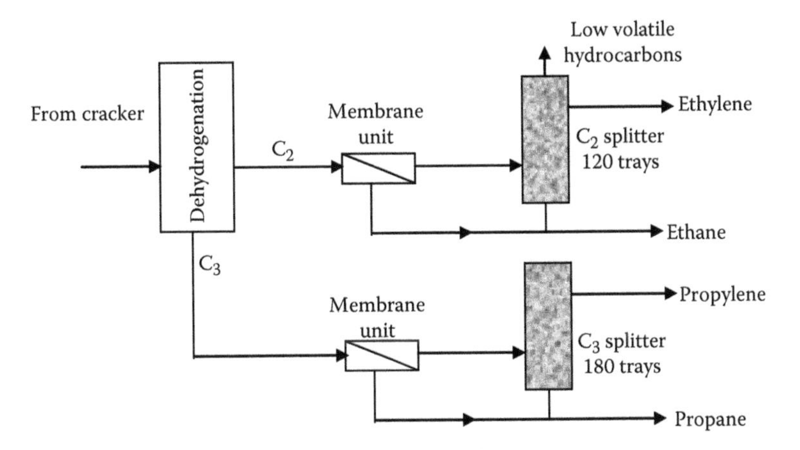


**FIGURE 5.2** Conventional separation of $C_2$ and $C_3$ olefin/paraffin mixture and possible implantation of membrane units in the process. (From Popescu, M., R. Popescu and C. Stratula. 2006. Metode fizico-chimice de tratare a poluantilor industriali atmosferici (in Romanian). Editura Academiei Romane. p. 199.)

explosive silver acetylide salt. Hence, future work in facilitated transport membranes for olefins/paraffin separation should be directed toward the development of chemically stable olefin carriers.

Positively polarized silver and gold nanoparticles have been demonstrated for use as stable olefin carriers for facilitated olefin transport membranes. The stabilized nanoparticles were responsible for the reversible interaction between the nanoparticles and olefin molecules and were employed to control the positive charge density of the surface (Sharma et al., 2003; Kang et al., 2006, 2008; Kim et al., 2004).

The new promising membrane materials for the olefin/paraffin separation seem to be the polyimides: the factor of up to 77 for the membranes with fluorine containing polyimides. Thus, different 4,4′-(hexafluoroisopropylidene) diphthalic anhydride (6FDA)–based polyimides were tested for the permeation and sorption properties of pure ethane, ethylene, propane, propylene and in order to determine separation factors for 50:50 olefin/paraffin feed mixtures. For the investigated polyimides, a preferred permeability of olefins compared to paraffins was found because of the smaller molecular dimensions of the olefins: the olefins show a shorter C=C distance (1.337 Å) compared to the paraffins, C–C (1.534 Å), and due to the $sp^2$ hybridization of the carbon atoms, the olefins are rigid "flat molecules" compared to the paraffins with $sp^3$ hybridization, where free C–C rotation is possible. The permeability of hydrocarbons in polyimides strongly depends on the shape of the penetration molecules; the permeability of olefins in 6FDA polyimides depends on the polymer structure, i.e., the free volume. Generally, with polyimides as membrane, olefins are sorbed preferentially; the reason for this preference is that the olefins show higher polarizability due to their $\pi$-electron system, compared to the paraffins and the separation process, which is mainly controlled by the diffusivity selectivity.

Different crosslinkable 4,4′-hexafluoro-isopropylidene diphthalic anhydride (6FDA) based copolyimides were synthesized using 3,5-diamino benzoic acid as one of the monomers providing a crosslinkable group and 15-crown-5-ether diamine suitable for creating facilitated transport sites. Crown ethers are well known for strong complexation of light metal ions but also complex formation with silver ions is possible. Thereby the ions are bonded into the cavity of the crown ether. The main advantage of this type of facilitated transport polymers is that no carrier medium is necessary which means "dry" membranes can be applied (Hess et al., 2006).

A new promising membrane was selected, for the further investigation of the morphological structure and permselectivities, using for separation binary mixtures of ethylene, ethane, propylene and propane at ambient temperature. The thin film of composite membrane was prepared by polyurethane (PU) dispersions with poly(dimethylsiloxane)/poly(ethylene glycol) (PDMS/PEG–based PU) and poly(vinylidene fluoride) (PVDF) substrate. The selectivities of propylene and propane to nitrogen were substantially improved, e.g., in a mixture containing 28% propylene and 72% nitrogen the selectivity of propylene to nitrogen reached 29.2 with a propylene permeance of 34.4 gas permeation unit (Jiang et al., 2008).

#### 5.2.1.2 Separation of $CO_2/CH_4$

In natural gas treatment applications, the feed gas usually is obtained directly from gas wells with a wide range of pressures (2–7 MPa) and compositions (5%–50% $CO_2$). The typical product is a $CH_4$-enriched residue stream containing less than 2% $CO_2$, which is sold as pipeline fuel. The residue stream is produced with essentially no pressure loss, while the $CO_2$-enriched permeate is produced at low pressure (0.1–0.5 MPa) but contains an appreciable amount of $CH_4$. The permeate stream often has some value as a low heating value fuel. As compared to natural gas treatment, the major difference in enhanced oil recovery application is that both permeate and residue streams have considerable value. The residue stream (higher than 98% $CH_4$) is used as a fuel, while the permeated stream (higher than 95% $CO_2$) is compressed and reinjected into the gas well (Chenar et al., 2008; Hao et al., 2008).

The mixed matrix composite membranes (MMC) constituted by two interpenetrating matrices of different materials offer the potential of combining the polymer processability with superior gas separation properties of rigid molecular sieving materials. The successful implementation of this membrane development lies on both the selection of polymeric matrix and inorganic filler, and the elimination of interfacial defects; for example, the addition of carbon molecular sieves (CMS) or activated carbon as the selective inorganic filler, into a polymeric matrix (polyetherimide, polyimide, acrylonitrile–butadiene–styrene (ABS), etc.). By using a two micro-mesoporous activated carbon (AC) in ABS copolymer, a new MMC was obtained for separation of $CO_2$ from $CH_4$. The micro-mesoporous activated carbons have different adsorptive capacity levels for polar and unsaturated compounds (i.e., $CO_2$, olefins) with respect to nonpolar and saturated ones (i.e., $CH_4$, paraffins). The increasing selectivities with ABS–AC could be partially explained considering the existence of a surface flux through the micro-mesoporous carbon media, with a mechanism of preferential surface diffusion of $CO_2$ (more adsorbable gas) over the $CH_4$ gas (less adsorbable). The high $CO_2$ productivity and selectivity of ABS–AC membranes can be attributed in part to (1) the intrinsic permselectivities of the ABS copolymer; (2) the selective gas capacity of AC adsorption for $CO_2$ gas and (3) the tight interfacial contact between the polymeric phase and filler phase. A future perspective should be devoted to obtaining a more homogeneous distribution of filler discrete entities ($< 3$ μm) inside of the polymeric matrix (Anson et al., 2004).

The size of micropores in the carbon membranes is in the 3–5 Å and the diameter of gas molecules are 3.3 and 3.8 Å for $CO_2$ and $CH_4$ respectively; the mixtures of these gases can be separated according to a molecular sieving mechanism (gas diffusivity and adsorption effects) (Ismail and David, 2001). Another MMC used polyethersulfone, polyphosphazene, copolymer ethylene vinyl acetate with vinyl acetate, etc with the micro-mesoporous carbon or microporous silica (Jha and Way, 2008; Ismail et al., 2008; Sadeghi et al., 2008; Li et al., 2008a, 2008b; Molyneux, 2008).

Nanoporous carbon (NPC) membranes show promise as a new generation of gas separation membranes suitable for $CO_2$ capture; the performances of separation depend on the pyrolysis temperature and operating temperature (Anderson et al., 2008; Muñoz et al., 2008). Understanding the mechanism of asymmetric membrane formation is very important in order to produce membranes with the expected morphology (Iqbal et al., 2008) and substructure defect-free (Kosuri and Koros, 2008).

#### 5.2.1.3 Removal of Volatile Organic Compounds

There are a number of conventional separation processes for the removal of volatile organic compounds (VOC) from the waste stream (absorption, adsorption, condensation, incineration, etc.), but vapor permeation through membranes offers significant opportunities of energy saving and reuse of the VOC, compared to the conventional VOC control processes, particularly if the VOC concentration is high. The recovery of VOC from waste or vent gases, for example, can be motivated in terms of environmental protection and economical aspect (Majundar et al., 2003; Obuskovic et al., 2003; Teplyakov et al., 2003; Daisley et al., 2006; Vu et al., 2003; Liu et al., 2006; Yeom et al., 2002a, 2002b, 2002c).

##### *5.2.1.3.1 Separation of Alcohols, Toluene, and Other Solvents*

Large- and small-scale manufacturing processes, in chemical, food, pharmaceutical, and petrochemical industries, fuel storage reservoirs, coating and finishing operations often release into the atmosphere air/$N_2$ streams containing large amounts of VOC. In two pilot scales, the VOC-containing feed to the emission from a batch reactor in a pharmaceutical plant or the air emissions from a paint booth were studied. From these, microporous polypropylene hollow fibers substrate having an ultrathin plasmapolymerized poly(dimethylsiloxane) (PDMS) skin on the outside surface of the fibers are used.

The pilot plant has a number of smaller reactors, which are usually operated in batch mode; Figure 5.3 is an overall scheme of the reactor assembly using the membrane unit. In a typical batch, a pilot reactor is charged with 120 L of solvent and operated at an elevated temperature, often near the boiling point of the solvent, under a layer of $N_2$. A substantial part of the solvent vapor is recovered from this stream and recycled back to the reactor as the stream is passed successively through two glass condensers. The exhaust stream leaving the second condenser still has 3,000–13,000 Vpm (0.3%–1.3%) of solvent; it is then discharged into a central exhaust line held at a pressure slightly

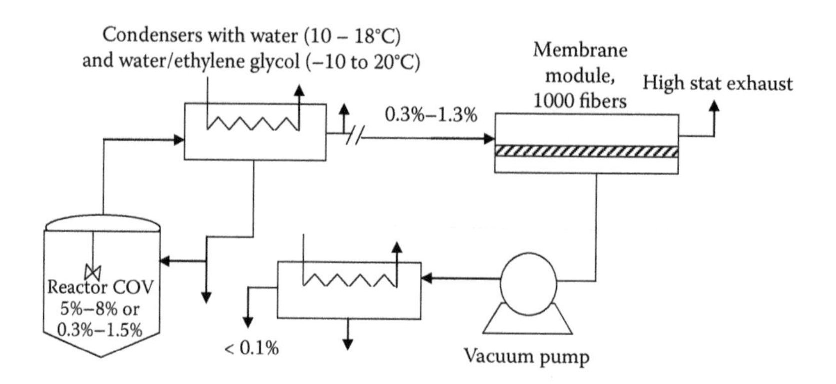


**FIGURE 5.3** Configuration for a reactor–condenser assembly with the membrane module. (From Popescu, M., R. Popescu and C. Stratula. 2006. Metode fizico-chimice de tratare a poluantilor industriali atmosferici (in Romanian). Editura Academiei Romane. p. 202.)

below 1 bar. Batch reactor operations give rise to variable emission rates over short periods; the VOC concentrations should be 5%–8% when the stream is drawn from condensers and then to the membrane unit. A highly VOC selective separation was performed with a thin layer of silicon oil, immobilized in part of the micropores, as a liquid in the hollow fiber membrane. The resulting benefits of having the thin immobilized liquid membrane (ILM) incorporated in the microporous structure were two to five times more VOC-enriched permeated since the nitrogen flux was drastically reduced and the separation factor was increased 5 to 20 times depending on the type of the VOC and the feed gas flow rate (Majundar et al., 2003; Obuskovic et al., 2003).

The VOC concentration in the gas stream was as high as 14% of methanol in the study batch reactor; other VOC emitted from the reactor were toluene 2.5% and ethyl acetate 4% (Figure 5.4). Very low levels of VOC in the range of 5–100 Vpm were involved in the paint booth emission study. More than 95% of the VOC present in the feed were successfully removed in each study. The influence of the effect of the polymer architecture on its permeability properties, PTMSP with other ratio of *cis-trans* form must be synthesized under well controlled conditions and then relevant films must be tested with another set of permeant probes. The maximal selectivity (~200) was obtained for toluene/$N_2$/air; a high ideal selectivity of vapor/vapor separation can be estimated for toluene/$H_2O$ and toluene/dimethylketone (Teplyakov et al., 2003).

The gas separation of VOC (toluene, methanol, acetone and methylene chloride) from $N_2$ feed gas at atmospheric pressure was studied using microporous polypropylene hollow fibers having ultrathin plasma-polymerized nonporous silicone skin on the outside surface. The operational mode in the membrane module was that of the feed gas flowing through the fiber bore (the nonskin side) and vacuum on the shell side.

The polyetherimide (PEI) membranes coated with polydimethylsiloxane (PDMS) are frequently used for removing acetone, ethyl acetate and ethanol from the air.

In particular, for the recovery of aromatic compounds, such as 1,2-benzisothiazolin-3-one (BIT) which show mass transfer rates through the membrane, a novel composite membrane was used; the membrane is composed of a thin nonporous PDMS selective layer coated on a microporous support layer cast from polyacrylonitrile, polyvinylidene fluoride, polyetherimide or polyphenylenesulfone (Daisley et al., 2006).

The mixed matrix film was composed of fine particles of high-selective carbon molecular sieves (CMS), 19 vol. %, dispersed within a glassy polyimide matrix; the membranes were used for the separation condensable impurity (toluene, 70 ppm) in $CO_2/CH_4$ = 10%/90% gas feeds, at 35°C for a period of up to 60 h. The adsorption surface of the toluene molecules onto the CMS particles may induce a slow relaxation leading to better packing at the interface between the carbon and matrix reflected by the protracted changes seen (Vu et al., 2003).

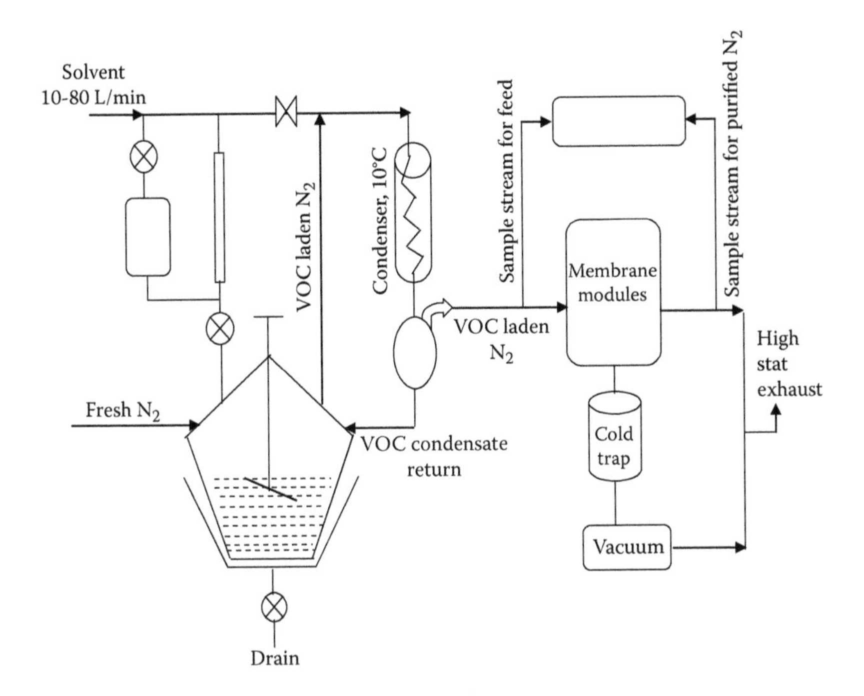


**FIGURE 5.4** Schematic diagram of a reactor emitting solvent vapor and the vapor permeation assembly. (From Popescu, M., R. Popescu and C. Stratula. 2006. Metode fizico-chimice de tratare a poluantilor industriali atmosferici (in Romanian). Editura Academiei Romane. P 382.)

A charged effluent with toluene is introduced by using a compressor into a membranous module. The feed retentate is recycled at the compressor inlet and permeate, drained with a vacuum pump, is introduced into a condenser, for toluene liquation, the selective permeated compound. The vacuum pomp assures the drainage of permeate, and a depressor in the membrane downstream represents the driving strength of the transfer. The evaluation was established for two toluene concentrations: 2 g/Nm$^3$ and 20 g/Nm$^3$ respectively, and provide a toluene recovery cost of 5.14 dollars per toluene kg and 0.8 dollars per toluene kg, respectively, and a treatment cost of 0.01 dollars/Nm$^3$ and 0.0016 dollars/Nm$^3$, respectively. The calculus demonstrates that the most important limit of membrane techniques is the pollutant concentration from the effluent. Thus, it is considered that for a concentration less than 1g/Nm$^3$, and efficiency over 95% the membrane techniques become expensive. Therefore, procedures should be found for pollutant enrichment before the treatment with the membrane (Le Cloirec, 1998).

#### *5.2.1.3.2 Recovery of Gasoline Vapors from the Polluted Air: Vaconocore System*

An important membrane vapor recovery system was the recovery of gasoline vapors from vent streams produced at large oil and gasoline terminals. During the transfer of hydrocarbons (HC) from tankers to holding tanks and then to trucks, off-gases are produced. The off-gas stream volume and vapor concentration vary widely, but the average emissions resulting from each transfer operation are large, in the range 0.03%–0.05% of the hydrocarbon transferred. The HC concentration of the emitted gas is generally quite high, in the range 10 to 30 vol. %, depending on the type of HC and type of transfer. Because the off-gas is an air–hydrocarbon mixture, the potential for formatting an explosive composition has to be considered in the design of the membrane vapor recovery system. The usual solution is to saturate the incoming feed mixture with additional HC vapor in a small contactor tower. This ensures that the feed to the compressor needed to operate the membrane unit is always comfortably above the upper explosion limit, regardless of the composition of the feed gas.

The fluid leaving the compressor is then a two phase mixture of gasoline containing dissolved vapors HC-saturated air (Figure 5.5). A phase separator, after the compressor, separates the HC liquid and gas phases. The vapor saturated gasoline is removed; the saturated vapor then passes to the membrane unit; the HC vapors are removed by using a HC-selective membrane. The HC enriched permeate is recycled to the front of the feed gas compressor; the HC of retentate contains 0.5 to 2% HC, mainly the light gases methane, ethane and propane. To meet air discharge regulations, this gas is usually sent to a final polishing step, most commonly a small, molecular sieve, a pressure swing absorption unit (PSA) which reduces the HC level to 0.2 to 0.5 vol. %.

In the last few years, the gasoline stations have installed small membrane systems to treat their tank vents. A flow scheme of this type of system is shown in Figure 5.6. Air from the gas station dispenser is collected and sent to the gasoline storage tank. When the pressure in the tank reaches a preset value, a pressure switch activates a small compressor that draws off excess vapor-laden air. A part of the HC vapors condense and are returned to the tank as liquid. The remaining HC permeate the membrane and are returned to the tank as concentrated vapor. Air, stripped of 95%–99% of HC, is vented. Typical systems are small, containing a single 1–2 $m^2$ membrane module. In addition to eliminating HC emissions, the unit pays for itself with the value of the recovered gasoline. For the separation of gasoline vapor, for example, hollow fiber composite membranes were used, comprising a thin layer of poly(ether block amide) supported on microporous poly(vinylidene fluoride) substrate; at ambient temperature, an overall permeate HC concentration as high as 95 wt. % was obtained. The membrane was found to be stable for gasoline vapor recovery during a 10-month period of testing (Le Cloirec, 1998; Liu et al., 2006).

#### *5.2.1.3.3 Recovery of Chlorinated Hydrocarbons*

The chlorinated hydrocarbons that were part of the homologous series of chloromethanes and chloroethanes were used as organic vapor and have been treated on the vapor permeation and separation of $VOC/N_2$ mixtures through a poly(dimethylsiloxane) (PDMS) membrane. The PDMS was composed of two parts: part A is mainly PDMS oligomers terminated with vinyl groups and part B is a mixture of Pt catalyst and PDMS oligomer with active hydrogens. Various chlorinated hydrocarbons (methylene chloride, chloroform, 1,2-dichloroethane and 1,1,2-trichloroethane) were used as

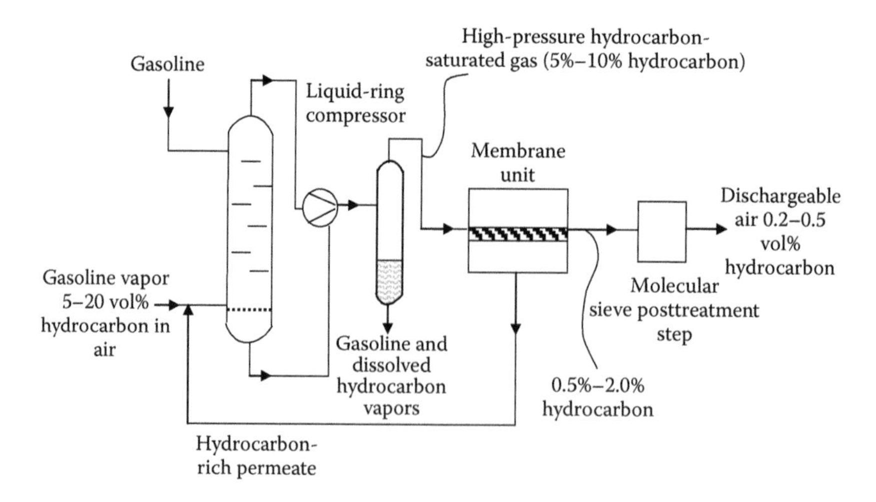


**FIGURE 5.5** Flow schematic of a gasoline vapor recovery system, using a combination of absorption and membrane separation to recover 98%–99% of the HC in the vent gas, followed by molecular sieve pressure swing absorption (PSA) unit to remove the final 1%–2% HC. (Baker, R.W.: *Membranes for Vapour/gas separation*, 2nd ed. 2006. p. 15. http://www.mtrinc.com/publications/MT01%20Fane%20Memb%20for%20VaporGas_Sep%202006%20Book%20Ch.pdf. Copyright Wiley-VCH Verlag GmbH & Co. KGaA. Reproduced with permission.)

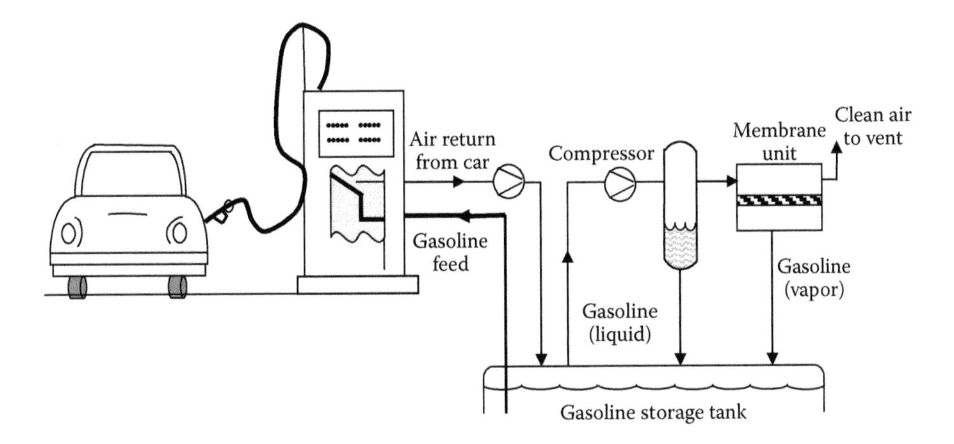


**FIGURE 5.6** Flow diagram of a membrane gasoline-vapor recovery unit suited to a retail gasoline station tank vent. (Baker, R.W.: *Membranes for Vapour/gas separation*, 2nd ed. 2006. p. 15. http://www.mtrinc.com/publications/MT01%20Fane%20Memb%20for%20VaporGas_Sep%202006%20Book%20Ch.pdf. Copyright Wiley-VCH Verlag GmbH & Co. KGaA. Reproduced with permission.)

VOC component; VOC in feed ranged from 0.3 to 1.5 vol. % and operating temperature 35°C–65°C. The permeation performance of the mixture as well as the selectivity toward VOC component showed a strong dependence of both VOC content on the feed and the condensability of VOC (Yeom et al., 2002a, 2002b, 2002c).

#### *5.2.1.3.4 Recycling of Vinyl Chloride Monomer*

The annual total world production of vinyl chloride monomer (VCM), which is approximately equal to PVC production, was about 26 million tons in 1995, for example. Most of the vinyl chloride that enters the environment comes from vinyl chloride manufacturing or processing plants, which release it into the air or into waste water. The purge gas is typically at 4 to 5 bar pressure and can contain much as 50 vol. % monomer (Le Cloirec, 1998).

EPA limits the amount that industries can release, because VCM is extremely volatile ($T_b = -13°C$), toxic (limit of exposure is 3.0 Vpm of air) and inflammable (explosive limits in air are 4–22 vol. %).

A typical process flow scheme to recover VCM is shown in Figure 5.7. Feed gas containing VCM and air is sent to the membrane system. The VCM-enriched permeate from the membrane system is

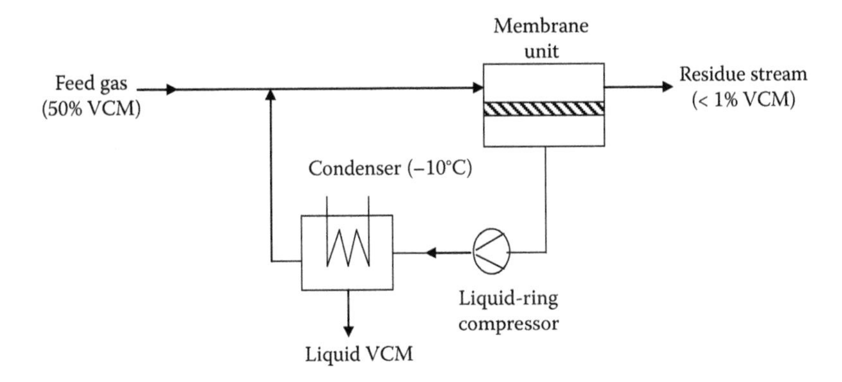


**FIGURE 5.7** Recovery of VCM in a polyvinylchloride plant. (Baker, R.W.: *Membranes for Vapour/gas separation*, 2nd ed. 2006. p. 15. http://www.mtrinc.com/publications/MT01%20Fane%20Memb%20for%20VaporGas_Sep%202006%20Book%20Ch.pdf. Copyright Wiley-VCH Verlag GmbH & Co. KGaA. Reproduced with permission.)

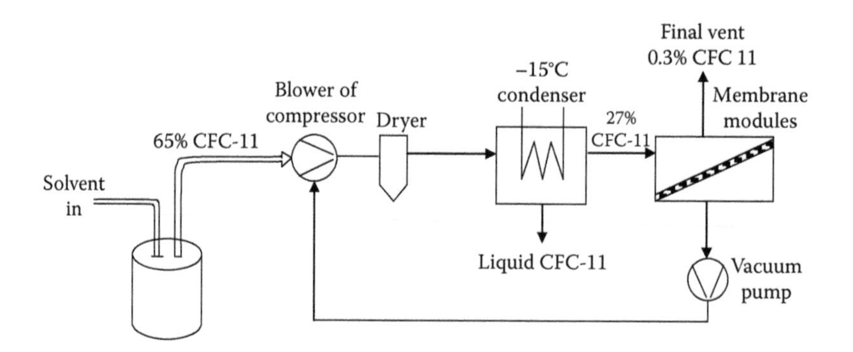


**FIGURE 5.8** A membrane system for recovering CFC. The system recovers 98% of the CFC contained in the feed stream. (From Simmons, V.L., Baker, R.W., Kaschemekat, J., et al. 1992. Membrane vapor separation systems for the recovery of halons and CFC. p. 295. http://www.bfrl.nist.gov/866/HOTWC/HOTWC2006/pubs/R0301015.pdf.)

compressed in a liquid-ring compressor and cooled to liquefy the VMC. A liquid-ring compressor is used because of the flammable nature of VCM. The non-condensable gases are mixed with the feed gas and returned to the membrane section. The residue stream is sent to the incinerator, where the remaining VCM is destroyed before venting the inerts. VCM recovery is more than 99%. The first unit of this type was installed by MTR (Membrane Technology and Research) in 1992.

#### *5.2.1.3.5 Recovery of CFC and Halons*

Chlorofluorocarbons (CFC, freon) are any of several compounds composed of carbon, fluorine, chlorine and hydrogen used mostly in refrigeration systems, as solvents and aerosol propellants. Halons are compounds in which the hydrogen atoms of a hydrocarbon have been replaced by bromine and other halogen atoms; very stable, they have long lifespans whose breakdown in the stratosphere causes ozone depletion. They are used for fire extinguishers (Le Cloirec, 1998).

The recovery of CFC by conventional technologies is difficult because of its low boiling point and high vapor pressure. Membrane Technology and Research Inc. has developed a halon and CFC recovery process based on membranes. The membrane used are: the polyimides, the poly(ethylene oxide), the poly(1-trimethylsilyl-1-propyne), etc. A flow diagram of a membrane system for the recovery of CFC 11 and CFC 113, for example, is shown in Figure 5.8; the air vented from the container as it fills is saturated with CFC vapor. This stream is sent to a dryer and then to a –15°C condenser, which reduces the CFC concentration to approximately 27%. Adding a membrane system reduces the vented CFC concentration from 27% to 0.3%. The sytem will pay for itself after only a few hundred hours of operation.

### 5.2.2 Applications of Pervaporation Processes

Pervaporation (PV) is a membrane process where the feed-stock is a liquid stream, while the permeate stream is a vapor; the feed is in the liquid phase and permeate is in the vapor phase.

Pervaporation separations by polymeric membranes have been studied for decades because of the potential industrial viability for separating azeotropic, close boiling, isomeric, or heat sensitive liquid mixtures.

Pervaporation processes are classified into three groups depending on the feed type and aim of the separation: hydrophilic pervaporation, hydrophobic pervaporation, and target organophilic pervaporation. The first is the most studied type; it is used for solving such problems as the dehydration of organic compounds (e.g., isopropanol, benzene, ethylene glycol, pyridine, acetic acid and extraction of water from mixtures, in particular, azeotropic solutions (for example, water–ethanol solutions). Hydrophobic pervaporation is also widely used: e.g., in waste water treatment, removal of

COV from drinking water, for the regeneration of organic solvents in the food industry and in biotechnology. Typical problems targeted with organophilic pervaporation are the separation of pairs of organic compounds such as benzene–cyclohexane, methanol-methyl-*tert*-butyl ether (MTBE) or ethanol-ethyl-*tert*-butyl ether (ETBE) and the separation of isomers (e.g., *n*-hexane-2,2-dimethylbutane; *m*-and *p*-xylene). Due to the complexity and high costs of the conventional processes, the membrane separation technology has been viewed in recent years as a viable alternative (Kim et al., 2002; Porter, 1990; Baker et al., 1991; Baker, 2004; Polyakov, et al., 2004; Matsui and Paul, 2002; Cunha et al., 2002; Vasudevan and Leland, 2007; Gimenes et al., 2007; Dubey et al., 2005; Huang et al., 2008; Zhang et al., 2007; Changand and Chang, 2004; Guan et al., 2006; Navajas et al., 2007; Kuhn et al., 2008; Liu et al., 2008; Castricum et al., 2008; Robert et al., 2008; Li et al., 2002; Guo et al., 2008; Sijbesna et al., 2008; Yahaya, 2008; Polyakov et al., 2003, 2004; Chen et al., 2008a; Li et al., 2008a, 2008b; Lin et al., 2008; Pandey et al., 2003; An et al., 2003, 2008; Lu et al., 2006; Peng et al., 2007a, 2007b; Luo et al., 2007; Lue et al., 2004; Tanaka et al., 2002; Mandal and Pangarkar, 2002, 2003; Ray and Ray, 2006a, 2006b; Khaye et al., 2005).

### 5.2.2.1 Hydrophilic PV

#### *5.2.2.1.1 Dehydration of Alcohols*

The energy efficient dehydration of low water content ethanol is a challenge for the sustainable production of fuel-grade ethanol; the ethanol forms with the water an azeotrope at 95.6 wt. % ethanol; the ethanol boils at 78.4°C, water at 100°C, but the azeotrope boils at 78.1°C.

Pervaporative membrane dehydration water/ethanol using a recently developed hydrophilic and hydrophobic membranes formulation showed excellent separation performance in dehydration. The selective membranes used are: organic membranes (poly(allylamine hydrochloride)/PVA) (Vasudevan and Leland, 2007), sericin/PVA (Gimenes et al., 2007), bacterial cellulose membranes impregnated with chitosan (CTSN–BCM) (Dubey et al., 2005), poly(thiol ester amide) (Huang et al., 2008); organic–inorganic membranes type hybrid ((PVA) with γ-aminopropyl-triethoxysilane (APTEOS) (Zhang et al., 2007) or type mixed matrix membranes (MMM) (copolymer of polysiloxane and phosphate ester cast on porous PVDF substrate) (Changand and Chang, 2004), a porous poly(acrylonitrile-co-methyl acrylate) intermediate layer and a polyphenylene sulfide (PPS) nonwoven fabrics substrate (Guan et al., 2006); inorganic membranes (mordenite tubular membranes (Navajas et al., 2007), DD3R zeolite membrane (Kuhn et al., 2008)).

For example, a schematic diagram of the system used to carry out the ethanol dehydration is shown in Figure 5.9. It consists of a two-chamber stainless steel module. The upper chamber holds the inlet and outlet ports for the ethanol–water feed liquid. The lower chamber is coupled to liquid

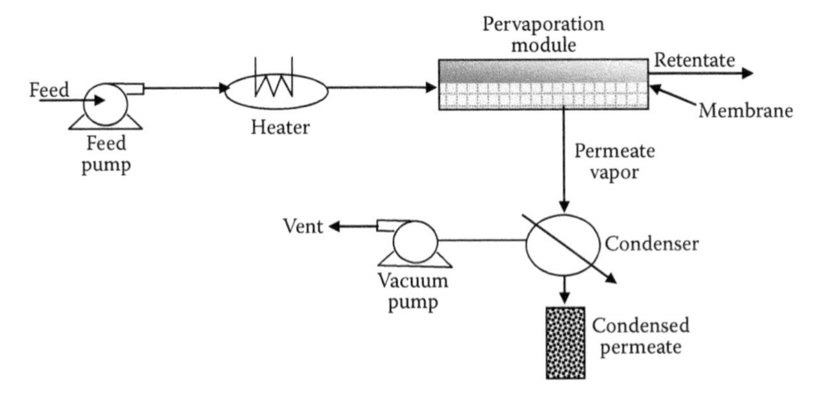


**FIGURE 5.9** Schematic diagram of a PV system. (Reprinted from *J. Memb. Sci.*, 306, Namboodiri, V.V, Vane, L.M., 209–215, p. 210, Copyright 2007, with permission from Elsevier.)

nitrogen traps and vacuum pump. The membrane is placed in between them, supported by a sintered stainless steel inlay. The active membrane area had 39 $cm^2$. The retentate stream was returned to the 20 L feed tank. Feed temperature was maintained with a constant temperature heat exchange bath. The feed flow rate through the membrane cell was kept constant at 2 L/min and this was determined to be sufficient to avoid liquid boundary layer effects on the mass transfer. Permeated vapor from the lower chamber was collected in a liquid nitrogen trap for time unit intervals and the mass of permeate collected was determined. The membrane showed a good stability over a 75 days period of continuous operation over a water concentration range of 0.5–42 wt. % (Vasudevan and Leland, 2007).

The composite membranes from poly(tetrafluoroethylene)/polyamide (PTFE/PA) are used to PV dehydration processes on a 70 wt. % isopropanol aqueous solution. The membranes exhibit good layer compatibility and stability due to the formation of covalent linkages between the PTFE substrate and PA layer. The membranes are stable under the pervaporation dehydration operations and show a high permeation flow of 1720 $g{\cdot}m^{-2}{\cdot}h^{-1}$ and a separation factor of 177 (Liu et al., 2008).

In the dehydration of *n*-butanol/water mixtures (95/5 wt. %) the following substances are used: organically hybrid linked 1,2-*bis*(triethoxysilyl)ethane (BTESE) on microporous molecular sieving (Castricum et al., 2008), ceramic-supported polyimide membranes (Robert et al., 2008) with a high separation factor of 4000 and 360, at 150°C, respectively.

##### *5.2.2.1.2 Dehydration of Benzene*

Three composite membranes of poly(vinyl alcohol)/poly(acrylonitrile) (PVA/PAN) which were crosslinked with formaldehyde, glutaraldehyde and maleic acid, respectively, were used for the dehydration of benzene at laboratory and pilot scale. In the pilot scale, the results explained that the water content could be reduced from 600 to 30 ppm at a temperature of about 70°C and downstream pressure of about 2000 Pa (Li et al., 2002).

##### *5.2.2.1.3 Dehydration of Ethylene Glycol*

The composite membrane of poly(vinyl alcohol)/poly(ethersulfone) (PVA/PES) was used for large-scale dehydration of the ethylene glycol/water mixture. The effect of interfacial crosslinking agent and hydrophilicity of the support layer on the interfacial adhesive strength and PV performance were investigated. The membrane exhibited a high separation factor of over 438, a high permeation flux of 427 $g{\cdot}m^{-2}{\cdot}h^{-1}$ for 80 wt. % ethylene glycol in the feed at 70°C and the required structural stability (Guo et al., 2008).

##### *5.2.2.1.4 Flue Gas Dehydration*

Composite hollow fiber membranes with a top layer of sulfonated poly(ether ether ketone) (SPEEK) were investigated and built into 20 fiber bundles. Coal-fired power plants produce electricity and in addition to that large volume flows of flue gas; to prevent condensation of the water vapor present in this flue gas stream, water has to be removed before emission into the atmosphere. The application of membrane technology for this separation is attractive due to the additional energy saving and the possible reuse of water (Sijbesma et al., 2008).

#### 5.2.2.2 Hydrophobic PV

##### *5.2.2.2.1 Separation of VOC (Toluene, BTEX, Aromatic/Aliphatic Hydrocarbons)*

A new composite membrane structure was tested and developed to improve the low permeability in typical composite membranes. Membrane and toluene (a representative VOC) permeation tests were accomplished to investigate the membrane structure. The tested membranes were polydimethylsiloxane (PDMS), as a top layer, and three different support layers: a non-woven fabric, a polysulfone (PSf) UF type membrane, and a non-woven fabric that was treated with a polyethylene glycol (PEG) solution. Thus, it can be said that the PEG treatment prevents the membrane resistance from increasing by clogging the macropores of the fabric. The membrane with PEG-treated non-woven fabric support layer had the best performance in toluene removal compared to various composite type membranes.

The toluene flux of the best membrane showed a maximum increase of 37.4% over a composite membrane with PSf and a 41.4% increase over that with nonwoven fabric (Kim et al., 2002).

The BTEX (benzene, toluene, ethylbenzene and xylene) are representative of an industrially significant family of aromatic hydrocarbon chemicals produced from industrial waste. A composite hollow fiber membrane module, employing hydrophobic microporous polypropylene hollow fibers, having a thin layer of cross-linkable vinyl-terminated silicone rubber, and polydimethylsiloxane (PDMS) coated on the inside diameter has been used for the separation of the BTEX mixture from water. An increase of the permeate pressure reduces the operating costs while the separation is enhanced; the permeation fluxes of both BTEX and water were found to increase with temperature and feed concentration (Yahaya, 2008).

For the separation of aromatic/aliphatic hydrocarbons crosslinked poly(methyl acrylate-*co*-acrylic acid) membranes and polyurethane dense membranes were used, for example. Copolymers of methyl acrylate and acrylic acid were synthesized to produce ionically crosslinked membranes using aluminum acetylacetonate for the separation of toluene/i-octane mixture at 100°C. A typical crosslinked membrane showed a normalized flux of 26 $kg{\cdot}\mu m{\cdot}m^{-2}{\cdot}h^{-1}$ and a selectivity of 13 for a 50/50 wt. % feed mixture (Matsui and Paul, 2002). The PU membranes were found to be selective toward aromatics in all systems studied (pentane, hexane, cyclohexane, isooctane, benzene, toluene, *m*-xylene). Both the swelling of polymer matrix and the permeate flux increase with aromatic weight fraction in the feed; the highest selectivity was achieved with a benzene/*n*-hexane mixture in the whole feed composition range (Cunha et al., 2002).

#### 5.2.2.2.2 *Separation of Mixtures of Chloromethanes*

The homologous series of chloromethanes is $CH_2Cl_2$, $CHCl_3$, and $CCl_4$. The process of their industrial production is the homogeneous or catalytic chlorination of methane; depending on the conditions of the process, the reaction products contain various quantities of all the three components. Currently, the process stream is being separated by rectification.

The PV of binary mixtures ($CH_2Cl_2$–$CHCl_3$, $CHCl_3$–$CCl_4$, $CH_2Cl_2$–$CCl_4$) through two amorphous copolymers of 2,2-bis-trifluoro-methyl-4,5-difluoro-1,3-dioxole and tetrafluoroethylene (Teflons AF) was studied at different temperatures, feed composition, and downstream pressure. AF Teflons possess several properties, especially for PV membranes: insolubility in common organic solvents (except for perfluorinated compounds), good chemical and thermal stability, excellent film forming properties, and low swelling tendency. The results show that the copolymer with the greater content of the dioxole comonomer and having the larger free volume (AF 2400, with a higher content (87%) of the dioxole component) is much more permeable; for certain mixtures the selectivity of PV through this material is similar to that of the less permeable copolymer (AF 1600, containing 65% of the dioxole component).

It can be presumed that hydrogen bonds by $CH_2Cl_2$ and $CHCl_3$ are formed; these compounds having "acid" hydrogen atoms are capable of hydrogen bonding in the processes of sorption and permeation in polymers. When temperature increases inter- and intra-molecular associates formed due to hydrogen bonding should dissociate. This effect should increase the permeation rate of $CH_2Cl_2$ and $CHCl_3$ but not $CCl_4$, thus increasing the selectivity of pervaporation (Polyakov et al., 2003, 2004).

#### 5.2.2.2.3 *Desulfurization of FCC Gasoline*

The composite membrane poly(dimethylsiloxane)/polyetherimide (PDMS/PEI) was used for the desulfurization of the cracked catalytic fluid (FCC gasoline); the organic sulfur compounds investigated are: thiophene, 2-methyl thiophene, 2,5-dimethyl thiophene, *n*-butyl mercaptan and *n*-butyl sulfide. The order of partially permeated flux and selectivity was: thiophene, 2-methyl thiophene, 2,5-dimethyl thiophene, *n*-butyl mercaptan and *n*-butyl sulfide (Chen et al., 2008a, 2008b).

PDMS–$Ni^{2+}Y$ zeolite hybrid membranes were produced and used for the pervaporation removal of thiophene from the model gasoline system. With the increase of $Ni^{2+}Y$ zeolite content, the permeation flux increased continuously, while the enrichment factor first increased and then decreased possibly due to the occurrence of defective voids within the organic–inorganic interface region. For example, the PDMS membrane containing 5.0 wt. % $Ni^{2+}Y$ zeolite exhibited the highest enrichment factor (4.84) with a permeation flux of 3.26 kg/($m^2$ h) for 500 ppm sulfur in feed at 30°C (Li et al., 2008a, 2008b).

The desulfurization of a thiophene mechanism of polyethylene glycol (PEG) membranes has been investigated by the study of solubility and diffusion behavior of typical gasoline components through PEG membranes with various crosslinking degrees. Crosslinking is an effective modification way to improve the overall performance of PEG membranes applied in gasoline desulfurization (Lin et al., 2008).

#### 5.2.2.3 Organophilic PV

##### *5.2.2.3.1 Separation of Benzene/Cyclohexane Mixture*

The separation of benzene/cyclohexane (Bz/Chx) mixtures is one of the most complicated processes in the petrochemical industry. As the difference in the boiling points of benzene and cyclohexane is only 0.6°C, and the interaction parameter between them is small, the liquid–vapor equilibrium curve of their binary liquid mixture is nearly a close-boiling system. Also, benzene and cyclohexane form azeotrope at 45 vol. % cyclohexane.

Thus, the separation of Bz and Chx is difficult by a conventional distillation process (azeotropic distillation and extractive distillation) because these components form close boiling point mixtures on the entire range of their compositions. The main advantage of PV is independent of relative volatilities of compounds and is not limited by the vapor–liquid equilibrium. The domination of sorption selectivity is due to the affinity between the double bonds of benzene molecules and the polar groups of the polymeric membrane; benzene has $\pi$ electrons that show a stronger affinity to polar molecules. Benzene's smaller size, smaller collision diameter, and planar shape enhance its diffusivity. The benzene had greater affinity toward PVA, for example, and the overall selectivity increased with increasing polymer density.

The comparison of performances of PV membranes for the separation of Bz/Chx showed a selectivity of 5–20 at the fluxes 1–10 kg·μm/$m^2$·h. The experiments indicated that a hybrid system (extractive distillation/PV) is more economical than either of the two processes alone.

Better results of separation Bz/Chx were obtained using modified membranes: aromatic polyamides with various aromatic diacids, PVA by acid-catalyzed acetalization (PVAc) (Pandey et al., 2003), binary blends PVC/EAC (ethylene-*co*-vinyl acetate copolymer) (An et al., 2003). A new hybrid membrane was prepared by filling carbon graphite (CG) into poly (vinyl alcohol) (PVA) and chitosan (CS) blending mixture, in hopes of improving the separation performance of the membranes by the synergistic effect of blending and filling (Lu et al., 2006), and membranes with carbon nanotube (poly(vinyl alcohol)/carbon nanotube hybrid membranes where the carbon nanotube was dispersed by using β-cyclodextrin (β-CD) (Peng et al., 2007a, 2007b) and nanocomposite membranes (PVA–CNT(CS)) prepared by incorporating chitosan-wrapped multiwalled carbon nanotube (MWNT) into poly(vinyl alcohol) (PVA) (Peng et al., 2007a, 2007b)).

An example of a PV experiment is shown in Figure 5.10. The effective surface area of the membrane in contact with the feed mixture is 28 $cm^2$. The permeation flux of benzene could be 61 g/$m^2$·h and the separation factor could be 41.2 at 333 K for Bz/Chx 50/50 wt. % mixtures (Peng et al., 2007).

Other types of membranes were prepared from blends of ethyl cellulose (EC) and hyperbranched-polyester (HBPE); the HBPE were employed as macromolecular crosslinking agents to enhance the pervaporation performance (Luo et al., 2007).

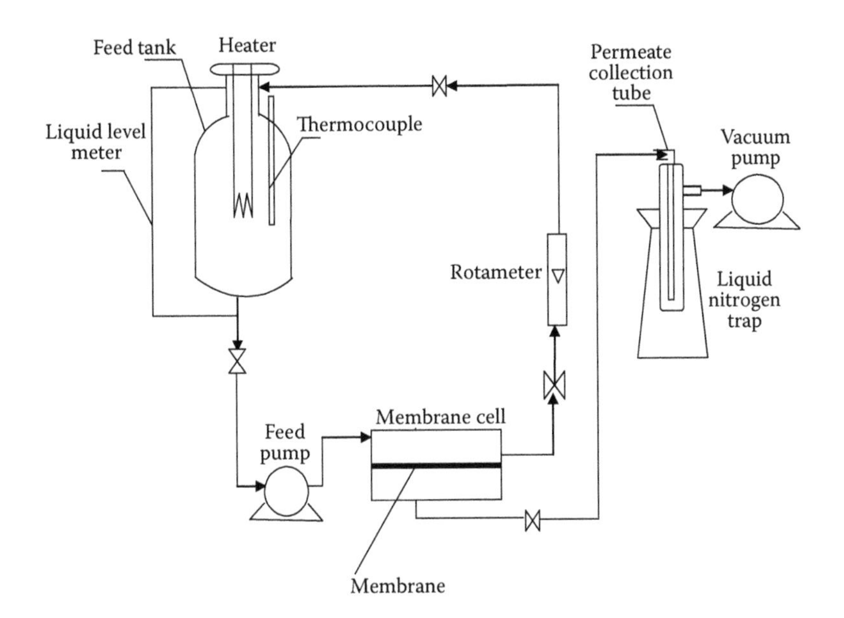


**FIGURE 5.10** Schematic diagram of the PV equipment. (Reprinted from *J. Memb. Sci.*, 297, Peng, F., C. Hu and Z. Jiang, 236–242, p. 238, Copyright 2007, with permission from Elsevier.)

The degree of swelling (DS) of membranes was investigated in the polyacrylonitrile-block-poly(methyl acrylate) (P(AN-b-MA)) block copolymer; it was found that the degree of swelling (DS) of membranes increased with increasing the MA content in the P(AN-b-MA) block copolymer and the benzene content in feed at 30°C (An et al., 2008).

A good results was obtained with a mixture of cation-exchange membranes containing copper ions (Cu (II)), (Lue et al., 2004) and crosslinked poly(ethylene oxide imide) segmented copolymer (PEO-PI) membranes (Tanaka et al., 2002), etc.

#### *5.2.2.3.2 Separation of Alcohols–Toluene/Benzene*

Methanol forms an azeotrope with toluene at 72.8 wt. % methanol. Pervaporative separation of methanol/toluene mixtures, over the entire range of 0%–100%, was obtained with hydrophilic membranes, for a selective separation of methanol (a polar compound) from its mixture with toluene/benzene (a non polar compound).

The hydrophilic membranes used are: cellulose acetate (CA), cellulose diacetate (CDA), cellulose triacetate (CTA), three blends of cellulose acetate and cellulose acetate butyrate (CBA), the copolymer of acrylonitrile and hydroxy ethyl methacrylate with three different compositions, the polyimide membrane, two blends of CTA with acrylic acid, poly(dimethylsiloxane) (PDMS), (Mandal and Pangarkar, 2002), polyacrylonitrile and four of its different copolymer membranes (acrylonitrile with 2-hydroxyethylmethacrylate, maleic anhydride, acrylic acid and methacrylic acid) (Mandal and Pangarkar, 2003); chitin/polyetherimide composite membranes had good pervaporation characteristics and were also found to be mechanically robust and stable to withstand the corrosive conditions.

#### *5.2.2.3.3 Separation of Toluene–Methanol*

For the pervaporative separation of a toluene–methanol mixture up to 10 wt. % of toluene in feed were used the hydrophobic membranes, for example, the natural rubber (NR) and poly(styrene-co-butadiene) rubber (SBR) has been crosslinked with sulfur and accelerator with three different doses of varied accelerator to sulfur ratios to obtain three crosslinked membranes from each of these two rubbers (NR-1, NR-2 and NR-3 and SBR-1, SBR-2 and SBR-3). Among these membranes, NR-1 and SBR-1 with highest crosslink density showed a maximum separation factor for toluene along with a good flux. It has also been found that for comparable crosslink densities SBR membranes showed a better separation factor than NR membranes (Ray and Ray, 2006a, 2006b).

The membranes with a good structure, crystallinity and thermal stability were the poly(ethylene terephthalate)-graft-polystyrene (PET-*g*-PST) membranes; grafted PET-*g*-PST membranes with degrees of grafting up to 35% were found to be better than the ungrafted PET for pervaporation of toluene/methanol system (Khaye et al., 2005).

### 5.2.3 Applications of Ion Exchange Membrane Processes

In ion exchange membranes, charged groups are attached to the polymer backbone of the membrane material. Electrodialysis is by far the largest use of ion exchange membranes, mainly to desalt brackish water or (in Japan) to produce concentrated brine. Long-term major applications for ion exchange membranes may be in the non-separation areas such as fuel cells (Baker, 2004). It has been anticipated that fuel cells for residential accommodations will be manufactured in the range 1–10 kW and portable fuel cells will be built in the sub-watt to 5 kW range; in world markets, within a much closer time frame, an estimated 550 GW of generating capacity will be added as per these projections (Smitha et al., 2005).

The $H_2/O_2$ fuel cell, commonly referred to as polymer electrolyte fuel cell (PEFC), and the direct methanol fuel cell (DMFC) are two types of fuel cells that use polymer electrolytes. DMFC have higher energy density but exhibit shortcomings such as (a) slower oxidation kinetics than PEFC below 100°C and (b) significant permeation of the fuel from the anode to the cathode resulting in a drop in efficiency of fuel utilization up to 50%.

The membrane material used can be classified into four categories: (1) perfluorinated ionomers, (2) hydrocarbons membranes, (3) sulfonated polyarylenes and (4) acid–base blends.

The perfluorinated sulfonic acid membranes have perfluorinated structures with attached sulfonic acid groups (Figure 5.11a) the Nafion® is the prominent polymer in the first category, the more "mature" membrane. Ultra-thin integral composite membranes consisting of a base material are preferably made of expanded PTFE that supports an ion exchange material such as, for instance,

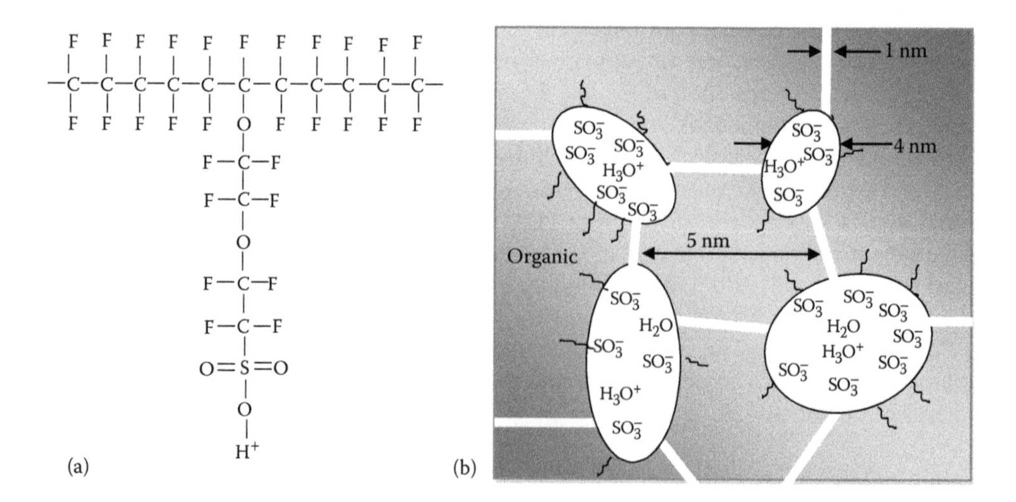


**FIGURE 5.11** (a) Chemical structure of Nafion; (b) schematic view of a cluster network model. (Reprinted from *J. Memb. Sci.*, 259, Sridhar, and A.A. Khan, 10–26, p. 15, 17, Copyright 2005, with permission from Elsevier.)

perfluorinated sulfonic acid resin; ion exchanges are separated from the fluorocarbon backbone thus forming spherical clusters (pores), which are connected by short narrow channels (Figure 5.11b). Much research has been conducted on the details of the transport of protons through the polymer matrix and novel methods of improving its properties, but led to the development of a sturdy inexpensive substitute to Nafion (Smitha et al., 2005; Yang et al., 2004; Chen et al., 2008b; Lee et al., 2008a; Holmberg et al., 2008; Tay et al., 2008; Nam et al., 2008; Kim et al., 2008). The perfluorinated polymer used most extensively has the trade name of Nafion, produced by DuPont. But the PEFC based on a perfluorinated membrane have typically been operated in a temperature range between 50°C–90°C. This temperature is a compromise between competing factors. A number of additions have been investigated to ascertain water retention at higher temperatures: silica, $TiO_2$, phosphoric acid, clorosulfonic acid, heteropolyacids (phosphotungstic acid, phosphomolybdenic acid, phosphotin acid), silicotungstic acid, thiophene (Smitha et al., 2005), zirconium phosphate (ZrP) (Yang et al., 2004; Chen et al., 2008b), formation of platinum nanoparticles in Nafion (Lee et al., 2008a, 2008b, 2008c), sulfonic acid functionalized zeolite beta (Holmberg et al., 2008), hybrid nanofiler containing silica (Tay et al., 2008), sulfonated poly(phenylmethyl silsesquioxane) (Nam et al., 2008), etc.

Hydrocarbon membranes are less expensive, commercially available and their structure permits the introduction of polar sites as pendant groups in order to increase the water uptake (Lee et al.; Norddin et al., 2008). Poly(vinyl alcohol) membranes are known to be good methanol barriers. The sulfonated polyarylenes are high temperature rigid polymers owing to the presence of inflexible and bulky aromatic groups (Bai and Ho, 2008; Lee et al., 2008c). The acid–base blends considered for fuel cell membranes involve the incorporation of an acid component into an alkaline polymer base to promote proton conduction (Mosa et al., 2008; Tripathi et al., 2008; Bello et al., 2008; Gasa et al., 2008; Fu et al., 2008).

Proton exchange membrane fuel cells (PEMFC) (Mosa et al., 2008) are studied as one of the new energy conversion devices. The current technology is based on perfluorosulfonic acid membranes as electrolytes, such as Nafion, which exhibit many specific properties: high proton conductivity, good mechanical and chemical stability. A new high-performance membrane PEMFC was prepared with different compositions in the systems $P_2O_5$–$ZrO_2$ and $SiO_2$–$P_2O_5$–$ZrO_2$ using the sol-gel technique. $SiO_2$ was incorporated for an improvement of the chemical and mechanical stability and $P_2O_5$ for higher proton conductivity; the membrane was doped with tungstophosphoric acid (PWA) to provide high proton conductivity.

The application of Nafion membranes to direct methanol fuel cells (DMFC) (Chen et al., 2008b; Tripathi et al., 2008; Bello et al., 2008), which uses methanol as fuel, causes the problem of methanol crossover membranes, thus lowering the DMFC performance. The incorporation of submicron particles of metal (IV) phosphates such as zirconium phosphate (ZrP) and tin phosphate (SnP) in polymer matrices tend to increase the proton conductivity of the polymers; with increasing proportion of metal particles, a notable degree of improvements in proton conductivity, especially at high temperature, can be anticipated (Smitha et al., 2005). The properties of two major polyelectrolyte membranes (NF–ZrP–I, NF–ZrP–d) controlling DMFC performance are proton resistance and methanol permeability of membranes. The NF–Zr–I membranes were prepared by inserting $ZrOCl_2$ into the NF (Nafion/PTFE) membrane after Nafion was annealed at 135°C; however, NF–Zr–d membranes were prepared by mixing $ZrOCl_2$ into Nafion solution before Nafion was annealed at 135°C. Thus, by morphology, NF–Zr–I membranes had lower methanol electro-osmosis and better DMFC performance than NF–Zr–d membranes. For NF–Zr–I to have better DMFC performance than NF–Zr–d, reason can be attributed to the process of hybridizing ZrP into NF membranes (Yang et al., 2004).

Phosphoric acid functional groups showed high intrinsic proton conductivity due to a high mobility of protonic charge carriers and a degree of self-dissociation in the polymer matrix (Tripathi et al., 2008). The composite membranes were obtained by embedding inorganic proton conducting material (tungstophosphoric acid) (TPA) into sulfonated poly(ether ether ketone) (SPEEK) polymer matrix (Bello et al., 2008).

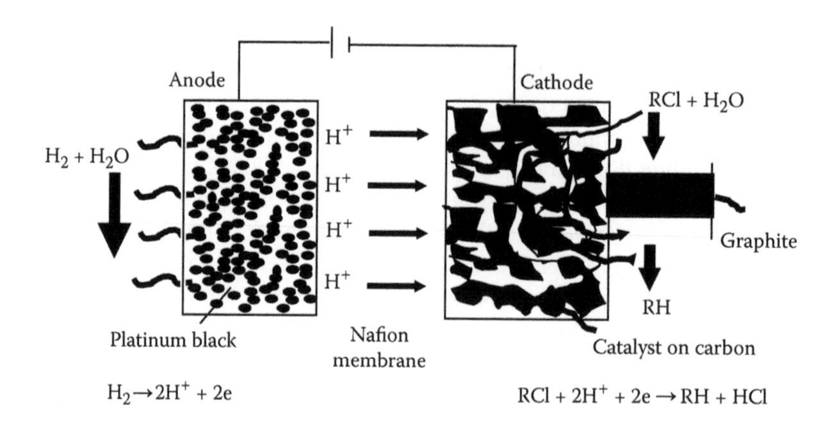


**FIGURE 5.12** Schematic representation of a modified fuel cell. (From Popescu, M., R. Popescu and C. Stratula. 2006. Metode fizico-chimice de tratare a poluantilor industriali atmosferici (in Romanian). Editura Academiei Romane, p. 383.)

Liu et al. treated gas-phase chlorinated solvents using a modified fuel cell (Figure 5.12). The electronegative character of the halogenated substituents is energetically poised to participate in reductive reactions of the form $RX + 2e + H^+ \rightarrow RH + X^-$. The reductive dehalogenation of gas-phase chlorinated alkanes ($CCl_4$, $CHCl_3$ and 1,1,1-trichloroethane), alkenes (perchloroethene (PCE) and trichloroethene (TCE)) was reduced in a modified fuel cell. Anode and cathode compartments are separated by a Nafion membrane. The $H_2$ was fed into the anode compartment and the RX (target compounds) was added to the cathode compartment in a $N_2$ stream at atmospheric pressure; RH is the product of hydrogenolysis processes.

### 5.2.4 Applications of Carrier-Facilitated Transport (Active Transport)

Carrier-facilitated transport involves supported liquid membranes containing a macroporous support in which the carrier (liquid phase) is immobilized by capillary force (Baker, 2004).

The transport species through these membranes is active because it is facilitated by the carrier that chemically combines with the permeant to be transported. In this case, the rate of chemical reaction is fast, compared to the rate of diffusion across the membrane, and the amount of material transported by carrier-facilitated transport is much larger than that transported by passive diffusion.

In carrier-facilitated transport membranes the active transport occurs at the membrane interphase, and depends on the type of reaction between the carrier and permeants. The transport of a component can be: facilitated transport (co-transport), where only one species is transported across the membrane by the carrier, or coupled transport (counter-transport) where the carrier agent couples the flow of two species (Baker, 2004; Araki and Tsukube, 1990). A schematic presentation of the active transport by a cation across the membrane is presented in Figure 5.13: by co-transport (Figure 5.13a) and by counter-transport (Figure 5.13b).

The approach to stabilizing facilitated transport membranes is to form multilayer structures in which the supported liquid –selective membrane is encapsulated between these layers of very permeable but non selective dense polymer layers. These membranes are proposed to be used in hydrometallurgy, biotechnology and in the treatment of industrial wastewater.

In this way, new separation engines were developed for the removal and recovery of metals ions and small organic compounds, from different solutions: active membranes, containing a carrier (extractant), a dense polymer layer and a support. In journals and books, these membranes are named: polymer inclusion membranes (PIMs) and activated composite membranes (ACMs).

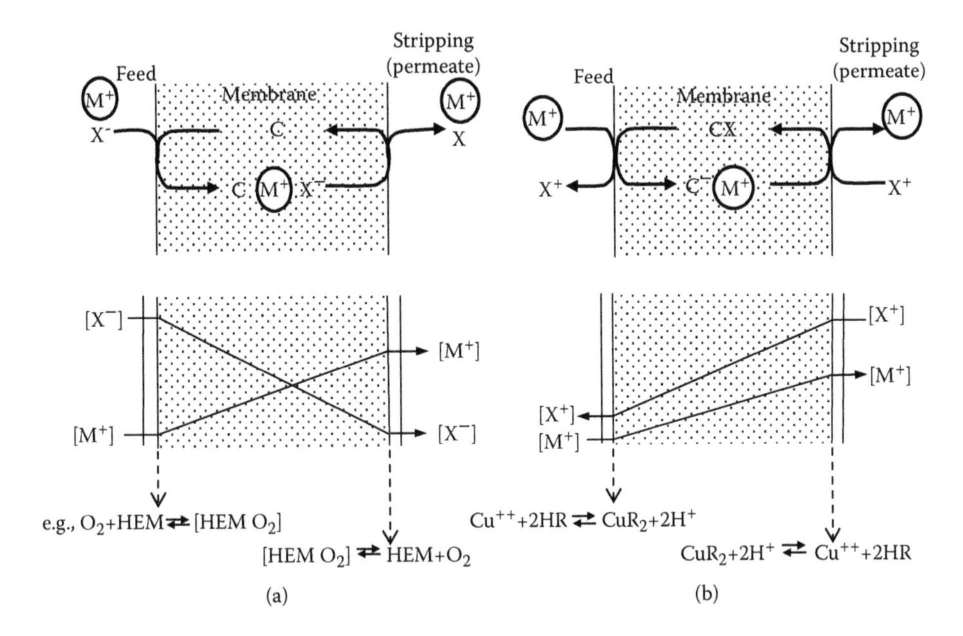


**FIGURE 5.13** Schematic presentation of carrier-facilitated transport (active transport) by (a) facilitated transport and (b) coupled transport. [M+], cation; [C], carrier; [X-], [X+], aqueous soluble coupled-transport ions; [HEM], hemoglobin. (Reprinted from *J. Memb. Sci.*, 281, Nghiem, L.D., P. Mornane, I.D. Potter, et al., 7–41, p. 32, Copyright 2006, with permission from Elsevier.)

Polymer inclusion membranes are membranes that contain a polymeric support such as cellulose triacetate (CTA) or poly (vinyl chloride) (PVC) to form a dense, flexible and stable film (Scindia et al., 2005), while activated composite membranes are membranes with a bi-layer structure consisting of a porous lower polymeric support (polysulfone) and a dense upper layer (polyamide) (Resina et al., 2008).

The most important applications of coupled transport are the removal or recovery of metals from aqueous solutions, such as contaminated ground water or industrial (hydrometallurgical) streams, and small organic solutes from different effluents. Thus, the first commercial applications of coupled transport are likely to be smaller plants installed in pollution control applications (Baker, 2004). It has been predicted that practical industrial applications of polymer inclusion membranes (PIMs) and activated composite membranes (ACMs) will be achieved in the near future. Below, some new laboratory research applications are presented.

### 5.2.4.1 Recovery of Precious Metals

#### *5.2.4.1.1 Recovery of Pt (VI)*

Platinum is a metal of high interest due to extensive anthropogenic uses such as in catalysis, electronic devices and biomedical devices (Hartlwy, 1991). The large demand of this metal next to its high price, and its high production cost from naturally occurring supplies have made the recovery of precious metal a feasible and cost effective alternative. Thus, the transport of platinum (IV) through PIM and ACM, with carrier Aliquat 336 (tricaprylmethylammonium chloride) from chloride media, is presented (Fontas et al., 2005; Resina et al., 2008). The transport of platinum (IV) has been studied through a PIM using a cellulose triacetate (CTA) membrane and 2-nitrophenyl octyl ether (NPOE) as plasticizer and through an ACM with a porous lower polymeric support prepared by phase inversion (polysulfones) and a dense upper layer made by interfacial polymerization (polyamide). To complete the investigation they used another type of membrane, named hybrid membrane (HM) containing carrier Alliquat 336 immobilized in a polymer matrix of CTA with a plasticizer, modified by the incorporation of an inorganic material (silanes) prepared by a sol-gel route. In the case of PIM the feed phase consisted of Pt (VI) in 0.5 M NaCl at pH 2, the receiving phase was 0.5 M $NaClO_4$, and the membrane working area was 8.5 $cm^2$. In the case of ACM and HM, the stripping phases contain anionic species 0.5 M $ClO_4^-$ and $NO_3^-$, respectively, and 12 $cm^2$ was the membrane working area. The authors proposed the following transport mechanism for PIM:

- Source phase:

$$PtCl_{6aq}^{2-} + 2(R_4N^+Cl^-)_{mem} \rightleftarrows (R_4N^+)_2PtCl_{6mem}^{2-} + 2Cl_{aq}^-$$

- Receiving phase:

$$(R_4N^+)_2PtCl_{6mem}^{2-} + 2ClO_{4aq}^- \rightleftarrows 2R_4N^+ClO_{4mem}^- \, PtCl_{6aq}^{2-}$$

and suggested that the driving force for the transport of Pt (VI) was provided by the high $ClO_4^-$ concentration gradient. It was found that the good flux of Pt (IV) was $13.3 \cdot 10^{-7}$ $mol \cdot m^{-2} \cdot s^{-1}$ when the concentration of the carrier in the PIM was 250 mM. The higher selectivity for Pt (VI) was showed, for a competition between Pd (II) and Pt (IV) cations and this was explained on the basis of the higher hydrophobicity of the Pt complex compared to the Pd complex making the transport faster for Pt than for Pd in the PIM.

In the case of ACM and HM, authors report that, comparing the results of both types of membrane with the same carrier concentration, 250 mM, the percentage of metal transported trough ACM was 77% and, for HM, 30%, and the percentage of Pt (IV) recovery was 56% for HM and 12% for ACM, after 24 h of experimentation. This difference is probably due to the higher diffusion resistance across the HM than for the ACM and the possibility of immobilizing higher amounts of carrier (1.41 mg $cm^{-2}$) in the HM matrix. Moreover, these membranes present higher selectivity toward Pt (IV) than Pd (VI). That could be explained by the fact that platinum (IV) has a larger ionic radius and higher oxidation state than palladium (II), and therefore, the higher number of ligands and the larger size of the complex make platinum species much more hydrophobic than those of palladium. Consequently, the extraction into the membrane phase is much faster for $PtCl_6^{2-}$ than $PdCl_4^{2-}$.

*5.2.4.1.2 Recovery of Au (III)*

The interest in the recovery of gold (III) in the processing of electronic wastes, particularly printed wire boards (PWBs), makes the transport of this precious metal using supported composite liquid membranes attractive for economical purposes (Veit et al., 2005; Faramarzi et al., 2004).

Argiropouls et al. have studied the extraction of Au (III) from hydrochloric acid solutions using a PIM, containing a PVC support and Aliquat 336 chloride as carrier ($R_4N^+Cl^-$). For the membrane containing 50% (w/w) Aliquat 336, gold was completely extracted from 2.5 M HCl solution containing 100 µg $L^{-1}$ of gold within 75 h. For the receiving phase they used a thiourea solution. This system was aimed at the recovery of gold from electronic scrap and was tested on a sample of scrap containing 96% (w/w) copper, 0.13% (w/w) gold and a low concentration of iron, dissolved in diluted Aqua Regia to give a 2.5 M HCl concentration. It was found that all the gold could be extracted by the membrane with a selectivity factor of $5 \cdot 10^5$ for gold over copper.

Another application for selective gold (III) separation is presented, using an organic–inorganic activated hybrid membrane containing Kelex 100 (7-(4-ethyl-1-methyloctyl)-8-hydroxyquinoline) as carrier (De San Miguel et al., 2008). The chloride media was used for gold separation and the permeabilities in the order of 0.05 $cm \cdot min^{-1}$ were determined for membranes with optimal composition (62.5% (w/w) poly(dimethylsiloxane) (PDMS), 31.25% (3-Aminopropyl) triethoxysilane (APTS), and 6.25% Kelex 100, using an aqueous 0.05 M mol·dm$^{-3}$ $Na_2S_2O_3$ strip solution. The Au (III) is extracted by the formation of $AuCl_4^-$ (R-$H_2Q^+$) species with an extraction equilibrium constant $10^{1.15}$. A selective separation of 10 ppm of Au (III) from 50 ppm of Cu (II) and 100 ppm of Zn (II) was possible; however, membranes could not be reused successfully for more than three cycles of 6 h due to mechanical instability. That can be explained by some accumulation of gold in the membrane because of complex properties matrix, which significantly reduces when Kelex 100 content increases from 5% (w/w).

#### 5.2.4.2 Recovery of Toxic Metals

##### *5.2.4.2.1 Recovery of Cr (VI)*

Because Cr (VI) is one of major toxic element present in the environment, a remarkable increase of the applications of composite liquid membranes in recovering this metal from residual aqueous solutions, for example generated by galvanic plants and mine waters, is observed. Chromium (VI) can be effectively removed from acidic chloride solutions using coupled transport across PIMs with tertiary amines ($R_3N$, where R = $n$-$C_6H_{13}$-$n$-$C_{12}H_{25}$) and Aliquat 336 as ions carrier, CTA or PVC as a support. The stripping phase used was 1 mol/L $NaNO_3$ (Scindia et al., 2005; Kozlowski, 2005). The highest fluxes of PIM transport were found for $n$-tributyl amines and Aliquat 336 as ion carrier. The variation of permeability coefficients of Cr (VI) were between 4.42 µm·s$^{-1}$ and 1.46 µm·s$^{-1}$ respectively at pH 2 and 8, in the feed solution, because the transport of ions across the membrane depends on their chemical form. The Cr (VI) permeation involves coupled-diffusion transport whereby Cr (VI) permeates from feed to stripping compartment and $NO_3^-$ permeates in the opposite direction. The diffusion mobility in the membrane phase is dependent on the nature of diffusing Cr (VI) species and membrane composition; the recovery of Cr (VI) from seawater samples was only 58% over a period of 10 h and, in the case of municipal water, the recovery factor was ≈ 99% for the same period of time.

In case of recovery of Cr (VI) from metallurgical wastewaters, ACMs were used (Melita and Popescu, 2008), containing Aliquat 336 as carrier; the pH of feed and stripping solutions was maintained between 4.00 and 5.50 to decrease the conversion probability of Cr (VI) to Cr (III) (Galán et al., 2005). The good separation of chromium was observed at 25 and 50 mg·L$^{-1}$ Cr (VI) respectively in feed solution. After 7 h treatment of the polluted water with ACM, 0.2 mg·L$^{-1}$ was the maximum unpolluted low admissible concentration and it could be used by the purification plant. A good enrichment factor, E (12.5 and 14.5) in the separation process was obtained in concordance with the calculated membrane capacity. Studies have related that the transport of Cr (VI) through porous polysulfone membranes is still often diffusion-controlled, but the morphology of the membrane affects the coupled transport of ($CrO_4^{2-}$) and ($Cr_2O_7^{2-}$) through the membrane.

##### *5.2.4.2.2 Recovery of Cd (II)*

Cadmium is known to be an environmental pollutant with toxic effects. In the human body it penetrates via inhalation of dust from industry, or orally by food. Cd accumulates in the kidneys and results in various diseases. But, despite its toxicity, Cd is widely used in metallurgy, electroplating, pigmenting and in the fabrication of Ni–Cd batteries. In the laboratory applications the PIMs for removal of Cd (II) were used. These membranes consisted of CTA, as the polymeric support, o-nitrophenyl octyl ether (NPOE) as the plasticizer and Aliquat 336 and D2EHPA (di-(2-ethylhexyl) phosphoric acid) as the carriers. Some of the aqueous samples used are collected from the Mediterranean Sea with the following characteristics: pH 8.2, conductivity 62 mS, 0.62 M $Cl^{-1}$, 0.68 M $Na^+$, 4325 mg·L$^{-1}$ $SO_4^{2-}$, 1415 mg·L$^{-1}$ $Mg^{2+}$ and $Ca^{2+}$, $K^+$, $HCO_3^-$ concentration was < 1000 mg·L$^{-1}$. These were enriched with the appropriate amount of Cd stock solution (from 2 to 30 mg·L$^{-1}$). Other real samples contain Cd from a Ni–Cd battery leaching and were brought up to the desired volume with the same acid and used as the feed solution. The extraction of metallic anions

($CdCl_3^-$ and $CdCl_4^{2-}$) by Aliquat 336 and metallic cations ($Cd^{2+}$) by D2EHPA is based on a coupled transport that involves the formation of an ion pair that can enter the membrane. The transport of Cd (II) trough membranes depends on the composition of the membrane, the concentration of the carrier, the type of plasticizer and its amount and the influence of the aqueous phase. The presence of $Cl^-$ in the feed phase is necessary for the extraction of cadmium by Aliquat 336, but similar results are obtained when metal is contained in both NaCl and HCl media, showing the efficiency of the PIM system when dealing with either acidic or saline solutions. In the case of D2EHPA, when the pH of the feed phase is increased the transport flux tends to decrease. This result is unexpected as classical (Baker, 2004; Pont et al., 2008; Vernet, 1991; Bertin and Averbeck, 2006; Kebiche-Senhadji et al., 2008).

This phenomenon was attributed to the formation of hydroxide forms of the metal less extractable with D2EHPA (Mellah and Benachour, 2006), the formation of emulsion, which depend both on pH, and on the concentration of the carrier (Alvarez-Salazar, 2005) or on a mechanism which involves the monomeric form (LH) or dimeric from ($LH_2$) of D2EHPA, to form the complexes $ML_2$ and $ML_2(LH)_2$. The existence of $LH_2$ is more probable in a range of pH lover than the pKa of D2EHPA, while the formation of the $ML_2$ species predominates in the opposite range (Kebiche-Senhadji et al., 2008). For a matrix with Cd (II) and a large amount of Ni (II), in feed solution, the PIM systems allowed the quantitative separation of Cd (II) after 400 min. of experimentation. In addition, the effect of the presence of Cu, Zn and Pb in feed solutions demonstrates that the membrane with D2EHPA carrier transports all the cations without any marked selectivity, while Aliquat 336 shows an improved selectivity for Cd (II) and Pb (II) as compared to Zn (II). The maximal Cd (II) recovery factors obtained in 8 h are 97.5% and 91.8% with D2EHPA and Aliquat 336, respectively.

#### 5.2.4.2.3 *Transport of Zn (II)*

In order to extract metal ions from hydrometallurgical effluents, D2EHPA is a carrier that in membrane processes presents a great selectivity toward $Zn^{2+}$, in relation to other metals such as $Cu^{2+}$, $Ni^{2+}$, $Ca^{2+}$, $Mn^{2+}$, $Fe^{3+}$, $Al^{3+}$ and $Cd^{2+}$ as it has been demonstrated in some membrane processes. On the other hand D2EHDTPA (di-(2-ethylhexyl) dithiophosphoric acid) is an analogous molecule to D2EHPA where two sulfur atoms substitute two oxygen atoms and its acidity is slightly stronger than that of D2EHPA, due to the named substitution. It has been shown that D2EHDTPA also shows a crossed transport between protons and metallic ions in a counter-transport system. One of the most utilized applications in the clean up effluent streams from the electroplating industry is, for example, the removal of Zn (II) with acidic carriers such as D2EHDTPA immobilized in ACMs and D2EHPA immobilized in PIMs. In both situations, the separation was achieved by pH control of the feed phase with 1.0 M HCl as the striping phase. Best selectivity for Zn (II) was achieved using the source phase at $pH < 2$, which was in accordance with the solvent extraction of D2EHPA for the metal ions. A novel type of PIM system, HM were prepared and used in applications. The metal ions used for selectivity experiments were $Cd^{2+}$, $Cu^{2+}$ and $Zn^{2+}$ that appear together in some industrial processes such as electroplating practices. Data were obtained by using ACM and PIM prepared with a concentration of 300 mM in D2EHPA, 230 mM in D2EHDTPA or, in other cases, including both carriers in the proportion 245:40 (mM) D2EHPA: D2EHDTPA. In the case of binary mixtures $Zn^{2+}$–$Cu^{2+}$, DE2HPA transferred about 43% of $Zn^{2+}$ from the feed to the stripping solution within 7 h of experiment, whereas almost all the $Cu^{2+}$ remained on the feed. In the case of recovery of $Zn^{2+}$ from a $Zn^{2+}$–$Cd^{2+}$ binary mixture the same carrier transported 63% of $Zn^{2+}$ in 7.5 h, while it was observed that there was no transport of $Cd^{2+}$. The tendency was reversed with D2EHDTPA, obtaining a major recovery for $Cd^{2+}$, 49% in 8 h while only 36% of $Zn^{2+}$ was recovered in the same period of time. The results obtained using a mixture of carriers contributed to consider that in the presence of D2EHPA most of the transport involves the $Zn^{2+}$, and almost 60% of the metal was recovered, whereas $Cd^{2+}$ was re-extracted in a 3% that was negligible (Resina et al., 2006; Macanas and Muñoz, 2005; Ulevicz et al., 2003; Filippou, 2004).

## REFERENCES

Alvarez-Salazar, G., A.N. Bautista-Flores, E. Rodriguez de San Miguel, M. Muhammed, and J. de Gyves. 2005. Transport characterization of a PIM system used for the extraction of Pb (II) using D2EHPA as carrier. *J. Memb. Sci.* 250: 247–257.

An, Q.F., J.W. Qian, H.B. Sun, et al. 2003. Compatibility of PVC/EVA blends and the pervaporation of their blend membranes for benzene/cyclohexane mixtures. *J. Memb. Sci.* 222(1/2): 113–122.

An, Q.F., J.W. Qian, Q, Zhao, et al. 2008. Polyacrylonitrile-block-poly(methyl acrylate) membranes 2: Swelling behavior and pervaporation performance for separating benzene/cyclohexane. *J. Memb. Sci.* 313(1/2): 60–67.

Anderson, C.J., S.J. Pas, G. Arora, et al. 2008. Effect of pyrolysis temperature and operating temperature on the performance of nanoporous carbon membranes. *J. Memb. Sci.* 322(1): 19–27.

Anson, M., J. Marchese, E. Garis, et al. 2004. ABS copolymer-activated carbon mixed matrix membranes for $CO_2/CH_4$ separation. *J. Memb. Sci.* 243: 19–28.

Araki, T., and H. Tsukube. 1990. *Liquid Membranes: Chemical Applications.* Boca Raton: CRC Press, Florida.

Bai, H., and W.S.W. Ho. 2008. New poly(ethylene oxide) soft segment-containing sulfonated polyimide copolymers for high temperature proton-exchange membrane fuel cells. *J. Memb. Sci.* 313(1/2): 75–85.

Baker, R.W. 2004. *Membrane technology and applications.* New York: John Wiley & Sons.

Baker, R.W. 2006. *Membranes for Vapour/gas separation*, 2nd ed. p. 15. http://www.mtrinc.com/publications/MT01%20Fane%20Memb%20for%20VaporGas_Sep%202006%20Book%20Ch.pdf.

Baker, R.W., E.L. Cussler, W. Eykamp, et al. 1991. *Membrane separation systems.* Noyes Data Corporation.

Basu, A., J. Akhtar, M. H. Rahman, et al. 2004. A Review of separation of gases using membrane systems. *Petroleum Science and Technology* 22(9/10): 1343–1368.

Bello, M., S.M.J. Zaidi, and S.U. Rahman. 2008. Proton and methanol transport behavior of SPEEK/TPA/MCM-41 composite membranes for fuel cell application. *J. Memb. Sci.* 322(1): 218–224.

Bertin, G., and D. Averbeck. 2006. Cadmium: cellular effects, modifications of biomolecules, modulation of DNA repair and genotoxic consequences—a review. *Biochimie* 88: 1549–1559.

Castricum, H.L., R. Kreiter, and H.M. van Veen. 2008. High-performance hybrid pervaporation membranes with superior hydrothermal and acid stability. *J. Memb. Sci.* 324(1/2): 111–118.

Changand, C.L., and M.S. Chang. 2004. Preparation of multi-layer silicone/PVDF composite membranes for pervaporation of ethanol aqueous solutions. *J. Memb. Sci.* 238(1/2): 117–122.

Chen, J., J. Li, R. Qi, et al. 2008a. Pervaporation performance of crosslinked polydimethylsiloxane membrane for deep desulfurization of FCC gasoline I. Effect of different sulphur species. *J. Memb. Sci.* 322(1): 113–121.

Chen, L.C., T.L. Yu, H.L. Lin, et al. 2008b. Nafion/PTFE and zirconium phosphate modified Nafion/PTFE composite membranes for direct methanol fuel cells. *J. Memb. Sci.* 307: 10–20.

Chenar, M.P., M. Soltanieh, and T. Matsuura. 2008. Application of Cardo-type polyimide (PI) and polyphenylene oxide (PPO) hollow fiber membranes in two-stage membrane systems for $CO_2/CH_4$ separation. *J. Emb. Sci.* 324(1/2): 85–94.

Cunha, V.S., M.L.L. Paredes, C.P. Borges, et al. 2002. Removal of aromatics from multicomponent organic mixtures by pervaporation using polyurethane. *J. Memb. Sci.* 206: 277–290.

Daisley, G.R., M.G. Dastgir, F.C. Ferreira, et al. 2006. Application of thin film composite membranes to the membrane aromatic recovery system. *J. Memb. Sci.* 268(1): 20–36.

De San Miguel, E.R., A.V. Garduño-Garcia, M.E. Nuñez-Gaytan, J. C. Aguilar, and J. de Gyves. 2008. Application of an organic–inorganic hybrid membrane for selective gold (III) permeation. *J. Memb. Sci.* 307: 1–9.

Dubey, V., L.K. Pandey, and C. Saxena. 2005. Pervaporative separation of ethanol/water azeotrope using a novel chitosan-impregnated bacterial cellulose membrane and chitosan–poly(vinyl alcohol) blends. *J. Memb. Sci.* 251(1/2): 131–136.

Faramarzi, M.A., M. Stagars, E. Pensini, W. Krebs, and H. Brandtl. 2004. Metal solubilization from metal containing solid material by cyanogenic *Chromobacterium violaceum. J. Biotechnol.* 113: 321–326.

Filippou, D. 2004. Innovative hydrometallurgical processes for the primary processing of Zn (II). 2005. *Miner. Process. Extr. Metall.* 25: 205–252.

Fontas, C., R. Tayeb, S. Tingry, M. Hidalgo, and P. Seta. 2005. Transport of platinum (VI) trough supported liquid membrane (SLM) and polymeric plasticized membrane (PPM). *J. Memb. Sci.* 263: 96–102.

Fu, R.Q., D. Julius, and L. Hong. 2008. PPO-based acid–base polymer blend membranes for direct methanol fuel cells. *J. Memb. Sci.* 322(2): 331–338.

Galán, B., D. Castañeda, and I. Ortiz. 2005. Removal and recovery of Cr (VI) from polluted ground water: a comparative study of ion-exchange technologies. *Water Res.* 39: 4317–4324.

Gasa, J.V., R.A. Weiss, and M.T. Shaw. 2008. Structured polymer electrolyte blends based on sulfonated polyetherketoneketone (SPEKK) and a poly(ether imide) (PEI). *J. Memb. Sci.* 320(1/2): 215–223.

Gimenes, M.L., L. Liu, and X. Feng. 2007. Sericin/poly(vinyl alcohol) blend membranes for pervaporation separation of ethanol/water mixtures. *J. Memb. Sci.* 295(1/2): 71–79.

Gonzo, E.E., M.L. Parentis, and J.C. Gottifredi. 2006. Estimating models for predicting effective permeability of mixed matrix membranes. *J. Memb. Sci.* 227(1/2): 46–54.

Guan, H.M., T.S. Chung, Z. Huang, et al. 2006. Poly(vinyl alcohol) multilayer mixed matrix membranes for the dehydration of ethanol–water mixture. *J. Memb. Sci.* 268(2): 113–122.

Guo, R., X. Fang, H. Wu, et al. 2008. Preparation and pervaporation performance of surface crosslinked PVA/PES composite membrane. *J. Memb. Sci.* 322(1): 32–38.

Hao, J., P.A. Rice, and S.A. Stern. 2008. Upgrading low-quality natural gas with $H_2S$- and $CO_2$-selective polymer membranes. Part II: Process design, economics, and sensitivity study of membrane stages with recycle streams. *J. Memb. Sci.* 320(1/2): 108–122.

Hartlwy, F.R. 1991. *Chemistry of the Platinum Group Metals: Recent Developments*. Elsevier: Amsterdam.

Hess, S., C.S. Bickel, and R.N. Lichtenthaler. 2006. Propene/propane separation with copolyimide membranes containing silver ions. *J. Memb. Sci.* 275(1/2): 52–60.

Holmberg, B.A., X. Wang, and Y. Yan. 2008. Nanocomposite fuel cell membranes based on Nafion and acid functionalized zeolite beta nanocrystals *J. Memb. Sci.* 320(1/2): 86–92.

Huang, S.H., W.L Lin, J. Liaw, et al. 2008. Characterization, transport and sorption properties of poly(thiol ester amide) thin-film composite pervaporation membranes. *J. Memb. Sci.* 322: 139–145.

Iqbal, M., Z. Man, H. Mukhtar, et al. 2008. Solvent effect on morphology and $CO_2/CH_4$ separation performance of asymmetric polycarbonate membranes. *J. Memb. Sci.* 318(1/2): 167–175.

Ismail, A.F., and L.I.B. David. 2001. A review on the latest development of carbon membranes for gas separation. *J. Memb. Sci.* 193: 1–18.

Ismail, A.F., T.D. Kusworo, and A. Mustafa. 2008. Enhanced gas permeation performance of polyethersulfone mixed matrix hollow fiber membranes using novel Dynasylan Ameo silane agent. *J. Memb. Sci.* 319(1/2): 306–312.

Jha, P., and J.D. Way. 2008. Carbon dioxide selective mixed-matrix membranes formulation and characterization using rubbery substituted polyphosphazene. *J. Memb. Sci.* 324(1/2): 151–161.

Jiang, X., J. Ding, and A. Kumar. 2008. Polyurethane-poly(vinylidene fluoride) (PU-PVDF) thin film composite membranes for gas separation. *J. Memb. Sci.* 323(2): 371–378.

Kang, S.W., D.H. Lee, J.H. Park, et al. 2008. Effect of the polarity of silver nanoparticles induced by ionic liquids on facilitated transport for the separation of propylene/propane mixtures. *J. Memb. Sci.* 322(2): 281–285.

Kang, S.W., J. Hong, J.H. Park, et al. 2008. Nanocomposite membranes containing positively polarized gold nanoparticles for facilitated olefin transport. *J. Memb. Sci.* 321(1): 90–93.

Kang, S.W., J.H. Kim, K. Char, et al. 2006. Nanocomposite silver polymer electrolytes as facilitated olefin transport membranes. *J. Memb. Sci.* 285(1/2): 102–107.

Kebiche-Senhadji, O., L. Mansouri, S. Tingry, P. Seta, and M. Benamor. 2008. Facilitated Cd(II) transport across CTA polymer inclusion membrane using anion (Aliquat 336) and cation (D2EHPA) metal carriers. *J. Memb. Sci.* 310: 438–445.

Khaye, M., M.M. Nasefand, and J. Mengual. 2005. Radiation grafted poly(ethylene terephthalate)–*graft*–polystyrene pervaporation membranes for organic/organic separation. *J. Memb. Sci.* 263(1/2): 77–95.

Kim, D.J., B.J. Chang, and J.H. Kim. 2008. Sulfonated poly(fluorenyl ether) membranes containing perfluorocyclobutane groups for fuel cell applications. *J. Memb. Sci.* 325(1): 217–222.

Kim, H.J., S.S. Nah, and R.B. Min. 2002. A new technique for preparation of PDMS pervaporation membrane for VOC removal. *Advances in Environmental Research*. 6: 255–264.

Kim, H.J., J. Won, Y.S. Kang. 2004. Olefin-induced dissolution of silver salts physically dispersed in inert polymers and their application to olefin/paraffin separation. *J. Memb. Sci.* 241(2): 403–407.

Kosuri, M.R., and W.J. Koros. 2008. Defect-free asymmetric hollow fiber membranes from Torlon®, a polyamide–imide polymer, for high-pressure $CO_2$ separations. *J. Memb. Sci.* 320(1/2): 65–72.

Kozlowski, C.A., and W. Walkowiak. 2005. Applicability of liquid membranes in chromium(VI) transport with amines as ion carriers. *J. Memb. Sci.* 266: 143–150.

Kuhn, J., K. Yajima, and T. Tomita. 2008. Dehydration performance of a hydrophobic DD3R zeolite membrane. *J. Memb. Sci.* 321(2): 344–349.

Le Cloirec, P. 1998. *Les composés organiques volatils (COV) dans l'environnement*. TEC & DOC Lavoisier: Paris.

Lee, C.H., C.H. Park, and Y.M. Lee. 2008a. Sulfonated polyimide membranes grafted with sulfoalkylated side chains for proton exchange membrane fuel cell (PEMFC) applications. *J. Memb. Sci.* 313(1/2): 199–206.

Lee, P.C., T.H. Han, D.O. Kim, et al. 2008b. In situ formation of platinum nanoparticles in Nafion recast film for catalyst-incorporated ion-exchange membrane in fuel cell applications. *J. Memb. Sci.* 322: 441–445.

Lee, W., H. Kim, and H. Lee. 2008c. Proton exchange membrane using partially sulfonated polystyrene-*b*-poly(dimethylsiloxane) for direct methanol fuel cell. *J. Memb. Sci.* 320(1/2): 78–85.

Li, B., D. Xu, Z. Jiang, et al. 2008a. Pervaporation performance of PDMS-$Ni^{2+}$Y zeolite hybrid membranes in the desulfurization of gasoline. *J. Memb. Sci.* 325(2): 293–301.

Li, J., C. Chen, B. Han, et al. 2002. Laboratory and pilot-scale study on dehydration of benzene by pervaporation. *J. Memb. Sci.* 203(1/2): 127–136.

Li, Y., T.S. Chung, and Y. Xiao. 2008b. Superior gas separation performance of dual-layer hollow fiber membranes with an ultrathin dense-selective layer. *J. Memb. Sci.* 325(1): 23–27.

Lin, L., Y. Kong, and Y. Zhang. 2008. Sorption and transport behavior of gasoline components in polyethylene glycol membranes. *J. Memb. Sci.* 325(1): 438–445.

Liu, Y.L., X. Feng, and D. Lawless. 2006. Separation of gasoline vapor from nitrogen by hollow fiber composite membranes for VOC emission control. *J. Memb. Sci.* 171(1/2): 114–124.

Liu, Y.L., C.H. Yu, and J.Y. Lai. 2008. Poly(tetrafluoroethylene)/polyamide thin-film composite membranes, via interfacial polymerization for pervaporation dehydration on an isopropanol aqueous solution. *J. Memb. Sci.* 315(1/2): 106–115.

Lu, L., H. Sun, F. Peng, et al. 2006. Novel graphite-filled PVA/CS hybrid membrane for pervaporation of benzene/cyclohexane mixtures. *J. Memb. Sci.* 281: 245–252.

Lue, S.J., F.J. Wang, and S.Y. Hsiaw. 2004. Pervaporation of benzene/cyclohexane mixtures using ion-exchange membrane containing copper ions. *J. Memb. Sci.* 240(1/2): 149–158.

Luo, Y., W. Xin, G. Li, et al. 2007. Pervaporation properties of EC membrane crosslinked by hyperbranched-polyester acrylate. *J. Memb. Sci.* 303(1/2): 183–193.

Macanas, J., and M. Muñoz. 2005. Mass transfer determining parameter in facilitated transport through di-(2-ethylhexyl)dithiophosphoric acid activated composite membranes. *J. Memb. Sci.* 534: 101–108.

Majumdar, S., D. Bhaumik, and K.K. Sirkar. 2003. Performance of commercial-size plasmapolymerized PDMS-coated hollow fiber modules in removing VOCs from $N_2$/air. *J. Memb. Sci.* 214(2): 323–330.

Mandal, S., and V.G. Pangarkar. 2002. Separation of methanol–benzene and methanol–toluene mixtures by pervaporation: effects of thermodynamics and structural phenomenon. *J. Memb. Sci.* 201(1/2): 175–190.

Mandal, S., and V.G. Pangarkar. 2003. Development of co-polymer membranes for pervaporative separation of methanol from methanol–benzene mixture—a solubility parameter approach. *Separation and Purification Technology* 30(2): 147–168.

Matsui, S., and D.R. Paul. 2002. Pervaporation separation of aromatic/aliphatic hydrocarbons by crosslinked poly(methyl acrylate-*co*-acrylic acid) membranes. *J. Memb. Sci.* 195: 229–245.

Melita, L., and M. Popescu. 2008. Removal of Cr (VI) from industrial water effluents and surface waters using activated composite membranes. *J. Memb. Sci.* 312: 157–162.

Mellah, A., and D. Benachour. 2006. The solvent extraction of zinc and cadmium from phosphoric acid solution by di-2-ethyl hexyl phosphoric acid in kerosene diluent. *Chem. Eng. Proc.* 45: 684–690.

Molyneux, P. 2008. Permeation of gases through microporous silica hollow-fiber membranes: Application of the transition-site model. *J. Memb. Sci.* 320(1/2): 42–56.

Mosa, J., A. Larramona, A. Durán, et al. 2008. Synthesis and characterization of $P_2O_5$–$ZrO_2$–$SiO_2$ membranes doped with tungstophosphoric acid (PWA) for applications in PEMFC. *J. Memb. Sci.* 307: 21–27.

Muñoz, D.M., E.M. Maya, J. Abajo, et al. 2008. Thermal treatment of poly(ethylene oxide)-segmented copolyimide based membranes: An effective way to improve the gas separation properties. *J. Memb. Sci.* 323(1): 53–59.

Nam, S.E., S.O. Kim, Y. Kang, et al. 2008. Preparation of Nafion/sulfonated poly(phenylsilsesquioxane) nanocomposite as high temperature proton exchange membranes. *J. Memb. Sci.* 322(2): 466–474.

Namboodiri, V.V, and L.M. Vane. 2007. High permeability membranes for the dehydration of low water content ethanol by pervaporation. *J. Memb. Sci.* 306(1/2): 209–15.

Navajas, A., R. Mallada, C. Téllez, et al. 2007. Study on the reproducibility of mordenite tubular membranes used in the dehydration of ethanol. *J. Memb. Sci.* 299(1/2): 166–173.

Nghiem, L.D., P. Mornane, I.D. Potter, et al. 2006. Extraction and transport of metal ions and small organic compound using polymer inclusion membranes (PIMs). *J. Memb. Sci.* 281: 7–41.

Norddin, M.N.A., A.F. Ismail, D. Rana, et al. 2008. Characterization and performance of proton exchange membranes for direct methanol fuel cell: Blending of sulfonated poly(ether ether ketone) with charged surface modifying macromolecule. *J. Memb. Sci.* 323: 404–413.

Obuskovic, G., S. Majumdar, and K.K. Sirkar. 2003. Highly VOC-selective hollow fiber membranes for separation by vapor permeation. *J. Memb. Sci.* 217(1/2): 99–116.

Pandey, L.K., C. Saxena, and V. Dubey. 2003. Modification of poly(vinyl alcohol) membranes for pervaporative separation of benzene/cyclohexane mixtures, *J. Memb. Sci.* 227(1/2): 173–182.

Paul, D.R., and Y.P. Yampl'skii. 1994. *Polymeric gas separation membranes*. Boca Raton: CRC Press.

Peng, F., C. Hu, and Z. Jiang. 2007a. Novel poly(vinyl alcohol)/carbon nanotube hybrid membranes for pervaporation separation of benzene/cyclohexane mixtures. *J. Memb. Sci.* 297(1/2): 236–242.

Peng, F., F. Pan, H. Sun, et al. 2007b. Novel nanocomposite pervaporation membranes composed of poly(vinyl alcohol) and chitosan-wrapped carbon nanotube. *J. Memb. Sci.* 300(1/2): 13–19.

Polyakov, A.M., L.E. Starannikova, and Y.P Yampolskii. 2004. Amorphous Teflons AF as organophilic pervaporation materials separation of mixtures of chloromethanes. *J. Memb. Sci.* 238(1/2): 21–32.

Polyakov, A.M., L.E. Starannikova, and Y.P. Yampolskii. 2003. Amorphous Teflons AF as organophilic pervaporation materials transport of individual components. *J. Memb. Sci.* 216(1/2): 241–256.

Pont, N., V. Salvadó, and C. Fontàs. 2008. Selective transport and removal of Cd from chloride solutions by polymer inclusion membranes. *J. Memb. Sci.* 318: 340–345.

Popescu, M., J.M. Blanchard, and J. Carre. 1998. *Analyse et traitement physicochimique des rejets atmosphériques industriels (émissions, fumées, odeurs, et poussières)*. TEC & DOC Lavoisier: Paris.

Popescu, M., R. Popescu, and C. Stratula. 2006. *Metode fizico-chimice de tratare a poluantilor industriali atmosferici (in roumanian)*. Editura Academiei Romane.

Porter, M.C. 1990. *Handbook of industrial membrane technology*. Noyes Publications.

Ray, S., and S.K. Ray. 2006a. Separation of organic mixtures by pervaporation using crosslinked rubber membranes. *J. Memb. Sci.* 207(1/2): 132–145.

Ray, S., and S.K. Ray. 2006b. Separation of organic mixtures by pervaporation using crosslinked and filled rubber membranes. *J. Memb. Sci.* 285(1/2): 108–119.

Resina, M., C. Fontàs, C. Palet, and M. Muñoz. 2008. Comparative study of hybrid and activated composite membranes containing Aliquat 336 for the transport of Pt (IV). *J. Memb. Sci.* 311: 235–242.

Resina, M., J. Macanas, J. de Gyves, and M. Muñoz. 2006. Zn(II), Cd(II) and Cu(II) separation through organic–inorganic Hybrid Membranes containing di-(2-ethylhexyl) phosphoric acid or di-(2-ethylhexyl) dithiophosphoric acid as carrier. *J. Memb. Sci.* 268: 57–64.

Robert, K., P.W. Damian, W.R. Charles, et al. 2008. High-temperature pervaporation performance of ceramic-supported polyimide membranes in the dehydration of alcohols. *J. Memb. Sci.* 319(1/2): 126–132.

Sadeghi, M., G. Khanbabaei, A.H. Dehaghani, et al. 2008. Gas permeation properties of ethylene vinyl acetate–silica nanocomposite membranes. *J. Memb. Sci.* 322(2):423–428.

Scindia, Y.M., A.K. Pandey, and A.V.R. Reddy. 2005. Coupled-diffusion transport of Cr(VI) across anion-exchange membranes prepared by physical and chemical immobilization methods. *J. Memb. Sci.* 249: 143–152.

Sharma, R.R., R. Agrawal, and S. Chellam. 2003. Temperature effects on sieving characteristics of thin-film composite nanofiltration membranes: pore size distributions and transport parameters. *J. Memb. Sci.* 223(1/2): 69–87.

Sijbesma, H., K. Nymeijer, R. Marwijk, et al. 2008. Flue gas dehydration using polymer membranes. *J. Memb. Sci.* 313(1/2): 236–276.

Simmons, V.L., Baker, R. W., Kaschemekat, J., et al. 1992. *Membrane vapor separation systems for the recovery of halons and CFC*. p. 295. http://www.bfrl.nist.gov/866/HOTWC/HOTWC2006/pubs/R0301015.pdf.

Smitha, B., S. Sridhar, and A.A. Khan. 2005. Solid polymer electrolyte membranes for fuel cell applications. *J. Memb. Sci.* 259(1/2): 10–26.

Tanaka, K., H. Kita, K. Okamoto, et al. 2002. Isotopic-transient permeation measurements in steady-state pervaporation through polymeric membranes. *J. Memb. Sci.* 197: 173–183.

Tay, S.W., X. Zhang, Z. Liu, et al. 2008. Composite Nafion® membrane embedded with hybrid nanofillers for promoting direct methanol fuel cell performance. *J. Memb. Sci.* 321(2): 139–145.

Teplyakov, V.V., D. Roizard, E. Favre, et al. 2003. Investigations on the peculiar permeation properties of volatile organic compounds and permanent gases through PTMSP. *J. Memb. Sci.* 220(1/2): 165–175.

Tripathi, B.P., A. Saxena, and V.K. Shahi. 2008. Phosphoric acid grafted bis (4-Y-aminopropyldiethoxysilylphenyl) sulfone (APDSPS)-poly(vinyl alcohol) cross-linked polyelectrolyte membrane impervious to methanol. *J. Memb. Sci.* 318(1/2): 288–297.

Ulevicz, M., W. Walkoviak, J. Gega, and B. Pospiech. 2003. Zinc (II) selective removal from other transition metal ions by solvent extraction and transport through polymer inclusion membrane with D2EHPA. *ARS Separation Acta* 2: 47–55.

Vasudevan, V.N., and M.V. Leland. 2007. High permeability membranes for the dehydration of low water content ethanol by pervaporation. *J. Memb. Sci.* 306(1/2): 209–215.

Veit, H.M., T.R. Diehl, A.P. Salami, J.S. Rodrigues, A.M. Bernardes, and J.A.S Tenório. 2005. Utilization of magnetic and electrostatic separation in the recycling of printed circuit boards. *Waste Manag.* 25: 67–74.

Vernet, J.P. 1991. *Heavy Metals in the Environment.* Elsevier: Amsterdam.

Vu, De Q., W.J. Koros, and S.J. Miller. 2003. Effect of condensable impurity in $CO_2/CH_4$ gas feeds on performance of mixed matrix membranes using carbon molecular sieves. *J. Memb. Sci.* 221: 233–239.

Yahaya, G.O. 2008. Separation of volatile organic compounds (BTEX) from aqueous solutions by a composite organophilic hollow fiber membrane-based pervaporation process. *J. Memb. Sci.* 319(1/2): 82–90.

Yang, C., S. Srinivasan, A.B. Bocarsly, et al. 2004. A comparison of physical properties and fuel cell performance of Nafion and zirconium phosphate/Nafion composite membrane. *J. Memb. Sci.* 237: 145–161.

Yeom, C.K., S.H. Lee, H.Y. Song, et al. 2002a. A characterization of concentration polarization in a boundary layer in the permeation of VOCs/$N_2$ mixtures through PDMS membrane, *J. Memb. Sci.* 205(1/2): 155–174.

Yeom, C.K., S.H. Lee, H.Y. Song, et al. 2002b. Vapor permeations of a series of VOCs/$N_2$ mixtures through PDMS membrane. *J. Memb. Sci.* 198: 129–143.

Yeom, C.K., S.H. Lee, H.Y. Song, et al. 2002c. Modeling and evaluation of boundary layer resistance at feed in the permeation of VOC/$N_2$ mixtures through PDMS membrane. *J. Memb. Sci.* 204(1/2): 303–322.

Zhang, Q.G., Q.L. Liu, Z.Y. Jiang, et al. 2007. Anti-trade-off in dehydration of ethanol by novel PVA/APTEOS hybrid membranes. *J. Memb. Sci.* 187(2): 237–245.

# 6 Treatment of Kraft Black Liquor Using Membrane-Based Separation Process

*Chiranjib Bhattacharjee and Dwaipayan Sen*

**CONTENTS**



## 6.1 INTRODUCTION

Fresh usable water, which is around 0.3% of the total water on earth, is one of the world's most valuable resources and essential to support human life and the environment. The majority of this surface water consumed by modern civilization is used to produce food, farming and different segments of other industries as process water. Already there is more wastewater generated and dispersed today than at any other time in the history of our planet, which results more than one out of six people lack access to safe drinking water, namely, 1.1 billion people, and more than two out of six lack adequate sanitation, namely, 2.6 billion people (Jong-Wook and Bellamy, 2004). 3900 children die every day from waterborne diseases (Jong-Wook and Bellamy, 2004). In reality, these figures could be much higher (Jong-Wook and Bellamy, 2004). The National Wild and Scenic Rivers System notes that it takes 1000 gallons of water to grow and process each pound of food per person in the United States per year. According to the Food Development Authority, the average person eats 1500 pounds of food each year. Therefore, approximately 1.5 million gallons of water is needed to process the food for just one person each year. This statistics shows that so much water is consumed as part of our everyday commodities. Apart from these, there are so many other segments where water consumption rate is also quite high. To fight the scarcity of fresh water, emphases are now given on the treatment and recycling of dispersed water or wastewater back into the process.

In this chapter, we will be focusing on the treatment of Kraft black liquor (KBL), effluent that is coming out of the paper and pulp industry. Paper manufacturing industry is a high capital, energy and water intensive industry. In India, around 905.8 million $m^3$ of water is consumed, whereas 695.7 million $m^3$ of wastewater is annually discharged by this sector (Trivedi et al., 1990). Here we are going to discuss about the treatment of KBL using the membrane separation technique and the consequent merits and demerits of the process. Over the last few decades, membrane-based separation techniques have gained considerable importance in chemical process industries. The increased interest through this technique has demonstrated numerous advantages that membrane separation systems have over many other separation systems such as centrifuges, bioreactors, vacuum filters and evaporators. Some of the advantages include a crystal clear filtrate and flexibility in separation quality based on filter media and materials compatibility due to a wide range of available membranes. Much of this technology was developed after World War II, yet several factors delayed upswing in the popularity of membrane separations until the 1970s. The challenges that limited the use of membrane filtration as a viable production scale separations alternative were unreliability, high cost and slow fluxes which ranged from 34 $L{\cdot}m^{-2}{\cdot}h^{-1}$ to 102 $L{\cdot}m^{-2}{\cdot}h^{-1}$. Putting all these problems aside, membrane separation process is the most efficient among all other processes (Bhattacharjee et al., 2008).

## 6.2 LITERATURE REVIEW ON KBL

Paper and pulp industry has some great importance in the world as because of the impact of this industry on the socio-economic development of a country. The first paper was produced around 2000 years ago by a Chinese person named Tsai Lun. As the days go on, demand for different forms of paper has increased resulting an increment in the paper production from 77,000,000 tons/year to 360,000,000 tons/year in between 1961 and 2004 (Ahonen et al., 2006). In India, production in chemical wood pulp has increased by around 160% in 20 years, from 1986 to 2006, especially in the export of printing and writing papers, where there is an astounding increase from 500 tons to 187,134 tons in the last 10 years, which reflects the growth of paper and pulp industry (FAOSTAT, 2006).

There are two methods by which one can produce the chemical pulp. One is called sulfate pulping (or "Kraft" pulping) and the other one is called sulfite pulping. Nowadays, 95% of the chemical pulp is produced by Kraft pulping (Ahonen et al., 2006). Benefit lies in Kraft pulping is that the cooking procedure can be done irrespective of the species of the wood, producing approximately 7 tons of black liquor from 1 ton of pulp. Though chemical pulping method is the most common and convenient method for producing pulp in paper industry, this process is a potential threat to the environment. One of these treatments is the application of bleaching agents to achieve the brightness of the paper. Several consecutive bleaching stages are employed in chemical pulping process. In the traditional bleaching procedure, several chlorine-containing chemicals are generally used in a proper bleaching sequence, either CEHDED or CEDED (C: chlorine, H: hypochlorite, D: chlorine dioxide, E: alkali to extract lignin between stages). Effluent that is coming out of any paper industry contains mainly these harmful chemicals along with chromium and other organic substances. This chapter deals with the treatment of paper mill effluent, i.e., KBL using ultrafiltration (UF), nanofiltration (NF) and reverse osmosis (RO) membrane separation processes.

UF is primarily a size exclusion based pressure driven membrane separation process where membranes are described by their nominal molecular weight cutoff (MWCO), which is defined as the smallest molecular weight species for which the membrane has more than 90% rejection. Several membrane characteristics are important in determining a membrane's suitability for separation applications. These overall characteristics can be summed up as porosity, morphology, surface properties, mechanical strength and chemical resistances. Especially the main difficulties that will be felt during any of the membrane separation processes, is the build-up of polarized layer resistance, which results in gradual flux decline. The later is commonly attributed to two different problems. One is membrane fouling (Ghaffour, 2004) and the other one is the concentration polarization (Pharoah et al., 2000). The first one is mainly due to adsorption of the solute at the membrane surface or simply due to pore plugging. These types of scenarios mainly attribute to permanent reduction in membrane performance. The second one is due to the creation of driving force in terms concentration in the reverse direction from the membrane surface. So one of the problems during membrane filtration is the declination of permeate flux with time. Mathematical expression for permeate flux can be given by the following osmotic pressure model equation,

$$J = \frac{\Delta P - \Delta\pi_m}{\mu R_m} \quad (6.1)$$

where $R_m$ is the membrane hydraulic resistance, $\Delta P$ is the hydrostatic pressure difference, $\Delta\pi_m = \pi_m - \pi_P = f(C_m)$ is the osmotic pressure difference, $\pi_P$ is the osmotic pressure on the permeate side, $C_m$ is the membrane concentration and $\mu$ is the viscosity of the feed solution. Osmotic effect is comparatively little in case of UF than RO.

Similarly, by using film theory model for concentration at the membrane surface the flux can be represented by (Bhattacharjee et al., 2005),

$$J = \frac{\Delta P - f[C_B \exp(J/K)]}{\mu R_m} \quad (6.2)$$

where $K$ is the mass transfer coefficient and $C_B$ is the bulk concentration of the rejected solute. The above equation often require iterative solution for determination of permeate flux.

### 6.2.1 Lignin: A By-Product from Kraft Process

KBL mainly contains various organic materials such as low molecular weight acids and the degradation products of the complex polyphenolic compounds, lignin, with several inorganic chemicals (Llamas et al., 2007). Table 6.1 (Wallberg et al., 2003) shows the typical composition of KBL. The main purpose for the treatment of KBL is to recover its lignin components. Lignin, which is a polysaccharide group of phenolic compounds, is basically a biofuel and can be separated out using UF and diafiltration. The possible structure of lignin is given in Figure 6.1. The most valuable product that can be obtained from lignin is lignosulfonates. This can be done in two-step reactions wherein the first step, delignification, is done by the acidic cleavage of the ether bonds along with the formation of electrophilic carbocations, which will again react with bisulfite ion to produce lignosulfonates, a by-product from the sulfite process. Figure 6.2 shows the reaction steps for lignosulfonates from lignin. Lignin is also dried and sold as an industrial chemical and used as a feedstock for polymeric products. An earlier study that evaluated lignin production from renewable and recycled streams showed that diversion of 10% of U.S. lignin can produce enough carbon fiber to replace half of the steel in all domestic passenger transport vehicles (Compere et al., 2001). Table 6.2 shows the applications of lignin as lignosulfonates in different segment of the industries.

**TABLE 6.1**
**Typical Composition of KBL**

| | Organic Compounds | | | | Inorganic Compounds | |
|---|---|---|---|---|---|---|
| **KBL Constituents** | **Ligneous Materials (Polyaromatic)** | **Saccharinic Acids (Hydroxy Acids)** | **Formic and Acetic Acids** | **Extractives** | **Sodium** | **Sulfur** |
| **Percentage (wt. %)** | 30–45 | 25–35 | 10 | 3–5 | 17–20 | 3–5 |

*Source:* Wallberg, O., et al., *Desalination*, 154, 187–199, 2003, with permission from Elsevier.

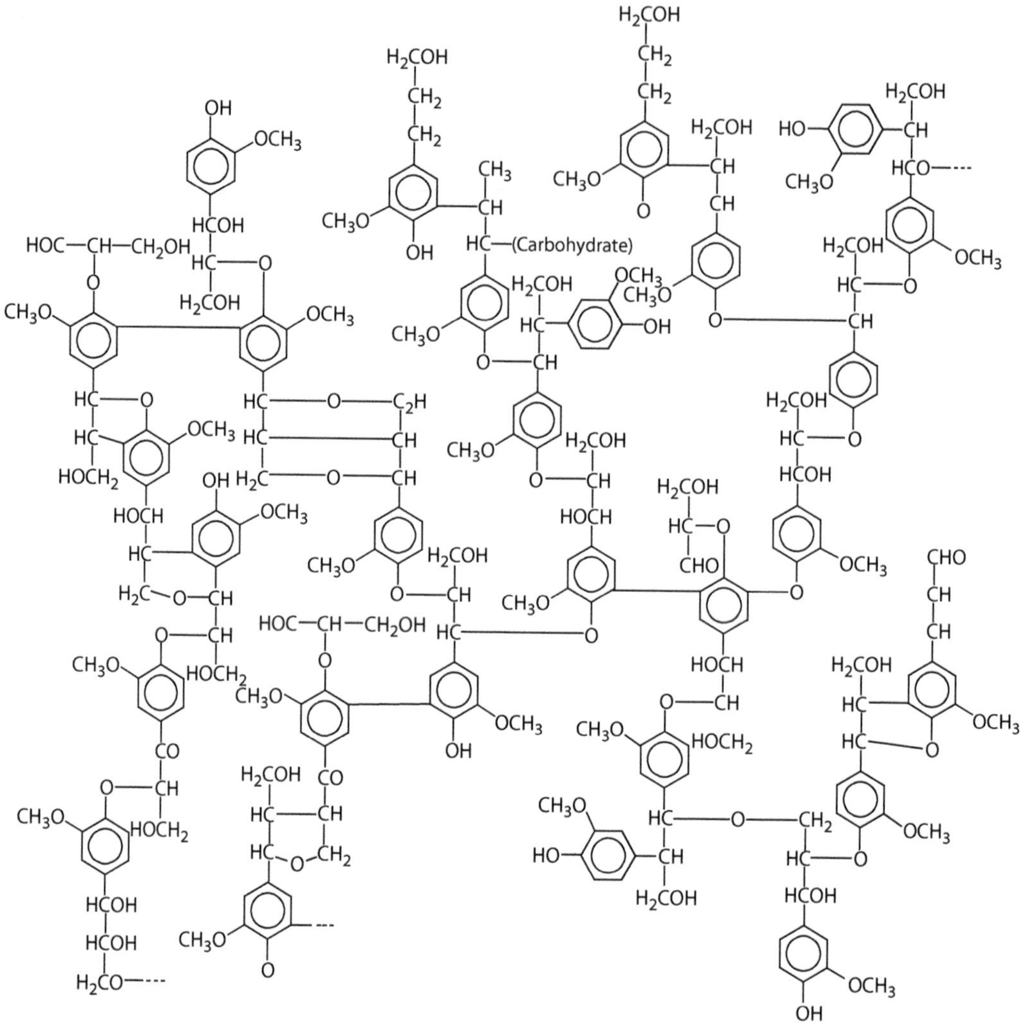

FIGURE 6.1 Probable molecular structure of lignin.

### 6.2.2 UF Process for the Treatment of KBL

UF of KBL is already gaining importance for the recovery of the valuable organics as well as to meet a part of the water requirements; the process, therefore, may finally tackle the problem of wastewater disposal as environmental regulations are becoming stringent day by day. However, one of the major drawbacks with the process of UF for treating KBL is the decline of flux with time, a

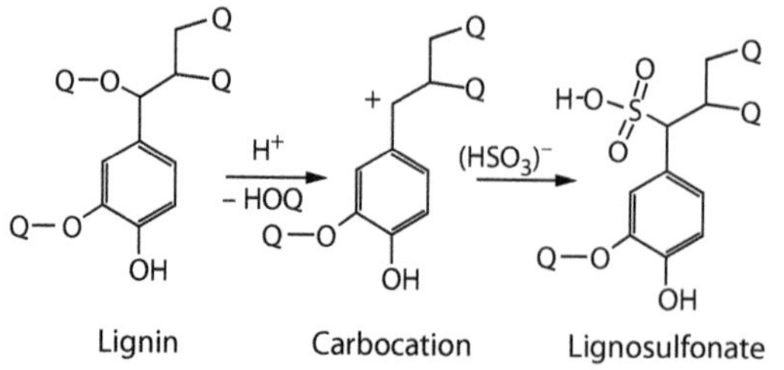

FIGURE 6.2 Reaction scheme for lignosulfonate from lignin.

**TABLE 6.2**
**Application of Lignin (Lignosulfonates) in Different Fields**

| Usage | Application |
|---|---|
| Lignin as binder | Soil stabilizer, plywood and particle board, coal briquettes, etc. |
| Lignin as dispersant | Leather tanning, pesticides and insecticides, oil drilling mud, etc. |
| Lignin as emulsifier | Asphalt emulsion, wax emulsion, etc. |
| Lignin as sequestrant | Water treatments for boilers and cooling systems, micronutrient systems, etc. |

typical characteristic of any pressure-driven membrane process. Such a decline is generally attributed to either gel formation, osmotic pressure retardation, or fouling of the membrane, resulting from reversible or irreversible pore plugging (Bhattacharjee et al., 2006). High dry solid contents are mainly attributed to these problems.

So far, UF has been employed for the treatment of KBL mainly for the following three purposes: (1) separation of lignin compounds from low molecular weight inorganics; (2) fractionation of high molecular weight lignin compounds; and (3) recovery of water (Bhattacharjee and Bhattacharya, 2006). Woerner and McCarthy (1984) suggested that to produce purified high molecular weight lignin, UF should be operated at low pressure and high alkalinity. Because of high biological oxygen demand (BOD), chemical oxygen demand (COD) loading and total solids of KBL, researchers have explored different UF membrane models for predicting the limiting flux value. Sridhar and Bhattacharya (1991) concluded that the phenomenon of concentration polarization layer deposition mainly attributes to the declination in the flux. Asymmetric cellulose acetate complex membrane with 5 kDa MWCO was used to handle black liquor with a concentration of up to 5 $g{\cdot}L^{-1}$. A permeate flux of about 37.8 $L{\cdot}m^{-2}{\cdot}h^{-1}$ at a pressure of 5.5 $kgf{\cdot}cm^{-2}$ was reported. Because of the formation of a concentration polarization layer, membrane modules having high shear creating capability are now being investigated for the treatment of KBL. These high-shear devices can be utilized to create intense turbulence on the membrane surface so as to reduce concentration polarization and enhance the permeation. In this regard, researchers have employed several membrane modules—one is the rotating disk membrane module (RDMM), and the other is the cross-flow membrane module.

#### 6.2.2.1 Treatment of KBL Using RDMM

Use of RDMM for treatment of KBL was investigated recently in detail by Bhattacharjee et al. (2005). For these purpose, raw materials had been collected from Supreme Paper Mills, Kolkata (India), which uses 90% rice straw and 10% jute straw for manufacturing writing grade paper. The total solid content of this liquor was typically 12–14 $g{\cdot}L^{-1}$. KBL contained organic constituents that are given in Table 6.1, along with 1% methanol and other inorganic compounds. Because of high solid contents, this black liquor was first taken for pretreatment before it was introduced in the membrane module to avoid fouling. It was first centrifuged with a speed of 12,500 rpm in order to reduce the total solid load. After centrifugation, the feed was subjected to vacuum microfiltration (MF) (Sartorius A.G., Göttingen, Germany), fitted with an oil-free portable vacuum pump (Sartorius A.G., Göttingen, Germany, model ROC 300 with moisture trap) and polyether sulfone (PES) membrane (47 mm diameter, pore size 0.45 μm). After pretreatment of KBL, permeate from MF unit was subjected to UF, using cellulose triacetate membrane of 5 kDa MWCO in RDMM (Figure 6.3) (Bhattacharjee et al., 2005). The module was equipped with two motors with speed controllers to provide rotation of the stirrer and membrane housing. Membrane and the stirrer were rotated in opposite direction to provide maximum shear in the vicinity of the membrane. UF process was performed with different transmembrane pressures (TMPs), membrane speed and stirrer speed. Table 6.3 (Bhattacharjee et al., 2005) describes the physical properties of KBL after different treatment stages. From the table, centrifugation stage shows a 4% drop in the total solid content because of the removal of lipids, grease and colloidal substances. After MF, there is a further increase in the

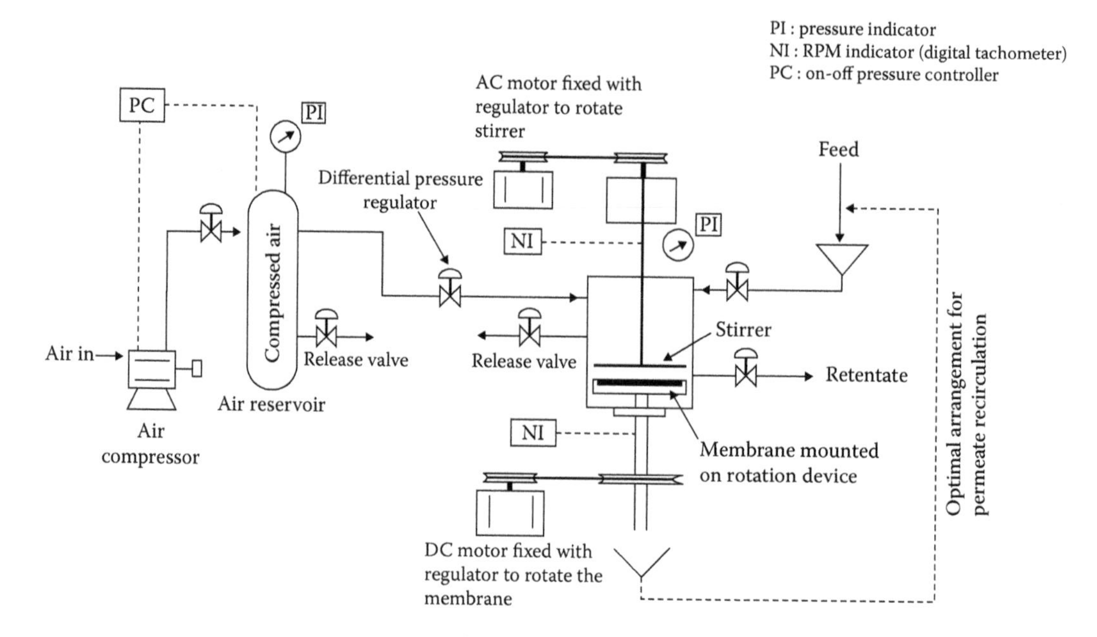


**FIGURE 6.3** Schematic diagram of RDMM setup. (Reprinted from *Sep Purif Technol*, 49, Bhattacharjee, C. and P. K. Bhattacharya, 281–290, Copyright 2006, with permission from Elsevier.)

quality of the effluent. Total solid content now shows a reduction of about 20%, and finally after UF, the reduction is 74% from the MF stage. Overall, the reduction was almost 80%, which reflects the efficacy of the UF in KBL treatment. Figure 6.4 shows the variation of permeate flux with varying TMPs, with a stirrer speed at 500 rpm, whereas Figure 6.5 shows steady state flux as a function of TMP. It is evident from the figure that the permeate flux increases almost linearly in fixed disk module until a TMP of 7 kgf·cm$^{-2}$ was applied. Beyond this pressure, the flux is seen to be nearly independent of pressure and falls into a mass-transfer controlled region. Increased pressure causes more flux, and accordingly, more solute is being carried toward the membrane. So at higher pressure, more solute rejection at the membrane is causing a gel/polarized layer to form on the membrane surface, and under this condition, the flux becomes almost independent of pressure (limiting

**TABLE 6.3**

**Improvements of the Different Wastewater Parameters (with Corresponding Statistical Error Values) and Its Comparison for Centrifuged BL, MF Permeate, and UF Permeate under Specified Conditions for RDMM**

| Physical Properties | Raw KBL | Centrifuged KBL | MF KBL Permeate | UF KBL Permeate (TMP 7 kgf·cm$^{-2}$) |
|---|---|---|---|---|
| Total solids (g·L$^{-1}$) | 12.34 | 11.79 | 9.50 | 2.50 |
| Turbidity (NTU) | 337 | 119 | 35.5 | 25 |
| BOD (mg·L$^{-1}$) | 6600 | 5850 | 3000 | 900 |
| COD (mg·L$^{-1}$) | 16,800 | 14,000 | 9800 | 3200 |
| pH (at 30°C) | 8.00 | 7.80 | 7.50 | 7.33 |
| Conductivity (mmoh·cm$^{-1}$) | 5.60 | 5.90 | 7.04 | 4.30 |
| Oil and grease content (mg·L$^{-1}$) | 564 | 150 | 100 | 30 |

*Source:* Bhattacharjee, C. and P. K. Bhattacharya, *Sep Purif Technol*, 49, 281–290, 2006, with permission from Elsevier.

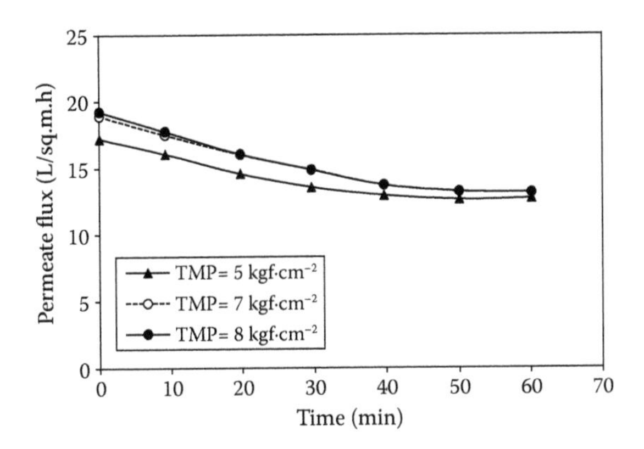


**FIGURE 6.4** Comparison of effect of TMPs on flux in stirred fixed disk membrane ($N_s$ = 500 rpm). (Reprinted from *Sep Purif Technol*, 49, Bhattacharjee, C. and P. K. Bhattacharya, 281–290, Copyright 2006, with permission from Elsevier.)

flux phenomenon), falling into the gel-layer controlled UF regime. In Figure 6.6, there is a variation of the permeate flux with the variation of the stirrer speed at a fixed TMP and membrane disk. The figure shows that there is a 33% increase in the initial permeate flux and a 54% increase in the steady state flux compared to nonrotating stirrer, which shows that stirrer motion enhances the feed velocity and shear stresses at the membrane surface resulting in higher permeate flux. Figure 6.7, which shows the variation of the permeate flux at different membrane speed, reflects the increase in initial permeate flux by 32% compared to the stationary membrane. Membrane rotation generates higher turbulence near the vicinity of the membrane surface that tends to shear-off deposited material, thereby reducing hydraulic resistance of the fouling layer and increasing flux. At higher rpm of membrane disk, solute particles responsible for pore plugging get disengaged from the pores, leaving relatively clear pores for passage of solvents and thereby giving higher flux. Membrane rotation is found to be more effective in reducing the concentration polarization as compared to stirrer speed.

The effect of TMPs on lignin rejection is depicted in Figure 6.8. The retention of lignin was more than 75% when operated at 8 kgf·cm$^{-2}$ TMP. At lower pressures, lower retention was found. As pressure increases, more liquid permeates through the membrane, leaving more solutes to retain, thereby constricting the pore opening and subsequently increasing rejection. Lignin purity, which was 36%–38% in the original black liquor (based on lignin concentration/total solid concentration), was found to improve to 48% in the UF retentate after carrying UF run for a period of 3 h using RDMM at 7 kgf·cm$^{-2}$ TMP with stirred speed of 500 rpm.

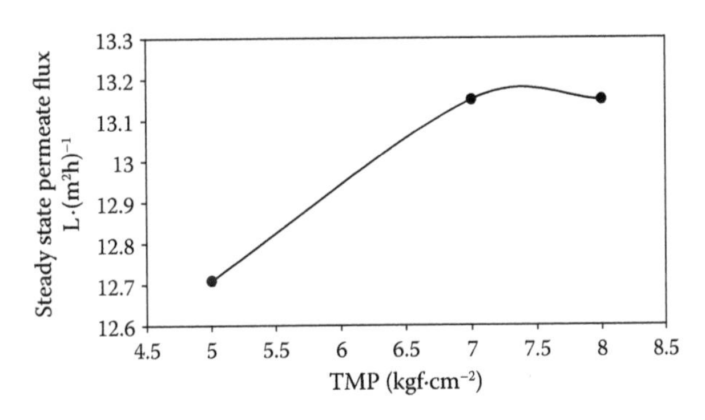


**FIGURE 6.5** Variation of steady state flux at different TMPs in stirred fixed disk membrane (Ns = 500 rpm). (Reprinted from *Sep Purif Technol*, 49, Bhattacharjee, C. and P. K. Bhattacharya, 281–290, Copyright 2006, with permission from Elsevier.)

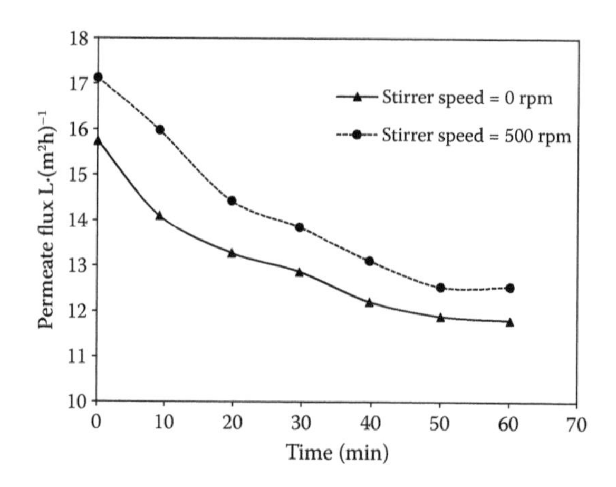


**FIGURE 6.6** Effect of stirrer speed on flux vs. time profile at a fixed membrane at a fixed TMP (TMP = 5 kgf·cm$^{-2}$). (Reprinted from *Sep Purif Technol*, 49, Bhattacharjee, C. and P. K. Bhattacharya, 281–290, Copyright 2006, with permission from Elsevier.)

BOD of permeate was found to get reduced by 70% when UF was carried out at a TMP of 7 kgf·cm$^{-2}$ with a stirrer speed of 500 rpm in the rotating disk membrane, whereas, COD was reduced by only 67%. In fact, at all the TMP level, BOD reductions were found to be more than the corresponding COD reduction on percentage basis. As the COD is caused by the presence of low molecular inorganic chemicals also, which might pass through the 5 kDa MWCO membrane, may give less %COD reduction compared to the corresponding %BOD reduction as most of the BOD comes from high molecular weight lignin/lignosulfonates. The effect of TMP on transient lignin concentration in the permeate at fixed membrane and stirrer speed has shown in Figure 6.9. As the TMP increases, more and more solutes get rejected by the membrane due to higher convective flow and result in higher polarized layer thickness. Moreover, increased pressure might cause compaction of the deposited solute layer, which could act as a secondary membrane, thereby reducing permeate concentration. Transient variation of the lignin concentration in permeate shows a decrease, followed by an increase after elapse of some time. Decrease of permeate concentration with time could be attributed due to gradual increase of polarized layer thickness, which could retard the solute transport in the vicinity of the membrane, thereby reducing permeate concentration. As the operation was carried out in batch mode, the retentate gets concentrated gradually, and after some time, this effect was found to dominate, which increases the permeate concentration. Thus, fractionation

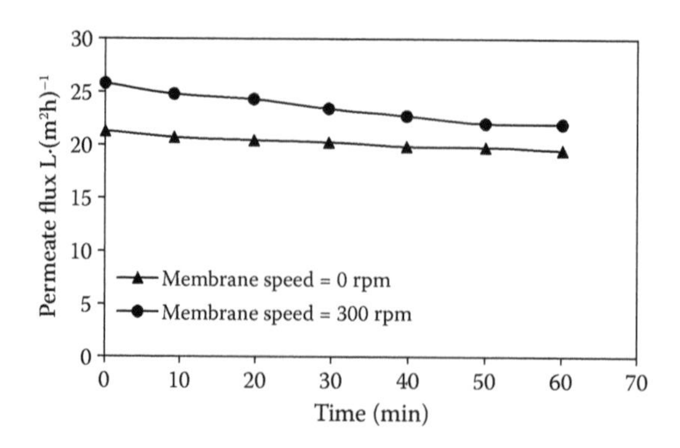


**FIGURE 6.7** Effect of membrane rotation speed on flux (TMP = 5 kgf·cm$^{-2}$, $N_s$ = 1000 rpm). (Reprinted from *Sep Purif Technol*, 49, Bhattacharjee, C. and P. K. Bhattacharya, 281–290, Copyright 2006, with permission from Elsevier.)

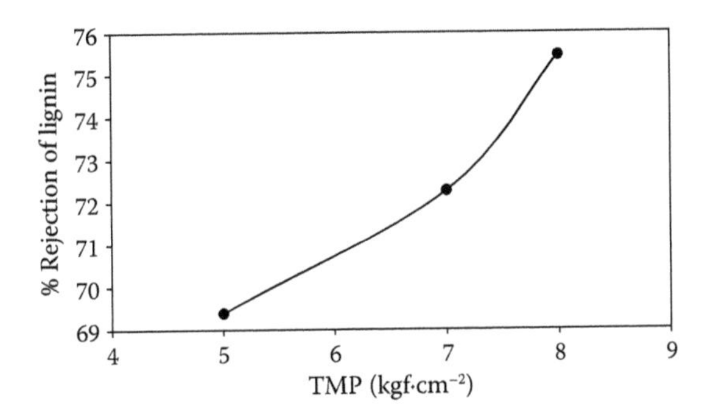


**FIGURE 6.8** Effect of TMPs on rejection ($N_m$ = 300 rpm, $N_s$ = 500 rpm). (Reprinted from *Sep Purif Technol*, 49, Bhattacharjee, C. and P. K. Bhattacharya, 281–290, Copyright 2006, with permission from Elsevier.)

of KBL using 5 kDa MWCO cellulose triacetate membrane is found to be quite useful in concentrating the lignin in the retentate as well as for improving the wastewater quality parameters.

#### 6.2.2.2 Treatment of KBL Using Cross-Flow Membrane Module

Another way to the treatment of KBL is routing the process through sedimentation followed by tertiary treatment such as adsorption. Another aspect of this tertiary treatment is the color removal of the liquor as the discharge of it in the river water results the prevention of the passage for the sunlight, which inhibits the growth of aquatic biota due to reduction in photosynthetic activity and primary production. Dilek and Gokcay (1994) reported 96% removal of COD from the paper machine, 50% from the pulping, and 20% for bleaching effluents by using alum as a coagulant. It is reported that, precipitation and chemical coagulation with lime, alum, aluminum chloride, or ferric sulfate results in good color removal (60–80%) from pulp and paper mill effluents. Murthy et al. (1991) reported a high removal of color by activated charcoal, fuller's earth, and coal ash. Shawwa et al. (2001) reported 90% removal of color, COD, and bleaching agents from bleached wastewater by the adsorption process, using activated coke as an adsorbent. Jonsson et al. (1996) reported on the treatment of paper coating color effluent by membrane filtration suggesting that the composition of the color had a significant influence on the performance. Membrane separation techniques were reported to be suitable for removing bleaching agents, COD, and color from pulp and paper mills (Zaidi et al., 1992; Afonso and Pinho, 1991).

Bhattacharjee et al. (2007) made a study on this tertiary treatment process having a route to cross-flow UF membrane separation process. Basically, in this study, two types of technique had

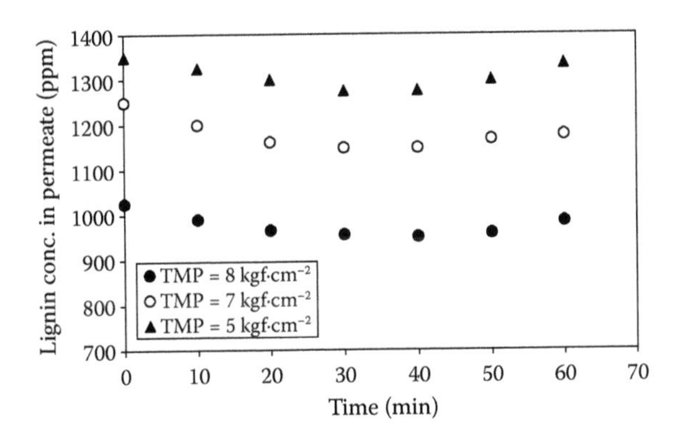


**FIGURE 6.9** Transient variation of lignin concentration in the permeate at different TMPs under fixed stirrer and membrane rotation speeds ($N_m$ = 300 rpm, $N_s$ = 500 rpm). (Reprinted from *Sep Purif Technol*, 49, Bhattacharjee, C. and P. K. Bhattacharya, 281–290, Copyright 2006, with permission from Elsevier.)

**TABLE 6.4**

**Comparison of BOD, COD, Total Solids, and Turbidity Values of Filtrate/Permeate Obtained after Adsorption and Membrane Processes for Cross-Flow UF Module**

| Parameters | Supernatant after Sedimentation Using Lime and Alum | Filtrate after Adsorption with 50 mg Activated Carbon/g Wastewater at pH 6.1 | Filtrate after Adsorption with 50 mg Activated Carbon/g Wastewater at pH 3.2 | Permeate from Cross-Flow Module at TMP = 2 $kgf \cdot cm^{-2}$ | Permeate from Cross-Flow Module at TMP = 2.5 $kgf \cdot cm^{-2}$ |
|---|---|---|---|---|---|
| TS, g/L | 7.64 ± 0.3 | 4.2 ± 0.15 | 3.1 ± 0.17 | 3.0 ± 0.14 | 2.9 ± 0.17 |
| BOD, mg/L | 4620 ± 22 | 820 ± 7.4 | 690 ± 7.2 | 750 ± 3.7 | 660 ± 3.3 |
| COD, mg/L | 12000 ± 55 | 3200 ± 14 | 2100 ± 12 | 3000 ± 14 | 2760 ± 18 |
| Turbidity, NTU | 235 ± 5 | 50 ± 2 | 36 ± 2 | 42 ± 2 | 28 ± 2 |
| Lignin concentration, g/L | 3.3846 ± 0.0034 | 0.6205 ± 0.0028 | 0.4680 ± 0.0054 | 0.5682 ± 0.0052 | 0.4742 ± 0.0016 |

*Source:* Bhattacharjee, S., et al., *Desalination*, 212, 92–102, 2007, with permission from Elsevier.

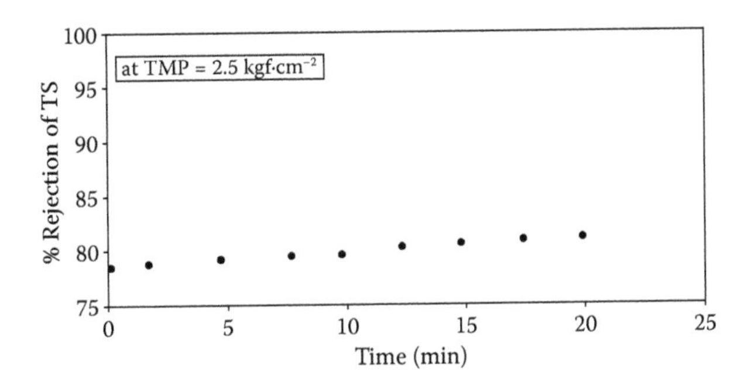


**FIGURE 6.10** Profile of percentage rejection based on total solids concentration against time at a TMP = 2.5 kgf·cm$^{-2}$. (Reprinted from *Sep Purif Technol*, 49, Bhattacharjee, C. and P. K. Bhattacharya, 281–290, Copyright 2006, with permission from Elsevier.)

been used for the treatment of KBL. One was adsorption on activated carbon and the other one was membrane separation. Main objective of this study was reduction of the color by removing total solid content and simultaneously making a reduction in COD as well as BOD. Initially 100 ml of KBL sample was taken in each of six glass cylinders. In this sample, 17.5% (wt/vol) (it was found to be the optimum dose at which water quality was better) alum solution was given and mixed at 160 rpm for coagulation. It was found that BOD removal was almost 65%. Table 6.4 (Bhattacharjee et al., 2007) shows a comparative study between adsorption and membrane separation technique for KBL treatment. It is evident from the table that in UF, almost 4% more amount of lignin was recovered compared to that in the activated carbon adsorption process. Figure 6.10 shows the rejection of total solids at 2.5 kgf·cm$^{-2}$ with time, which is almost 80% and which is quite effective to meet the objective of this treatment scheme.

De and Bhattacharya (1996) had made a similar study on cross-flow UF separation for KBL, where two different types of cross-flow module were used. One is low rejecting UF (LRUF) membrane (obtained from Permionics, India) and the other one is the high rejecting UF (HRUF) thin film composite membrane (obtained from Hydranautics India Ltd.). Here they had applied LRUF at 7 kgf·cm$^{-2}$ followed by HRUF at 1.8 kgf·cm$^{-2}$ which results in almost 91% rejection.

### 6.2.3 NF Process for the Treatment of KBL

In all the above discussions, we have confined ourselves in UF process for KBL whereas NF nowadays is emerging as an equally competitive route to the UF. In case of NF chances of fouling of the membrane is more pronounced. To avoid these unusual circumstances generally UF followed by NF are used for the treatment of KBL. In a study by Mänttäri et al. (1997), it was found that the reduction of total organic carbon (TOC), COD and total solids in UF was 50%–60% whereas the reduction was more than 80% in case of NF. Dafinov et al. (2005) had made a study on UF/NF application where they had used tubular ceramic membrane with a skin layer of $TiO_2$ and $ZrO_2$ on a support made up of α-alumina. First the effluent was treated with 5 kDa and 15 kDa membrane and the collected permeate was again treated with 1 kDa (NF range) membrane. But main difficulties with the NF are the percentage of separation. It was found that apart from lignin the dry matter retention was around 13% as because of the bindings with the retained lignin, which made the separation or purity percentage on the lower side.

### 6.2.4 RO Process for the Treatment of KBL

Another pressure driven membrane process that is used for the treatment of KBL is RO. Yalcin et al. (1999) had applied both UF and RO processes for the treatment of KBL. This study shows that almost 90%–95% of the coloring agents, total solids and other organic and inorganic agents can be removed by RO alone. Three types of membranes were used for this study: UF, RO with brackish water (BW) and RO with seawater (SW). First, the effluent was treated by UF membrane followed by RO (BW and SW) membrane separately. 98%–99% of COD, color and conductivity were removed by the application of both these processes simultaneously. Another aspect of the reverse osmosis for KBL is the treatment of the stripped condensate coming out from the evaporator in bleach plant. RO application on stripped condensate reduces the toxicity of the mills effluent (Genco and Heiningen, 2000).

## 6.3 CONCLUSION

From the above discussions, it can be understood that membrane separation has an important role to play in the treatment of KBL. UF, NF and RO are the three pressure-driven membrane separation techniques that have been used extensively for treatment of KBL. Especially UF is most frequently used membrane separation process in the treatment of KBL, as main disadvantage lying with NF and RO is the membrane fouling. But as the removal efficiency of COD, conductivity, and TDS are quite high for both NF and RO, these are now increasingly used in combination with UF. UF plant for the treatment of KBL installed at the Borregaard Industries (Sapsborg, Norway) with a membrane (20,000, PCIPUI20) surface area of 1120 $m^2$ is still under operation since 1981. It was reported that the plant handles 50 $m^3 \cdot h^{-1}$ of feed liquor with 12% solids producing a retentate stream of 16 $m^3 \cdot h^{-1}$ and 22% solid (Liu et al., 2004), which proves the widespread application of UF for KBL treatment.

## REFERENCES

Afonso, M. D., and M. N. Pinho. 1991. Membrane separation processes in pulp and paper production. *Filtration & Separation* 28: 42–44.

Ahonen, A., I. Bodin, M. Dinets, M. Gadin, F. Staub, and T. Åström. 2006. *Hazards in paper and pulp industries—from an engineering insurance perspective.* Paper presented at IMIA Conference, Boston.

Bhattacharjee, C., and P. K. Bhattacharya. 2006. Ultrafiltration of black liquor using rotating disk membrane module. *Separation and Purification Technology* 49: 281–290.

Bhattacharjee, C., P. Sarkar, and D. Sen. 2008. Treatment of casein whey using polyether sulphone ultrafiltration membrane in cross-flow membrane module. *The Journal of The Institution of Engineers* 88: 39–43.

Bhattacharjee, C., P. Sarkar, S. Datta, B. B. Gupta, K. Bhattacharya. 2006. Parameter estimation and performance study during ultrafiltration of Kraft black liquor. *Separation and Purification Technology* 51: 247–257.

Bhattacharjee, S., S. Datta, and C. Bhattacharjee. 2005. Studies on Water Quality Parameters during Ultrafiltration of Kraft Black Liquor. *The Journal of The Institution of Engineers* 86: 1–7.

Bhattacharjee, S., S. Datta, and C. Bhattacharjee. 2007. Improvement of wastewater quality parameters by sedimentation followed by tertiary treatments. *Desalination* 212: 92–102.

Compere, A. L., W. L. Griffith, C. F. Leitten, and S. Petrovan. 2004. *Improving the fundamental properties of Lignin-based carbon fiber for transportation applications.* In Proceedings of the 36th International SAMPE Technical Conference, San Diego, CA.

Dafinov, A., J. Font, and R. Garcia-Valls. 2005. Processing of black liquors by UF/NF ceramic membranes. *Desalination* 173: 83–90.

De, S., and P. K. Bhattacharya. 1996. Flux prediction of black liquor in cross flow ultrafiltration using low and high rejecting membranes. *Journal of Membrane Science* 109: 109–123.

Dilek, F. B., and C. F. Gokcay. 1994. Treatment of effluents from hemp-based pulp and paper industry—I. Waste characterization and physico-chemical treatability. *Water Science and Technology* 29: 161–164.

FAOSTAT. 2006. *Statistical Database of the Food and Agricultural Organization of the United Nations.*

Genco, J. M., and A. V. Heiningen. 2000. *Comparative Analysis of XL-2 Projects (Third Report for IP XL-2 Project).* A Technical report by Pulp and Paper Process Development Center, University of Maine, Department of Chemical Engineering.

Ghaffour, N. 2004. Modeling of fouling phenomena in cross-flow ultrafiltration of suspensions containing suspended solids and oil droplets. *Desalination* 167: 281–291.

Jong-Wook, L., and C. Bellamy. 2004. *Meeting the MDG drinking water and sanitation target (A mid-term assessment of progress).* WHO/UNICEF Joint Monitoring Programme on Water Supply and Sanitation Report.

Jonsson, A. S., C. Jonsson, M. Teppler, P. Tomani, and S. Wännström. 1996. Treatment of paper coating color effluents by membrane filtration. *Desalination* 105: 263–276.

Liu, G., Y. Liu, J. Ni, H. Si, and Y. Qian. 2004. Treatability of Krafi spent liquor by microfiltration and ultrafiltration. *Desalination* 160: 131–141.

Llamas, P., T. Dominguéz, J. M. Vargas, J. Llamas, J. M. Franco, and A. Llamas. 2007. A novel viscosity reducer for Kraft process black liquors with a high dry solids content. *Chemical Engineering and Processing* 46: 193–197.

Mänttäri, M., J. Nuortila-Jokinen, and M. Nyström. 1997. Evaluation of nanofiltration membranes for filtration of paper mill total effluent. *Filtration & Separation* 34: 275–280.

Murthy, B. S. A., T. A. Sihorwala, H. V. Tilwankar, and D. J. Killedar. 1991. Removal of colour from pulp and paper mill effluents by sorption technique—a case study. *Indian Journal of Environmental Protection* 11: 360–362.

Pharoah, J. G., N. Djilali, and G. W. Vickers. 2000. Fluid mechanics and mass transport in centrifugal membrane separation. *Journal of Membrane Science* 176: 277–289.

Shawwa, A. R., D. W. Smith, and D. C. Sego. 2001. Color and chlorinated organics removal from pulp mills wastewater using activated petroleum coke. *Water Research* 35: 745–749.

Sridhar, S., and P. K. Bhattacharya. 1991. Limiting flux phenomena in ultrafiltration of Kraft black liquor. *Journal of Membrane Science* 57: 187–206.

Trivedi, R. C., R. M. Bhardwaj, B. Sengupta, S. Sharma, K. Mehrotra, and C. P. Singh. 1990. *Water quality in India—Status and Trend.* CPCB MINARS Publication, Report No. 20.

Wallberg, O., A. S. Jonsson, and R. Wimmerstedt. 2003. Fractionation and concentration of Kraft black liquor lignin with ultrafiltration. *Desalination* 154: 187–199.

Woerner, D. L., and J. L. McCarthy. 1984. Ultrafiltration of Kraft black liquor. *AIChE Symposium Series.* 80: 25–33.

Yalcin, F., I. Koyuncu, I. Oztürk, and D. Topacik. 1999. Pilot scale UF and RO studies on water reuse in corrugated board industry. *Water Science and Technology* 40: 303–310.

Zaidi, A., H. Buisson, S. Sourirajan, and H. Wood. 1992. Ultra- and nano-filtration in advanced effluent treatment schemes for pollution control in the pulp and paper industry. *Water Science and Technology* 25: 263–276.

# 7 Treatment of Refinery Wastewater Using Membrane Processes

*Muhammad H. Al-Malack and Muhammad Muhitur Rahman*

## CONTENTS



## 7.1 INTRODUCTION

The occurrence of oil-containing wastewater and the corresponding contamination of water sources by oil began with the production and utilization of petroleum and its products. Before the introduction of the wastewater treatment and reuse schemes, it was a common practice to discharge the wastes into water bodies or bare surfaces. However, strict regulations, increased hauling costs, and environmental concerns made oily wastewater treatment a prominent issue for most industries.

The type and concentration of pollutants in a given refinery's effluent depend on the chemical make-up of the crude oil and the processes used to make the final products. Refineries use large amounts of water in the refining process and as a cooling agent. This water picks up waste oil and impurities from the refining process. Some impurities are in the crude oil itself, such as heavy metals, sulfide, and phenols, while others are created during the refining process, such as cyanide, dioxins, and furans. All of these chemicals can be toxic to aquatic life at very low concentrations. The major problem of oily wastewater is associated with its suitable disposal. The refinery wastewater has been marked as one of the key environmental pollutants with a great effect on the biodiversity. The reclamation and reuse of such oily wastewaters are needed especially in the oil-producing arid regions because of water scarcity. Although there are several methods used in treating oily wastewaters, investigation for improving the plant performance in terms of better effluent quality and cost effectiveness is continuing.

## 7.2 GENERATION AND CHARACTERISTICS OF REFINERY WASTEWATER

### 7.2.1 Generation

Wastewater generation is one of the major environmental problems that are encountered in most industries. In petroleum refineries, large volumes of water are used, especially for cooling systems. Furthermore, surface water runoff and sanitary wastewater are also generated. The quantity of wastewater generated and their characteristics depend on the process configuration. As a general guide, approximately 3.3 to 5 $m^3$ of wastewater per ton of processed crude oil are generated when cooling water is recycled. Wisjunpratpo and Kardena (2000) reported that 130 $m^3$ wastewater per hour was discharged by one of the refineries located in West Java, Indonesia. In 1952, Giles estimated that oil refineries used 5400 mgd of water (based on national crude-oil production of 7 million barrels per day). This volume of water represented about 20% of the total industrial consumption in the United States and slightly less than 50% of municipal needs. Although the sheer volume of water from oil refineries looms as so large a problem, the American Petroleum Institute reported that 80% to 90% of the total water used by the average refinery is for cooling purposes only and is not contaminated except through leaks in the lines.

### 7.2.2 Characteristics

To evaluate the oil-containing wastewater and to properly select the treatment methods and procedures for monitoring and operating treatment facilities, the clear understanding of the chemical and physiochemical composition of oils in wastewater is essential. Generally, petroleum refineries generate polluted wastewater containing biological oxygen demand (BOD) and chemical oxygen demand (COD) levels of approximately 150–250 mg/l and 300–600 mg/l, respectively; phenol levels of 20–200 mg/l; oil levels of 100–300 mg/l in desaltered water and up to 5,000 mg/l in tank bottoms; benzene levels of 1–100 mg/l; benzo[*a*]pyrene levels of 1–100 mg/l; and heavy metal levels of 0.1–100 mg/l for chromium and 0.2–10 mg/l for lead. Al Zarooni and Elshorbagy (2006) characterized Al Ruwais refinery wastewater. They reported that average values of BOD and COD in the CPI effluent were 130 ppm and 420 ppm, respectively. TPH and PAHs were reported to be 650 ppm and 56 ppm, respectively. They also reported TKN and sulfate values of 60 ppm and 200 ppm, respectively. Phenols and PCBs were also detected at levels of 21.7 ppm and 10 to 3900 ppb, respectively. Other investigators reported similar and different pollutants such as cyanide and chlorides (Lingbo et al., 2005; Gulyas and Reich, 2000; Viero et al., 2008; Xiangling et al., 2005). Weston (1950) gave the sanitary characteristics of typical refinery wastes.

## 7.3 UNIT OPERATIONS AND PROCESSES USED IN TREATMENT OF REFINERY WASTEWATER

Refinery wastewaters often require a combination of different treatment methods to remove oil and other contaminants before discharge. A typical system may include sour water striping, gravity separation of oil and water, dissolved air floatation, biological treatment, and clarification. A final polishing step using filtration, activated carbon, or chemical treatment may also be required. Several investigators, including Galil et al. (1997), Murray and Malone (1997), Kenawy et al. (1997), Carriere et al. (1997), Watanabe et al. (1991), Matthews and Choi (1987), Roques and Aurelle (1986), Pushkarev et al. (1983), Viraraghavan and Mathavan (1990), Hamia et al. (2007), Lavallee and Nadreau (1997), Gher et al. (1993), Valade et al. (1996), Al-Muzaini, et al. (1994), Moursy et al. (2003), El-Naas and Makhlouf (2008), Tyagi et al. (1993), Jou and Huang (2003), Tang and Lu (1993), Stepnowski et al. (2002), Coelho et al. (2006), and Sun et al. (2008) reported the use of a combination of different processes in treating oily wastewater. The application of membrane processes in refinery wastewater treatment will be fully detailed and discussed in the subsequent sections.

## 7.4 USE OF MEMBRANES IN REFINERY WASTEWATER TREATMENT

### 7.4.1 Theoretical Background of Membrane Processes

In membrane processes, both dead-end and cross-flow filtration are implemented (Figure 7.1). The decline in flux or permeation rate is believed to be the major hindrance of the wide implementation of membrane processes such as microfiltration, ultrafiltration, nanofiltration, and reverse osmosis in the water and wastewater treatment industry. This decline in flux rate is attributed to the formation of a dynamic or secondary membrane on top of the primary membrane.

Darcy's Law can be applied in describing the filtration rate of cross-flow microfiltration. The resistance includes explicitly the contribution of cake and filter medium:

$$v = \frac{k\Delta P}{\mu l}, \tag{7.1}$$

where $v$ is the liquid velocity, $k$ is the proportionality constant or Darcy's law permeability of the medium, $\Delta P$ is the pressure drop across the media, $l$ is the media thickness, and $\mu$ is the liquid viscosity.

$$\frac{l}{k} = R_{\mathrm{m}} + R_{\mathrm{c}}, \tag{7.2}$$

where $R_{\mathrm{m}}$ is the resistance caused by filter media and $R_{\mathrm{c}}$ is the resistance caused by cake. Thus, when filtering a suspension containing a wide range of particles and colloids using a microporous membrane at a constant pressure, the filtration flux, $J$, can be expressed by the following resistance equation:

$$J = \frac{\Delta P}{\mu R_{\mathrm{t}}}, \tag{7.3}$$

where $R_{\mathrm{t}}$ is the total resistance to the flux

$$= R_{\mathrm{m}} + R_{\mathrm{c}}.$$

At a constant applied pressure and, for a given solution, constant viscosity (at a given temperature and composition), the filtration flux is inversely proportional to the total resistance. The total resistance consists of the resistance caused by the filter media ($R_{\mathrm{m}}$), which is the product of the pore size, the pore density, and the pore depth, and the resistance due to cake formation, which is the product of internal colloidal fouling, concentration polarization, and formation of the gel layer.

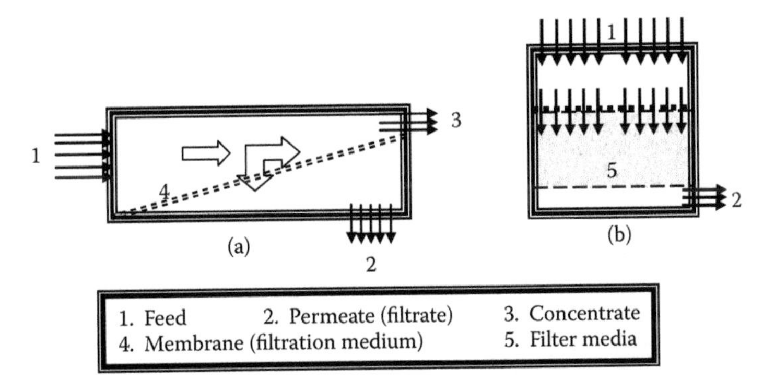


**FIGURE 7.1** Schematic diagrams of (a) cross-flow filtration and (b) dead-end filtration.

### 7.4.2 Fouling Mechanisms of Membrane Processes

Fouling mechanisms, as a result of dynamic membrane formation, were divided into three categories by Tanny (1978). Class I dynamic membranes are formed when filtering suspensions where the particles have a particle size greater than the pore size of the membrane. This phenomenon is known as concentration polarization.

Class II dynamic membranes are created when filtering dilute suspensions of colloidal particles of particle size much smaller than the pore size of the membrane. In this case, the flux decline mechanism was found to behave according to an internal pore clogging phenomenon rather than cake build-up on the membrane surface. Visvanathan and Ben Aim (1989) and other investigators (Tanny et al., 1979; Hermia, 1982; Grace, 1956) reported similar results. All investigators reported the following fouling model, which represents the decrease in permeate volume:

$$\frac{t}{V} = \frac{1}{Q_0} + \frac{k_1 t}{2}, \tag{7.4}$$

where $V$ is the permeate volume, $t$ is the filtration time, $Q_0$ is the initial flux rate, and $k_1$ is the filtration constant.

After some time, the colloidal particles will be brought up to the membrane surface, and the flux behavior will proceed in accordance with the following classical cake filtration model (Murkes and Carlsson, 1988):

$$\frac{t}{V} = \frac{1}{K_1}(V - 2V_{\mathrm{f}}), \tag{7.5}$$

where $V_{\mathrm{f}}$ is the volume of permeate that produces a hydraulic resistance equal to that of the membrane and $K_1$ is the cake filtration constant. A different form of the cake filtration model was reported by Visvanathan and Ben Aim (1989) and was as follows:

$$\frac{t}{V} = \frac{1}{Q_0} + \frac{K_1 V}{2}. \tag{7.6}$$

Al-Malack and Anderson (1996) investigated the formation of dynamic membranes with cross-flow microfiltration. They concluded that dynamic membrane formation obeys the standard law of filtration in the first few minutes of membrane formation (15 min). As time passes, the dynamic membrane formation was found to proceed according to the classical cake filtration model. Tanaka et al. (1997) investigated the characteristics in cross-flow microfiltration using different yeast suspensions. They reported that the experimental steady-state flux agreed well with that calculated by the following equation, which is used in dead-end filtration:

$$J = \frac{\Delta P}{\mu(R_{\mathrm{m}} + R_{\mathrm{c}})}, \tag{7.7}$$

where $J$ is the permeate flux, $\Delta P$ is the pressure difference across the membrane, $\mu$ is the liquid viscosity, $R_{\mathrm{m}}$ is the membrane hydraulic resistance, and $R_{\mathrm{c}}$ is the cake hydraulic resistance.

Class III dynamic membranes are formed when filtering polymers or polyelectrolyte molecules of equal size to the membrane size.

With respect to fouling control and mitigation, McAlexander and Johnson (2003) investigated the mitigation of membrane fouling via backpulsing for commercial kerosene, distilled kerosene, and the refinery fluid. They reported that backpulsing improved recovery rates by 30% over a 15-h period, but it could not prevent the occurrence of some irreversible fouling. Nghiem and Schäfer (2006) investigated the cause of membrane fouling at a small water recycling plant using a hollow-fiber microfiltration system. Results obtained from this study indicate that the membrane was fouled by a mixture of colloids and organic matters, enhanced by the presence of multivalent cations.

### 7.4.3 Performance of Membranes

The performance of membranes is primarily a function of the operating parameters, and is measured by the filtrate flow rate (flux) and its quality. Cross-flow velocity, transmembrane pressure, temperature, pore size of the membrane and concentration of suspended solids in the feed were reported to affect the performance of membrane processes (Forman et al., 1990; Wakeman and Tarleton, 1991; Bhave et al., 1988; Hazlett et al., 1987; Anderson et al., 1986).

In recent years, a new technique has been discovered using different types of membranes to treat oil-containing wastewater. Fan and Wang (2000) conducted a study where inorganic ceramic membranes were used to treat oil-containing wastewater. They reported that, using this technique, the oil concentration in the permeate was lower than 5 mg/l, which represented a removal efficiency of more than 95%. In another study conducted by Ting and Chi-Sheng (1999), porous ceramic membranes were used to separate oil from an oil/water emulsion. The separation was performed using a cross-flow ultrafiltration process. They reported that the separation of oil from water was attributed to the capillary force on the hydrophilic surface of the membrane. The separation was found to be optimum at a transmembrane pressure of 300 kPa and a feed rate of 6 ml/min with back flush, using 0.02 μm membranes. The oil content was reduced from 2 to 0.4 wt. % achieving a separation factor of 5. Ceramic membranes were also used by Wang and Fan (2000), and Mueller et al. (1997) to treat oily wastewater. These investigators reported results that are in agreement with those reported by Ting and Chi-Sheng (1999).

Other investigators, including Santos and Wiesner (1997) and Reed et al. (1997), reported the use of ultrafiltration membranes in treating oily wastewater. In order to improve its performance in treating oily wastewater, ultrafiltration membranes were also reported to be used in combination with other processes such as ozonation (Chang et al., 2001), coagulation (Tansel et al., 2001), wetlands (Reed et al., 1998), vibration (Vigo et al., 1993), and reverse osmosis (Krug and Attard, 1990). The authors reported that the membranes exhibited 100% rejection of suspended solids and produced effluents with the oil and lubricant content of less than 10 ppm. The retention of oil and lubricant in the RO process was found to be 100%. Li et al. (2006) developed a hydrophilic hollow fiber ultrafiltration (UF) membrane for oil–water separation. The experimental results showed that oil retention of the membrane was over 99% and oil concentration in the filtrate was below 10 mg/l. It was concluded that the cellulose hollow fiber UF membrane developed in the study was not only resistant to fouling but also tolerant to a wide pH range (1–14). Teodosiua et al. (1999) investigated the possibilities of using ultrafiltration as a pretreatment for reverse osmosis to treat refinery effluent. Experimental results indicated that average removal efficiencies of 98% for turbidity and TSS and 30% for COD were achieved. They reported that the best chemical cleaning agents were found to be citric acid and sodium hydroxide. Zhang et al. (2005) investigated the effectiveness of ultrafiltration (UF) technology for treatment of refinery wastewater using powdered activated carbons (PACs) and coagulant. They reported that the UF unit performance could be significantly improved by simultaneously adding 15 mg/l of PACs and 0.8 ml/l of HCA into the system. The removal rates of TOC in the wastewater were over 99%. Fratila-Apachitei et al. (2001) studied the treatment of refinery and petrochemical effluents using two UF membranes that had different molecular weight cut-offs (50 and 150 kDa). They reported that the two membranes performed very differently, where the 150-kDa membrane showed a very fast flux decline (20% in 2 min) requiring frequent backwashing, whereas in the case of the 50-kDa membrane, 20% flux decline was reached in 20 min. Cai et al. (2000) utilized the formation of $MnO_2$ dynamic membrane to treat diatomite mineral processing wastewater and oil refinery wastewater. They reported stable performance and over 98% of turbidity removal rate. In addition, 5 wt. % HCI solution was used for cleaning the $MnO_2$ membrane. Lia et al. (2006) used a modified tubular UF module to purify oily wastewater from an oil field. They reported that retentions of chemical oxygen demand and total organic carbon were more than 90% and 98%, respectively. The addition of nano-sized alumina particles was reported to improve the performance of the membrane. Zhong et al. (2003) investigated combination of flocculation and microfiltration for the treatment of refinery wastewater. They recommended operation conditions for pilot and industrial application as transmembrane pressure of 0.11 MPa, and cross-flow velocity of 2.56 m/s. In a study conducted by Zhang et al. (2008) where a combination of demulsification and reverse osmosis (RO) was used to treat high-strength and stable oil/water emulsion. The pilot-scale experiment results showed that the removal efficiency of COD, turbidity, and oil was 99.96%, 100%, and 100%, respectively, in 30–50 min at 80°C–90°C with demulsifier dosage of 0.1% (w/v), for the demulsification process, and at 35°C–40°C with a driving pressure of 3.6 MPa and the flow rate of 1.5–1.6 $m^3/h$ for the RO process.

### 7.4.4 Membrane Bioreactors

The configuration of membrane along with activated sludge commonly known as membrane bioreactor (MBR) is of two types (Figure 7.2). The first one is MBRs with internal submerged membrane filtration (SM-MBR) where the membrane filtration is carried out directly in the activated sludge tank. Another is MBR with external membrane filtration (CF-MBR) where the membrane filtration is carried out outside the activated sludge tank. The concentrate, that is the retained activated sludge, is returned to the activated sludge tank.

Scholz and Fuchs (2000) examined the feasibility of applying a CF-MBR to treat surfactant containing oil water emulsion. Trials in an MBR with a high activated sludge concentration of up to 48 g/l showed that oily wastewater containing surfactants was biodegraded with high efficiency. The average removal of COD and TOC during the experiment was 94%–96% for fuel oil, and 97%–98% for lubricating oil respectively at a hydraulic retention time (HRT) of 13.3 h.

Seo et al. (1997) also investigated the effect of HRT on the biodegradability of oil, where at an HRT of more than 10 days, the removal efficiency of oil was found to be more than 90%. The performance of the cross-flow MBR was also investigated by Daubert et al. (2003), Sutton et al. (1992) and Gaines et al. (2000).

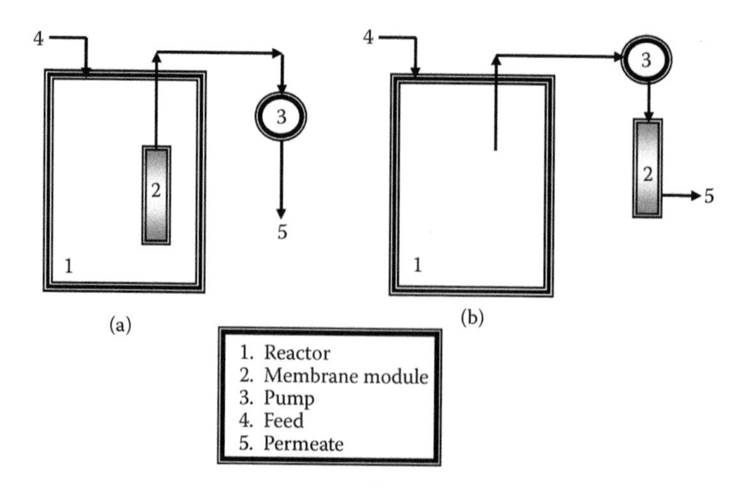


**FIGURE 7.2** Schematic diagrams of (a) submerged MBR and (b) MBR.

#### 7.4.4.1 Kinetics of MBR Process

The basic equation that describes the interaction between the growth of microorganisms and utilization of the growth limiting substrate in the activated sludge process are based on the Monod (1949) equations. The Monod model is still the most commonly and widely used model for the study of the biokinetic coefficients.

- Microbial death and viability
- Microbial decay

Zhang et al. (2002) used a combinational approach with considering HRT as an evaluation index to discuss factors, such as maximum specific removal rate (K), saturation constant ($K_s$), maintenance coefficient (m), maximum specific growth rate ($\mu_m$) and observed yield coefficient ($Y_{obs}$). He reported values of K and $K_s$ for petrochemical wastewater, as 0.185 and 154.2, respectively. In another study, Fan et al. (1998) reported a coefficient of COD removal k, for petrochemical wastewater between 0.017 to 0.080 l/(mg·d). Tellez et al. (1995) evaluated the biokinetic coefficients of New Mexico oilfield produced water. Using respirometric techniques for determination of the biokinetic constants, values of 1.37 mg/l and 0.136 $h^{-1}$ were obtained for $K_s$ and $\mu_{max,}$ respectively. Changes in cell yield were also evident, however, yields increased from 0.41 to 0.69 mg biomass/mg total *n*-alkane. According to Suman Raj and Anjaneyulu (2005), typical values of half velocity constant ($K_s$), yield coefficient (Y) and endogenous decay coefficient ($k_d$) in industrial wastewater varies within a range of 850 to 5200 mg/l, 0.3 to 0.72 mg/mg, and 0.05 to 0.18 $day^{-1}$, respectively. Rahman et al. (2005) determined the Biokinetic coefficients for the treatment of refinery wastewater. They reported the values of saturation constant ($K_s$), specific growth rate (μ), yield coefficient (Y) and endogenous decay coefficient ($k_d$) for MLSS 3000 mg/l as 659.45 mg COD/l, 1.2 $day^{-1}$, 0.222 mg/mg, 0.09 $day^{-1}$ respectively.

#### 7.4.4.2 Performance of MBR

Rahman and Al-Malack (2005) used a cross-flow MBR in treating wastewater discharged by a petroleum refinery at MLSS concentrations of 5000 and 3000 mg/l. The results of the investigation showed that a COD removal efficiency of more than 93% was obtained at both MLSS values. The study also showed that hydraulic retention time did not have a significant effect on the system's performance. The relationship between permeate flux (J) and cross-flow velocity (V) was found to be best described by a power relationship ($J = kV^n$) where constants $k$ and $n$ were affected by MLSS concentration. The cleaning mechanism investigation showed that cleaning the membrane with an acidic detergent, with a pH value of about 1.5, produced the best results. Xianling et al. (2005) carried out aerobic biotreatment of the refinery wastewater in a 170 $m^3$ pilot-scale gas–liquid–solid three-phase flow airlift loop bioreactor (ALR) with a low ratio of height to diameter, in which biological membrane replaced the activated sludge. Under the optimum operation condition, the effluent COD and $NH_4$–N were lower than 100 and 15 mg/l, respectively for more than 40 days. Furthermore, this pilot-scale airlift loop bioreactor generated only one-third of the sludge waste compared to the traditional activated sludge process. Viero et al. (2008) investigated the treatment of a refinery wastewater (oily stream) using a submerged membrane bioreactor (SMBR) operating with constant permeate flux. The membrane had a key role in the process, since it improved COD and TOC removal efficiencies by 17 and 20%, respectively. Elmaleh and Ghaffor (1999) observed aggregation processes of hydrocarbons on the bacterial floes. They found the process leading to larger particles with an optimal hydrocarbons/biological solids ratio. This induces a significant flux increase up to 150 l/h·$m^2$. The progressive fouling was limited by use of helical baffles introduced in the filtration element operated at 0.5 bar. The application of submerged membrane and submerged MBR in treatment of refinery wastewaters was also reported by Bu-Naiyan et al. (2006) and Conner et al. (2008), respectively.

## 7.5 ECONOMY OF MEMBRANE PROCESS

The use of membrane systems has experienced exponential growth over the last few years due to their ability to treat water and wastewater that meet stringent compliance standards. With their modular design and sophisticated automation, membrane plants are now being built to provide safety and flexibility with minimal operator intervention. Another driving force for the popularity of membranes is that their costs have never been lower. Lower costs are stimulated by increased competition, increased demand that favor scale of economics for manufacturing, and more efficient process operation. According to an American Water Works Association Research Foundation (AWWARF) survey in 2002, the capital cost for low-pressure membrane system was in the range of \$0.18–\$0.23/gallon per day (gpd) and was expected to lower to \$0.15–\$0.20/gpd in coming decades (Lozier and Jacangelo, 2002). Further reduction in cost would largely come from either the development of high flux, low fouling membranes, or more efficient way to operate membrane systems. Aileen and Kim (2007) reported that permeate flux and transmembrane pressure, which correlate to permeate viscosity and total resistance, are directly related to the cost of membrane processes. MBRs typically operate at higher biomass concentrations than conventional biological treatment processes. The advantage that this provides is increased volumetric loading and decreased sludge production, which in turn lowers capital investment costs for civil works and reduces sludge disposal costs. Biomass concentration also influences energy costs. A reduced oxygen transfer rate is associated with higher biomass concentrations, so energy cost for aeration increases accordingly. Aeration accounts for a significant portion of energy costs in the operation of MBR systems. Thus, optimizing the oxygen supply can favorably affect operating costs. Yoon et al. (2004) reported a comparative economical assessment between MBR and the combined biological and chemical processes (CBCP). The CBCP showed capital investment of US\$ 204/m$^3$ lower than that of the MBR, and operating cost was estimated at US\$ 0.048/m$^3$ less than that of the MBR. The major factor largely attributing to less economical for the MBR was originated from both membrane price and its life expectancy.

## 7.6 MEMBRANE FUTURE PERSPECTIVE

Due to the increasing stringency in environmental standards and scarcity in water resources, the implementation of alternative advanced water and wastewater treatment technologies are being encouraged and enforced. In recent years, membrane processes, namely, microfiltration (MF), ultrafiltration (UF), reverse osmosis (RO), nanofiltration (NF), dialysis, and electrodialysis have been established as very efficient techniques for the separation of solids from water and wastewater samples. The aforementioned processes are now extensively being employed to produce drinking water from seawater and brackish water, to recover precious products from industrial effluents, and to perform various concentration and purification missions in the municipal and industrial effluents. Due to the high costs of membrane processes and shortcomings in membrane performance represented by rapid flux decline, the implementation of membrane processes did not spread widely in recent years. As the renewable sources of fresh water dwindle and become inadequate, however, and as the governmental policies of water recycling and reuse become more enforceable, the utilization of membrane techniques in water and wastewater industries is anticipated to have a growing pattern because of their clear technical and sometimes economical advantages when compared to conventional and competing water and wastewater treatment processes. New membranes with better separation characteristics and improved thermal, chemical, and mechanical properties are being developed. Novel system designs as well as complete new process concepts, such as MBRs, have been developed on a laboratory or pilot plant scale. The facilitated transport through liquid membranes offers another important technology for the selective removal of heavy metal ions and organic solvents from industrial effluents. The processes have been demonstrated in the laboratory and are now becoming available on an industrial scale. The future advancements of the technical and economical aspects of membrane processes is anticipated to play a very significant role in the treatment of municipal and industrial wastewaters, particularly, in countries suffering of water shortages.

## REFERENCES


Aileen, N.L.N., and Albert, S.K. 2007. A mini-review of modeling studies on membrane bioreactor (MBR) treatment for municipal wastewaters. *Desalination* 212: 261–281.

Al Zarooni, M., and Elshorbagy, W. 2006. Characterization and assessment of Al Ruwais refinery wastewater. *J. Hazard. Mater.* A136: 398–405.

Al-Malack, M.H. and Anderson, G.K. 1996. Formation of dynamic membranes with crossflow microfiltration. *J. Membr. Sci.* 112: 287.

Al-Muzaini, S., Khordagui, H., and Hamouda, M.F. 1994. Removal of VOC from refinery and petrochemicals wastewater using dissolved air floatation. *Water Sci. Tech.* 30: 79–90.

Anderson, G.K., James, A., and Saw, C.B. 1986. Crossflow Microfiltration–A Membrane for Biomass Retention in Anaerobic Digestion. *4th World Filtration Congress, Belgium.*

Bhave, R.R., and Fleming, H.L. 1988. Removal of Oily Contaminants in Wastewater with Microporous Alumina Membranes. *AIChE Symposium Series* 84, no. 261: 19–27.

Bu-Naiyan, M.H., Al-Aama, S., Al-Malack, M.H, Balubaid, A., and Al-Abeesy, S. 2006. Treatment of Oily Wastewater using Immersed Membrane Process at the Riyadh Petroleum Refinery. *A paper presented at the third UAE–Japan Symposium: Sustainable GCC Environmental and Water Resources.* Abu Dhabi, UAE.

Cai, B., Ye, H., and Yu, L. 2000. Preparation and separation performance of a dynamically formed $MnO_2$ membrane. *Desalination* 128: 247–256.

Carriere, P., Reed, B., Lin, W., Roark, G. and Viadero, R. 1997. Pilot-scale treatment of an Al-manufacturer oily wastewater. *Proc. Technomic Publ Co Inc, Lancaster, PA, USA*, 587–596.

Chang, I.-S., Chung, C.-M., and Han, S.-H. 2001. Treatment of oily wastewater by ultrafiltration and ozone. *Desalination* 133(3): 225–232.

Coelho, A., Antonio V.C., Marcia, D., and Sant'Anna Jr., G.L. 2006. Treatment of petroleum refinery sourwater by advanced oxidation processes. *J. Hazard. Mater.* B137: 178–184.

Conner, W., Liu, J., and Yee, J. 2008. Comparative Evaluation of MBR/RO and PAC Enhanced MBR/RO Treatment of Refinery Oily Wastewater. *A paper presented at the fourth Joint KFUPM-JCCP Environment symposium: GCC Environment and Sustainable Development.* Dhahran, Saudi Arabia.

Daubert, I., Muriel, M.B., Claude, M., et al. 2003. Why and how membrane bioreactors with unsteady filtration conditions can improve the efficiency of biological processes. *Ann. N.Y. Acad. Sci.* 984: 420–435.

Elmaleh S. and Ghaffor, N. 1996. Upgrading oil refinery effluents by cross-flow ultrafiltration. *Water Sci. Technol.* 34(9): 231–238.

El-Naas, M.H., and Makhlouf, S. 2008. A spouted bed bioreactor for the biodegradation of phenols in refinery wastewater. *J. Biotechnol.* 136S: S647–S677.

Fan, S., and Wang, J. 2000. Treatment of oil-containing wastewater with an inorganic membrane. *J. Dalian. Univ. Technol.* 40(1): 61–63.

Fan, Y., Wang, J., and Jiang, Z. 1998. Test of membrane bioreactor for waste water treatment of a petrochemical complex. *J. Environ. Sci. (China)* 10(3): 269–275.

Forman, S.M., DeBernardez, E.R., Feldberg, R.S., and Swartz, R.W. 1990. Crossflow filtration for the separation of inclusion bodies from soluble proteins in recombinant *Escherichia coli* cell lysate. *J. Membr. Sci.* 48(1): 263–279.

Fratila-Apachitei, L.E., Kennedy, M.D., Linton, J.D., Blumed, I., and Schippers, J.C. 2001. Influence of membrane morphology on the flux decline during dead-end ultrafiltration of refinery and petrochemical waste water. *J. Membr. Sci.* 182: 151–159.

Gaines, F.R., Dunn, W.G., Del V. M., Manning, J.G., and Mark R. 2000. Annual Conf. and Exposition on Water Quality and Wastewater Treatment, 73rd, Anaheim, CA, 2810–2836.

Galil, N., Zeira, B.D., and Rebhun, M. 1997. Characteristics of hydrocarbons removals in wastewater from an integrated oil refinery, Technion, Israel. *Proc. of the 52nd Industrial Waste Conference.*

Gher, R., Swartz, C., and Offringa, G. 1993. Removal of triahalomethane precursors from eutrophic water by dissolved air floatation. *Water Res.* 27: 41–49.

Giles, R.N. 1952. A rational approach to industrial waste disposal problems. *Sewage Ind. Wastes* 24: 1495.

Grace, H.P. 1956. Structure and performance of filter media. *AlChe J.* 2: 307.

Gulyas, H., and Reich, M. 2000. Organic compounds at different stages of a refinery wastewater treatment plant. *Water Sci. Technol.* 32(7): 119–126.

Hamia, M.L., Al-Hashimi, M.A., and Al-Dooric, M.M. 2007. Effect of activated carbon on BOD and COD removal in dissolved air flotation unit treating refinery wastewater. *Desalination* 216: 116–122.

Hazlett, J.D., Kutowy, O., and Tweddle, T.A. 1987. Processing of crude oils with polymeric ultrafiltration membranes. *AIChE Symposium Series* 272:101–107.

Hermia, J.V. 1982. Constant pressure blocking filtration laws—application to power-law non-Newtonian fluids. *Trans. Inst. Chem. Eng.* 60: 183.

Jou, C.G., and Huang, G.C. 2003. A pilot study for oil refinery wastewater treatment using a fixed-film bioreactor. *Adv. Environ. Res.* 7: 463–469.

Kenawy, F.A., Kandil, M.E., Fouad, M.A., and Abo-arab, T.W. 1997. Environmental protection from refinery oily waste water effluents by using coalescers. *Proc. of Middle East Oil Show. Society of Petroleum Engineers (SPE), Richardson, TX, USA*, 363–370.

Krug, T.A., and Attard, K.R. 1990. Treating oily waste water with reverse osmosis. *Water Pollut. Contr.* 128(5): 16–18.

Lavallee, H.C., and Nadreau J. 1997. Dissolved air floatation system use increases for secondary clarification. *Pulp & Paper* 71: 99–101.

Li, H.J., Yi, M.C., Qin, J.J., et al. 2006. Development and characterization of antifouling cellulose fiber UF membranes for oil–water separation. *J. Membr. Sci.* 279: 328–335.

Lia, Y.S., Yan, L., Xianga, C.B., and Hong, L.J. 2006. Treatment of oily wastewater by organic–inorganic composite tubular ultrafiltration (UF) membranes. *Desalination* 196: 76–83.

Lingbo L., Song Y., Congbi, H., and Guangbo, S. 2005. Comprehensive characteristics of oil refinery effluent-derived humic substance using various specific approaches. *Chemosphere* 60: 467–476.

Lozier, J., and Jacangelo, J. 2002. Where Are We Headed?—The Future of Membrane Treatment. *Proceedings of Water Quality Technology Conference*, Seattle, WA.

Matthews, R.R., and Choi, M.S. 1987. Produced Water Treating Equipment: Recent Field Tests. *American Inst of Chem Eng, National Meeting 1987. AIChE* 3c: 17.

McAlexander, B.L., and Johnson, D.W. 2003. Backpulsing fouling control with membrane recovery of LNAPL. *J. Membr. Sci.* 227: 137–158.

Monod, J. 1949. Growth of bacterial cultures. *Ann. Rev. Microbiol.* 3: 371–394.

Moursy, A.S., and Abo El-Ela, S.E. 2003. Treatment of oily refinery wastes using a dissolved air flotation process. *Environ. Inter.* 7(4): 267–270.

Mueller, J., Cen, Y., and Davis, R.H. 1997. Crossflow microfiltration of oily water. *J. Membr. Sci.* 129(2): 221–235.

Murkes, J. and Carlsson, C. G. 1988. *Crossflow Filtration.* John Wiley & Sons, New York.

Murray, B.G. and Malone, K.S. 1997. Prioritizing wastewater reduction and recycling. *Environ. Technol.* 7(7): 40–43.

Nghiem, L.D., and Schäfer, A.I. 2006. Fouling autopsy of hollow-fibre MF membranes in wastewater reclamation. *Desalination* 188: 113–121.

Pushkarev, V.V., Yuzhaninov, A.G., and Men, S.K. 1983. *Treatment of Oil-Containing Wastewater*, 6. Allerton Press, Inc., New York.

Rahman, M.M. and Al-Malack, M. H. 2006. Performance of cross flow membrane bioreactor (CF-MBR) when treating refinery wastewater. *Desalination* 191: 16–26.

Rahman, M.M., Al-Malack, M.H., and Bukhari, A.A. 2005. Kinetics of CF-MBR Processes in the Treatment of Refinery wastewater. *Proc. of Workshop of SAWEA 2005.*

Reed, B.E., Carriere, P., Lin, W., Roark, G., and Viadero, R. 1998. Oily wastewater treatment by ultrafiltration: Pilot-scale results and full-scale design. *Pract. Period. Hazard. Toxic Radioact. Waste Manage.* 2(3): 100–107.

Reed, B.E., Lin, W., and Viadero, R. Jr. 1997. Oil-based lubricant/coolant treatment by high-shear rotary ultrafiltration. *Proc of 52nd Ind. Waste Conf.* 429–440.

Roques, H. and Aurelle, Y. 1986. Recent developments in the treatment of oily effluents. *Water Sci. Technol.* 18(9): 91–103.

Santos, S.M., and Wiesner, M.R. 1997. Ultrafiltration of water generated in oil and gas production. *Water Environ. Res.* 69(6): 1120–1127.

Scholz, W., and Fuchs, W. 2000. Treatment of oil-contaminated wastewater in a membrane bioreactor, *Water Res.* 34(14): 3621–3629.

Seo, G.T., Lee, T.S., Moon, B.H., Choi, K.S., and Lee, H.D. 1997. Membrane separation activated sludge for residual organic removal in oil wastewater. *Water Sci. Technol.* 36(2): 275–282.

Stepnowski, P., Siedlecka, E.M., Behrend, P., and Jastorff, B. 2002. Enhanced photo-degradation of contaminants in petroleum refinery wastewater. *Water Res.* 36: 2167–2172.

Suman Raj, D.S., and Anjaneyulu, Y. 2005. Evaluation of biokinetic parameters for pharmaceuticals wastewater using aerobic ox integrated with chemical treatment. *Process Biochem.* 40: 165–175.

Sun, Y., Zhang, Y., and Quan, X. 2008. Treatment of petroleum refinery wastewater by microwave-assisted catalytic wet air oxidation under low temperature and low pressure. *Sep. Purif. Technol.* 62: 565–570.

Sutton, P.M., and Mishra, P.N. 1992. The membrane biological reactor for industrial wastewater treatment and bioremediation. *International Symposium on the Implementation of Biotech in Industrial Waste Treatment and Bioremediation*, Grand Rapids: 175–191.

Tanaka, T., Tsuneyoshi, S., Kitazawa, W., and Nakanishi, K. 1997. Characteristics in crossflow filtration using different yeast suspension. *Sep. Sci. Technol.* 32(11): 1885.

Tang, S. and Lu, X. 1993. The use of *Eichhornia crassipesto* cleanse oil-refinery wastewater in China. *Ecol. Eng.* 2(3): 243–251.

Tanny, G.B. 1978. Dynamic membranes in ultrafiltration and reverse osmosis. *Sep. Purif. Meth.* 7(2): 183.

Tanny, G.B., Strong, D.K., Presswood, W.G. and Meltzert, T.H. 1979. The adsorptive retention of *Pseudomonas diminuta* by membrane filters. *J. Parenter. Drug Assoc.* 33: 40.

Tansel, B., Regula, J. and Shalewitz, R. 2001. Evaluation of ultrafiltration process for treatment of petroleum contaminated waters. *Water Air Soil Pollut.* 126(3–4): 291–305.

Tellez, G.T., Nirmalakhandan, N., and Gardea-Torresdey, J.L. 1995. Evaluation of biokinetic coefficients in degradation of oilfield produced water under varying salt concentrations. *Water Res.* 29(7): 1711–1718.

Teodosiua, C.C, Kennedy, M.D., van Straten, H.A., and Schippers, J.C. 1999. Evaluation of secondary refinery effluent treatment using ultrafiltration membranes. *Water Res.* 33(9): 2172–2180.

Ting, J., and Chi-Sheng, J. 1999. Cross-flow ultra filtration of oil/water emulsions using porous ceramic membranes. *J. Chinese Inst. Chem. Eng.* 30(3): 207–214.

Tyagi, R.D., Tran, F.T., and Chowdhury, A.K.M.M. 1993. Biodegradation of petroleum wastewater in a rotating biological contactor with PU foam attached to the disks. *Water Res.* 27(1): 91–99.

Valade, M.T., Edzwald, J.K., Tobiason, J.E., Dahlquist, J., Hedberg, T., and Amato, T. 1996. Particle removal by floatation: pretreatment effects. *J. AWWA* 88(12): 35–47.

Viero, A.F., De Meloa, T.M., Torres, A.P.R., and Ferreira, N.R. 2008. Effects of long-term feeding of high organic loading in a SMBR treatment oil refinery wastewater. *J. Membr. Sci.* 319: 223–230.

Vigo, F., Uliana, C., Ravina, F., Lucifredi, A., and Gandoglia, M. 1993. Vibrating ultrafiltration module Performance in the 50–1000 Hz frequency range. *Sep. Science Tech.* 28(4): 1063–1076.

Viraraghavan, T., and Mathavan, G.N. 1990. Treatment of oily waters using peat. *Water Pollut. Res. J. Can.* 25(1): 73–90.

Visvanathan, C., and Ben Aim, R. 1989. Studies on colloidal membrane fouling mechanisms in crossflow microfiltration. *J. Membr. Sci.* 45(2): 3.

Wakeman, R.J., and Tarleton, E.S. 1991. Colloidal fouling of microfiltration during the treatment of aqueous feed streams. *Desalination and Water Reuse Proc. of the 12th International Symposium.*

Wang, L., and Fan, L. 2000. Treating oil-containing restaurant wastewater with anaerobic-oxidation-sedimentation bioreactor. *J. Chem. Eng. Chinese Univ.* 14(4): 358–362.

Watanabe, K., Yamanouchi, H., Ueta, Y., and Nagata, O. 1991. Present situation and problems related to marine oil–water separation techniques. *Water Sci. Technol.* 23(1–3): 319–328.

Weston, R.F. 1950. Separation of oil refinery wastewater. *Ind. Eng. Chem.* 42: 607.

Wisjunpratpo and Kardena, E. 2000. Biotreatment of natural oil and gas industry wastes. *Proceedings of Sixth AEESEAP Triennial Conference,* Kuta, Bali, Indonesia.

Xianling L., Jianping, W., Qing Y., and Xueming, Z. 2005. The pilot study for oil refinery wastewater treatment using a gas–liquid–solid three-phase flow airlift loop bioreactor. *Biochem. Eng. J.* 27: 40–44.

Yoon, T.I., Lee, H.S., and Kim, C.G. 2004. Comparison of pilot scale performances between MBR and hybrid conventional waste water treatment systems. *J. Membr. Sci.* 242: 5–12.

Zhang, H., Fang, S., Ye C., et al. 2008. Treatment of waste filature oil/water emulsion by combined demulsification and reverse osmosis. *Sep. Purif. Technol.* 63: 264–268.

Zhang, J.C., Wang, Y.H., Song, L.F., et al. 2005. Feasibility investigation of refinery wastewater treatment by combination of PACs and coagulant with ultrafiltration. *Desalination* 174: 247–256.

Zhang, S., Houten, R., Eikelboom, D.H., Jiang, Z., Fan, Y., and Wang, J. 2002. Determination and discussion of hydraulic retention time in MBR system. *J. Environ. Sci.* 14(4): 501–507.

Zhong, J., Sun, X., and Wang, C. 2003. Treatment of oily wastewater produced from refinery processes using flocculation and ceramic membrane filtration. *Sep. Purif. Technol.* 32: 93–98.

# 8 Advanced Oxidation Process and Nanofiltration for Treatment of Textile Plant Effluent

## *A Brief Review*

*Phalguni Banerjee and Sirshendu De*

**CONTENTS**



## 8.1 INTRODUCTION

Textile industries generate huge quantities of wastewater from various steps in the dyeing and finishing processes (Desphande, 2001). Nearly 2% of the total worldwide production of dyes is discharged directly as aqueous effluent. Further, almost 10% are lost during the textile dyeing process. The efficiency of application of color in the form of dyes onto textile fibers is not a very efficient process (Nigam et al., 1996). The efficiency of color application varies depending on the method of dye delivery. Consequently, the wastewater produced by the textile industry is generally colored as a result of unfixed dyes (Shu et al., 2005). Certain classes of dyes like reactive dyes are generally found at relatively high concentrations in wastewater. This is a result of their lower ability of fixation to fibers like cotton and viscose. These wastewater streams are often found to be biorecalcitrant due to the presence of various additional chemicals like fixation agents, bleaching agents and surfactants. Large amounts of salts are added to the dye baths for improving dye fixation. Salt concentrations at high levels of up to 100 g/L are necessary for reactive dyes (Mauersberger et al., 1948). In addition to salts, caustic solution of about 23% NaOH is also added to improve properties like fiber strength, shrinkage resistance, luster, and dye affinity. The escalating problem of synthetic dyes is that due to their recalcitrant nature, they are a major contributor to pollution in the aquatic environment, which has led to the enactment of strict environmental regulations (Hessel et al., 2007).

Dyes are generally synthetic organic compounds. They contain many different functional groups and heavy metals. Therefore, dye-containing effluent is highly toxic to the aquatic organisms, which can inhibit photosynthesis. Ingestion of wastewater containing dyes can lead to several health hazards like severe headache, skin irritation and acute diarrhea (Meier et al., 2000).

The guidelines, recommendations and propositions that have been formulated for ensuring the protection of water resources are many. The costs related to water use and wastewater treatment form a considerable portion of the operational expenses of companies worldwide. In France, for example, a sample of the colored effluent is diluted by a factor of 30 and if it is found that there is no visible coloration after dilution, the effluent is considered to be complying with the norm. The discharge limits for release into the environment are biological oxygen demand (BOD) less than 30 mg/L; chemical oxygen demand (COD) less than 250 mg/L in many countries worldwide; suspended solids (SS) less than 50 mg/L (Hessel et al., 2007). Therefore, protecting the environment from pollutants is matter of great concern worldwide.

Due to high costs of treatment of wastewater discharged at sewage works, the dyehouses are in many cases forced to have their own wastewater treatment facilities. The color and chemical composition of textile effluents are subject to many variations due to daily routines and seasonal effects. Thus, a combination of techniques is considered more suitable to achieve effective treatment of colored wastewater.

## 8.2 CONVENTIONAL METHODS FOR TREATMENT OF WASTEWATER

Hypersaline effluents are often found to be recalcitrant to biological treatment. Biological treatment systems sometimes become poisoned due to the chemicals present in these effluents. Thus, they cannot be sustained for organics with concentrations of 1% or more (Lefebvre et al., 2006). Thus, as a result, a suitable combination of physical and chemical treatment methods becomes necessary to eliminate organic matter as well as the salts from the effluents. Solar evaporation is one of the low-cost techniques, which reduces the volume of the effluent and thereby concentrates the salts and organic content of saline effluent (Moletta et al., 2006). Coagulation–flocculation is also used as a pretreatment of hypersaline effluent to remove their colloidal COD (Ellouze et al., 2003). However, coagulation–flocculation is not efficient for salt removal. The conventional processes that are used to remove color due to dyes from wastewater thus include chemical coagulation, flocculation, flotation (Kabil et al., 1994), sedimentation (Vallero et al., 2005), physical and chemical adsorption (McKay et al., 1984), advanced oxidation processes like ozonation, photocatalytic oxidation (Kitis et al., 1984 and Kuo et al., 1992), Fenton's process, photo-Fenton oxidation, ultraviolet irradiation, and electrochemical oxidation (Mantzavinos et al., 2004). They have been reported in the literature as effective means for treatment of synthetic and actual textile effluent.

The limitations associated with all these conventional processes are also very significant. For example, ozonation is very expensive process, and it must be continuously produced during operation due to its instability (Contreras et al., 2001). Advanced oxidation processes also involve chemicals like hydrogen peroxide, which are expensive. Although advanced oxidation is found to be effective against many organics, sometimes it cannot oxidize certain organics (Bigda, 1995). Biological treatments are also not suitable always. This is found especially when there are intermittent waste flows or the wastes contain some substances, which are toxic to biological growth. Sometimes there are nonbiodegradable impurities in the effluent containing wastewater.

Activated carbon, clay minerals, zeolites, metal oxides, agricultural wastes, biomass and polymeric materials are used as adsorbents in adsorption (Nouri et al., 2002; Oliveira et al., 2004; Wu et al., 2004; Atia et al., 2003). Due to large surface area, porous structure, high surface reactivity, inertness and thermal stability, activated carbon has found worldwide application in water treatment. Activated carbon has been reported to have a utility over a broad pH range. Although it has been applied for removal of many organic compounds, application of activated carbon is also associated with many disadvantages. Activated carbon is expensive. Its regeneration is also an expensive process. Its powdery form is difficult to separate from the effluent. Moreover, adsorption is a slow process and it transfers pollutants from one phase to another instead of eliminating them from the environment. Additionally, the numerous composite adsorbents that have been developed for water purification by adsorption and catalytic oxidation processes have their own advantages and disadvantages (Jiuhui, 2008). The process of chemical precipitation has its limitation that it cannot reduce the contaminant far enough to meet the water-quality standards. Thus, there is a great need to find effective and economical methods for the removal of pollutants. This has resulted in the search for unconventional methods and materials that might be useful for treatment of wastewater (Bambang et al., 2007). Electrochemical water or wastewater technologies require large capital investment and expensive electricity supply. This has made the applications of such technologies limited. The application of supercritical water oxidation for treating industrial wastes has also met with only limited success. Supercritical water oxidation is an oxidation process and it occurs in water above thermodynamic critical point of a mixture, i.e., $T_c$ = 374°C and $P_c$ = 22.1 MPa (Bambang et al., 2007). Under these conditions, water acquires unique properties, which are used to destroy hazardous wastes. Supercritical water is used as a reaction medium. The unique solvating properties provide enhanced solubility to organic pollutants like chlorinated hydrocarbons, which may be otherwise insoluble. Thus, it allows a single-phase reaction of aqueous wastes and provides an environment free of interphase mass transfer limitations, faster reaction kinetics, and an increased selectivity to complete oxidation. The limitations of supercritical water oxidation are, requirement of high operating pressure, possibility of reactor plugging due to salt formation and sometimes a corrosive behavior and other higher processing costs (Bambang et al., 2007; Hodes et al., 2004; Tester et al., 1993). Organics can also be destroyed by incineration. But the process is economically feasible only when the wastewater effluent streams contain relatively large concentrations of organics. Incineration must be carried out at very high temperature as much as 900°C–1100°C, and with 100% to 200% excess air. Thus, incineration is not really an efficient process (Bambang et al., 2007). Advanced oxidation process (AOP) is an oxidation process, which generates hydroxyl radicals in sufficient quantity to effect water treatment. To achieve a high performance in AOP, Fenton's reagent (consisting of a combination of $FeSO_4{\cdot}7H_2O$ and hydrogen peroxide) is used. Ozone and UV radiation enhances the performance of Fenton's reagent and is therefore also sometimes used for AOP. Hydroxyl radicals degrade the organic hazardous dyes to $CO_2$ and $H_2O$. This is because hydroxyl radical is a strong oxidizing agent. In literature, Fenton and Fenton-like reactions are found to be generally efficient for decolorizing and detoxifying textile effluent. Hydroxyl radicals are generated by the reaction between $H_2O_2$ and ferrous ions. The slow regeneration of $Fe^{3+}$ to $Fe^{2+}$ is the rate-determining step of the reaction. The rate of dye degradation is fast in the beginning due to high initial concentration of $Fe^{2+}$. However, subsequently the rate drastically drops due to the fall in the concentration of $Fe^{2+}$ (Kuo et al., 1992; Kusic et al., 2007; Lunar et al., 2000).

## 8.3 MEMBRANE SEPARATION PROCESSES

In view of the above limitations of conventional techniques, membrane based processes can offer attractive alternatives. Membrane materials come in very diverse ranges depending on chemical composition and physical structure. The fundamental importance of membrane separation is the mechanism by which separation is achieved. Membranes are categorized as dense or porous on this basis. Reverse osmosis (RO), electrodialysis (ED) and nanofiltration (NF) are considered to be under the category of dense membranes. Dense membranes effect separation by the physical and chemical interactions between the permeating constituents and the membrane material. The separations achieved by dense membrane are highly selective. On the other hand, porous membranes are considered to cause separation by sieving, which is a mechanical process. Thus, porous membranes are mechanically close to the conventional filtration methods. Ultrafiltration separates colloidal and dissolved macromolecular species. The ability of membranes to reject solute materials is defined by the molecular weight cut-off (MWCO) in Daltons (i.e., the relative molecular weight) of the rejected solute (Bellona et al., 2004).

Membrane processes like reverse osmosis, ultra-low pressure reverse osmosis and nanofiltration have found increasingly widespread applications in water treatment. They are also utilized for the purpose of reclaiming water or recycling water and other valuable components, and thereby achieving higher process efficiency (Fusaoka et al., 2001; Mohammad et al., 2002; Drewes et al., 2003). Membrane processes often achieve high and efficient removals of constituents such as dissolved solids, organic carbon, inorganic ions, salts and organic compounds (Braghetta et al., 1997).

NF and UF membranes are composed of a thin membrane skin that acts as the porous strainer. They contain a thicker layer below the thin membrane, which gives mechanical support to the membrane to withstand high pressure required during these processes. The membrane skin usually carries a negative charge, which minimizes the adsorption of negatively charged solutes present in membrane feedwater. Therefore, it reduces membrane fouling and increases the rejection of dissolved salts (Braghetta et al., 1997). Thus, the membrane technologies have become a viable alternative to the wastewater treatment for the recovery of valuable components as well as reclamation of water. Numerous studies involving application of pressure-driven membrane processes like nanofiltration, ultrafiltration and reverse osmosis in the treatment of process wastewater generated in various steps of textile dyeing and finishing processes have been carried out and are reported in literature (Treffry et al., 1983; Porter et al., 1984). Nanofiltration (NF) is less energy intensive compared to reverse osmosis. Therefore, NF has been widely used for the treatment of industrial effluents (Yu et al., 2001). NF has the ability to retain dyes and other lower-molecular-weight organic compounds (MWCO: 200 to 1000). In NF, salt (as NaCl) can easily pass through the membrane skin. The salt rich permeate thus generated from NF can be reused in the preparation of dye baths used in textile industries, and thus, it minimizes the process water requirement. However, the major operational problem encountered in membrane separation is caused due to membrane fouling. This leads to flux decline as a result of solute particles accumulating over the membrane surface due to transmembrane pressure drop. This is known as concentration polarization. This offers resistance against the solvent flux. The membrane fouling also depends on the concentration of the organics in the feedwaters. Higher concentration of organics increases the membrane fouling in general. Majority of flux decline due to concentration polarization is recovered after membrane cleaning. Therefore, concentration polarization is mostly reversible in nature. However, sometimes it is not possible to remove the adsorbed solute from membrane. This type of fouling is known as irreversible fouling.

## 8.4 OBJECTIVE OF THE STUDY

The overall objective of the work presented here is to remove the dyes and effectively recover saline water from a textile effluent containing two dyes Cibacron Red (CR) and Cibacron Black (CB) by combination of AOP and NF. The present study combined these two operations, with an aim to take advantages of both the processes. The study consisted of two schemes: Scheme 1 is a combination of AOP as the first operation and is followed by NF. Scheme 2 is a two-step application of NF. Hydrodynamic studies using turbulent promoters are also carried out to enhance the permeate flux.

## 8.5 EXPERIMENTAL

The textile effluent used in the present study was collected from a textile dye house, Singhal Brothers, located in Kolkata, India. The effluent contained a mixture of two reactive dyes: reactive red (Cibacron Red, molecular weight 855.5 g/gmol) with a concentration of 78 mg/L and reactive black (Cibacron Black, molecular weight 924.5 g/gmol) with a concentration of 165 mg/L. The pH of the effluent was 12.4, salt content (as NaCl) was 78.6 g/L, COD was 4840 mg/L and total solids content (TS) was 85 g/L. In subsequent sections, the two dyes (reactive red and reactive black) are denoted by CR and CB, respectively.

### 8.5.1 Membrane

A composite membrane (MWCO = 400) consisting of polyamide skin over polysulfone support supplied by Genesis Membrane Sepratech Pvt. Ltd., Mumbai was used in the Nanofiltration (NF) experiments. The permeability of the membrane was found to be $2.1 \times 10^{-11}$ $m^2 s/kg$.

Figure 8.1 is a schematic diagram of the NF cell. The experimental setup in NF consisted of a rectangular cross-flow cell made of stainless steel. The cell comprised of a top and bottom flange and the bottom flange was grooved to provide permeate flow. A porous stainless steel plate placed over the lower plate provided mechanical support to the membrane. Two neoprene rubber gaskets were placed above the membrane. The length of the flow channel was 25.8 cm, and the width was 4.8 cm. The height of the flow channel was 0.76 mm. There was an effluent storage tank and feed was pumped into the membrane channel with the help of a reciprocating pump. The retentate stream was passed back to the feed channel. The flow rate of the effluent through the membrane cell was measured with the help of a rotameter. Nanofiltration with turbulent promoters was carried out by using 16 equispaced thin copper wires, of diameter 19 mm that was placed between the two gaskets. The spacing between the promoters was about 14 mm.

AOP study was carried out in batches of 500 ml in absence of direct sunlight at 30°C with stirring speed of 50 rpm. The duration of each experiment was 3 h. The concentration of the dyes in the effluent was measured every 30 min. The reactions were carried out at atmospheric pressure.

A UV spectrophotometer (make: Genesis 2, Thermospectronic, USA) was used to measure the absorbance of dye. The absorbance value was calibrated with known concentrations of the dye. Spectrophotometric analysis during AOP was extensively used in the literature [27]. The maximum absorbance value of CR was at 523 nm and that for CB was 598 nm.

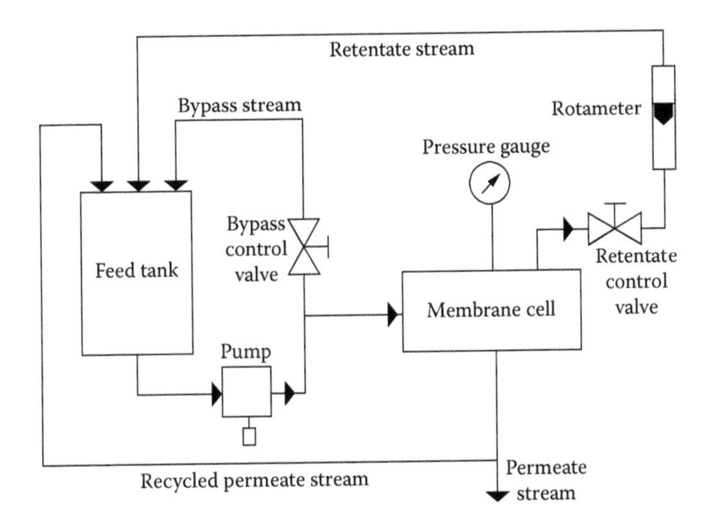


**FIGURE 8.1** Schematic diagram of cross-flow NF membrane module.

## 8.6 RESULTS AND DISCUSSION

### 8.6.1 Scheme 1: AOP Followed by NF

The first part of the study is pretreatment with AOP as the first step followed by NF. It is observed that dye degradation for both dyes is substantial in the first 30 min and gradual thereafter. Initially due to the presence of high concentration of $Fe^{2+}$, the production of hydroxyl radicals is higher. But during the process, $Fe^{2+}$ is converted to $Fe^{3+}$. The regeneration to $Fe^{2+}$ is a slow process and hence becomes the rate-determining step after the fast initial phase of dye degradation. The Cibacron Red concentration after 30 min is degraded by 61% for $FeSO_4{\cdot}7H_2O$ concentration of 417 and 3240 mg/L of $H_2O_2$. Beyond 30 min, the degradation of dyes becomes extremely slow.

By increasing the $H_2O_2$ concentration, moderate increase in the rate of dye degradation is observed. For example, at 417 mg/L of $FeSO_4{\cdot}7H_2O$, the degradation in Cibacron Red dye concentration after 30 min increased from 61% to 78% when $H_2O_2$ concentration increases to 4320 mg/L from 3240 mg/L. Similarly, the degradation in Cibacron Black dye concentration after 30 min increased from 70% to 86%.

Ferrous sulfate reacts with hydrogen peroxide to generate hydroxyl radicals, which in turn degrades the dye according to the following reactions (Neyens et al., 2001):

$$Fe^{2+} + H_2O_2 \rightarrow Fe^{3+} + OH^* + OH^- \tag{8.1}$$

$$RH + OH^* \rightarrow H_2O + R^* \tag{8.2}$$

From Equation 8.1, it can be observed that the amount of the generation of hydroxyl radicals depends upon the concentration of both ferrous sulfate and hydrogen peroxide. In this study, the molar ratio of ferrous ions to hydrogen peroxide is much smaller than 1.0. Therefore, increase in the concentration of ferrous ions has more pronounced effect in generation of hydroxyl radicals compared to the increase in the concentration of hydrogen peroxide. The rate of dye degradation is thus more in case of increase in concentration of $FeSO_4{\cdot}7H_2O$.

During AOP, it is observed that dye concentration of Cibacron Red drops to 15 mg/L, and Cibacron Black falls to 23 mg/L within the first 30 min for a solution having an initial concentration of 556 mg/L of $FeSO_4{\cdot}7H_2O$ at a concentration of 2160 mg/L hydrogen peroxide in this study. After 30 min, the rate of dye decomposition reduces significantly. This composition of Fenton's reagent is thus considered to be the most suitable among all the compositions studied. Therefore, the composition of Fenton's reagent selected for the rest of the experiments is 556 mg/L $FeSO_4{\cdot}7H_2O$ and 2160 mg/L $H_2O_2$, and the duration of AOP is selected as 30 min. Subsequently, it is observed that the effluent after AOP treatment contains a large quantity of sludge. Thus, the supernatant of AOP-treated effluent is microfiltered and subjected to NF.

Thus, the feed to NF had the following characteristics: CB concentration is 23 mg/L, CR concentration is 15 mg/L, pH was 8.5, conductivity is 11.8 S/m, total solids is 78 g/L, COD is 151 mg/L and NaCl is 76.8 g/L. Durations of NF experiments are about 0.5 h. NF is carried out for laminar flow regimes. The steady state permeate flux values for all the operating cross-flow velocities studied are shown in Figure 8.2. The water flux using distilled water is also shown.

It is observed from Figure 8.2 that the permeate flux increases with pressure almost linearly in the range of operating pressure studied herein. This is due to the increased driving force. At a particular pressure, the flux values vary in a narrow range for various cross-flow velocities. For example, at 689 kPa pressure, flux varies between about $7.7 \times 10^{-6}$ to $9.7 \times 10^{-6}$ $m^3/m^2s$ and at 828 kPa, this range is about $1.06 \times 10^{-5}$ $m^3/m^2s$ to $1.11 \times 10^{-5}$ $m^3/m^2s$. It may be observed from the figure that pure water flux is about 1.8 times the permeate flux at highest cross-flow velocity (0.91 m/s), at 552 kPa and about 1.7 times at 828 kPa. This decline in flux (when compared to pure water flux) is due to a phenomenon known as concentration polarization. In the membrane flow channel, solute particles

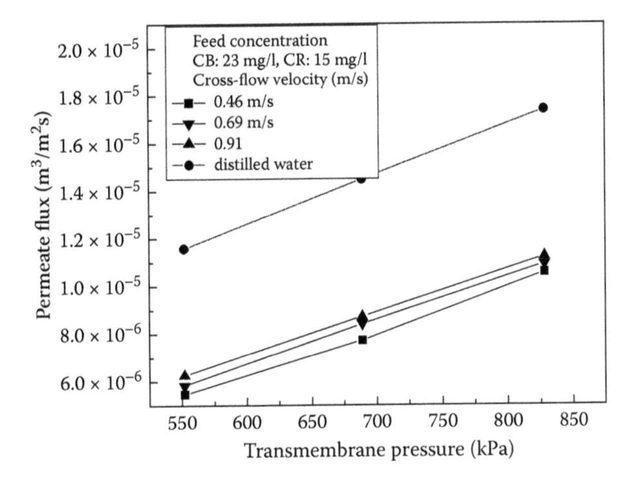


**FIGURE 8.2** Permeate flux variation in NF of scheme 1.

accumulate over the membrane interface, causing a decrease in flux due to reduction in driving force. However, as evident from this figure, permeate flux increases (although marginally) with cross-flow velocity. This is due to reduction in concentration polarization by the sweeping action of the cross-flow over the membrane surface.

Variation of observed retention of the dyes with the operating conditions after NF in scheme 1 is shown in Figure 8.3. Concentration of both dyes is almost zero (observed dye retention = 1.0) up to 689 kPa pressures for all cross-flow velocities except 0.46 m/s. Even at 828 kPa, permeate concentration of both the dyes are well below 1 mg/L (when $R_{0,1} > 0.96$ and $R_{0,2} > 0.93$) at cross-flow velocities above 0.69 m/s. But at higher pressure (e.g., at 828 kPa) and lower cross-flow velocity, i.e., 0.46 m/s, permeate concentration of both dyes is above 1.0 mg/L (i.e., when $R_{0,1} < 0.96$ and $R_{0,2} < 0.93$).

It may be noted here that higher pressure and lower cross-flow velocity induce increased concentration polarization. At higher pressure, more solutes are convected to the membrane surface, and the sweeping action of the cross-flow becomes less at lower cross-flow velocity, leading to severe concentration polarization. Under this condition, more solutes diffuse through the membrane and thereby, increasing the dye concentration in permeate. Therefore, although the effect of cross-flow velocity is marginal in the permeate flux (refer to Figure 8.2), it has a significant effect on the

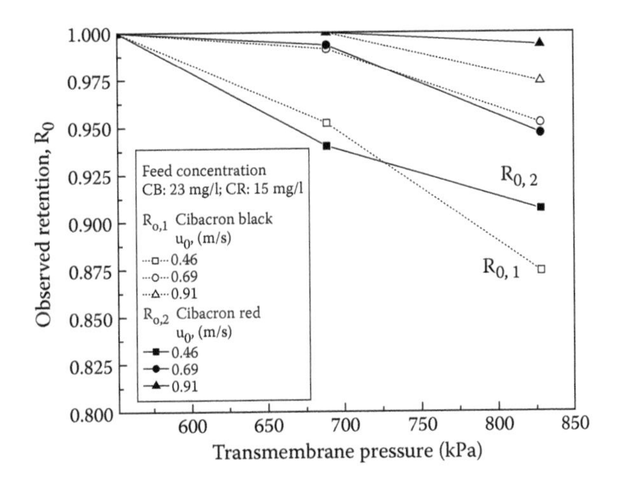


**FIGURE 8.3** Observed retention of the dyes in NF of scheme 1.

**TABLE 8.1**
**Characterization of Effluent at the End of NF in Scheme 1**

| S. No. | Pressure × $10^5$ (Pa) | Cross-Flow Velocity (m/s) | Conductivity (S/m) | Fraction NaCl Recovered | TDS (g/L) | COD (mg/L) |
|---|---|---|---|---|---|---|
| 1 | 5.52 | 0.46 | 11.61 | 0.975 | 75.7 | 62.0 |
| 2 | 5.52 | 0.69 | 11.51 | 0.962 | 75.0 | 61.0 |
| 3 | 5.52 | 0.91 | 11.38 | 0.947 | 74.2 | 60.0 |
| 4 | 6.89 | 0.46 | 11.69 | 0.985 | 76.2 | 66.0 |
| 5 | 6.89 | 0.69 | 11.61 | 0.975 | 75.7 | 63.2 |
| 6 | 6.89 | 0.91 | 11.51 | 0.963 | 75.0 | 63.0 |
| 7 | 8.28 | 0.46 | 11.75 | 0.992 | 76.6 | 69.0 |
| 8 | 8.28 | 0.69 | 11.69 | 0.985 | 76.2 | 67.0 |
| 9 | 8.28 | 0.91 | 11.60 | 0.974 | 75.6 | 64.5 |

permeate concentration. Hence, from the range of operating conditions studied here, 828 kPa pressure and 0.91 m/s cross-flow velocity is selected as suitable for this step of nanofiltration.

Characterization of the effluent coming out of the NF step in scheme 1 is presented in Table 8.1, as function of operating conditions.

It is observed from the table that the range of various properties of the effluent is as follows: COD from 60 to 69 mg/L, sodium chloride recovered from 95% to 99.2%. pH of the original effluent is about 12.4, primarily due to presence of NaOH. However, after AOP in step 1 of scheme 1, pH reduces to around 8.5. This is because of the acidic nature of Fenton's reagent. Original effluent has a COD value of 4840 mg/L and the COD of the final treated effluent is about 62 mg/L. Also, on an average, about 97% NaCl is recovered in the final treated effluent. Therefore, this stream can be recycled back to the upstream unit as a source of makeup water after addition of NaOH only. A small quantity of make-up NaCl is required as almost 97% NaCl is recovered.

### 8.6.2 Scheme 2: Two-Step NF

The second part of the study is a two-step NF, under laminar flow regime. The steady state values of permeate flux for all operating conditions after NF in step 1 are shown in Figure 8.4. This figure

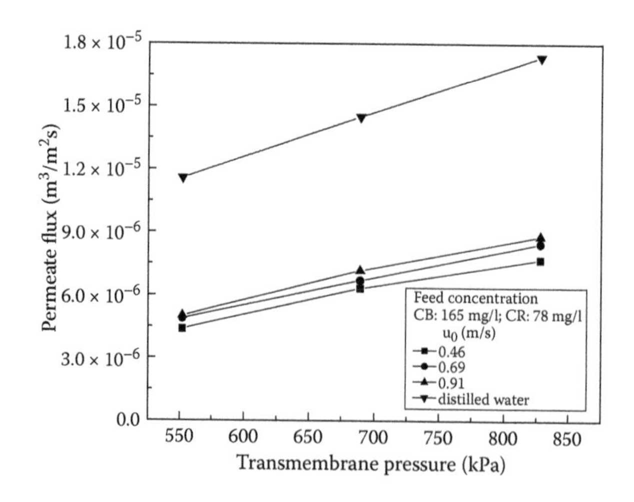


**FIGURE 8.4** Permeate flux variation in first step NF of scheme 2.

shows the usual trend that flux increases almost linearly with pressure and increases marginally with cross-flow velocity (in the range of operating conditions studied herein). The reason for this behavior is already discussed.

The noticeable observation from this figure is that the flux values are quite lower in this case, compared to Figure 8.2. For example, at 689 kPa pressure, range of permeate flux is from about $6.3 \times 10^{-6}$ to $7.3 \times 10^{-6}$ $m^3/m^2s$ (at cross-flow velocities of 0.46 and 0.91 m/s respectively) and that at 828 kPa, is from $7.22 \times 10^{-6}$ to $8.33 \times 10^{-6}$ $m^3/m^2s$. This is due to the effect of feed concentration. In Figure 8.2, the feed concentration of dyes to NF is already lowered to 23 mg/L for CB and 16 mg/L for CR after degradation by AOP. On the other hand, in Figure 8.4, the original dye effluent was subjected to first stage NF with dye concentration as 78 mg/L for CR and 165 mg/L of CB. Higher concentration in feed favors concentration polarization. As feed concentration is more, more solutes accumulate on membrane surface, leading to an increase in osmotic pressure. This decreases the effective driving force, resulting in lower permeate flux. Therefore, flux decline with respect to pure water flux is more in this case.

Variations of permeate concentration of dyes with operating condition after first stage NF in scheme 2 is shown in Figure 8.5. It was observed that CB concentration in permeate varies from 5 mg/L ($R_{0,1} = 0.96$) to 13.1 mg/L ($R_{0,1} = 0.92$) and that of CR varies between 6 mg/L ($R_{0,1} = 0.92$) to 18 mg/L ($R_{0,2} = 0.78$). As feed concentration of dyes is more, concentration polarization is more severe as discussed earlier. Hence, diffusional flux of the dyes through the membrane leads more to enhanced permeate concentration. It can be observed from Figure 8.5 that the observed retention of CB varied between 0.92 to 0.98, and that of CR varied between, 0.78 to 0.93.

There are also two trends, which are obvious here. Firstly, permeate concentration increases (and observed retention decreases) with pressure significantly up to 689 kPa and gradually thereafter. Convective flux of solutes increases with pressure leading to deposition of more dyes on the membrane surface. This increases diffusional flux of solutes through the membrane, and thereby, the permeate concentration increases. With further increase in the pressure, backward diffusion of solutes from membrane surface to the bulk also increases due to concentration gradient and hence rate of diffusion of dyes through the membrane slows down. Hence, concentration of the dyes in the permeate increases but at slower rate with pressure. Secondly, at a fixed pressure, permeate dye concentration decreases with cross-flow velocity (i.e., observed retention increases). Membrane surface concentration of dyes becomes lower due to more sweeping action at higher cross-flow velocity

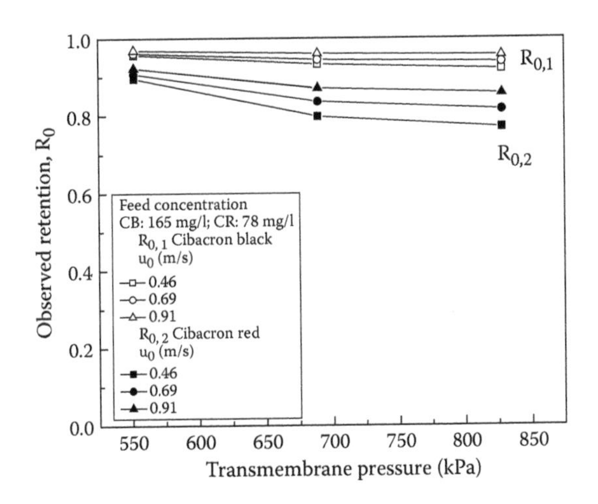


**FIGURE 8.5** Observed retention of the dyes in first step NF of scheme 2.

**TABLE 8.2**
**Characterization of Effluent at the End of Step 1, Scheme 2**

| S. No. | Pressure × $10^5$ (Pa) | Cross-Flow Velocity (m/s) | Conductivity (S/m) | Fraction NaCl Recovered | TDS (g/L) | COD (mg/L) |
|---|---|---|---|---|---|---|
| 1 | 5.52 | 0.46 | 11.4 | 0.97 | 74.9 | 94.0 |
| 2 | 5.52 | 0.69 | 11.3 | 0.95 | 73.4 | 88.0 |
| 3 | 5.52 | 0.91 | 11.1 | 0.948 | 72.9 | 83.0 |
| 4 | 6.89 | 0.46 | 11.4 | 0.98 | 75.4 | 113.0 |
| 5 | 6.89 | 0.69 | 11.32 | 0.969 | 74.6 | 104.0 |
| 6 | 6.89 | 0.91 | 11.28 | 0.963 | 74.2 | 94.0 |
| 7 | 8.28 | 0.46 | 11.54 | 0.987 | 76.1 | 121.0 |
| 8 | 8.28 | 0.69 | 11.45 | 0.98 | 74.4 | 108.0 |
| 9 | 8.28 | 0.91 | 11.39 | 0.970 | 74.7 | 97.0 |

and hence diffusional flux of the dyes through the membrane is less. Hence, permeate concentration of dyes is less (observed retention is more) at higher cross-flow velocity. The characteristics of the effluent after the first stage NF is given in Table 8.2.

As observed from Figure 8.5, the concentration of both dyes are much above 1 mg/L ($R_{0,1} < 0.99$ and $R_{0,2} < 0.987$) in the first step NF of scheme 2, the effluent generated from first step of NF is subjected to one more stage of NF so that permeate dye concentration in the final effluent becomes further lowered. Therefore, about 10 L of the permeate from first stage of NF is collected at 828 kPa pressure and 0.91 m/s cross-flow velocity with concentration of dyes about 7 mg/L ($R_{0,1} = 0.957$) of CB and 11 mg/L ( $R_{0,2} = 0.86$) of CR. This effluent is subjected to second stage NF. Variation of permeate flux of second stage NF with operating conditions is shown in Figure 8.6. It is observed from Figure 8.6, that flux values are now higher than those of first stage NF. This is because feed concentration of second stage NF is much less compared to that of first stage NF, and thus the extent of concentration polarization is less in the second stage. For example, at 689 kPa pressure, flux varies in the range of $7.8 \times 10^{-6}$ to $9.1 \times 10^{-6}$ $m^3/m^2s$, and at 828 kPa, it is from $1.03 \times 10^{-5}$ to $1.17 \times 10^{-5}$ $m^3/m^2s$.

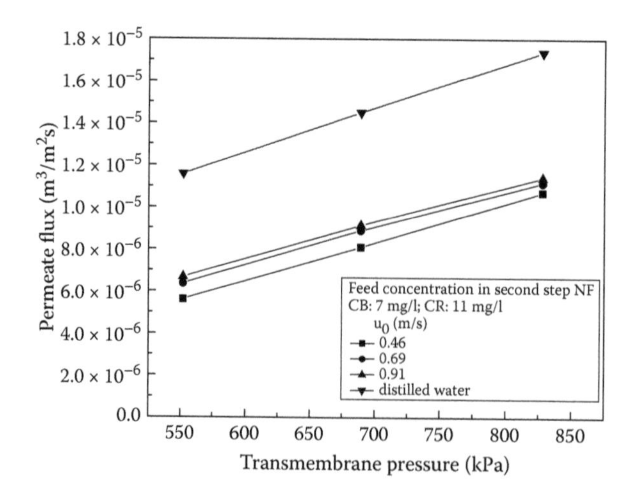


**FIGURE 8.6** Permeate flux variation in second step NF of scheme 2.

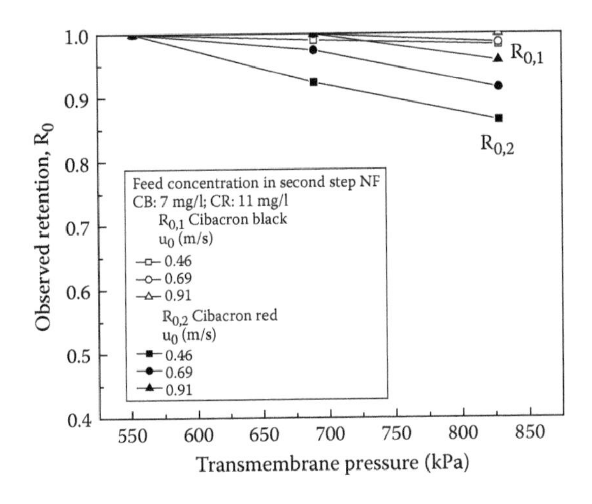


**FIGURE 8.7** Observed retention of the dyes in second step NF of scheme 2.

Variation of observed retention of the dyes in the second stage NF with operating conditions is shown in Figure 8.7.

It is found that for operating pressure above 689 kPa and cross-flow velocity of 0.46 m/s, CR concentration in the permeate exceeds 1.0 mg/L ($R_{0,1} < 0.85$). For 0.91 m/s cross-flow velocity, at all the operating pressure, concentration of both dyes in permeate is virtually zero ($R_{0,1} = 1.0$; $R_{0,2} = 1.0$). Hence 828 kPa and 0.91 m/s cross-flow velocity may be selected as operating condition for second stage NF.

Characteristics of permeate out of second stage of NF in scheme 2 is presented in Table 8.3. Since, in scheme 2, no AOP is involved, almost all NaOH is recovered in permeate stream. Recovery of NaCl is also in the range of 92% to 94% with respect to the original effluent. COD of the treated effluent is in the range of 64 to 71 mg/L, which is much lower compared to the actual feed (4840 mg/L).

Therefore, permeate from second stage NF can be recycled completely (even without make up NaOH as compared to scheme 1) resulting into almost no make-up chemicals, thus lowering the operating cost of the process. Dyes recovered from the retentate can also be recycled.

**TABLE 8.3**
**Characterization of Effluent at the End of Step 2 NF, Scheme 2**

| S. No. | Pressure × $10^5$ (Pa) | Cross-Flow Velocity (m/s) | Conductivity (S/m) | Fraction NaCl Recovered | TDS (g/L) | COD (mg/L) |
|---|---|---|---|---|---|---|
| 1 | 5.52 | 0.46 | 11.21 | 0.959 | 71.4 | 64.0 |
| 2 | 5.52 | 0.69 | 11.12 | 0.948 | 70.8 | 64.0 |
| 3 | 5.52 | 0.91 | 11.03 | 0.937 | 69.9 | 63.0 |
| 4 | 6.89 | 0.46 | 11.36 | 0.977 | 72.9 | 67.0 |
| 5 | 6.89 | 0.69 | 11.29 | 0.969 | 72.3 | 64.5 |
| 6 | 6.89 | 0.91 | 11.25 | 0.964 | 72.0 | 63.5 |
| 7 | 8.28 | 0.46 | 11.42 | 0.985 | 73.5 | 71.5 |
| 8 | 8.28 | 0.69 | 11.36 | 0.977 | 72.9 | 66.0 |
| 9 | 8.28 | 0.91 | 11.30 | 0.970 | 72.4 | 64.5 |

### 8.6.3 Hydrodynamic Studies

The application of higher cross-flow velocities results in further lowering of surface concentration of solutes. This leads to significant increases in the permeate flux and also the reduction of dye concentrations in permeate.

In the transition flow regime, NF was conducted at two cross-flow velocities, viz., 1.83 and 2.28 m/s, corresponding to Reynolds numbers 2781 and 3465, respectively. The permeate flux obtained was found to be significantly higher compared to the laminar flow regime. For example in scheme 2, step 1 NF, at 550 kPa pressure, permeate flux is $6.05 \times 10^{-6}$ $m^3/m^2s$ at a cross-flow velocity of 2.28 m/s, whereas at 1.83 m/s (laminar flow regime), it is only $5.04 \times 10^{-6}$ $m^3/m^2s$. At 828 kPa, the fluxes at the same velocities are $1.08 \times 10^{-5}$ $m^3/m^2s$ and $8.3 \times 10^{-6}$ $m^3/m^2s$, respectively.

Similarly, in the step 2 NF, where the concentration of dyes were much lower, the permeate flux in transition regime is found to be higher than that at laminar flow regime. Thus, at a velocity of 2.28 m/s and a pressure of 550 kPa, the flux is $7.22 \times 10^{-6}$ $m^3/m^2s$, which is higher than $6.8 \times 10^{-6}$ $m^3/m^2s$, the value of flux at 1.83 m/s. Thus, the fluxes in the transition flow regime are much higher than that in the laminar flow regime. For example, at 828 kPa, and for a velocity of 2.28 m/s, the flux is 23% higher than that at a velocity of 0.91 m/s (Reynolds number = 1383) in scheme 1, first step NF.

Similarly, dye concentrations in the permeate were significantly less in the transition flow regime. For example, in scheme 2, first step NF, at 828 kPa and cross-flow velocity of 2.28 m/s, the CB and CR concentrations are 5.8 and 8.0 mg/L, respectively, which is much smaller than that at 828 kPa and velocity of 0.91 m/s, where CB and CR concentrations are 7.2 and 11.2 mg/L respectively. The salt recovery remained significantly high and almost close to that in the laminar flow regime.

In the turbulent flow regime, the NF was conducted at three cross-flow velocities, viz., 2.74, 3.2, and 3.65 m/s. Corresponding Reynolds numbers were 4165, 4864 and 5548, respectively. The permeate flux was significantly higher than even the transition regime. For example in the step 1 NF of scheme 2, at 828 kPa pressure and velocity of 3.65 m/s, the permeate flux is $1.26 \times 10^{-5}$ $m^3/m^2s$, which is higher than the flux at the velocity of 2.28 m/s, which is found to be $1.18 \times 10^{-5}$ $m^3/m^2s$ (transition flow regime). Similar trends are also observed in the Step 2 NF of scheme 2 as well as step 2 NF of scheme 1. Thus, the fluxes in the turbulent flow regime are much higher than that in the laminar flow regime. For example, at 828 kPa, and a velocity of 3.65 m/s the flux is 37% higher than at cross-flow velocity of 0.91 m/s (Reynolds number = 1383), in step 1 NF in scheme 2.

Similarly, the dye concentrations in permeate are found to be less than in the turbulent regime than in the transition regime. For example, at 828 kPa and velocity of 3.65 m/s, the CB and CR concentrations were 4.1 and 6.4 mg/L, respectively, which was found to be less than that of the laminar and transition regime as reported earlier.

### 8.6.4 Turbulent Promoters

Utilization of turbulent promoters in the laminar flow regime results in enhanced permeate flux. Thus, for different flow conditions the increase in the permeate flux is between 4.4% and 22.7% as compared to the laminar flow regime in first step NF of scheme 2. The permeate concentration of the dyes also reduced significantly by about 6% to 34% with respect to laminar regime CR and by about 3% to 20% for CB. The permeate flux is enhanced due to more turbulence created by the promoters.

## 8.7 CONCLUSIONS

Two schemes are tested for the treatment of textile plant containing two dyes (CR and CB). In scheme 1, AOP is used followed by NF. In scheme 2, two stage NF is employed.

In scheme 1, composition of Fenton's reagent selected is 556 mg/L ferrous sulfate and 2160 mg/L of hydrogen peroxide as the AOP step. In AOP step at the end of 30 min, CR concentration reduces to 15 mg/L and CB concentration reduces to 23 mg/L. And after AOP, NF is applied in step 2 to the AOP-treated effluent. It is then found that dye concentrations of both CR and CB, further fall below 1 mg/L in permeate stream at pressures of 552 kPa and 689 kPa, at all cross-flow velocities studied in this work.

In scheme 2, 828 kPa pressure and 0.91 m/s cross-flow velocity are selected in the first step NF. In the second step of this scheme, both dyes are removed completely at all operating pressures when the cross-flow velocity is 0.91 m/s. As far as removal of dyes was concerned, both schemes showed comparable performance. But, in case of recovery of chemicals (sodium chloride and sodium hydroxide), performance of scheme 2 is found better than scheme 1. In scheme 1, some amount of NaOH is neutralized due to the acidic nature of Fenton's reagent and final treated salt-rich stream had a pH value of 8.6. In order to recycle this stream, make up sodium hydroxide should be added. On the other hand, in scheme 2, final permeate had a pH about 12.3, close to the feed pH, indicating almost complete recovery of sodium hydroxide and this stream could be recycled thus lowering the operating cost. Recovery of sodium chloride was also almost same in both the schemes. Use of turbulent promoters resulted in flux enhancement of 4.2 to 10% in scheme 1. Similarly, it was observed that there was 5 to 22.7% increment in flux in step 1 (NF) of scheme 2. Moreover, a significant reduction in dye concentration was also observed in both the schemes.

## REFERENCES


Atia, A.A., A.M. Donia, S.A. Abou-El-Enein, and A.M. Yousif. 2003. Studies on uptake behavior of copper(II) and lead(II) by amine chelating resins with different textural properties. *Sep. Purif. Technol.* 33: 295–301.

Bambang, V., and K. Jae-Duck. 2007. Supercritical water oxidation for the destruction of toxic organic wastewaters: A review. *J. Environ. Sci.* 19: 513–522.

Bellona, C., J.E. Drewes, P. Xu, and G. Amy. 2004. Factors affecting the rejection of organic solutes during NF/RO treatment—a literature review. *Water Res.* 38: 2795–2809.

Bigda, R. 1995. Consider Fenton's chemistry for wastewater treatment. *Chem. Eng. Program.* 91: 62–66.

Braghetta, A., F.A. Digiano, and W.P. Ball. 1997. Nanofiltration of natural organic matter: pH and ionic strength effects. *J. Environ. Eng.* 123(7): 628–40.

Contreras, S., M. Rodríguez, E. Chamarro, S. Esplugas, and J. Casado. 2001. Oxidation of nitrobenzene by $O_3$/UV: The influence of $H_2O_2$ and Fe(III). Experiences in a pilot plant. *Water Sci. Technol.* 44: 39–46.

Desphande, S.D. 2001. Ecofriendly dyeing of synthetic fibres. *Indian J. Fibre Textile Res.* 26: 136–142.

Drewes, J., M. Reinhard, and P. Fox. 2003. Comparing microfiltration-reverse osmosis and soil-aquifer treatment for indirect potable reuse. *Water Res.* 37: 3612–3621.

Ellouze, E., R.B. Amar, and B.A.B. Salah. 2003. Coagulation–flocculation performances for cuttlefish effluents treatment. *Environ. Technol.* 24 (11): 1357–1366.

Fusaoka, Y., T. Inoue, M. Murakami, and M. Kurihara. 2001. Drinking water production using cationic and anionic charged nanofiltration membranes. *Proceedings of the AWWA Membrane Technology Conference*, San Antonio, TX.

Hessel, C., C. Allegre, M. Maisseu, F. Charbit, and P. Moulin. 2007. Review: Guidelines and legislation for dye house effluents. *J. Environ. Manag.* 83: 171–180.

Hodes, M., P.A. Marrone, and G.T. Hong. 2004. Salt precipitation and scale control in supercritical water oxidation: Part A. Fundamental and research. *J. Supercrit. Fluid.* 29: 265–288.

Jiuhui, Q.U. 2008. Research progress of novel adsorption processes in water purification: A review. *J. Environ. Sci.* 20: 1–13.

Kabil, M.A., and S.E. Ghazy. 1994. Separation of some dyes from aqueous solutions by floatation. *Sep. Sci. Technol.* 29: 2533–2539.

Kitis, M., C.D. Adams, and G.T. Daigger. 1999. The effects of Fenton's reagent pretreatment on the biodegradability of nonionic surfactants. *Water Res.* 33: 2561–2568.

Kuo, W.G. 1992. Decolourising dye wastewater with Fenton's reagent. *Water Res.* 26: 881–886.

Kusic, H., A.L. Bozic, and N. Koprivanac. 2007. Fenton type processes for minimization of organic content in coloured wastewaters. Part I. Processes optimization. *Dyes Pigments* 74: 380–387.

Lefebvre, O., and R. Moletta. 2006. Treatment of organic pollution in industrial saline wastewater: A literature review. *Water Res.* 40: 3671–3682.

Lefebvre, O., N. Vasudevan, M. Torrijos, K. Thanasekaran, and R. Moletta. 2006. Anaerobic digestion of tannery soak liquor with an aerobic post-treatment. *Water Res.* 40 (7): 1492–1500.
Lunar, L., D. Sicilia, S. Rubio, D. Perez-Bendito, and U. Nickel. 2000. Degradation of photographic developers by Fenton's reagent: Condition optimization and kinetics for metal oxidation. *Water Res.* 34: 1791–1802.
Mantzavinos, D., and E. Psillakis. 2004. Enhancement of biodegradability of industrial wastewaters by chemical oxidation pre-treatment. *J. Chem. Technol. Biotechnol.* 79: 431–454.
Mauersberger, H.R. 1948. *Textile Fibers: Their Physical, Microscopical and Chemical Properties.* New York: John Wiley & Sons Inc.
McKay, G. 1984. Analytical solution using a pore diffusion model for the adsorption of basic dye on silica. *Am. Inst. Chem. Eng. J.* 30: 692–697.
Meier, J., T. Melin, and L.H. Eilers. 2000. Nanofiltration and adsorption on powdered adsorbent as process combination for the treatment of severely contaminated waste water. *Desalination* 146: 361–366.
Mohammad, A., and N. Ali. 2002. Understanding the steric and charge contributions in NF membranes using increasing MWCO polyamide membranes. *Desalination* 147: 205–212.
Neyens, E., and J. Baeyens. 2003. A review of classic Fenton's peroxidation as an advanced oxidation technique. *J. Hazard. Mater.* 98: 33–50.
Nigam, P., I.M. Banat, D. Singh, and R. Marchant. 1996. Microbial process for the decolorization of textile effluent containing azo, diazo and reactive dyes. *Process Biochem.* 31: 435–442.
Nouri, S., F. Haghseresht, and G.Q.M. Lu. 2002. Comparison of adsorption capacity of p-cresol & p-nitrophenol by activated carbon in single and double solute. *Adsorption* 8: 215–223.
Oliveira, L.C.A., D.I. Petkowicz, A. Smaniotto, and S.B.C. Pergher. 2004. Magnetic zeolites: A new adsorbent for removal of metallic contaminants from water. *Water Res.* 38: 3699–3704.
Porter, J., and G. Goodman. 1984. Recovery of hot water, dyes and auxiliary chemicals from textile wastestreams. *Desalination* 49: 185–192.
Shu, L., T.D. Waite, P.J. Bliss, A. Fane, and V. Jegatheesan. 2005. Nanofiltration for the possible reuse of water and recovery of sodium chloride salt from textile effluent. *Desalination* 172: 235–243.
Tester, J.W., H.R. Holgate, and F.J. Armellini. 1993. Supercritical water oxidation technology: A review of process development and fundamental research. In: *Emerging Technologies in Hazardous Waste Management III* (ed., D.W. Tedder and F.G. Pohland.). Washington, DC: American Chemical Society. Chapter 3.
Treffry-Goatley, K., C. Buckley, and G. Groves. 1983. Reverse osmosis treatment and reuse of textile dyehouse effluents. *Desalination* 47: 313–320.
Vallero, M.V.G., G. Lettinga, and P.N.L. Lens. 2005. High rate sulfate reduction in a submerged anaerobic membrane bioreactor (SAMBaR) at high salinity. *J. Membr. Sci.* 253 (1–2): 217–232.
Wu, R.C., J.H. Qu, and H. He. 2004. Removal of azo-dye Acid Red B (ARB) by adsorption and combustion using magnetic $CuFe_2O_4$ powder. *Appl. Catal. B* 48: 49–56.
Yu, S., C. Gao, H. Su, and M. Liu. 2001. Nanofiltration used for desalination and concentration in dye production. *Desalination* 140: 97–100.

# 9 Micellar-Enhanced Ultrafiltration and Its Applications

*Chandan Das and Mihir Kumar Purkait*

## CONTENTS

## 9.1 INTRODUCTION

Application of different classes of surfactants is increasing nowadays in a number of technological areas. The main areas are (1) biotechnology, (2) electronic printing, (3) high technology electronic ceramics, (4) magnetic recording, (5) microelectronics, (6) nonconventional energy production, (7) novel pollution control, and (8) novel separation techniques. Among them, use of surfactant in separation processes is a major area of research in surfactant and separations science. Surfactant-based separations have a number of potential advantages over traditional methods. They often are low-energy intensive processes because large temperature or phase changes are not needed for separations. At times, surfactants offer improved selectivity over a conventional solvent system. An additional advantage of micellar systems is that they are compatible with electrochemical detection while conventional organic solvents are not. Surfactants are often environmentally innocuous and of low toxicity, so the leakage of a small amount of surfactant into an aqueous process stream from the separation may be tolerable, in contrast to toxic solvents from liquid–liquid extraction. Surfactant-based separation of toxic chemicals is often getting attractive nowadays.

The traditional methods for treatment of metal ion–containing waste are mainly adsorption (Verma et al., 2006; Xu et al., 2002), ion exchange (Tenório and Espinosa, 2001), precipitation (Dean et al., 1972), and flocculation (Rout et al., 2006). The most popular method is adsorption, but this is quite a slow process due to mass transfer limitations. These techniques are quite energy intensive. Desired separations to a satisfactory level are often unattainable in conventional separations based on adsorption. A rate-governed separation process, such as reverse osmosis (RO), is more useful and is already recognized as an efficient technique for the separation of several inorganic and organic compounds. Compared to nanofiltration, ultrafiltration, and microfiltration, relatively "dense" membranes are used in the RO process. Permeability of RO membranes is very low, and thus to get the desired throughput (permeate flux), high operating pressure is required. Hence, these processes are energy intensive with high capital and operating costs. Therefore, a modified membrane separation process, micellar-enhanced ultrafiltration (MEUF), can prove to be a competitive alternative wherein the operating pressure requirement is low compared to RO and a membrane of higher permeability can be used (Purkait et al., 2004a, 2004b). In MEUF, small pollutants are bound (for ionic pollutants) or solubilized (for organic pollutants) in large surfactant micelles, which can be separated by ultrafiltration membranes with a larger pore size. Separation of metal ions using MEUF is a field of active research during the last 15 years (Scamehorn and Harwell, 1989; Baek and Yang, 2004; Purkait et al., 2006). Applications of MEUF for removal of various metal cations include $Cd^{2+}$ (Kim et al., 2006), $Ni^{2+}$ (Yurlova et al., 2002), $Cs^{+}$, $Sr^{2+}$, $Mn^{2+}$ (Juang et al., 2003), $Pb^{2+}$ (Gzara, 2000), $Cu^{2+}$ (Liu and Li, 2005), $Al^{3+}$ (Hankins et al., 2005), $Cr^{3+}$ (Aoudia et al., 2003), etc. But most of these studies are for a single-component system. An industrial effluent may contain more than one ionic pollutant. Scant literature data are available for treatment of binary mixture of metal ions (e.g., $Sr^{2+}$ and $Mn^{2+}$, $Sr^{2+}$ and $Zn^{2+}$, etc.) using MEUF (Juang et al., 2003).

Dissolved organic substances (DOS) are also another source of water pollution. Removal of trace amounts of DOS from aqueous stream using MEUF is reported (Dunn et al., 1985, 1987). Several industrial effluents (including coal refining, textiles, dyes, and synfuel processing) contain unacceptable concentrations of DOS as well as multivalent metal ions (e.g., heavy metals). Generally, two-stage separations are required for effective removal of DOS and heavy metals. MEUF has the potential to allow simultaneous removal of DOS and heavy metals and can combine two unit operations into one. Little research work on simultaneous removal of DOS and metal ions is reported (Dunn et al., 1985; Witek et al., 2006). Simultaneous removal of phenol or ortho cresol and zinc or nickel ions using MEUF is reported by Dunn et al. (1985). Witek et al. (2006) reported simultaneous separation of phenolic compounds along with chromium ion. In these works, only the feasibility of the separation of the solutes is reported. However, viability of the MEUF process depends on the permeate flux, which is the throughput of the system. Apart from that, variation of the permeate flux and solute retention with the operating conditions and recovery of surfactant are also important design criteria.

Removal of anionic pollutants like, chromate, nitrate, and permanganate using MEUF is also reported (Baek and Yang, 2004; Purkait et al., 2005; Baek et al., 2003). Few references for treatment of binary ionic mixture using MEUF are available (Baek and Yang, 2004; Purkait et al., 2005; Kim et al., 2006; Juang et al., 2003; Baek et al., 2003; Akita et al., 1999). Again, these are for same kinds of ionic pollutants, i.e., either binary mixture of cations (Kim et al., 2006; Juang et al., 2003; Akita et al., 1999) or anions (Baek and Yang, 2004; Purkait et al., 2006; Baek et al., 2003). Use of MEUF to treat wastewater containing both cationic and anionic pollutants is not available in the literature. A system of mixed micelles containing cationic as well as anionic surfactants is required for this purpose. Micelles of cationic surfactants solubilize the anionic pollutants, and those of anionic surfactants solubilize the cationic pollutants. MEUF using a mixed micellar system is reported in the literature for removal only of single cations (Aoudiaet al., 2003; Rathman and Scamehorn, 1984). However, in all these studies, mixed micelles are generated from an ionic surfactant and a nonionic surfactant. The primary aim of all these works was to reduce the critical micellar concentration on the ionic surfactant by the nonionic surfactant, so that the solubilization capacity of ionic micelles increases. Simultaneous separation of both cations (copper ions) and anions (permanganate ions) using MEUF has been attempted with mixed micelles of sodium dodecyl sulfate (SDS) and cetylpyridinium chloride (CPC) (Das et al., 2008a). The effects of operating conditions, namely, transmembrane pressure drop, cross-flow rate, and composition of the feed (having both pollutants) are studied based on the retention characteristics of each solute and permeate flux of the system. The performance is compared with the single-solute system in the same range of operating conditions.

## 9.2 MEMBRANE

A thin-film composite (TFC) membrane with polyamide skin of molecular weight cutoffs 1 kDa and 5 kDa was used for the MEUF of dye and metal ions, respectively. The membranes were supplied by M/s, Permionics Membranes Pvt. Ltd., Gorwa, Vadodara, India. The membranes were hydrophilic in nature and used without any further treatment. Membrane permeabilities were measured using distilled water and were found to be $3.57 \times 10^{-11}$ m/Pa s and $3.62 \times 10^{-11}$ m/Pa s, respectively.

### 9.2.1 Unstirred Batch Cell

The unstirred batch experiments are conducted in a 50-ml-capacity filtration cell (MILLIPORE, model 8050, USA). Inside the cell, a circular membrane is placed over a base support. The membrane diameter is 4.45 cm, and the effective membrane area is 13.4 cm$^2$. The maximum allowable pressure is 518 kPa. The permeate is collected from the bottom outlet of the cell. The cell is pressurized using nitrogen. The schematic of the experimental setup is shown in Figure 9.1 (Purkait et al., 2004b).

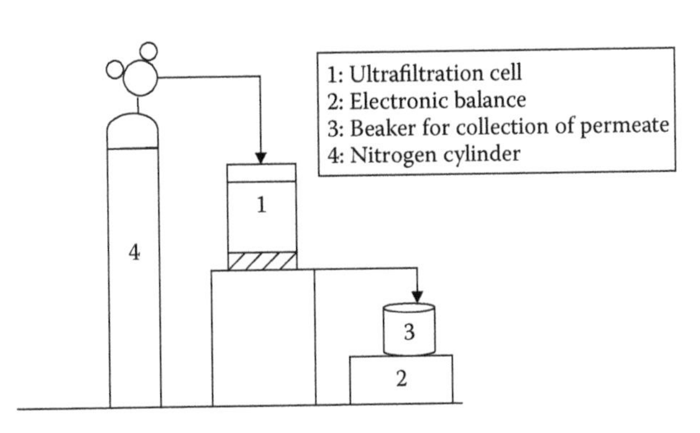


**FIGURE 9.1** Batch experimental setup. (Reprinted from *Sep. Purif. Technol.*, 37(1), Purkait, M.K., et al., 81–92, Copyright 2004, with permission from Elsevier.)

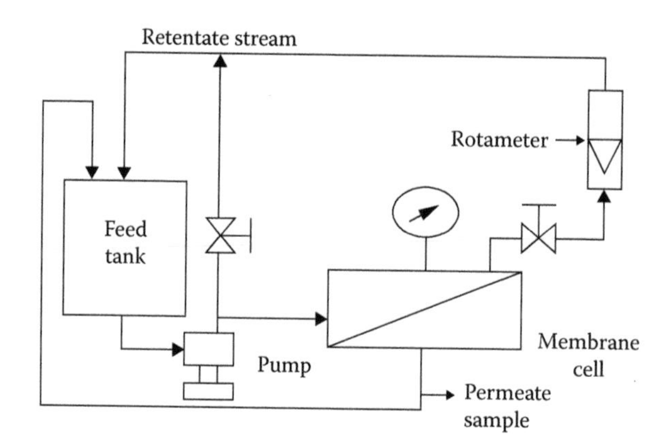


**FIGURE 9.2** Cross-flow setup. (Reprinted from *J. Colloid. Interf. Sci.* 285 (1), Purkait, M.K. et al., 395–402, Copyright 2005, with permission from Elsevier.)

### 9.2.2 Cross-Flow Cell

A rectangular cross-flow cell, made of stainless steel, was designed and fabricated. A schematic of the experimental setup was shown in Figure 9.2. The effective length of the membrane was $37.2 \times 10^{-2}$ m and width was $5.0 \times 10^{-2}$ m. The channel height depended on the thickness of the gasket and after tightening of the flanges was found to be $3.5 \times 10^{-3}$ m. The micellar solutions of different solutes were placed in a stainless steel feed tank of $10 \times 10^{-3}$ m$^3$ capacity. A plunger pump was used to feed the micellar solution into the cell. The retentate stream as well as permeate stream were recycled to maintain a constant concentration in the feed tank. A bypass from the pump delivery to the feed tank was provided. The two valves in the bypass and the retentate lines were used to vary the pressure and the flow rate through the cell, independently (Purkait et al., 2004a).

## 9.3 SELECTION OF FEED SURFACTANT CONCENTRATIONS

To select the suitable feed surfactant concentrations, two sets of experiments were conducted. The first set of experiment was conducted using CPC, and that of for SDS in second set using surfactant solutions more than critical micellar concentration (CMC).

### 9.3.1 Selection of CPC Concentration in Feed

The effects of feed CPC concentration (above CMC) on the permeate flux and permeate concentration are shown in Figure 9.3 at the end of batch experiment and at 345 kPa pressure (Purkait et al., 2006). It is evident from the figure that the permeate flux decreases sharply with CPC concentration. This is because at a higher CPC concentration, the gel type layer of micellar aggregates offers resistance against the solvent flux. From Figure 9.3, it may also be observed that the permeate concentration of CPC remains below its CMC value (0.322 kg/m$^3$) up to a feed concentration of 10.0 kg/m$^3$. The increase of permeate CPC concentration becomes gradual when feed CPC concentration increases from 10.0 to 20.0 kg/m$^3$, but beyond that, the permeate concentration increases sharply. This increase in permeate concentration is due to the concentration polarization over the membrane surface, which increases the convective transport of CPC molecules along with smaller-sized micelles to the permeate side and thereby increases the permeate concentration.

### 9.3.2 Selection of SDS Concentration in Feed

The feed anionic surfactant (SDS) concentration was determined by conducting six experiments using only SDS solution in the concentration range of 5 to 40 kg/m$^3$ (5, 10, 15, 20, 30, and 40 kg/m$^3$)

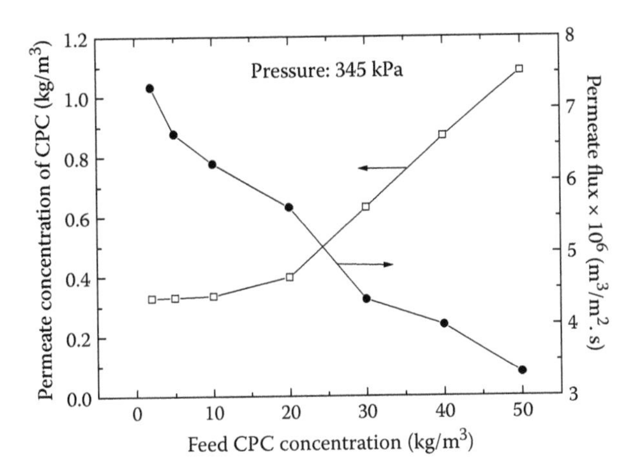


**FIGURE 9.3** Variation of permeate flux and permeate concentration of CPC with feed CPC concentration at the end of experiment and at 345 kPa pressure (>CMC). (Reprinted from *J. Hazard. Mater.* 136(3), Purkait, M.K., et al., 972–977, Copyright 2006, with permission from Elsevier.)

at 276 kPa pressure and cross-flow rate of $1.67 \times 10^{-5}$ m³/s (Das et al., 2008a, 2008b). Variations of permeate flux and permeate concentration for each of the surfactants are shown in Figure 9.4. It is observed from the figure that the permeate flux decreases and permeate concentration increases with surfactant concentration in the feed. During filtration, surfactant micelles form a gel layer over the membrane surface. Thickness of this gel layer increases with surfactant concentration resulting in further decrease in permeate flux. Permeate concentration of surfactant increases with feed concentration. As surfactant concentration increases, permeate concentration may even go beyond the CMC level due to the leakage of smaller-sized micelles. Therefore, feed surfactant concentration is selected such that permeate concentration is around the level of CMC. Based on these criteria and from the results presented in Figure 9.4, feed SDS concentrations are selected as 25 kg/m³ for (CMC = 2.33 kg/m³).

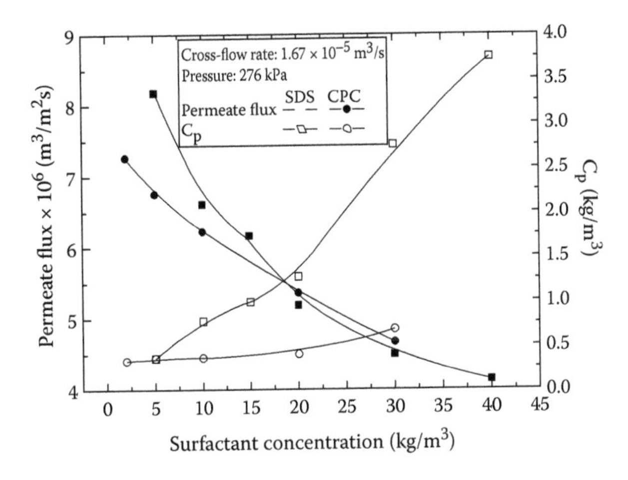


**FIGURE 9.4** Selection of surfactant concentration. (Reprinted from *Chem. Eng. J.*, 144, Das, C., et al., 35–41, Copyright 2008, with permission from Elsevier.)

**TABLE 9.1**
**Experimental Conditions for MEUF**

| Surfactant | Solute | Mode of Operation | Feed Surfactant Conc. (kg/m³) | Feed Solute Conc. (kg/m³) | Pressure Drop (kPa) | Cross-Flow Rate (L/h) |
|---|---|---|---|---|---|---|
| CPC | Eosin | Batch | 5.0, 10.0, 20.0, 25.0 | 0.01 | 276, 345, 414 | — |
| CPC | Eosin | Batch | 10.0 | 0.005, 0.20, 0.30, 0.40 | 276 | — |
| CPC | Eosin | Cross-flow | 10.0 | 0.005, 0.01, 0.02, 0.03, 0.04 | 276 | 30 |
| | | | 10.0 | 0.01 | 345, 414 | 30 |
| | | | 10.0 | 0.02 | 276 | 45, 75 |
| | | | 5.0, 20.0, 25.0 | 0.10 | 276 | 30 |
| SDS | Cu | Cross-flow | 5.0, 10.0 | 4.0 | 276 | 60 |
| | | | 15.0, 20.0 | | | |
| | | | 25.0, 35.0 | 0.05, 1.0, 2.0, 2.5, 3.0, 3.5, 4.0 | 276 | 60 |
| | | | 25 | | | |
| SDS | BN | Cross-flow | 25 | 0.04, 0.09, 0.24, 0.4, 0.5 1.0 | 276 | 60 |

## 9.4 OPERATING CONDITIONS

Filtration experiments are carried out with only solutes in aqueous solution and the mixture of surfactant and different solutes with different concentrations in an unstirred batch cell as well as cross-flow cell. Experiments are designed to observe the effects of the concentrations of surfactants, solutes and the transmembrane pressure drop on the permeate flux and the retention of both the solutes and surfactants. The various operating conditions for batch and cross-flow experiments are shown in Table 9.1.

## 9.5 ANALYSIS

Feed and permeate concentrations of surfactant (CPC) and solutes (eosin and beta naphthol) are measured by a UV spectrophotometer (make: Thermo Spectronic, USA; model: GENESYS 2). Copper present in various samples were estimated by Orion Aplus™ Benchtop Ion Meter (supplied by M/s, Thermo Electron Corporation, Beverly, MA, USA) using ion-specific electrodes. SDS concentration was determined by a two-phase titration according to Epton (1948). The titrant was benzothonium chloride (often called hyamine 1622), a cationic surfactant; the indicator was an acidic mixture of a cationic dye (dimindium bromide) and an anionic dye (disulfine blue VN). The titration was carried out in a water chloroform medium. The pink color of the chloroformic phase was discharged at the end point; the chloroform layer became gray, and with one further drop of the hyamin, the chloroform layer turned to blue. SDS concentration was determined using the following equation:

$$\text{SDS concentration} = \frac{a \times \text{molar concentration of hyamine} \times 288.38}{5 \text{ ml of sample}} \quad (9.1)$$

where $a$ was the volume (ml) of hyamine 1622 required for titration.

## 9.6 MEUF OF EOSIN DYE

This section describes the removal of eosin dye, a toxic chemical, using MEUF. The first part of this section explains the effects of transmembrane pressure drop and different combinations of surfactant and eosin dye concentrations on the permeate flux and retention characteristic of both the dye and surfactant in the batch cell. Variation of flux and retention in the cross-flow cell at different experimental condition are discussed in the second part.

### 9.6.1 Unstirred Batch Cell

#### 9.6.1.1 Variation of Eosin Retention and Permeate Flux during UF of Eosin Solution in Absence and in Presence of Surfactant

The variation of eosin retention with and without surfactant is presented in Figure 9.5 for different feed pressures (276, 345 and 414 kPa) and at dye concentration of $10 \times 10^{-3}$ kg/m$^3$. In both cases, it is observed that the observed retention decreases with time (Purkait et al., 2004b). As the filtration progresses, more solutes will be deposited on the membrane surface leading to an increase in the membrane surface concentration (concentration polarization). This results in an increase in the convective transport of the solutes to the permeate side, thereby increasing the permeate concentration and subsequently, decreasing the observed retention with time. For higher operating pressure, at the same time instant, retention is low. This is due to the fact that at higher operating pressure, the convective transport of the solutes through the membrane is high leading to higher value of the permeate concentration thus lowering the value of observed retention. It can be observed from the figure that for dye solution without surfactant, the observed retention varies from 10% to 26% at the end of the operation under different operating pressures. When surfactant is used, the retention of the dye has been significantly increased, as shown in the figure to about of 68% to 74%, at the end of the experiment. This clearly indicates that the dyes are solubilized within the surfactant micelles, which are subsequently rejected by the ultrafiltration membrane.

The variation of the permeate flux with and without surfactants is shown in Figure 9.6 (Purkait et al., 2004b). Two trends can be observed from this figure. First, in both cases, the permeate flux declines over the time of operation and second, the permeate flux is more at higher operating pressure. As discussed earlier, due to concentration polarization, the membrane surface concentration increases with time. This increases the osmotic pressure at the membrane solution interface

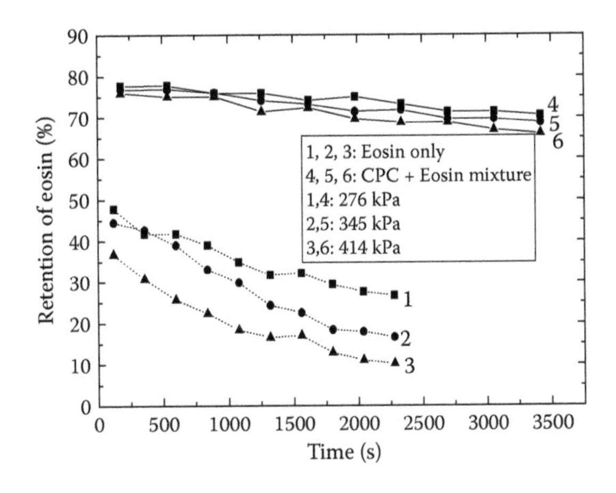


**FIGURE 9.5** Variation of observed retention of eosin with time at different operating pressure differences. Feed eosin concentration is $10 \times 10^{-3}$ kg/m$^3$ and CPC concentration is 10 kg/m$^3$. (Reprinted from *Sep. Purif. Technol.*, 37(1), Purkait, M.K., et al., 81–92, Copyright 2004, with permission from Elsevier.)

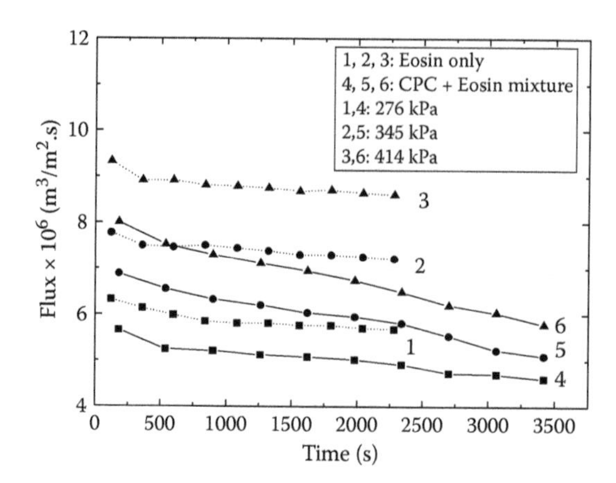


**FIGURE 9.6** Variation of the permeate flux with time at different operating pressure differences. Feed eosin concentration is $10 \times 10^{-3}$ kg/m$^3$ and CPC concentration is 10 kg/m$^3$. (Reprinted from *Sep. Purif. Technol.*, 37(1), Purkait, M.K., et al., 81–92, Copyright 2004, with permission from Elsevier.)

and therefore reduces the driving force for the permeating solution. This leads to a decline of the permeate flux with the time of operation. For example, in the case of filtration of eosin only, the permeate flux decreases from about $9.3 \times 10^{-6}$ m$^3$/m$^2$s (at 180 s) to $8.6 \times 10^{-6}$ m$^3$/m$^2$s (at 2200 s) at 414 kPa. This indicates flux drop of about 7.5% during 40 min of operation. For the case of MEUF of eosin and CPC mixture, the permeate flux drops from $8.0 \times 10^{-6}$ m$^3$/m$^2$s (at 180 s) to $5.8 \times 10^{-6}$ m$^3$/m$^2$s (at 3400 s.) at 414 kPa. Therefore, the flux of the micelle containing mixture drops about 27% during 57 min of operation. At the same instance of time, increase in operating pressure simply increases the driving force across the membrane; thus, flux is more at higher pressure. For example, in case of filtration of only eosin, at the end of the experiment, the permeate flux is $5.7 \times 10^{-6}$ m$^3$/m$^2$s at 276 kPa pressure and it is $8.6 \times 10^{-6}$ m$^3$/m$^2$s at 414 kPa. This indicates about 51% increase in flux when pressure increases from 276 to 414 kPa. On the other hand, in case of MEUF of eosin and CPC mixture, the permeate flux increases from $4.6 \times 10^{-6}$ m$^3$/m$^2$s to $5.8 \times 10^{-6}$ m$^3$/m$^2$s at the end of the experiment while pressure increases from 276 to 414 kPa, indicating about 26% increase in flux.

It is also evident from the figure, that the permeate flux is more for only dye solution compared to that with the surfactants. By addition of surfactants above the critical micellar concentration, the surfactant micelles form aggregates, generating a deposited layer over the membrane surface. This increases the resistance against the solvent flux through the membrane. This results in a decrease in the permeate flux compared to that of the dye solution alone. For example, at 276 kPa pressure, the permeate flux is about $4.6 \times 10^{-6}$ m$^3$/m$^2$s at the end of the experiment for eosin and surfactant mixture, whereas, for eosin only it is about $5.7 \times 10^{-6}$ m$^3$/m$^2$s.

#### 9.6.1.2 Effects of Operating Pressure and Feed CPC Concentration on the Permeate Flux and Observed Retention of Eosin

The effects of the operating pressure and different dosage of feed CPC concentration for a fixed eosin concentration of $10 \times 10^{-3}$ kg/m$^3$ on the permeate flux are presented in Figure 9.7. Figure 9.7 represents the flux data at the end of experiment. The figure shows that the permeate flux decreases with feed CPC concentration at a fixed operating pressure. For example, at 276 kPa pressure flux decreases from $5.8 \times 10^{-6}$ m$^3$/m$^2$s to about $3.5 \times 10^{-6}$ m$^3$/m$^2$s when CPC concentration increases from 5 to 25 kg/m$^3$. This trend may be explained by the increase in resistance against the solvent flux due to micellar aggregates as described in the preceding section. At the constant CPC

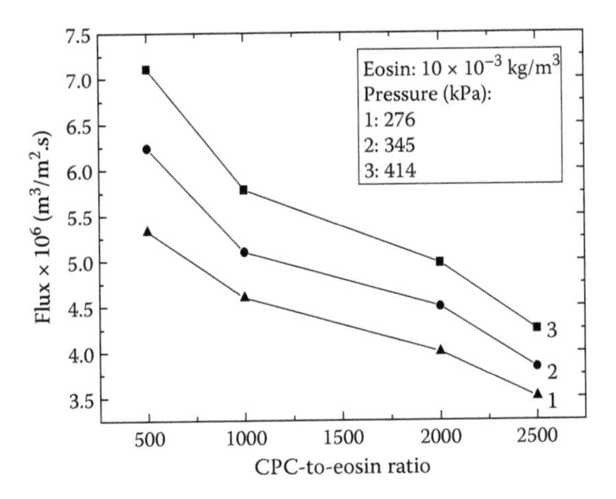


**FIGURE 9.7** Variation of permeate flux with CPC-to-eosin ratio at different pressures at the end of experiment. Feed eosin concentration is $10.0 \times 10^{-3}$ kg/m$^3$. (Reprinted from *Sep. Purif. Technol.*, 37(1), Purkait, M.K., et al., 81–92, Copyright 2004, with permission from Elsevier.)

concentration, flux increases with pressure. For example, at CPC-to-eosin ratio of 2500 (i.e., feed CPC concentration 25 kg/m$^3$), flux increases from $3.5 \times 10^{-6}$ m$^3$/m$^2$s to $4.25 \times 10^{-6}$ m$^3$/m$^2$s while pressure increases from 276 to 414 kPa. This occurs due to increase in the effective driving force (Purkait et al., 2004b).

The effect of operating pressure on the observed retention as a function of CPC-to-eosin concentration ratio at the end of the experimental duration is shown in Figure 9.8 (Purkait et al., 2004b). From the figure, it can be observed that for a fixed pressure, eosin retention increases with CPC concentration. With increase in surfactant concentration, the micelle concentration in the solution increases. This results in more solubilization of eosin in CPC micelles, thereby increasing the retention of eosin. For a fixed CPC-to-eosin ratio, the retention of eosin increases with decrease in operating pressure. This may be due to the fact that at higher operating pressure, micelles may become

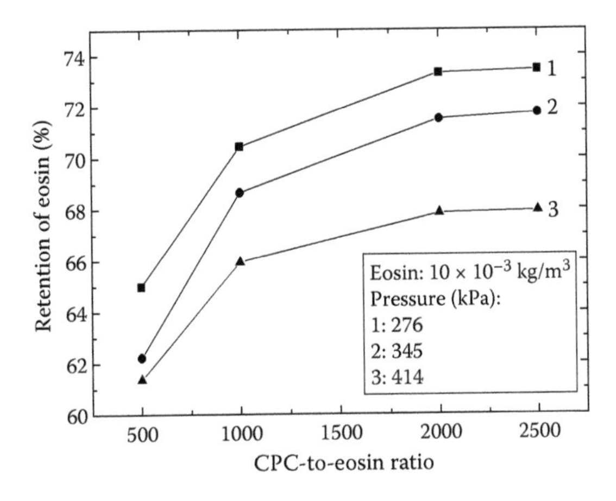


**FIGURE 9.8** Variation of eosin retention with CPC-to-eosin ratio at different pressures after 1 h of experiment. Feed concentrations are eosin: $10.0 \times 10^{-3}$ kg/m$^3$ and CPC: 5, 10, 20, and 25 kg/m$^3$. (Reprinted from *Sep. Purif. Technol.*, 37(1), Purkait, M.K., et al., 81–92, Copyright 2004, with permission from Elsevier.)

compact and therefore solubilization capability of the micelles decreases (Dunn et al., 1987; Syamal et al., 1995). Therefore, less amount of dye is solubilized in the micelles at higher operating pressure and the permeate concentration is more leading to a decrease in observed retention of eosin at higher pressure. It can also be observed from Figure 9.8 that at a fixed pressure, the increase of eosin retention is fast for lower CPC-to-eosin concentration ratio and is gradual as this ratio increases. Beyond a ratio of 2000, the increase in eosin retention is marginal. Therefore, lower operating pressure and a ratio of feed concentration of CPC to eosin of 2000 may be considered as optimum for maximum removal of eosin.

#### 9.6.1.3 Effects of Feed Eosin Concentration on the Observed Retention of Dye at Fixed CPC Concentration

In this case, the surfactant concentration is kept fixed at 10 kg/m$^3$ and eosin concentration is varied as $5 \times 10^{-3}$, $10 \times 10^{-3}$, $20 \times 10^{-3}$, and $30 \times 10^{-3}$ and $40 \times 10^{-3}$ kg/m$^3$. The experiments are conducted at a relatively low pressure of 276 kPa (as lower pressure is a favorable operating condition for higher retention of eosin as shown in Figure 9.9 (Purkait et al., 2004b). In Figure 9.9, the variation of retention of eosin with CPC-to-eosin ratio at the end of 1 h of operation at 276 kPa pressure is presented for both the combinations of surfactant and dye mixture. This figure reveals important information regarding the ratio of CPC to dye to obtain maximum solubilization of the dye in the surfactant micelles. Curve 1 in the figure indicates that the dye concentration is fixed and CPC concentration is gradually increased, whereas, the curve 2 indicates that the CPC concentration is fixed and the ratio is varied by varying the concentration of dyes. Although the CPC-to-eosin ratio is in the same range in both the cases, the retention of dye is more in the first case compared to that in the second case. In the first case, since CPC concentration is increased, concentration of micelles increases and more dye will be solubilized in the micelle. In the later case, the micelle concentration is fixed (as the surfactant concentration is kept constant) and dye concentration is gradually increased. Since the concentration of micelle is constant in this case, the solubilization capacity of dye is also constant. Hence further increase in dye concentration only results in an increase of unsolubilized dye concentration which will be increasing the permeate concentration and hence reduces the observed retention of dyes. In essence, it can be concluded from this figure that in order to determine an optimum surfactant-to-dye ratio to get maximum removal of the dye, the ratio should be changed by varying the surfactant concentration (keeping dye concentration fixed) to calculate the concentration of CPC required for a given dye concentration.

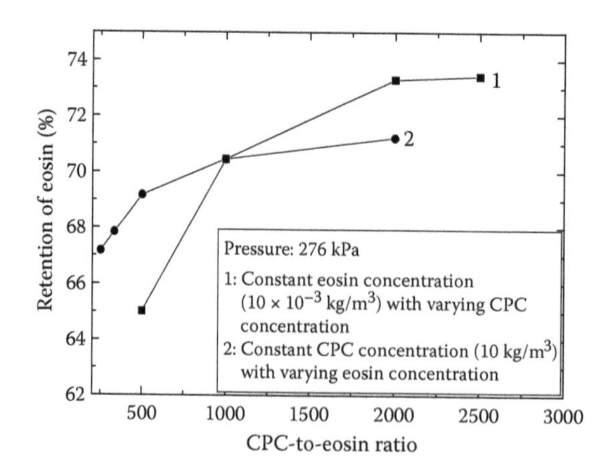


**FIGURE 9.9** Variation of eosin retention with CPC-to-eosin ratio at 276 kPa. (Reprinted from *Sep. Purif. Technol.*, 37(1), Purkait, M.K., et al., 81–92, Copyright 2004, with permission from Elsevier.)

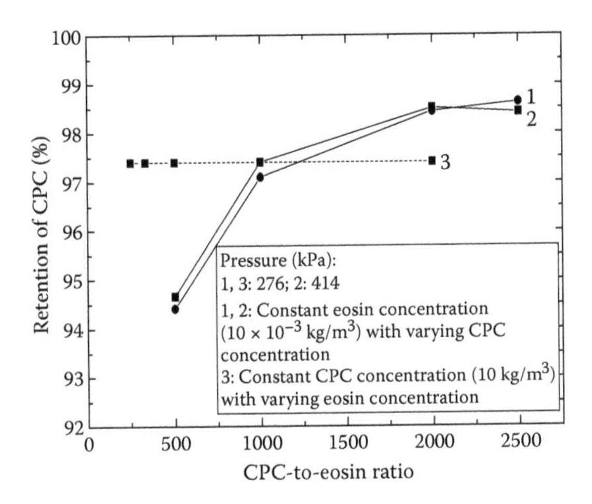


**FIGURE 9.10** Variation of CPC retention with CPC-to-eosin ratio. (Reprinted from *Sep. Purif. Technol.*, 37(1), Purkait, M.K., et al., 81–92, Copyright 2004, with permission from Elsevier.)

The retention of CPC at the end of the experimental run is presented in Figure 9.10 for both the combinations of CPC and dye mixture at different pressures (Purkait et al., 2004b). It is observed from the figure that for the case of constant eosin concentration ($10 \times 10^{-3}$ kg/m$^3$), retention of CPC is in the range of 94% to 98% with varying CPC-to-surfactant ratio. There is an increasing trend of observed retention of CPC with CPC-to-eosin ratio. Since all the feed concentrations of surfactants are much above CMC, the micelles formed are retained by the membrane and the free surfactants at the concentration of CMC pass to the permeate stream. Hence, with the increase in feed CPC concentration ($C_0$), permeate CPC concentration ($C_p$) remains around the CMC (0.322 kg/m$^3$) value and therefore, the ratio $C_p/C_0$ decreases and observed retention increases with CPC-to-eosin ratio. In case of constant CPC and varying dye concentrations, a constant CPC retention of about 97.5% is obtained. This is because the permeate surfactant concentration remains at CMC and the observed retention of CPC remains unaltered although CPC-to-eosin ratio has been increased by addition of more dyes at constant feed CPC concentration.

Once the concept of the separation of dye in an unstirred batch cell using MEUF is successfully validated, the mixture is subjected to a steady state cross-flow ultrafiltration, as this should substantially improve flux and retention. A detailed parametric study is also conducted to observe the effects of the operating conditions on the permeate flux and observed retention.

#### 9.6.1.4 Effect of the Feed CPC Concentration on Permeate Flux and the Retention of Both Dye and CPC

The effect of feed CPC-to-dye concentration ratio on the retention and permeate flux is presented in Figure 9.11. The concentration ratio of CPC to dye is varied by changing the CPC concentration at a fixed dye concentration of $10 \times 10^{-3}$ kg/m$^3$. The figure shows that the retention of both CPC and eosin increases with feed CPC concentration. But beyond the ratio of 2000, the increase in retention becomes gradual. For example, when CPC-to-eosin ratio increases from about 500 to 2000, the retention increases from 69.6% to 77.8% but retention becomes 78.2% when the ratio increases to 2500. Figure 9.11 also shows that the permeate flux decreases with feed CPC concentration at constant dye concentration. For example, flux decreases from $6.4 \times 10^{-6}$ m$^3$/m$^2$s to about $4.5 \times 10^{-6}$ m$^3$/m$^2$s when CPC concentration is increased from 5 to 25 kg/m$^3$ (CPC-to-eosin ratio increases from 500 to 2500). This is due to the resistance offered by the deposited layer of the micellar aggregates over the membrane surface (Purkait et al., 2006).

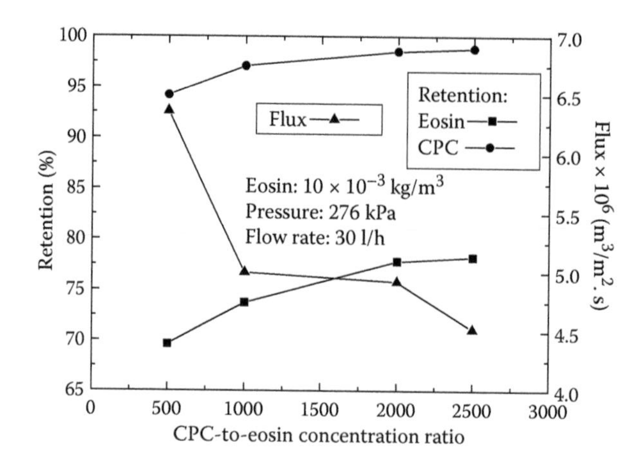


**FIGURE 9.11** Effect of the feed CPC concentration on permeate flux and the retention of both dye and CPC. (Reprinted from *Sep. Purif. Technol.*, 37(1), Purkait, M.K., et al., 81–92, Copyright 2004, with permission from Elsevier.)

#### 9.6.1.5 Effect of the Feed Dye Concentration on the Permeate Flux and Retention of Both Dye and CPC

The effect of dye concentration in feed on the retention of dye and permeate flux is shown in Figure 9.12 (Purkait et al., 2006). In this case, CPC-to-dye ratio is varied by changing the dye concentration keeping CPC concentration constant at 10 kg/$m^3$. It may be observed from the Figure 9.12 that the flux and retention increase marginally with CPC-to-dye ratio. Beyond the ratio of 2000, the retention remains almost unchanged. On the other hand, the variation of the permeate flux is marginal with CPC-to-dye concentration ratio. It may be noted that the major contribution of resistance against the solvent flux comes from the CPC micelles. Since CPC concentration remains constant and the solute concentration varies over a range between 4 and 40 × $10^{-3}$ kg/$m^3$, the permeate flux remains almost unaltered.

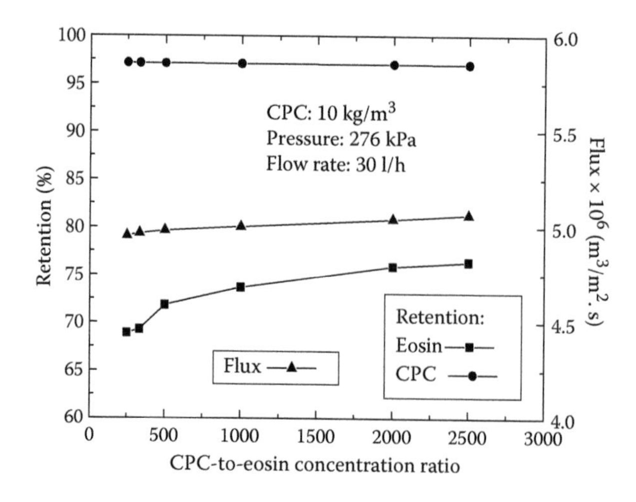


**FIGURE 9.12** Effect of the feed dye concentration on the permeate flux and retention of both dye and CPC. (Reprinted from *Sep. Purif. Technol.*, 37(1), Purkait, M.K., et al., 81–92, Copyright 2004, with permission from Elsevier.)

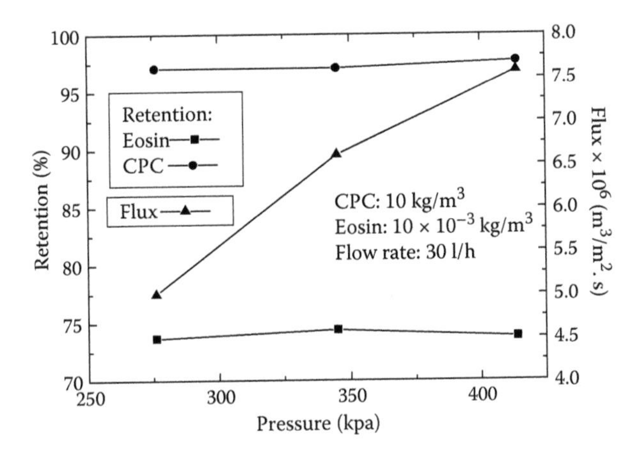


**FIGURE 9.13** Effect of pressure on the observed retention and the permeate flux. (Reprinted from *J. Hazard. Mater.* 136(3), Purkait, M.K., et al., 972–977, Copyright 2006, with permission from Elsevier.)

#### 9.6.1.6 Effect of Pressure Drop on the Observed Retention of Dye and Permeate Flux

Variation of the dye retention with the applied pressure is shown in Figure 9.13 (Purkait et al., 2006). It may be observed from the figure that the retention of dye and CPC remains almost independent of pressure. The effect of the operating pressure on the permeate flux has also been presented in Figure 9.13 using 10 kg/m$^3$ of CPC and 10 × 10$^{-3}$ kg/m$^3$ of eosin. The figure shows that the flux increases with pressure almost linearly within the pressure range. This occurs due to an increase in the effective driving force.

#### 9.6.1.7 Effect of Cross-Flow Rate on the Observed Retention of Dye and Permeate Flux

The effects of cross-flow rate on permeate flux and observed retention of eosin dye is presented in Figure 9.14 (Purkait et al., 2006). It may be observed from the figure that (1) the retention of both dye and CPC remains almost unchanged with cross-flow rate, which shows the independency of the extent of solubilization with cross-flow rate and (2) marginal increase in flux value with cross-flow

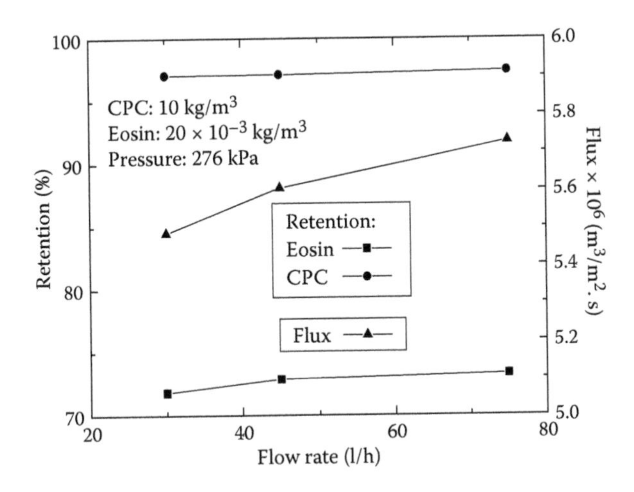


**FIGURE 9.14** Effect of cross-flow rate on the observed retention and the permeate flux. (Reprinted from *J. Hazard. Mater.* 136(3), Purkait, M.K., et al., 972–977, Copyright 2006, with permission from Elsevier.)

rate. This is due to the fact that the resistance offered by the deposited layer of the micellar aggregates is slightly lowered with the increase in cross-flow rate considered herein.

## 9.7 MEUF OF METAL ION AND ORGANIC POLLUTANT

In this section, application of MEUF is investigated to treat multicomponent solutes from aqueous stream. A binary mixture of copper and beta naphthal (BN) has been studied (Das et al., 2008c). Sodium dodecyl sulfate (SDS) is taken as the anionic surfactant. An organic polyamide membrane of MWCO 5000 has been used in a continuous cross-flow ultrafiltration unit. Suitable feed surfactant concentration is experimentally found out for the selected membrane. Effects of various operating conditions, e.g., feed concentration of solutes, transmembrane pressure drop and cross-flow rate on the observed retention and permeate flux are studied in detail. A two-step chemical treatment process, available in literature, involving calcium chloride and sodium carbonate is tested for recovery of surfactant from the permeate as well as from the retentate stream (Scamehorn and Harwell, 1989). Optimum consumption of chemicals is also found out. A schematic representation of counterion binding of metal ions and the consequent filtration by MEUF is presented in Figure 9.15 (Das et al., 2008c).

## 9.8 MEUF OF BINARY MIXTURE OF $Cu^{2+}$ AND BN

In this section, separation characteristics of a mixture of metal ion ($Cu^{2+}$) and organic pollutant (beta naphthol) are investigated using MEUF. The experimental system, membrane, procedure, etc. are already described earlier.

Six compositions in binary mixture are selected for MEUF study. The compositions are selected such that both the solutes vary in lower to higher range. The concentrations (in $kg/m^3$) of solutes are $Cu^{2+}$: BN = 0.05:1; 1:0.5; 2:0.4; 3:0.24; 3.5:0.09; 4:0.04 (refer to Table 9.1). Variation of permeate

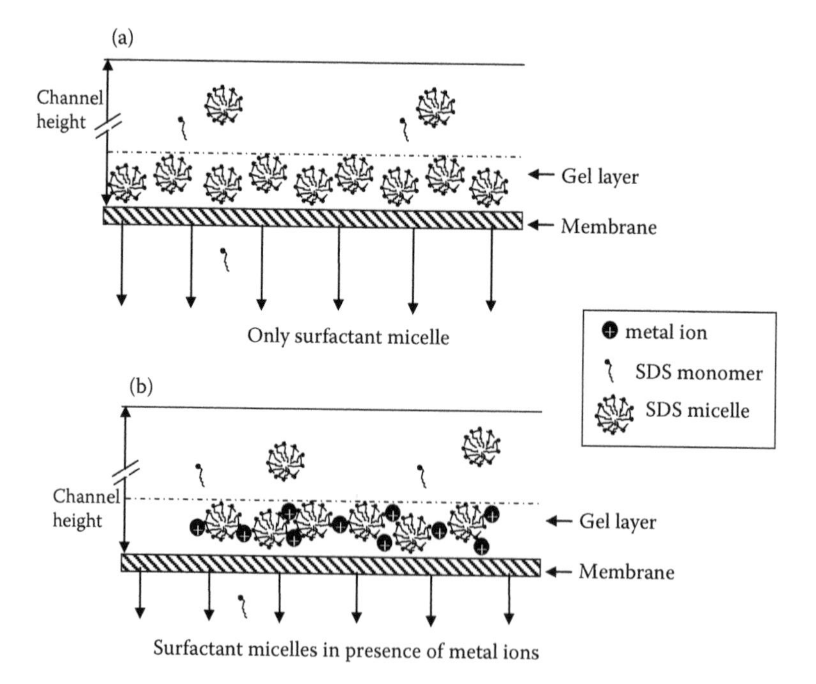


**FIGURE 9.15** A schematic representation of solubilization of organics and metal ions and filtration by MEUF. (Reprinted from *Colloid. Surf. A*, 318, Das C., et al., 125–133, Copyright 2008c, with permission from Elsevier.)

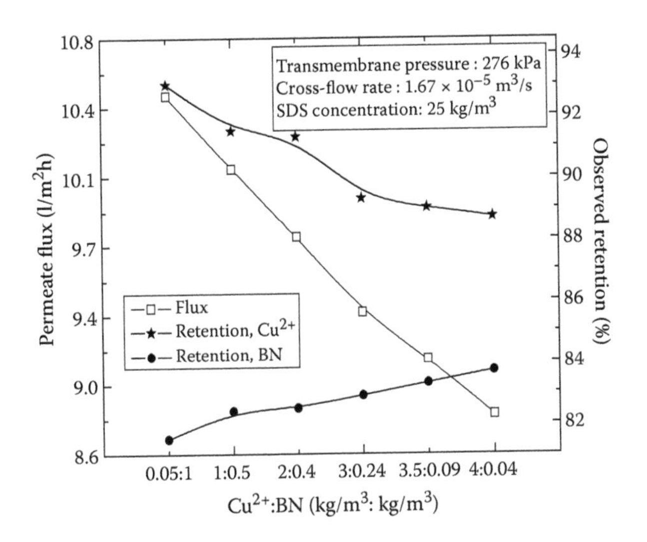


**FIGURE 9.16** Effect of feed concentration on permeate flux and solute retention in Cu BN mixture.

flux and observed retention for these six compositions of mixture are plotted in Figure 9.16. It is observed from this figure that the permeate flux decreases as the copper concentration increases. The decrease in flux is about 16% in the concentration range considered herein. Interestingly, it may be noted that almost same amount of flux decline occurs for pure copper system in the same concentration range. Therefore, the flux decline in the mixture of copper and beta naphthol is dominated by the presence of copper, the reason for which is discussed earlier. It is clear from the figure that the retention of copper ions decreases from about 93% to 89% as the concentration of copper increases in the feed. Similarly, the retention of BN decreases from about 84% to about 82% as the BN concentration increases in the feed. The reasons have already been discussed earlier. Interestingly, it may be noted that the retention of both the solutes are slightly less in the mixture (compared to single-solute system). For example, for a mixture of (3:0.24 kg/m$^3$) solute, the retention of copper is about 90% whereas that for BN is 83%. At the same conditions and feed concentration, the retention in case of single-solute system, is about 92% for both solutes. In copper beta naphthol system, copper ions are quickly adsorbed in the outer surface of oppositely charged micelles. This may cause steric hindrance to the solubilization of beta naphthol within micellar core. Therefore, retention of beta naphthol decreases slightly compared to a pure component system. Solubilization of copper is favored over that of BN. As observed in the case of copper calcium mixture, in case of copper beta naphthol system also, the solute retentions are found to be independent of operating conditions and the permeate flux increases marginally with cross-flow rate and operating pressure drop in the range studied here.

## 9.9 CONCLUSIONS

In this section, the major conclusions of the present chapter are summarized.

### 9.9.1 Ultrafiltration of Surfactant Solutions

A wide range of feed CPC and SDS concentrations are used. Significant flux enhancement is achieved in cross-flow experiments. From the experimental results, it may be conclude that it is better to select 10 kg/m$^3$ of CPC and 25 kg/m$^3$ of SDS solutions for the MEUF experiments subject to get optimum (1) surfactant retention and (2) permeate flux.

### 9.9.2 MEUF of Eosin Dye

MEUF of eosin dye in CPC solution has been studied. The feed CPC concentration ranges from 5 to 25 kg/m$^3$ and eosin concentration ranges from 5 to 40 × 10$^{-3}$ kg/m$^3$ with pressure varying from 276 to 414 kPa. Flux and retention characteristics of both CPC and eosin at different experimental conditions have been investigated and analyzed. The retention of dye without using surfactant is only 10% at a typical feed dye concentration of 10 × 10$^{-3}$ kg/m$^3$ whereas under the same operating pressure (276 kPa), retention increases up to 73.4% using surfactant micelles. The maximum retention of eosin is obtained at a surfactant-to-eosin ratio of 2000. The permeate flux decline for the batch system is analyzed using a resistance in series model. Various resistances are quantified, compared and their dependence on the operating conditions, e.g., the operating pressure and feed surfactant to CPC ratio have been investigated. During the cross-flow experiments, retention of eosin increased to about 78%. The cross-flow rate improves the permeate flux to about 10% to 18.5% without altering the dye retention.

### 9.9.3 MEUF of Binary Mixture of $Cu^{2+}$ and BN

Simultaneous separation of copper and beta naphthol from their aqueous binary mixture was studied using MEUF. A wide range of concentration of both solutes was studied. The concentration of beta naphthol was from 0.04 to 1 kg/m$^3$, and that of copper was from 0.05 to 4 kg/m$^3$. Retention of copper was found to be in the range of 93% to 89%, and that for beta naphthol was from 81% to 84%. These retentions were slightly less than those of pure component systems. Retention values were found to be independent of MEUF operating conditions, e.g., cross-flow rate and transmembrane pressure drop. The surfactants SDS and CPC concentrations in permeate are close to CMC level. The retention of copper ion is from 97.6% to 99.99% using 25 kg/m$^3$ of SDS and is independent of presence of other two anionic pollutants. The extent of solubilization differs from solute to solute depending on their ionic character, which also influences their retention and flux during MEUF. Solubilization increases with increase in ionic character.

## ACKNOWLEDGMENTS

The research outputs presented in this chapter were carried out at Prof. Sirshendu De's Laboratory at Indian Institute of Technology Kharagpur. Hence, the authors thankfully acknowledge the contribution of Prof. De.

## REFERENCES

Aoudia, M., N. Allal, A. Djennet, and L. Toumi. 2003. Dynamic micellar enhanced ultrafiltration: Use of anionic (SDS)–nonionic (NPE) system to remove $Cr^{3+}$ at low surfactant concentration. *J. Membr. Sci.* 217: 181–192.

Akita, S., L.P. Castillo, S. Nii, K. Takahashi, and H. Takeuchi. 1999. Separation of Co(II)/Ni(II) via micellar-enhanced ultrafiltration using organophosphorus acid extractant solubilized by nonionic surfactant. *J. Membr. Sci.* 162: 111–117.

Baek, K., B.K. Kim, and J.W. Yang. 2003. Application of micellar enhanced ultrafiltration for nutrients removal. *Desalination* 156: 137–144.

Baek, K., and J.W. Yang. 2004. Cross-flow micellar-enhanced ultrafiltration for removal of nitrate and chromate: Competitive binding. *J. Hazard. Mater.* B108: 119–123.

Das, C., P. Maity, S. DasGupta, and S. De. 2008a. Separation of cation–anion mixture using micellar-enhanced ultrafiltration in a mixed micellar system. *Chem. Eng. J.* 144: 35–41.

Das C., S. DasGupta, and S. De. 2008b. Simultaneous separation of mixture of metal ions and aromatic alcohol using cross flow micellar-enhanced ultrafiltration and recovery of surfactant. *Sep. Sci. Technol.* 43: 71–92.

Das C., S. DasGupta, and S. De. 2008c. Prediction of permeate flux and counterion binding during cross-flow micellar-enhanced ultrafiltration. *Colloid. Surf. A.* 318: 125–133.

Dean, J.G., F.L. Bosqui, and K.H. Lanouette. 1972. Removing heavy metals from wastewater. *Environ. Sci. Technol.* 6: 518–522.
Dutt, G.B. 2005. Comparison of microenvironments of aqueous sodium dodecyl sulfate micelles in the presence of inorganic and organic salts: A time-resolved fluorescence anisotropy approach. *Langmuir* 21: 10391–10397.
Dunn Jr., R.O., J.F. Scamehorn, and S.D. Christian. 1985. Use of micellar-enhanced ultrafiltration to remove dissolved organics from aqueous stream. *Sep. Sci. Technol.* 20: 257–284.
Dunn Jr., R.O., J.F. Scamehorn, and S.D. Christian. 1987. Concentration polarization effects in the use of micellar-enhanced ultrafiltration to remove dissolved organic pollutants. *Sep. Sci. Technol.* 22: 763–789.
Dunn Jr., R.O., J.F. Scamehorn, and S.D. Christian. 1989. Simultaneous removal of dissolved organics and divalent metal cations from water using micellar-enhanced ultrafiltration. *Colloids and Surfaces.* 35: 49–56.
Epton, S.R. 1948. A new method for the rapid titrimetric analysis of sodium alkyl sulphates and related compounds. *Trans. Faraday Soc.* 44: 226–230.
Gzara, L., A. Hafiane, and M. Dhahbi. 2000. Removal of divalent lead cation from aqueous streams using micellar-enhanced ultrafiltration. *Rev. Sci. Eau.* 13(3): 289–304.
Hankins, N., N. Hilal, O.O. Ogunbiyi, and B. Azzopardi. 2005. Inverted polarity micellar enhanced ultrafiltration for the treatment of heavy metal polluted wastewater. *Desalination* 185: 185–202.
Juang, R.S., Y.Y. Xu, and C.L. Chen. 2003. Separation and removal of metal ions from dilute solutions using micellar-enhanced ultrafiltration. *J. Membr. Sci.* 218: 257–267.
Kim, H., K. Baek, J. Lee, J. Iqbal, and J.W. Yang. 2006. Comparison of separation methods of heavy metal from surfactant micellar solutions for the recovery of surfactant. *Desalination* 191: 186–192.
Liu, K., and C.W. Li. 2005. Combined electrolysis and micellar enhanced ultrafiltration (MEUF) process for metal removal. *Sep. Purif. Technol.* 43: 25–31.
Mazer, N.A., G.B. Benedek, and M.C. Carey. 1976. An investigation of the micellar phase of sodium dodecyl sulfate in aqueous sodium chloride solutions using quasielastic light scattering spectroscopy. *J. Phys. Chem.* 80: 1075–1085.
Purkait, M.K., S. DasGupta, and S. De. 2004a. Resistance in series model for micellar enhanced ultrafiltration of eosin dye. *J. Colloid Interface. Sci.* 270: 496–506.
Purkait, M.K., S. DasGupta, and S. De. 2004b. Removal of dye from wastewater using micellar-enhanced ultrafiltration and recovery of surfactant. *Sep. Purif. Technol.* 37(1): 81–92.
Purkait, M.K., S. DasGupta, and S. De. 2006. Micellar enhanced ultrafiltration of eosin dye using hexadecyl pyridinium chloride, *J. Hazard. Mater.* 136(3): 972–977.
Rout, T.K., D.K. Sengupta, G. Kaur, and S. Kumar. 2006. Enhanced removal of dissolved metal ions in radioactive effluents by flocculation. *Int. J. Miner. Process.* 80(2–4): 215–222.
Rathman, J.F., and J.F. Scamehorn. 1984. Counterion binding on mixed micelles. *J. Phys. Chem.* 88: 5807–5816.
Scamehorn, J.F., and J.H. Harwell. (Eds.). 1989. *Surfactant Based Separation Processes*, Surfactant Science Series, Vol. 33, New York: Marcel Dekker, Inc.
Syamal, M., S. De, and P.K. Bhattacharya. 1995. Phenol solubilization by cetylpyridinium chloride micelles in micellar enhanced ultrafiltration. *J. Membr. Sci.* 37: 99–107.
Tenório, J.A.S., and D.C.R. Espinosa. 2001. Treatment of chromium plating process effluents with ion exchange resins. *Waste Manage.* 21: 637–642.
Verma, A., S. Chakraborty, and J.K. Basu. 2006. Adsorption study of hexavalent chromium using tamarind hull-based adsorbents. *Sep. Purif. Technol.* 50: 336–341.
Witek, A., A. Koltuniewicz, B. Kurczewski, M. Radziejowska, and M. Hatalski. 2006. Simultaneous removal of phenols and $Cr^{3+}$ using micellar-enhanced ultrafiltration process. *Desalination* 191: 111–116.
Xu, Y.H., T. Nakajimi, and A. Ohki. 2002. Adsorption and removal of arsenic (V) from drinking water by aluminum-loaded Shirasu-zeolite. *J. Hazard. Mater.* B92: 275–287.
Yurlova, L., A. Kryvoruchko, and B. Kornilovich. 2002. Removal of Ni(II) ions from wastewater by micellar-enhanced ultrafiltration. *Desalination* 144: 255–260.

# 10 Membrane Hybrid Systems in Wastewater Treatment

*Saravanamuthu Vigneswaran, Hok-Yong Shon, and Jaya Kandasamy*

## CONTENTS



## 10.1 INTRODUCTION

Membrane technology has been increasingly applied in water and wastewater treatment. Membrane filtration involves separation of dissolved, colloidal, and particulate constituents from a pressurized fluid using microporous materials. Membranes are categorized into four main groups based on the size of their pores, namely, reverse osmosis (RO), nanofiltration (NF), ultrafiltration (UF), and microfiltration (MF).

Microfiltration with the largest pore size of 0.1–10 μm is commonly used to separate suspended particulates, large colloids, and bacteria; hence, it is suitable for the treatment of water that has high turbidity and low color or organics content (Schafer, 2001). Similarly, ultrafiltration (0.001–0.1 μm) can exclude macromolecules and fine colloidal suspensions such as proteins, dyes, and bacteria, but the removal of dissolved organics is limited. MF and UF can also be used as pretreatment for NF and RO processes (Schafer, 2001).

Nanofiltration with smaller pore sizes of between 15 and 30 Å is employed for water softening and the removal of disinfection by-products (DBPs). NF can reduce 60%–80% of the hardness and more than 90% of color-causing substances (Cheryan, 1998). Reverse osmosis with the smallest pore size (5–15 Å) is widely used as a polishing treatment to remove salts and low-molecular-weight

(MW) pollutants from water and wastewater. In general, the selection of a suitable membrane is dependent on the targeted materials to be removed and the quality of treated effluent.

In the membrane filtration process, the removal efficiency is significantly influenced by the nature of the solution, including MW distribution, charge groups, and polarity (e.g., hydrophobicity or hydrophylicity) of the compounds. The interactions between organic matter and the membranes are either adsorption, which easily leads to membrane fouling, or rejection due to electrostatic and/or steric exclusion (Hillis, 2000). Other factors that can affect removal efficiency are pH, ionic strength, and calcium content in the solution (Amy and Cho, 1999).

Previous studies showed that UF can reduce effluent organic matter (EfOM) by 40%–60% whereas NF can produce 80% removal (Ernst et al., 2000). Snyder et al. (2001) found that polar and charged compounds were easier to be separated by UF and NF compared with less polar and neutral compounds. RO can effectively remove >90% of low-MW micromolecules (Schafer et al., 2001; Yoon et al., 2006). Combined membrane processes have also been studied; for instance, an MF and UF system was able to reduce 75% of DOC while an MF and NF system was able to remove in excess of 90% (Jarusutthirak et al., 2002).

The cost of membrane treatment is adversely proportional to the size of contaminants to be removed. In other words, the smaller the molecules, the harder it is for them to be removed, thus higher costs. Therefore, MF and UF have attracted more attention than NF and RO, as they require relatively lower cost and lower pressure for operation. In fact, UF and MF coupled with physicochemical and/or biological processes can achieve superior pollutant removal in an economical manner.

## 10.2 PRINCIPLES OF MEMBRANE PROCESSES

### 10.2.1 Filtration Mechanisms

In the filtration process, the membrane functions as a thin semipermeable barrier that allows some certain components in the feed stream to permeate freely while retaining the passage of other components. The transport of the permeable components through the membrane is driven by the pressure. As a result, the two components are completely separated, and the product stream is relatively free of unwanted impermeable constituents. The characteristics and comparison of different membranes are presented in Table 10.1.

### 10.2.2 Membrane Filtration Modes

Membrane filtration is operated in two main modes known as dead-end and cross-flow (Figure 10.1). In dead-end filtration, the feed is passed through the membrane perpendicularly where the solids

**TABLE 10.1**
**Characteristics of Membranes**

| Membrane Operation | Driving Force | Mechanism or Separation | Molecular Weight Cut-Off Range (Da) | Pore Size Range (μm) | Operating Pressure (kPa) |
|---|---|---|---|---|---|
| MF | Pressure or vacuum | Sieve | >100,000 | 0.1–10 | 20–640 |
| UF | Pressure | Sieve | >2,000–100,000 | 0.01–0.1 | 60–4700 |
| NF | Pressure | Sieve + solution/ diffusion + exclusion | 300–1000 | 0.001–0.01 | 1490–4680 |
| RO | Pressure | Solution/diffusion + exclusion | 100–200 | <0.001 | 2500–7500 |

*Source:* Adapted from Stephenson et al. (2000).

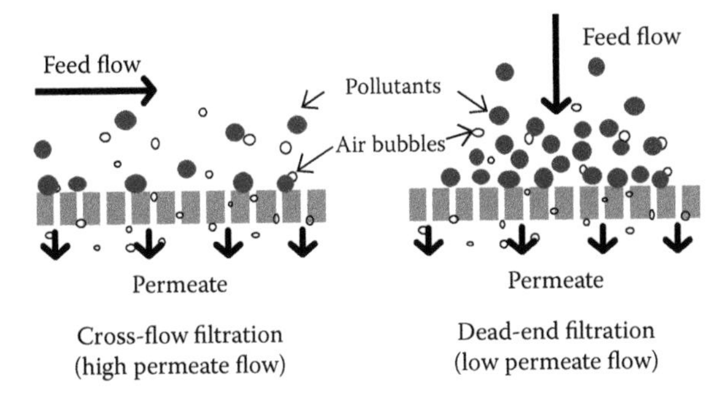


**FIGURE 10.1** Membrane filtration modes.

and targeted pollutants are retained and accumulate on the membrane surface. As the layer gets thicker with filtration time, the permeation through the membrane will experience greater resistance, and the flux gradually decreases.

In the cross-flow operation, the feed flows parallel to the membrane surface. Only a small part of the water permeates through the membrane under the applied pressure while the rest is recirculated into the feed water. The accumulation of the contaminants on the membrane surface is substantially reduced in cross-flow filtration as the flow sweeps a part of the deposits over the surface. Therefore, the permeate flux can be maintained at a higher level than in dead-end filtration. However, cross-flow filtration requires higher pressure and energy to maintain the flow at a relatively high speed.

The reduction of the flux during the filtration caused by the buildup of filter cake is also referred to as membrane fouling. Once fouling occurs, the membrane needs to be cleaned by means of physical and chemical cleaning or backward flushing to remove contaminants attached to the surface or trapped inside the pores of the membrane. After cleaning, the membrane can be operated for another filtration period. In comparison, cross-flow mode can attain a more stable flux and longer filtration time while requiring less cleaning time than in the dead-end mode.

### 10.2.3 Submerged Membrane System

Membrane filtration unit can be mounted externally or within the reactor. In the submerged membrane system, the membrane module is mounted and immersed in a reactor. The solid–liquid separation is governed by a suction pump that is connected to withdraw the permeate out of the membrane. Air scour is also supplied from the bottom of the reactor to provide oxygen as well as generate turbulence along the surface of the membrane. Aeration is an important factor used to control concentration polarization and membrane fouling of the MF submerged membranes.

Compared with the external membrane system, the submerged (internal) system involves more membrane area per unit volume and involves a lower transmembrane pressure (TMP). Moreover, the submerged system reduces space requirements and minimizes installation and operating costs. The submerged membrane concept was initially developed for membrane bioreactors (MBRs). However, it is now widely applied in water and wastewater treatment and other hybrid membrane processes.

### 10.2.4 Membrane Fouling Control and Pretreatment

Membrane fouling is the major challenge in membrane technology. It can be controlled by altering the operating conditions (i.e., operating under critical flux and pressure), by adding antifoulant chemicals to control membrane scaling, and by pretreatment (Mulder, 1996). A brief summary of the membrane fouling control is presented in Table 10.2. Modification of membrane properties, transformation of module processes, effective cleaning procedures, compaction, and pretreatment

**TABLE 10.2**
**Various Methods to Reduce Membrane Fouling**

| Pretreatment | Membrane Properties | Module Process | Cleaning | Compaction |
|---|---|---|---|---|
| 1. Heat treatment | 1. Narrow pore size | 1. Decrease of concentration polarization | 1. Hydraulic | 1. Mechanical deformation of polymeric membrane matrix |
| 2. pH adjustment | 2. Hydrophobic/ hydrophilic charge | 2. Increase of mass transfer coefficient (high flow velocities) | 2. Mechanical | 2. Low pressure |
| 3. Addition of complexing | 3. Preadsorption of the membrane | 3. Turbulence promoters | 3. Chemical | |
| 4. Chlorination | | | 4. Electric | |
| 5. Adsorption | | | | |
| 6. Flocculation | | | | |
| 7. Pre-MF | | | | |
| 8. Pre-UF | | | | |

*Source:* Adapted from Shon (2006).

are some of the methods used to reduce membrane fouling. Among the different methods, pretreatment is one of the most effective ways to reduce membrane fouling.

Pretreatment is considered the most effective alternative to reduce membrane fouling and membrane lifetime as it can simply be added to the existing membrane application. As such, pretreatment is the first choice in reducing fouling in membrane installations. When pretreatment is applied, membrane material, module type, feed-water quality, recovery ratio, and final water quality must be carefully incorporated to select an appropriate pretreatment method.

Pretreatment can be classified into four groups: physical, chemical, biological, and electrical strategies (Table 10.3). Pretreatment can remove soluble salts (hardness), colloids (silt, Fe, Al and silica), solids (suspended solid and particulate organics), biological material, oxidizing agents, and dissolved organic matter.

Pretreatment can be selected depending on feed water quality, membrane performance, and permeate water quality. The conventional pretreatment, such as a media filter, does not provide sufficient pretreatment to the membrane process. Thus, a combination of biological and physicochemical processes such as coagulation–flocculation, sedimentation, conventional filtration, oxidation, and adsorption processes is necessary to control the membrane fouling.

#### 10.2.4.1 Physical Pretreatment

Pretreatment methods in which the application of physical forces predominates are known as physical unit operations. Physical pretreatments consist of media filtration, cartridge filter, MF, UF and shock treatment. Flocculation is also partially involved in this category when the aggregated floc after flocculation is settled down. Physical pretreatments mostly remove colloids and suspended

**TABLE 10.3**
**Classification of Physical, Chemical, Biological, and Electrical Pretreatment**

| Treatment | Pretreatment |
|---|---|
| Physical pretreatment | Adsorption, media filtration, cartridge filter, MF, UF, UV pretreatment |
| Chemical pretreatment | pH adjustment, scale inhibitor, ion exchange, lime softening, flocculation, chlorination, dechlorination |
| Biological pretreatment | MBR, media biofiltration, conventional biological treatment |
| Electrical pretreatment | Electromagnetic treatment |

solids, which are one of the major membrane foulants. In addition, dissolved organic matter can be partially removed by physical pretreatment such as UF or activated carbon.

##### 10.2.4.1.1 Adsorption

Adsorption is predominantly a physical phenomenon where organic matter is attracted to the surface of adsorbent namely activated carbon by intermolecular forces of attraction. Small size (less than 10 nm) organic foulant can be removed by adsorption. Adsorption by activated carbon is mainly caused by van der Waals force and electrostatic force between organic matter and activated carbon. In principle, any porous solid can be an adsorbent, however for an efficient and economical adsorption process, the adsorbent must have a large surface, long life, and a well-defined micro-crystalline structure to possess high adsorption selectivity and capacity. The main factors that affect the adsorption of organics are (1) the characteristics of adsorbent: surface area, particle size, and pore structure; (2) the characteristics of adsorbate: solubility, molecular structure, ionic or neutral nature; and (3) the characteristics of the solution: pH, temperature, presence of competing organic and inorganic substances.

##### 10.2.4.1.2 Media Filter

A media filter, which is applied to remove colloids and particulate solids, is one of the most widely used pretreatment processes. A bed of sand, crushed granite, granular activated carbon or other materials can be used to filter contaminated waters. The removal of membrane organic foulant occurs at the initial filtration stage. Once microbial communities become established on the filter, the media filter functions as a biofilter. As such, a physical filter is always associated with biofiltration in the long-term operation.

##### 10.2.4.1.3 Cartridge Filter

A cartridge filter is a simple option to pretreat water before membrane filtration. The cartridge filter can remove large particulate organic matter (more than 5 μm). Generally, string-wound polypropylene filters that remove 1- to 5-μm particles are used. It is applied when contaminations in the influent are lower than 100 ppm. Surface or depth-type filter can be utilized. Surface filters that are made of thin materials such as papers, woven wire and cloths remove particles at the surface of the cartridge filter. On the other hand, depth-type filters remove particles by the thickness of the medium. Cartridge filters can be easily disposable and replaced.

##### 10.2.4.1.4 MF and UF

Pretreatment using MF and UF is considered as an advanced pretreatment to NF/RO. MF and UF have become a preferred pretreatment option due to their compactness, consistent pretreated water quality and convenient operation. MF is the membrane process with the largest pores. It can be used to filter suspended particulates, large colloids, bacteria, and organics. Since the pore size of MF is relatively large, air backflush or permeate backwash can be used to clean the deposits from the pores and surface of the membrane. Physical sieving is the major rejection mechanism in MF. The deposit or cake on the membrane also acts as a self-rejecting layer, and thus, MF can retain even smaller particles or solutes than its pore size. UF removes virus and results in partial removal of color. It enables the concentration, purification, and fractionation of macromolecules such as proteins, dyes, and other polymeric materials. It is widely used in the industrial wastewater treatment, where recycling of raw materials, products, and by-products is of primary concern.

##### 10.2.4.1.5 UV Irradiation

Pretreatment of UV irradiation can be used to disinfect the feed water to membrane application. While germicidal UV wavelengths range from 200 to 300 nm, 260 nm is the most effective wavelength for disinfection (Bolton and Hooper, 2004). UV light can be absorbed into the DNA of a microorganism to rupture the cross-bonds between the two strands of DNA, or RNA in the case of virus. UV is now considered as the most effective disinfectant for the control of *Cryptosporidium*, which cannot be removed by chlorine at practical dosages. Further, disinfection by $Cl_2$ produces disinfectant by-products (DBP). The advantages of UV pretreatment compared to the other chemical disinfectants such as chlorine, chloramine, chlorine dioxide, and ozone are (1) no chemical requirement, (2) lower cost, (3) reduced hazards compared to chlorine gas, (4) minimal space requirement, (5) short residence time, (6) no DBP formation, and (7) no residual disinfection, which is good for membrane preservation. The disadvantage is that periodic lamp cleaning is required.

#### 10.2.4.2 Chemical Pretreatment

Chemical pretreatment means that the removal or conversion of contaminants is brought about by the addition of chemicals or by other chemical reactions. Flocculation, pH adjustment, scale inhibitor, ion exchange, lime softening, chlorination, and dechlorination are the most common chemical methods used as pretreatment.

##### *10.2.4.2.1 Flocculation*

The process of overcoming the repulsive barrier and allowing aggregation to occur is called flocculation. Flocculation removes most colloids that carry a negative charge in water and wastewater, but a colloidal dispersion does not have a net electrical charge. The primary charges on the particles are counterbalanced by charges in the aqueous phase, resulting in an electrical double layer at every interface between the solid and the water. The forces of diffusion and electrostatic attraction spread the charge around each particle in a diffuse layer. Repulsive electrical forces and attractive van der Walls forces interact between particles in the solution, producing a potential barrier that prevents aggregation.

##### *10.2.4.2.2 pH Adjustment*

Pretreatment of pH adjustment is used when feed water has a high salt concentration. Hardness and alkalinity in the feed water lead to scaling on the membrane. As such, acid is added to maintain carbonates in their soluble carbonic acid form.

##### *10.2.4.2.3 Scale Inhibitors*

Scale inhibitors known as antiscalants interfere with the formation of all scales on the membrane surface, which starts when some soluble salts exceed their solubility and precipitate. Compared to acid pretreatment, which can only protect calcium carbonate and calcium phosphate scales, antiscalants additionally inhibit sulfate and fluoride scales and disperse colloids and metal oxides. Special products also exist to inhibit silica formation.

##### *10.2.4.2.4 Ion Exchange*

Pretreatment of ion exchange is used to remove small size of organic matter, as it is more economical than activated carbon, carbonaceous resins or metal oxides if on-site regeneration of ion exchanger is performed. Waters containing small organic matter are difficult to treat by the flocculation processes, as the smaller molecules and uncharged species are less effectively removed by flocculation. Since organic foulant has negative charge at neutral pH, basic anion exchanger is commonly used.

##### *10.2.4.2.5 Lime ($Ca(OH)_2$) Softening*

Pretreatment of lime ($Ca(OH)_2$) softening converts the $Ca^{2+}$ and $Mg^{2+}$ compounds in water into calcium carbonate ($CaCO_3$) and magnesium hydroxide ($Mg(OH)_2$). To generate $CaCO_3$ and $Mg(OH)_2$, the pH of the water needs to be increased by dosage with lime. $Ca^{2+}$ and $Mg^{2+}$ compounds can be removed at pH 9.0–9.5 and pH 10.0–0.5, respectively. Silica concentration is also reduced during a lime–soda softening process.

##### *10.2.4.2.6 Chlorination*

Chlorination pretreatment is used to reduce membrane biofouling, which is the most problematic foulant in membrane operation. Chlorine is added as a disinfectant to remove microorganisms that generate membrane biofouling. Chlorine is a strong oxidizer. It kills bacteria by stripping electrons off the structures in bacteria cell walls, entering into the cells and then breaking down enzymes and other molecules once inside the cell. Chlorine reacts to form a pH-dependent equilibrium mixture of chlorine, hypochlorous acid, and hydrochloric acid.

##### *10.2.4.2.7 Dechlorination*

Dechlorination treatment is the process of removing residual chlorine after a pretreatment of chlorination. This needs to be done before the membrane filtration stage as the chlorine used as a pretreatment will irreversibly damage the thin film composite membranes. Sulfur dioxide ($SO_2$) and sodium (meta) bisulfate ($Na_2S_2O_5$) are most commonly used. When sulfur dioxide or sulfite salts are added in chlorinated water, aqueous sulfur compounds in the +4 oxidation (S (IV)) state are generated. The S (IV) species, such as the sulfite ion ($SO_3^{2-}$), reacts with both free and combined forms of chlorine (WEF, 1996).

#### 10.2.4.3 Biological Pretreatment

Biological pretreatment brought by microbial activity is often used with biofiltration and MBR. Biological pretreatment can break down the biodegradable organic compounds and some nutrients into carbon dioxide and compost.

##### *10.2.4.3.1 MBR Technology*

Pretreatment of MBR technology is emerging as one of the most promising technologies in a biological wastewater treatment plant due to recent technical innovations and significant cost reductions. MBR is a porous membrane (MF or UF). MBR pretreatment produces high-quality effluent in terms of organic foulants. MBR can be easily associated with conventional processes. As such, for wastewater reuse, MBR can be combined with RO (MBR followed by RO).

##### *10.2.4.3.2 Biofiltration*

Biofiltration pretreatment is one of the most common processes used in membrane filtration. In addition to providing filtration, it is a pollution control technology using microorganisms to degrade biodegradable inorganic and organic compounds. A physical filtration at the initial stage of sand filtration, dual-media filter and granular activated carbon (GAC) filter, converts to biofiltration with time. Any type of filter with attached biomass on the filter media may be defined as a biofilter. It can be the trickling filter in the water treatment plant, or horizontal rock filter in a polluted stream, or GAC or sand filter in a water treatment plant. Biofilters have been successfully used for pretreatment to membrane filtration.

##### *10.2.4.3.3 Electrical Pretreatment*

Electrical and electromagnetic pretreatment utilizes electrical charging and electromagnetic force. Electrical power uses water forces with oppositely charged ions to move in the opposite directions. This results in the formation of microscopic nuclei, which cause calcium carbonate to precipitate within the water. Then, electromagnetic power can be applied and water is passed through a magnetic field. This leads to a change of the electrical properties of the water. This also separates agglomerations or clusters of water molecules. The electric and electromagnetic water pretreatment can improve the dissolution capacity of water. Electric and electromagnetic pretreatment removes sparingly soluble salts, colloidal particles and suspended solids. However, this pretreatment is not often used in membrane filtration.

## 10.3 MEMBRANE HYBRID SYSTEMS

Low-pressure membrane filtration, such as MF and UF, presents many advantages over conventional water treatment processes including superior and consistent product water, less chemical use, smaller spatial requirements and easier siting and expansion (Adham et al., 1995). Moreover, MF has the ability to integrate with other technologies such as biological processes, flocculation, and adsorption. Recently, photocatalysis has also attracted increasing attention in order to improve the treatment and achieve the desired water quality.

### 10.3.1 Membrane Bioreactor

MBR, an innovative technology, is a combination of activated sludge and membrane separation processes into a single process where suspended solids and microorganisms are separated from the treated water by membrane filtration. The entire biomass is confined within the system, providing both perfect control of the sludge age for the microorganisms in the reactor and the disinfection of the influent.

The optimum design of MBR process is very complex since it is dependent on many factors including feed characteristics, mixed liquor suspended solids (MLSS) concentration, sludge retention time (SRT), hydraulic retention time (HRT), operational flux, membrane material cost, energy consumption, and sludge treatment and disposal and their interrelation (Stephenson et al., 2000). Some of the advantages and disadvantages of aerobic MBR process are discussed here.

The advantages of MBR process are

- High-quality effluent, free from solids
- The ability to disinfect without the need for chemicals
- Complete independent control of HRT and SRT without the need to select for flocculent microorganisms or biofilms
- Reduced sludge production compared to other aerobic processes
- Process intensification through high biomass concentrations
- Treatment of recalcitrant organic fractions and improved stability of processes such as nitrification and
- Ability to treat high-strength wastes

Biomass separation MBR employs MF or UF modules for biomass retention. The membrane module can either be placed in external circuit to the bioreactor or submerged into the bioreactor, as shown in Figures 10.2a and 10.2b, respectively.

The membranes used in biomass separation MBRs are asymmetric with a dense top layer or skin of 0.1 to 0.5 in thickness and a supporting thick sublayer (Visvanathan et al., 2000). The skin can be placed either on the outside of the membrane called outer skinned membrane (OSM) or inside the membrane called inner skinned membrane (ISM) (AWWARF et al., 1996). This top layer eventually defines the characteristics of membrane separation.

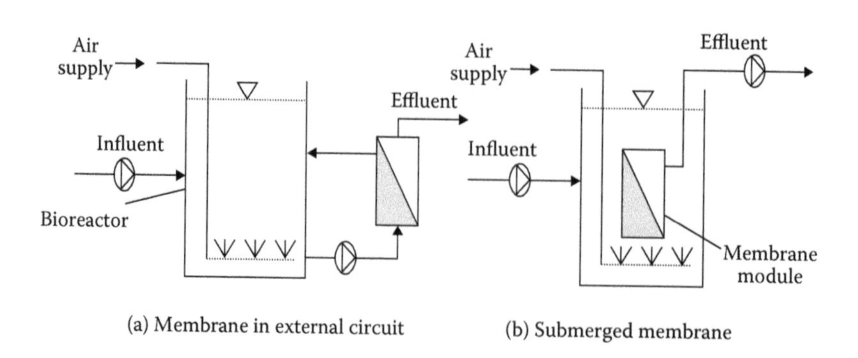


**FIGURE 10.2** Biomass separation MBR (a) membrane in external circuit and (b) submerged membrane.

Membrane in external circuit system, also known as recirculated MBR, is independent of the bioreactor. It can be operated with either outer- or inner-skinned membranes. In this system, the feed enters the bioreactor where it contacts biomass. This mixture is then pumped around a recirculation loop containing the membrane module where the permeate is discharged and the retentate is returned to the bioreactor. The TMP and the cross-flow velocity for membrane operation are both generated from a pump.

Submerged membrane system requires outer-skinned membranes and is independent of recirculation loop as the separation occurs within the bioreactor itself. In this system, the pressure across the membrane can only be applied by suction through the membrane or by TMP derived from the hydraulic head of the water above the membrane. Therefore, the power requirement for operation is generally lower than the membrane in external circuit system, which requires cross-flow membrane filtration. Fouling control is achieved by scouring of the membrane surface with aeration. The movement of bubbles close to the membrane surface generates the necessary liquid shear intensity.

Sidestream membrane configuration has been employed at full-scale domestic and industrial wastewater treatment plants worldwide. Since early 1990s, the submerged membrane configuration due to its lower energy demand has emerged as a more suitable option to serve small populations. Judd et al. (2001) found that the energy demand for a submerged system permeate product was twice as energy efficient (2 kWh/m$^3$) as the sidestream system (3.9 kWh/m$^3$). Its application from lab-scale to full-scale setup has progressed very rapidly due to the demand for decentralized treatment and high-energy-efficiency units. Kubota (flat sheet) and Zenon (hollow fiber) are both commercially available submerged systems and the most significant manufacturers in terms of growth and total installed area. Table 10.4 presents the difference in energy consumption and other operational parameters for the two MBR configurations.

Energy consumption arises from power requirements for pumping feed water, recycling of retentate, permeate suction and aeration. According to Table 10.5, there is a substantial difference between energy consumption of the two MBR operating systems. For example, submerged membrane systems do not require retentate recycle and some do not require permeate suction (operated under hydraulic head). Aeration is utilized in significantly different ways for the two MBR configurations. In the sidestream configuration, aeration is supplied to the bioreactor by fine bubble aerators, which are highly efficient for supplying oxygen to the biomass. On the other hand, coarse

**TABLE 10.4**
**Membrane Configuration, Operating Parameters, and Energy Consumption for MBR Systems**

| Process | Submerged | | | | Sidestream | | | |
|---|---|---|---|---|---|---|---|---|
| Membrane[a] | P&F | P&F | HF | HF | T | T | HF | HF |
| Material[b] | PS | PE | PE | PE | PS | C | C | PS |
| Pore size (1 μm)/MWCO (KDa) | 0.4 | 0.4 | 0.1 | 0.1 | 50 | 300 | 0.1 | 0.1 |
| Surface area (m$^2$) | 0.24 | 0.96 | 2 | 4 | 2.6 | 0.08 | 1.1 | 0.39 |
| TMP (bar) | 0.1 | 0.3 | 0.13 | 0.15 | 5 | 2 | 2 | 2.75 |
| Permeate flux (L/m$^2$/h) | 7.9 | 20.8 | 8 | 12 | 170 | 175 | 77 | 8.3 |
| Cross-flow velocity (m/s) | 0.5 | 0.3–0.5 | – | – | 1–2 | 3 | 1.5–3.5 | – |
| Energy, permeate (kWh/m$^3$) | – | 0.013 | 0.005 | 0.23 | 0.17 | 9.9 | 32 | 0.045 |
| Energy, aeration (kWh/m$^3$) | 4.0 | 0.009 | 0.140 | 70.00 | 0.52 | 2.8 | 9.1 | 10 |
| Total energy consumption (kWh/m$^3$) | 4.0 | 0.022 | 0.145 | 70.23 | 0.69 | 12.7 | 41.1 | 10.045 |

*Source:* Adapted from Khan (2008) and Gander et al. (2000).

[a] P&F, plate and frame; HF, hollow fiber; T, tubular.

[b] PS, polysulfone; PE, polyethylene; C, ceramic.

**TABLE 10.5**
**Performance Comparison between Activated Sludge Process (ASP) and MBR**

| Process | SRT (d) | HRT (h) | COD Removal (%) | TSS Removal (%) | Ammonia N Removal (%) | Reference |
|---|---|---|---|---|---|---|
| Sidestream MBR | 30 | – | 99 | 99.9 | 99.2 | Stephenson et al. (2000) |
| Submerged MBR | 0.25 | 3 | 97.3 | 99.9 | 40.9 | Ng and Hermanowicz (2005) |
| | 0.5 | 3 | 97.5 | 99.9 | 44.6 | |
| | 2.5 | 3 | 98.4 | 99.9 | 40.9 | |
| | 5 | 6 | 98.2 | 99.9 | 99.9 | |
| ASP | 20 | – | 94.5 | 60.9 | 98.9 | Stephenson et al. (2000) |
| ASP | 0.2 | 3 | 77.5 | 75.7 | 41.1 | Ng and Hermanowicz (2005) |
| | 5 | 3 | 78.7 | 76.0 | 34.4 | |
| | 0.5 | 6 | 83.0 | 81.7 | 344 | |
| | 2.5 | 6 | 93.8 | 94.7 | 99.5 | |
| | 5 | | | | | |

*Source:* Khan (2008).

bubble aeration is used in submerged systems generating the cross-flow as well as scouring to the membrane and providing oxygen to the MBR biomass. Coarse bubble aerators are less efficient than fine bubble aerators for supplying oxygen to the biomass but have the advantage of lower cost. It is due to this low cost of aeration and pumping that the total energy consumption tends to be lower in submerged reactor as compared to sidestream systems. In some operations, coarse and fine bubble aerators are used in combination (Le-Clech et al., 2003; Germain et al., 2005).

The membrane performance in a MBR system is characterized by the rejection, normally expressed as removal efficiency, of the respective concentrations of the target contaminants in the feed and by the permeability i.e., flux per unit pressure. Removal of particles, including biological and nonbiological colloids and macromolecules, is achieved by sieving and adsorption. One of the main advantages of MF and UF membranes is the significant ability to disinfect, by rejection of both bacteria and viruses, resulting in an effluent free from pathogenic microorganisms. The rejection is further improved with time due to the buildup of the dynamic membrane.

A major factor preventing the widespread application of MBRs is capital and operating cost, the most significant of which is the cost of the membrane. The actual costs of membrane and the fractional importance of membrane replacement on the overall costs are both decreasing. For example, costs for the Kubota system have decreased from US\$400 per $m^2$ in 1992 to US\$100 per $m^2$ in 2000. More importantly, the relative fraction of the overall cost associated with membrane replacement has decreased from ca. 54% to ca. 9% over the same period (Khan, 2008).

Studies have also attempted to identify the maximum size at which MBRs are competitive with traditional process. For a flow of 100 $m^3$/d MBR plant, a reduction of 20% in membrane investment costs and 18% in operating costs would be required to make it economically competitive. However, there is no doubt that the maximum competitive size of MBR plants is increasing and will continue to do so.

### 10.3.2 Membrane Hybrid System Coupled with Physicochemical Processes

Fouling is considered as the major technical problem when operating a membrane system in water and wastewater treatment. Fouling is due to the deposition and accumulation of primarily effluent organic matter (EfOM) on the surface and inside pore entrances of the membrane resulting in flux loss and higher energy requirements. Experience has shown that pretreatment of the feed water can significantly enhance the performance of membrane systems. Various processes are available for pretreatment including coagulation/flocculation, ion exchange, UV/ozone and adsorption by activated carbon. The employment of these methods aims to reduce particulates and dissolved organic matter, increase rejection, and minimize biological fouling of the membrane. Although pretreatment leads to an increase in additional cost, it will be recovered over the life of the plants in terms of extending the life of the membrane and improving the performance of the system.

#### 10.3.2.1 Flocculation–Membrane Hybrid System

MF with a large pore size can only remove a very small amount of dissolved organic compounds, especially trace organic compounds, taste and odor-causing materials, and synthetic organic chemicals. In order to increase their rejection by the membrane, the dissolved organics need to be aggregated into particulates. This can be done by adding coagulants and/or flocculants such as alum, aluminum sulfate and ferric chloride. Consequently, coagulation has been employed widely as a pretreatment option for MF/UF.

In-line flocculation–microfiltration is a method of reducing the degree of internal clogging. Since the chances for internal clogging are higher in microfiltration, this method is generally employed in MF systems. From the laboratory scale studies, a significant improvement in the performance of cross-flow microfiltration was observed when this process was incorporated with an in-line flocculation arrangement (Vigneswaran et al., 1991). The in-line flocculation is used to achieve two objectives: eliminating the penetration of colloidal particles into the membrane pores and modification of deposit characteristics.

The mechanism involved in this process is simple (Figure 10.3). Internal clogging caused by colloidal particles reduces the permeate flux and life span of the membrane (Figure 10.3a). To avoid this, the membrane can be selected with a lower pore size range (Figure 10.3b). By this means, the internal clogging phenomenon can be controlled. But higher pressure has to be applied and lower flux will result in the membranes with smaller pore diameter. On the other hand, flocculation of the suspension, makes the particle size bigger (microfloc formation). Thus, in-line CFMF can be used to remove particles and associated pollutants with lower pressure and higher flux (Figure 10.3c).

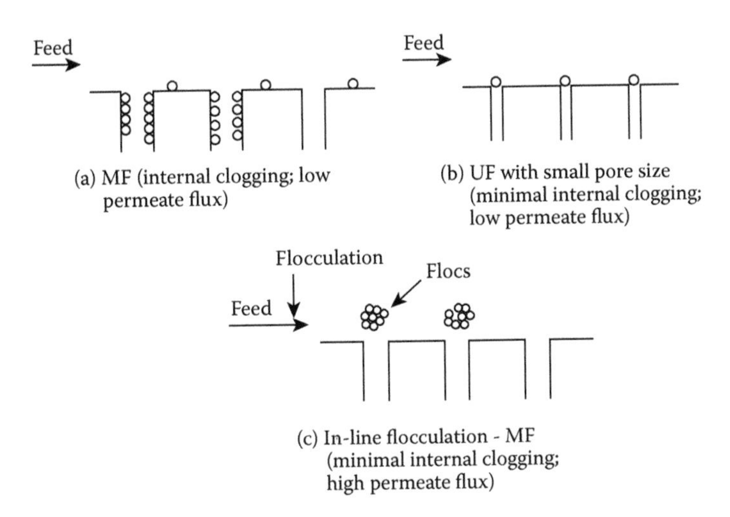


**FIGURE 10.3** Advantage of flocculation before cross-flow filtration (adapted from Vigneswaran et al., 1991). A combined system of flocculation as a pretreatment can reduce suspended solids, and phosphorus while reducing fouling.

Flocculation as a pretreatment to membrane filtration improves the removal of hydrophobic compounds, resulting in the increase of permeate flux. According to Chen et al. (2007), natural organic matter (NOM) removal by the membrane after coagulation increased by 20%, with 14.3% for the hydrophobic fraction and 2.5% for hydrophilic fraction, respectively. Correspondingly, the flux was reported to recover completely after backwash for each filtration cycle compared with gradual 20% flux drop when membrane was used alone (Chen et al., 2007). However, the performance of flocculation/membrane combined system is significantly dependent on the nature of the organic matters, coagulant dosage and floc characteristics leading to a process that is difficult to control (Shon et al., 2005).

Flocculation is also one of the most effective pretreatment methods to remove effluent organic matter (EfOM). Abdessemed and Nezzal (2002) observed that flocculation with $FeCl_3$ removed 77% of COD in wastewater. They also found that flocculation increased the permeate flux by 46.6%. Shon et al. (2005) showed that $FeCl_3$ flocculation removed 68% of organic matter (in terms of DOC) from the biologically treated wastewater. The majority of organics removed were the ones with the large MW.

The detailed effect of $FeCl_3$ flocculation was investigated as a pretreatment to UF in treating synthetic wastewater containing synthetic organic matter by Shon (2006). The effect of flocculant dose was studied in terms of organic removal and membrane flux decline. UF with optimum dose of $FeCl_3$ (68 mg/L) did not experience any flux decline during the whole operation of 6 h (Figure 10.4). The preflocculation with a smaller dose of 20 mg/L of $FeCl_3$ led to a sharp flux decline in the UF (more than 35% in 6 h). To understand the phenomenon of the flux decline of UF, the MW ranges of organic matter removed by different doses of $FeCl_3$ and by the posttreatment of UF were studied (Figure 10.5). The deviant crease circle shows that the pretreatment of flocculation with reduced $FeCl_3$ dose (less than the optimum dose) is possible as an adequate pretreatment to minimize the flux decline and to obtain high organic removal. In the study of Shon (2006), a dose of 50 mg $L^{-1}$ of $FeCl_3$ was sufficient to run the UF with no (or minimal) flux decline and high DOC removal. Specific experiments need to be conducted with particular wastewater at the time of operation to determine the suitable dose of $FeCl_3$.

#### 10.3.2.2 Adsorption–Membrane Hybrid System

Combining activated carbon adsorption with membrane filtration MF is another promising option to improve DOC removal efficiency. In the hybrid system, the small molecular species not usually

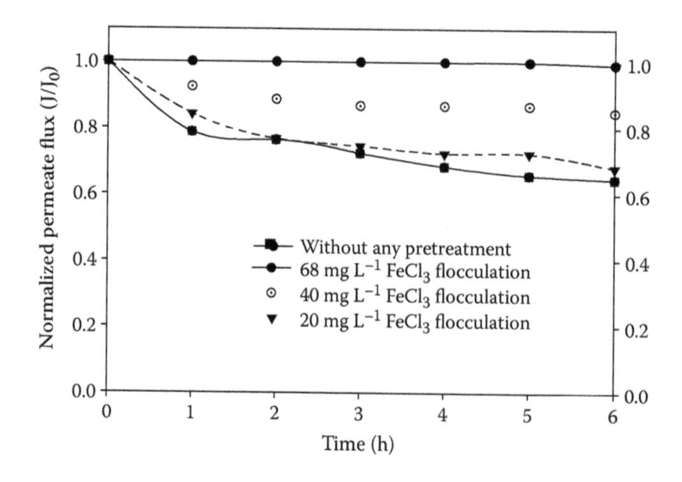


**FIGURE 10.4** Temporal variation of filtration flux of UF after a pretreatment of flocculation at different $FeCl_3$ doses (NTR 7410 UF membranes, $J_0$ = 1.84 m/d at 300 kPa; cross-flow velocity = 0.5 m/s; MWCO of 17,500 Da; Reynolds number: 735.5; shear stress: 5.33 Pa).

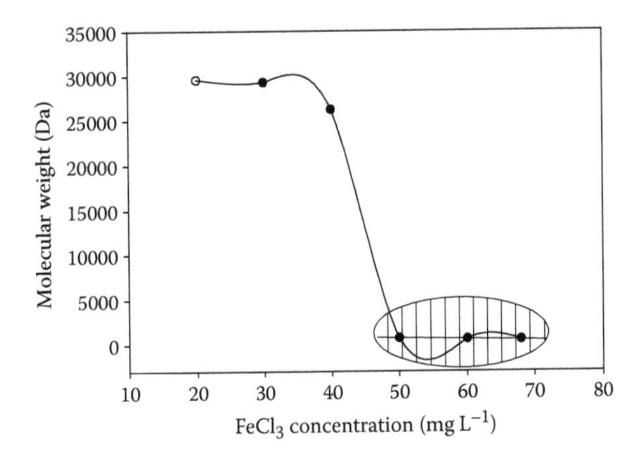


**FIGURE 10.5** Correlation between the $FeCl_3$ concentrations and the corresponding weight-averaged MW values in the flocculated effluent.

rejected by the membrane alone are adsorbed onto the activated carbon, which subsequently can be separated easily by the MF process. As a result, organic compounds are retained from the treated water. There are two types of activated carbon available namely powdered activated carbon (PAC) and granular activated carbon (GAC). Between the two, PAC is preferable due to its large surface area per unit volume and its affinity for a wide range of dissolved organics (Guo et al., 2005).

In the combined system, PAC can be dosed continuously to the membrane reactor or dosed at the beginning of the membrane filtration cycle. Previous studies showed that PAC preadsorption contributed to higher dissolved organic carbon (DOC) and disinfection by-products (DBPs) removal, flux enhancement and fouling reduction; thus prolonging the continuous filtration time (Lebeau et al., 1998; Guo et al., 2005). One of the hybrid system known as submerged membrane adsorption hybrid system (SMAHS) integrate the entire treatment activity including adsorption/biodegradation, liquid–solid separation and sludge accumulation into one single unit, which can offset the disadvantage of the large equipment size and space requirement (Guo et al., 2005) (Figure 10.6).

In addition, SMAHS has several advantages. The system can be operated consistently for a long period with a very low energy (0.2 kWh/m$^3$) and without any major sludge problems. Using a dose of 10 g/L of PAC added at the beginning of the experiment, Guo et al. (2005) obtained approximately 84% of TOC removal for a 15-day operation. A long-term study conducted showed that addition of 10 g/L of PAC at the beginning of the operation and replacement of 1% of PAC on a daily basis led

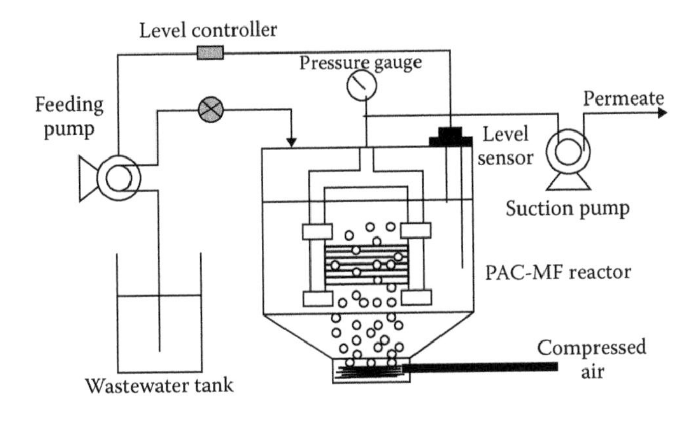


**FIGURE 10.6** Experimental setup of SMAHS (adapted from Guo et al., 2005).

to a consistent TOC removal of more than 90%. The PAC dose was found to be less than 20 mg/L when one considers an operation period of 50 days.

In Chatel-Gerard, France, a new plant was needed to replace the existing direct filtration plant. The raw water source was a groundwater, which, because of surface water contact, suffered occasional (and significant) increases in turbidity, color, TOC, coliforms, and atrazine. For this purpose, a PAC/UF system was found to be suitable and economical. The plant capacity is 600 $m^3$/d, which, with an average flux of 100 L/$m^2$·h. This plant used six Aquasource modules, each with a 70-$m^2$ surface area. The PAC dose varies between 10 and 15 mg/L, backwashing is done once per hour, and chemical cleaning is done once per year. Plant data show that the average atrazine is kept below the European standard of 0.1 μg/L, with excellent removal of turbidity, coliforms, and odor (Lebeau et al., 1998).

The performance of ultrafiltration (UF) was studied in terms of normalized permeate flux ($J/J_0$) with and without adsorption pretreatment. The pretreatment of PAC adsorption helped in the reduction of flux decline. For example, the reduction in the $J/J_0$ for UF after a pretreatment of PAC adsorption (with 1 g/L) was from 1 to 0.71 after 6 h of operation (i.e., 29% decline). The decline with no pretreatment was 0.66 (i.e., 34% decline). Adsorption using larger PAC dosages (0.5 g/L) removed the majority of the relatively smaller MW of organic matter (200 Da to 600 Da). However, the majority of the relatively larger MW organic matter could not be removed by adsorption alone. The PAC used had a pore radius from 1 to 5 nm with mean radius of 1.8 nm. The observed removal of a portion of a large MW organics by PAC may have been due to adsorption onto the larger pores, or onto the outer surface of PAC. The MW distribution results are consistent with the trend in the flux decline.

#### 10.3.2.3 Photocatalysis–Membrane Hybrid System

Semiconductor photocatalysis can remove waterborne pollution by sensitizing refractory and persistent organic pollutants, oxidizing microbial cells and deactivating viruses.

The hybrid photocatalysis/membrane system has attracted growing attention due to the potential synergy of both processes (Figure 10.7). MF not only aids the separation of suspended catalysts but also improves the effluent quality. In return, semiconductor photocatalysis helps to reduce membrane fouling by photomineralizing foul-causing compounds (Shon et al., 2008). As a result, photocatalysis partially enhances the consistent performance of MF.

The photocatalytic reactor and membrane module can be integrated either externally or internally. In the external system, the degradation of organic compounds mainly occurs in the photoreactor and the photocatalysts that are subsequently confined by means of membrane filtration. In the internal system, UV light and membrane module are embedded in a single reactor where the photosensitization and filtration of organic matter can take place simultaneously.

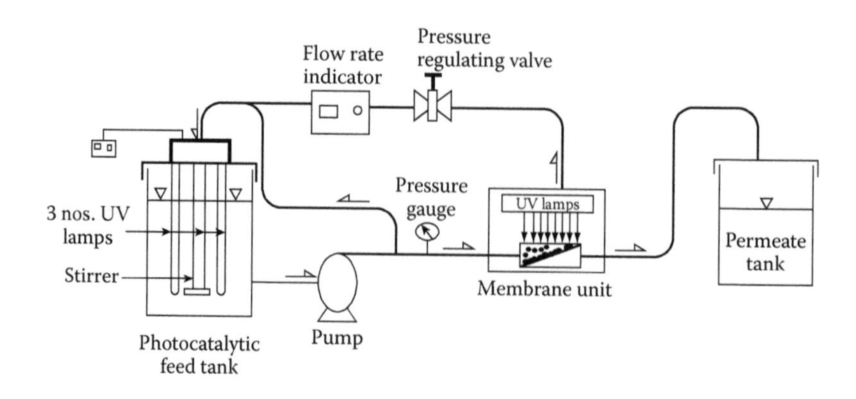


**FIGURE 10.7** Schematic drawing of photocatalysis cross-flow membrane filtration unit.

In comparison, the single system is more advanced in terms of compactness and footprint. However, the design and operation of the membrane photoreactor are difficult to optimize due to limited knowledge on the dual interaction between UV photocatalysis and membrane. There is also a concern that UV radiation produces a large amount of oxidizing hydroxyl radicals and heat and this may affect the performance and the life span of the membrane (Azague et al., 2007). In addition, the selectivity of more stable membrane leads to higher installation and operation costs; thus, the system become less economically viable.

The separated setup of photocatalysis/membrane system is relatively straightforward and allows each reactor to be operated and optimized independently. In other words, coupling the two reactors can simply take advantage of the two processes with little modification (Molinari et al., 2002). Moreover, various types of commercial membranes can be used in the hybrid system. In particular, submerged MF is the most attractive because of its low pressure and low cost requirements (Molinari et al., 2002).

Several studies have been conducted to evaluate the efficiency of the photocatalysis and membrane integrated systems for water purification and wastewater reclamation and reuse. The retention of $TiO_2$ slurry by means of different membrane types has been adapted and assessed (Sappideen, 2000; Molinari et al., 2002). The results of those studies revealed an optimal range of $TiO_2$ dose of 0.1–1.0 g/L and optimum membrane pore size of 0.2 μm gained the highest efficiencies for this specific coupling system.

Molinari et al. (2002) carried out the investigation of different system configurations using 4-nitrophenol as the pollutant. The authors reported a better degradation up to 80% obtained after 5–6 h of operation by the suspended catalyst reactor compared with 51% of the catalyst deposited on membrane system (Molinari et al., 2002).

Previous studies (Erdei et al., 2008; Ho, 2008) indicated that the effectiveness of the photocatalysis/membrane integrated system was relatively dependent on pH values (optimal range of 4.5–6.0), catalyst doses, UV lamp surface area, and the initial feed concentrations. However, most of the studies were performed with single compounds or drinking water with low concentration of NOM (TOC = 2–3 mg/L). A recent study on photocatalysis–membrane hybrid system for organic removal from biologically treated sewage effluent showed superior DOC degradation of more than 80% by the hybrid system. Moreover, it was demonstrated that photosensitization with $TiO_2$/UV could effectively reduce membrane fouling and enhance the permeate flux of the submerged membrane reactor. This method was also successful in recovering and recycling the $TiO_2$ (Shon et al., 2008).

#### 10.3.2.4 Biofiltration–Membrane Hybrid System

Biofilter is used to remove majority of organic matter. Biofilter is a conventional filter with biomass attached onto the filter media as a biofilm, where the organics are adsorbed on it and biodegraded by the microorganisms. The biofiltration process is economical and environmentally friendly in treating wastewater of relatively small volume. Previous studies on biofilter have shown that it could remove organics and nutrients in significant quantities and produce high-quality effluent (Chaudhary et al., 2003; Yang et al., 2001; and Sakuma et al., 1997).

Shon (2006) investigated UF performances with and without different pretreatments in terms of organic removal efficiency (DOC) and flux decline (Table 10.6). The flux decline is due to membrane fouling which depends on the composition of the feed and hydrodynamic conditions. In the present experiments, the hydrodynamic conditions were fixed to a predetermined value. Thus, the flux decline is mainly related to the feed composition. The feed composition is influenced by the pretreatment. The operation of UF membranes can be improved by pretreatment. For example, The UF NTR 7410 filtration (MWCO 17,500 Da) without pretreatment resulted in rapid filtration flux decline with time. When the large MW was removed by flocculation, the rate of decline of flux was minimized. The PAC adsorption or GAC biofilter alone as pretreatment reduced the flux up to 32% and 43%, respectively. The DOC removal after the pretreatment of GAC biofilter was the highest up to 79%. This may be due to different roles of pretreatment in removing the MW distribution of EfOM.

**TABLE 10.6**
**DOC Removal and Flux Decline with and without Pretreatment with UF NTR 7410 Membrane at 18 h Operation (MWCO 17,500 Da; $J_0$ = 3.01 m/d at 300 kPa; Cross-Flow Velocity = 0.5 m/s)**

| Pretreatment | DOC Removal (%) | Flux Decline (%) |
|---|---|---|
| Without pretreatment | 40 | 71 |
| PAC adsorption | 74 | 32 |
| $FeCl_3$ flocculation | 63 | 20 |
| GAC biofiltration | 79 | 43 |

### 10.3.2.5 Ion Exchange – Membrane Hybrid System

Magnetic ion exchange (MIEX®) resin can effectively remove dissolved organic matter from biologically treated sewage effluent (secondary effluent) and produce high-quality water. It removes the hydrophilic substances that are not normally removed by flocculation or adsorption. Especially when MIEX contactor is used as pretreatment for a submerged membrane hybrid system, higher effluent quality and longer operation time can be achieved (Zhang et al., 2006). Experiments also showed that the submerged membrane process with MIEX pretreatment can achieve superior organic removal (Zhang et al., 2006). The MIEX process has mainly been used as a batch process, which requires a large area for accommodating both contact tank and settling tank in the treatment process. In a study by Zhang et al. (2008), a fluidized bed reactor was used as pretreatment for a submerged membrane to a semicontinuous process.

Figure 10.8 shows the total DOC removal efficiency of the submerged membrane hybrid system with fluidized MIEX contactor as pretreatment. The total DOC removal efficiency by the MIEX reactor–membrane hybrid system can be as high as 75%–85% with a pretreatment of MIEX reactor of 100 ml (80.3 cm depth during fluidization) of MIEX resin in the fluidized column. Submerged membranes alone could only remove an average of 35% of DOC from the wastewater. Although MIEX as a pretreatment was efficient in removing DOC, the use of membranes as posttreatment is essential to remove the suspended solids and the carry-over resins.

Figure 10.9 shows the TMP development during 8 h of operation of the submerged membrane hybrid system with fluidized bed MIEX as pretreatment. With the MIEX pretreatment, the submerged membrane system resulted in only a marginal increase of TMP. The TMP development was

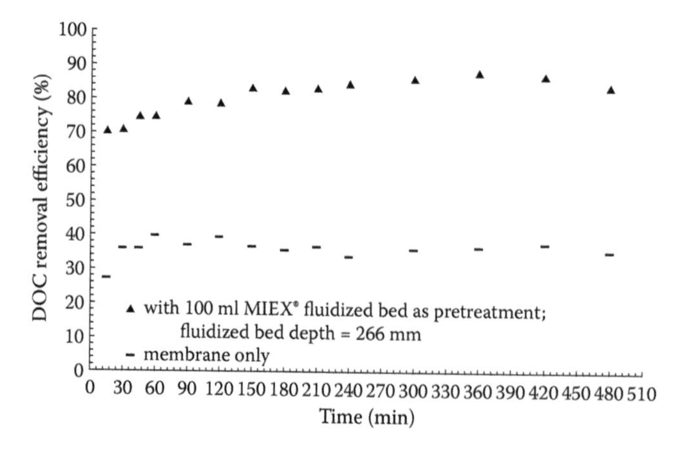


**FIGURE 10.8** DOC of the wastewater = 10 mg/L; permeate flux = 96 L/m²/h; backwash rate = 240 L/m²/h; membrane backwash duration = 1 min; membrane backwash frequency = 1 h.

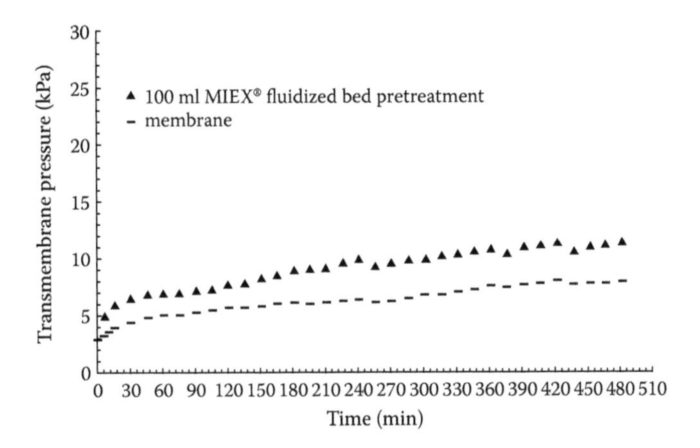


**FIGURE 10.9** TMP profile of submerged membrane hybrid system with fluidized bed MIEX contactor as pretreatment (DOC of the wastewater = 10 mg/L; permeate flux = 96 L/m²/h; backwash rate = 240 L/m²/h; membrane backwash duration = 1 min; membrane backwash frequency = 1 h).

less than 10 kPa during 8 h of operation. With a higher amount of MIEX and a longer contact time, the TMP development was slightly higher, up to 19 kPa at the end of 8 h of operation. This might have been caused by the overflow of the small MIEX particles from the fluidized bed reactor into the membrane tank and their accumulation in the submerged membrane tank. The escape of MIEX resin into the membrane tank was visually observed during the 8-h filtration operation.

## 10.4 CONCLUSION

This chapter highlights the benefits of different membrane hybrid systems such as MBRs, membrane adsorption, hybrid systems, etc, in reducing the membrane fouling and producing water superior in quality. In a similar way, pretreatments such as biosorption, photocatalysis and ion exchange can also reduce the membrane fouling.

## REFERENCES

Adham, S.S., Jacangelo, J.G., and Laine, J.M. 1995. Low-pressure membrane assessing integrity. *JAWWA* 87: 62–76.

Abdessemed, D., and Nezzal, G. 2002. Treatment of primary effluent by coagulation–adsorption–ultrafiltration for reuse. *Desalination* 152(13): 367–373.

Amy, G.L., and Cho, J. 1999. Interactions between natural organic matter (NOM) and membranes: Rejection and fouling. *Water Sci. Technol.* 40(9): 131–139.

AWWARF, LdE and WRC. 1996. *Water Treatment Membrane Processes.* USA: McGraw-Hill, New York.

Azrague, K., Aimar, P., Benoit-Marquie, F., and Maurette, M.T. 2007. A new combination of a membrane and a photocatalytic reactor for the depollution of turbid water. *Appl. Catal. B Environ.* 72: 197–206.

Bolton G.L., and Hooper J.M. 2004. Selection of a UV disinfection system for Busselton water. *Water J.* 31 (February): 22–25.

Chaudhary, D. S., Vigneswaran, S., Ngo, H. H., Shim, W. G., and Moon, H. 2003. Granular activated carbon (GAC) biofilter for low strength wastewater treatment. *Environ. Eng. Res. KSEE* 8(4): 184–192.

Chen, Y., Dong, B.Z., Gao, N.Y., and Fan, J.C. 2007. Effect of coagulation pretreatment on fouling of an ultrafiltration membrane. *Desalination* 204(1–3): 181–188.

Cheryan, M. 1998. *Ultrafiltration and Microfiltration Handbook.* Lancaster, PA: Technomic.

Ernst, M., Sachse, A., Steinberg, C.E., and Jekel, M. 2000. Characterization of the DOC in nanofiltration permeates of a tertiary effluent. *Water Res.* 34(11): 2879–2886.

Erdei, L. 2008. *Membrane Hybrid Systems and Fouling Index.* Thesis, Doctor of Philosophy, Faculty of Engineering, University of Technology Sydney, Australia.

Gander, M., Jefferson, B., and Judd, S. 2000. Aerobic MBRs for domestic wastewater treatment: A review with cost considerations. *Sep. Purif. Technol.* 18: 119–130.

Germain, E., Stephenson, T., and Pearce, P. 2005. Biomass characteristics and membrane aeration: Toward a better understanding of membrane fouling in submerged membrane bioreactors (MBRs). *Biotechnol. Bioeng.* 90: 316–322.

Guo, W.S., Shim, W.G., Vigneswaran, S., and Ngo, H.H. 2005. Effect of operating parameters in a submerged membrane adsorption hybrid system: Experiments and mathematical modeling. *J. Membr. Sci.* 247(1): 65–74.

Guo, W.S., Vigneswaran, S., Ngo, H.H., Nguyen T.B.V., Ben Aim R., 2006. Influence of bioreaction on a long-term operation of a submerged membrane adsorption hybrid system. *Desalination* 191(1–3): 92–99.

Hillis, P. 2000. *Membrane Technology in Water and Wastewater Treatment.* Cambridge, UK: Royal Society of Chemistry.

Ho, D. 2008. *A Photocatalysis Hybrid System for Effluent Organic Matter (EfOM) Removal from Biologically Treated Sewage Effluent (BTSE).* Thesis, Master of Engineering, Faculty of Engineering, University of Technology Sydney, Australia.

Jarusutthirak, C., Amy, G., and Croué, J.-P. 2002. Fouling characteristics of wastewater effluent organic matter (EfOM) isolates on NF and UF membranes. *Desalination* 145: 247–255.

Judd, S. J., Le-Clech, P., Taha, and T., Cui Z. F. 2001. Theoretical and experimental representation of a submerged membrane bio-reactor system. *Membr. Technol.* 135(7): 4–9.

Lebeau, T., Lelievre, C., Buisson, H., Cleret, D., Van de Venter, L.W., and Cote, P. 1998. Immersed membrane filtration for the production of drinking water: Combination with PAC for NOM and SOCs removal. *Desalination* 117(2): 219–231.

Le-Clech, P., Jefferson, B., and Judd, S. 2003. Impact of aeration, solids concentration and membrane characteristics on the hydraulic performance of a membrane bioreactor. *J. Membr. Sci.* 218: 117–129.

Khan, S. J. 2008. *Influence of Hydrodynamic and Physico-Chemical Approaches on Fouling Mitigation in a Membrane Bioreactor*, PhD Thesis (EV-08-03), School of Environment, Resources and Development, Asian Institute of Technology, Bangkok, Thailand.

Molinari, R., Borgese, M., Drioli, E., Palmisano, L., and Schiavello, M. 2002. Hybrid processes coupling photocatalysis and membranes for degradation of organic pollutants in water. *Catal. Today* 75: 77–85.

Mulder, M. 1996. *Basic Principle of Membrane Technology.* New York, Kluwer Academic, Springer.

Ng, H.Y., Hermanowicz, S.W. 2005. Membrane bioreactor operation at short solids retention time: Performance and biocharacteristics. *Water Res.* 39: 981–992.

Sakuma, H., Maki, Y., Tanaka, T., and Kabuto, T. 1997. Nitrogen removal by biological filter using floating media. *Proceedings of 6th IAWQ Asia-Pacific Regional Conference, Seoul, 1997*: 283–289.

Sappideen, S. 2000. *Hybrid Photocatalytic Oxidation of Organics over $TiO_2$.* PhD Thesis, University of New South Wales.

Schafer, A. 2001. *Natural organics removal using membranes: Principles, performance and cost.* Lancaster, PA: Technomic Publishing.

Shon, H.K., Vigneswaran, S., Ben Aim, R., Ngo, H.H., Kim, In S., and Cho, J. 2005. Influence of flocculation and adsorption as pretreatment on the fouling of ultrafiltration and nanofiltration membranes: Application with biologically treated sewage effluent. *Environ. Sci. Technol.* 39(10): 3864–3871.

Shon, H.K. 2006. *Ultrafiltration and Nanofiltration Hybrid Systems in Wastewater Treatment and Reuse.* Thesis, Doctor of Philosophy, Faculty of Engineering, University of Technology, Sydney, Australia.

Shon, H.K., Phuntsho, S., and Vigneswaran, S. 2008. Effect of photocatalysis on the membrane hybrid system for wastewater treatment. *Desalination* 225: 235–248.

Snyder, S.A., Villeneuve, D.L., Snyder, E.M., and Giesy, J.P. 2001. Identification and quantification of estrogen receptor agonists in wastewater effluents. *Environ. Sci. Technol.* 35: 3620–3625.

Stephenson, T., Judd, S., Jefferson, B., and Brindle, K. 2000. *Membrane Bioreactors for Wastewater Treatment.* UK: IWA Publishing.

Vigneswaran, S. 1991. *Application of Microfiltration for Water and Wastewater Treatment.* ENSIC, Asian Institute of Technology, No. 31, June 1991.

Visvanathan, C., Aim, R.B., and Parameshwaran, K. 2000. Membrane separation bioreactors for wastewater treatment. *Crit. Rev. Environ. Sci. Technol.* 30: 1–48.

Water Environment Federation (WEF). 1996. *Operation of Municipal Wastewater Treatment Plants*, 5th edition, MOP 11. Alexandria, VA: Water Environment Federation.

Yang, L., Chou, L., and Shieh, W. K. 2001. Biofilter treatment of aquaculture water for reuse applications. *Water Res.* 35: 3097.

Yoon, S. H., and Collins, J.H. 2006. The novel flux enhancing method for membrane bioreactor (MBR) process using polymer. *Desalination* 191(1): 52–61.

Zhang, R., Vigneswaran, S., Ngo, H.H., and Nguyen, H. 2006. Magnetic ion exchange (MIEX) resin as a pre-treatment to a submerged membrane system in the treatment of biologically treated wastewater. *Desalination* 192: 296–302.

Zhang, R., Vigneswaran, S., Ngo, H.H., and Nguyen, H. 2008. Fluidized bed magnetic ion exchange (MIEX) as pre-treatment process for a submerged membrane reactor in wastewater treatment and reuse. *Desalination* 227(1–3): 85–93.

# 11 Membrane Emulsification
## *Current State of Affairs and Future Challenges*

*Karin (C.G.P.H.) Schroën, Cees J. M. van Rijn, and Remko M. Boom*

**CONTENTS**



### 11.1 INTRODUCTION

For the dispersion of one liquid in another, various methods are available such as high-pressure homogenization, rotor–stator systems, and ultrasound treatment, and they all have their specific pros and cons. Besides these known methods, some emerging technologies are becoming of age and are expected to move into the emulsification field soon. Various forms of membrane emulsification have been reported since Nakashima and Shimizu filed their first Japanese patent in 1986 (Nakashima and Shimizu, 1986); among others, cross-flow emulsification and premix emulsification have been used for various applications. The newest development in emulsion technology is the use of microtechnological devices; and although most are still operated at small scale, they are very useful tools to investigate emulsion formation in detail.

This chapter gives a brief introduction on various emulsification techniques including membrane emulsification, and they will be compared on various aspects. Next, a number of membrane emulsification techniques will be discussed in detail together with the various applications for which they can be used. In the last sections, microtechnological devices will be shortly touched upon, in relation to droplet formation, and the challenges that need to be tackled to mature this technology and membrane emulsification toward practical application.

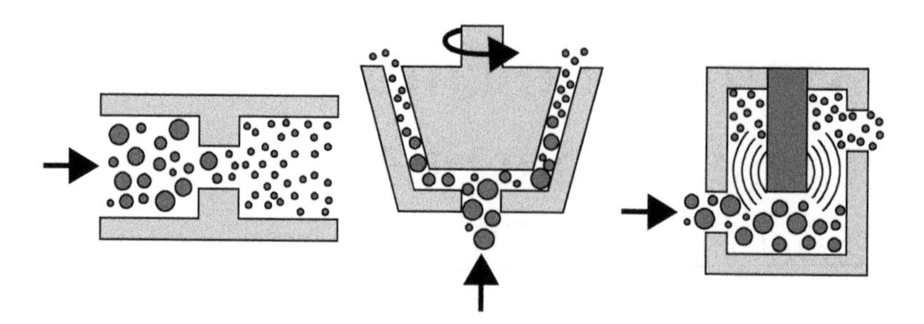

FIGURE 11.1 Schematic representation of classic emulsification methods. From left to right: high-pressure homogenizer, rotor–stator system, and ultrasound. (From van der Zwan, E. A., et al., *Am. Inst. Chem. Eng. J.*, 54, 2190–2197, 2008. With permission.)

## 11.2 COMPARISON OF EMULSIFICATION TECHNIQUES

In many industries, oil and water need to be mixed intimately in order to obtain a fine emulsion. Some examples of emulsions are paints, spreads, sauces, cosmetic crèmes, pharmaceutical ointments etc. Since a (macro) emulsion is intrinsically instable, it needs to be stabilized by components that adhere to the formed interface, therewith also reducing the interfacial tension. Mostly, the emulsification process is started with a mixture of oil and water, which is already a coarse emulsion. Upon passage of the emulsification device, the coarse emulsion is broken up into smaller droplets, and quite often, this process needs to be repeated to obtain the desired droplet size with a sufficiently narrow dispersion in droplet size. The methods of choice for industrial emulsification are high-pressure homogenization, rotor–stator systems, and ultrasound treatment, and they are schematically depicted in Figure 11.1. For high-pressure homogenization, the mixture is pushed through a small orifice at high velocity; in the rotor–stator system, the mixture is sheared between two walls; and during ultrasound treatment, cavitation results in rupture of droplets into smaller ones.

Membrane emulsification can be carried out as premix emulsification or as cross-flow emulsification (schematically depicted in Figure 11.2). In premix emulsification again a coarse emulsion is used, which is broken up upon passage through the membrane; also here repeated passage is used to meet the droplet size specifications. For cross-flow emulsification, one of the phases is dispersed into the other, via a membrane. Once a droplet is formed on top of the membrane, it is sheared off (mostly by the cross-flowing liquid) and carried away. Unlike the other emulsification methods mentioned in this introduction, no repeated passage can be used: the process conditions need to be such that this one-step method renders the correct droplet size.

When comparing the methods, various aspects can be considered, but this chapter will concern only those aspects that we consider most relevant, such as productivity, energy usage, droplet size,

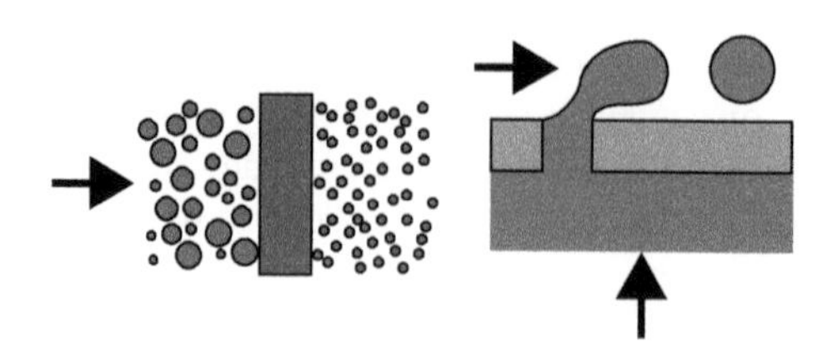

FIGURE 11.2 Schematic representation of premix membrane emulsification, and cross-flow emulsification. (From van der Zwan, E. A., et al., *Am. Inst. Chem. Eng. J.*, 54, 2190–2197, 2008. With permission.)

**TABLE 11.1**
**Comparison of Various Emulsification Methods**

| Method | Productivity | Energy Usage | Droplet Size | Monodispersity | Fouling |
|---|---|---|---|---|---|
| High pressure homogenization | ++++ | ++++ | Micron to submicron | + | 0 |
| Rotor–stator | +++ | ++++ | Microns | + | 0 |
| Ultrasound | +++ | ++++ | Micron to submicron | + | 0 |
| Premix membrane emulsification | ++ | ++ | Micron to submicron | ++ | – – – |
| Cross-flow membrane emulsification | + | + | Micron to submicron | +++ | – |
| Microtechnological emulsification | 0 | + | Micron to submicron | ++++ | – |

monodispersity of the emulsion, and sensitivity to fouling (Table 11.1). The top three entries in Table 11.1 are the classic technologies, and they are characterized through their high productivity albeit at high-energy consumption. The dispersity in droplet sizes that they yield is considered the reference against which any new technology has to compete, and is therefore indicated with one +. In general, the classic methods are not sensitive to fouling, unlike methods based on membranes and microtechnological devices. The membrane emulsification techniques have much lower energy consumption compared to the classic technologies, and maybe even more importantly, they can render emulsions with considerably narrower droplet size distributions. For as far as the actual droplet size is concerned, the membrane emulsification techniques are comparable with the classic technologies except for the rotor–stator systems which give larger droplets. The last entry in the table relates to microtechnological devices, which are known for their extremely narrow droplet size distributions, and their energy consumption is expected to be comparable to cross-flow membrane emulsification, and therewith considerably lower than in the classic technologies. Detailed information on the emulsification methods can be found in the four cited works of the group of Professor Schubert—Karbstein and Schubert (1995), Schröder and Schubert (1999), Behrend and Schubert (2001), and Lambrich and Schubert (2005)—and in Gijsbertsen-Abrahamse (2003) and Eisner (2007).

Although the last three entries in Table 11.1 are very attractive from the point of view of energy consumption, their productivity is not high enough to warrant large-scale practical application, albeit that for specific shear-sensitive products they are relevant because these products cannot be made otherwise. Good reviews that show the options of membrane emulsification in detail are available in literature. We recommend the work of Joscelyne and Trägårdh (2000; general review), Charcosset et al. (2004; general review), Vladisavljevic and Williams (2005; overview with many products), van der Graaf et al. (2005a; double emulsions), and Charcosset (2009; specific for food).

## 11.3 CHARACTERIZATION OF MEMBRANE EMULSIFICATION

It is clear from what is stated above that membrane emulsification could have a great future, if the productivity can be increased, and ideally, while the monodispersity is preserved. In order to reach this objective, thorough understanding of the droplet formation mechanism is needed. Various approaches for this can be found in literature, such as visualization, derivation of scaling relations and modeling. Interesting results are available for both cross-flow and premix emulsification, and below we give a selection of these findings.

### 11.3.1 Cross-Flow Membrane Emulsification

There are a number of techniques that we qualify as cross-flow membrane emulsification. The first one is the classical form that is depicted in Figure 11.2, where one phase is pressurized through a membrane; the cross-flowing continuous phase shears of the droplets and carries them away from the surface. The technique was first introduced by Nakashima and co-workers in Japan as early as 1986 (Nakashima and Shimizu, 1986), and later publications became available (e.g., Nakashima et al., 1991, for basic emulsions; Nakashima et al., 2000, for more complex systems). The technology did not stay an academic exercise; Morinaga Company experimented with membrane emulsification for food production (Katoh et al., 1996), and launched a product (a low fat spread) based on it.

At the end of the previous century and the beginning of this century, membrane emulsification was picked up by various scientific groups, e.g., the groups of Professor Trägårdh, e.g., Joscelyne and Trägårdh (1999) and Rayner and Trägårdh (2002); Professor Schubert, e.g., Schröder and Schubert (1999) and Lambrich and Schubert (2005); and our own group, e.g., Abrahamse et al. (2001) and van der Graaf et al. (2004). Further, we like to mention traveling scientist Dr. Goran Vladisavljevic, who worked within the research groups of Schubert (e.g., Vladisavljevic and Schubert, 2002, 2003a, 2003b; Vladisavljevic et al., 2004) and currently works at Loughborough University (Vladisavljevic and Williams 2005). The work of all these scientists gave a sound basis for membrane emulsification technology and kept it in the center of attention.

To generate the required shear for droplet detachment, different designs have been proposed in literature. Just to name a few, Stillwell et al. (2007) investigated a stirred cell in which a membrane was mounted at the bottom of a vessel and a stirrer was used to put the continuous phase in motion; however, the resulting emulsions were rather polydisperse due to the differences in shear across the membrane. Eisner-Schadler (Schadler and Windhab, 2006; Eisner, 2007), Aryantia et al. (2006), and Yuan et al. (2008) used a different approach by rotating the (metal) membranes to shear off the droplets. This results in better control over droplet size. When compared to regular cross-flow emulsification, the droplets are much larger. This is due to the technical limitations of the metal membranes that cannot be produced with the same pore sizes as other membranes.

Cross-flow emulsification is mostly used for oil-in-water emulsions or products that start as an oil-in-water emulsion (see also the premix emulsification section for a description of the preparation of some other products, or Vladisavljevic and Williams (2005) for an overview of products), and practically all references mentioned here refer to this case. For oil-in-water emulsification, a hydrophilic membrane is used; the to-be-dispersed phase should ideally not wet the membrane, otherwise the oil will stick to the membrane and form large droplets. A topic that is much less studied is emulsification of water in oil. In that case, a hydrophobic membrane is needed, which should remain hydrophobic during processing in spite of the presence of surface-active components. This latter effect is not very easy to achieve since surface-active components have a great preference for surfaces such as the membrane represents. When adsorbed on a hydrophobic surface, the components are expected to make the surface hydrophilic, and emulsification may be unsuccessful unless an effective layer is applied that does not change wettability (too much) (see, e.g., Schroën et al., 1993, for effects of protein adsorption). In the work of Vladisavljevic et al. (2002) on water in oil emulsification with polypropylene hollow fibers, also indications for wettability changes can be found. The researchers describe how pre-wetting of the membrane by oil reduces the droplet size, compared to pre-wetting with water (at least a factor of 2 in droplet diameter). Further, they also noted that the fluxes decreased strongly in time. Both effects may be indications for wettability changes taking place. Because of this, emulsification of water in oil is much more complex than of oil in water. Hydrophilic membranes are usually not strongly influenced by adsorbing species, and remain hydrophilic.

#### 11.3.1.1 Force Balance Analysis

Although the designs vary, all cross-flow technologies have the common characteristics that one phase is pressurized through a membrane and that the droplets are sheared off after a certain time, which determines their size. The first ones to derive equations to describe this cross-flow emulsification were Peng and Williams (1998), and their approach was followed by many others. Peng and Williams start from the forces acting on the forming droplet, the interfacial tension force, which keeps the droplet connected to the membrane pore, the shear force that tries to remove the droplet. In the case of big droplets, also the buoyant and inertia forces need to be considered; however, for emulsion droplets that are below 10 μm, only the interfacial tension and shear force are relevant, and buoyancy and inertia can be neglected.

The interfacial tension and shear forces can be summarized in a capillary number defined as:

$$Ca_c = \frac{\eta_c v_c}{\gamma_{ow}}. \tag{11.1}$$

In this capillary number, which is defined for the continuous phase, $\eta_c$ is the dynamic viscosity of the continuous phase, $v_c$ is the characteristic average velocity of the continuous phase, and $\gamma_{ow}$ is the interfacial tension between oil and water phase. At high Ca, the resulting droplets will be small because of the high shear (which may be a result of high continuous phase velocity or high viscosity or both) and/or low interfacial tension, and obviously at low Ca, the droplets are big because of opposite reasons.

In the work of van Rijn (2004; figure 15, page 368), the equation was used as a scaling relation for classic and microtechnological membranes to relate the droplet size to the Ca number. Rayner and Trägårdh (2002) used the capillary number directly to predict the droplet size. Although a reasonable description is obtained, the authors also noted that there is room for improvement, and the role of operating parameters needs to be elucidated further. Also, Abrahamse et al. (2002) started from the force balance when analyzing droplet formation with microsieves (discussed in more detail in the visualization section) and identified various factors that influenced the droplet size beyond what was expected from the basic force balance. This was confirmed by Lepercq-Bost et al. (2008), who used a ceramic membrane and found similar effects.

#### 11.3.1.2 Visualization

Although the force balance as proposed by Peng and Williams (1998) has contributed to better understanding of membrane emulsification, it is clear that observations that show more details of droplet formation are needed to come to a better understanding. Membranes have a pore size distribution and have a rough surface, which can make visualization hard (see, e.g., Christov et al., 2002). Therefore, Abrahamse et al. (2002) did pioneering work with microsieves which have uniform pores and a flat surface (van Rijn and Elwenspoek, 1995; van Rijn, 2004) (Figure 11.3). These devices are made through micromachining technology, which allows precise positioning of the pores and design of the pore geometry. In spite of the fact that these devices are very thin, they have considerable strength due to a layer of silicon nitride (van Rijn et al., 1997).

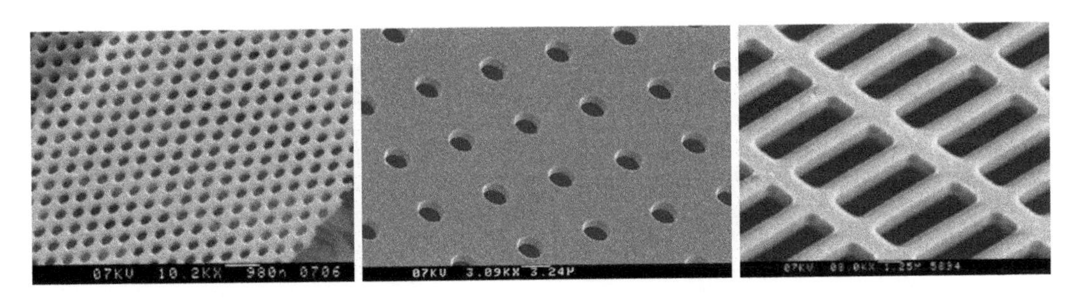


**FIGURE 11.3** Various microsieve designs. (From Aquamarijn Microfiltration B.V. With permission.)

In the work of Abrahamse et al. (2002), it was shown that the forming emulsion droplets influenced each other, which lead to premature droplet snap off, resulting in polydisperse emulsions. Further, pore activation was found to be low and dependent on droplet formation at neighboring pores and the hydraulic resistance of the membrane (as elaborated on in Gijsbertsen-Abrahamse et al., 2004), and therewith, it was clear that for monodisperse emulsions, not only the process conditions should be controlled but that also the membrane morphology plays a major role. This was confirmed in a paper by Vladisavljevic et al. (2007), who polished an SPG membrane to enable visualization. From this they could conclude, that although the membrane had a porosity of approximately 55%, the actual percentage of droplet forming pores was only 1% irrespective of the applied transmembrane pressure. This implied that the largest pores were activated first, while the neighboring smaller pores could not be activated. This was explained by the authors because of their much higher activation pressure, and in that respect, the available number of pores was used rather inefficiently. An alternative explanation is that a pore on activation experiences a pressure drop, as the maximum Laplace pressure does not have to be sustained. The resulting pressure drop inhibits the surrounding, connected pores from become active as well. The last explanation implies that low pore activation is intrinsic and not necessarily related to the polydispersity of the pores in the membrane. However, the findings could be used to design membranes/pores with better activation behavior as described by Gijsbertsen-Abrahamse et al. (2004). The pores should not communicate, their resistance should be high, and they should be placed at a certain distance from each other, in order to get uniform pore activation (see also Figure 11.4a) and monodisperse emulsions, and these can be achieved with microsieves.

#### 11.3.1.3 Computer Modeling

Because the basic force balance did not predict the experimental studies sufficiently accurate, some groups resolved to computer modeling. Abrahamse et al. (2001) reported their first results obtained by CFD and could, for example, show that the interfacial tension and the surface properties, as reflected in the contact angle, are very relevant for the droplet size. If the contact angle is such that the to-be-dispersed phase wets the membrane, emulsification will even be lost completely. Rayner et al. (2004) used the Surface Evolver (a package calculating the interfacial shape under static conditions) under conditions for which the force balance does not hold, and they could predict the droplet size with an average error of 8%, which is an improvement compared to using the force balance only. Within the group of Professor Drioli, the force balance models were adjusted to also include, among others, contact angle effects (De Luca and Drioli, 2006), and further torque balance models have been suggested in De Luca et al., 2008, but unfortunately, this did not lead to accurate predictions of the droplet size.

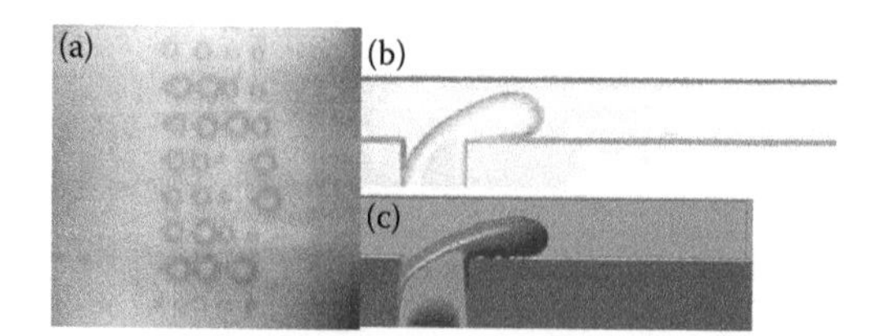


**FIGURE 11.4** (a) Top view of a microsieve with droplets forming from the pores. (From van der Graaf, S., *Membrane emulsification: Droplet formation and effects of interfacial tension*, PhD thesis, Wageningen University, 2006. With permission.) (b) Side view of a T-shaped junction with the oil pressurized from the bottom channel into the horizontal channel with the cross-flowing continuous phase. (From van der Graaf, S., et al., *Colloid. Surface. A*, 266, 106–116, 2005. With permission.) (c) Result from Lattice Boltzmann simulation for conditions as in the image above. (From van der Graaf, S., et al., *Langmuir*, 22, 4144–4152, 2006. With permission.)

Although the modeling results are once again a step forward in understanding membrane emulsification, the questions regarding the actual mechanism remain (also discussed further in the microtechnology section). Connected to this, the influence of the (dynamic) interfacial tension on droplet formation is still uncharted territory. As early as 1998, the importance of the dynamic interfacial tension was investigated by Schröder et al. (1998) through the bursting membrane technique, and it could be concluded that for conditions typical for membrane emulsification, a dynamic interfacial tension value is needed, which is between the equilibrium value that represents a surfactant saturated surface, and the value of an empty interface. The only work we know of that used dynamic interfacial tension values is by van der Graaf et al. (2004), where droplet volume tensiometry was used to extrapolate interfacial tension values to the conditions applied for droplet formation on a microsieve. In a way, it is not surprising that information on dynamic interfacial tension values is not available, given the lack of methods that allow analysis on the typical time scales relevant to membrane emulsification, which are orders of magnitude shorter than in standard interfacial tension measurement devices.

### 11.3.2 Premix Membrane Emulsification

Cross-flow membrane emulsification gives very monodisperse emulsions at low disperse phase fraction. When higher concentrations are targeted, the continuous phase needs to be recycled over the membrane and in general this leads to loss of monodispersity, either through action of droplets that shear off forming droplets, or side effects such as a pump that damages the produced emulsion. Therefore, premix emulsification is an interesting option for emulsions with higher disperse phase fractions. Unlike cross-flow emulsification, premix emulsification starts with an emulsion that needs to be dispersed further. Due to the applied transmembrane pressure, large droplets pass through constrictions, i.e., the pores of the membrane and break-up into smaller ones. In specific cases, phase transition (reversal) can be induced by the properties of the membrane, and highly concentrated emulsions (>90%) can be obtained. This is nicely illustrated in the work of Suzuki et al. (1993, 1999), who used a hydrophobic PTFE membrane to disperse the continuous water phase of the premix into the oil phase that wetted the membrane and subsequently became the continuous phase. But mostly, phase transition by the membrane is not targeted because this option only works for specific conditions.

When premix emulsification is used, it is mostly not used as a single stage operation but is applied repeatedly, as is the case in the classic emulsification techniques discussed previously. Away from the regular emulsions described for example by Ribeiro et al. (2005), surprisingly enough premix emulsification is also reported for preparation of vulnerable multiple emulsions starting from coarser multiple emulsions. Two early papers by Shima et al. (2004) and Vladisavljevic et al. (2004) show that premix emulsification can effectively be used to reduce the size of the droplets while also maintaining high encapsulation efficiency (>90%), in the case of the work of Shima and coworkers. For more information on the preparation of multiple emulsions, including membrane emulsification, please see the overview paper by Muschiolik (2007).

Further, premix membrane emulsification has been used for the production of (biodegradable) particles, and in this case, the various time scales that play a role need to be tuned very carefully (Sawalha et al., 2008b). A polymer solution (sometimes the monomer is used which is polymerized later) is mixed with another liquid in which the solvent can dissolve readily, but the polymer cannot. During emulsification, phase separation sets in and the polymer starts to solidify, and this should not take place faster than the actual emulsification process, otherwise the membrane will block. Some recent examples can be found in the work of Zhou et al. (2008) who prepared agarose beads of around 5 μm, and Lv et al. (2009) who reported chitosan beads that are in the nanometer range.

From solid polymeric beads, it is only a small step to hollow particles, which can be applied as ultrasound contrast agents. In this case, an additional liquid is added to the polymer solution, and this liquid is chosen such that it is not soluble in the continuous phase and remains inside the solidifying polymer droplet. Later this liquid can be removed, e.g., by freeze-drying. For the specific case of polylactide, much information was gathered by Sawalha (2009). A scaling relation was reported by Sawalha et al. (2008a), which relates, e.g., number of passes, interfacial tension, and applied pressure to the size of the particles. Further, the importance of the choice of the continuous phase is discussed by Sawalha et al. (2008b), which enables tuning of the time scales that are relevant to premix membrane emulsification. Kooiman et al. (2009) demonstrated that it is even possible to load hollow particles with an oil-soluble drug if two extra liquids are added to the polymer solution prior to premix emulsification. One of the liquids can be removed by freeze-drying, and forms the hollow core of the particle that can be activated by ultrasound, while the other liquid contains the drug that initially remains inside the particle, but can be released at increased ultrasound levels at which the bubble bursts.

The products discussed here for premix emulsification, and many others, can be produced by cross-flow membrane emulsification as well, and in some cases also by microtechnological devices. For an extensive review of the various options that are available, we like to refer to the work of Vladisavljevic and Williams (2005), and more specifically their figure 1.

When comparing cross-flow emulsification and premix emulsification, much more seems to be known about cross-flow emulsification. Many parameter studies have been carried out, but it has not been possible to bring the various results for either technique together in a comprehensive framework or model, which is not surprising given the numerous parameters that play a role, and the complexity of the process. One way to shed some light is to try to isolate as many parameters as possible, and the ideal tool to do so is microtechnology, which allows detailed observation in specifically designed structures, and well-defined conditions. Since this new development is in our opinion essential to bring membrane emulsification to maturity, we have dedicated the next section to this topic, and chose to discuss only the devices that are most closely related to membrane emulsification.

## 11.4 MICROTECHNOLOGY

In one of the previous sections, we presented results obtained with a microsieve: a microtechnological device with uniform pore size. It was used to visualize cross-flow emulsification, and helped to understand among others, pore activation. Although microsieves are very useful tools, unfortunately, they only allow observation from the top as described in van der Graaf et al. (2006; see Figure 11.4a), where a side view would allow better understanding of the droplet formation mechanism. Therefore, so-called T-shaped junctions were designed. The continuous phase flows through the horizontal channel in Figure 11.4, and the to-be-dispersed phase is pressurized through the vertical channel to form droplets at the T-junction. The dimensions of the device are chosen such that visualization can be carried out easily; the situation at the junction is expected to closely resemble the situation during cross-flow emulsification. Details can be found in the work of van der Graaf et al. (2005b), typical images of the microsieve and the T-shaped junction can be found in Figure 11.4.

From comparison of the images obtained by high-speed imaging, and Lattice Boltzmann simulations, new insights in droplet formation were obtained (van der Graaf, 2006). The force balance turned out to be more complex than originally presumed. Droplet formation seems to take place in two parts. In stage one, the droplet obtains a certain size at which the droplet is taken with the flow of the continuous phase. This droplet size can be estimated with a force balance. The droplet starts to move, even though it is still connected to the pore mouth with a neck and is therefore not yet detached. As long as the neck connects the droplet to the pore, it will grow further in size until the neck breaks. Because of this, the droplet size is not only a function of the applied shear rate, but also of the applied oil flow rate ("transmembrane pressure"), which makes droplet formation less simple than originally assumed. This can explain the differences in reported results. It is only fair to mention that in their original work, Peng and Williams (1998) indicated that there were limitations to the use of the force balance. They identified non-instantaneous droplet release as one of them.

To our knowledge, microtechnological devices have been used only by van der Zwan et al. (2006) to visualize premix emulsification. Not surprisingly, various break-up mechanisms were found in the constrictions that were investigated, such as snap-off due to localized shear, break-up due to interfacial tension effects, and break-up due to steric hindrance. Further, the interaction of the droplets leads to extensive droplet break-up, either in the investigated structure or in the layer of droplets that accumulates in front of the structure. Overall, premix emulsification turned out to be a complex process that was hard to quantify.

Within the field of microtechnology, other emulsification devices have been proposed as well, and most of these use spontaneous droplet generation. As an example to explain the principle, we use emulsification with a terrace structure, which is depicted in Figure 11.5 (the cover plate is invisible); in general, the term microchannel is used for this type of device. The to-be-dispersed phase is introduced through an inlet channel onto a shallow, wider area called the terrace. On the terrace, the phase assumes a disk shape due to the presence of the cover plate, and this disk grows until it reaches the end of the terrace. Now the liquid can move into the deeper part of the structure and assume a shape that is more spherical, and thus thermodynamically favorable. Once the size of the developing droplet exceeds a certain value, a droplet will be formed.

The microchannel was first introduced by Kikuchi et al. (1992) for the observation of blood cells, and later, it was adjusted for emulsion formation by Kawakatsu et al. (1997), again within the group of Professor Nakajima. Here, the technology was further investigated, which resulted in many papers by Sugiura and coworkers published between 2000 and 2004 (Sugiura et al., 2000, 2001a, 2001b, 2002a, 2002b, 2002c, 2002d, 2004a, 2004b). The technology was taken a step further within the same group, by Kobayashi and co-workers, who worked on a scale-up of the microchannels, resulting in straight-through emulsification devices. A selection of publications between 2002 and 2008 can be found in the reference list, which also includes some modeling studies (Kobayashi et al., 2002, 2004, 2005a, 2005b, 2005c, 2005d, 2006, 2007). In general, it can be said that the size of the droplet is determined by the dimensions of the terrace and the flow rate of the to-be-dispersed phase, *not* by the continuous phase and this makes this technology rather robust.

From the experimental work and the modeling studies carried out in Professor Nakajima's group a good understanding of microchannel emulsification was obtained, however not all parameters could be covered, and that was one of the reasons for our group to use different modeling approaches, to eventually be able to chart the window of operation of microchannels even better. van der Zwan et al. (2009b) used the lattice Boltzmann method, and were able to model the full emulsification process, including various aspects that were relevant for a geometric analysis of the system, derived shortly after that (van der Zwan et al., 2009a). This latter study yielded a design map based on an adapted capillary number that covers flow velocity, viscosity, and interfacial tension, but which also comprises the dimensions of the terrace. With this design map, the droplet diameter can be predicted over a wide range of conditions. Van Dijke et al. (2008) used CFD simulations to derive an

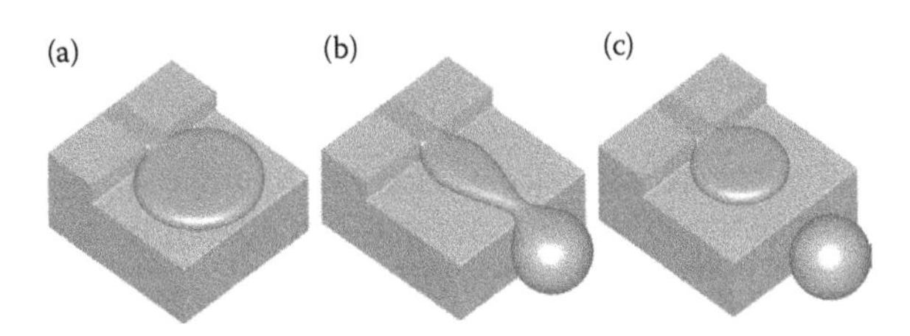


**FIGURE 11.5** Spontaneous droplet formation in terrace based systems. (a) The to-be-dispersed phase is pushed through the channel on the left onto the terrace where it forms a disk. As soon as the disk reaches the end of the terrace (b), the liquid can flow in a deeper part and assume a spherical shape. As soon as the size of the developing droplet exceeds a certain size, a droplet will be formed (c). (Reprinted with permission from van der Zwan, E., et al., *J. Colloid Interf. Sci.*, 335, 112–122. Copyright 2009b, American Chemical Society.)

analytical model that uses a flux criterion from which microchannel emulsification can be described for an even wider range of conditions, and this approach seems to have the best predictive value.

The work done in Professor Nakajima's group, plus the models derived by van der Zwan and van Dijke, show different aspects of microchannel emulsification and together they make that this (microtechnological) emulsification technology is currently understood best. Whether this also enhances the chances of microchannel emulsification, is discussed in the next section in which some future challenges are presented together with concluding remarks.

## 11.5 CONCLUDING REMARKS AND FUTURE CHALLENGES

When comparing the classic emulsification methods with membrane/microdevice emulsification, it is immediately clear that in spite of their high energy usage, and the polydispersity of the produced emulsions, classic methods are still very much in use. Various issues need to be resolved for the membrane/microdevice-related methods. However, given the current speed of the developments including detailed understanding through visualization and modeling, and the versatility of the technology in respect to specialty products that cannot or can hardly be produced with classic technology, it is expected that emulsification with membranes and microdevices will become a valuable new addition to the world of emulsification. An example of new products that become possible through the new methods is the production of capsules produced by layer-by-layer adsorption (Sagis et al., 2008), for which a monodisperse starting emulsion is needed to warrant uniform loading of the capsules.

When comparing the various shear-based emulsification methods like cross-flow emulsification and premix emulsification with spontaneous emulsification methods, it needs to be mentioned, that the rate at which droplets are formed from one pore are very different. In shear-based emulsification, the droplet formation rate is orders of magnitude higher than in spontaneous emulsification methods. However, the emulsions prepared by the latter methods are much more monodisperse than those produced by membrane emulsification, and that is expected to be an important "selling point" for microchannel technology for specific products. When combining these two conclusions, it would only be logical to work with shear-based microdevices that combine the best of both worlds: monodisperse emulsions at high production rates. An illustrative example is the microsieve for which results were discussed in the main text.

When looking further into the future, the shear-based Y-junction design that was investigated by Steegmans et al. (2009) could become of interest because the droplet size can be controlled more easily, i.e., by the continuous phase flow only. This could make them the preferred option over T-shaped junctions and membranes, for which both the oil and water phases need to be controlled. Further, it is known from the work of Nisisako et al. (2002) and Nisisako and Torii (2008) that shear-based microdevices can be scaled up (to some extent). Besides, microdevices do not necessarily have to be produced in silicon-related materials: some have been produced in polymer via various methods (e.g., Martynova et al., 1997; Roberts et al., 1997; Duffy et al., 1998; Liu et al., 2004; Narasimhan and Papautsky, 2004; Vogelaar et al., 2005), enabling mass production and hence mass parallelization of the devices. It is expected that this will bring down the price of microdevices, which, together with ease of operation and scalability, will be important in the further development of emulsification with microdevices.

An area, which is still relatively uncharted though relevant for any emulsification technology, is the influence of surfactants on interfaces. This can be reflected in the effect of the dynamic interfacial tension on droplet formation (e.g., van der Graaf et al., 2004), on product stability (e.g., Vladisavljevic and Schubert 2003a; Vladisavljevic et al., 2008), but also in wettability changes of the surface as discussed previously for water in oil emulsions. Regarding dynamic interfacial tension, it is expected that microtechnological devices will allow measurement of this parameter on scales that are relevant for any emulsification method (unpublished results of Steegmans and coworkers). This may well give a refreshing perspective on scaling studies. Within modeling, dynamic interfacial tension is still a complex topic, but progress is made using the lattice Boltmann method as is described in the work of van der Sman and van der Graaf (2006). Regarding wettability changes, surface modification can be used, and tailored to the specific surface that needs protecting and to the components from which it needs to be protected. For example for polypropylene membranes, an effective coating existing of block copolymers is reported which acts against protein adsorption (Schroën et al., 1993). For microtechnological devices based on silicon nitride and silicon carbide, modification methods have very recently become available, as is discussed in the work of Arafat et al. (2004, 2007) and Rosso et al. (2009). A range of molecules can be covalently attached to these surfaces, including some with, for example, protein-repelling functionalities.

If specific modification methods are not (yet) available and fouling (or difficult cleaning) is a mayor issue, as is often the case in premix membrane emulsification, the glass bead system presented by van der Zwan and et al. (2008) is an interesting alternative. It uses premix emulsification by passage through a layer of glass beads instead of a membrane, which was originally proposed as a model system for premix membrane emulsification. In practice, it turned out that the glass bead system is effective in emulsification, while it also could easily be cleaned by re-suspension of the particles, re-assembled, and re-used. The glass bead system is a creative way of using the knowledge obtained in membrane emulsification studies to come to solutions, and it is our firm belief that serendipity applied in this way will contribute to breakthroughs in the emulsification field. Also new (hybrid) developments in the field of microdevices should be mentioned in this respect. For example, Gu et al. (2008) and Malloggi et al. (2008) applied electrowetting in a microfluidic device, to speed up and control emulsion droplet generation. The valve system presented by Abate et al. (2009) for flow-focusing devices could also be interesting for scale-up given the formation frequencies that are in the kilohertz range. These are only a few examples, but it is clear that there are many options waiting to be applied for emulsion production.

## ACKNOWLEDGMENTS

We are indebted to Anneke Abrahamse for her microsieve work, Sandra van der Graaf for her work with T-junctions, Eduard van der Zwan for his premix and direct emulsification work, Hassan Sawalha for his work on ultrasound contrast agent preparation, Francisco Rossier for his encapsulation work, Koen van Dijke for his work on spontaneous emulsification, and Maartje Steegmans for her work on shear-based emulsification in Y-junctions.

## REFERENCES

Abate, A.R., M.B. Romanowsky, J.J. Agresti, and D.A. Weitz. 2009. Valve-based flow focusing for drop formation. *Appl. Phys. Lett.* 94: 023503.

Abrahamse, A.J., R. van Lierop, R.G.M. van der Sman, A. van der Padt, and R.M. Boom. 2002. Analysis of droplet formation and interactions during cross-flow membrane emulsification. *J. Membr. Sci.* 204: 125–137.

Abrahamse, A.J., A. van der Padt, R.M. Boom, and W.B.C. de Heij. 2001. Process fundamentals of membrane emulsification: Simulation with CFD. *Am. Inst. Chem. Eng. J.* 47: 1285–1291.

Arafat, A., K. Schroën, L. de Smet, E. Sudhölter, and H. Zuilhof. 2004. Tailor-made functionalization of silicon nitride surfaces. *J. Am. Chem. Soc.* 126: 8600–8601.

Arafat, A., M. Giesbers, M. Rosso, E.J.R. Sudhölter, C.G.P.H. Schroën, R.G. White, et al. 2007. Covalent biofunctionalization of silicon nitride surfaces. *Langmuir* 23: 6233–6244.

Aryantia, N., R.A. Williams, R. Houa, and G.T. Vladisavljevic. 2006. Performance of rotating membrane emulsification for O/W production. *Desalination* 200: 572–574.

Behrend, O., and H. Schubert. 2001. Influence of hydrostatic pressure and gas content on continuous ultrasound emulsification. *Ultrason. Sonochem.* 8: 271–276.

Charcosset, C. 2009. Preparation of emulsions and particles by membrane emulsification for the food processing industry. *J. Food Eng.* 92: 241–249.

Charcosset, C., I. Limayem, and H. Fessi. 2004. The membrane emulsification process—a review. *J. Chem. Technol. Biotechnol.* 79: 209–218.

Christov, N.C., D.N. Ganchev, N.D. Vassileva, N.D. Denkov, K.D. Danov, and P.A. Kralchevsky. 2002. Capillary mechanisms in membrane emulsification: Oil-in-water emulsions stabilized by Tween 20 and milk proteins. *Colloid. Surface. A* 209: 83–104.

De Luca, G., F.P. Di Maio, A. Di Renzo, and E. Drioli. 2008. Droplet detachment in cross-flow membrane emulsification: Comparison among torque- and force-based models. *Chem. Eng. Process.* 47: 1150–1158.

De Luca, G., and E. Drioli. 2006. Force balance conditions for droplet formation in cross-flow membrane emulsifications. *J. Colloid Interf. Sci.* 294: 436–448.

Duffy, D.C., J.C. McDonald, O.J.A. Schueller, and G.M. Whitesides. 1998. Rapid prototyping of microfluidic systems in poly(dimethylsiloxane). *Analytic. Chem.* 70: 4974–4984.

Eisner, V. 2007. Emulsion processing with a rotating membrane (ROME). Dissertation ETH Zürich, number 17153.

Gijsbertsen-Abrahamse, A.J., A. van der Padt, and R.M. Boom. 2004. Status of cross-flow membrane emulsification and outlook for industrial application. *J. Membr. Sci.* 230: 149–159.

Gijsbertsen-Abrahamse, A.J. 2003. Membrane emulsification: Process principles. PhD thesis, Wageningen University.

Gu, H., F. Malloggi, S.A. Vanapalli, and F. Mugele. 2008. Electrowetting-enhanced microfluidic device for drop generation. *Appl. Phys. Lett.* 93: 183507.

Joscelyne, S.M., and G. Trägårdh. 2000. Membrane emulsification—a literature review. *J. Membr. Sci.* 169: 107–117.

Joscelyne, S.M., and G. Trägårdh. 1999. Food emulsions using membrane emulsification: Conditions for producing small droplets. *J. Food Eng.* 39: 59–64.

Karbstein, H., and H. Schubert. 1995. Developments in the continuous mechanical production of oil-in-water macro-emulsions. *Chem. Eng. Process.* 34: 205–211.

Katoh, R., Y. Asano, A. Furuya, K. Sotoyama, and M. Tomita. 1996. Preparation of food emulsions using a membrane emulsification system. *J. Membr. Sci.* 113: 131–135.

Kawakatsu, T., Y. Kikuchi, and M. Nakajima. 1997. Regular-sized cell creation in microchannel emulsification by visual microprocessing method. *J. Am. Oil Chemist. Soc.* 74: 317–321.

Kikuchi, Y., K. Sato, H. Ohki, and T. Kaneko. 1992. Optically accessible microchannels formed in a single-crystal silicon substrate for studies of blood rheology. *Microvasc. Research.* 4: 226–240.

Kobayashi, I., X.F. Lou, S. Mukataka, and M. Nakajima. 2005a. Preparation of monodisperse water-in-oil-in-water emulsions using microfluidization and straight-through microchannel emulsification. *J. Am. Oil Chemist. Soc.* 82: 65–71.

Kobayashi, I., S. Mukataka, and M. Nakajima. 2004. CFD simulation and analysis of emulsion droplet formation from straight-through microchannels. *Langmuir* 20: 9868–9877.

Kobayashi, I., S. Mukataka, and M. Nakajima. 2005b. Effects of type and physical properties of oil phase on oil-in-water emulsion droplet formation in straight-through microchannel emulsification, experimental and CFD studies. *Langmuir* 21: 5722–5730.

Kobayashi, I., S. Mukataka, and M. Nakajima. 2005c. Novel asymmetric through-hole array microfabricated on a silicon plate for formulating monodisperse emulsions. *Langmuir* 21: 7629–7632.

Kobayashi, I., S. Mukataka, and M. Nakajima. 2005d. Production of monodisperse oil-in-water emulsions using a large silicon straight-through microchannel plate. *Industr. Eng. Chem. Res.* 44: 5852–5856.

Kobayashi, I., M. Nakajima, K. Chun, Y. Kikuchi, and H. Fujita. 2002. Silicon array of elongated through-holes for monodisperse emulsion droplets. *Am. Inst. Chem. Eng. J.* 48: 1639–1644.

Kobayashi, I., K. Uemura, and M. Nakajima. 2006. CFD study of the effect of a fluid flow in a channel on generation of oil-in-water emulsion droplets in straight-through microchannel emulsification. *J. Chem. Eng. Japan* 39: 855–863.

Kobayashi, I., K. Uemura, and M. Nakajima. 2007. Formulation of monodisperse emulsions using submicron-channel arrays. *Colloid. Surface. A* 296: 285–289.

Kooiman, K., M.R. Böhmer, M. Emmer, H.J. Vos, C. Chlon, W.T. Shi, et al. 2009. Oil-filled polymer microcapsules for ultrasound-mediated delivery of lipophilic drugs. *J. Control. Rel.* 133: 109–118.

Lambrich, U., and H. Schubert. 2005. Emulsification using microporous systems. *J. Membr. Sci.* 257: 76–84.

Lepercq-Bost, E., M.-L. Giorgi, A. Isambert, and C. Arnaud. 2008. Use of the capillary number for the prediction of droplet size in membrane emulsification. *J. Membr. Sci.* 314: 76–89.

Liu, H., M. Nakajima, and T. Kimura. 2004. Production of monodispersed water-in-oil emulsions using polymer microchannels. *J. Am. Oil Chemist. Soc.* 81: 705–711.

Lv, P.-P., W. Wei, F.-L. Gong, Y.-L. Zhang, H.-Y. Zhao, J.-D. Lei, et al. 2009. Preparation of uniformly sized chitosan nanospheres by a premix membrane emulsification technique. *Indust. Eng. Chem. Res.* 48(19): 8819–8828.

Malloggi, F., H. Gu, A.G. Banpurkar, S.A. Vanapalli, and F. Mugele. 2008. Electrowetting—a versatile tool for controlling microdrop generation. *Euro. Phys. J. E* 26: 91–96.
Martynova, L., L.E. Locascio, M. Gaitan, G.W. Kramer, R.G. Christensen, and W.A. MacCrehan. 1997. Fabrication of plastic microfluid channels by imprinting methods. *Analytic. Chem.* 69: 4783–4789.
Muschiolik, G. Multiple emulsions for food use. 2007. *Curr. Opin. Colloid Interf. Sci.* 12: 213–220.
Nakashima, T., and M. Shimizu. 1986. Porous glass from calcium alumino boro-silicate glass. *Ceram. Japan* 21: 408.
Nakashima, T., M. Shimizu, and M. Kukizaki. 1991. Membrane emulsification by microporous glass. *Key Eng. Mater.* 61–62: 513.
Nakashima, T., M. Shimizu, and M. Kukizaki. 2000. Particle control of emulsion by membrane emulsification and its applications. *Adv. Drug Deliv. Rev.* 45: 47–56.
Narasimhan, J., and I. Papautsky. 2004. Polymer embossing tools for rapid prototyping of plastic microfluidic devices. *J. Micromech. Microeng.* 14: 96–103.
Nisisako, T., and T. Torii. 2008. Microfluidic large-scale integration on a chip for mass production of monodisperse droplets and particles. *Lab Chip* 8: 287–293.
Nisisako, T., T. Torii, and T. Higuchi. 2002. Droplet formation in a microchannel network. *Lab Chip* 2: 24–26.
Peng, S.J., and R.A. Williams. 1998. Controlled production of emulsions using a crossflow membrane. Part I: Droplet formation from a single pore. *Trans IChemE* 76: 894–901.
Rayner, M., and G. Trägårdh. 2002. Membrane emulsification modelling: How can we get from characterisation to design? *Desalination* 145: 165–172.
Rayner, M., G. Trägårdh, C. Trägårdh, and P. Dejmek. 2004. Using the surface evolver to model droplet formation processes in membrane emulsification. *J. Colloid Interf. Sci.* 279: 175–185.
Ribeiro, H.S., L.G. Rico, G.G. Badalato, and H. Schubert. 2005. Production of O/W emulsions containing astaxanthin by repeated premix membrane emulsification. *J. Food Sci. E* 70: 117–123.
Roberts, M.A., J.S. Rossier, P. Bercier, and H. Girault. 1997. UV laser machined polymer substrates for the development of microdiagnostic systems. *Analytic. Chem.* 69: 2035–2042.
Rosso, M., M. Giesbers, A. Arafat, K. Schroën, and H. Zuilhof. 2009. Covalently attached organic monolayers on SiC and SixN4 surfaces: Formation using UV light at room temperature. *Langmuir* 25: 2172–2180.
Sagis, L.M.C., R. de Ruiter, F.J. Rossier Miranda, J. de Ruiter, K. Schroën, A. van Aelst, et al. 2008. Polymer microcapsules with a fiber-reinforced nanocomposite shell. *Langmuir* 24: 1608–1612.
Sawalha, H.I.M. 2009. Polylactide microcapsules and films: Preparation and properties. PhD thesis, Wageningen University.
Sawalha, H., Y. Fan, K. Schroën, and R. Boom. 2008a. Preparation of hollow polylactide microcapsules through premix membrane emulsification—effects of nonsolvent properties. *J. Membr. Sci.* 325: 665–671.
Sawalha, H., N. Purwanti, A. Rinzema, K. Schroën, and R. Boom. 2008b. Polylactide microspheres prepared by premix membrane emulsification—effects of solvent removal rate. *J. Membr. Sci.* 310: 484–493.
Schadler, V., and E.J. Windhab. 2006. Continuous membrane emulsification by using a membrane system with controlled pore distance. *Desalination* 189: 130–135.
Schröder, V., O. Behrend, and H. Schubert. 1998. Effect of dynamic interfacial tension on the emulsification process using microporous, ceramic membranes. *J. Colloid Interf. Sci.* 202: 334–340.
Schröder, V., and H. Schubert. 1999. Production of emulsions using microporous, ceramic membranes. *Colloid. Surface. A* 152: 103–109.
Schroën, C.G.P.H., M.C. Wijers, M.A. Cohen-Stuart, A. van der Padt, and K. van't Riet. 1993. Membrane modification to avoid wettability changes due to protein adsorption in an emulsion/membrane bioreactor. *J. Membr. Sci.* 80: 265–274.
Shima, M., Y. Kobayashi, T. Fujii, M. Tanaka, Y. Kimura, S. Adachi, and R. Matsuno. 2004. Preparation of fine W/O/W emulsion through membrane filtration of coarse W/O/W emulsion and disappearance of the inclusion of outer phase solution. *Food Hydrocolloids* 18: 61–70.
Steegmans, M.L.J., C.G.P.H. Schroën, and R.M. Boom. 2009. Characterization of emulsification at flat microchannel Y junctions. *Langmuir* 25: 3396–3401.
Stillwell, M.T., R.G. Holdich, S.R. Kosvintsev, G. Gasparini, and I.W. Cumming. 2007. Stirred cell membrane emulsification and factors influencing dispersion drop size and uniformity. *Industr. Eng. Chem. Res.* 46: 965–972.
Sugiura, S., M. Nakajima, S. Iwamoto, and M. Seki. 2001a. Interfacial tension driven monodispersed droplet formation from microfabricated channel array. *Langmuir* 17: 5562–5566.
Sugiura, S., M. Nakajima, N. Kumazawa, S. Iwamoto, and M. Seki. 2002a. Characterization of spontaneous transformation-based droplet formation during microchannel emulsification. *J. Phys. Chem. B* 106: 9405–9409.

Sugiura, S., M. Nakajima, and M. Seki. 2002b. Effect of channel structure on microchannel emulsification. *Langmuir* 18: 5708–5712.

Sugiura, S., M. Nakajima, and M. Seki. 2002c. Prediction of droplet diameter for microchannel emulsification. *Langmuir* 18: 3854–3859.

Sugiura, S., M. Nakajima, and M. Seki. 2002d. Preparation of monodispersed emulsion with large droplets using microchannel emulsification. *J. Am. Oil Chemist. Soc.* 79: 515–519.

Sugiura, S., M. Nakajima, and M. Seki. 2004a. Prediction of droplet diameter for microchannel emulsification: Prediction model for complicated microchannel geometries. *Industr. Eng. Chem. Res.* 43: 8233–8238.

Sugiura, S., M. Nakajima, J.H. Tong, H. Nabetani, and M. Seki. 2000. Preparation of monodispersed solid lipid microspheres using a microchannel emulsification technique. *J. Colloid Interf. Sci.* 227: 95–103.

Sugiura, S., M. Nakajima, H. Ushijima, K. Yamamoto, and M. Seki. 2001b. Preparation characteristics of monodispersed water-in-oil emulsions using microchannel emulsification. *J. Chem. Eng. Japan* 34: 757–765.

Sugiura, S., M. Nakajima, K. Yamamoto, S. Iwamoto, T. Oda, M. Satake, et al. 2004b. Preparation characteristics of water-in-oil-in-water multiple emulsions using microchannel emulsification. *J. Colloid Interf. Sci.* 270: 221–228.

Suzuki, K., K. Hayakawa, and Y. Hagura. 1999. Preparation of high concentration O/W and W/O emulsions by the membrane phase inversion emulsification using PTFE membranes. *Food Sci. Technol. Res.* 5: 234–238.

Suzuki, K., I. Shuto, and Y. Hagura. Application of membrane emulsification method for preparing food emulsions and emulsion characteristics, developments in food engineering. *Proceedings 6th International Congress on Engineering of Food*, Chiba. Japan, 1993, p. 167.

van der Graaf, S. 2006. Membrane emulsification: Droplet formation and effects of interfacial tension. PhD thesis, Wageningen University.

van der Graaf, S., T. Nisisako, C.G.P.H. Schroën, R.G.M. van der Sman, and R.M. Boom. 2006. Lattice Boltzmann simulations of droplet formation in a T-shaped microchannel. *Langmuir* 22: 4144–4152.

van der Graaf, S., C.G.P.H. Schroën, R.G.M. van der Sman, and R.M. Boom. 2004. Influence of dynamic interfacial tension on droplet formation during membrane emulsification. *J. Colloid Interf. Sci.* 277: 456–463.

van der Graaf, S., C.G.P.H. Schroën, and R.M. Boom. 2005a. Preparation of double emulsions by membrane emulsification—a review. *J. Membr. Sci.* 251: 7–15.

van der Graaf, S., M.L.J. Steegmans, R.G.M. van der Sman, C.G.P.H. Schroën, and R.M. Boom. 2005b. Droplet formation in a T-shaped microchannel junction: A model system for membrane emulsification. *Colloid. Surface. A* 266: 106–116.

van der Sman, R.G.M., and S. van der Graaf. 2006. Diffuse interface model of surfactant adsorption onto flat and droplet interfaces. *Rheol. Acta* 46: 3–11.

van der Zwan, E., K. Schroën, K. van Dijke, and R.M. Boom. 2006. Visualization of droplet break-up in premix membrane emulsification using microfluidic devices. *Colloid. Surf. A* 277: 223–229.

van der Zwan, E.A., C.G.P.H. Schroën, and R.M. Boom. 2008. Premix membrane emulsification by using a packed layer of glass beads. *Am. Inst. Chem. Eng. J.* 54: 2190–2197.

van der Zwan, E., K, Schroën, and R. Boom. 2009a. A Geometric model for the dynamics of microchannel emulsification. *Langmuir* 25: 7320–7327.

van der Zwan, E., R. van der Sman, K. Schroën, and R.M. Boom. 2009b. Lattice Boltzmann simulations of droplet formation during microchannel emulsification. *J. Colloid Interf. Sci.* 335: 112–122.

van der Zwan, E. 2008. Emulsification with microstructured systems. Process principles. PhD thesis Wageningen University.

van Dijke, K.C., K. Schroën, and R.M. Boom. 2008. Microchannel emulsification: From computational fluid dynamics to predictive analytical model. *Langmuir* 24: 10107–10115.

Van Rijn, C.J.M. 2004. Nano and micro engineered membrane technology. *Membrane Science and Technology Series 10*. Elsevier. ISBN 0444514899, 9780444514899, 384 p.

Van Rijn, C.J.M., and M.C. Elwenspoek. 1995. Micro filtration membrane sieve with silicon micro machining for industrial and biomedical applications. *Proc. Inst. Electr. Electron. Eng.* 29: 83–87.

Van Rijn, C.J.M., M. van der Wekken, W. Nijdam, and M. Elwenspoek. 1997. Deflection and maximum load of microfiltration membrane sieves made with silicon micromachining. *J. Microelectromech. Syst.* 6: 48–54.

Vladisavljevic, G.T., I. Kobayashi, and M. Nakajima. 2008. Generation of highly uniform droplets using asymmetric microchannels fabricated on a single crystal silicon plate: Effect of emulsifier and oil types. *Powder Technol.* 183: 37–45.

Vladisavljevic, G.T., I. Kobayashi, M. Nakajima, R.A. Williams, M. Shimizu, and T. Nakashima. 2007. Shirasu porous glass membrane emulsification: Characterisation of membrane structure by high-resolution x-ray microtomography and microscopic observation of droplet formation in real time. *J. Membr. Sci.* 302: 243–253.

Vladisavljevic, G.T., and H. Schubert. 2002. Preparation and analysis of oil-in-water emulsions with a narrow droplet size distribution using Shirasu-porous-glass (SPG) membranes. *Desalination* 144: 167–172.

Vladisavljevic, G.T., and H. Schubert. 2003a. Influence of process parameters on droplet size distribution in SPG membrane emulsification and stability of prepared emulsion droplets. *J. Membr. Sci.* 225: 15–23.

Vladisavljevic, G.T., and H. Schubert. 2003b. Preparation of emulsions with a narrow particle size distribution using microporous-alumina membranes. *J. Dispers. Sci. Technol.* 24: 811–819.

Vladisavljevic, G.T., M. Shimizu, and T. Nakashima. 2004. Preparation of monodisperse multiple emulsions at high production rates by multi-stage premix membrane emulsification. *J. Membr. Sci.* 244: 97–106.

Vladisavljevic, G.T., S. Tesch, and H. Schubert. 2002. Preparation of water-in-oil emulsions using microporous polypropylene hollow fibers: Influence of some operating parameters on droplet size distribution. *Chem. Eng. Process.* 41: 231–238.

Vladisavljevic, G.T., and R.A. Williams. 2005. Recent developments in manufacturing emulsions and particulate products using membranes. *Adv. Colloid Interf. Sci.* 113: 1–20.

Vogelaar, L., R.G.H. Lammertink, J.N. Barsema, W. Nijdam, L.A.M. Bolhuis-Versteeg, C.J.M. van Rijn, et al. 2005. Phase separation micromolding: A new generic approach for microstructuring various materials. *Small* 1: 645–655.

Yuan, Q., R. Houa, N. Aryantia, R.A. Williams, S. Biggs, S. Lawson, H. Silgram, M. Sarkar, and R. Birch. 2008. Manufacture of controlled emulsions and particulates using membrane emulsification. *Desalination* 224: 215–220.

Zhou, Q.-Z., L.-Y. Wang, G.-H. Ma and Z.-G. Su. 2008. Multi-stage premix membrane emulsification for preparation of agarose microbeads with uniform size. *J. Membr. Sci.* 322: 98–104.

# 12 Emerging Membrane Technologies and Applications for Added-Value Dairy Ingredients

*Laurent Bazinet, David P. Lachkar, and Yves Pouliot*

**CONTENTS**



## 12.1 INTRODUCTION

Milk is a constituent of many food products. The functionality of the various components in milk could be used more effectively if they were available separately. Fractionation of milk components enables a more constant quality of products (for example, cheese) and the development of new products such as bioactive peptides. Therefore, milk fractionation will lead to a more efficient and diverse use of milk. During the past 25 years, dairy products and dairy processing have seen many changes. The most important change that has probably influenced the dairy industry is the development of membrane technology. Membrane separation technology seems to be a logical choice for the fractionation of milk, because many milk components can be separated according to their sizes and/or charges.

## 12.2 BASICS ON MEMBRANE PROCESSES

### 12.2.1 Filtration

#### 12.2.1.1 Definition

Membrane processes are used to concentrate or fractionate a liquid. The separation process is based upon the selective permeability of one or more of the liquid constituents though a membrane in order to obtain two liquids with different composition: retentate and permeate. The retentate, which contains the solute(s), which is concentrated, is the part of the liquid retained by the membrane. The permeate, which is the part of the liquid that passes the membrane, will be deficient in the solute(s), which is concentrated. The mechanisms governing mass transport in different membrane processes vary as a function of the membrane type, process conditions, and equipment configuration. Generally, the driving force used to achieve the desired hydrodynamic flow though the membrane is a hydrostatic pressure gradient (Rosenberg, 1995; Bird, 1996).

#### 12.2.1.2 Filtration System

The pressure-driven separation technologies using membranes in the dairy industry include microfiltration (MF), ultrafiltration (UF), nanofiltration (NF), and reverse osmosis (RO). Table 12.1 summarizes the filtration spectrum of membrane processes applied to milk constituents.

### 12.2.2 Electrodialysis

#### 12.2.2.1 Definition

Electrodialysis (ED) is an electrochemical process based on the migration of electrically charged ionic species through perm-selective membranes, under an electric field. This electric field is created by the application of a potential difference between two electrodes immersed in an aqueous solution enriched in mineral or organic ionic species (Leitz, 1986).

In a conventional ED system, anion-exchange membranes (AEM) and cation-exchange membranes (CEM) are arranged in an alternating pattern. A frame spacer equipped with a grill serving both to support the membrane and to promote turbulence is placed between the AEM and the CEM to allow circulation of a solution between them. Each solution compartment is connected by an external circuit to a separate reservoir to allow recirculation of the solutions.

#### 12.2.2.2 Membranes

Under the influence of an electrical field, cations migrate toward the cathode. They leave the dilution compartment by crossing the CEM and are then retained in the concentration compartment because they cannot pass the AEM. Simultaneously, the anions migrate toward the anode. They leave the

**TABLE 12.1**
**Filtration Spectrum Available for the Separation of Milk Constituents**

| Filtration Process | Pore Size | Separation Domain |
|---|---|---|
| MF | >0.1 μm | Somatic cells, bacteria, spores, fat globules, casein micelles |
| UF | 1–500 nm | Soluble proteins, caseinomacropeptide |
| NF | 0.1–1 nm | Indigenous peptides, salts (divalent cations) |
| RO | <0.1 nm | Salts (monovalent cations), lactose |

*Source:* Pouliot, Y., *Internat. Dairy J.*, 18, 735–740, 2008. With permission.

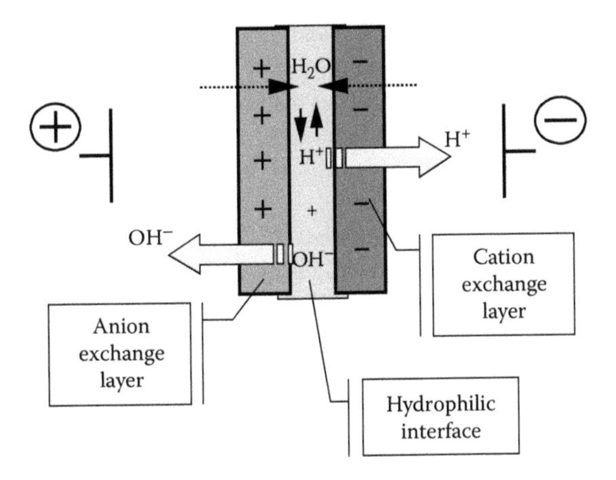


**FIGURE 12.1** Schematic representation of a bipolar membrane. (Adapted from Bazinet, L., *CRC Critical Review in Food Science and Nutrition*, 45, 307–326, 2005. With permission.)

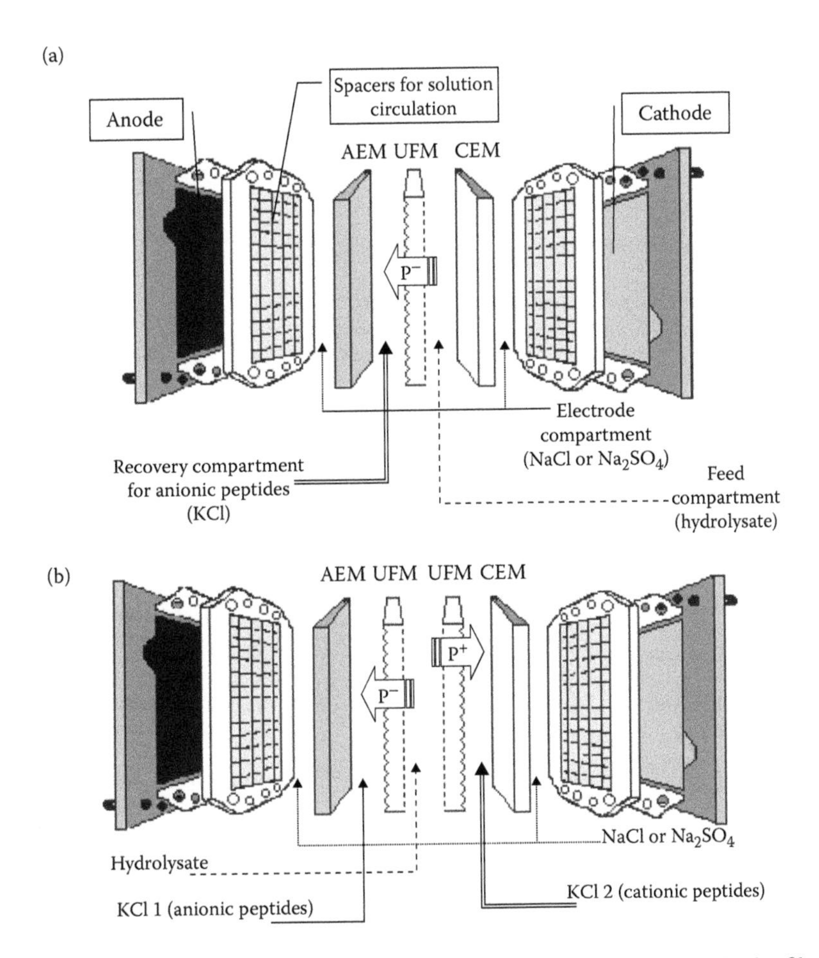


**FIGURE 12.2** Electrodialysis with ultrafiltration membrane cell configuration. UFM, ultrafiltration membrane; AEM, anion-exchange membrane; CEM, cation-exchange membrane. (a) With one UFM for separation of anionic peptides. (b) With two UFMs for simultaneous separation of anionic and cationic peptides. (Adapted from Bazinet, L., and Firdaous, L., *Recent Patents Biotechnol.*, 3, 61–72, 2009. With permission.)

dilution compartment, pass through the AEM, and are retained in the concentration compartment, as they cannot go through the CEM (Bazinet, 2005).

Bipolar membranes appeared commercially at the end of the 1980s. Bipolar membranes have the property to dissociate water molecules under an electric field and to generate a flow of protons $H^+$ at the cationic interface and a flow of hydroxyl ions $OH^-$ at the anion-exchange interface (Figure 12.1) (Pourcelly and Bazinet, 2008).

Electrodialysis with ultrafiltration membrane (EDUF), a technique developed recently by Bazinet et al. (2005a), has demonstrated very high selectivity for the separation of organic charged biomolecules. In this system, one or more ultrafiltration membrane is stacked as a molecular barrier in a conventional electrodialysis cell (Figure 12.2). EDUF couples size exclusion capabilities of UF or porous membranes with the charge selectivity of electrodialysis (ED).

## 12.3 APPLICATIONS TO ADDED-VALUE DAIRY INGREDIENTS

The applications of membrane processes in dairy processing can be classified into three main areas (Figure 12.3), namely: (1) alternatives to some unit operations such as centrifugation, evaporation, debacterization and demineralization, (2) means to resolve separation issues such as defatting of whey, protein recovery and separation, milk fat globule fractionation (Goudédranche et al., 2000), or recycling of solutions and spores removal, and (3) tools to create new dairy products such as UF cheeses (Ras, Pavé d'Affinois, Domiati, etc.), extended shelf life milk, whey-based beverages and textured milk products. This simple classification highlights the versatility the membrane processes have acquired over the years and their wide range of applications in the dairy industry.

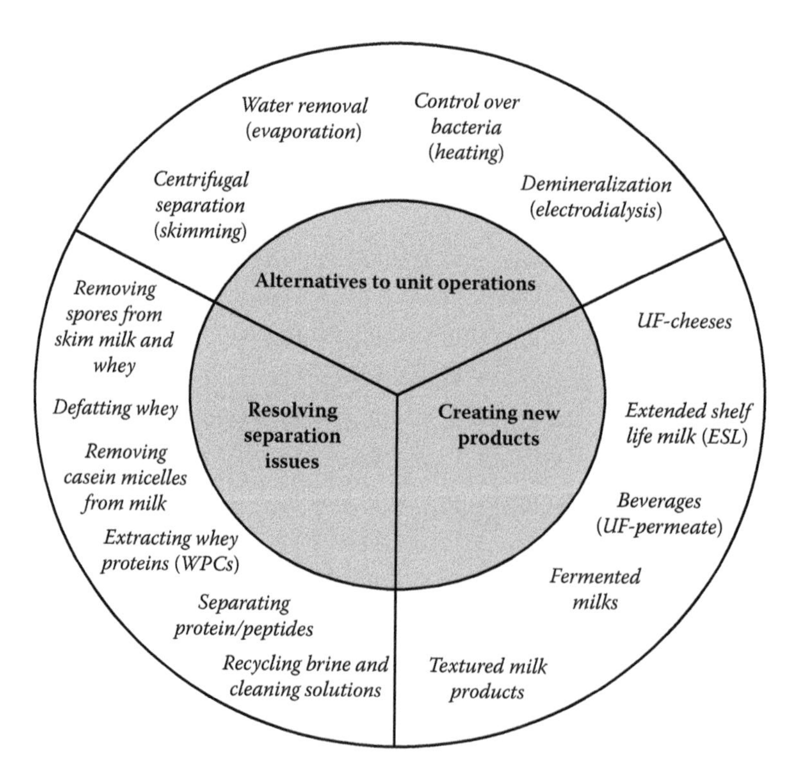


**FIGURE 12.3** Membrane processes in the dairy industry: a look at the applications. (From Pouliot, Y., *Internat. Dairy J.*, 18, 735–740, 2008. With permission.)

## 12.3.1 Proteins

### 12.3.1.1 Caseins

Whole milk contains 2.6% (w/w) casein, but the volume fraction of the micelles can be as much as 10% of the milk. Besides casein, the micelles consist of calcium, phosphate and water. On the outside, they have a "hairy" κ-casein surface. These "hairs" (glycomacropeptide, or GMP) cause steric hindrance and prevent aggregation of the casein micelles in milk. The diameter of the micelles is between 20 and 300 nm with an average of about 110 nm.

#### *12.3.1.1.1 UF/MF Combination*

The native casein micelles can be concentrated from skim milk by ceramic membranes with pore sizes between 0.05 and 0.2 μm. When whole or skim milk is circulated along a MF membrane with a pore size diameter of 0.1–0.2 μm, a microfiltrate with a composition close to that of sweet whey is obtained (Fauquant et al., 1988). This microfiltrate is crystal clear, sterile and can be claimed without any viral particle (Gautier et al., 1994) if the downstream equipment prevents recontamination. The MF retentate is an enriched milk solution of native micellar casein. Excepting the pore size of the MF membrane, engineering characteristics of the technology are the same as those used for bacteria removal of milk, i.e., use of a device leading to a uniform transmembrane pressure (UTP), a retentate velocity of at least 7 $m{\cdot}s^{-1}$, and a running temperature of 50°C–55°C. As proposed by Pierre et al. (1992) and by Schuck et al. (1994), purified micellar casein can be prepared according to this application of MF by following a process with three main steps: (1) concentration of the MF retentate up to a volumetric reduction factor (VRF) of 3–4; (2) diafiltration (DF) of this MF retentate with 4 diavolumes of RO treated water; and (3) concentration of the diafiltered MF retentate up to a VRF of 6–7.

Many variants of this process can be envisaged such as use of UF instead of MF at step (c), DF at lower or higher VRF or uses of saline or acid solutions for DF instead of pure water.

#### *12.3.1.1.2 Coupling Conventional ED with Chemical Acidification*

In this industrial procedure, the skim milk is acidified in an ED cell by batch processing to a pH ranging from 4.9 and 5.0 (Laiterie Triballat, 1979). The pH decrease of the solution is due to the generation of protons by dissociation of water molecules at the interface of ion-exchange membranes. This first step completed, the skim milk is separated from the ED cell circuit and the acidulation is completed with dilute acid until pH 4.6 is reached. Afterward, the curd is classically prepared by separation from the whey, washing, pressing, and drying to produce casein powder.

#### *12.3.1.1.3 SAFIR Process*

Conventional ED was used for the preparation of acid caseinates from fresh skim milk (Bolzer, 1985). The milk was circulated in a three-compartment ED cell to undergo a stoechiometric exchange of protons against milk constitutive cations. A main reservoir containing caseinate at pH 2.5 is used as pH control of the process. A known quantity of milk is added to the acid caseinate to reach a pH of 3.5. The mixture is then circulated in an ED cell, of which the outlet pH is about 1.8 to 2.4. Part of this very acid caseinate is used to maintain the pH of the control pH reservoir, while the remainder is supplied by pipe to the production process.

#### *12.3.1.1.4 Bipolar Membrane Electroacidification*

The skim milk was circulated in a three-pair-cell configuration, and the bipolar membrane electroacidification (BMEA) was performed in batch process using a current density of 20 $mA/cm^2$ (Bazinet et al., 1999). The pH of the solution was lowered from pH 6.6 to 4.4. Lowering the pH to 4.4 allowed a precipitation of 80%–85% of the total protein. The remaining proteins were whey proteins soluble at pH 4.4. This technology allowed the production of bovine milk casein isolates with a similar or higher protein content than those of commercial isolates (97% vs. 95% on a dry basis).

### 12.3.1.2 Whey Proteins

#### *12.3.1.2.1 Separation by UF*

Fractionation of the main whey protein, first attempted in the mid-1980s when Pearce (1988) proposed exploiting the low heat stability of calcium-free α-lactalbumin (α-La), has recently undergone significant improvements. α-La is a calcium metalloprotein containing one mole of ionic calcium per mole of protein. Removal of calcium from α-La by adjusting pH to about 3.8 or by the addition of a sequestering agent such as citric acid or sodium citrate (Bramaud et al., 1995) results in much reduced thermal stability compared to the native protein. Subsequent heating around 55°C for a limited period of time leads to a reversible and partially denatured form, which undergoes aggregation. This property of α-La is used in two processes developed by Pearce (1983) and Maubois et al. (1987). The first one uses whey concentrated by ultrafiltration to about 12% total solids. pH is adjusted to 4.2 and heated at 65°C for 5 min to cause aggregation of α-La. During this treatment, both immunoglobulins and the serum albumin coprecipitate with α-La. Separation of the precipitate is performed in a continuous desludging clarifier (Pearce, 1983). The supernatant is diafiltered through a 50-kDa cutoff UF membrane to yield purified β-lactoglobulin (β-Lg). In the second process (Maubois et al., 1987), the thermocalcic aggregation process prior to UF concentration up to volume reduction ration (VRR) 30, clarifies whey. A pH value of 3.8 and heating at 55°C for 30 min are used to allow coprecipitation of α-La, immunoglobulins and bovine serum albumin. Separation of a highly purified (95%) soluble β-Lg is carried out by centrifugation or by MF (pore size 0.1 μm) with a diafiltration step. The sediment or MF retentate yields a 70%-purity α-La fraction. Further improvement of the process and the purity of α-La fraction are to be expected from a novel cascade of separations (Muller, 1996): prepurification step by UF of defatted clarified whey (Muller et al., 1999); precipitation step carried out on the ultrafiltrate. The purity of the final fractions was 83%–97% for α-La and 0.98–0.96 for β-Lg. The coproducts have higher value (WPI 99) (Gésan-Guiziou et al., 1999) than those obtained with the process based on α-La precipitation directly from whey (Bramaud et al., 1997).

#### *12.3.1.2.2 Coupling Ultrafiltration and ED with pH Chemical Adjustment*

The method for separation of enriched β-Lg and α-La fractions from whey ultrafiltered was developed by Amundson and Watanawanichakorn (1982) and Slack et al. (1986). The pH is adjusted at pH 4.65 with concentrated HCl or NaOH before ED demineralization, to extract low molecular ions such as $Na^+$, $K^+$, $Ca^{2+}$ and $Mg^{2+}$. The pH of the demineralized concentrate is readjusted to pH 4.65, if necessary, with 0.1 N HCL or NaOH, and a precipitate containing mainly β-Lg is formed. This precipitate is separated from the α-La enriched fraction by centrifugation. However, with a maximum limiting 90% volume reduction of fresh acid whey (about 2.9% protein) by UF and demineralization by ED at pH 4.65, only 33.51% of the acid whey proteins were recovered (Bazinet, 2005).

#### *12.3.1.2.3 Coupling ED, pH Chemical Adjustment, and Thermal Treatment*

In their process, Stack et al. (1995) sought to provide an efficient integrated process for treating whey for the recovery of its constituents, particularly a considerably pure β-Lg fraction, an enriched α-La fraction, and lactose. The raw whey is treated by ED to reduce its mineral content, and then a cation-exchanger is used to remove calcium particularly. Next, the whey is subjected to a heat treatment at a temperature ranging between 71°C and 98°C for 50 to 95 s. The β-Lg remains soluble in a large extent. In the next stage, the pH of the liquor is adjusted to a pH of between 4.3 and 4.7 at a temperature less than 10°C and then heated to a temperature between 35°C and 54°C for 1 to 3 h. The α-La component of this whey liquor is flocculated at these temperatures (Bazinet, 2005). Stack et al. (1995) did not mention the yield of recovering for both whey protein fractions.

##### 12.3.1.2.4 *Bipolar Membrane Electroacidification*

A whey protein isolate (WPI) solution at different concentration was circulated in a one pair cell configuration on the cationic layer of the bipolar membrane. BMEA was carried out in batch process using a current density of 20 mA/cm$^2$, and the pH of the solution was lowered from pH 6.9 to 4.6. A 10% protein concentration was demonstrated as the best combination of electrodialytic parameters and protein recovery: separation of a 95.3% pure β-Lg fraction at a 53.4% yield of recovering. The β-Lg-enriched fraction contained only 2.7% of α-La for a 98% total protein purity (Bazinet et al., 2004a, 2004b).

#### 12.3.1.3 Minor Compounds

##### 12.3.1.3.1 *Lactoferrin*

Bovine lactoferrin (LF), an 80 kDa iron-binding glycoprotein, has been reported to have important nutraceutical and biological properties. However, with an average concentration of 0.1% of LF in whey, the large scale use of LF requires a purification process combining good selectivity, high extraction yield and low cost of production.

*12.3.1.3.1.1 Separation by MF* Nowadays, high-purity LF is obtainable on a laboratory scale using microfiltration affinity purification (Chen and Wang, 1991). In this process, a combination of affinity chromatography with membrane filtration was used to isolate lactoferrin and immunoglobulin G from cheddar cheese whey. Heparin Sepharose, protein G Sepharose, and protein G bearing Streptococcal cells were used as adsorbents to form affinity complexes with target proteins. These could be retained by a 0.2-μm MF membrane and separated from unbound whey proteins. Lactoferrin with 95% purity and 92% iron-binding capacity, and immunoglobulin G with 90% purity and 86% activity could be obtained with reasonable yields.

*12.3.1.3.1.2 Separation by Electromicrofiltration* Electrically enhanced membrane filtration (EMF) consists in superimposing an electrical field to a conventional membrane filtration unit. In EMF, the electrical field acts as an additional driving force to the transmembrane pressure (PT). Brisson et al. (2007) showed the potential of EMF for separation of more complex protein mixture such as cheese whey LF using MF membranes with 0.5 μm of pore size diameters.

*12.3.1.3.1.3 Separation of Lactoferrin by Electrodialysis with Ultrafiltration Membrane (EDUF)* To overcome the disadvantages of low selectivity and solution contamination of pressure-driven and chromatographic processes, EDUF was tested by Ndiaye et al. (2010). A 0.1% LF solution at pH 3.0 was treated by EDUF with an ultrafiltration membrane having a 500-kDa molecular weight cutoff (MWCO). EDUF treatments were carried out using 200 mL of LF solutions in a batch process using a constant voltage difference of 20 V. A migration rate going up to 46% was obtained after 4 h of treatment. Thus, it appears that the EDUF process could allow the separation of large proteins, such as LF, from dairy solutions (Ndiaye et al., 2010).

##### 12.3.1.3.2 *Growth Factors*

The fractionation of growth factors from milk or colostrum is challenged by a number of important factors, namely, the range of molecular weight (MW) and isoelectric point (pI) of growth factors relative to that of other constituents, and the occurrence of interactions between constituents. Since casein micelles are present in milk and colostrum, the main approach used to separate growth factors from colostrum has been to remove casein by acid or rennet treatment but others have made successful attempts at MF using diluted colostrum. In the case of whey, Francis et al. (1995) clearly demonstrated that UF membranes with a 3- to 100-Da MWCO could not successfully concentrate the mitogenic activity of whey. Nevertheless, Hossner and Yemm (2000) achieved the separation of IGF-I and IGF-II on a 30,000 Da UF membrane by performing DF at pH 8.0, which allowed the passage of IGF-II to the permeate due to its pI of 7.5. Maubois et al. (2003) proposed a combination of acidification and heat treatment to precipitate a TGF-β2-rich fraction from native whey. This precipitated material was further concentrated using a 0.1 μm MF membrane and was characterized by a predominant content in α-La, containing also 15% of the initial protein and 70% of the initial TGF-β2. Akbache et al. (2009) also demonstrated that UF/DF concentration of whey obtained from microfiltration of milk can potentially be used to produce growth factors extracts with high contents in TGF-β2 and low contents in IGF-I.

### 12.3.2 Bioactive Peptides

#### 12.3.2.1 Casein Peptides

##### *12.3.2.1.1 Glycomacropeptide*

GMP is a heterogeneous peptide of 64 amino acids formed by casein cleavage using chymosin during cheese production. The enzyme promotes rupture of the Phe105–Met106 bond of κ-casein, unstabilizing and denaturating the casein micelle, forming two peptides. One of the peptides, named para-κ-casein contains the residues 1 to 105 and stays in the casein micelle. The other peptide, known as GMP, is formed by the residues 106 to 169, which carry all of the sugars in casein. This peptide is also known as caseinomacropeptide (CMP) or casein-derived peptide (CDP).

There are currently two main methods to separate GMP from whey proteins using ultrafiltration. The first one is described by Kawasaki et al. (1993) and reveals the heterogeneity of GMP regarding its molecular mass with pH variation. The authors made CMP isolation from a WPC solution acidified to pH 3.5, by removing whey proteins by UF, the CMP being present in the permeate. Membranes with a higher or lower cut off can be used, making it possible to increase operational flow. The second method is described by Martin-Diana et al. (2002) and is based on the heat stability of CMP compared to other whey proteins. This process includes acidification, heating and ultrafiltration of milk whey to obtain powdered CMP with a protein content of 75%–79%. The authors recovered CMP from the whey protein concentrates of cow, ewe and goat milks, in levels between 71% and 76% and purity of 75% to 90%. Using this technique, the heat treatment at 90°C for 1 h causes complete denaturation and aggregation of whey proteins. They can be removed by centrifugation at 5200 g, 4°C for 15 min. The supernatant containing CMP can be concentrated by ultrafiltration with a 10-kDa membrane. The disadvantage of this method is that part of the whey proteins lose their functionality due to denaturation.

##### *12.3.2.1.2 Phosphopeptides*

Pierre et al. (1992) demonstrated that native phosphocaseinate can be separated from raw milk by tangential membrane microfiltration with a pore diameter of 0.2 μm followed by purification though water diafiltration.

##### *12.3.2.1.3 $\alpha_{s2}$-Casein Peptide Separation by Electromembrane Filtration (EMF) Technology*

Bargeman et al. (2002) used a specially designed filtration module to isolate positively charged peptides from an $\alpha_{s2}$-casein ($\alpha_{s2}$-CN) hydrolysate. After 4 h of EMF, the $\alpha_{s2}$-CN f (183–207) was the most important peptide in the permeate and was enriched from 7.5% of the total protein components in the feed to 25% in the permeate. In a further series of experiments, the authors observed that a higher transport rate of $\alpha_{s2}$-CN f (183–207) could be obtained by increasing the hydrolysate (feed) concentration, the applied potential difference and by decreasing the conductivity of the feed solution. The limits of this technology are probably the fact that it needs the use of a special designed module and that the membrane surface is limited by the flat design of the module.

#### 12.3.2.2 Whey Protein Peptides (Mainly β-Lactoglobulin)

Beta-lactoglobulin (β-Lg), the major whey protein, can release different bioactive peptidic sequences by enzymatic hydrolysis (Pihlanto-Leppälä et al., 1998; Bazinet and Firdaous, 2009). However, higher functionality would require the purification of these bioactive peptides from the protein hydrolysate. Among others, β-Lg f (142–148) has been reported to be the most potent antihypertensive sequence.

##### *12.3.2.2.1 Separation by UF*

A two-step ultrafiltration process can be used to obtain whey peptide fractions (Turgeon and Gauthier, 1990). In this study, whey protein concentrate (WPC) and heat-treated WPC (90°C, pH 2.5, 10 min) were hydrolyzed with trypsin and chymotrypsin in an ultrafiltration reactor (PM 30). The total hydrolysates were further fractionated, resulting in mixtures of polypeptides characterized by a high content of peptides larger than 5000 Da and fractions made up of amino acids and small peptides rich in small components (~2000 Da).

##### *12.3.2.2.2 Separation by NF*

Tsuru et al. (1994) showed that several nanofiltration membranes can be used to separate amino acids and peptides on the basis of charge interaction with the membranes since most of them contain charged functional groups. A membrane SG13 with an MWCO between 1000 and 5000 Da was shown to be the best to separate peptides from whey protein enzymatic hydrolysates (Pouliot et al., 1999). Groleau et al. (2004) showed that a G-10 NF membrane with an MWCO of 2.5 kDa is more selective for the β-Lg tryptic hydrolysate separation of basic peptides of MW <1000 Da than for larger negatively charged peptides. Chay Pak Ting et al. (2007) showed that the same kind of NF membrane has the best potential for separating the acidic, neutral, and basic peptides in whey proteins hydrolysates composed mainly by β-LG peptides because acidic peptides are completely retained at pH 9.

##### *12.3.2.2.3 Separation of β-Lg Bioactive Peptides by EDUF*

Poulin et al. (2006) were the first to evaluate the feasibility of separating basic/cationic and acid/anionic peptides from a β-Lg hydrolysate using one or several 20 kDa UF membrane stacked in ED cell. The authors demonstrated that EDUF was a selective method for bioactive peptide separation since among a total of 40 peptides in the raw hydrolysate, only 13 were recovered in the separated adjacent solutions (KCl 1 and KCl 2). In the best conditions, a total migration of 29% was obtained for an antihypertensive peptide (β-Lg f (142–148)) after only 90 min of processing with a relative concentration of 30% (Poulin et al., 2007).

##### *12.3.2.2.4 Separation of Lactoferrin Bioactive Peptides*

Using the same special flat design EMF module as previously described in section 3.2.1.3, Bargeman et al. (2000) isolated cationic peptides from a LF hydrolysate with a 10-kDa MWCO polysulfone flat-sheet membrane. Among the seven fractions revealed by HPLC analysis, they identified the potent bactericidal peptide Lfcin-B (LF f (17–42)). However, the Lfcin-B transport rate reached was relatively low: a reduction of the Lfcin-B concentration in the concentrate less than 15% was obtained after 180 min of treatment.

### 12.3.3 Lipids

#### 12.3.3.1 Separation of MFGM by MF

Separation of milk fat in small and large globules was recently proposed and patented by Goudédranche et al. (2000) through the use of special ceramic MF membranes, which causes very low damage to the native globular membrane. It was claimed that milks containing as fat only globules with a diameter lower than 2 mm led to smoother and finer textural characteristic dairy products: cream, liquid milk, cheese etc. than those made from reference milks or from milks containing large fat globules with the exception of butter. The observed results were ascribed to the ability of fat globule membrane components to bind water and to the difference in triglycerides composition.

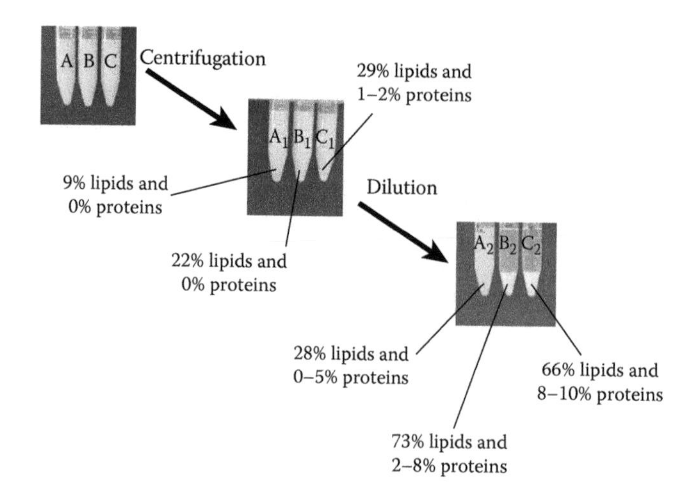


**FIGURE 12.4** Effect of bipolar membrane electroacidification (sample B) and coupled effect of electrodialysis demineralization and bipolar membrane electroacidification (sample C) on WPC (sample A) lipid and protein separation. (Adapted from Lin Teng Shee, F., et al., *J. Agri. Food Chem.*, 55, 3985–3989, 2007. With permission.)

#### 12.3.3.2 Separation of Phospholipids by BMEA

Lin Teng Shee et al. (2005) used BMEA technology to separate phospholipids from whey products. A 54% increase in precipitation rate was observed for BMEA technology after centrifugation (sample B1) in comparison with centrifugation alone (sample A1). Products A1, B1, and C1, which were the supernatants coming from the centrifugation of products A, B, and C (Figure 12.4), presented similar levels of total solids and proteins. Furthermore, the effect of the ionic strength on precipitation rates of fats was demonstrated by water dilution of the WPC samples after electroacidification/centrifugation (B2) and by coupling demineralization/electroacidification/centrifugation (C2) (Bazinet et al., 2005b; Lin Teng Shee et al., 2007). It appeared that the acidification by BMEA followed by a dilution resulted in the respective reduction of lipids of 73% and 66% for products B2 and C2. This new process would have two advantages, the production of a phospholipids-enriched fraction, which could be used in cosmeceutics and nutraceuticals, and a purified (demineralized and delipided) and more valuable protein fraction after concentration of the whey (Lin Teng Shee et al., 2007).

## 12.4 CONCLUSION

This chapter focused on the principal uses of membrane technologies for processing dairy products. But, some things could be developed in the future to improve these technologies. For example, combining filtration and electrodialysis could be studied because these techniques have always been studied in separate ways. Using different electric fields can also be studied because they can enhance the migration of targeted molecules and prevent membranes fouling. Moreover, the interactions between bioactive molecules and membranes should be understood because these interactions have an influence on mass transfer and fouling. Finally, new membrane materials could allow the developing of new applications.

## REFERENCES

Akbache, A., É. Lamiot, O. Moroni, S. Turgeon, S.F. Gauthier, and Y. Pouliot. 2009. Use of membrane processing to concentrate TGF-β2 and IGF-I from bovine milk and whey. *Journal of Membrane Science* 326:435–440.

Amundson, C.H., and S. Watanawanichakorn. 1982. Production of enriched protein fractions of β-lactoglobulin and α-lactalbumin from cheese whey. *Journal of Food Processing and Preservation* 6:55–71.

Bargeman, G., M. Dohmen-Speelmans, I. Recio, M. Timmer, and C. Van den Horst. 2000. Selective isolation of cationic amino acids and peptides by electro-membrane filtration. *Lait* 80:175–185.

Bargeman, G., J. Houwing, I. Recio, G.H. Koops, and C. Van den Horst. 2002. Electro-membrane filtration for the selective isolation of bioactive peptides from an $\alpha_{s2}$-casein hydrolysate. *Biotechnology and bioengineering* 80:599–609.

Bazinet, L. 2005. Electrodialytic phenomena and their applications in the dairy industry: A review. *CRC Critical review in Food Science and Nutrition* 45:307–326.

Bazinet, L., J. Amiot, J.-F. Poulin, A. Tremblay, and D. Labbé. 2005a. Process and system for separation of organic charged compounds. World patent no. 082495A1.

Bazinet, L., F. Ling Teng Shee, P. Angers, and W. Ben Ounis. 2005b. Method for extracting lipids from biological solutions. Demande de Brevet PCT/CA2005/000577.

Bazinet, L., and L. Firdaous. 2009. Membrane processes and devices for separation of bioactive peptides. *Recent Patents on Biotechnology* 3:61–72.

Bazinet, L., D. Ippersiel, and B. Mahdavi. 2004a. Fractionation of whey protein by bipolar membrane electro-acidification. *Innovative Food Sciences and Emerging Technology* 5:17–25.

Bazinet, L., D. Ippersiel, and B. Mahdavi. 2004b. Effect of conductivity adjustment on the separation of whey protein by bipolar membrane electroacidification. *Journal of Agricultural and Food Chemistry* 52:1980–1984.

Bazinet, L., F. Lamarche, D. Ippersiel, and J. Amiot. 1999. Bipolar membrane electro-acidification to produce bovine milk casein isolate. *Journal of Agricultural and Food Chemistry* 47:5291–5296.

Bird, J. 1996. The application of membrane systems in the dairy industry. *Journal of the Society of Dairy Technology* 49:16–23.

Bolzer, R. 1985. Installation and Process for the Preparation of Acid Caseinates. U.S. patent no. 4,559,119.

Bramaud, C., P. Aimar, and G. Daufin. 1995. Thermal isoelectric precipitation of alpha-lactalbumin from a whey protein concentrate: Influence of protein-calcium complexation. *Biotechnology and Bioengineering* 47:121–130.

Bramaud, C., P. Aimar, and G. Daufin. 1997. Optimization of a whey protein fractionation process based on the selective precipitation of alpha-lactalbumin. *Lait* 77:411–423.

Brisson, G., M. Britten, and Y. Pouliot. 2007. Electrically-enhanced crossflow microfiltration for separation of lactoferrin from whey protein mixtures. *Journal of Membrane Science* 297:206–216.

Chay Pak Ting, B.P., S.F. Gauthier, and Y. Pouliot. 2007. Fractionation of β-lactoglobulin tryptic peptides using spiral wound nanofiltration membranes. *Separation Science and Technology* 42:2419–2433.

Chen, J.P., and C.H. Wang. 1991. Microfiltration affinity purification of lactoferrin and immunoglobulin G from cheese whey. *Journal of Food Science* 56:701–706.

Fauquant, J., J.-L. Maubois, and A. Pierre. 1988. Microfiltration du lait sur membrane minérale. *Technique Laitière* 1028:21–23.

Francis, G.L., G.O. Regester, H.A. Webb, and F.J. Ballard. 1995. Extraction from cheese whey by cation-exchange chromatography of factors that stimulate the growth of mammalian cells. *Journal of Dairy Science* 78:1209–1218.

Gautier, M., A. Rouault, S. Méjean, J. Fauquant, and J.-L. Maubois. 1994. Partition of *Lactococcus lactis* bacteriophage during the concentration of micellar casein by tangential 0.1 μm pore size microfiltration. *Lait* 74:419–423.

Gésan-Guiziou, G., G. Daufin, M. Timmer, D. Allersma, and C. Van der Horst. 1999. Process steps for the preparation of purified fractions of alpha-lactalbumin and beta-lactoglobulin from whey protein concentrates. *Journal of Dairy Research* 66:225–236.

Goudédranche, H., J.J. Fauquant, and J.-L. Maubois. 2000. Fractionation of globular milk fat by membrane microfiltration. *Lait* 80:93–98.

Groleau, P.E., J.-F. Lapointe, S.F. Gauthier, and Y. Pouliot. 2004. Effect of aggregating peptides on the fractionation of β-LG tryptic hydrolysate by nanofiltration membrane. *Journal of Membrane Science* 234:121–129.

Hossner, K.L., and R.S. Yemm. 2000. Improved recovery of insulin-like growth factors (IGFs) from bovine colostrum using alkaline diafiltration. *Biotechnology and Applied Biochemistry* 32:161–166.

Kawasaki, Y., H. Kawakami, M. Tanimoto, S. Dosako, A. Tomizawa, M. Kotake, and I. Nakajima. 1993. pH-Independent molecular weight changes of κ-casein glycomacropeptide and its preparation by ultrafiltration. *Milchwissenschaft* 48:191–195.

Laiteries Triballat. 1979. Procédé et installation pour la préparation de la caséine à partir du lait et produits ainsi obtenus. French Patent no. 2,428,626.

Leitz, F.B. 1986. Measurements and control in electrodialysis. *Desalination* 59:381–401.

Lin Teng Shee, F., P. Angers, and L. Bazinet. 2005. Precipitation of cheddar cheese whey lipids by electrochemical acidification. *Journal of Agricultural and Food Chemistry* 53:5635–5639.

Lin Teng Shee, F., P. Angers, and L. Bazinet. 2007. Delipidation of a whey protein concentrate by electroacidification with bipolar membranes (BMEA). *Journal of Agricultural and Food Chemistry* 55:3985–3989.

Martin-Diana, A.B., M.J. Fraga, and J. Fontecha. 2002. Isolation and characterization of caseinomacropeptide from bovine, ovine and caprine cheese whey. *European Food Research and Technology* 214:282–286.

Maubois, J.-L., J. Fauquant, P. Jouan, and M. Bourtourault. 2003. Method for obtaining a TGF-beta enriched protein fraction in activated form, protein fraction and therapeutic applications. Patent PCT/WO 03/006500.

Maubois, J.-L., A. Pierre, J. Fauquant, and M. Piot. 1987. Industrial fractionation of main whey proteins. *Bulletin of the International Dairy Federation* 212:154–159.

Muller, A. 1996. Procédé d'obtention d'Alpha-Lactalbumine de haute Pureté: Étapes élémentaires du fractionnement des Protéines du Lactosérum et mise en Cascade. Ph.D. Dissertation, École Nationale Supérieure Agronomie, Rennes, France.

Muller, A., A. Daufin, and B. Chaufer. 1999. Ultrafiltration modes of operation for the separation of alpha-lactalbumin from acid casein whey. *Journal of Membrane Science* 153:9–21.

Ndiaye, N., Y. Pouliot, L. Saucier, L. Beaulieu, and L. Bazinet. 2010. Electroseparation of bovine lactoferrin from model and whey solutions. *Separation and Purification Technology* 74(1):93–99.

Pearce, R.J. 1983. Thermal separation of β-lactoglobulin and α-lactalbumin in bovine cheddar cheese whey. *Australian Journal of Dairy Technology* 38:144–149.

Pearce, R.J. 1988. Fractionation of whey proteins. *Bulletin of the International Dairy Federation* 212:150–153.

Pierre, A., J. Fauquant, Y. Le Graet, M. Piot, and J.-L. Maubois. 1992. Préparation de phosphocaséinate natif par microfiltration sur membrane. *Lait* 72:461–474.

Pihlanto-Leppälä, A., T. Rokka, and H. Korhonen. 1998. Angiotensin I converting enzyme inhibitory peptides derived from bovine milk proteins. *International Dairy Journal* 8:325–331.

Poulin, J.-F., J. Amiot, and L. Bazinet. 2006. Simultaneous separation of acid and basic bioactive peptides by electrodialysis with ultrafiltration membrane. *Journal of Biotechnology* 123:314–328.

Poulin, J.-F., J. Amiot, and L. Bazinet. 2007. Improved peptide fractionation by electrodialysis with ultrafiltration membrane: influence of ultrafiltration membrane stacking and electrical field strength. *Journal of Membrane Science* 299:83–90.

Pouliot, Y. 2008. Membrane processes in dairy technology—from a simple idea to worldwide panacea. *International Dairy Journal* 18:735–740.

Pouliot, Y., M.C. Wijers, S.F. Gauthier, and L. Nadeau. 1999. Fractionation of whey protein hydrolysates using charged UF/NF membranes. *Journal of Membrane Science* 158:105–114.

Pourcelly, G., and L. Bazinet. 2008. Developments of bipolar membrane technology in food and bio-industries. In *Handbook of Membrane Separations: Chemical, Pharmaceutical and Biotechnological Applications*, ed. A.K. Pabby, A.M. Sastre, and S.S.H. Rizvi, 582–633. New York: Marcel Dekker Inc.

Rosenberg, M. 1995. Current and future applications of membrane processes in the dairy industry. *Trends in Food Science & Technology* 6:12–19.

Schuck, P., M. Piot, S. Méjean, Y. Le Graet, J. Fauquant, G. Brulé, and J.-L. Maubois. 1994. Déshydratation des laits enrichis en caséine micellaire par microfiltration, comparaison des proprietés des poudres obtenues avec celles d'une poudre de lait ultra-propre. *Lait* 74:47–63.

Slack, A.W., C.H. Amundson, and C.G. Hill. 1986. Production of enriched β-lactoglobulin and α-lactalbumin whey protein fractions. *Journal of Food Processing and Preservation* 10:19–30.

Stack, F.M., M. Hennessy, D. Mulvihill, and B.T. O'Kennedy. 1995. World Patent no. 9534216-C1.

Tsuru, T., T. Shutou, S.I. Nakao, and S. Kimura. 1994. Peptide and amino acid separation with nanofiltration membranes. *Separation Science and Technology* 29:971–984.

Turgeon, S.L., and S.F. Gauthier. 1990. Whey peptide fractions obtained with a two-step ultrafiltration process: Production and characterization. *Journal of Food Science* 55:106–110,157.

# 13 Membrane-Based Separation Process for Juice Processing

*P. Rai and Sirshendu De*

## CONTENTS

## 13.1 INTRODUCTION

Membrane processes such as ultrafiltration (UF), microfiltration (MF), and reverse osmosis (RO) have been widely applied to the dairy, food, and beverage industry after the discovery of asymmetric membranes by Loeb and Sourirajan in the early 1960s. The major application of membrane technology in fruit juice processing is clarification and concentration.

The traditional method of juice clarification involves a number of steps involving centrifugation to remove suspended particle in juice, depectinization of juice, fining treatment to remove haze, and, finally, filtration by the diatomaceous earth to remove the fining agents. The major advantages of membrane-based processes over traditional processing are retention of enzymes by the membrane, savings in time, removal of fining agents (diatomaceous earth, gelatin, and bentonite), extraction efficiency of 95% to 97% (Swientek, 1983). Requirement of heat treatment of the juice is eliminated, and the juice may be cold-sterilized using aseptic processing and packaging.

Fruit juices are usually concentrated by multistage vacuum evaporation to reduce the storage volume and shipping costs and increase shelf life, but there is loss of fresh juice flavors, color, and taste due to the thermal effects. Membrane technology is a most promising alternative in maintaining the flavor of fresh juice. In this review, the application of membrane technology in fruit juice processing is discussed.

## 13.2 CLARIFICATION OF FRUIT JUICE

Application of UF for the clarification of fruit and vegetable juices has been extensively studied during the last 25 years. Heatherbell et al. (1977) introduced UF to clarify apple juice and were able to obtain a stable clear product. UF has also been investigated for the clarification of pear, orange, lemon, starfruit, kiwifruit, guava, pineapple, apple, and passion fruit juice (Kirk et al., 1983; Capannelli et al., 1984; Sulaiman et al., 1998; Wilson et al., 1983; Chan et al., 1982; Jiraratananon et al., 1997; De Bruijn et al., 2003; Jiraratananon et al., 1996). Juice clarification is an essential step before some treatments such as removing polyphenolic compounds, bitterness, tartness, and acids with adsorbent resins; deacidification by electrodialysis (ED); recovery of natural color substances; and concentration by membrane technologies (Carabasa et al., 1998; Goloubev et al., 1989; Laflamme et al., 1993; Bailey et al., 2000). The efficiency of these postclarification treatment processes is increased considerably by the removal of suspended solids.

The juice can be clarified by depectinization, followed by clarification by UF or MF (Cheryan, 1998). Pretreatment methods reduce the particulate matter in the juice, which improves the flux and allows the attainment of higher concentration factors.

### 13.2.1 Pretreatment of Juice for Clarification

Juice contains high concentration of pectin and other constituents such as cellulose, hemicellulose, and protein. These make the juice highly viscous, causing difficulty in the subsequent clarification process by UF and MF. Therefore, there is a need for the pretreatment of juice before the clarification step. The most common method for pretreatment of juice is enzymatic treatment by pectinase.

#### 13.2.1.1 Optimizing Enzyme Treatment of Juice

Pectins make the juice clarification process difficult because of their fiber-like structure and because they form a highly viscous gel-type layer on the membrane surface during filtration. Pectinase hydrolyzes pectin and causes pectin protein complexes to flocculate. The resulting juice has a much lower amount of pectins and lower viscosity, which is advantageous for the subsequent filtration process.

Depectinization directly leads to an improvement of the permeate flux during UF. Pectinase has been used for depectinization of tangerine juice before clarification at constant temperature (Chamchong et al., 1991). The commercial pectic enzyme was added to plums in varying concentration at constant temperature and time (Chang et al., 1995). Viquez et al. (1981) tested six commercial enzymes for clarification of banana juice at different enzyme concentrations for 2 h at 45°C. Depectinization by pectinase treatment depends upon time, enzyme concentration, temperature, and pH. So to make the enzymatic treatment more efficient, the determination of optimum condition of these process variables is important. Rai et al. (2004) used response surface methodology to optimize the enzyme treatment of mosambi juice (*Citrus sinensis* (L.) Osbeck). The optimum depectinization conditions are viscosity (minimum), clarity (high), and AIS (minimum). The process variables were determined using contour plots, and it was found to be as follows: time, 99.27 min; temperature, 42°C; and enzyme concentration, 0.0004% w/v.

#### 13.2.1.2 Alternative Pretreatment Methods

With pectinase being a costly method, alternate pretreatment methods to substitute pectinase have been investigated. Apart from pectin, the other factors affecting the filtration of juice are protein, fiber, suspended solid, etc. The suspended solids responsible for fouling are insoluble pectin, cellulose, hemicellulose, and lignin (Kirk et al., 1983). Therefore, apart from the pectin, other constituents also influence the permeate flux (Girard et al., 2000). Along with enzyme treatment, various treatments that have been researched upon are protease for removal of protein, centrifugation of juice, centrifugation of depectinized juice, and use of various fining agents (gelatin and bentonite), alone or in combination with enzymes. Pectinase used during UF of apple juice accounts for almost 25% to 30% of the total processing cost (Porter, 1990). Meyer et al. (2001) recommended that precentrifugation of juice is done during clarification of juice. The review for juice extraction suggested the combination of centrifugation and UF for juice extraction (Beveridge, 1997). The high-speed centrifuge (5000–10,000 g) is used for the separation of micron-sized particle, and that will finally be helpful during UF of juice (Cheryan, 1998).

Rai et al. (2006b) studied the efficacy of various pretreatment processes in comparison to pectinase treatment to find out the possibility of the substitution of pectinase for juice pretreatment. Based on the steady state permeate flux and cost of pretreatment, it was found that the pretreatment cost by centrifugation is about four times less compared to enzymatic treatment.

### 13.2.2 Selection of Membrane

Selection of membrane with proper molecular weight cutoff (MWCO)/pore size is important before going for clarification by UF and MF. Therefore, a membrane with appropriate MWCO should be selected to treat the pretreated juice.

Fouling that occurs during filtration is due to concentration polarization and adsorption inside the pores by complete, intermediate, and partial pore blocking (Field et al., 1995). The effect of each of these fouling mechanisms on flux decline depends on factors such as membrane material and pore size, operating conditions, and solute content and its particle size distribution (Bai et al., 2002).

Todisco et al. (1996) studied the effect of the fouling on permeate flux during clarification of orange juice by MF (0.3 μm). The permeate flux decline followed the cake filtration mechanism during low velocity (Re = 5000) and complete pore blocking at higher velocity (Re = 15,000). During the filtration of pineapple juice using MF (0.01 μm) and UF (100 K), it was found that during the first few minutes there was complete pore blocking followed by cake filtration (De Barros et al., 2003).

Rai et al. (2006a) selected the membrane for clarification of depectinized mosambi juice. The steady state permeate flux obtained during filtration was found to be independent of MWCO of various UF membranes. The flux obtained during MF was found to be less than all the UF membranes studied. Although complete or partial pore blocking might occur during initial stage of the filtration, the growth of the gel-type layer over the membrane surface by the higher molecular weight solutes dictates the long-term flux decline. A membrane of MWCO 50 kDa was found suitable for the clarification of depectinized mosambi juice.

### 13.2.3 Role of Pectin during Clarification by UF

Quantification of various parameters that affect the permeate flux is important to make the clarification process efficient. Pectin is a well-known gelling agent (Thakur et al., 1997), and its concentration in fruit juice is up to 1 wt. % (Sulaiman et al., 2001). Rai et al. (2005a) conducted a systematic parametric study of aqueous solutions of sucrose (only), pectin (only), a mixture of sucrose and pectin (synthetic juice), and depectinized mosambi juice in unstirred UF to observe the behavior of the permeate flux and permeate concentration. It was observed that the pectin present in the solution forms a gel-type layer over the membrane surface. The gel layer of pectin caused the rapid decline of the permeate flux for pectin–sucrose mixture. The sucrose, which was freely permeable through the membrane, diffused slowly through the gel of pectin, and the permeate concentration of sucrose increased gradually almost up to the level of the feed concentration. UF runs of the enzymatically treated juice exhibited a decline in the permeate flux due to the development of the leftover (after enzyme treatment) pectin gel on the membrane surface.

### 13.2.4 Modeling during Clarification of Juice

#### 13.2.4.1 Quantification of Flux Decline of Depectinized Juice

The membrane fouling can be quantified by rate of flux decline (Kelly et al., 1993), fouling time (De Bruijn et al., 2003), relative flux as the ratio of the steady state and the pure water flux (Mänttäri et al., 2000), fouling coefficients from logarithmic and exponential models during filtration (Kuo et al., 1983), fouling layer resistance (Girard et al., 1999), modified fouling index (Boerlage et al., 2002) and measurement of deposit thickness (Hamachi et al., 1999).

The flux decline and gel resistance during unstirred UF of enzymatically treated mosambi were analyzed by gel filtration theory (Rai et al., 2005b). The compressibility factor of gel layer was found to be 0.37 and the calculated gel layer thickness varies from 4.12 μm to 74.1 μm at different operating pressures at the end of experiment.

#### 13.2.4.2 Modeling of Sucrose Permeation through a Pectin Gel

Fruit juices contain mixtures of low molecular weight (LMW) and high molecular weight (HMW) solutes. The LMW compounds (MW<1 kDa) consist of sugars, organic acids, amino acids, pigments, vitamins, etc. and HMW components consists of proteins, enzymes, pectic substances, etc. During filtration of fruit juice by UF and MF, the nonpermeating class of solutes (consists of HMW) tends to form a gel on the membrane surface and LMW solutes pass though the gel and thereafter, through membrane. Since, sucrose is freely permeable through the high-MWCO UF membrane, the permeate sucrose concentration continuously increases and after a certain time of filtration, it approaches the feed concentration (Rai et al., 2005a). These data for sucrose diffusivity through the pectin gel is important for design of membrane modules and evaluation of the filtration performance.

Rai et al. (2006c) mathematically modeled the transport of the sucrose through the pectin gel and model was solved numerically and validated by experimental data. The hindered diffusion coefficient of sucrose was determined for various operating conditions by optimizing the experimental permeate concentration profile. It was found that the diffusion coefficient is reduced by a factor of 2 to 6 for various synthetic juice compositions.

#### 13.2.4.3 Modeling of Performance of a Filtration Unit at Steady State

A series of experiments using synthetic juice (sucrose and pectin mixture) as well as depectinized mosambi juice were conducted in a stirred UF cell under continuous mode of operation (Rai et al., 2007b). A model based on gel layer filtration was formulated to quantify the transient flux decline by modifying an already available model (De et al., 1997). The mass transfer coefficient was calculated from the steady state flux data, and a new pressure-dependent Sherwood number was proposed to estimate the mass transfer coefficient for such systems.

#### 13.2.4.4 Resistance in Series Model

In resistance in series model, the flux decline is due to irreversible membrane fouling (Wiesner et al., 1999) and reversible fouling (concentration polarization) over the membrane surface in addition to the membrane resistance (De et al., 1997). Vladisavljević et al. (2003) used the resistance in series model during UF of depectinized apple juice to study the variation of total resistance (sum of membrane and fouling resistance) with the feed flow rate. The variation of fouling resistance was studied with the filtration time and the effects of operating pressure and cross-flow velocity during clarification of apple juice by UF (De Bruijn et al., 2003).

Rai et al. (2006d) proposed a resistance in series model to quantify the flux decline during UF of depectinized mosambi juice. A systematic method was outlined to quantify the time variation of each constituent resistance, namely, the membrane resistance, adsorption, pore plugging and reversible fouling resistance. The contribution of fouling resistance was 60% to 74%, adsorption resistance was 4% to 6%, pore plugging resistance was from 9% to 27%, and membrane resistance was from 8% to 10% of the overall total resistance.

### 13.2.5 Flux Enhancement during Clarification

Higher permeate flux during clarification of fruit juice is important to make the membrane technology commercially viable for the fruit juice industry. Generally, two approaches are adopted to enhance the permeate flux: in the first, there are various pretreatments methods (using pectinase, pectinase plus amylase, fining by gelatin, bentonite, silica etc.), and in the second approach, hydrodynamic conditions are altered in the membrane flow chamber itself.

#### 13.2.5.1 Effect of Various Pretreatment Methods on Permeate Flux

The increase in permeate flux was found to be significant when depectinized apple juice was centrifuged at 4000 × g prior to UF using 20- and 50-kDa membranes in plate and frame modules (Sheu et al., 1987). The fining treatments have been reported prior to UF, and the flux was found to be higher (Youn et al., 2004). The fining agents most often used to clarify juices are gelatin and bentonite (Amerine et al., 1976). Apart from gelatin and bentonite, other fining agents such as tannin, casein, silica gel, and egg albumin also have some effect on clarifying fruit juices, but are seldom used (Chan et al., 1992).

Rai et al. (2007a) studied the effect of various pretreatment processes on the performance of UF of mosambi juice at fixed pressure and stirring speed. Various pretreatment methods investigated were centrifugation, fining by gelatin, fining by bentonite, fining by bentonite followed by gelatin, enzymatic treatment using pectinase, enzymatic treatment followed by centrifugation, and enzymatic treatment followed by fining using bentonite. The enzymatic treatment followed by bentonite was found to be the best, and permeate flux was 23 L/m$^2$h, which corresponded to about 77% increase in flux compared to only enzymatic treatment.

#### 13.2.5.2 Change in Hydrodynamic Conditions under Continuous Cross-Flow Mode

Various techniques have been used to minimize flux decline by altering the hydrodynamic conditions in flow channel. The devices suggested for promoting turbulence are paddle mixers, static mixers, kenic (series of helical elements) mixers, rods with rings, glass beads, moving balls, detached strips, and mesh-like spacers. These can be placed near the membrane surface to promote turbulent flow (Cheryan et al., 1998). The studies have reported the use of backflushing/washing and pulsed flow for improving flux during fruit juice clarification (Ben Amar et al., 1990). Rai et al. (2010) reported the effects of various hydrodynamic conditions (laminar flow, laminar flow with turbulent promoters, flow under transient, and turbulent regime) on the permeate flux during UF of synthetic and depectinized mosambi juice in cross-flow filtration.

#### 13.2.5.3 Effect of External Electric Field on Permeate Flux

The use of external electric field reduces the concentration polarization near the membrane surface by applying external d.c. electric field across the membrane in case of filtration of charged particles. The polarity of the electric field is such that electrically charged particles move away from the membrane surface by electrophoretic migration (Moulik et al., 1967; Mullon et al., 1985).

Sarkar et al. (2008) studied the effect of external electric field on the enhancement of permeate flux during clarification of mosambi (*C. sinensis* (L.) Osbeck) juice by UF, using a 50-kDa-MWCO flat-sheet polyethersulfone membrane in cross-flow UF under laminar flow conditions for a wide range of operating conditions. The permeate flux increased from 8.65 to 11.75 L/m$^2$h (i.e., 35.8% flux enhancement) at a cross-flow velocity of 0.12 m/s and a transmembrane pressure of 360 kPa by applying electric field of 400 V/m. Theoretical models, based on the use of integral method for boundary layer analysis was developed for the prediction of permeate flux. It was found to work well for the UF of the model synthetic juice because values of physical properties were known and well characterized (Sarkar et al., 2008).

### 13.2.6 Production of Cold-Sterilized Juices

The shelf life of juice processed by UF and MF is very much important to make these processes successful. The maintenance of aseptic environment during filtration and packaging is essential for longer shelf life of juice. The major problems during storage are nonenzymatic browning and loss of ascorbic acid. Using UF and MF with aseptic processing and packaging, a cold-sterilized juice could potentially be produced. Several parameters influence the rate of microbial spoilage, enzymatic degradation, chemical changes and flavor deterioration. These include hygiene during packaging, temperature of storage, oxygen exposure, light exposure and interactions with the package itself. Thus, the packaging process, the choice of packaging materials and the storage conditions will all have an impact on shelf life and product quality.

Heatherbell et al. (1977) filtered apple juice (having microorganism concentration $3.8 \times 10^4$ cfu/mL) with a 50-kDa MWCO polysulfone membrane to a twofold concentration. The permeate had low counts of microorganism (<1 cfu/mL), while the concentration of microorganism in retentate reached $1 \times 10^5$ cfu/mL. No evidence of spoilage was detected in aseptically bottled product after storage at room temperature for 6 months. The membrane sterilization applications rely on MF with an absolute rating, but for the UF membranes, the probability of contamination depends on the system and the associated nominal rating. All yeasts and molds and most bacteria are retained by MF membranes of 0.45 μm, but due to the development of a gel layer formation during filtration, coarser MF membranes of up to 0.8 μm may potentially allow the production of commercially stable filtrates. The shelf life study of apple juice clarified using MF and UF polymeric membranes is reported (Girard et al., 1999).

Good quality assurance practices must be followed in aseptic filling operation. Reid et al. (1990) describe the set up for aseptic filtration and filling of beer into polyethylene terephthalate containers. The bottle rinsing, filling, and capping operations were maintained in a filtered air environment to minimize contamination, and sterile water was used for the bottle cleaning.

The clarified mosambi juice was packaged and stored at both room and refrigerated temperature. The ultrafiltered juice stored in amber colored and transparent vials for 1 month, and its quality parameters (total soluble solid, pH, clarity, color, and ascorbic acid) were monitored at 5 days interval (Rai et al., 2008). The juice stored at room temperature was spoiled in 3 days. The soluble solid and pH of juice stored at refrigerated temperature in both the vials did not change with time. The degradation kinetics of ascorbic acid and nonenzymatic browning followed the zero-order and first-order kinetics, respectively, for both the storage vials.

## 13.3 CONCENTRATION OF FRUIT JUICE

The solids content of the juice is increased from 10% to 12% up to 65%–75% by weight during the concentration of fruits and vegetables juice. During concentration, significant amount of water is removed which is beneficial in reducing transport, packaging, and storage cost with greater stability of the concentrates (Ramteke, 1993). Concentration also assists in preventing microbial spoilage of the juice (Downes, 1990).

The commonly used technologies for juice concentration are evaporation and freeze concentration. The development of energy recovery system in evaporative concentration reduces energy consumption and it becomes comparable to other concentration techniques. But use of evaporator increases the capital cost by up to 300% (Addison, 1986).

Concentration by evaporation is associated with high temperature (more than 50°C) leading to the loss of most of the aroma compounds present in juice and the aroma profile undergoes an irreversible change (Maccarone et al., 1996). Juice concentration by evaporation is usually coupled with aroma stripping and the recovered aroma is added back to the concentrated juice. The aroma stripping process has drawbacks like high energy consumption, change of sensory attributes (color, taste and aroma), loss of nutrients (vitamin C) and transferring only 40%–65% of the total volatiles into the aroma concentrate (Lazarides et al., 1990). This problem can be reduced by vacuum evaporation process, but significant loss of aroma of juice due to evaporation is unavoidable, resulting in poor sensory attributes of the concentrated juice (Yu et al., 1986). Freeze concentration process gives concentrated juice of good quality since it operates at low temperature (van Niestelrooij, 1998) but its use is limited due to too high investment cost and final concentration up to 45%–50% only (Nguyen, 2000).

The membrane-based technologies are an alternative to the conventional process of juice concentration. It can be done at room temperature, which results in a low energy consumption and production of high quality juices. The different membrane techniques to concentrate juice are RO, direct osmosis (DO), membrane distillation (MD) and osmotic distillation (OD) (Courel, 1999; Girard et al., 2000; Medina et al., 1988; Vaillant et al., 2001; Cussler, 1984; Merlo et al., 1986; Merlo et al., 1986; Ishii et al., 1981; Watanabe et al., 1982; Pepper et al., 1985; Gherardi et al., 1986).

### 13.3.1 Reverse Osmosis

The juice concentration by RO is used in food industry for last 30 years, however, the concentration above 30% (total solid) is not economical due to severe fouling and reduction in permeate flux (Hogan et al., 1998; Nguyen, 1991). Membrane compaction during RO can affect the porous substructure of the membrane and it offers additional resistance to permeation during concentration (Leightell, 1972; Peri et al., 1973). In industry, it is practice to concentrate the juice in range of 42 to 65 Brix, so RO is used as pre concentration technique.

### 13.3.2 Direct Osmosis

The removal rate of water is proportional to the difference in osmotic pressures of the osmotic solution and the juice during DO. As the juice becomes more concentrated, its osmotic pressure increases and therefore, the rate of water removal decreases. The suitability of DO for higher concentration requires more study. An evaluation at industrial scale needs to be examined. Life of membrane is also important issue. DO requires high investment costs and high energy consumption during operation.

### 13.3.3 Membrane Distillation

MD involves thermally driven transport of water vapor through a porous hydrophobic membrane and its driving force is proportional to the partial pressure difference generated by temperature or a pressure difference. MD has been shown in laboratory trials for concentrating juice to higher levels than RO (Jiao et al., 1992). The loss of volatile components and heat degradation occur due to the heat requirement of the feed stream in order to maintain gradient of vapor pressure (Barbe et al., 1998).

### 13.3.4 Osmotic Distillation

In OD process, the energy consumption is less than RO and MD. The OD process operates at low temperature, which avoids degradation of heat-sensitive components, and overcome the drawbacks of RO and MD (Kunz et al., 1996).

#### 13.3.4.1 Effect of Operating Conditions on OD Flux

Clarified passion fruit juice was concentrated on an industrial scale using calcium chloride as a electrolyte at the concentration of 5.45 (M) and temperature 30°C, the evaporation flux decreased from 0.73 kg $h^{-1}$ $m^{-2}$ at 30 g TSS/100 g to 0.55 kg $h^{-1}$ $m^{-2}$ at 60 g TSS/100 g (Vaillant et al., 2001). There was a 25% decrease in flux due to an increase in concentration, and flux decline became more severe at 40 g TSS/100 g, which was the break-even point of the viscosity with respect to concentration. It is the point where the juice viscosity increases exponentially and the evaporation flux decreases steadily, even though brine concentration is either increasing or remains constant. Similar observations were made during concentration for sucrose solutions and orange juice (Courel et al., 2000; Cassano et al., 2003).

Bui et al. (2003) studied the effect of operating parameters such as temperature, feed concentration, brine and feed cross-flow velocity during concentration of glucose solution by OD. Feed concentration and brine velocity have significant effect on OD flux, but the importance of these factors depends on the range of feed concentration. The brine velocity is more important at lower feed concentration range. In the lower feed concentration range, the flux rate is higher. Hence, more severe polarization at the membrane surface on the brine side occurs, and consequently the flux rate is more dependent on the brine velocity at lower feed concentration range. The effect of the feed velocity on flux is not significant at lower feed concentration but significant at high feed concentration. The effect of temperature, which is exponential in nature, has strongest effect on OD flux. This is mainly due to the addition of temperature gradient as driving force and reduction of solution viscosity.

Bui et al. (2005) studied the effect of temperature and concentration polarization on the flux during concentration of glucose solution by OD. It was found that temperature polarization was less dependent on the operating conditions in comparison to the concentration polarization. The concentration and temperature polarization in OD were found to contribute up to 18% to the flux reduction. The flux reduction due to concentration polarization was found to be larger than the one due to temperature polarization.

Concentration of camu-camu juice up to 600 g $kg^{-1}$ was obtained using OD. Nutritional quality of the concentrated juice was very similar to the original one in terms of vitamin C content (losses below 5%) and flux was close to 10 kg $h^{-1}$ $m^{-2}$ (Rodrigues et al., 2004).

#### 13.3.4.2 Effect of OD on the Retention of Aroma during Concentration

The quantitative headspace GC analysis of 20–35 volatile components showed a loss of about 32% of the volatile components in orange and about 39% in passion fruit juices during concentration by MF followed by OD (Shaw et al., 2001).

Barbe et al. (1998) studied the effect of nine types of hydrophobic microporous membranes on the retention of a range of volatile organic species during OD of the model aqueous solutions, gordo grape juice and orange juice. The gas chromatography–mass spectrometry headspace analyses of the feed materials coupled with scanning electron microscopy and image analyses of the membranes indicated that the membranes with large surface pore diameters will be beneficial in industrial applications for retention of volatile flavor and fragrance components.

Ali et al. (2003) studied the transfer kinetics of four aroma compounds by GC during the concentration of a sucrose solution in a semi-industrial pilot plant by OD. The loss of aroma compounds decreased considerably by reducing circulation velocity, temperature of the solution and was found to be less than evaporation process.

In OD, the gradient of vapor pressure of water across the hydrophobic membrane is maintained by lowering the vapor pressure on the downstream side relative to that on the upstream side by using a concentrated brine stripper. But in case of MD, the requirement to heat the feed stream to maintain the water vapor pressure gradient, leads to a significant loss of organic volatiles. As in the case of MD, OD is not limited by the osmotic pressure of the feed. The pilot plant trials of grape and orange juice concentrated to about 70 Brix with both the feed and stripper streams maintained at ambient temperature (Johnson et al., 1989; Sheng et al., 1991).

#### 13.3.4.3 Osmotic Solutions and Its Management

The important properties to be considered for stripping solution are high water solubility, high surface tension (to get high penetration pressure), negligible volatility (to avoid counterdiffusion toward the juice and loss during regeneration), nontoxic, non corrosive and of low cost. The widely used stripping agents are solution of sodium and calcium chloride. Due to its low solubility, NaCl has limited driving force and hence it is not used for industrial purpose. Calcium chloride is more effective due to its high solubility but it is corrosive. Among the inorganic stripping solution, $CaCl_2$ is more effective. The most effective stripping solution is the potassium salts of ortho and pyrophosphoric acid. These have low equivalent weights, high water solubilities and steep positive temperature coefficients of solubility (Deblay, 1995; Hogan et al., 1998; Lefebvre, 1992; Michaels, 1998). Normally they are present in biological fluids and thus safe for food if present in low amount.

To overcome the problems associated with the use of brines, Celere et al. (2003) tried the organic stripping solution propylene glycol (PG) and Glycerol with good extractive power. These organics are easy to handle and result in comparable fluxes with advantages related to the absence of corrosion and scaling.

The problem with commercial application of OD is the management of the diluted osmotic solution and its concentration. Due to corrosion and scaling, the regeneration of exhausted brines becomes expensive. To make the process economical, it is essential to reuse it several times before it is removed from the process. Beaudry et al. (1990) and Herron et al. (1994) reported the use of an evaporator to reconcentrate the diluted osmotic solution. Apart from evaporation, other methods are solar ponds, RO, and pervaporation to reconcentrate osmotic solution (Thompson, 1991).

#### 13.3.4.4 Integrated Membrane Processes

The potential for concentrating fruit juice by integrated membrane processes for higher concentration and flux has been tried for industrial production. The effect of prefiltration by UF on concentration of grape juice using OD has been investigated using membranes with various pore diameters (Bailey et al., 2000). UF using membranes with pore diameters of 100 micrometer or less resulted in increase of OD flux in comparison to juice without UF. This is due to reduction in the viscosity of the concentrated juice–membrane boundary layer due to removal of protein.

Fresh kiwifruit juice, after enzymatic treatment, was clarified by UF followed by concentration using OD. The final concentration of TSS was higher than 60 Brix (at 25°C) with average evaporation fluxes of 1 $kg/m^2h$ using a calcium chloride dihydrate at 60 w/w% as stripping solution (Cassano et al., 2004). The concentrated juice had almost same total antioxidant activity and vitamin C content was preserved.

Cassano et al. (2003) used the integrated membrane processes for the concentration of citrus (orange and lemon) and carrot juice by using UF–RO–OD. The RO process was used to preconcentrate the permeate coming from the UF step up to 15–20 g TSS/100 g and final concentration by OD up to 60–63 g TSS/100 g at an average throughput of about 1 $kg/m^2h$. The juice obtained after concentration using OD preserved its total antioxidant content, color and aroma.

A pilot scale process involving MF followed by OD was used to concentrate orange and passion fruit juice (Shaw et al., 2001). The juice was concentrated threefold to 33.5 and 43.5 Brix, respectively.

Melon juice obtained from fruits discarded by exporters was concentrated using MF followed by OD. The process resulted into two products, a clarified concentrate of melon juice with its quality intact and a retentate (after MF) enriched in pro vitamin A, which can be used as raw material to extract beta-carotene or as functional drinks. The concentration of juice was 550 g TSS/kg with water evaporation flux in range of 0.70 to 0.57 kg/m$^2$h (Vaillanta et al., 2005).

The Wingara Wine group, an Australian Company has developed an integrated membrane process consisting of UF–RO–OD (of capacity 50 L/h) and a single stage brine evaporator for fruit juice concentration. The pre concentration by RO or UF–RO resulted in concentration of 65-70 Brix with quality comparable to that of OD alone, but at significant reduction in processing cost (Hogan et al., 1998). RO was used as preconcentration step to concentrate 18 Brix juice to about 30 Brix. It reduced the amount of water to be removed by OD to deliver a 68 Brix concentrate by about 56%. This reduces the evaporator capacity to concentrate brine solution by one-half and the membrane area for OD by over 30%, leading to a decrease in capital and operating costs.

#### 13.3.4.5 Pilot Plants and Industrial Applications

Several pilot plants have been established in Australia (Thompson, 1991; Johnson et al., 1989; Durham et al., 1994), France and industrial installations are also coming up in the United States and in Europe (Sheng, 1993). Orange, apple and grape juice were concentrated using plate and frame modules of membrane area 1 m$^2$ and flux 0.02 to 2.8 L/m$^2$h, depending on the nature and concentration of fruit juice (Sheng, 1993). The Australian Syrinx Research Institute concentrated fruit juice up to 75% solid using a spiral wound membrane and the flux varied between 5 and 10 L/m$^2$h. In Australia, wine was made from reconstituted concentrate grape and grape juice was concentrated using a pilot plant having a feed rate of approximately 80 to 100 L/h to produce approximately 20 to 25 L/h of 68 Brix concentrate (Thompson, 1991).

Concentration of fruit and vegetable juice has been successfully demonstrated at pilot facilities in Melbourne and Mildura, Australia. The Melbourne plant contains two 19.2 m$^2$ Liqui-Cel modules, along with UF and RO equipment and a brine evaporator to concentrate 50 L/h of juice to 65%–70% solid by weight. This plant contains twenty-two 4 in. × 28 in. Liqui-Cel modules (total interfacial area: 425 m$^2$) operated at 30°C–35°C and two atmosphere. Color, flavor and aroma retention were good but there is problem in module cleaning (Hogan et al., 1998; Sirkar, 1995; Sirkar, 1997).

## 13.4 ELECTRODIALYSIS APPLICATIONS

Electrodialysis is an electrically driven membrane separation process that is capable of separating, concentrating, and purifying selected ions from aqueous solution (as well as some organic solvents). The process is based on the property of ion exchange membranes to reject anions or cations selectively. ED can remove salts from food, dairy, and other products, waste streams and other solutions, as well as concentrated salts, acids or bases. The system is a useful tool to remove unwanted total dissolved solids that can build up in product streams.

Cation-selective membranes consist of sulfonated polystyrene, while anion-selective membranes consist of polystyrene with quaternary ammonia. Sometimes pretreatment is necessary before ED can take place. Suspended solids with a diameter that exceeds 10 μm need to be removed, or else they will plug the membrane pores. There are also substances that are able to neutralize a membrane, such as large organic anions, colloids, iron oxides and manganese oxide. These destabilize the selective property of the membrane. Pretreatment methods, which aid the prevention of these effects, are active carbon filtration (for organic matter), flocculation (for colloids) and filtration techniques.

### 13.4.1 Deacidification of Juices

Tropical fruit juices are appreciated for their intense aroma and flavor but reduction of its acidity is essential for its use in other food products (Bhatia et al., 1979; Johnson et al., 1985; Couture et al., 1992; Scott, 1995).

Various methods have been attempted for deacidification of fruit juice are conventional chemical method based on precipitation of calcium citrate using calcium hydroxide or calcium carbonate, ion-exchange (IE) and ED processes using homopolar or bipolar membranes. However, the application of the precipitation method has limitations regarding the legislation of some countries and that consumers' preference of natural products (Vera et al., 2003a). Among the various deacidification methods investigated, such as calcium salts precipitation and ion-exchange resins, ED presented great advantages because it produces good quality juice, without the use of chemicals and reduces the effluent production (Vera et al., 2003b).

ED has shown satisfactory performances for deacidification of various fruit juice such as orange (Goboulev et al., 1989), grape, pineapple (Adhikary et al., 1983) and passion fruit, castilla mulberry, naranjilla, and araza (Vera et al., 2007b). Moreover, the sensorial characteristics of fruit juices were only slightly modified (Vera et al., 2007a). The use of bipolar membranes (BM) for the deacidification has additional advantages. Since the electrohydrolysis of water produced $H^+$ and $OH^-$ ions, which can be used for the deacidification, and the production of organic acids (Bailly, 2002), leading to an economical and environmental benefits (Chuanhui et al., 2006).

### 13.4.2 Control of Enzymatic Browning

The demand for cloudy or unclarified apple juice is more and it is increasing (Hervé, 1997). It contains significant quantities of suspended pulp that supplies dietary fiber and important nutrients. However, it is very difficult to produce superior quality juice since cloudy apple juice is very sensitive to enzymatic browning because it contains considerable quantities of polyphenols and polyphenol oxidase (PPO) (Lea, 1990).

Zemel et al. (1990) reported that temporarily lowering the pH of apple juice to 2.0 and then adjusting it back to the same pH would irreversibly inhibit PPO activity and stabilize juice color. They added hydrochloric acid concentrate and caustic soda to the juice to adjust its pH. However, the treatment resulted in the formation of salts, which affect the flavor of the juice.

Tronc et al. (1997, 1998) demonstrated the feasibility of acidifying cloudy apple juice and returning the pH to its initial value without altering the flavor by using ED with bipolar and cationic membranes. However, the process was too lengthy, and the time required was more than 1 h, the energy required was too high (197 kWh/$m^3$ of juice), and it required the addition of exogenous KCl to the juice to reach pH 2.0.

Lam Quoc et al. (2006) demonstrated that a combination of acidification (by ED with bipolar membranes) with mild heat treatment increases the rate of inactivation of the PPO and PME. The best combination was mild heat treatment at 45°C for a juice acidified at pH 2.0, which reduced the time required to inactivate the enzymes to only 5 min. The color of the juice treated by combining ED and mild heat treatment was clearer than that of the juice treated by ED only, and its opalescence was more stable.

### 13.4.3 Demineralization of Juices

The alkali metal cations were suspected of being highly melassigenic; they hold sugar in the molasses and prevent it from being recovered during crystalline white sugar manufacturing. The order of melassigenic effect among the alkali cations follow the order K>Na>Ca>Mg (Khattabi et al., 1996). The methods tried in the sugar manufacturing industry to eliminate melassigenic ions are ion-exchange resins, synthetic adsorbents, coagulants, and membranes.

Removal of melassigenic ions for beet sugar solutions has been carried out using an improved ED stack equipped with high-alkali, temperature- and organic-resistant membranes and new spacers offering perfect mechanical stability at up to 60°C (Elmidaouia et al., 2006). Operating at temperatures higher than 40°C is obligatory to minimize the growth of microorganisms. The results obtained for the three treated solutions (juice, syrup, and mother liquor B) seriously encourage the use of ED in removing melassigenic ions from these solutions (Elmidaouia et al., 2004). The quality of the three solutions was improved, and an important gain in purity was observed. During all the period of tests, no contamination and no fouling were observed. Following this preliminary study, it is recommended that ED be introduced to obtain clear beet juice.

## 13.5 PERVAPORATION APPLICATION

Pervaporation is a membrane separation process in which the components from a liquid mixture permeate selectively through a dense membrane. The affinity between the permeant and the polymer material that constitutes the membrane, as well as its mobility through the membrane matrix, is responsible for its transport. These parameters define the membrane selectivity. The driving force in the pervaporation process is a chemical potential gradient obtained by reduction of partial pressure on the permeate side, which can be accomplished, for example, by applying vacuum or blowing a sweep inert gas on this side.

Aroma is responsible for the odor and taste of fruit juice. Aroma compounds can be aldehydes, alcohols, ketones, esters, lactones, etc. and are present in typically ppm or ppb levels. Conventional processes such as solvent extraction, flash distillation, and adsorption are the techniques primarily used for aroma recovery during fruit juice concentration by evaporation, which may cause undesirable changes in the flavor profiles and thus overall quality (Nisperos-Carriedo, 1990).

Pervaporation can be employed to recover the aroma compounds preserving the molecule integrity (due to the low thermostability of aroma compounds), having an efficient extraction process (high selectivity) and respecting the environment (avoiding solvent use, low energy consumption) (Shepherd et al., 2002; Karlsson et al., 1995). Number of studies has been conducted to evaluate the performance of the pervaporation process for the recovery of aroma compounds from apple, grape, orange, and kiwi fruit juices (Rajagopalan et al., 1995; Bengtsson et al., 1989; Shepherd et al., 2002; Cassano et al., 2004). Despite all favorable results obtained at laboratory scale, aroma recovery by pervaporation has not yet achieved industrial scale, and further studies on the subject are needed for experimental data on pilot-scale processing of real juices and/or industrial streams (such as condensate from juice evaporators) (Cristina et al., 2006).

## 13.6 FUTURE AREA OF RESEARCH

Fundamental issues such as interactions between juice compositions and the membrane materials and subsequent membrane fouling should be investigated. The centrifugation study should be conducted in details for substitution of enzyme treatment. The low cost aseptic packaging machine should be developed for packaging of clarified fruit juice. The details of composition of fruit juice in terms of soluble solid, suspended solid, colloidal particles in terms of amount and their particle size should be determined. The membrane bioreactor should be tried for clarification of fruit juice. Calculation of the extra energy involved in turbulent promoters, changing feed flow from laminar to turbulent and applying electric field for flux enhancement.

The serious drawback of the OD process is low flux, and therefore, efforts should be directed to optimize the operating conditions in order to attain industrially competitive fluxes. To make the process economically viable, there is a need to handle and reuse the diluted brine solution after concentration. To overcome the problems associated with the use of brines, search for easy-to-handle organic extractants with good extractive power should be made. There is a problem of corrosion and scaling during the regeneration of exhausted brines by evaporator, so the economic feasibility of other concentration techniques like solar evaporation, pervaporation, ED, etc. should be tried. A detailed study of integrated membrane technology for fruit and vegetable juices by using different combinations of UF, MF, RO, and OD should be done. The optimization of different processes in the integrated system to get a desired concentration and economics is required. To make the OD process cost effective, studies on developments of new membranes that are highly selective, permeable, and stable in long-term application for juice processing are needed. The process engineering, including module design, process design, and optimization, of the OD process in detail is required.

The combination of ED with bipolar membranes and heat treatment can be considered as a simple and efficient stabilization method applicable to fruit juices. However, to fully address the commercial viability of the ED process for stabilization of cloudy apple juice, a combination of both the acidification and pH readjustment steps in one ED operation will have to be carried out.

Although favorable results have been obtained at the laboratorial scale, aroma recovery by pervaporation has not achieved an industrial level. The experimental data on pilot-scale processing of real juice and condensate from juice evaporators should be generated with an assessment of the sensory quality of the final product. There is a need for comprehensive models for process design and simulation.

## 13.7 CONCLUSION

The use of membrane technology in fruit juice processing will increase with time and the consumer demand for juice of high quality. The major issue in juice clarification is high permeate flux to make the process commercially viable. Although the cost of OD in concentration of fruit and vegetable juice is higher than the existing technology, its ability to concentrate juice, as compared with evaporation, and to preserve the original quality of the juice makes it more attractable. The OD process overcomes the drawbacks of RO and MD in concentrating juice: RO is prone to high osmotic pressure limitation and fouling, whereas in MD, loss of volatile components and heat degradation occur due to the heat requirement for the feed stream to maintain the gradient of pressure of water vapor; on the other hand, OD will be free from such problems if it operates at low temperature. The OD is more costly than conventional processes such as evaporation and RO for concentration of juice. However, its efficacy to remove water from other LMW solutes (volatile or nonvolatile) to yield concentrates of superior quality is quite advantageous. OD has a unique ability to produce highly concentrated solutions of heat-sensitive products with minimal deterioration. Hybrid integrated membrane processes that involve the use of pressure-driven membrane preconcentration, such as UF and RO, promise to reduce the overall costs of OD concentration and broaden its applicability.

## REFERENCES

Addison, W. 1986. Recent trend in fruit juice concentration, fruit and vegetable for processing. *Acta Horticulture* 194: 241–247.

Adhikary, S.K., Harkare, W.P., and K.P. Govindan. 1983. Deacidification of fruit juices by electrodialysis. *Indian Journal of Technology* 21: 120–123.

Ali, F., Dornier, M., Duquenoy, A., et al. 2003. Evaluating transfers of aroma compounds during the concentration of sucrose solutions by osmotic distillation in a batch-type pilot plant. *Journal of Food Engineering* 60: 1–8.

Amerine, M.A., Berg, H.W., and W.V. Cruess. 1976. *The technology of wine making*. Westport: AVI Publishing Co.

Bailey, A.F.G., Barbe, A.M., Hogan, P.A., et al. 2000. The effect of ultrafiltration on the subsequent concentration of grape juice by osmotic distillation. *Journal of Membrane Science* 164: 195–204.

Bailly, M. 2002. Production of organic acids by bipolar electrodialysis: Realizations and perspectives. *Desalination* 144: 157–162.

Bai, R.B., and H.F. Leow. 2002. Microfiltration of activated sludge wastewater—the effect of system operation parameters. *Separation and Purification Technology* 29: 189–198.

Barbe, A.M., Bartley, J.P., Jacobs, A.L., et al. 1998. Retention of volatile organic flavour/fragrance components in the concentration of liquid foods by osmotic distillation. *Journal of Membrane Science* 145: 67–75.

Beaudry, E.G., and K.A. Lampi. 1990. Osmosis concentration of fruit juices. *Fluessiges Obst* 57: 652–656, 663–664.
Ben Amar, R., Gupta, B.B., and M.Y. Jaffrin. 1990. Apple juice clarification using mineral membranes: Fouling control by backwashing and pulsating flow. *Journal of Food Science* 55: 1620–1625.
Bengtsson, E., Tragardh, G., and B. Hallstrom. 1989. Recovery and concentration of apple juice aroma compounds by pervaporation. *Journal of Food Engineering* 10: 65–71.
Beveridge, T. 1997. Juice extraction from apples and other fruits and vegetables. *Critical Review Food Science and Nutrition* 37: 449–469.
Bhatia, A.R., Dang, R.L., and G.S. Gaur. 1979. Deacidification of apple juice by ion exchange resins. *Indian Food Packer* (January–February): 15–19.
Boerlage, S.F.E., Kennedy, M.D., Dickson, M.R., et al. 2002. The modified fouling index using ultrafiltration membranes (MFI-UF): Characterisation, filtration mechanisms and proposed reference membrane. *Journal of Membrane Science* 197: 1–21.
Bui, A.V., Nguyen, Minh H., and J. Muller. 2003. A laboratory study on glucose concentration by osmotic distillation in hollow fibre module. *Journal of Food Engineering* 63: 237–245.
Bui, A.V., Nguyen, Minh H., and J. Muller. 2005. Characterisation of the polarisations in osmotic distillation of glucose solutions in hollow fibre module. *Journal of Food Engineering* 68: 391–402.
Capannelli, G., Bottino, A., Munari, S., et al. 1994. The use of membrane processes in the clarification of orange and lemon juices. *Journal of Food Engineering* 21: 473–483.
Carabasa, M., Ibarz, A., Garza, S., et al. 1988. Removal of dark compounds from clarified juices by adsorption processes. *Journal of Food Engineering* 37: 25–41.
Cassano, A., Jiao, B., and E. Drioli. 2004. Production of concentrated kiwifruit juice by integrated membrane process. *Food Research International* 37:139–148.
Cassano, A., Drioli, E., Galaverna, G., et al. 2003. Clarification and concentration of citrus and carrot juices by integrated membrane processes. *Journal of Food Engineering* 57: 153–163.
Celere, M., and C. Gostoli. 2003. Osmotic distillation with propylene glycol, glycerol and glycerol–salt mixtures. *Journal of Membrane Science* 229:159–170.
Chamchong, H., and A. Noomhorm. 1991. Effect of pH and enzymatic treatment on microfiltration and ultrafiltration of tangerine juice. *Journal of Food Process Engineering* 14: 21–34.
Chan, W.Y., and B.H. Chang. 1992. Production of clear guava nectar. *International Journal Food Science and Technology* 27: 435–441.
Cheryan, M. 1998. *Ultrafiltration and Microfiltration Handbook.* Lancaster, PA: Technomic Publishing Company.
Chuanhui, H., and X. Tongwen. 2006. Electrodialysis with bipolar membranes for sustainable development. *Environmental Science and Technology* 40: 5233–5243.
Chang, T., Siddiq, M., Sinha, N.K., et al. 1995. Commercial pectinases and the yield and quality of Stanley plum juice. *Journal of Food Processing and Preservation* 19: 89–101.
Courel, M. 1999. Mass transfer study in osmotic evaporation: Application to fruit juice concentration. PhD dissertation, University of Montpellier, Montpellier, France.
Courel, M., Dornier, M., Herry, J.M., et al. 2000. Effect of operating conditions on water transport during the concentration of sucrose solutions by osmotic distillation. *Journal of Membrane Science* 170: 281–289.
Couture, R., and R. Rouseff. 1992. Debittering and deacidifying sour orange (*Citrus aurantium*) juice using neutral and anion exchange resins. *Journal of Food Science* 57: 380–384.
Cristina, C.P., Cláudio, P.R. Jr., Ronaldo, N., et al. 2006. Pervaporative recovery of volatile aroma compounds from fruit juices: A review. *Journal of Membrane Science* 274: 1–23.
Cussler, E.L. 1984. *Diffusion: Mass transfer in fluid systems.* Cambridge: Cambridge University Press.
De, S., and P.K. Bhattacharya. 1997. Modeling of ultrafiltration process for a two component aqueous solution of low and high (gel-forming) molecular weight solutes. *Journal of Membrane Science* 136: 57–69.
De, S., Bhattacharjee, S., Bhattacharya, P.K., et al. 1997. Generalized integral and similarity solutions for concentration profiles for ultrafiltration. *Journal of Membrane Science* 130: 99–121.
Deblay, P. 1995. Process for at least partial dehydration of an aqueous composition and devices for implementing the process. U.S. Patent 5,382,365, 26 January.
De Barros, S.T.D., Andrade, M.G., Mendes, E.S., et al. 2003. Study of fouling mechanism in pineapple juice clarification by ultrafiltration. *Journal of Membrane Science* 215: 213–224.
De Bruijn, J.P.F., Venegas, A., Martínez, J.A., et al. 2003. Ultrafiltration performance of carbosep membranes for the clarification of apple juice. *Lebensmittel Wissenschaft und Technologie* 36: 397–406.
Downes, J.W. 1990. Equipment for extraction and processing of soft and pome fruit juices. In *Production and Packaging of Non-carbonated Fruit Juices and Fruit Beverages*, ed. D. Hicks, 158–181. Glasgow: Blackie.

Durham, R.J., and M.H. Nguyen. 1994. Hydrophobic membrane evaluation and cleaning for osmotic distillation of tomato puree. *Journal of Membrane Science* 72: 53–72.

Elmidaoui, A., Chay, L., Tahaikt, M., et al. 2004. Demineralisation of beet sugar syrup, juice and molasses using an electrodialysis pilot plant to reduce melassigenic ions. *Desalination* 165: 435.

Elmidaouia, A., Chaya, L., Tahaikta, M., et al. 2006. Demineralisation for beet sugar solutions using an electrodialysis pilot plant to reduce melassigenic ions. *Desalination* 189: 209–214.

Field, R.W., Wu, D., Howell, J.A., et al. 1995. Critical flux concept for microfiltration fouling. *Journal of Membrane Science* 100: 259–272.

Girard, B., and L. Fukumoto. 1999. Apple juice clarification using microfiltration and ultrafiltration polymeric membranes. *Lebensmittel Wissenschaft und Technologie* 32: 290–298.

Girard, B., and L. Fukumoto. 2000. Membrane processing of fruit juices and beverages: A review. *Critical Review in Food Science and Nutrition* 40: 91–157.

Gherardi, S., Bazzarini, R., Trifiro, A., et al. 1986. Preconcentration of tomato juice by reverse osmosis. *Industria Conserve* 61: 115–119.

Goloubev, V.N., and B. Salem. 1989. Traitement à l'électrodialyse du jus d'orange. *Indust. Agric. Alimen.* 106: 175–177.

Hamachi, M., Cabassud, M., Davin, A., et al. 1999. Dynamic modelling of crossflow microfiltration of bentonite suspension using recurrent neural network. *Chemical Engineering Process* 38: 203–210.

Heatherbell, D.A., Short, J.L., and P. Strubi. 1997. Apple juice clarification by ultrafiltration. *Confructa* 22: 157–169.

Herron, J.R., Beaudry, E.G., Jochums, C.E., Medina, L.E. 1994. U.S. Patent No. 5,281,430, 25 January.

Hervé, M. 1997. Les axes de recherches sur les jus. *Process* 1124: 82–83.

Hogan, P.A., Canning, R.P., Peterson, P.A., et al. 1998. A new option: Osmotic distillation. *Chemical Engineering Progress*: 49–61.

Ishii, K., Konomi, S., Kojima, K., et al. 1981. Development of a tomato juice concentration system by reverse osmosis. In *Synthetic Membranes: Vol. II—Hyperfiltration and Ultrafiltration Uses*, ed. A.F. Turbak, ACS Symposium Series, Washington, DC: 1–16.

Jiao, B., Molinari, R., Calabro, V., et al. 1992. Application of membrane operations in concentrated citrus juice processing. *Agro-Industry Hi-tech* 3: 19–27.

Jiraratananon, R., and A. Chanachai. 1996. A study of fouling in the ultrafiltration of passion fruit juice. *Journal of Membrane Science* 111: 39–48.

Jiraratananon, R., Uttapap, D., and C. Tangamornsuksun. 1997. Self-forming dynamic membrane for ultrafiltration of pineapple juice. *Journal of Membrane Science* 129: 135–143.

Johnson, R.L., and B.V. Chandler 1985. Ion exchange and adsorbent resins for removal of acids and bitter principles from citrus juices. *Journal of Science and Food Agriculture* 36: 480–484.

Johnson, R.A., Valks, R.H., and M.S. Lefebvre. 1989. Osmotic distillation—A low temperature concentration technique. *Australian Journal Biotechnology* 3: 206–207.

Karlsson, H.O.E., Loureiro, S., and G. Trägårdh. 1995. Aroma compound recovery with pervaporation. *Journal of Food Engineering* 26: 177–191.

Kelly, S.T., Opong, W.S., and A.L. Zydney. 1993. The influence of protein aggregates on the fouling of microfiltration membranes during stirred cell filtration. *Journal of Membrane Science* 80: 175–187.

Kirk, D.E., Montgomery, M.W., and M.G. Kortekaas. 1983. Clarification of pear juice by hollow fiber ultrafiltration. *Journal of Food Science* 48: 1663–1666.

Khattabi, M.O. E1, Alaoui Hafidi, My, R., and E1 Midaoui, A. 1996. Reduction of melassigenic ions in cane sugar juice by electrodialysis. *Desalination* 107: 149–157.

Kuo, K.P., and M. Cheryan. 1983. Ultrafiltration of acid whey in a spiral-wound unit: Effect of operating parameters on membrane fouling. *Journal of Food Science* 48: 1113–1118.

Kunz, W., Benhabiles, A., and R. Ben-Kim. 1996. Osmotic evaporation through macroporous hydrophobic membranes: A survey of current research and applications. *Journal of Membrane Science* 121: 25–36.

Laflamme, J., and R. Weinand. 1993. New developments by the combination of membrane filtration and adsorption technology. *Fruit Processing* 9: 336–342.

Lam Quoc, A., Mondor, M., Lamarche, F., et al. 2006. Effect of a combination of electrodialysis with bipolar membranes and mild heat treatment on the browning and opalescence stability of cloudy apple juice. *Food Research International* 39: 755–760.

Lazarides, H.N., Iakovidis, A., and H.G. Schwartzberg 1990. Aroma loss and recovery during falling film evaporation. In *Engineering and Food. Vol. 3. Advanced Processes*, ed. W.E.L. Spiess, and H. Schubert, 96–105. Amsterdam: Elsevier Applied Science Publishers.

Lea, A.G.H. 1990. Apple juice. In *Production and Packaging of Non-carbonated Fruit Juices and Fruit Beverages*, ed. D. Hicks, 182–225. New York: Van Nostrand.

Leightell, B. 1972. Reverse osmosis in the concentration of food. *Process Biochemistry* 3: 41.

Lefebvre, M.S.M. 1992. Osmotic distillation process and semipermeable barriers therefore. U.S. Patent 5,098,566, 24 March.

Maccarone, E., Campisi, S., Cataldi Lupo, M.C., et al. 1996. Thermal treatments effects on the red orange juice constituents. *Industria Bevande* 25: 335–341.

Mänttäri, M., Puro, L., Nuortila-Jokinen, J., et al. 2000. Fouling effects of polysaccharides and humic acid in nanofiltration. *Journal of Membrane Science* 165: 1–17.

Medina, B.G., and A. Garcia. 1988. Concentration of orange juice by reverse osmosis. *Journal of Food Process Engineering* 10: 217–230.

Merlo, C.A., Rose, W.W., Pedersen, L.D., et al. 1986. Hyperfiltration of tomato juice during long term high temperature testing. *Journal of Food Science* 51: 395–398.

Merlo, C.A., Rose, W.W., Pedersen, L.D., et al. 1986. Hyperfiltration of tomato juice: Pilot plant scale high temperature testing. *Journal of Food Science* 51: 403–407.

Meyer, A.S., Koser, C., and A.N. Jens. 2001. Efficiency of enzymatic and other alternative clarification and fining treatments on turbidity and haze in cherry juice. *Journal of Agriculture and Food Chemistry* 49: 3644–3650.

Michaels, A.S. 1998. Methods and apparatus for osmotic distillation. U.S. Patent 5,824,223, 20 October.

Moulik, S.P., Cooper, F.C., M. Bier. 1967. Forced-flow electrophoretic filtration of clay suspensions: Filtration in an electric field. *Journal of Colloid and Interface Science* 24: 427–432.

Mullon, C., Radovich, J.M., and B.A. Behnam. 1985. Semi-empirical model for electroultrafiltration–diafiltration. *Separation Science and Technology* 20: 63–72.

Nguyen, M.H. 1991. New strategies for food product concentration by membrane technology. *Chemical Engineering in Australia* 16: 10–11.

Nguyen, H. M. 2000. Alternatives to evaporation prior to drying: Developments and applications of freeze concentration and membrane processes. In *Proceeding of Food Engineering Conference—Melbourne, Australia*, March 2000.

Nisperos-Carriedo, M.O., and P.E. Shaw. 1990. Comparison of volatile flavor components in fresh and processed orange juices. *Journal of Agriculture and Food Chemistry* 38:1048.

Pepper, D., Orchard, A.C.J., and A.J. Merry. 1985. Concentration of tomato juice and other fruit juices by reverse osmosis. *Desalination* 53: 157–166.

Peri, C., Battisti, P., and D. Setti. 1973. Solute transport and permeability characteristics of reverse osmosis membranes. *Lebensmittel Wissenschaft und Technologie* 6: 127.

Porter, M.C. 1990. *Handbook of Industrial Membrane Technology.* Park Ridge: Noyes.

Rai, P., Majumdar, G.C., DasGupta, S., et al. 2004. Optimizing pectinase usage in pretreatment of mosambi juice for clarification by response surface methodology. *Journal of Food Engineering* 64: 397–403.

Rai, P., Majumdar, G.C., DasGupta, S., et al. 2005a. Understanding ultrafiltration performance with mosambi juice in an unstirred batch cell. *Journal of Food Process Engineering* 28: 166–180.

Rai, P., Majumdar, G.C., DasGupta, S., et al. 2005b. Quantification of flux decline of depectinized mosambi (*Citrus sinensis* (L.) Osbeck) juice using unstirred batch ultrafiltration. *Journal of Food Process Engineering* 28: 359–377.

Rai, P., Majumdar, G.C., Sharma, G., et al. 2006a. Effect of various cutoff membranes on permeate flux and quality during filtration of mosambi (*Citrus sinensis* (L.) Osbeck) juice. *Food Bioproduct Processing* 84: 213–219.

Rai, P., Majumdar, G.C., Jayanti, V.K., et al. 2006b. Alternate pretreatment methods to substitute enzyme treatment for clarification of mosambi juice using ultrafiltration. *Journal of Food Process Engineering* 29: 202–218.

Rai, P., Majumdar, G.C., DasGupta, S., et al. 2006c. Modeling of sucrose permeation through a pectin gel during ultrafiltration of depectinized mosambi (*Citrus sinensis* (L.) *Osbeck*) Juice. *Journal of Food Science* 71: E87–E94.

Rai, P., Majumdar, G.C., DasGupta, S., et al. 2006d. Resistance in series model for ultrafiltration of mosambi (*Citrus sinensis* (L.) *Osbeck*) juice in a stirred continuous mode. *Journal of Membrane Science* 283: 116–222.

Rai, P., Majumdar, G.C., DasGupta, S., et al. 2007a. Effect of various pretreatment methods on permeate flux and quality during ultrafiltration of mosambi juice. *Journal of Food Engineering* 78: 561–568.

Rai, P., Majumdar, G.C., DasGupta, S., et al. 2007b. Modeling of permeate flux decline of synthetic fruit juice and mosambi juice (*Citrus sinensis* (L.) *Osbeck*) in stirred continuous ultrafiltration. *Lebensmittel Wissenschaft und Technologie* 40: 1765–1773.

Rai, P., Majumdar, G.C., DasGupta, S., et al. 2008. Storage study of ultrafiltered mosambi (*Citrus sinensis* (L.) *Osbeck*) juice. *Journal of Food Processing and Preservation* 32: 923–934.

Rai, P., Majumdar, G.C., DasGupta, S., et al. 2010. Flux enhancement during of ultrafiltration of mosambi (*Citrus sinensis* (L.) *Osbeck*) juice. *Journal of Food Process Engineering* 33: 554–567.
Rajagopalan, N., and M. Cheryan. 1995. Pervaporation of grape juice aroma. *Journal of Membrane Science* 104: 243–250.
Ramteke, R.S. 1993. Methods for concentration of fruit juices: A critical evaluation. *Journal of Food Science and Technology* 30: 391–402.
Reid, G.C., Hwang, A., Meisel, R.H., et al. 1990. The sterile filtration and packaging of beer into polyethylene terephthalate containers. *Journal of American Society Brewer Chemistry* 48: 85–91.
Rodrigues, R.B., Menezes, H.C., Cabral, L.M.C., et al. 2004. Evaluation of reverse osmosis and osmotic evaporation to concentrate camu-camu juice (*Myrciaria dubia*). *Journal of Food Engineering* 63: 97–102.
Sarkar, B., DasGupta, S., and S. De. 2008. Effect of electric field during gel-layer controlled ultrafiltration of synthetic and fruit juice. *Journal of Membrane Science* 307: 268–276.
Sarkar, B., DasGupta, S., and S. De. 2008. Cross-flow electro-ultrafiltration of mosambi (*Citrus sinensis* (L.) Osbeck) juice. *Journal of Food Engineering* 89: 241–245.
Scott, K. 1995. *Handbook of Industrial Membranes*. Amsterdam: Elsevier Advance Technology.
Shaw, P.E., Lebrun, M., Dornier, M., et al. 2001. Evaluation of concentrated orange and passionfruit juices prepared by osmotic evaporation. *Lebensmittel Wissenschaft und Technologie* 34: 60–65.
Sheng, J., Johnson, R.A., and M.S. Lefebvre 1991. Mass and heat transfer mechanisms in the osmotic distillation process. *Desalination* 80: 113–121.
Sheng, J. 1993. Osmotic distillation technology and its applications. *Australian Chemical Engineering Conference* 3: 429–432.
Shepherd, A., Habert, A.C., and C.P. Borges. 2002. Hollow fiber modules for orange juice aroma recovery using pervaporation. *Desalination* 148: 11–114.
Sheu, M.J., and R.C. Wiley. 1984. Influence of reverse osmosis on sugar retention in apple juice concentration. *Journal of Food Science* 49: 304.
Sheu, M.J., Wiley, R.C., and D.V. Schlimme. 1987. Solute and enzyme recoveries in apple juice clarification using ultrafiltration. *Journal of Food Science* 52: 732–736.
Sirkar, K.K. 1995. Membrane separations: Newer concepts and applications for the food industry. In *Bioseparation Processes in Foods*, ed. R.K. Singh and S.S.H. Rizvi, 353–356. New York: Marcel Dekker.
Sirkar, K.K. 1997. Membrane separation technologies: Current developments. *Chemical Engineering Communication* 157: 145–184.
Sulaiman, M.Z., Sulaiman, N.K., and L.Y. Shih. 1998. Limiting permeate flux in the clarification of untreated starfruit juice by membrane ultrafiltration. *Journal of Chemical Engineering* 68: 145–148.
Sulaiman, M.Z., Sulaiman, N.K., and M. Shamel. 2001. Ultrafiltration studies on solutions of pectin, glucose and their mixtures in a pilot scale crossflow membrane unit. *Journal of Chemical Engineering* 84: 557–563.
Swientek, R.J. 1986. Ultrafiltration expanding role in food and beverage processing. *Food Process* 47: 71–83.
Thakur, B.R., Singh, R.K., and A.K. Handa. 1997. Chemistry and uses of pectin. *Critical Reviews in Food Science and Nutrition* 37: 47–73.
Thompson, D. 1991. The application of osmotic distillation for the wine industry. *The Australian Grape Grower & Wine Maker* April: 11–14.
Todisco, S., Pena, L., Drioli, E., et al. 1996. Analysis of the fouling mechanism in microfiltration of orange juice. *Journal of Food Processing and Preservation* 20: 453–466.
Tronc, J.S., Lamarche, F., and J. Makhlouf. 1997. Enzymatic browning inhibition in cloudy apple juice by electrodialysis. *Journal of Food Science* 62: 75–78.
Tronc, J.S., Lamarche, F., and J. Makhlouf. 1998. Effect of pH variation by electrodialysis on the inhibition of enzymatic browning in cloudy apple juice. *Journal of Agriculture Food Chemistry* 46: 829–833.
Vaillant, F., Jeanton, E., Dornier, M., et al. 2001. Concentration of passion fruit juice on an industrial pilot scale using osmotic evaporation. *Journal of Food Engineering* 47: 195–202.
Vaillanta, F., Cissea, M., Chaverri, M., et al. 2005. Clarification and concentration of melon juice using membrane processes. *Innovative Food Science and Emerging Technologies* 6:213–220.
Van Niestelrooij, M. 1998. Freeze concentration and its application in the fruit juice industry. *Fruit Processing* 11: 460–462.
Vera, E., Ruales, J., Dornier, M., et al. 2003a. Comparison of different methods for deacidification of clarified passion fruit juice. *Journal of Food Engineering* 59: 361–367.
Vera, E., Ruales, J., Dornier, M., et al. 2003b. Deacidification of the clarified passion fruit juice using different configurations of electrodialysis. *Journal of Chemical Technology and Biotechnology* 78: 918–925.
Vera, E., Sandeaux, J., Persin, F., et al. 2007a. Deacidification of clarified tropical fruit juices by electrodialysis. Part I. Influence of operating conditions on the process performances. *Journal of Food Engineering* 78: 1427–1438.

Vera, E., Sandeaux, J., Persin, F., et al. 2007b. Deacidification of clarified tropical fruit juices by electrodialysis. Part II. Characteristics of the deacidified juices. *Journal of Food Engineering* 78: 1439–1445.

Viquez, F., Lastreto, C., and R.D. Cooke. 1981. A study of the production of clarified banana juice using pectinolytic enzymes. *Journal of Food Technology* 16: 115–125.

Vladisavljević, G.T., Vukosavljević, P., and B. Bukvić. 2003. Permeate flux and fouling resistance in ultrafiltration of depectinized apple juice using ceramic membranes. *Journal of Food Engineering* 60: 241–247.

Watanabe, A., Ohtani, T., and S. Kimura. 1982. Performance of dynamically formed Zr(IV)-PAA membrane during concentration of tomato juice. *Nippon Nogeikagaku Kaishi* 56: 339.

Wiesner, M.R., and S. Chellam. 1999. The promise of membrane technology. *Environmental Science and Technology* 33: 360–366.

Wilson, E.L., and D.J.W. Burns. 1983. Kiwifruit juice processing using heat treatment techniques and ultrafiltration. *Journal of Food Science* 48:1101–1105.

Youn, K.S., Hong, J.H., Bae, D.H., et al. 2004. Effective clarifying process of reconstituted apple juice using membrane filtration with filter-aid pretreatment. *Journal of Membrane Science* 228: 179–186.

Yu, Z.R., and B.H. Chiang. 1986. Passion fruit juice concentration by ultrafiltration and evaporation. *Journal of Food Science* 51: 1501–1515.

Zemel, G.P., Sims, C.A., Marshall, M.R., et al. 1990. Low pH inactivation of polyphenol oxidase in apple juice. *Journal of Food Science* 55: 562–565.

# 14 Fouling in Membrane Processes Used for Water and Wastewater Treatment

*Matthias Kraume and Fangang Meng*

## CONTENTS



## 14.1 INTRODUCTION

Water scarcity is among the most serious crises facing the world as a result of poor water management and climate change. Of a population of 6.75 billion, 1.1 billion lack access to drinking water, and 2.6 billion people are short of adequate water for sanitation. To cope with the situation of water scarcity, investigators and engineers have been attempting to develop treatment methods of every sorts, which aim to eliminate the micropollutants in water bodies (i.e., wastewater treatment process), or to increase water supplies via the safe reuse of wastewater and efficient desalination of seawater and brackish water (Shannon et al., 2008). Among all the processes used for water and wastewater treatment, membrane technology is a very attractive, reliable, and friendly process at present. Membranes are selective barriers that allow some types of matter to pass through while retaining others. Permeability and selectivity are two basic parameters of the membrane, and they are strongly dependent on membrane pore size and pore size distribution. The membrane-based processes used for water and wastewater treatment include direct membrane filtration (e.g., microfiltration, ultrafiltration, nanofiltration, and reverse osmosis) and hybrid membrane processes (e.g., membrane bioreactors and coagulation-coupled membrane filtration). The implementation of membrane processes strongly depends on water components and end-user requirements.

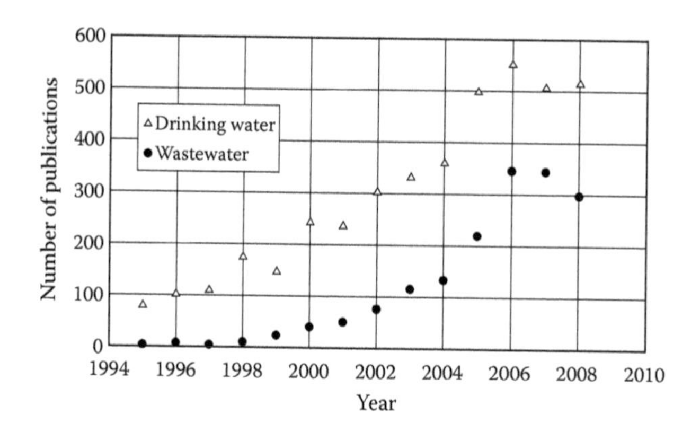


**FIGURE 14.1** Number of publications per year on fouling related papers over the last decade for membrane-based drinking water production and MBR wastewater treatment. (Data from Google Scholar.)

However, there are several issues that have not yet been fully understood, and they still are a significant obstacle toward the broad application of membrane processes. One of the issues is the understanding and mitigating of membrane fouling, which is inevitably associated with membrane processes (Chen et al., 2004b). Throughout the last years, a strongly increasing number of investigations per year were performed (see Figure 14.1) in order to obtain more detailed information on membrane fouling or to look for efficient strategies for fouling control in drinking water production and wastewater treatment. But these investigations were of different focus, and therefore, it is highly desirable to summarize and compare the results obtained during the last few years. To complement the current knowledge on membrane fouling, this chapter is focused on two issues: fouling behavior (see Section 14.2) and fouling control strategies (see Section 14.3).

## 14.2 FOULING BEHAVIOR

### 14.2.1 Definition of Membrane Fouling

Membrane fouling can be defined as the undesirable deposition and accumulation of particulate matter, microorganisms, colloids, and solutes on membranes. As shown in Figure 14.2, membrane fouling can be attributed to both membrane pore clogging and cake deposition on membranes which is usually the predominant fouling component (Lee et al., 2001). Membrane fouling is a very complicated phenomenon and results from multiple causes. Particle sizes of the pollutants in wastewater or drinking water may strongly affect fouling mechanisms in a membrane filtration system. If the size of foulants is comparable with the diameter of the membrane pores (i.e., colloids), or smaller than the pore size (i.e., solutes), adsorption at the internal pore surfaces and pore blocking may occur. However, if foulants (i.e., sludge flocs and colloids) are much larger than the membrane pores,

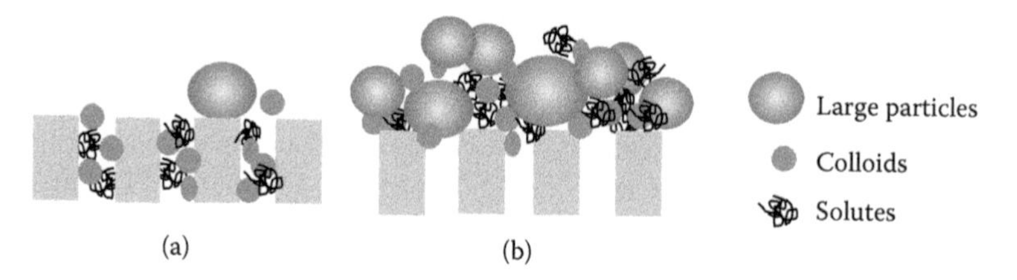


**FIGURE 14.2** Schematic illustration showing membrane fouling: (a) pore blocking and (b) cake layer. (Reprinted from *Water Research*, 43(6), Meng, F., et al., 1489–1512, Copyright 2009, with permission from Elsevier.)

they tend to form a cake layer on the membrane surface. Membrane fouling results in a reduction in permeate flux or an increase in transmembrane pressure (TMP) depending on the operation mode. Generally, the intensity of membrane fouling results from the following mechanisms:

1. Adsorption of solutes or colloids within/on membranes
2. Formation of a cake layer on the membrane surface
3. Detachment of foulants mainly caused by shear forces
4. The spatial and temporal changes of the foulant composition during the long-term operation (e.g., the change of bacterial community and biopolymer components in cake layer)

In fact, the occurrence of membrane fouling also strongly depends on the membranes used. Microfiltration (MF) and ultrafiltration (UF) membranes are usually used to reject large particles such as flocs and colloids, so MF/UF fouling is mainly caused by the deposition of flocs and colloids on membranes. In contrast, the fouling of nanofiltration (NF) and reverse osmosis (RO) membranes originally results from concentration polarization and is followed by cake fouling. From the viewpoint of fouling components, membrane fouling can be classified into three major categories: biofouling, organic fouling, and inorganic fouling. A fundamental understanding of the formation of membrane foulants will help to develop more effective approaches for fouling control.

### 14.2.2 Biofouling, Organic Fouling, and Inorganic Fouling

#### 14.2.2.1 Biofouling

Biofouling refers to the deposition, growth and metabolism of bacteria cells on the membranes, which has aroused a significant concern in membrane filtration processes (Pang et al., 2005; Wang et al., 2005). Biofouling may start with the deposition of individual bacteria on the membrane surface, after which the cells multiply and form a cake layer. Many researchers suggest that soluble microbial products (SMP) or extracellular polymeric substances (EPS) released by bacteria play important roles in formation of biological foulants and cake layer on membrane surfaces (Flemming et al., 1997; Judd, 2006; Ramesh et al., 2007).

The deposition of bacteria cells can be visualized by techniques such as scanning electron microscopy (SEM), confocal laser scanning microscopy (CLSM) and atomic force microscopy (AFM), direct observation through the membrane (DOTM) technique. (Vrouwenvelder et al., 1998) reported that the organisms on NF or RO membranes were found to be metabolically active in three pilot plants which are used for drinking water production, and the cell concentration on the membranes was quantified as $2.1 \times 10^8$ cells/cm$^2$. The combination of biomass related parameters such as adenosine triphosphate (ATP) and total cell account seems to be a suitable method for identification of NF or RO biofouling, whereas plate counts were found not to be appropriate to characterize biofouling (Vrouwenvelder et al., 2008). Zhang et al. (2006a) used a DOTM to observe the interactions between the bioflocs and the membrane surface. The images showed that the bioflocs could move across the membrane surface by rolling and sliding. More recently, CLSM has become a powerful approach for characterization of membrane biofouling, which can not only identify the deposited cell, but also present the 3D structure of the fouling layer. Yun et al. (2006) characterized the biofilm structure and analyzed its effect on membrane permeability in MBR for dye wastewater treatment. They found that membrane filterability was closely associated with the structural parameters of the biofilms (i.e., porosity, biovolume). The visualization of biofouling using these techniques is helpful for understanding of the proceeding of floc/cell deposition and the microstructure or architecture of the cake layer.

In addition, a few investigations have been performed to study the microbial community structures and microbial colonization on the membrane surfaces in membrane process (Chen et al., 2004a; Jinhua et al., 2006; Zhang et al., 2006b; Miura et al., 2007; Hwang et al., 2008). The microbial community structures can be investigated using microbiology methods such as polymerase chain reaction denaturing gradient gel electrophoresis (PCR-DGGE) and fluorescence in situ hybridization (FISH). Hörsch et al. (2005) studied biofouling of UF and NF membranes for drinking water treatment by using FISH. It was found that the bacterial composition of the primary fouling layer showed a dominance of *γ-proteobacteria*, and the mature fouling layer, which is similar to the population structure of raw water, was mainly composed of *α-proteobacteria* and *β-proteobacteria*. This result indicated that the *γ-proteobacteria* attached to membranes readily and formed a primary biofilm, and were then colonized by other bacteria. Miura et al. (2007) studied the microbial communities in a full-scale submerged MBR used to treat real municipal wastewater delivered from the primary sedimentation basin of a municipal wastewater treatment facility over 3 months. They also reported that the microbial communities on membrane surfaces were quite different from those in the suspended biomass. In this study, their FISH and 16S rRNA gene sequence analysis, however, revealed that the *β-proteobacteria* probably played a major role in the development of the mature biofilms, which led to severe irremovable membrane fouling. Jinhua et al. (2006) reported that *γ-proteobacteria* more selectively adhered to and grew on membrane surface than other microorganisms, and the deposited cells had higher surface hydrophobicity than suspended sludge. Hwang et al. (2008) analyzed the biofilm structure in a laboratory-scale MBR treating synthetic wastewater and came to the conclusion that three factors affect the pattern of the TMP increase due to fouling: (1) the increasing thickness of the biocake resulting from the deposition of bacterial flocs from the bulk phase, (2) the reduction of biocake porosity arising from the compression effect exerted on the biocake by the increasing external pressure (i.e., the TMP), and (3) the generation of EPS like polysaccharides inside the biocake layer.

The high shear stress induced by aeration can select the deposition of cells. Some cells can be detached easily by the shear stress, but others still deposit on the membrane surface tightly. The selective deposition of the cell relies on the affinity of cells to membranes. And, due to the anoxic condition in the cake layer, the temporal change of microbial community structure may take place. We can see that some specific bacteria in water bodies or activated sludge play an important role in membrane biofouling. The fundamental understanding of deposition behavior of bioflocs/cells and mechanisms of cell attachment on membranes will be crucial for the development of appropriate biofouling control strategies.

#### 14.2.2.2 Organic Fouling

Organic fouling in membrane processes refers to the deposition of natural organic matter (NOM) or SMP, which is mainly composed of proteins, humic acids and polysaccharides, on the membranes. NOM is known as the main fouling-causing substances in water treatment, and SMP are generally recognized as the main foulants in MBRs. Due to the small size, NOM or SMP can be deposited onto the membranes more readily due to the permeate flow, as the back transport to the bulk phase due to lift forces is weaker in comparison to large particles (e.g., colloids aggregates and sludge flocs). For NOM, humic substances are the most detrimental foulants, which can cause severe irreversible fouling through pore blocking and membrane adsorption. But, some investigations also showed that polysaccharides having large particle size are of high importance for membrane fouling when UF membranes are used for water recycling (Laabs et al., 2006; Zheng et al., 2009). Of high interest is that hydrophilic NOM (or nonhumic substances) might be a more significant foulant. Fan et al. (2001) reported that hydrophilic neutrals were the most important membrane foulants when compared with hydrophobic acids and transphilic acids. In recent years, algogenic organic matter (AOM), which is produced by algae in lakes or rivers, has been paid more and more attention to. AOM is mainly composed of hydrophilic polysaccharides and hydrophobic proteins having a negative zeta potential (Henderson et al., 2008). AOM compounds cannot be well rejected by NF membranes due to the low molecular weight and relatively low hydrophobicity (Her et al., 2004). For a given MBR process, Metzger et al. (2007) reported that the upper cake layer consisted of a porous, loosely bound cake layer with a similar composition to the biomass flocs. The intermediate cake layer, which consisted of equal parts of SMP and biomass aggregates, featured a high concentration of polysaccharides. The lower layer, representing the irremovable fouling fraction and predominantly consisting of SMP, features a higher concentration of strongly bound proteins. This study revealed the spatial distribution of SMP on the membrane surface.

In order to figure out the detailed information on the deposited biopolymers, identification of these substances is indispensable. Fourier transform infrared (FTIR) spectroscopy, three-dimensional excitation-emission matrix fluorescence spectroscopy (EEM), and high-performance size exclusion chromatography (HP-SEC) are powerful analytical tools for investigation of the organic fouling. Through the FTIR spectra, the major components of NOM foulants were identified as proteins and polysaccharides (Lee et al., 2004; Her et al., 2008). The EEM is often used to characterize protein-like and humic-like substances in water bodies or in membrane foulants (Liu et al., 2007; Kimura et al., 2009). With the help of EEM it was demonstrated that protein-like substances, organic colloids and humic-like substances are of high importance for membrane fouling in both MBR and water purification. A study by Teychene et al. (2008) showed that the deposition of soluble compounds was very important for MBR fouling whereas the impact of the colloidal fraction (poorly presents in supernatant) was less; and the results of HP-SEC and EEM analyses revealed the important role of protein-like substances (polypeptides) in MBR fouling. Figure 14.3 shows a typical HP-SEC chromatogram of sludge supernatant from a pilot-scale MBR. It can be seen that the SMP in sludge supernatant was mainly composed of biopolymers (i.e., polysaccharides and proteins), humic substances and low-molecular-weight substances, in which biopolymers are of high importance to MBR fouling due to their larger size. An early study by Rosenberger et al. (2006) demonstrated that polysaccharides had higher fouling propensity because most of the polysaccharides in the supernatant could be retained on the membrane surface due to the large size. Consequently, the biopolymers are of high importance for MF and UF processes. Whereas humic acids and the low-molecular-weight substances play a significant role in NF and RO processes. In addition to the molecular size, the deposition of SMP or NOM on membranes strongly depends on their affinity to membranes. Therefore, the development of membrane fouling depends on both membrane properties and foulant species.

### 14.2.2.3 Inorganic Fouling

Inorganic fouling or membrane scaling refers to the formation of deposits of inorganic scales such as $CaCO_3$, $CaSO_4$, $BaSO_4$, $Mg(OH)_2$, silica and so on, which results from the increased concentration of one or more species beyond their saturation limits and they ultimately precipitate onto the membranes (Al-Amoudi et al., 2007). Inorganic fouling can often be observed in NF or RO filtration of drinking water due to concentration polarization on the membrane surface, e.g., the precipitation

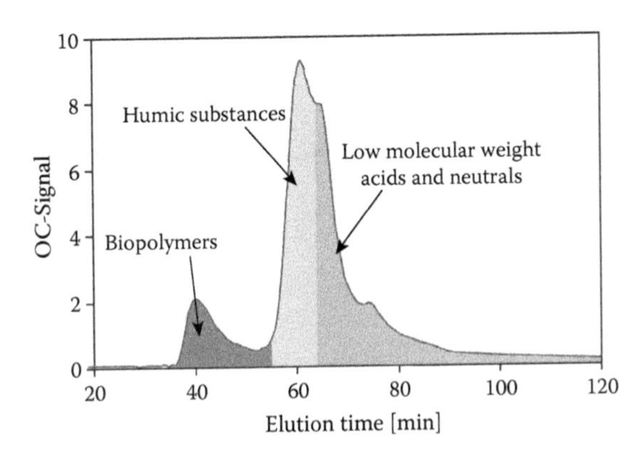


**FIGURE 14.3** A typical HPLC–SEC chromatogram of sludge supernatant from a pilot-scale MBR. (Reprinted from *Water Research*, 41(17), Haberkamp, J., et al., 3794–3802, Copyright 2007, with permission from Elsevier.)

of $BaSO_4$ and $CaCO_3$ on membranes (Van der Hoek et al., 2000). Al and Ca were also detected in membrane deposits during filtration of sand-filtered ozonated water (Her et al., 2007). Sometimes, the inorganic fouling can be ignored. For example, it was reported that the contribution of inorganic substances to membrane foulants was less than 15% in a pilot-scale NF plant for the treatment of surface water (Speth et al., 1998). It was also found that in the presence of calcium, alginate resulted in severe fouling due to the complexation between them (Jermann et al., 2007).

Recently, inorganic fouling has also been reported in MBR processes. Ognier et al. (2002) reported a severe $CaCO_3$ fouling in a pilot MBR with a ceramic ultrafiltration membrane module. In this study, the synthetic wastewater was prepared with hard tap water ($Ca^{2+}$ = 120 mg/L and $Mg^{2+}$ = 8 mg/L). They found that the high alkalinity of the activated sludge (pH = 8–9) could cause the precipitation of $CaCO_3$. Lyko et al. (2007) also found that metal substances (e.g., $Fe^{3+}$) were a more significant contributor to membrane fouling than biopolymers. Sometimes, the fouling caused by inorganic scaling is not easy to be eliminated even by chemical cleaning (You et al., 2006). These findings indicate that inorganic fouling has become more and more important in membrane processes. But the understanding of inorganic fouling is still not clear. The investigation of the limiting concentration of metal ions in the feedwater that can lead to inorganic fouling will be of great interest, since the chemical component of water or wastewater is in close relation with the formation of precipitation.

The inorganic fouling can form through two ways (see Figure 14.4): chemical precipitation and biological precipitation. During membrane filtration, there is a great variety of cations and anions such as $Ca^{2+}$, $Mg^{2+}$, $Al^{3+}$, $Fe^{3+}$, $CO_3^{2-}$, $SO_4^{2-}$, $PO_4^{3-}$, and $OH^-$. Concentration polarization will lead to even higher concentration of concentrated salts at the membrane surface, especially for NF and RO membranes, so that saturation concentrations are reached although the bulk concentration is still below the standard level. Carbonates are one kind of the predominant salts in inorganic fouling. Carbonates of metals such as Ca, Mg, and Fe cause an increase of the membrane scaling potential (You et al., 2005).

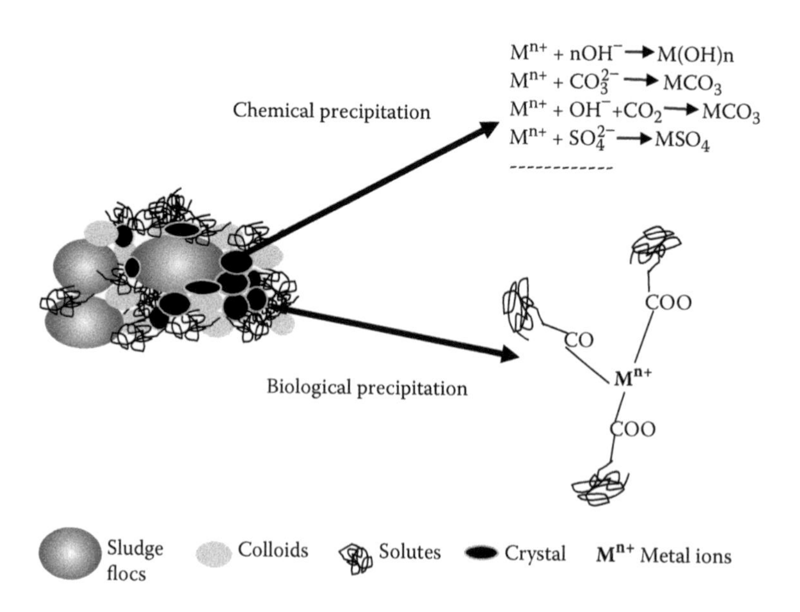


**FIGURE 14.4** Schematic illustration of the formation of inorganic fouling in MBRs. (Reprinted from *Water Research*, 43(6), Meng, F., et al., 1489–1512, Copyright 2009, with permission from Elsevier.)

Biological precipitation is another contribution to inorganic fouling. The biopolymers contain ionizable groups such as $COO^-$, $CO_3^{2-}$, $SO_4^{2-}$, $PO_4^{3-}$, and $OH^-$. The metal ions can be easily captured by these negative ions. In some cases, calcium and acidic functional groups (R-COOH) can form complexes and build a rigid organic matter network that may exacerbate flux decline (Costa et al., 2006). When the metal ions approach the membranes, they could be caught by the attached bio-cake layer via complexation and charge neutralization and thereby accelerate membrane fouling. Bridging between deposited biopolymers and metal ions further enhanced the compactness of the fouling layer (Hong et al., 1997). The synergistic interactions between different kinds of foulants (e.g., bacterial clusters, colloids, macromolecules, and inorganic elements) could result in faster and more substantial foulant deposition on the membrane surface.

Despite the fact that inorganic fouling is a troublesome phenomenon in membrane processes, it is possible to avoid or limit inorganic fouling by pretreatment of feedwater and/or chemical cleaning. But, the presence of a small quantity of calcium can benefit the membrane permeation in some membrane processes due to its positive effect on sludge or colloid/sludge floccubility (Kim et al., 2006). Optimum concentrations of inorganic matter can improve a membrane filtration process. As inorganic fouling can result in severe irremovable fouling, chemical cleaning is more effective than physical cleaning in the removal of inorganic precipitation. Chemical cleaning agents such as EDTA might efficiently remove inorganic substances from the membrane surface. EDTA can form a strong complex with $Ca^{2+}$; biopolymers associated with $Ca^{2+}$ ions are replaced by EDTA through a ligand exchange reaction (Al-Amoudi et al., 2007). As a result, the metal ions in fouling layers can be removed by chemical agents such as EDTA, and then inorganic fouling can be controlled.

## 14.3 CONTROL STRATEGIES

Even though membrane fouling is very complex and it is an inevitable phenomenon during membrane filtration, five strategies are applied to control fouling: pretreatment of feedwater, improvement of hydrodynamic conditions, subcritical flux operation, implementation of physical/chemical cleaning, and development of low-fouling membranes.

### 14.3.1 Pretreatment of Feedwater

A commonly used approach to control membrane fouling is the implementation of pretreatment (e.g., dosage of coagulants and/or adsorbents) to decrease the level of solutes and colloids in the solution or to enhance the flocculation ability of colloids or flocs. The dosage of coagulants and/or adsorbents can not only enhance membrane permeability but also improve membrane effluent quality significantly. Evidence suggested that the compounds of dissolved organic carbon that are preferentially removed by coagulation are the higher-molecular-weight, more hydrophobic, and more acidic compounds (Randtke, 1988; Dennett et al., 1996; White et al., 1997), indicating that the coagulation might be an efficient method to eliminate fouling-causing substances. A recent study indicated that the dosage of a small amount of $Fe^{3+}$ (0.05 mmol/L) could help remove biopolymer concentration by more than 30%, whereas humic substances, low-molecular-weight acids, and total DOC decreased by less than 10% (Haberkamp et al., 2007). As shown in Figure 14.5, with increasing coagulant (ferric chloride) dosage, more organic matter could be removed. It must be addressed that different coagulants might have a distinct role in the elimination of NOM. Guo et al. (2005) observed that $FeCl_3$ was capable of removing large-sized organic compounds, whereas powdered activated carbon (PAC) reduced mainly the smaller-sized organic substances, and 0.4 mmol/L $FeCl_3$ can yield a higher permeate flux than 2 g/L PAC. More recently, Cai et al. (2008) reported that heated aluminum oxide particles (HAOPs) can remove more NOM substances and NOM fractions with high specific ultraviolet absorbance ($SUV_{254}$) than alum or $FeCl_3$. But different membranes (e.g., UF and MF) might have a different fouling behavior for a given feedwater and coagulant. A

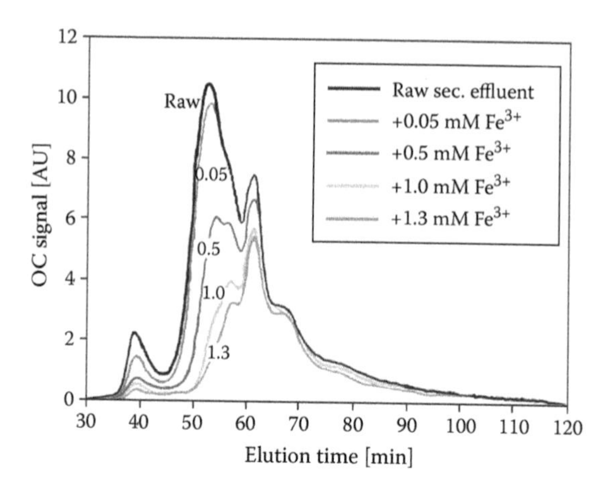


**FIGURE 14.5** LC-OCD chromatograms of secondary effluent before and after coagulation using different dosing of ferric chloride (each sample threefold diluted). (Reprinted from *Water Research*, 41(17), Haberkamp, J., et al., 3794–3802, Copyright 2007, with permission from Elsevier.)

recent investigation, for example, reported that UF membranes fouled readily during filtration of river water whereas less fouling occurred after coagulation. In contrast, for MF membranes, more severe fouling occurred after coagulation (Howe et al., 2006). It was attributed to the increase of large-sized flocs (>1 μm) after coagulation, which seemed to be more significant foulants for MF membranes than for UF membranes.

In recent years, coagulation or flocculation also has been used in MBRs to decrease the level of SMP in sludge supernatant or enhance the flocculation ability of sludge flocs. The addition of powdered activated carbon (PAC) is a simple and convenient method for fouling control. The PAC can not only be incorporated into the bioflocs forming biologically activated carbon (BAC) (Ng et al., 2006), but also adsorb biopolymers in the sludge suspension. The addition of PAC to MBRs provides a solid support for biomass growth, and hence reduces floc breakage (Hu et al., 2007). Moreover, the BAC flocs in MBRs are very stable and dense, which can help to prevent particle accumulation on the membranes. The coagulants can remove SMP by charge neutrality and bridging (Wu et al., 2006). Addition of optimum calcium can induce lower SMP concentration, lower hydrophobicity, lower concentration of filamentous bacteria and better flocculation, which result in a reduced cake resistance and pore blocking resistance (Kim et al., 2006). Attempts have also been made to use alum, ferric chloride, and chitosan as coagulants or filter aids (Ji et al., 2008; Song et al., 2008; Tian et al., 2008; Zhang et al., 2008b). Zhang et al. (2008b) reported that the addition of ferric chloride at the optimal concentration could reduce both SMP with MW > 10 kDa in the supernatant and the fraction of small particles (sludge flocs) in the range of 1–10 μm. In general, careful dosing must be applied so that all of the added substance is bound in flocs and no residual remains in the liquid phase. The improvement of membrane flux or sludge filterability also depends on the coagulant used. Ji et al. (2008) found that the membrane fouling rate of MBRs was in the order of control MBR without coagulant > Al2(SO4)3 added MBR > chitosan added MBR > polymeric ferric sulfate (PFS) added MBR. In addition, cationic polymeric chemicals were found to be favorable due to their steady and successful performance in fouling control (Koseoglu et al., 2008). It has been reported that polymeric coagulants could supply more positive charges and longer chain molecules, so that they had a better effect on filterability enhancement of sludge suspensions than monomeric coagulants, while excess addition of polymeric coagulant led to "colloidal restabilization" (Wu et al., 2006). Recently, special coagulants, so-called membrane fouling reducers (MFR) or membrane flux enhancers (MFE), have been developed to reduce membrane fouling in MBRs (Guo et al., 2008; Koseoglu et al., 2008). The addition of MFR can lead to the flocculation of activated sludge. SMP is also entrapped by the microbial flocs during the course of the flocculation, leading to an increase in the concentration of bound EPS (Hwang et al., 2007). A systematic study on the influence of 30 flux enhancing chemicals showed that neither the biological activity nor the oxygen transfer coefficient is significantly influenced by most additives.

### 14.3.2 Optimization of Hydrodynamic Conditions

One of the approaches used for fouling control during membrane operation is the improvement of hydrodynamic conditions (i.e., shear stress, cross-flow velocity) at the membrane surface. As expected, improvement of hydrodynamic conditions could lead to an increase in turbulence and mass transfer coefficient and then weaken concentration polarization.

Air sparging has been shown to significantly enhance the performance of some membrane processes owing to the beneficial effect of bubbling to control concentration polarization and cake deposition (Cui et al., 2003). It is likely that air-sparging technique is often used in MBR processes (Chang et al., 2002a; Psoch et al., 2005a, 2006, 2008; Guglielmi et al., 2008), because the MBRs themselves need aeration to provide dissolved oxygen as substrate for the microorganisms and for the biomass suspension. For a given air-sparging MBR, air bubbles are injected into the lumen or outside of membranes to enhance permeate flux. After being injected into a membrane module, coarse bubbles are formed and moving upward due to buoyancy. The bubble motion generates a variety of flow patterns, depending on the ratio of gas and liquid flow rates, the module geometry, interfacial tension and inclination. In Figure 14.6, common flow patterns inside a vertical tube are shown graphically. The main factor characterizing the flow pattern is the void fraction $\varepsilon$ in the pipe, which depends directly on the gas and liquid phase velocities. The flow pattern, as classified according to the value of $\varepsilon$, is defined as follows (Cabassud et al., 2001; Chang et al., 2002a):

1. Bubble flow ($\varepsilon < 0.2$): air bubbles are dispersed in the liquid phase.
2. Slug flow ($0.2 < \varepsilon < 0.9$): flow comprises alternate "slug" of gas and liquid.

In a tubular filtration module, a slug flow pattern was found to be most effective for the improvement of membrane permeation (Li et al., 1997; Mercier et al., 1997). As described by Psoch et al. (2005b), under slug flow conditions water and air slugs remove the fouling layer by inducing shear stress. Additionally, a thin water film flows downward due to the displacement caused by the rising gas slugs (Cabassud et al., 2001). The turbulence caused by slug flow reduces the concentration polarization strongly (Bellara et al., 1996; Ducom et al., 2003). These turbulent movements are to some extent capable of removing foulants from the membranes.

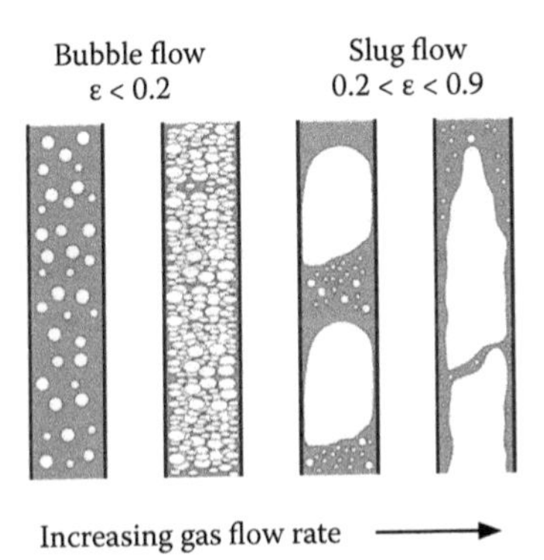


**FIGURE 14.6** Different bubble shapes in a two-phase vertical membrane tube.

Recently, there has been a strong trend toward the use of air sparging in submerged MBRs where air is generally injected outside the fibers rather than inside the lumens of the fibers (Chang et al., 2002a; Ghosh, 2006). The air sparging is very efficient to enhance critical flux in membrane filtration of sludge suspension (Yu et al., 2003). A long-term investigation by Psoch et al. (2005b) showed that air sparging could significantly increase the permeate flux even over several weeks of operation. They found that if the superficial liquid and gas velocities are fairly close to each other, the highest flux values are achievable, and suggested that the optimum air flow rate amounts to 80% of the liquid flow rate. It suggests that the ratio of gas and liquid flow rates is an important factor affecting the performance of air-sparging MBRs.

In low-pressure membrane processes, such as MBRs, the bubble size also plays a significant role for hydrodynamic conditions and energy demand. Fane et al. (2005) compared the effect of two nozzle sizes, 0.5 and 1.0 mm in diameter, on bubble size and membrane fouling. The larger nozzle could result in bigger bubbles than those resulting from the smaller nozzle. However, the fouling control, characterized by dTMP/dt, was noticeably improved using the smaller nozzle with the smaller bubbles. A more recent study by Prieske et al. (2008), however, showed that the smaller bubble size (1 mm) induces a slower circulation velocity compared to large bubbles (2 and 3 mm) due to a smaller gas hold-up difference between the riser and the downcomer. The authors concluded that larger bubbles seem to be more efficient for air scour of the membrane surface because the resulting drag and lift forces on the membranes are much higher due to higher circulation velocities. It has also been reported that fouling reduction increased with the air flow rates up to a given value, and beyond this flow rate, no further enhancement was achieved (Ndinisa et al., 2006a). In addition, under certain conditions, intermittent airflow can achieve better fouling control than continuous filtration, and it also reduces energy requirements.

Membrane module design also has significant impacts on the enhancement of membrane permeation by air sparging. Ghosh (2006) designed two different membrane modules for air-sparging MBR, which are shown in Figure 14.7a (type 1) and Figure 14.7b (type 2). The results showed that the type 1 membrane module was able to better utilize the positive effects of air sparging because a type 1 membrane module could give a more defined airlift-induced flow within the shell. In addition, in the type 1 membrane module, the gas bubbles were confined in close proximity of the

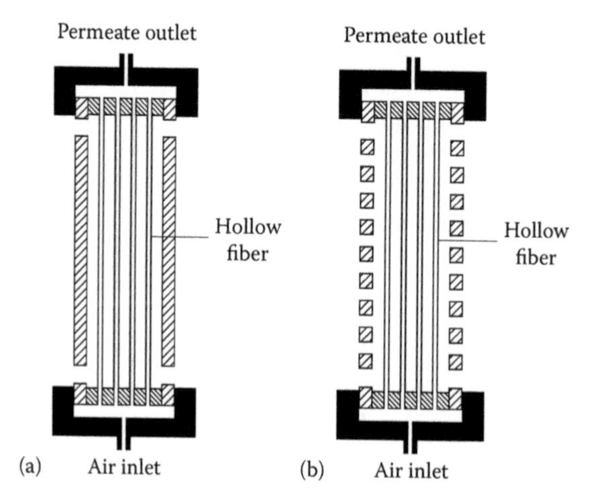


**FIGURE 14.7** Two approaches of air sparging. The modules had a length of 200 mm, an inner diameter of 25 mm, and an outer diameter of 30 mm. Each module was made of 52 hollow fibers having an effective filtration length of 180 mm. The modules were equipped with a Perspex shell (a) and perforated Perspex shell (b). (Reprinted from *Journal of Membrane Science*, 274(1–2), Ghosh, R., 73–82, Copyright 2006, with permission from Elsevier.)

hollow fiber membranes during their residence within the shell; in the type 2 module, however, some bubbles escaped through the holes at different locations on the shell. Mayer et al. (2006) compared various aeration devices for air sparging in cross-flow tubular membrane filtration and showed that complex aeration systems with multiple orifices injecting air homogeneously into the feed flow worked best. However, when the cross-flow filtration unit is operated at a fixed cross-flow velocity, simple tube air-sparging devices may be more practical. The distribution of the gas phase, bubble size, and distance between the bubbles, which depend on the ratio of gas and liquid flow rates and the membrane module, is therefore important for fouling control.

There are still some issues that need to be considered during the operation of air-sparging MBRs. The sludge suspension is a very complex system, so sludge characteristics should be taken into consideration when air sparging is applied to enhance membrane permeation. For example, higher air-sparging intensity will lead to floc breakage, which may lead to intensified fouling.

To promote turbulence close to the membrane surface, Broussous et al. (2000) developed a new ceramic tubular membrane with a helical relief stamp at the membrane surface. Two flow structures were found: a central flow and helical secondary flow between the helical relief, which can mitigate concentration polarization and cake fouling. For hollow fiber membranes, a lateral sway imposed on the hollow fiber can increase the pseudo-steady-state permeate flux by mechanically disturbing the foulant layer formed on membranes (Berube et al., 2006).

For submerged hollow fibers using aeration, such as the membrane bioreactor, the preferred fiber orientation should be vertical rather than horizontal though the overall effect of fiber orientation on membrane permeation is smaller than that of aeration (Chang et al., 2002c). The size of the gap between the submerged flat-sheet membranes is also important for two-phase flow and fouling control (Ndinisa et al., 2006a). As the gap was increased from 7 mm to 14 mm, the fouling became worse and the degree of fouling reduction by two-phase flow decreased by at least 40% based on suction pressure rise (dTMP/dt). Moreover, fiber movement and fouling control are influenced by fiber tightness with significantly improved performance for slightly loose fibers (Wicaksana et al., 2006).

The interaction of two-phase flow with membranes is a complex issue. During the last few years, analysis for the hydrodynamics and the fluid flow pattern adjacent to the membrane have been investigated by using computational fluid dynamics (CFD) mathematical modeling and simulation (Ahmad et al., 2005). The multiphase flow simulation by CFD technique can provide microscopic understanding of the fouling mechanism, and it has been proven to be a powerful tool to aid in membrane module design (Li et al., 2006). Ndinisa et al. (2006b) studied the fouling in a submerged flat sheet MBR using two-phase flow characterization and CFD simulation. It was found that the flux enhancement by the increasing bubble size was primarily due to an increase in the overall shear stress on the membranes and to more turbulence generated by the gas phase. Buetehorn et al. (2007) studied the instantaneous fiber displacement within a single hollow fiber bundle by using CFD simulation and computer tomography (CT). It was observed that the hollow fibers displacement varied with hollow fiber length. Rios et al. (2007) used CFD simulations based on the volume of fluid (VOF) approach to evaluate the shear stress distributions near the membrane surface for different air velocities. The results showed that the broadness of the stress distribution as well as the shear stresses increased significantly with rising air velocities.

In brief, enhancement of hydrodynamic condition is one of the effective approaches to mitigate membrane fouling. But, the hydrodynamic condition has close relation with aeration intensity, bubble size, membrane module configuration, MLSS concentration and sludge viscosity etc. Therefore, the hydrodynamic condition in membrane processes is very complex; and optimization of membrane module and aeration combined with CFD modeling and simulation might be helpful for the future enhancement of hydrodynamic condition.

### 14.3.3 Subcritical Flux Operation

Operation below the critical flux is an efficient approach to avoid severe fouling including removal and irreversible fouling within a given filtration system. Field et al. (1995) introduced the critical flux concept, operation below the critical flux concept is called subcritical flux or nonfouling operation and is expected to cause little irreversible fouling. For a short-term membrane filtration, when the permeate flux is set below the critical flux, the TMP remains stable and fouling is removable. In contrast, when it exceeds the critical flux, the TMP increases and might lead to a TMP jump. As a matter of fact, for a long-term operation of membrane processes, irreversible fouling can occur even if they are operated below the critical flux (Guglielmi et al., 2007). Ognier et al. (2004) reported that despite the initial choice of subcritical flux filtration conditions, gradual fouling was seen to develop which, after long periods of operation without intermediary membrane regeneration, proved to be hydraulically irremovable. The critical flux value depends on membrane characteristics, operating conditions (i.e., aeration intensity, temperature), and sludge characteristics.

Even though strictly speaking it does not exist in MBR, critical flux is often used as a practical guideline for plant design and operation. It is acknowledged that significant differences above and below a clearly defined flux exist, so to distinguish between high- and low-fouling rates or to indicate the point of significant change the sustainable flux has been introduced (Bacchin et al., 2006). It can be defined as the flux above which the rate of fouling is economically and environmentally unsustainable (Bacchin et al., 2006) or for which TMP increases gradually at an acceptable rate, such that chemical cleaning is not necessary (Le-Clech et al., 2006). Obviously, such a value can only be assessed in long-term trials. It might be therefore be more accurate to speak of apparent sustainable flux when values are obtained from short-term laboratory trials. The decision for any one value is rather arbitrary but will be below the so-called critical flux.

Subcritical flux operation has also been used to control NF and RO membranes fouling (Lisitsin et al., 2005; Zhang et al., 2008c; Vrouwenvelder et al., 2009). A reduction in operation flux can cause a significant change in the scaling mechanism (Lisitsin et al., 2005). However, a recent study performed by Vrouwenvelder et al. (2009) found that subcritical flux operation is not a suitable approach to control biofouling of spiral wound RO and NF membranes used for drinking water treatment. In addition, most of the recent investigations on the determination of critical flux are based on ex-situ measuring, which cannot offer the real fouling propensity. Recently, an in-situ method was developed by de la Torre et al. (2008), which can provide more reliable information about critical flux than ex-situ methods. Huyskens et al. (2008) developed an on-line measuring method, which was used to evaluate the reversible and irreversible fouling propensity of MBR mixed liquor in a reproducible way. These studies imply that it is possible to develop on-line or in-situ method to determine critical flux or reversible/irreversible fouling. It is also of high interest to develop a unified measuring method and an appropriate test device. On the other hand, it should be pointed out that the critical flux is usually determined by using flux-step experiments, which can not reflect the fouling behavior in long-term operating membrane processes (Kraume et al., 2009). During initial filtration, colloids, solutes and microbial cells pass through and precipitate inside the membrane pores. But, during the long-term operation of membrane filtration, the deposited cells multiple and yield EPS, which can clog the pores and forms a strongly attached fouling layer. The occurrence of membrane fouling is a very complex process. Thus, how to predict and use the critical flux concept is of great significance for membrane processes. Further discussion of critical flux can be found in recent review articles by (Pollice et al., 2005; Bacchin et al., 2006).

### 14.3.4 Implementation of Physical and Chemical Cleaning

To recover membrane permeability, membrane cleaning including physical and chemical cleaning procedures can be conducted after short- and long-term operation of membrane filtration. Physical cleaning or chemical cleaning can remove membrane foulants from membranes via relaxation, backwashing, hydraulic scouring and chemical reactions. Physical cleaning based on backflushing or just water rinsing is usually used for the removal of reversible fouling whereas chemical cleaning is applied for the removal of irreversible fouling in membrane processes. Figure 14.8 shows the typical development of transmembrane pressure (TMP) during long-term operation of full-scale MBRs

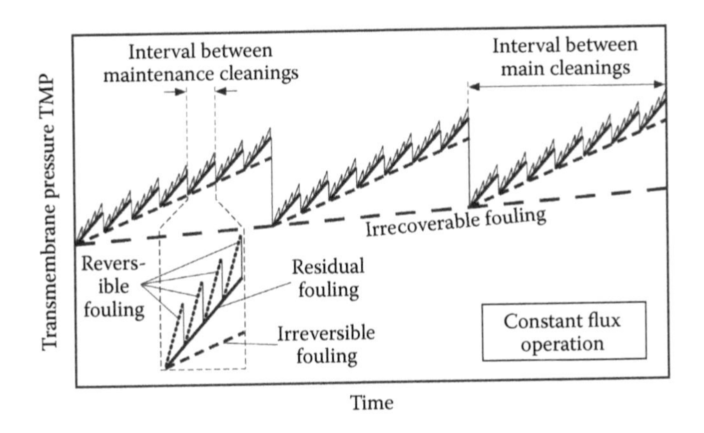


**FIGURE 14.8** Schematic illustration showing different cleaning methods for flux recovery during long-term operation of full-scale MBRs. (Reprinted from *Desalination*, 236, Kraume, M., et al., 94–103, Copyright 2009, with permission from Elsevier.)

including different cleaning procedures. The physical cleaning by backflushing and the chemical maintenance cleaning is performed in situ to remove cake fouling or residual fouling. When the membrane permeability is no longer sufficient, an intensive chemical cleaning is implemented (Kraume et al., 2009). It can be seen that most of the foulants that had accumulated on the membranes during long-term operation can be removed by chemical cleaning. After chemical cleaning, only a very small irrecoverable fouling remains, which cannot be removed by any approaches. Compared with physical cleaning, the mechanisms of chemical cleaning are very complicated. Weis et al. (2003) reported that chemical cleaning can recover membrane permeation in three ways: (1) foulants may be removed directly, (2) morphology of foulants may be changed, and (3) surface chemistry of the fouling layer may be altered. The cleaning agents can reduce the cohesion forces between the foulants themselves and the adhesion between the foulants and the membrane surface (Zondervan et al., 2007). For example, the EDTA cleaning agent can reduce $Ca^{2+}$ in membrane foulants via a ligand exchange reaction (Li et al., 2004).

Based on the feedwater composition and foulant composition, chemical agents of different sorts such as NaOH, NaClO, $H_2O_2$, EDTA, and HCl have been proposed. Several factors such as temperature, pH, concentration of the chemical agents, and contact time have been observed to impact cleaning efficiencies. In general, acid cleaning is effective in elimination of inorganic fouling or scaling for the membranes. In contrast, caustic and oxidizing agents are effective in removal of organic fouling and biofouling. Li et al. (2004) found that 1 mM EDTA and 10 mM SDS could complete remove membrane foulants, and a higher pH value enhanced flux recovery. They also found that the increase of SDS concentration would significantly increase cleaning efficiency. Nigam et al. (2008), however, found that a lower concentration of caustic solution (0.2%) was observed to provide the best performance compared with higher concentrations (0.3% and 0.5%). The higher concentration solutions probably enhanced the swelling of the foulants which consequently resulted in additional blockage of the pores and thus the permeate flux recovery was not as high. Figure 14.9 compares the performance of different chemical agents used for cleaning UF membranes fouled by surface water. It can be seen that caustic and oxidizing agents have higher instant cleaning rate and overall cleaning efficiencies. Sometimes, to fully recover membrane flux, combined cleaning protocols might be required. The combined acid/base chemical cleaning was found to be the best for oil contaminated membranes (Al-Obeidani et al., 2008). Even though membrane cleaning is a rapid and simple method to recover membrane permeation, the additional manpower utilization and energy consumed during cleaning can increase the overall cost of water treatment (Al-Amoudi et

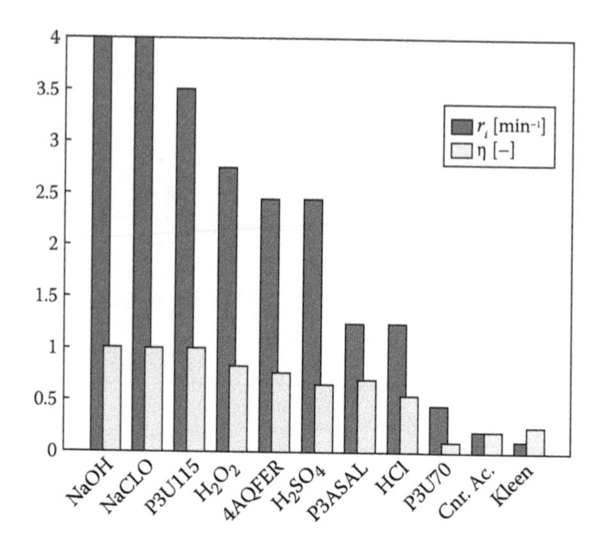


**FIGURE 14.9** Overall instant cleaning rate $r_i$ [1/min] and overall cleaning efficiency η [–] for different cleaning agents (0.05 mol/L) in membrane process used for surface water filtration. (Reprinted from *Journal of Membrane Science*, 304, Zondervan, E. and Roffel, B., 40–49, 2007, with permission from Elsevier.)

al., 2007). Therefore, the implementation of membrane cleaning and the cleaning frequency should be balanced with the overall energy consumption.

Up to now all measures to regain permeability (increased air sparging, backflushing, relaxation, chemical cleaning) are neither optimized not controlled, i.e., they are carried out before they are necessary, thereby wasting energy, permeate or chemicals, or too late. Currently, the application of mechanistic models which describe filtration and fouling mechanisms are tested for the development of model-based control strategies for membrane operation that reduce energetic expenditure and increase filtration efficiency (Drews et al., 2009).

### 14.3.5 Development of Antifouling Membranes

Membrane characteristics including pore size, porosity, surface charge, roughness, and hydrophilicity/hydrophobicity, etc., have certain impacts on membrane performance, especially on membrane fouling. The determination of suitable membrane pore sizes has been extensively investigated in the 1990s. Pore size distribution is likely to be one of the parameters affecting membrane performance. A narrow pore size distribution is preferred to control membrane fouling in most membrane processes.

The membrane materials always show different fouling propensity due to their different pore size, morphology and hydrophobicity. For example, the composite membrane having a dense layer over the support material is usually more resistant to membrane fouling than symmetric membrane having a homogeneous structure (Chae et al., 2008). Chae et al. (2008) reported that the cake layer of the composite membrane contained relatively large amounts of organics and inorganics compared to that of a symmetric membrane during membrane filtration of drinking water, but it was effectively removed during the backwashing period. It is due to the fact that the dense layer on the composite membrane can retain foulants and prevent them from penetrating into membrane pores. It also has been reported that a polyvinylidene fluoride (PVDF) membrane is superior to a polyethylene (PE) membrane in terms of prevention of irreversible fouling in MBRs used for treatment of municipal wastewater (Yamato et al., 2006). The fouling behavior of the membrane used is determined by the affinity between foulants (e.g., NOM/EPS/SMP) and membranes. Zhang et al. (2008a) studied the affinity between EPS and three polymeric ultrafiltration membranes, and observed that the affinity of the three membrane were orderly polyacrylonitrile (PAN) < PVDF < polyethersulfone (PES). It suggests that among these membranes the PAN membrane is more fouling-resistant. Inorganic membranes, such as aluminum, zirconium, and titanium oxide, show superior hydraulic, thermal, and chemical resistance. But, these inorganic membranes are not the preferred option for large-scale water/wastewater treatment because of the high membrane costs. In addition, inorganic membranes can induce severe inorganic fouling (i.e., struvite formation) due to ligand exchange reaction as well as surface charge effect (Kang et al., 2002). So, the inorganic membranes might be used in some special applications such as high-temperature wastewater treatment (Zhang et al., 2006c).

In general, membrane fouling occurs more readily on hydrophobic membranes than on hydrophilic ones because of hydrophobic interaction between membrane foulants and membrane materials. As a result, much attention has been given to reduce membrane fouling by modifying hydrophobic membranes to relative hydrophilic (Yu et al., 2005a, 2005b). This chapter is focused on the novel and significant findings in membranes or filters used for water and wastewater treatment. The detailed discussion on the impacts of membrane material on membrane fouling can be found in recent review articles (Chang et al., 2002b; Le-Clech et al., 2006).

The main objective of new membrane material development is to reduce the high investment costs for the membrane modules and/or to enhance and maintain membrane flux. Buonomenna et al. (2008) recently reported that the plasma modified NF membranes exhibited favorable performance in terms of fouling control and rejection improvement during membrane filtration of aqueous solution containing organic compounds. To improve the antifouling property of membranes in an MBR for wastewater treatment, the membranes were subjected to surface modification by $NH_3$ and $CO_2$ plasma treatment by Yu et al. (2005a, 2005b, 2008). The water contact angle reduced significantly after $NH_3$ and $CO_2$ plasma treatment due to the fact that $-NH_2$ groups and $-COOH$ groups were grafted on the membrane surface. Fouling rates of the $NH_3$ and $CO_2$ plasma-treated membranes were 50% lower than those of the unmodified membranes. Although the plasma treatment processes has many advantages, e.g., a very shallow modification depth compared with other modification techniques, it still has drawbacks. For example, the chemical reactions of the plasma treatment are rather complex, so the surface chemistry of the modified surface is difficult to understand in detail and thus, currently it is not possible to extend plasma treatment on the large scale (Yu et al., 2007).

To overcome these disadvantages of plasma treatment, Yu et al. (2007) applied the surface graft polymerization method to improve the membrane permeation. In the study, the surface modification of polypropylene microporous membranes was accomplished by UV irradiation in aqueous acrylamide solutions. The contact angle data showed that the hydrophilicity of the modified membrane increased strongly with the increase of the grafting degree. Akbari et al. (2006) also found that the modified UF membranes, based on UV-photo grafting of sodium p-styrene sulfonate and [2-(acryloyloxy)ethyl] ammonium chloride, can be successfully sued for the treatment of dye effluents, and these modified membranes were far less sensitive to fouling. Even though the modified membrane showed better filtration ability than the unmodified membrane, this method has the disadvantage of applying high-energy methods, such as UV irradiation, plasma treatment, gamma irradiation, and chemical reaction, resulting in an increase in membrane cost (Asatekin et al., 2006).

Recently, a self-assembly technique, which is one of the simplest and most effective methods to fabricate a thin film on the membrane surface, was employed for fabricating a fouling resistance membrane in MBR (Asatekin et al., 2006). In this study, commercial polyvinylidene fluoride ultrafiltration membranes (PVDF UF) were coated with the amphiphilic graft copolymer polyvinylidene fluoride–graft–polyoxyethylene methacrylated (PVDF-g-POEM), to create thin-film composite nanofiltration membranes (TFC NF). The new TFC NF membranes exhibited no irremovable fouling in a 10-day dead-end filtration of bovine serum albumin, sodium alginate, and humic acid at concentrations of 1000 mg/L. The antifouling properties of the TFC NF membranes were attributed to both the nanoscale dimensions of the hydrophilic channels through the coating, which greatly restrict the size of permeate species, and the unique properties of polyethylene oxide (PEO), which can resist the adsorption of EPS on the membrane surface. To improve antifouling property of UF membranes in water treatment, nanosilver was incorporated into polysulfone UF membranes by Zodrow et al. (2009), and the modified UF membranes were observed to increase membrane hydrophilicity and consequently reduce membrane fouling. Previously, $TiO_2$-embedded polymeric membranes have also been prepared by a self-assembly process (Bae et al., 2005a, 2005b). The surface of a $TiO_2$-embedded membrane can be more hydrophilic than that of an unmodified membrane due to the higher affinity of metal oxides to water. The investigation showed that the flux of the modified membrane was improved by nearly 100% (Bae et al., 2005b). The $TiO_2$ nanoparticles embedded in the membrane have shown antibacterial property, especially when illumination is available (Kwak et al., 2001). These investigations confirmed that a self-assembly technique can be successfully used to modify the membranes for fouling control.

Blending of conventional hydrophobic membrane materials with amphiphilic polymers has been proven as a powerful approach. Recently, for example, a PVDF membrane was modified by blending with amphiphilic hyperbranched-star polymer (Zhao et al., 2007). As shown in Figure 14.10, an amphiphilic hyperbranched-star polymer (HPE-g-MPEG) with about 12 hydrophilic arms in each molecule was synthesized by grafting methoxy polyethylene glycol (MPEG) to the hyperbranched polyester (HPE) molecule using terephthaloyl chloride (TPC) as the coupling agent and blended with PVDF membrane to fabricate porous membranes via a phase inversion process. It was found that the modified membrane showed higher protein resistance, higher flux, and better flux recovery than the pure PVDF membrane. Figure 14.11 gives the comparison of flux between pure PVDF membrane and modified PVDF membrane.

Next-generation membranes should aim at developing of economical, high-flux, nonfouling membrane before viable MBR processes can be achieved (Shannon et al., 2008). The nonfouling or low-fouling membrane should have much narrower pore size distributions, stronger hydrophilicity, and larger porosity than the currently used membranes. At this point, the microsieve membrane, which has very uniform pore size (Brans et al., 2006; Ning Koh et al., 2008), can provide a useful alternative for the development of narrow pore size distribution membranes. On the other hand, the nanotechnology might be of interest for the development of strong hydrophilic membranes.

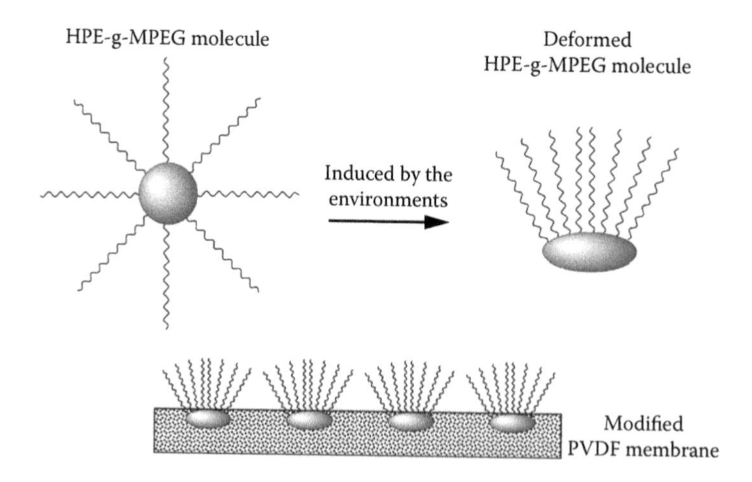


**FIGURE 14.10** Deformation of the HPE-g-MPEG molecule during the immersion precipitation process and the sketch of molecular conformation in the final membrane. (Reprinted with permission from Zhao, Y.-H., et al., 2007, 5779–5786. Copyright 2007 American Chemical Society.)

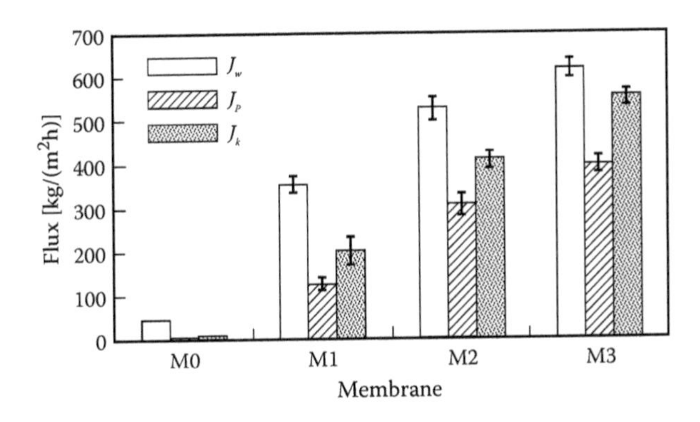


**FIGURE 14.11** Permeation fluxes of pure water and BSA solution through the pure PVDF membrane and membranes with an HPE-g-MPEG/PVDF ratio of (M0) 0, (M1) 1/10, (M2) 2/10, and (M3) 3/10, where $J_w$ is pure water flux, $J_p$ is BSA solution flux, and $J_k$ is pure water flux of BSA-fouled membrane after washing with DI water. (Reprinted with permission from Zhao, Y.-H., et al., 2007, 5779–5786. Copyright 2007 American Chemical Society.)

## 14.4 CONCLUDING REMARKS

In this chapter, membrane fouling in membrane processes used for water and wastewater treatment was addressed. Depending on the specific membrane process, membrane fouling can be classified into three types, which are biofouling, organic fouling, and inorganic fouling. These three types of fouling usually occur at the same time and are combined with each other. The synergistic interactions between different kinds of fouling result in faster and more substantial membrane flux decline. To mitigate membrane fouling effectively, a number of methods have been developed by investigators and engineers. Dosing of coagulant and/or adsorbents can decrease the level of solutes and colloids in the solution or enhance the flocculation ability of flocs, which can not only increase membrane permeability but improve membrane effluent quality significantly as well. Air sparging is an efficient strategy to control concentration polarization and remove cake layers. Further enhancement of hydrodynamic conditions is one of the effective approaches to mitigate membrane fouling. Particularly, the combination of aeration and membrane module design with CFD simulations might be helpful for the improvement of hydrodynamic conditions. Operation below the critical flux is one of the effective approaches to avoid severe fouling within a given filtration system. The membrane filtration period can be prolonged as a subcritical flux is imposed. Membrane modification has also provided a potential alternative for fouling control. The implementation of membrane cleaning, including physical cleaning and chemical cleaning, can recover membrane permeation significantly, but these cleaning methods/agents are of different efficiencies.

## ACKNOWLEDGMENT

Fangang Meng gratefully acknowledges the financial support of the Alexander von Humboldt (AvH) Foundation.

## REFERENCES

Ahmad, A. L., Lau, K. K., Bakar, M. Z. A. and Shukor, S. R. A. 2005. Integrated CFD simulation of concentration polarization in narrow membrane channel. *Computers & Chemical Engineering* 29(10), 2087–2095.

Akbari, A., Desclaux, S., Rouch, J. C., Aptel, P. and Remigy, J. C. 2006. New UV-photografted nanofiltration membranes for the treatment of colored textile dye effluents. *Journal of Membrane Science* 286(1–2), 342–350.

Al-Amoudi, A. and Lovitt, R. W. 2007. Fouling strategies and the cleaning system of NF membranes and factors affecting cleaning efficiency. *Journal of Membrane Science* 303(1–2), 4–28.

Al-Obeidani, S. K. S., Al-Hinai, H., Goosen, M. F. A., Sablani, S., Taniguchi, Y. and Okamura, H. 2008. Chemical cleaning of oil contaminated polyethylene hollow fiber microfiltration membranes. *Journal of Membrane Science* 307, 299–308.

Asatekin, A., Menniti, A., Kang, S., Elimelech, M., Morgenroth, E. and Mayes, A. M. 2006. Antifouling nanofiltration membranes for membrane bioreactors from self-assembling graft copolymers. *Journal of Membrane Science* 285(1–2), 81–89.

Bacchin, P., Aimar, P. and Field, R. W. 2006. Critical and sustainable fluxes: Theory, experiments and applications. *Journal of Membrane Science* 281(1–2), 42–69.

Bae, T.-H. and Tak, T.-M. 2005a. Effect of $TiO_2$ nanoparticles on fouling mitigation of ultrafiltration membranes for activated sludge filtration. *Journal of Membrane Science* 249(1–2), 1–8.

Bae, T.-H. and Tak, T.-M. 2005b. Preparation of $TiO_2$ self-assembled polymeric nanocomposite membranes and examination of their fouling mitigation effects in a membrane bioreactor system. *Journal of Membrane Science* 266(1–2), 1–5.

Bellara, S. R., Cui, Z. F. and Pepper, D. S. 1996. Gas sparging to enhance permeate flux in ultrafiltration using hollow fibre membranes. *Journal of Membrane Science* 121(2), 175–184.

Berube, P. R. and Lei, E. 2006. The effect of hydrodynamic conditions and system configurations on the permeate flux in a submerged hollow fiber membrane system. *Journal of Membrane Science* 271(1–2), 29–37.

Brans, G., Kromkamp, J., Pek, N., Gielen, J., Heck, J., van Rijn, C. J. M., van der Sman, R. G. M., Schroen, C. G. P. H. and Boom, R. M. 2006. Evaluation of microsieve membrane design. *Journal of Membrane Science* 278(1–2), 344–348.

Broussous, L., Schmitz, P., Boisson, H., Prouzet, E. and Larbot, A. 2000. Hydrodynamic aspects of filtration antifouling by helically corrugated membranes. *Chemical Engineering Science* 55(21), 5049–5057.

Buetehorn, S., Koh, C. N., Wintgens, T., Melin, T., Volmering, D. and Vossenkaul, K. 2007. Investigating hydrodynamics in submerged hollow-fibre membrane filtration units in municipal wastewater treatment using computational fluid dynamics (CFD). In: Lesjean, B. (ed.) *2nd IWA National Young Water Professionals Conference*, Berlin/Germany, 89–96.

Buonomenna, M. G., Lopez, L. C., Davoli, M., Favia, P., d'Agostino, R. and Drioli, E. 2008. Polymeric membranes modified via plasma for nanofiltration of aqueous solution containing organic compounds. *Microporous and Mesoporous Materials* 120(1–2), 147–153.

Cabassud, C., Laborie, S., Durand-Bourlier, L. and Laine, J. M. 2001. Air sparging in ultrafiltration hollow fibers: Relationship between flux enhancement, cake characteristics and hydrodynamic parameters. *Journal of Membrane Science* 181(1), 57–69.

Cai, Z., Kim, J. and Benjamin, M. M. 2008. NOM removal by adsorption and membrane filtration using heated aluminum oxide particles. *Environmental Science and Technology* 42(2), 619–623.

Chae, S.-R., Yamamura, H., Ikeda, K. and Watanabe, Y. 2008. Comparison of fouling characteristics of two different poly-vinylidene fluoride microfiltration membranes in a pilot-scale drinking water treatment system using pre-coagulation/sedimentation, sand filtration, and chlorination. *Water Research* 42(8–9), 2029–2042.

Chang, I.-S. and Judd, S. J. 2002a. Air sparging of a submerged MBR for municipal wastewater treatment. *Process Biochemistry* 37(8), 915–920.

Chang, I. S., Clech, P. L., Jefferson, B. and Judd, S. 2002b. Membrane fouling in membrane bioreactors for wastewater treatment. *Journal of Environmental Engineering* 128(11), 1018–1029.

Chang, S., Fane, A. G. and Vigneswaran, S. 2002c. Experimental assessment of filtration of biomass with transverse and axial fibres. *Chemical Engineering Journal* 87(1), 121–127.

Chen, C. L., Liu, W. T., Chong, M. L., Wong, M. T., Ong, S. L., Seah, H. and Ng, W. J. 2004a. Community structure of microbial biofilms associated with membrane-based water purification processes as revealed using a polyphasic approach. *Applied Microbiology and Biotechnology* 63(4), 466–473.

Chen, J. C., Li, Q. and Elimelech, M. 2004b. In situ monitoring techniques for concentration polarization and fouling phenomena in membrane filtration. *Advances in Colloid and Interface Science* 107(2–3), 83–108.

Costa, A. R., de Pinho, M. N. and Elimelech, M. 2006. Mechanisms of colloidal natural organic matter fouling in ultrafiltration. *Journal of Membrane Science* 281(1–2), 716–725.

Cui, Z. F., Chang, S. and Fane, A. G. 2003. The use of gas bubbling to enhance membrane processes. *Journal of Membrane Science* 221(1–2), 1–35.

de la Torre, T., Lesjean, B., Drews, A. and Kraume, M. 2008. Monitoring of transparent exopolymer particles (TEP) in a membrane bioreactor (MBR) and correlation with other fouling indicators. *Water Science and Technology* 58(10), 1903–1909.

Dennett, K. E., Amirtharajah, A., Moran, T. F. and Gould, J. P. 1996. Coagulation: Its effect on organic matter. *Journal of the American Water Works Association* 88129–142.

Drews, A., Arellano-Garcia, H., Schoeneberger, J., Schaller, J., Wozny, G. and Kraume, M. 2009. Model-based recognition of fouling mechanisms in membrane bioreactors. *Desalination* 236, 224–233.

Ducom, G. and Cabassud, C. 2003. Possible effects of air sparging for nanofiltration of salted solutions. *Desalination* 156(1–3), 267–274.

Fan, L., Harris, J. L., Roddick, F. A. and Booker, N. A. 2001. Influence of the characteristics of natural organic matter on the fouling of microfiltration membranes. *Water Research* 35(18), 4455–4463.

Fane, A. G., Yeo, A., Law, A., Parameshwaran, K., Wicaksana, F. and Chen, V. 2005. Low pressure membrane processes — doing more with less energy. *Desalination* 185(1–3), 159–165.

Field, R. W., Wu, D. and Howell, J. A. 1995. Critical flux concept for microfiltration fouling. *Journal of Membrane Science* 100(3), 259–272.

Flemming, H. C., Schaule, G., Griebe, T., Schmitt, J. and Tamachkiarowa, A. 1997. Biofouling—the Achilles heel of membrane processes. *Desalination* 113(2–3), 215–225.

Ghosh, R. 2006. Enhancement of membrane permeability by gas-sparging in submerged hollow fibre ultrafiltration of macromolecular solutions: Role of module design. *Journal of Membrane Science* 274(1–2), 73–82.

Guglielmi, G., Chiarani, D., Judd, S. and Andreottola, G. 2007. Flux criticality and sustainability in a hollow fibre submerged membrane bioreactor for municipal wastewater treatment. *Journal of Membrane Science* 289, 214–248.

Guglielmi, G., Chiarani, D., Saroj, D. P. and Andreottola, G. 2008. Impact of chemical cleaning and air-sparging on the critical and sustainable flux in a flat sheet membrane bioreactor for municipal wastewater treatment. *Water Science and Technology* 57(12), 1873–1879.

Guo, W. S., Vigneswaran, S. and Ngo, H. H. 2005. Effect of flocculation and/or adsorption as pretreatment on the critical flux of crossflow microfiltration. *Desalination* 172(1), 53–62.

Guo, W. S., Vigneswaran, S., Ngo, H. H., Kandasamy, J. and Yoon, S. 2008. The role of a membrane performance enhancer in a membrane bioreactor: A comparison with other submerged membrane hybrid systems. *Desalination* 231(1–3), 305–313.

Haberkamp, J., Ruhl, A. S., Ernst, M. and Jekel, M. 2007. Impact of coagulation and adsorption on DOC fractions of secondary effluent and resulting fouling behaviour in ultrafiltration. *Water Research* 41(17), 3794–3802.

Henderson, R. K., Baker, A., Parsons, S. A. and Jefferson, B. 2008. Characterisation of algogenic organic matter extracted from cyanobacteria, green algae and diatoms. *Water Research* 42(13), 3435–3445.

Her, N., Amy, G., Chung, J., Yoon, J. and Yoon, Y. 2008. Characterizing dissolved organic matter and evaluating associated nanofiltration membrane fouling. *Chemosphere* 70(3), 495–502.

Her, N., Amy, G., Park, H. R. and Song, M. 2004. Characterizing algogenic organic matter (AOM) and evaluating associated NF membrane fouling. *Water Research* 38(6), 1427–1438.

Her, N., Amy, G., Plottu-Pecheux, A. and Yoon, Y. 2007. Identification of nanofiltration membrane foulants. *Water Research* 41(17), 3936–3947.

Hong, S. and Elimelech, M. 1997. Chemical and physical aspects of natural organic matter (NOM) fouling of nanofiltration membranes. *Journal of Membrane Science* 132(2), 159–181.

Hörsch, P., Gorenflo, A., Fuder, C., Deleage, A. and Frimmel, F. H. 2005. Biofouling of ultra- and nanofiltration membranes for drinking water treatment characterized by fluorescence in situ hybridization (FISH). *Desalination* 172(1), 41–52.

Howe, K. J., Marwah, A., Chiu, K.-P. and Adham, S. S. 2006. Effect of coagulation on the size of MF and UF membrane foulants. *Environmental Science and Technology* 40(24), 7908–7913.

Hu, A. Y. and Stuckey, D. C. 2007. Activated carbon addition to a submerged anaerobic membrane bioreactor: Effect on performance, transmembrane pressure, and flux. *Journal of Environmental Engineering* 133(1), 73–80.

Huyskens, C., Brauns, E., Van Hoof, E. and De Wever, H. 2008. A new method for the evaluation of the reversible and irreversible fouling propensity of MBR mixed liquor. *Journal of Membrane Science* 323(1), 185–192.

Hwang, B.-K., Lee, W.-N., Park, P.-K., Lee, C.-H. and Chang, I.-S. 2007. Effect of membrane fouling reducer on cake structure and membrane permeability in membrane bioreactor. *Journal of Membrane Science* 288(1–2), 149–156.

Hwang, B. K., Lee, W. N., Yeon, K. M., Park, P. K., Lee, C. H., Chang, I. S., Drews, A. and Kraume, M. 2008. Correlating TMP increases with microbial characteristics in the bio-cake on the membrane surface in a membrane bioreactor. *Environmental Science & Technology* 42(11), 3963–3968.

Jermann, D., Pronk, W., Meylan, S. and Boller, M. 2007. Interplay of different NOM fouling mechanisms during ultrafiltration for drinking water production. *Water Research* 41(8), 1713–1722.

Ji, J., Qiu, J., Wong, F.-S. and Li, Y. 2008. Enhancement of filterability in MBR achieved by improvement of supernatant and floc characteristics via filter aids addition. *Water Research* 42(14), 3611–3622.

Jinhua, P., Fukushi, K. and Yamamoto, K. 2006. Bacterial community structure on membrane surface and characteristics of strains isolated from membrane surface in submerged membrane bioreactor. *Separation Science and Technology* 41(7), 1527–1549.

Judd, S. 2006. *The MBR Book*. Elsevier Science and Technology, Amsterdam.

Kang, I.-J., Yoon, S.-H. and Lee, C.-H. 2002. Comparison of the filtration characteristics of organic and inorganic membranes in a membrane-coupled anaerobic bioreactor. *Water Research* 36(7), 1803–1813.

Kim, I. S. and Jang, N. 2006. The effect of calcium on the membrane biofouling in the membrane bioreactor (MBR). *Water Research* 40(14), 2756–2764.

Kimura, K., Naruse, T. and Watanabe, Y. 2009. Changes in characteristics of soluble microbial products in membrane bioreactors associated with different solid retention times: Relation to membrane fouling. *Water Research* 43(4), 1033–1039.

Koseoglu, H., Yigit, N. O., Iversen, V., Drews, A., Kitis, M., Lesjean, B. and Kraume, M. 2008. Effects of several different flux enhancing chemicals on filterability and fouling reduction of membrane bioreactor (MBR) mixed liquors. *Journal of Membrane Science* 320(1–2), 57–64.

Kraume, M., Wedi, D., Schaller, J., Iversen, V. and Drews, A. 2009. Fouling in MBR—What use are lab investigations for full scale operation? *Desalination* 236, 94–103.

Kwak, S.-Y., Kim, S. H. and Kim, S. S. 2001. Hybrid organic/inorganic reverse osmosis (RO) membrane for bacterial anti-fouling. 1. Preparation and characterization of $TiO_2$ nanoparticle self-assembled aromatic polyamide thin-film-composite (TFC) membrane. *Environmental Science and Technology* 35, 2388–2394.

Laabs, C. N., Amy, G. L. and Jekel, M. 2006. Understanding the size and character of fouling-causing substances from effluent organic matter (EfOM) in low-pressure membrane filtration. *Environmental Science and Technology* 40(14), 4495–4499.

Le-Clech, P., Chen, V. and Fane, T. A. G. 2006. Fouling in membrane bioreactors used in wastewater treatment. *Journal of Membrane Science* 284(1–2), 17–53.

Lee, J., Ahn, W.-Y. and Lee, C.-H. 2001. Comparison of the filtration characteristics between attached and suspended growth microorganisms in submerged membrane bioreactor. *Water Research* 35(10), 2435–2445.

Lee, N., Amy, G., Croue, J. P. and Buisson, H. 2004. Identification and understanding of fouling in low-pressure membrane (MF/UF) filtration by natural organic matter (NOM). *Water Research* 38(20), 4511–4523.

Li, Q. L. and Elimelech, M. 2004. Organic fouling and chemical cleaning of nanofiltration membranes: Measurements and mechanisms. *Environmental Science & Technology* 38(17), 4683–4693.

Li, Q. Y., Cui, Z. F. and Pepper, D. S. 1997. Effect of bubble size and frequency on the permeate flux of gas sparged ultrafiltration with tubular membranes. *Chemical Engineering Journal* 67(1), 71–75.

Li, Y.-L., Chang, T.-H., Wu, C.-Y., Chuang, C.-J. and Tung, K.-L. 2006. CFD analysis of particle deposition in the spacer-filled membrane module. *Journal of Water Supply: Research and Technology—AQUA* 55(7–8), 589–601.

Lisitsin, D., Hasson, D. and Semiat, R. 2005. Critical flux detection in a silica scaling RO system. *Desalination* 186(1–3), 311–318.

Liu, R., Lead, J. R. and Baker, A. 2007. Fluorescence characterization of cross flow ultrafiltration derived freshwater colloidal and dissolved organic matter. *Chemosphere* 68, 1304–1311.

Lyko, S., Al-Halbouni, D., Wintgens, T., Janot, A., Hollender, J., Dott, W. and Melin, T. 2007. Polymeric compounds in activated sludge supernatant—Characterisation and retention mechanisms at a full-scale municipal membrane bioreactor. *Water Research* 41(17), 3894–3902.

Mayer, M., Braun, R. and Fuchs, W. 2006. Comparison of various aeration devices for air sparging in crossflow membrane filtration. *Journal of Membrane Science* 277(1–2), 258–269.

Meng, F., Chae, S. R., Drews, A., Kraume, M., Shin, H. S. and Yang, F. 2009. Recent advances in membrane bioreactors (MBRs): Membrane fouling and membrane material. *Water Research* 43(6), 1489–1512.

Mercier, M., Fonade, C. and Lafforgue-Delorme, C. 1997. How slug flow can enhance the ultrafiltration flux in mineral tubular membranes. *Journal of Membrane Science* 128(1), 103–113.

Metzger, U., Le-Clech, P., Stuetz, R. M., Frimmel, F. H. and Chen, V. 2007. Characterisation of polymeric fouling in membrane bioreactors and the effect of different filtration modes. *Journal of Membrane Science* 301(1–2), 180–189.

Miura, Y., Watanbe, Y. and Okabe, S. 2007. Membrane biofouling in pilot-scale membrane bioreactors (MBRs) treating municipal wastewater: Impact of biofilm formation. *Environmental Science & Technology* 41(2), 632–638.

Ndinisa, N. V., Fane, A. G. and Wiley, D. E. 2006a. Fouling control in a submerged flat sheet membrane system: Part I—Bubbling and hydrodynamic effects. *Separation Science and Technology* 41(7), 1383–1409.

Ndinisa, N. V., Fane, A. G., Wiley, D. E. and Fletcher, D. F. 2006b. Fouling control in a submerged flat sheet membrane system: Part II—Two-phase flow characterization and CFD simulations. *Separation Science and Technology* 41(7), 1411–1445.

Ng, C. A., Sun, D. and Fane, A. G. 2006. Operation of membrane bioreactor with powdered activated carbon addition. *Separation Science and Technology* 41(7), 1447–1466.

Nigam, M. O., Bansal, B. and Chen, X. D. 2008. Fouling and cleaning of whey protein concentrate fouled ultrafiltration membranes. *Desalination* 218, 313–322.

Ning Koh, C., Wintgens, T., Melin, T. and Pronk, F. 2008. Microfiltration with silicon nitride microsieves and high frequency backpulsing. *Desalination* 224(1–3), 88–97.

Ognier, S., Wisniewski, C. and Grasmick, A. 2002. Characterisation and modelling of fouling in membrane bioreactors. *Desalination* 146(1–3), 141–147.

Ognier, S., Wisniewski, C. and Grasmick, A. 2004. Membrane bioreactor fouling in sub-critical filtration conditions: A local critical flux concept. *Journal of Membrane Science* 229(1–2), 171–177.

Pang, C. M., Hong, P., Guo, H. and Liu, W. T. 2005. Biofilm formation characteristics of bacterial isolates retrieved from a reverse osmosis membrane. *Environmental Science and Technology* 39(19), 7541–7550.

Pollice, A., Brookes, A., Jefferson, B. and Judd, S. 2005. Sub-critical flux fouling in membrane bioreactors—a review of recent literature. *Desalination* 174(3), 221–230.

Prieske, H., Drews, A. and Kraume, M. 2008. Prediction of the circulation velocity in a membrane bioreactor. *Desalination* 231(1–3), 219–226.

Psoch, C. and Schiewer, S. 2005a. Critical flux aspect of air sparging and backflushing on membrane bioreactors. *Desalination* 175(1), 61–71.

Psoch, C. and Schiewer, S. 2005b. Long-term study of an intermittent air sparged MBR for synthetic wastewater treatment. *Journal of Membrane Science* 260(1–2), 56–65.

Psoch, C. and Schiewer, S. 2006. Anti-fouling application of air sparging and backflushing for MBR. *Journal of Membrane Science* 283(1–2), 273–280.

Psoch, C. and Schiewer, S. 2008. Long-term flux improvement by air sparging and backflushing for a membrane bioreactor, and modeling permeability decline. *Desalination* 230(1–3), 193–204.

Ramesh, A., Lee, D. J. and Lai, J. Y. 2007. Membrane biofouling by extracellular polymeric substances or soluble microbial products from membrane bioreactor sludge. *Applied Microbiology and Biotechnology* 74, 699–707.

Randtke, S. J. 1988. Organic contaminant removal by coagulation and related process combinations. *Journal of the American Water Works Association* 80, 40–56.

Rios, N., Nopens, I. and Vanrolleghem, P. 2007. Hydrodynamic CFD simulation of a two-phase flow in a single tube of an ultrafiltration membrane for a side-stream membrane bioreactor. In: Lesjean, B. (ed.) *2nd IWA National Young Water Professionals Conference*, Berlin/Germany, 113–120.

Rosenberger, S., Laabs, C., Lesjean, B., Gnirss, R., Amy, G., Jekel, M. and Schrotter, J. C. 2006. Impact of colloidal and soluble organic material on membrane performance in membrane bioreactors for municipal wastewater treatment. *Water Research* 40(4), 710–720.

Shannon, M. A., Bohn, P. W., Elimelech, M., Georgiadis, J. G., Marieas, B. J. and Mayes, A. M. 2008. Science and technology for water purification in the coming decades. *Nature* 452, 301–310.

Song, K.-G., Kim, Y. and Ahn, K.-H. 2008. Effect of coagulant addition on membrane fouling and nutrient removal in a submerged membrane bioreactor. *Desalination* 221(1–3), 467–474.

Speth, T. F., Summers, R. S. and Gusses, A. M. 1998. Nanofiltration foulants from a treated surface water. *Environmental Science and Technology* 32(22), 3612–3617.

Teychene, B., Guigui, C., Cabassud, C. and Amy, G. 2008. Toward a better identification of foulant species in MBR processes. *Desalination* 231(1–3), 27–34.

Tian, J.-Y., Liang, H., Li, X., You, S.-J., Tian, S. and Li, G.-B. 2008. Membrane coagulation bioreactor (MCBR) for drinking water treatment. *Water Research* 42(14), 3910–3920.

Van der Hoek, J. P., Hofman, J. A. M. H., Bonne, P. A. C., Nederlof, M. M. and Vrouwenvelder, H. S. 2000. RO treatment: Selection of a pretreatment scheme based on fouling characteristics and operating conditions based on environmental impact. *Desalination* 127(1), 89–101.

Vrouwenvelder, H. S., Van Paassen, J. A. M., Folmer, H. C., Hofman, J. A. M. H., Nederlof, M. M. and Van Der Kooij, D. 1998. Biofouling of membranes for drinking water production. *Desalination* 118(1–3), 157–166.

Vrouwenvelder, J. S., Manolarakis, S. A., van der Hoek, J. P., van Paassen, J. A. M., van der Meer, W. G. J., van Agtmaal, J. M. C., Prummel, H. D. M., Kruithof, J. C. and van Loosdrecht, M. C. M. 2008. Quantitative biofouling diagnosis in full scale nanofiltration and reverse osmosis installations. *Water Research* 42(19), 4856–4868.

Vrouwenvelder, J. S., van Paassen, J. A. M., van Agtmaal, J. M. C., van Loosdrecht, M. C. M. and Kruithof, J. C. 2009. A critical flux to avoid biofouling of spiral wound nanofiltration and reverse osmosis membranes: Fact or fiction? *Journal of Membrane Science* 326(1), 36–44.

Wang, S., Guillen, G. and Hoek, E. M. V. 2005. Direct observation of microbial adhesion to membranes. *Environmental Science and Technology* 39(17), 6461–6469.

Weis, A., Bird, M. R. and Nystrom, M. 2003. The chemical cleaning of polymeric UF membranes fouled with spent sulphite liquor over multiple operational cycles. *Journal of Membrane Science* 216(1–2), 67–79.

White, M. C., Thompson, J. D., Harrington, G. W. and Singer, P. C. 1997. Evaluating criteria for enhanced coagulation compliance. *Journal of the American Water Works Association* 89, 64–77.

Wicaksana, F., Fane, A. G. and Chen, V. 2006. Fibre movement induced by bubbling using submerged hollow fibre membranes. *Journal of Membrane Science* 271(1–2), 186–195.

Wu, J., Chen, F., Huang, X., Geng, W. and Wen, X. 2006. Using inorganic coagulants to control membrane fouling in a submerged membrane bioreactor. *Desalination* 197(1–3), 124–136.

Yamato, N., Kimura, K., Miyoshi, T. and Watanabe, Y. 2006. Difference in membrane fouling in membrane bioreactors (MBRs) caused by membrane polymer materials. *Journal of Membrane Science* 280(1–2), 911–919.

You, H. S., Huang, C. P., Pan, J. R. and Chang, S. C. 2006. Behavior of membrane scaling during crossflow filtration in the anaerobic MBR system. *Separation Science and Technology* 41(7), 1265–1278.

You, H. S., Tseng, C. C., Peng, M. J., Chang, S. H., Chen, Y. C. and Peng, S. H. 2005. A novel application of an anaerobic membrane process in wastewater treatment. *Water Science and Technology* 51(6–7), 45–50.

Yu, H.-Y., Hu, M.-X., Xu, Z.-K., Wang, J.-L. and Wang, S.-Y. 2005a. Surface modification of polypropylene microporous membranes to improve their antifouling property in MBR: NH3 plasma treatment. *Separation and Purification Technology* 45(1), 8–15.

Yu, H.-Y., Liu, L.-Q., Tang, Z.-Q., Yan, M.-G., Gu, J.-S. and Wei, X.-W. 2008. Mitigated membrane fouling in an SMBR by surface modification. *Journal of Membrane Science* 310(1–2), 409–417.

Yu, H.-Y., Xie, Y.-J., Hu, M.-X., Wang, J.-L., Wang, S.-Y. and Xu, Z.-K. 2005b. Surface modification of polypropylene microporous membrane to improve its antifouling property in MBR: CO2 plasma treatment. *Journal of Membrane Science* 254(1–2), 219–227.

Yu, H.-Y., Xu, Z.-K., Lei, H., Hu, M.-X. and Yang, Q. 2007. Photoinduced graft polymerization of acrylamide on polypropylene microporous membranes for the improvement of antifouling characteristics in a submerged membrane-bioreactor. *Separation and Purification Technology* 53(1), 119–125.

Yu, K., Wen, X., Bu, Q. and Xia, H. 2003. Critical flux enhancements with air sparging in axial hollow fibers cross-flow microfiltration of biologically treated wastewater. *Journal of Membrane Science* 224(1–2), 69–79.

Yun, M. A., Yeon, K. M., Park, J. S., Lee, C. H., Chun, J. and Lim, D. J. 2006. Characterization of biofilm structure and its effect on membrane permeability in MBR for dye wastewater treatment. *Water Research* 40(1), 45–52.

Zhang, G., Ji, S., Gao, X. and Liu, Z. 2008a. Adsorptive fouling of extracellular polymeric substances with polymeric ultrafiltration membranes. *Journal of Membrane Science* 309(1–2), 28–35.

Zhang, H.-F., Sun, B.-S., Zhao, X.-H. and Gao, Z.-H. 2008b. Effect of ferric chloride on fouling in membrane bioreactor. *Separation and Purification Technology* 63(2), 341–347.

Zhang, J., Chua, H. C., Zhou, J. and Fane, A. G. 2006a. Factors affecting the membrane performance in submerged membrane bioreactors. *Journal of Membrane Science* 284(1–2), 54–66.

Zhang, K., Choi, H., Dionysiou, D. D., Sorial, G. A. and Oerther, D. B. 2006b. Identifying pioneer bacterial species responsible for biofouling membrane bioreactors. *Environmental Microbiology* 8(3), 433–440.

Zhang, S., Yang, F., Liu, Y., Zhang, X., Yamada, Y. and Furukawa, K. 2006c. Performance of a metallic membrane bioreactor treating simulated distillery wastewater at temperatures of 30 to 45°C. *Desalination* 194(1–3), 146–155.

Zhang, Y. P., Chong, T. H., Fane, A. G., Law, A., Coster, H. G. L. and Winters, H. 2008c. Implications of enhancing critical flux of particulates by AC fields in RO desalination and reclamation. *Desalination* 220(1–3), 371–379.

Zhao, Y.-H., Zhu, B.-K., Kong, L. and Xu, Y.-Y. 2007. Improving hydrophilicity and protein resistance of poly(vinylidene fluoride) membranes by blending with amphiphilic hyperbranched-star polymer. *Langmuir* 2007(23), 5779–5786.

Zheng, X., Ernst, M. and Jekel, M. 2009. Identification and quantification of major organic foulants in treated domestic wastewater affecting filterability in dead-end ultrafiltration. *Water Research* 43(1), 238–244.

Zodrow, K., Brunet, L., Mahendra, S., Li, D., Zhang, A., Li, Q. and Alvarez, P. J. J. 2009. Polysulfone ultrafiltration membranes impregnated with silver nanoparticles show improved biofouling resistance and virus removal. *Water Research* 43(3), 715–723.

Zondervan, E., and Roffel, B. 2007. Evaluation of different cleaning agents used for cleaning ultra filtration membranes fouled by surface water. *Journal of Membrane Science* 304, 40–49.

# 15 Fundamentals of Liquid Membrane

*Aloke Kumar Ghoshal*

## CONTENTS



## 15.1 INTRODUCTION

The increasing environmental concern has enforced to suppress the discharge of industrial wastes into the environment across the world. There has been a nutritional interest and the need for an efficient and cost effective process either for removal of toxic species or for the selective extraction and recovery of valuable species. Solvent extraction technology came into existence in the late 1940s. With the ever-increasing needs of separation engineering and the continuing evolution of new extractants, solvent extraction technology developed rapidly and has played a pinnacle role in diversified engineering fields. In the last few decades, the compounding of solvent extraction with other processes has created a number of new separation processes, which are solvent tinctured resins, extraction chromatography, electrostatic extraction, supercritical extraction and membrane-based extraction. These membrane-based separation techniques are emerging as promising avenue in economizing energy compared to formal and traditional separation processes like packed bed absorption, adsorption, distillation, cryogenic separation, etc. Membrane-based technology has incurred appreciable attention because of its low energy consumption, capability of integration with other processes, easy scaling up and adjustable characteristics of the membrane. Membrane separation processes, in general, differ based on size (*viz.*, microfiltration, ultrafiltration, nanofiltration), affinity (*viz.*, reverse osmosis, pervaporation, gas separation), charge (*viz.*, dialysis, electrodialysis), and chemical nature (*viz.*, carrier-mediated separation) of the separated particles (Mulder, 1991; Pinnau and Freeman, 1999). It also differs based on the membranes materials such as solid membrane and liquid membrane separation (Mulder, 1991). The major issues related to the membrane separation processes are separation goal, nature of species to be retained, nature of species to be transported through membrane, minor or major species of feed solution transported through membrane, driving forces, mechanism for transport and phase of feed and permeate streams.

Table 15.1 summarizes the principal features of various membrane separation processes. Although the separations based on solid membrane (organic or polymeric and inorganic), such as ultrafiltration, reverse osmosis, and electrodialysis, are more stable, they face the problems of low flux rate, membrane fouling, low selectivity, and need of large-sized equipment. The above issues are the driving forces in developing superior membranes. Liquid membrane (LM), which employs liquid as membrane material is an outcome of such development. An immiscible liquid can serve as a membrane between two liquids or gas phases. Thus, LM, in general, is a homogenous, nonporous, thin film of organic liquid interposed between two aqueous phases of different compositions. The aqueous phases are miscible with each other but the organic phase is immiscible with either of them. In a liquid, the diffusion coefficient and the solubility are different for different solutes, and the measure of the permeability is the product of these two terms. Because the diffusion coefficients in the liquids are typically orders of magnitude higher than in polymers and inorganic membranes, a higher flux can be obtained, and therefore, a liquid can yield selective permeability and separation. Separations based on LM have the advantages of being able to generate a large surface area in small-sized equipment, simultaneous extraction and stripping in a single step, and high separation factor and selectivity. The attractiveness of LM separation lies in the fact that diffusivities of solute in liquids are several orders ($10^{-6}$ to $10^{-5}$ $cm^2{\cdot}s^{-1}$) of magnitude higher than solid membranes (above $10^{-8}$ $cm^2{\cdot}s^{-1}$) (Treybal, 1980). One of the benefits of using a liquid membrane is that a carrier is often used to enhance the transport of solute through the membrane phase, thus making the liquid membrane highly selective. The carrier often makes a complex that is transported through the liquid membrane phase. Although the value of diffusion coefficient for the complex is less than that for smaller species diffusing directly, increase in distribution coefficient ($m$) to a larger order makes the transport process facilitated with higher rate of transport. Thus, liquid membranes are relatively high in efficiency and, as such, are being looked into for industrial applications. The principal domains of applications for the LM technique are hydrometallurgy, treatment of wastewater from the chemical industry, gas separation, medical application, biotechnology, etc.

However, liquid membranes require stability in order to be effective, and if they are pushed out of the pores or ruptured in some way due to pressure differentials or turbulence, then they just do not work.

## 15.2 TRANSPORT MECHANISM THROUGH LM

The transport mechanisms in LM can be categorized in two major types: passive transport and active transport. The passive transport that follows solution diffusion mechanism involves the transport of selective species from one phase to the other due to gradient in chemical or electrical potential. The basic transport mechanism can be altered by adding a specific carrier agent or by providing external electric or photo impulses. The active transport involves the driving force provided by the chemical reactions such as oxidation and reduction to transport the solute from feed to receiving phase. LM transport mechanisms can again be divided into two categories depending on the use of carriers. These are ordinary or free diffusive transport and carrier-mediated or facilitated transport. The ordinary or free diffusive transport, the simplest solution diffusion mechanism, is usually followed during the permeation of solute through the liquid membrane due to concentration gradient (Ellinghorst et al., 1989). Transport of component $j$ from feed phase to membrane phase occurs by higher solubility or diffusivity of solute $j$ in the membrane phase. The mass transfer rate in this case is low and depends on the solubility of the solute in the organic as well as receiving/strip phases. In carrier-mediated transport, a carrier agent is added to the membrane phase to increase the rate of mass transfer or efficiency of the LM separation. Therefore, it is also termed as facilitated transport.

**TABLE 15.1**
**Principal Features of Various Membrane Separation Processes**

| Separation Process | Separation Goal | Nature of Species Retained | Nature of Species Transported | Minor/major Species Transported | Driving Force | Mechanism for Transport | Phase of Feed |
|---|---|---|---|---|---|---|---|
| Gas permeation | Stream enriched in particular component | Larger species | Smaller species | Either | Concentration gradient | Solution diffusion | Gaseous |
| Pervaporation | Same as above | Same as above | More soluble/smaller/ more volatile | Preferably minor species | Concentration gradient/ temperature gradient | Solution diffusion | Liquid feed, gaseous permeate |
| Dialysis | Macrosolute solution free of microsolute | >0.02 μm retained | Microsolute, smaller solute | Minor species | Concentration gradient | Sieving, hindered diffusion in micropore | Liquid |
| Electrodialysis | Solution free of microions | Co-ions, macroions, and water retained | Microionic species | Minor ionic species | Electrical potential gradient | Counterion transport via ion exchange membranes | Liquid |
| Reverse osmosis | Solvent free of all solutes | 1-Å to 10-Å microsolute species | Solvent, species retained may be electrolytic or volatile | Major species solvent | Hydrostatic pressure gradient vs. osmotic pressure gradient | Preferential sorption | Liquid |
| Ultrafiltration | Solution free of macrosolutes, macrosolute fractionation | 10-Å to 200-Å macrosolute species | Solution of microsolutes | Major solvent, minor microsolutes | Hydrostatic pressure gradient vs. osmotic pressure gradient | Sieving | Liquid |
| Microfiltration | Solution free of particles, gas-free particles | 0.02-μm to 10-μm particles | Solution/gas free of particles | Major solvent, minor microsolutes | Hydrostatic pressure gradient | Sieving | Liquid or gas |
| Emulsion liquid membrane | Stream/stream enriched or depleted in a particular species | Generally not size-selective except in host–guest chemistry | Species with high solubility in liquid membrane | Minor species | Concentration gradient, pH gradient | Solution diffusion, facilitated transport | Liquid feed, emulsion containing permeate |

The carrier should be soluble only in the membrane phase and should have the ability to form a complex reversibly with a specific solute. Various mechanisms that take place in carrier-mediated transport through LM are cation transport, anion transport, and neutral guest transport. Each of them is briefly discussed below.

### 15.2.1 Cation Transport

Figures 15.1 and 15.2 show cation transport which take place in symport (cotransport, where transport of the two species occurs in the same direction) and antiport (countertransport, where transport of the two species occurs in the opposite direction) configurations respectively. In the symport configuration, a carrier channelizes the guest along with cotransported anion together across the membrane (Figure 15.1). This takes place in several steps, as detailed below.

- Diffusion of guest cation from the bulk of the feed phase to the feed–membrane interface (Phase I)
- Dissolution of solute in the LM
- Formation of the guest cation–carrier complex at the feed solution–membrane interface
- Diffusion of the formed complex through the LM
- Decomplexation at the membrane-receiving/strip phase interface
- Release of solute from the membrane phase to the receiving/strip phase (Phase II)
- Diffusion of free carrier back to the feed membrane interface
- Diffusion of solute from membrane–strip interface to the bulk strip phase

Removal of $Cu^{+2}$ through LM using 2-hydroxy-5-dodecylbenzaldehyde (2H5DBA) as a carrier is a typical example of such type of transport. The carrier contains hydroxyl and aldehydic groups as ion-binding groups, thereby facilitating the transport. The complexation (extraction) and decomplexation (stripping) reactions at feed/membrane and membrane/strip, respectively, are as follows (Monlinari et al., 2006):

Extraction reaction: $$Cu^{+2}_{(aq)} + 2HA_{(org)} \Leftrightarrow CuA_{2(org)} + 2H^{+}_{(aq)}$$

Stripping reaction: $$CuA_{2(org)} + 2H^{+} \Leftrightarrow 2HA + Cu^{+2},$$

where HA is the acidic extractant.

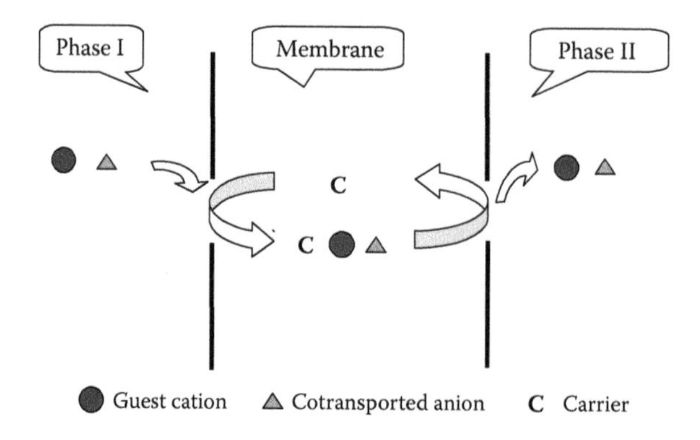


**FIGURE 15.1** Cationic symport.

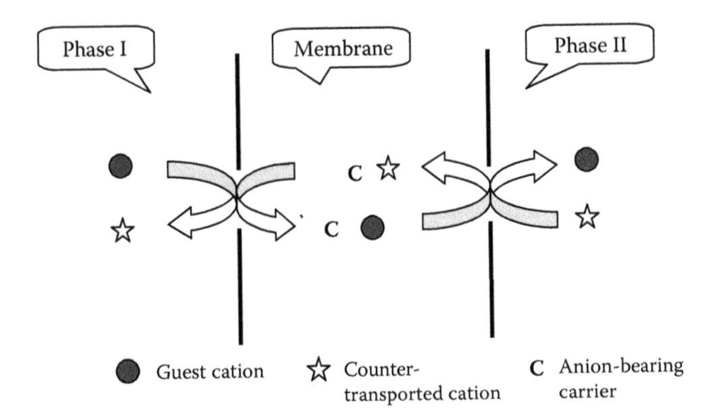


**FIGURE 15.2** Cationic antiport.

In the case of the antiport configuration (Figure 15.2), an anionic carrier is used, and the transport steps are slightly different, as mentioned below.

- Diffusion of guest cation from the bulk of the feed phase to the feed–membrane interface (Phase I).
- Dissolution of solute in the LM.
- Formation of the guest cation–carrier complex at the feed solution–membrane interface.
- Diffusion of the formed complex through the LM.
- Decomplexation at the membrane-receiving/strip phase interface.
- Release of solute from the membrane phase to the receiving/strip phase (Phase II).
- Formation of the countertransported ion–carrier complex at the strip solution–membrane interface.
- The carrier complex along with the countertransported ion diffuses back across the membrane.

Transport of cobalt (II) ions through a liquid membrane using D2EHPA (di(2-ethylhaxyl) phosphoric acid) as carrier (denoted as HR in the reaction) and $H^+$ as counterions is a typical example of such type of transport (Leon and Guzman, 2005). The carrier diffuses from the bulk membrane phase to the feed/membrane interface, where Co (II) ions are exchanged for protons as shown in the reaction below.

$$Co^{+2}_{(aq)} + 2(HR)_{2(org)} \Leftrightarrow CoR_2(HR)_{2(org)} + 2H^+$$

The Co (II) carrier complex thus formed diffuses through the membrane to the membrane/strip interface, where, protons are again exchanged for Co (II) ions by reverse reaction. Co (II) is released into the strip phase and the carrier is regenerated to begin a new separation cycle (Figure 15.3). The transport mechanism is therefore a coupled counterion transport mechanism with Co (II) and $H^+$ moving in the opposite direction.

### 15.2.2 Anion Transport

Anion symport transport steps are similar to cation symport as discussed earlier with guest is an anion in place of being a cation (Figure 15.4). Typical example of cotransport is extraction of

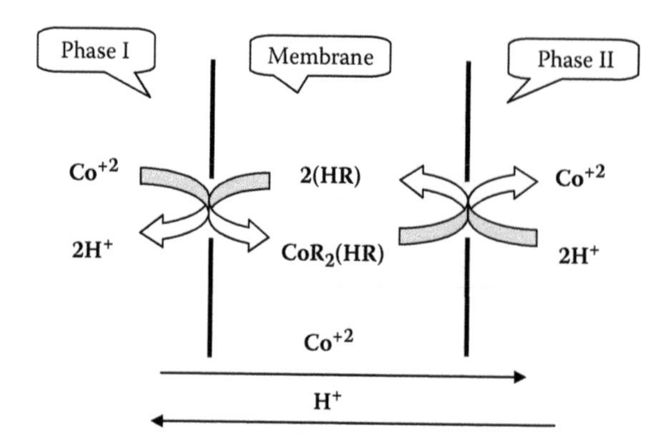


**FIGURE 15.3** Diagram of the facilitated transport of Co(II) ions using D2EHPA as carrier and $H^+$ as counterions.

mercury (II) ions through a liquid membrane in presence of the carrier, TOA (Tri-octylamine, a basic extractant) (Figure 15.5). The transport mechanisms for transport of mercury (II) as discussed below have been proposed by Li et al., 1996.

The transport mechanism follows the following steps:

1. Complexation of Hg (II) by chloride ions in the feed solution:

$$Hg^{+2} + nCl^- \leftrightarrow HgCl_n^{2-n}$$

2. Reaction of $R_3N$ (taken as TOA) at the feed/membrane interface with HCl in the feed solution:

$$R_3N_{(org)} + H^+Cl^-_{(aq)} \Leftrightarrow R_3NHCl_{(org)}$$

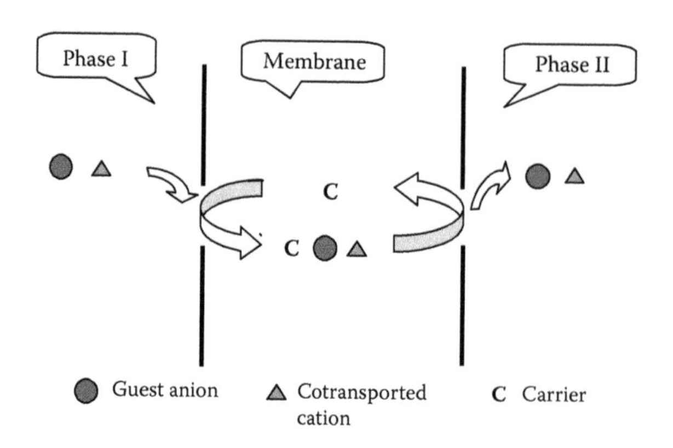


**FIGURE 15.4** Anionic symport.

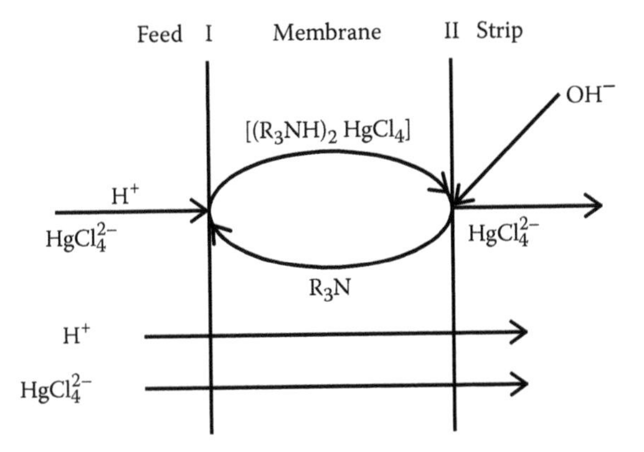


**FIGURE 15.5** Cotransport mechanism for mercury transport.

3. In the feed, $HgCl_4^{2-}$ exchange with $Cl^-$ of $R_3NH^+Cl^-$ in the membrane phase:

$$2R_3NH^+Cl^-_{(org)} + HgCl_4^{2-} \Leftrightarrow (R_3NH)_2HgCl_{4(org)} + 2Cl^-_{(aq)}$$

4. The stripping reaction:

$$(R_3NH)_2HgCl_{4(org)} + OH^- \Leftrightarrow R_3N_{(org)} + HgCl^{2-}_{4(aq)} + 2H_2O$$

Anion antiport (countertransport) transport is schematically represented in Figure 15.6. Countertransport mechanism has been demonstrated by Kontturi et al., 1990 while studying on the separation of lignosulfonate (LS) through LM. The reaction at the feed/membrane interface follows a two-step mechanism.

At feed/membrane interface,

$$R_3N + HCl \Leftrightarrow R_3NH^+Cl^-$$

$$R_3NH^+Cl^- + LSNa_n \Leftrightarrow R_3NHLSNa_{(n-1)} + NaCl,$$

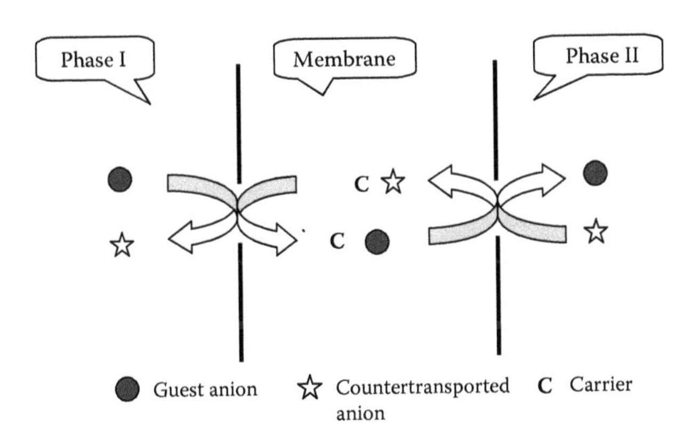


**FIGURE 15.6** Anionic antiport.

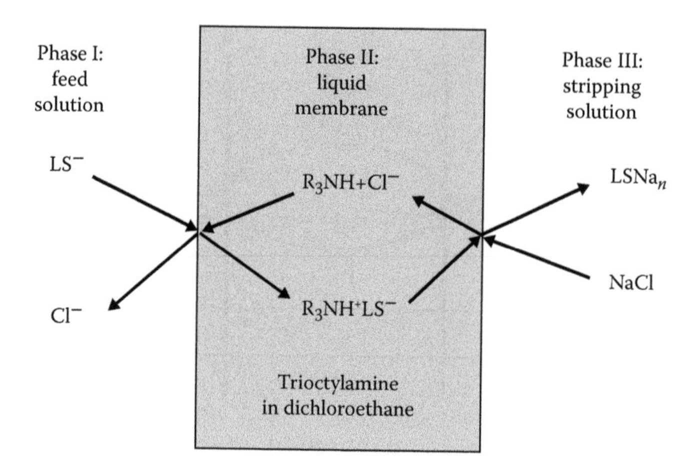


**FIGURE 15.7** Countertransport mechanism for lignosulfate transport.

where $R_3N$ is trioctylamine and $LSNa_n$ is sodium lignosulfonate.

Reaction at the membrane/strip interface is

$$R_3NHLSNa_{(n-1)} + NaCl \Leftrightarrow R_3NH^+Cl^- + LSNa_n.$$

A schematic representation of the countertransport mechanism during separation of lignosulfonate (LS) through LM is shown in Figure 15.7.

### 15.2.3 Neutral Guest Transport

In this mechanism, usually, a neutral guest molecule is transported between two organic phases through a water liquid membrane with a water-soluble carrier, as shown in Figure 15.8. Neutral species/molecules are channelized via symport configuration. A typical example of such transport is the use of water-soluble carrier cyclophane in the transport of molecules like pyrene and biphenyl (Diederich and Dick, 1984). Methylated carriers (e.g., methylated cyclodextrins) are proved to be effective for the transport of neutral azobenzene derivatives through water liquid membrane (Harada and Takahashi, 1987). Transport of oxygen, carbon dioxide, and other gaseous molecules with the help of carrier through liquid membranes is the example of such transport phenomenon.

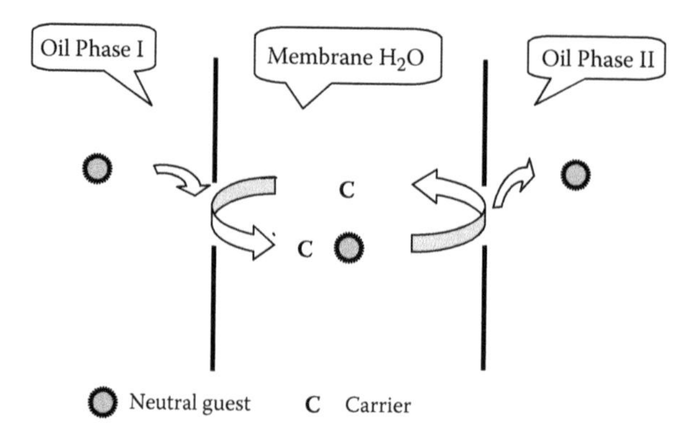


**FIGURE 15.8** Transport of neutral guest.

## 15.3 IMPORTANT ISSUES REGARDING LM-BASED SEPARATION

### 15.3.1 Choice of Solvent

Selection of a right solvent (or membrane phase) is the prime issue in all kinds of LM-based separation processes. The solvent should be chosen in such a way that it is immiscible with the aqueous phase, has low viscosity and volatility and at the same time it should have a high distribution coefficient (Mulder, 1991; Noble and Way, 1987). The performance of various solvents should be examined by estimating distribution coefficients ($m$) of the solute of concern in solvents. The organic solvents are generally flammable, volatile, toxic and their use leads to environmental and safety risks. Therefore, it is essential to look for environment friendly solvent for LM applications. The room temperature ionic liquids (RTILs) are considered as green solvent due to their nonflammable nature and negligible vapor pressure. Since, these ionic liquids are generally expensive and their toxicity data are not available (Venkateswaran and Palanivelu, 2006), vegetable oils such as palm oil, sunflower oil and coconut oil being nonhazardous, naturally occurring, easily available, nontoxic, low cost, easily biodegradable and renewable sources is considered as ideal replacement for the conventional volatile organic diluents. Chakrabarty et al. in their recent publication (Chakrabarty et al., 2010b) studied the feasibility of using coconut oil as a diluent for the removal of mercury through supported liquid membrane (SDLM). Coconut oil and TOA were used as diluent and extractant respectively for the separation of mercury from its aqueous solution. Feasibility of various polymeric support materials such as PE (polyethylene), PTFE (polytetrafloroethylene), PVDF (polyvinyldenefluoride), and Nylon 6, 6, and the stability of their SLM were tested. Effects of various fundamental parameters that affect the transport of mercury through the SLM, such as feed pH, strip phase concentration, carrier concentration, feed concentration, and feed-to-strip ratio, were investigated. They have finally reported that SLM configuration "PVDF–coconut oil–TOA" proves promising in the separation of mercury, that the carrier has an insignificant effect on the transport of mercury in the range of experiments conducted because about 91% of mercury could be removed without a carrier, and that the performance of coconut oil as diluent was better than that of organic solvents such as dichloroethane and heptane. Table 15.2 presents typical equilibrium distribution coefficient values for LS in the various organic solvents investigated (Chakrabarty et al., 2009a). It is observed that while separation of LS is negligible when $n$-heptane, hexane, and toluene are present, considerable separation occurs in 1,2-dichloroethane, followed by carbon tetrachloride and chloroform.

### 15.3.2 Two-Phase Equilibrium Study

Two-phase equilibrium study is an important step in order to select a suitable LM of high separation factor, i.e., high distribution coefficient. Various process parameters such as temperature, pH and carrier concentration do have significant impact on the separation factor. Therefore, a preliminary

**TABLE 15.2**
**Equilibrium Distribution Coefficient ($m$) of LS in Various Solvents**

| Experimental Condition | Solvent Used | Distribution Coefficient ($m$) |
|---|---|---|
| Concentration of LS in feed: 100 mg·L$^{-1}$; feed phase pH: 2.0 | Dichloroethane | 21.5 |
| Carrier (TOA) concentration: 0.8% (v/v) | Carbon tetrachloride | 8.5 |
| Temperature: 30°C | Chloroform | 1.25 |
| Stirring speed: 100 rpm | $n$-Heptanes | 0.584 |
| Duration of stirring: 6 h | Toluene | 0.368 |
| Feed phase volume: 25 ml | Hexane | 0.247 |
| Membrane phase volume: 25 ml | | |

investigation in the form of a two-phase study is essential to decide on the membrane liquid and carrier (if any) and other optimum physicochemical conditions such as temperature, pH, and carrier concentration.

pH plays an important role in the LM-based separation processes. Effect of pH is dependent on the types of solute, solvent, carrier and also the mechanism of transport. The two-phase equilibrium experiments in presence of carrier TOA revealed that maximum transport of lignosulfonate (LS), a polydispersed macromolecule occurred at pH 2 (Chakrabarty et al., 2009a). This was because of the fact that an acidic condition was necessary for the protonation of the carrier TOA, which eventually enhances the transport of LS into the membrane phase. At higher pH value, the distribution of LS decreased due to incomplete protonation of TOA at the feed/membrane interface. At much lower pH, i.e., in a highly acidic condition, LS molecules with higher molar mass on average are not ionized, and increase in chloride ion at lower pH inhibits the formation of amine complex at equilibrium.

Increase in temperature is expected to increase the equilibrium distribution coefficient, and a similar trend was observed in the case of equilibrium distribution coefficient of LS in the LM (Chakrabarty et al., 2009a). This is due to the fact that increase in temperature favors feed/membrane interface reaction. However, the distribution coefficient rises sharply above the optimum temperature 313 K because of the fact that the effective charge of LS drops to zero above 313 K (Kontturi et al., 1990; Kontturi, 1988) and carrier-mediated transport becomes ineffective.

Presence of carrier increases the equilibrium distribution coefficient. Once the carrier is added to the membrane phase, the solute reacts with the carrier and results in formation of a complex in the membrane phase. Thus, the mass transfer through the feed/membrane interface is enhanced as well as controlled by the chemical reaction at the interface that yields a high distribution coefficient. The maximum distribution coefficient depends on the saturation capacity of the membrane phase for the complex. However, the optimum carrier concentration would vary depending on the initial feed phase concentration of the solute, the ratio of volumes of feed phase to membrane phase, and other physicochemical properties such as temperature and pH. The equilibrium distribution coefficient of LS reaches a maximum at 2% (v/v) TOA. The following section discusses elaborately about carriers in LM separation (Chakrabarty et al., 2009a).

### 15.3.3 Carriers in LM Separation

Carriers being added in trace quantity to the membrane liquid play an important role in selective separation and concentration of specific solutes with high flux of transport. Carriers should have the following characteristic features (Kamiński and Kwapiński, 2000):

- Quick binding and release of particular solute(s)
- Selective and reversible binding with a component in the solution
- Nonaffinity with a solvent
- Noncoalescence
- Nontoxicity

Carriers are categorized into acidic, basic, and neutral primarily on the basis of their functional groups. Therefore, careful integration of a suitable carrier into the liquid membrane phase is very important to enhance the effective solubility of the species to be transported in the membrane phase. Carriers used in various LM processes are listed in Table 15.3.

Acidic carriers are most effective for extracting the cations by exchanging the cations for their protons. The carriers have COOH, P(OH), $SO_3H$ or chelating groups. Some of the commonly used acidic carriers for liquid membrane cation extraction include chelating agents such as β-diketones, hydroxyoximes, 8-hydroxyquinolines, and alkyl phosphorous compounds. These chelating carriers form coordination complex with positively charged metal ions and help in selective separation of

**TABLE 15.3**
**Carriers Used in Various LM Processes**

| Acidic Carrier | Extracted Material | References | Basic Carrier | Extraction Of | References | Neutral Carrier | Extracted Material | References |
|---|---|---|---|---|---|---|---|---|
| Acetylacetone | Cr (III) | Sekine et al., 1988 | Primene JMT $[(C_8H_{17})NH_2]$ | Mo(VI) | Neková and Schrotterova, 1999 | Di-benzo-18-crown-6 | $Na^+$, Ur(VI) | Goyette et al., 2003; Mohite et al., 2001 |
| Benzoylacetone | Ur (VI) | Rajan and Shinde, 1996 | Trioctylamine $[(C_8H_{17})_3N]$ | Cr(VI) | Kumbasar, 2008 | Di-cyclohexyle-18-crown-6 | Hg (II) | Jabbari et al., 2001 |
| β-Hydroxime | Cu (II) | Sengupta et al., 2007 | Trinonylamine $[(C_9H_{19})_3N]$ | Lignosulfonate | Kontturi et al., 1990 | Cryptana (2,2,1) $[N_2O_5]$ | Ag(I), Cu(II) and Zn(II) | Arous and Kerdjoudj, 2004 |
| D2EHPA[di(2-ethylhaxyl) phosphoric acid] | Ag (I), Cu (II) | Gherrou et al., 2002 | Trilaurylamine | Lignosulfonate | Kontturi et al., 1990 | Cryptand(2,2,2) $[N_2O_6]$ | Ag(I), Cu(II) and Zn(II) | Arous and Kerdjoudj, 2004 |
| Cyanex-272 ($R_2POH$) | Co (II) | Swain et al., 2007 | Alamine 336 $[R_3N(R:C_8–C_{10})]$ | Citric acid | Yordanov and Boyadzhiev, 2004 | | | |
| Trioctylphosphinoxide | As (V) | Perez et al., 2007 | Aliquat 336 $[CH_3(C_8C_{17})_3N]$ | Cd (II) | Lv et al., 2007 | | | |
| | | | Trioctylamine $[(C_8H_{17})_3N]$ | Hg (II) | Shaik et al., 2010; Chakrabarty et al., 2010a, 2010b, 2010c | | | |
| | | | Trioctylamine $[(C_8H_{17})_3N]$ | LS | Chakrabarty et al., 2009a, 2009b; Zhou, 1996 | | | |

the metals. Alkylphosphorous compounds often called as "liquid cation exchangers" are also widely used because they are cheap and their metal salts are more soluble in organic phase than metal chelates. A typical example of acidic carrier-mediated transport is presented below, where copper is extracted from its aqueous solution with D2EHPA (Gherrou et al., 2002).

$$Cu^{2+} + 2C_{16}H_{34}O_3P(OH) \rightarrow C_{16}H_{34}O_3P(OCu) + 2H^+$$

The basic carriers are used for the extraction of the anionic metal complexes such as $Cd(CN)_4^{-2}$, $Cr_2O_7^{-2}$, and $V_3O_9^{-3}$. High molecular weight primary, secondary, and tertiary amines and the quaternary alkylammonium ions are typical examples of basic carriers. These carriers are regarded as "liquid anion exchangers." A typical example of basic carrier-mediated transport is the transport of cadmium ion (Cd (II)) in the form of $CdCl_4^{-2}$ from the aqueous acidic solution by the carrier trioctylamine (TOA). TOA (also denoted as $R_3N$ here) accepts a proton to form a positively charged species, and Cd (II) is extracted as $(R_3NH_2)CdCl_4$ as shown below (Warey, 2007).

$$CdCl_{4(aq)}^{2-} + 2H_{(aq)}^{+} + 2R_3N_{(org)} \rightarrow (R_3NH_2)CdCl_{4(org)}$$

Neutral carriers are used to extracts uncharged metal complexes in aqueous solutions, where the metal species is coordinated with two different types of ligands, i.e., a water-soluble anion and an organic-soluble electron-donating functional group. Organic phosphoryl compounds and macrocyclic molecules are very common neutral carriers. Phosphoryl compounds are used for the recovery of the actinides and lanthanides. Macrocyclic ligand carriers such as crown ethers are used for the separation of alkali, alkaline earth metals, and trivalent lanthanides. A typical example of such type of carrier-mediated transport is given below, where silver ion is cotransported along with the nitrate

**TABLE 15.4**
**LM System Employed in the Extraction of Anions**

| Material Extracted | Carrier | Diluent | Stripping Agent | Reference |
|---|---|---|---|---|
| $HgCl_4^{2-}$ | Trioctylamine (TOA) | Toluene | NaOH | Li et al., 1996 |
| $HgCl_4^{2-}$ | TOA | Dichloroethane | NaOH | Shaik et al., 2010; Chakrabarty et al., 2010a, 2010c |
| $HgCl_4^{2-}$ | TOA | Coconut oil | NaOH | Chakrabarty et al., 2010b, 2010c |
| $Hg(pic)^-$ | Tetrathia-12-crown-4 | Chloroform | EDTA | Shamsipur et al., 2002 |
| $Hg(pic)^-$ | Bis-calixarene-nitrile derivative | Chloroform | $H_2O$ | Alpoguz et al., 2002 |
| $Cr_2O_7^{2-}$ | Cyanex-923 | Xylene | NaCl | Alguacil et al., 2005 |
| $Cr_2O_7^{2-}$ | Tri-*n*-butylphosphate | Hexane | NaOH | Muthuraman et al., 2009 |
| Lignosulfonate | TOA | Decanol | NaOH | Kontturi et al., 1990 |
| Lignosulfonate | Trilaurylamine | Decanol | NaOH | Kontturi et al., 1990 |
| Lignosulfonate | TOA | Dichloroethane | NaOH | Chakrabarty et al., 2009a, 2009b, 2010a, 2010c |
| $NO_3^-$ | TOA | Kerosene | NaOH | Lin and Long, 1997 |

*Note:* pic, picrate.

anion from the aqueous solution by the neutral carrier di-cyclohexanone-18-crown-6 (Bachiri et al., 1996).

$$Ag^{+}_{(aq)} + NO^{-}_{3(aq)} + DC18C6 \rightarrow Ag[DC18C6]NO_{3(org)}$$

Various combinations of LM systems employed in the extraction of different materials in anionic form are presented in Table 15.4.

## 15.4 LIQUID MEMBRANE CONFIGURATIONS

Liquid membranes, based on configuration, can be broadly classified into three types: bulk liquid membrane (BLM), emulsion liquid membrane (ELM), and immobilized liquid membrane (ILM), also called supported liquid membrane (SLM). General descriptions of the various types of LM configurations are presented below.

### 15.4.1 Bulk Liquid Membrane

Bulk liquid membrane (BLM) is the simplest of all LMs. It usually consists of aqueous feed and stripping phases separated by a water-immiscible liquid membrane phase (Chakrabarty, 2010c; Shaik et al., 2010). All three phases are placed in a tank with bulk quantities as shown in Figure 15.9. Two aqueous phases are separated with a solid barrier (e.g., a glass wall) and the membrane phase shares an interface with both these aqueous phases. The liquid membrane in Figure 15.9 is heavier to aqueous phases and is placed at the bottom. The aqueous phases are separated by a glass plate. Two stirrers are used on either sides rotating at fairly slow speeds, in the range of 100 to 300 rpm. High speed of the stirrers can lead to mixing between the aqueous phases due to turbulent mixing. Desired solute is transported from the feed phase to the strip phase through the membrane liquid. The membrane phase can also contain the carrier in the case of a study on carrier-mediated transport. In contrary to the present case, as shown in Figure 15.9, the LM, when lighter than the aqueous phases, is placed above the aqueous phases. The separating plate is extended till the bottom and is tightly fixed to ensure no leakage of fluid from one aqueous phase to the other. BLM is used to understand the basic principles of liquid membrane and for laboratory-scale experiments. It is generally not upgraded to a pilot/commercial scale because of its low mass transfer surface area to

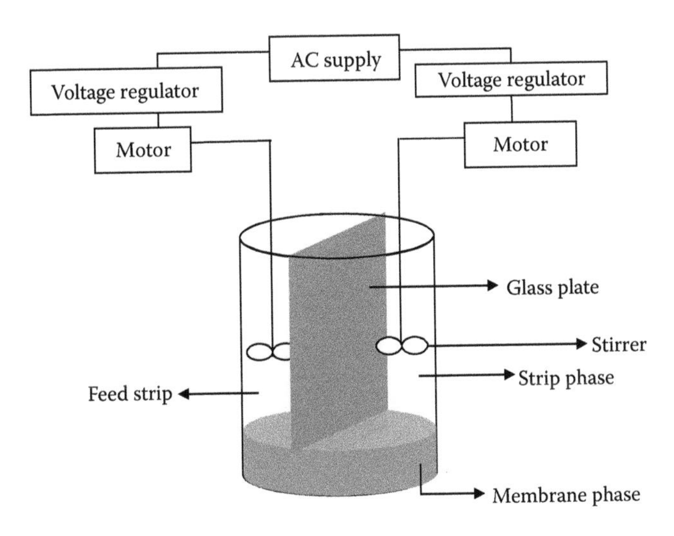


**FIGURE 15.9** Bulk liquid membrane.

volume ratio. Various important equilibrium and kinetic parameters useful for mass transfer studies can also be estimated through BLM experiments.

### 15.4.2 Emulsion Liquid Membrane

The emulsion liquid membrane (ELM) also called as surfactant liquid membrane or liquid surfactant membrane, are essentially double emulsions, i.e., water/oil/water (W/O/W) or oil/water/oil (O/W/O), where ions from the external or source phase are transported to the internal or stripping phase with or without a carrier in the membrane phase. Emulsions are of two types: O/W (oil dispersed in water) and W/O (water dispersed in oil) emulsions. Emulsion liquid membranes are usually prepared by first forming an emulsion between two immiscible phases and then dispersing it into a third (continuous) phase by agitation for extraction.

In the case of W/O/W emulsion, the organic membrane phase containing the solvent and the emulsifier segregates the aqueous phases but allows selective transport of solute across it with or without the help of the carrier. The emulsion drop size, in general, varies from 10 μm to 1 mm, and the internal strip droplet size varies from 1.0 to 10 μm (Chakrabarty, 2010c). The size of the globules depends on the characteristics and concentration of the surfactant in the emulsion, the viscosity of the emulsion, and the intensity of the mode of mixing. The transport rate through this membrane is very high due to the large surface area of the ELM per unit source phase volume. Concentrations in the receiving phase are also increased by a large factor due to the lower ratio of source phase volume to receiving phase volume. High diffusive flux and low energy consumption, simple configuration/equipment, zero disposals, recycling of material, operational simplicity, and high efficiency are the several other advantages of ELM-based process.

The membrane phase in ELM generally contains some surfactants and additives as stabilizing agents to develop a stable emulsion. To form a stable emulsion, the surfactant should be chosen carefully. The selection of an appropriate surfactant depends on the hydrophylic–lipophylic balance (HLB) which is defined as the percentage of hydrophilic functional groups in the surfactant molecule divided by five. Surfactants with a low HLB are more lipid loving and thus tend to make a water-in-oil emulsion while those with a high HLB are more hydrophilic and tend to make an oil-in-water emulsion. The diameter of the internal droplets in emulsion should be in the range of 1–3 μm to achieve good emulsion stability (Zhou, 1996). The schematic representation is shown in Figure 15.10.

The following set of guidelines has been suggested by Kopp (Noble and Way, 1987) for the formation of stable water-in-oil emulsions.

- Concentration of surfactant soluble in organic phase should be in the range of 0.1 to 5 wt. %.
- Viscosity of organic phase should be in the range of 30 to 1000 mPa.s.

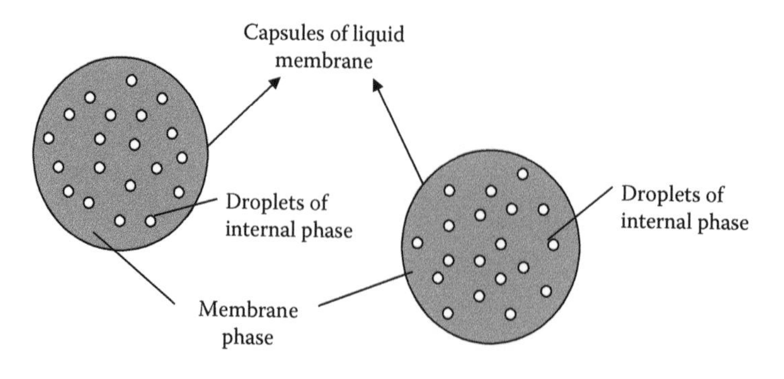


**FIGURE 15.10** Schematic of an emulsion liquid membrane system.

- Volume ratio of the internal phase to membrane phase should be in the range of 0.2 to 2.
- Volume ratio of internal phase to continuous external phase should be in the range of 0.2 to 0.05.
- Volume ratio of continuous phase to emulsion phase should be in the range of 1 to 40.
- Surfactant HLB value should be in the range of 6 to 8.

Figure 15.11 presents a schematic of a typical ELM-based continuous separation process. The entire process consists of four unit operations (Zhou, 1996): Emulsification: the membrane phase and internal phase are mixed and form water-in-oil emulsion. Mass transport operation: the W/O emulsion droplets are dispersed in the external phase to make W/O/W emulsion. Mass transport of the solute takes place from external phase to the internal phase through the oil (O) phase. Settling: the droplets of the emulsion and the external phase are separated in a settling tank. Breaking: the emulsion of the membrane phase and the internal phase with transported solute in it is then demulsified and separated using an electrical emulsification apparatus. The solute is obtained as a concentrated aqueous solution, and the membrane phase is recycled and used again in the emulsification operation.

Various applications of ELM-based separation include, among others, removal of toxic substances from wastewater; extraction of organic acids; recovery of pharmaceutical products and biomedicals; extraction of chlorophenol from aqueous solutions; extraction of nickel, zinc, and mercury from industrial effluents; extraction of organic chemicals from dye effluents; extraction of phenolics from wastewater; separation and purification of fermentation products; amino acid separation and purification; desalination of seawater; and novel drug delivery systems.

The major problems associated with ELM are the following:

- Parameters like ionic strengths, pH, etc. that influence the emulsion stability must be controlled.
- The emulsion must be formulated to withstand the shear generated by mixing during the extraction.
- Demulsification to get back the receiving and membrane phases is a difficult task, since in order to make the emulsion stable, work against the ease of breaking it backed down.

The effectiveness of ELMs is a result of two facilitated mechanisms: type 1 and type 2 facilitation (Zhou, 1996). In type 1 facilitation, the reaction in the internal phase of the ELM effectively maintains zero solute concentration through the minimization of the diffusing solute species in the

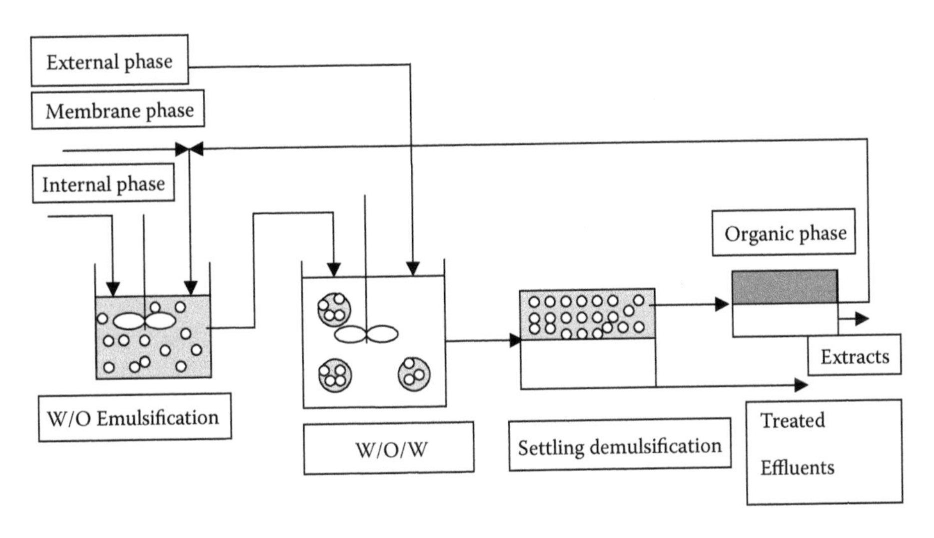


**FIGURE 15.11** Schematic diagram of the emulsion liquid membrane extraction process.

internal phase. Extraction of phenol form aqueous phase is a very good example. In type 2 facilitation, also called carrier facilitated transport, the diffusing species is carried across the membrane phase by incorporating a "carrier" compound (complexing agent or extractant) in the membrane phase. The procedure and the mechanism are illustrated below for such type through an example where extraction of strontium ion takes place with the help of di(2-ethylhexyl) phosphoric acid (D2EHPA) as carrier agent (Zhou, 1996).

The mechanism involves the following steps:

1. Mass transfer of strontium ion ($Sr^{+2}$) from the external bulk phase to the external interface between the external phase and membrane phase through the mass transfer film resistance.
2. $Sr^{+2}$ reacts with the carrier or extractant (D2EHPA), hereafter denoted as HA, present in the membrane phase at the interface and forms a membrane phase soluble complex $SrR_2$:

$$Sr^{+2}_{external} + 2HR_{interface(ext./memb.)} \rightarrow SrR_{2interface(ext./memb.)} + 2H^{+}_{external}.$$

3. Transport of $SrR_2$ takes place through the membrane phase to the interface between the membrane phase and the internal phase:

$$SrR_{2interface(ext./memb.)} \rightarrow SrR_{2interface(memb./int.)}.$$

4. $SrR_2$ reacts with hydrogen ions from the internal phase at the membrane/internal phase interface releasing $Sr^{+2}$ from the membrane phase into the internal phase:

$$SrR_{2interface(memb./int.)} + 2H^{+}_{internal} \rightarrow 2HR_{interface(memb./int.)} + Sr^{+2}_{internal}.$$

5. The extractant HR diffuses back from the membrane/internal interface to the external/membrane interface:

$$HR_{interface(memb./int.)} \rightarrow HR_{interface(ext./memb.)}.$$

This facilitated transfer of strontium ions occurs when the pH value of the internal phase is lower in value than the external phase.

### 15.4.3 Supported Liquid Membrane

Supported liquid membrane (SLM) essentially consists of an organic liquid immobilized in small pores of a support, hence called as immobilized liquid membrane. The solvent/diluent acts as the membrane phase in the case of ordinary diffusive transport. For carrier-mediated transport, the membrane phase is prepared by dissolving the carrier into the solvent/diluent. The porous support material serves as a framework or supporting layer for the membrane phase. It can be inorganic or organic (polymer), with suitable chemical properties and mechanical stability. The porosity of such support materials should be high in order to obtain a high permeation rate or flux. Besides porosity, permeation rate also depends upon the membrane thickness because the flux is inversely proportional to the membrane thickness (Mulder, 1991). Therefore, support should be as thin as possible to obtain high flux but at the same time should be mechanically stable.

According to the geometry of the supports, SLM is generally of three types: flat sheet, hollow fiber, and spiral wound. Flat-sheet SLM (Figure 15.12) uses support material in sheet form; it is simple and of low cost, but it requires a large space. The small volumes of organic components in the membrane and extraction in a single technological step provide the advantage of using expensive carriers. The SLM process is of high separation factor, easy scale up, low energy requirements, low

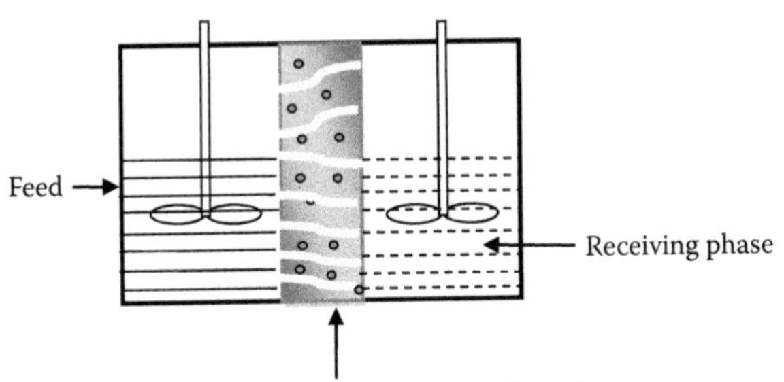


**FIGURE 15.12** Schematic diagram of a flat-sheet-supported liquid membrane.

capital and operating costs, etc. Figure 15.13 shows a typical setup of a flat-sheet SLM (Chakrabarty et al., 2009b, 2010a, 2010b, 2010c).

Hollow-fiber SLM (HFSLM) is like a shell and tube heat exchanger (Figure 15.14), with a high mass transfer area per unit volume (Winston Ho and Sirkar, 2001). Large numbers of thin porous fibers in the shape of tubes (ID: 0.5–1.0 mm) (Warey, 2007) are placed in an outer shell. The pores of the fibers are filled with the membrane phase. The feed phase and strip phase are separately passed through either fiber and shell side or vice versa. Carrier in the membrane phase helps faster transport of solute from feed phase to strip phase. Following are the several inherent advantages offered by HFSLM:

- The surface area and membrane thickness provide rapid transportation.
- The source/receiving phases are more easily recoverable than the emulsion system.
- The entire source and receiving phase are not in contact with the membrane at any given instant.
- Leakage and contamination are easily contained.

There are a few problems associated with this system, which are as follows:

- Very hydrophobic membrane solvents are required to maintain integrity.
- Hollow-fiber system must be cleaned between uses.
- Pore fouling and formation of cakes in filters often occur due to surface effects and particles in the system.
- High capital costs.

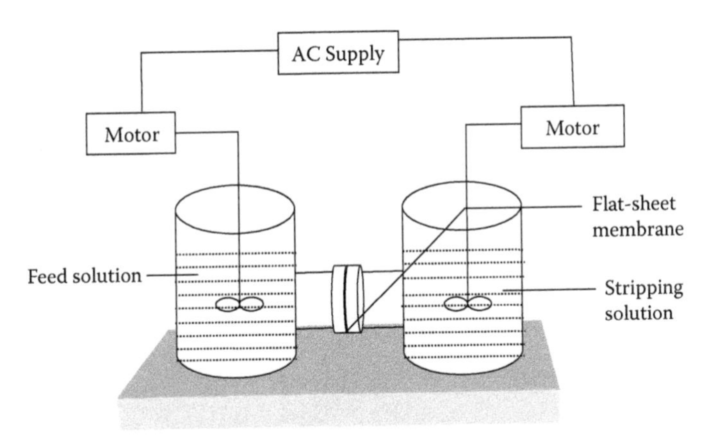


**FIGURE 15.13** Schematic representation of a supported liquid membrane setup.

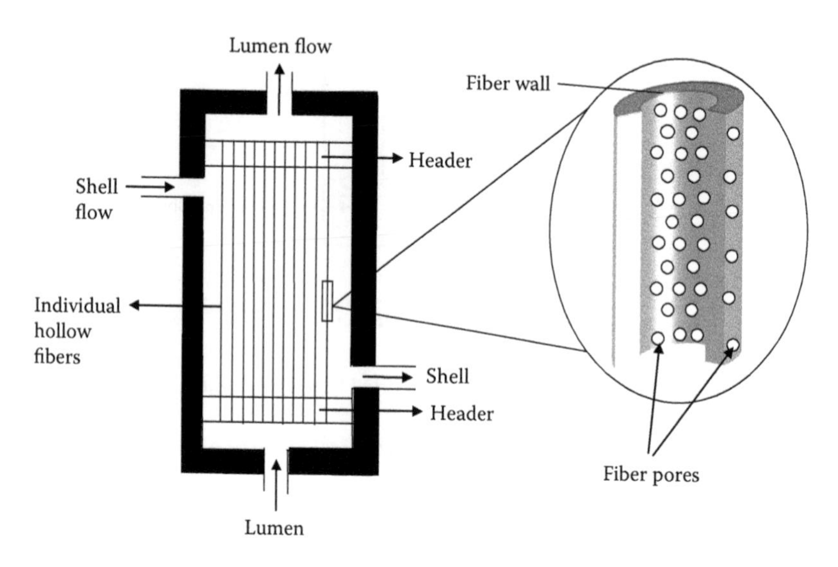


**FIGURE 15.14** Hollow-fiber-supported liquid membrane.

To avoid the problems due to low stability of organic liquid phase in the pores, in another mode of operation, aqueous and organic solutions flow continuously, with both phases coming into contact through the pores of the fiber wall. Thus, only one unit operation, either extraction or back-extraction, is realized in one module, and it is necessary to use two modules for the whole process. Pressure difference is also applied to avoid phase entrainment.

In order to carry out both extraction and recovery simultaneously, contained liquid membranes (CLM), where the extracting organic solvent is contained in the interstices of two sets of microporous hollow fibers (Figure 15.15), are used. The feed and strip solutions pass through the fiber lumens. The shell side is filled with the membrane liquid connected to an external membrane liquid reservoir at a pressure higher than the pressure of both feed and strip solutions. However, it suffers from problems like pore fouling, cleaning of fibers between uses, high pressure drop, and high cost. A similar configuration is the two-hollow-fiber SLM, where the source phase passes through one channel of hollow fiber and the receiving phase with another channel (Figure 15.16). Such configuration has advantages such as the possibility to use solvents with lower hydrophobicities, ease of renewal of solvent and carrier, high transport rate, and minimum leakage and contamination. However, the associated demerits are that the transport rate strongly depends on the stirring of the membrane phase, that the formation of boundary layer slows down the transport rate, fouling, and high capital cost.

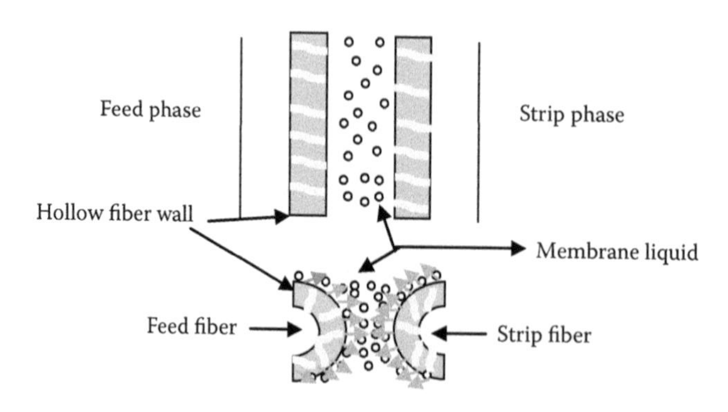


**FIGURE 15.15** Schematic of HFCLM.

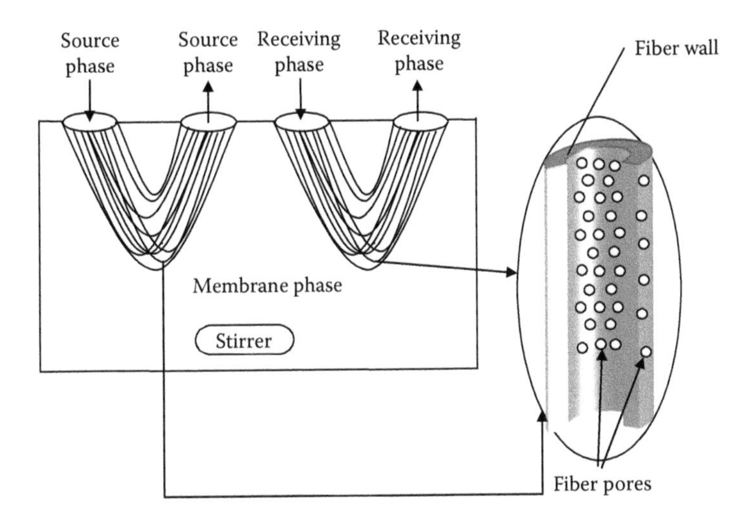


**FIGURE 15.16** Two-hollow-fiber-supported liquid membrane.

The spiral wound membrane (Figure 15.17) (Chakrabarty, 2010c) is like a plate and frame system wrapped around a central perforated collection pipe in a similar fashion to a sandwich roll (Winston Ho and Sirkar, 2001). It is made of four layers such as membrane, feed channel, another membrane, and strip channel, which forces all the separated materials toward the perforated tube in the center. The feed flows axially through the cylindrical module parallel along the central pipe whereas the permeate flows radially toward the central pipe. This type of SLM is an intermediate between flat-sheet and hollow-fiber membrane in terms of stability and surface area per unit volume.

The SLM possess many potential advantages, e.g., low capital, operating, maintenance, and energy cost; high separation factor; low extractant consumption; and easy to scale up. However, the major drawback of SLM is the instability of the liquid film due to the gradual loss of membrane solution by dissolution and spontaneous emulsification during the operation. The membrane instability makes it difficult to be implemented for large-scale applications.

The major factors responsible for the degradation of membrane stability are as follows:

- Continuous wetting of the pores in the membrane support by the aqueous phase.
- Rupture of membrane pores due to pressure difference.
- Common solubility of species from the aqueous phase and liquid membrane phase.

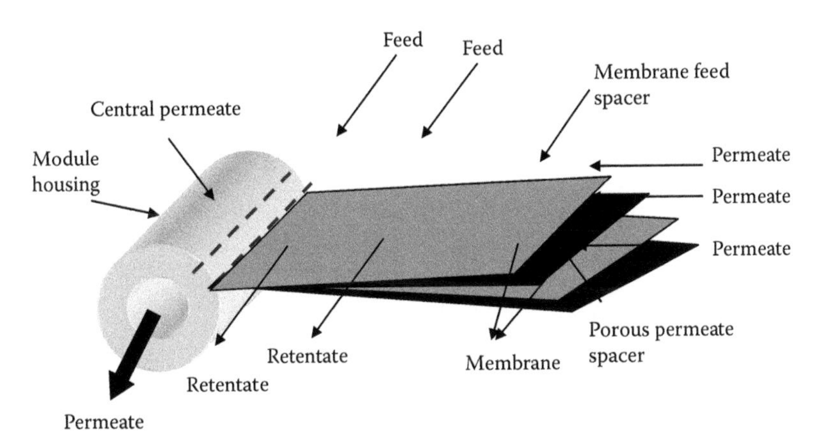


**FIGURE 15.17** Schematic of spiral wound SLM.

- Emulsion formation in the liquid membrane phase.
- Blockage of membrane pores by precipitation of a carrier complex.
- Stability can also be affected by the polymeric support and its pore radius.
- Organic solvent used in the liquid membrane and interfacial tension between the aqueous phase and the membrane phase.

### 15.4.4 Electrostatic Pseudo Liquid Membrane

Electrostatic pseudo liquid membrane (ESPLM) (Winston Ho and Sirkar, 2001) was developed in the year 1988 and has shown effectiveness in extracting metals from aqueous solutions. The setup (Zhong-Mao, 1990) uses the principle of electrostatics. According to the principle, a water drop inside the continuous oil phase will decompose into droplets due to application of a high electrical field (above critical field strength) (Zhong-Mao, 1990). A schematic representation of the ESPLM process is shown in Figure 15.18.

The setup is a reaction tank filled with organic solution containing extractant. The upper part of reaction tank is divided into two parts namely; extraction cell and a stripping cell with a baffle plate, and in the bottom part a divider is provided which divide the extraction settler and stripping settler from each other. The extracting and stripping cells are in connection with its extractor settler and stripping settler. The organic solution flows freely by the baffled channels and the aqueous solutions

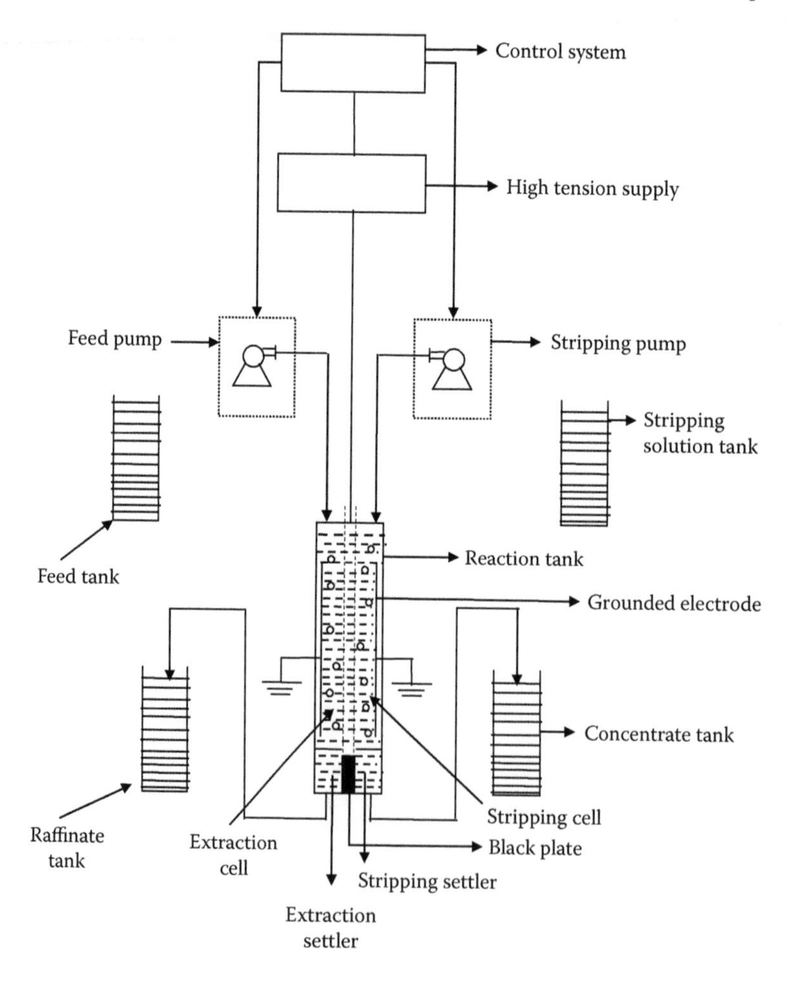


**FIGURE 15.18** Schematic representation of electrostatic pseudo liquid membrane.

are completely separated from being immiscible from one another. Pair of electrodes is hopped one in each side of the extraction and stripping cells.

High voltage electrostatic field is employed over the extraction and stripping cells at the same time when the solutions added to respective cells are disseminated into legion of droplets in the continuous organic phase. The aqueous droplets containing the solute are extracted into the continuous oil phase. The complex formed in the continuous phase moves by its own concentration gradient and diffuses across the baffle plate into the stripping cell. The extracted species are stripped into the aqueous stripping droplets in the stripping cell and the extractant diffuses back to the extraction cell through the baffle plate by its own concentration gradient.

These aqueous droplets in the extraction and stripping cells settles down due to gravity in the continuous oil phase, and thereby provides countercurrent extraction and stripping respectively. These aqueous droplets depart from the electrical field as they move forward at the respective settlers, where they disunite from the continuous oil phase, and the raffinate and concentrate are then collected.

Although it is a simple process, with easy operation, high flux, negligible leakage and swelling, and low energy consumption, it is effective only when the continuous phase is an organic solution (Winston Ho and Sirkar, 2001).

### 15.4.5 Hollow-Fiber Renewal Liquid Membrane

Hollow-fiber renewal liquid membrane (HFRLM) setup (Zhongqi et al., 2007) is based on the surface renewal theory. It integrates the advantages of fiber membrane extraction, liquid film permeation and other liquid membrane processes. HFRLM setup (Figure 15.19) consists of hydrophobic hollow fibers, the pores of which are occupied with the organic phase. The lumen side is mainly the aqueous phase in which organic phase droplets are made uniformly dispersed. Because of the wetting affinity of the organic phase and hydrophobic fibers, a thin organic film is developed in the lumen side of the fibers. The shear force of the flowing fluid in the lumen side causes the organic microdroplets to peel off the surface of the liquid membrane layer, which is again filled up by the organic droplets present in the aqueous phase of the lumen side. Thus, the renewal of the membrane liquid takes place, which reduces the mass transfer resistance in the aqueous boundary layer in the lumen side and consequently accelerates the mass transfer rate. The additional organic phase in the lumen side helps renewal of the membrane liquid, maintenance of the continuity of the liquid membrane layer and replenishment of the loss of the membrane liquid.

Merits of the HFRLM are as follows (Zhongqi et al., 2007):

1. Increased mass transfer rate because of the renewal effect of liquid membrane and high membrane surface area.
2. Long-time stability.
3. Expensive solvents can be used because of low extractant consumption.

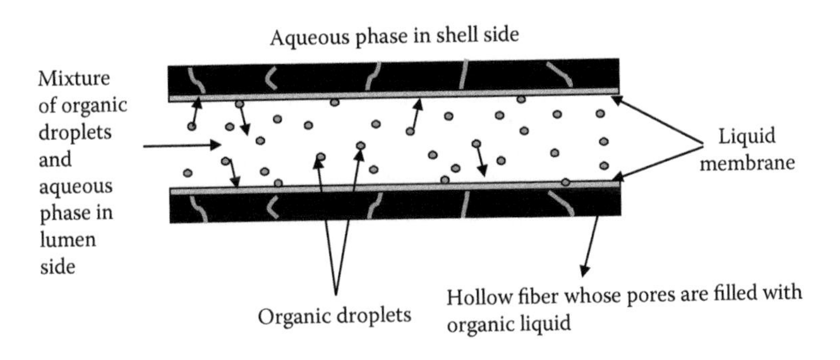


**FIGURE 15.19** Principle of HFRLM.

4. Low energy consumption; low capital, maintenance, and operation costs.
5. Easy to scale up due to the modular hollow fiber device and easy to operate.
6. This technique can be applied for simultaneous extraction and stripping processes using either w/o/w systems or o/w/o systems, with fiber being either hydrophobic or hydrophilic.

## REFERENCES

Alguacil, F.J., A.G. Coedo, and M.T. Dorado. 2005. Transport of chromium (VI) through a Cyanex-923-xylene flat sheet supported liquid membrane. *Hydrometallurgy* 5: 51–56.

Alpoguz, H.K., S. Memon, M. Ersoz, and M. Yilmaz. 2002. Transport of $Hg^{2+}$ through bulk liquid membrane using bis-calix[4]arene nitril derivative as carrier: Kinetic analysis. *New J. Chem.* 26: 477–480.

Arous, O., A. Gherrou, and H. Kerdjoudj. 2004. Removal of Ag(l), Cu(II) and Zn(Il) ions with a supported liquid membrane containing cryptands as carrier. *Desalination* 161: 295–303.

Bachiri, A.E., A. Hag, and M. Burgard. 1996. Recovery of silver nitrate by transport across a liquid membrane containing dicyclohexano-18-crown-6 as a carrier. *J. Membr. Sci.* 121: 159–168.

Chakrabarty, K., K. Vamsee Krishna, P. Saha, and A.K. Ghoshal. 2009a. Extraction and recovery of lignosulfonate from its aqueous solution using bulk liquid membrane. *J. Membr. Sci.* 330: 135–144.

Chakrabarty, K., P. Saha, and A.K. Ghoshal. 2009b. Separation of lignosulfonate from its aqueous solution using supported liquid membrane. *J. Membr. Sci.* 340: 84–91.

Chakrabarty, K., P. Saha, and A.K. Ghoshal. 2010a. Simultaneous separation of mercury and lignosulfonate from aqueous solution using supported liquid membrane. *J. Membr. Sci.* 346: 37–44.

Chakrabarty, K., P. Saha, and A.K. Ghoshal. 2010b. Separation of mercury from its aqueous solution through supported liquid membrane using environmentally benign diluents. *J. Membr. Sci.* 350: 395–401.

Chakrabarty, K. 2010c. Liquid membrane based technology for removal of pollutants from wastewater. Ph.D Dissertation, Indian Institute of Technology Guwahati.

Diederich, F. and K. Dick. 1984. A new water-soluble macrocyclic host of the cyclophone type: Host–guest complexation with aromatic guests and in aqueous solution and acceleration of the transport areas through an aqueous phase. *J. Am. Chem. Soc.* 106: 3024.

Ellinghorst, G., B. Goetz, A. Niemoeller, H. Scholz, H.E.A. Brueschke, and G. Tusel. 1989. Method for production of solution diffusion membranes and their application for pervaporation. U.S. Patent no. 4865743, September 12.

Gherrou, A., H. Kerdjoudj, R. Molinari, and E. Drioli. 2002. Removal of silver and copper ions from acidic thiourea solutions with a supported liquid membrane containing D2EHPA as carrier. *Sep. Purif. Tech.* 28: 235–244.

Goyette, M.L., T.L. Longin, R.D. Noble, and C.A. Koval. 2003. Selective photofacilitated transport of sodium ions through liquid membranes: Key factors in experimental design, transport results and comparison with a mathematical model. *J. Membr. Sci.* 212: 225–235.

Harada, A. and S. Takahashi. 1987. Transport of neutral azobenzene derivatives by methylated cyclodextrin. *J. Chem. Soc. Chem. Commun.* 527–528.

Jabbari, A., M. Esmaeili, and M. Shamsipur. 2001. Selective transport of mercury as $HgCl_4^{2-}$ through a bulk liquid membrane using $K^+$-dicyclohexyl-18-crown-6 as carrier. *Sep. Purif. Tech.* 24: 139–145.

Kamiński, W. and W. Kwapiński. 2000. Applicability of liquid membranes in environmental protection. *Polish J. Environ. Stud.* 9: 37–43.

Kontturi, A.K. 1988. Diffusion coefficient and effective charge number of lignosulfonate. *J. Chem. Soc, Faraday Trans.* 84: 40–43.

Kontturi, A.K., K. Kontturi, P. Niinikoski, and G. Sundholm. 1990. Extraction of a polyelectrolyte using SLM–II extraction and fraction of lignosulfonate. *Acta Chem. Scandinavica* 44: 883–891.

Kumbasar, R.A. 2008. Selective separation of chromium (VI) from acidic solutions containing various metal ions through emulsion liquid membrane using trioctylamine as extractant. *Sep. Purif. Tech.* 64: 56–62.

Leon, G. and M.A. Guzman. 2005. Kinetic study of the effect of carrier and stripping agent concentrations on the facilitated transport of cobalt through bulk liquid membranes. *Desalination* 184: 79–87.

Li, Q., Q. Liu, and X. Wei. 1996. Separation study of mercury through an emulsion liquid membrane. *Talanta* 43: 1837–1842.

Lin, C.C., and R.L. Long. 1997. Removal of nitric acid by emulsion liquid membrane: Experimental results and model prediction. *J. Membr. Sci.* 134: 33–45.

Lv, J., Q. Yang, J. Jiang, and T.S. Chung. 2007. Exploration of heavy metal ions transmembrane flux enhancement across a supported liquid membrane by appropriate carrier selection. *Chem. Eng. Sci.* 62: 6032–6039.

Mohite, B.S., S.G. Mane, and S.M. Sawant. 2001. Solvent extraction of uranium (VI) using dibenzo-18-crown-6 in nitrobenzene from ammonium thiocyanate medium. *J. Radioanal. Nucl. Chem.* 249: 613–616.

Monlinari, R., T. Poerio, and P. Argurio. 2006. Selective removal of $Cu^{+2}$ versus $Ni^{+2}$, $Zn^{+2}$, and $Mn^{+2}$ by using a new carrier in a supported liquid membrane. *J. Membr. Sci.* 280: 470–477.

Mulder, M. 1991. *Basic Principles of Membrane Technology*. Kluwer Academic Publishers, Dordrecht.

Muthuraman, G., T.T. Teng, C.P. Leh, and I. Norli. 2009. Use of bulk liquid membrane for the removal of chromium (VI) from aqueous acidic solution with tri-*n*-butyl phosphate as a carrier. *Desalination* 249: 884–890.

Neková, P., and D. Schrotterova. 1999. Extraction of Mo (VI) by primene JMT. *Solv. Extract. Ion Exch.* 17: 163–175.

Noble, R.D., and J.D. Way. 1987. *Liquid Membrane Technology*. National Bureau of Standards, Centre for Chemical Engineering, Boulder, CO.

Perez, M.M.E., J.A.R. Aguilera, T.I. Sauce, M.P. Gonzalez, R. Navarro, and M.A. Rodriguez. 2007. Study of As(V) transfer through a supported liquid membrane containing trioctylphosphine oxide (Cyanex-921) as carrier. *J. Membr. Sci.* 302: 119–126.

Pinnau, I., and B.D. Freeman. 1999. *Membrane Formation and Modification*. ACS Symposium Series; American Chemical Society: Washington DC.

Rajan, M., and V.M. Shinde. 1996. Synergistic extraction and separation study of uranium (VI). *J. Radioanal. Nucl. Chem.* 203: 169–176.

Sekine, T., K. Inaba, and H. Aikawa. 1988. Rate of solvent extraction of Cr (III) with acetylacetone into carbon tetrachloride. *Bull. Chem. Soc. Jpn.* 61: 1131–1134.

Sengupta, B., M.S. Bhakhar, and R. Sengupta. 2007. Extraction of copper from ammoniacal solutions into emulsion liquid membranes using LIX 84. *Hydrometallurgy* 89: 311–318.

Shaik, A.B., K. Chakrabarty, P. Saha, and A.K. Ghoshal. 2010. Separation of Hg (II) from its aqueous solution using bulk liquid membrane. *Industr. Eng. Chem. Res.* 49: 2889–2894.

Shamsipur, M., M. Hossein, and M.G. Azimi. 2002. Highly selective and efficient transport of mercury(II) ion across a bulk liquid membrane containing tetrathia-12-crown-4 as a specific ion carrier. *Sep. Purif. Tech.* 27: 155–161.

Swain, B., J. Jeong, J. Chun, L. Gae, and H. Lee. 2007. Extraction of Co(II) by supported liquid membrane and solvent extraction using Cyanex-272 as an extractant: A comparison study. *J. Membr. Sci.* 288: 139–148.

Treybal, R.E. 1980. *Mass Transfer Operation*. 3rd Ed. McGraw Hill Publishers, Singapore.

Venkateswaran, P. and K. Palanivelu. 2006. Recovery of phenol from aqueous solution by supported liquid membrane using vegetable oils as liquid membrane. *J. Hazard. Mater.* B131: 146–152.

Warey, P.B. 2007. *New Research on Hazardous Materials*. Nova Publishers, New York.

Winston Ho, W.S. and K.K. Sirkar. 2001. *Membrane Handbook*. Kluwer Academic Publishers, Dordrecht.

Yordanov, B., and L. Boyadzhiev. 2004. Pertraction of citric acid by means of emulsion liquid membranes. *J. Membr. Sci.* 238: 191–197.

Zhong-Mao, G. 1990. A new liquid membrane technology—electrostatic pseudo liquid membrane. *J. Membr. Sci.* 52: 77–88.

Zhongqi, R., W. Zhang, Y.M. Liu, Y. Dai, and C. Cui. 2007. New liquid membrane technology for simultaneous extraction and stripping of copper (II) from wastewater. *Chem. Eng. Sci.* 62: 6090–6101.

Zhou, D.W. 1996. Emulsion liquid membrane removal of arsenic and strontium from wastewater an experimental and theoretical study. Ph.D Dissertation. New Jersey Institute of Technology.

# 16 Applications and Advances with Supported Liquid Membranes

*Michael E. Vilt and W. S. Winston Ho*

## CONTENTS

## 16.1 SUPPORTED LIQUID MEMBRANES

This work focuses on the use of traditional SLMs and SLMs with strip dispersion (Ho, 2003). This section will discuss the basic principles of supported liquid membranes (SLMs) and review a variety of applications that exist for SLMs. Other processes related to SLMs, such as contained SLM (Sirkar, 1996), stagnant sandwich liquid membrane (SSwLM) (Molinari et al., 2005), and ionic liquid SLM (Scovazzo et al., 2002) are important separation processes but detailed discussion of these are out of this chapter's primary scope.

Supported liquid membrane (SLM), also referred to as immobilized liquid membrane (ILM), is a term generally used to describe a liquid membrane separation process in which a liquid membrane phase is embedded in a solid porous support and placed between a feed and receiving phase. Traditionally, the liquid membrane phase is an organic solution and is embedded in the pores of a hydrophobic microporous polymer support. The liquid membrane solution typically contains a carrier dissolved in an organic solvent, and is immiscible with the aqueous feed and receiving solutions. When the loaded support is placed between the feed and receiving solutions, as shown in Figure 16.1, an SLM is formed. Transport of a target species in the feed solution can then occur by carrier-facilitated transport, which will be described later. The configuration of the SLM is described by what kind of polymeric support is being used. The two main configurations are the flat-sheet SLM (FSSLM) and the hollow-fiber SLM (HFSLM). Experimental studies usually start with a system that consists of flat-sheet membrane in a two-compartment cell, as shown in Figure 16.2 (Hassoune et al., 2008). FSSLM systems may also include pump induced flow of the feed and receiving solutions. HFSLMs are normally operated under flow conditions and are used to demonstrate potential industrial applications, as this configuration provides high surface area per module volume. Industrial use of SLMs has been hampered by the problem of instability.

### 16.1.1 Mass Transfer Mechanism

Typically, SLMs transport species from the feed solution to the receiving solution by Type 2 facilitation or carrier-facilitated transport and consists of the following steps (de Gyves and Rodriguez de San Miguel, 1999):

1. Target species diffusion from the bulk phase across the aqueous diffusional layer to the membrane interface
2. Formation of carrier-target species complex in the membrane phase at the interface

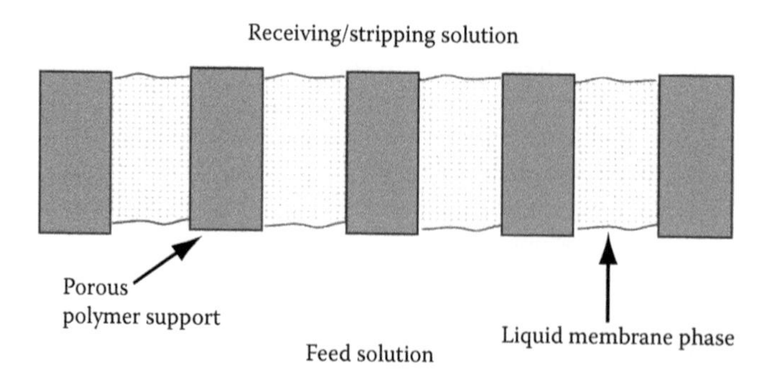


**FIGURE 16.1** Schematic diagram of supported liquid membrane.

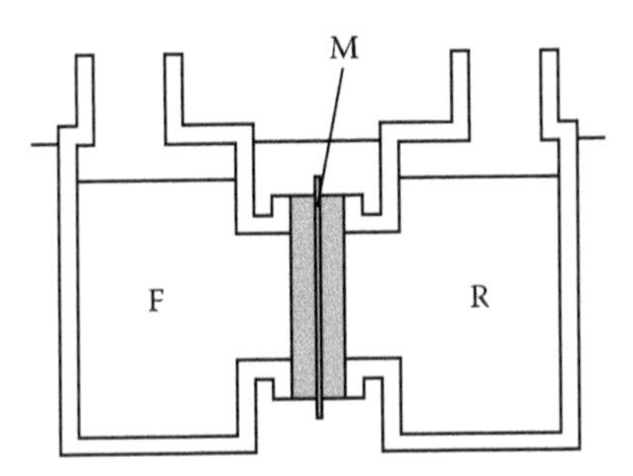


**FIGURE 16.2** Typical two-compartment cell for flat sheet supported liquid membrane experiments, where F, M, and R are the feed, membrane, and receiving solutions, respectively. (Adapted from Hassoune, H., et al., 2008. *J. Membr. Sci.* 315: 180–186.)

3. Diffusion of the complexed carrier across the membrane
4. The release of the target species at the strip interface and regeneration of the carrier
5. The diffusion of the released target species from the membrane-strip interface to the bulk strip phase
6. The return of the carrier across the membrane

Carrier-facilitated transport can be further classified into coupled cotransport and coupled countertransport (de Gyves and Rodriguez de San Miguel, 1999). When a neutral or basic carrier is used in the SLM process cotransport takes place. In cotransport, counter ions along with the target species in the feed solution form a carrier complex in the liquid membrane phase and are both transferred to the strip solution. Countertransport occurs when an acidic carrier loses a proton to the feed solution during the formation of the target species–carrier complex. The carrier is regenerated when the target species is released into the strip solution while simultaneously gaining a proton from the strip solution. Additional steps, specific to the transport mechanism, can also be included for:

7. Diffusion of cospecies or counterspecies across the aqueous diffusional layer
8. Carrier complexation with cospecies or counterspecies

## 16.2 APPLICATIONS

SLMs have been used for a variety of applications. Major applications include the removal and recovery of metals, the management of nuclear waste, and the recovery of biological and biochemical species. Other applications have included the removal of textile dyes from wastewater, the concentration of analytical samples, and controlled crystal growth.

### 16.2.1 Metal Removal and Recovery

The use of SLMs for metal separation has been extensive. A detailed review, completed in 1999 by de Gyves and Rodriguez de San Miguel is available. Table 16.1 displays recent SLM systems for some of the most commonly studied metals. Other metals, which have also been studied, include aluminum (Berends and Witkamp, 2001), arsenic (Perez et al., 2007; Prapasawat et al. 2008), gold (Kumar and Sastre, 2000; Gherrou and Kerdjoudj, 2002; Alguacil et al., 2005), lithium (Bansal et al., 2005), molybdenum (Yassine, 2000; Marchese et al., 2004), palladium (Fontas et al., 2003; Uheida et al., 2004; Zaghbani et al., 2007), platinum (Bhandare and Argekar, 2002b; Fontas et al., 2005), rhodium (Fontas et al., 2000), and silver (Shamsipur et al., 2003).

**TABLE 16.1**
**Supported Liquid Membranes for the Removal/Recovery of Selected Metals**

| Target Metal Ions | Configuration | Feed | Strip | Carrier/Diluent | Reference |
|---|---|---|---|---|---|
| Bi (III) | FSSLM | 2M $H_2SO_4$/0.5 M HCl | DI water | Cyanex 921/kerosene | Reyes-Aguilera et al., 2008 |
| Bi (III) | FSSLM | 0.1 M HCl | 2.5 M $HNO_3$ | Cyanex 301/4-chloroa-cetophenon | Madaeni and Zand, 2005 |
| Cd (II) | FSSLM | 0.1–5 M HCl | DI water | Cyanex 923/Solvesso 100 (Exxon) | Alonso et al., 2006 |
| Cd (II) | FSSLM | 0.1 M HCl, 0.4 M NaCl | 0.50 M $CH_3COONH_4$ | TOA/xylene | He and Ma, 2000a |
| Cd (II) | FSSLM | 0.1 M HCl, 0.4 M NaCl | 0.50 M $CH_3COONH_4$ | N235/xylene | He and Ma, 2000b |
| Cd (II) | FSSLM | Orthophosphoric acid | 4 M HCl | Cyanex 302/kerosene | Urtiaga et al., 2000 |
| Cd (II) | FSSLM | 3 M NaCl | 0.04 M EDTA, 1M NaCl | TBP/cyclohexane | Nowier et al., 2000 |
| Cd (II) | HFSLM | pH = 5 by acetate buffer | pH = 1.8 | D2EHPA/kerosene | Marchese and Campderros, 2004 |
| Cd (II) | FSSLM | pH = 3–6, sulfate | 0.9 M $H_2SO_4$ | D2EHPA/kerosene | Tripathy et al., 2002 |
| Cd (II) | FSSLM | 2.7 M HCl, 1 M $H_3PO_4$ | DI water | Cyanex 923/xylene | Alguacil and Navarro, 2001 |
| Cd (II) | FSSLM | 1 M HCl | 0.001 M EDTA | Aliquat 336/Kelfex 100/LIX 54/ kerosene | Lv et al., 2007 |
| Cd (II), Pb (II) | FSSLM | pH = 8.5, 0.1 M NaCl | pH = 3, 0.1 M NaCl | Kelex 100/kerosene | Aguilar et al., 2001 |
| Cd (II), Zn (II) | FSSLM | HCl–$CH_3COONa$ buffer | 1.26 M $H_2SO_3$ | D2EHPA/kerosene | Swain et al., 2004a |
| Cd (II), Zn (II) | FSSLM | $Cl^-$, $NO^{3-}$, $SO_4^{2-}$, $SCN^-$, $ClO^{3-}$, $CH3COO^-$ | 1.26 M $H_2SO_4$ | TOPS-99/kerosene | Swain et al., 2006 |
| Cd (II), Zn (II) | FSSLM | pH = 2–5, sulfate | 0.5 M $H_2SO_4$ | D2EHPA, PC88A/kerosene | Juang et al., 2004a |
| Cd (II), Zn (II) | FSSLM | 0.1 M HCl | 0.01 M HCl | Aliquat 336/kerosene, *n*-decanol | Juang et al., 2004b |
| Cd (II), Pb (II), Zn (II), Ag (I) | FSSLM | pH = 8 by TEA and Tricine | pH = 3 by $HNO_3$ or HCl | lasalocid A/NPOE | Canet et al., 2002 |
| Cd (II), Co (II), Cu (II), Fe (III), Ni (II), Pb (II) | FSSLM | pH = 3 by $HNO_3$ | 2 M $HNO_3$ | TOPO, D2EHPA, DOPPA/kerosene, DOTDDA/dihexylether | Belkhouche et al., 2006 |

| | | | | | |
|---|---|---|---|---|---|
| Co (II) | FSSLM | Acetate buffer | 0.1 M $H_2SO_4$ | Cyanex 272/kerosene | Swain et al., 2007a |
| Co (II) | FSSLM | 0.1–1 M HCl | NaOH | TEA/cyclohexanone | Bukhari et al., 2004 |
| Co (II) | HFSLM | pH 5 | 0.1M HCl | D2EHPA/kerosene | Prakorn et al., 2006 |
| Co (II), Li (I) | FSSLM | pH = 4.00–6.75, acetate buffer | 0.1 M $H_2SO_4$ | Cyanex 272/kerosene | Swain et al., 2007b |
| Co (II), Mn (II) | FSSLM | pH = 2.5–6.5, sulfate | 0.5 M $H_2SO_4$ | D2EHPA/Exxsol D100 | Alguacil, 2002 |
| Co (II), Ni (II) | HFSLM | pH = 4–6 | $H_2SO_4$ | HEH(EHP)/kerosene | Lee et al., 2004 |
| Co (II), Ni (II) | HFSLM | pH = 6 | 1.5 M $H_2SO_4$ | D2EHPA/kerosene | Jeong et al., 2003 |
| Co (II), Ni (II) | FSSLM | pH = 1–6, sulfate | 1 M $H_2SO_4$ | D2EHPA, Cyanex 272, 301, 302/ kerosene | Gega et al., 2001 |
| Cu (II) | HFSLM | Ammoniacal wastewater | 2 M $H_2SO_4$ | LIX54/kerosene | Yang and Kocherginsky, 2006 |
| Cu (II) | FSSLM/ HFSLM | Ammoniacal wastewater | 2 M $H_2SO_4$ | LIX54, LIX84/kerosene | Yang et al., 2006 |
| Cu (II) | FSSLM | $Ni^{2+}$,$Zn^{2+}$,$Mn^{2+}$ | pH = 2.2–3 | 2H5DBA/kerosene | Molinari et al., 2004 |
| Cu (II) | FSSLM | pH = 3.5, sulfate | pH = 1.5 | D2EHPA/*n*-decane | Molinari et al., 2005 |
| Cu (II) | FSSLM | $Cu^{+2}$, $Zn^{+2}$, $Co^{+2}$, $Cd^{+2}$, $Ni^{+2}$ | 2.3 M $H_2SO_4$ | LIX 984/kerosene | Ata, 2005 |
| Cu (II) | FSSLM | pH = 1–3 | 1.8 M $H_2SO_4$ | Acorga M5640/Iberfluid | Alguacil et al., 2002 |
| Cu (II) | FSSLM | $Ca^{2+}$,$Cd^{2+}$, $Pb^{2+}$, $Zn^{2+}$, $Ni^{2+}$ | pH = 1.0, 1% $HNO_3$ | LIX-84/kerosene | Cooper et al., 2004 |
| Cu (II) | HFSLM | pH = 2.8–3.2, mine water | 2-4 M $H_2SO_4$ | LIX-860/Kermac 500-T | Valenzuela et al., 2002 |
| Cu (II) | FSSLM | 0.1–0.5 M $(NH_4)_2CO_3$ | 1.8 M $H_2SO_4$ | LIX 973N/Iberfluid | Alguacil, 2001 |
| Cu (II) | FSSLM | pH = 1–3 | 1.8 M $H_2SO_4$ | MOC-55 TD/Iberfluid | Alguacil et al., 2001 |
| Cu (II) | HFSLM | pH = 1.5–3 | 1.5 M $H_2SO_4$ | PC88A, LIX84/kerosene | Lee et al., 1999 |
| Cu (II) | FSSLM | pH = 4 | 0.25 M $H_2SO_4$ | D2EHPA/coconut oil | Venkateswaran et al., 2007 |
| Cu (II), Fe (III) | FSSLM | pH = 2, sulfate | 0.5 M $H_3PO_4$ | D2EHPA/1-decanol | Zhang and Gozzelino, 2003 |
| Cu (II), Zn (II) | FSSLM | pH = 1.5–5 | 0.9 M $H_2SO_4$ | D2EHPA/kerosene | Sarangi and Das, 2004 |
| Cu (II), Zn (II), Co (II), Ni (II) | FSSLM | Synthetic leach liquor | 0.9 M $H_2SO_4$ | LIX 84, D2EHPA, Cyanex 272/ kerosene | Parhi and Sarangi, 2008 |
| Cu (II), Zn (II) | FSSLM | $Cl^-$, $NO^{3-}$, $SO_4^{-2}$, $ClO_3^-$, $CH_3COO^-$ | | D2EHPA/kerosene | Swain et al., 2004b |

*(continued)*

**TABLE 16.1 (Continued)**
**Supported Liquid Membranes for the Removal/Recovery of Selected Metals**

| Target Metal Ions | Configuration | Feed | Strip | Carrier/Diluent | Reference |
|---|---|---|---|---|---|
| Cu (II), Cr (VI), Zn (II) | FSSLM/ HFSLM | pH = 2.5 | 2 M $H_2SO_4$ | LIX 984N/kerosene | Yang et al., 2001 |
| Cu (II),Ag (I) | FSSLM | Acidic thiourea medium | DI water | DB18C6/chloroform | Gherrou et al., 2001a |
| Cu (II), Ag (I) | FSSLM | Acidic thiourea medium | pH = 0–2, $HNO_3$ | D2EHPA/chloroform | Gherrou et al., 2002a |
| Cu (II), Ag (I), Zn (II) | FSSLM | Nitrate | DI water | DB18C6, DA18C6/chloroform | Gherrou et al., 2001b |
| Cu (II), Ag (I), Zn (II) | FSSLM | Nitrate | DI water | DB18C6/chloroform | Gherrou et al., 2002b |
| Cu (II), Ag (I), Zn (II) | FSSLM | Nitrate | DI water | macrobicyclic polyethers/chloroform | Arous et al., 2004 |
| Cr (III) | FSSLM/ HFSLM | pH = 5.5, MES buffer | pH = 0, $HNO_3$ | Lasalocid A/NPOE | Tayeb et al., 2007 |
| Cr (III) | FSSLM | 1.5 M $Na_2SO_4$ | 1.5 M $H_2SO_4$ | TEA/cyclohexanone | Chaudry et al., 2007 |
| Cr (VI) | FSSLM | 0.5 M HCl | hydrazine sulfate | Cyanex 923/cumene | Alguacil and Alonso, 2003 |
| Cr (VI) | FSSLM | 0.1 M HCl | 0.1 M NaOH | TOA/toluene | Kozlowski and Walkowiak, 2005 |
| Cr (VI) | FSSLM | pH = 1–5 | 0.5 M NaOH | TBP | Venkateswaran and Palanivelu, 2005 |
| Cr (VI) | FSSLM | 0.5 M HCl | 0.5 M NaCl | Cyanex 923/xylene | Alguacil et al., 2000 |
| Cr (VI) | FSSLM | pH = 3.5–5.6, $Cr^{3+}$ | 2 M NaCl | Aliquat 336/NPOE, THF | Choi and Kim, 2005 |
| Cr (VI) | FSSLM | pH = 3 | 0.1–2.0 M NaOH | Aliquat 336/toluene | Park et al., 2004 |
| Fe (III) | FSSLM | 0.2 M HCl | 6 M HCl | Acylisoxazolones/xylene | Buonomenna et al., 2002 |
| Fe (III), Cu (II), Ni (II) | FSSLM | pH = 2, 1 M NaCl | 0.01 and 2 M HCl | Alamine 336, LIX 84/kerosene | Gill et al., 2000 |
| Hg (II) | FSSLM | 1 M HCl | 0.11 M NaSCN | Cyanex 471X/kerosene | Bhandare and Argekar, 2002a |
| Hg (II) | FSSLM | $Hg(NO_3)_2$ or $HgCl_2$ | DI water | calix[4]arene derivative/NPOE | Alpoguz et al., 2007 |

| | | | | | |
|---|---|---|---|---|---|
| Hg (II), Ag (I) | FSSLM | 0.001 M picric acid | 0.04 M $Na_2S_2O_3$, 0.025 M EDTA | PhenS2O, TT12C4/NPOE | Shamsipur et al., 2006 |
| Hg (II), As (III) | HFSLM | HCl | NaOH | TOA/toluene | Sangtumrong et al., 2007 |
| Ni (II) | FSSLM/ HFSLM | Nickel bath rinse solutions | 1.5M $H_2SO_4$ | LIX 860, Cyanex 302/kerosene | Van de Voorde et al., 2004 |
| Ni (II) | FSSLM | pH = 2–5 | 0.25–1.0 M $H_2SO_4$ | Acorga M5640/Exxsol D100 | Alguacil et al., 2006 |
| Zn (II) | FSSLM | $Co^{2+}$, $Ni^{2+}$, $Cu^{2+}$, $Pb^{2+}$, $Cd^{2+}$, $Hg^{2+}$, $Fe^{3+}$, $Cr^{3+}$ | 0.014 M L-cysteine | DC18C6/NPOE | Shamsipur et al., 2000 |
| Zn (II) | FSSLM | pH = 2.5, $Fe^{3+}$, $Ca^{2+}$, $Mn^{2+}$ | 3.5 M HCl | D2EHPA/kerosene | Ata and Colak, 2005 |
| Zn (II) | FSSLM | pH = 0.5–3.0, $Fe^{3+}$, $Ca^{2+}$, $Mn^{2+}$ | 0.5–3.5 M HCl | D2EHPA/kerosene | Ata et al., 2004 |
| Zn (II) | FSSLM | sulfate, $Co^{2+}$ | 0.5 M $H_2SO_4$ | D2EHPA/Exxsol D100 | Alguacil and Alonso, 2005 |
| Zn (II) | FSSLM | 1 M HCl | 1 M NaCl | Cyanex 923/Solvesso 100 (Exxon) | Alguacil and Martinez, 2001 |

*Note:* 2H5DBA, 2-hydroxy-5-dodecylbenzaldehyde; AcorgaM5640, 2-hydroxy-5-nonylbenzaldehyde oxime; Alamine 336, tri-octyl/decyl amine; Aliquat 336, tricaprylmethylammonium chloride; Cyanex 272, bis(2,4,4-trimethylpentyl)phosphinic acid; Cyanex 301, disooctyldithiophosphinic acid; Cyanex 302, bis(2,4,4-trimethylpentyl)thiophosphinic acid; Cyanex 471X, triisobutylphosphine sulfide; Cyanex 921, tri-*n*-octylphosphine oxide; Cyanex 923, phosphine oxides mixture; D2EHPA, bis(2-ethylhexyl)phosphoric acid; DA18C6, diaza-18-crown-6; DB18C6, dibenzo-18-crown-6; DC18C6, dicyclohexano-18-crown-6; DOPPA, di[p-(1,1′,3,3′-tetramethylbutyl)phenyl] phosphoric acid; DOTDDA, 1,12-(4,9-dioxatrisdecanyl)diphosphonic acid; HEH (EHP), (2-ethylhexyl)phosphonic acid mono-2-ethylhexyl ester; LIX 54, 1-phenyldecanone-1,3-diones; LIX 84, 2-hydroxy-5-nonylacetophenone; LIX 860, 5-dodoecylsalicylaldoxime; LIX 973N, mixture of 5-nonylsalicylaldoxime and 2-hydroxy-5-nonylacetophenone oxime; LIX 984N, mixture of 5-nonylsalicylaldoxime and 2-hydroxy-5-nonyl-acetophenone oxime; MOC-55 TD, 5-dodecylsalicylaldoxime; N235, tricapryl amine; NPOE, *o*-nitrophenyl octyl ether; PC-88A, 2-ethylhexyl phosphonic acid mono-2-ethylhexyl ester; PhenS2O, mixed aza-thioether crown containing a 1,10-phenanthroline subunit; TBP, tri-*n*-butyl phosphate; TEA, triethanolamine; THF, tetrahydrofuran; TOA, tri-*n*-octylamine; TOPO, trioctylphosphine oxide; TT12C4, tetrathia-12-crown-4.

### 16.2.2 Nuclear Waste Management

The removal of radioactive materials from nuclear waste streams has been accomplished with SLMs. Since reviews covering this area are available (Shukla et al., 1996; Mohapatra and Manchanda, 2003), recent developments will be briefly discussed.

#### 16.2.2.1 Uranium

The carrier Cyanex 272 (bis(2,4,4-trimethylpentyl)phosphinic acid) achieved selective separation of uranium (VI) from phosphate waste containing Am and Pu (Sawant et al., 2003). Aromatic diluents had higher extraction rates of $UO_2C1^{3-}$ and $UO_2C1_4^{2-}$ compared with polar diluents with the extractant Alamine 336 (tri(octyl/decyl) amine), possibly due to unfavorable hydrogen bonding with the latter (Lakshmi et al., 2004). Selective separation of uranium from a mixture of other fission products was demonstrated with the carrier Aliquat 336 (trioctylmethylammonium chloride) (Mohapatra et al., 2006a).

Selective uranium transport from feed solutions containing irradiated thorium was accomplished with the carrier di(2-ethylhexyl) isobutyramide (D2EHIBA) (Shailesh et al., 2008). The highest uranium transport was observed with the diluents *tert*-butylbenzene and diethylbenzene, due to the combination of low viscosity and low acid uptake from the feed solution.

Low uranium transport was reported when the carrier TOPO (tri-*n*-octylphosphine oxide) was studied for uranium separation from a phosphoric acid feed solution (Singh et al., 2008). The addition of nitric acid to the feed solution increased uranium transport and over 90% recovery was obtained when the phosphoric acid concentration was reduced to 0.001 M. The recovery of uranium with the carrier TBP (tri-*n*-butyl phosphate) in HFSLMs has also been reported (Ura et al., 2006; Ramakul et al., 2007).

#### 16.2.2.2 Plutonium

An FSSLM containing the carrier tri-*iso*-amyl phosphate (TAP) achieved over 90% extraction from a feed solution of 4 M $HNO_3$ with a Pu concentration of 1 mg/L (Shukla et al., 1998). A two-stage SLM process was able to treat analytical waste from the plutonium-uranium extraction (PUREX) process from an initial Pu concentration of 470 mg/L to a final concentration of 43 mg/L (Kedari et al., 1999). This system used the carrier 2-ethylhexylphosphonic acid (KSM-17, equivalent to PC 88A), dissolved in dodecane, and was supported on polypropylene PP flat sheets.

The recovery of plutonium from nitrate solutions was studied in both FSSLM and HFSLM systems with the carrier Cyanex 923 (a phosphine oxides mixture) and it was concluded that 1 M $HNO_3$ in the receiving solution provided the highest recovery (Sawant et al., 2004). An HFSLM system with an organic membrane solution composed of TBP in dodecane obtained a plutonium recovery of ~70% from a feed solution that contained various fission products (Rathore et al., 2001). This work was extended and was able to achieve simultaneous removal of plutonium and uranium (Rathore et al., 2004).

#### 16.2.2.3 Americium

An FSSLM process achieved over 90% recovery of americium when using a mixture of the carriers TOPO and 3-phenyl-4-benzoyl-5-isoxazolone (HPBI) (Mohapatra et al., 1999). Other studies have investigated the transport of americium using the carrier dimethyldibutyltetradecyl-1,3-malonamide (DMDBTDMA) (Sriram and Manchanda, 2002). The recommended diluent for DMDBTDMA was dodecane based on its physicochemical properties, diffusion coefficient, and distribution ratio. A continuous test was conducted for 20 days in a FSSLM with *N,N,N′,N′*-tetraoctyl-3-oxapentane diamide (TODGA) for the recovery of americium, in which the degree of recovery was consistently over 95% (Ansari et al., 2006). Additional studies with TODGA for the removal of trivalent actinides/lanthanides from high-level waste solutions have been completed (Ansari et al., 2008). A derivative of TODGA, tetra(2-ethylhexyl) diglycolamide (TEHDGA), is obtained by the substitution of the octyl group for a 2-ethylhexyl group, and was studied in an FSSLM system (Panja et al., 2008).

#### 16.2.2.4 Cesium

Selective cesium extraction was obtained with the carriers 1,3-dipropyloxycalix[4]arene crown ether and 1,3-dipropyloxycalix[4]arene dibenzo crown ether (Kim et al., 2001). Cesium was recovered with an efficiency of 80% in an SLM system composed of the carrier di-*t*-butylbenzo 18-crown-6 (DTBB18C6) dissolved in nitrobenzene and toluene and a receiving solution of distilled water (Mohapatra et al., 2004). An FSSLM process recovered cesium with an efficiency of ~90% with a calix[4]-bis-2,3-naphtho-crown-6 (CNC) concentration of $5 \times 10^{-4}$ M in a solvent mixture of 80% 2-nitrophenyl octyl ether and 20% *n*-dodecane (Raut et al., 2008).

#### 16.2.2.5 Strontium

A solvent mixture composed of a 1:1 ratio benzene-octanol obtained the highest strontium transport in an SLM with the carrier di-*tert*-butylcyclohexano 18-crown-6 (DTBCH18C6) (Mohapatra et al., 2006b). This result did not agree with batch extraction studies, suggesting that the diluents viscosity played an important role in this process. An SLM with 4,4′(5′)di-*tert*-butyl-cyclohexano-18-crown-6 (DtBuCH18C6) was recently reported for strontium recovery (Rawat et al., 2007). Strontium transport was slow due to cotransport of nitric acid but was increased with the addition of $Al(NO_3)_3$ to the feed.

#### 16.2.2.6 Cerium

Cerium was recovered with an efficiency of 99.8% in an FSSLM composed of the carrier octyl(phenyl)-*N*,*N*′-diisobutylcarbamoylmethyl phosphine oxide (CMPO), the modifier tributyl phosphate (TMP), and the diluent dodecane (Teramoto et al., 2000a). Teramoto et al. (2000b) extended this study in a plate-and-frame SLM module and stability was limited to a period of 5 hours. To address the stability problem, small amounts of organic membrane solution were added to the strip solution as demonstrated by Ho et al. (1999) that could replace any membrane solution lost in the support, providing stable SLM operation. Asymmetric high-density polyethylene (HDPE) hollow fibers were compared against commercial available PP fibers as supports in HFSLMs with CMPO (Fu et al., 2004a). The HDPE fibers obtained higher cerium recovery due to larger pore size and increased porosity at the inner surface. A mathematical model for cerium transport with CMPO was proposed in another study by Fu et al. (2004b).

### 16.2.3 Recovery of Antibiotics, Organic Acids, Amino Acids, and Sugars

The recovery of many biological and biochemical species have been accomplished with the use of SLMs. These include antibiotics (penicillin), organic acids (citric and lactic acid), amino acids (phenylalanine), and sugars.

#### 16.2.3.1 Antibiotics

The majority of SLM systems for antibiotic recovery have focused on the recovery of penicillin G. Marchese et al. (1989) first transported penicillin G (benzylpenicillin) through an SLM using a flat-sheet polytetrafluoroethylene (PTFE) support. The extraction of penicillin G and penicillin V with subsequent enzymatic hydrolysis to 6-aminopenicillanic acid was studied subsequently (Tsikas et al., 1992). When polypropylene hollow fibers were used as a support, *N*-lauryl-*N*-trialkylmethylamine (Amberlite LA-2) demonstrated superior stability in long-term studies compared with 1-methyltrioctylammonium chloride (Adogen 464) (Tsikas et al., 1992). Amberlite LA-2, a secondary amine, has been the most common carrier used in penicillin recovery with SLMs. Investigation of an SLM with Amberlite LA-2 in 1-decanol as the membrane phase with a PP flat-sheet support determined that the mass transfer of penicillin G was controlled by both the aqueous film layer and the liquid membrane phase (Lee et al., 1993). This work was extended to show the strong transport competition between penicillin G and phenylacetic acid when both are in the feed solution (Lee et al., 1994). Amberlite LA-2 was also used as a carrier for determination of the transport mechanism and rates of transport of penicillin G through an SLM (Juang et al., 1998). Another study utilizing SLMs included the use of hexafluorophosphates of 1-*n*-butyl-, 1-*n*-hexyl- and 1-*n*-octyl-3-methyl imidazolium, and trioctylmethylammonium chloride (TOMAC) as ionic liquids in a polyvinylidene fluoride (PVDF) SLM (Matsumoto et al., 2007). The ionic liquids in this study, however, did not improve extraction or stripping compared with conventional solvent extraction with butyl acetate. Cephalosporin C (CPC) extraction through a PP SLM containing the carrier Aliquat 336 was reported (Ghosh et al., 1995). When increasing the carrier concentration, the transport of CPC increased until a maximum was reached. After this point, increasing carrier concentration decreased transport due to a large decrease in diffusion coefficients in the membrane phase, resulting from an increase in viscosity.

The removal of other drugs not classified as antibiotics has also been of interest. For example, the removal of diclofenac, a nonsteroidal anti-inflammatory drug, from wastewaters is of interest, since it is not removed during standard water treatment procedures (Molinari et al., 2006). An SLM system composed of TBP dissolved in *n*-decane obtained a higher diclofenac flux value compared with an SSwLM, 2.4 versus 1.7 mmol/h-$m^2$; however, the long-term stability of the SSwLM was ~2.5 times higher.

### 16.2.3.2 Organic Acids

The recovery of citric acid from a fermentation broth with a PP support was demonstrated by Friesen et al. with the carrier trilaurylamine (Friesen et al., 1992). Citric acid extraction with the carrier Alamine 336 was studied with the solvents heptane and xylene, in which heptane led to instability (Sirman et al., 1990). Systems composed of xylene gave acceptable performance and the limiting transport step was determined to be diffusion across the membrane. In another study, tri-*n*-octylamine (TOA) and TOA salts were compared for citric acid extraction, and it was found that all the TOA salts studied increased citric acid transport, with the citrate salt having the highest enhancement (Juang and Chen, 1997). The extraction reaction of citric acid with TOA can be classified into two reaction regions: fast and slow. It was reported that in an SLM with a PTFE support that the reaction is part of the slow reaction region (Park et al., 2006). SLM systems for citric acid recovery that were subjected to a thermal gradient were proposed by Rockman et al. (1997). In an SLM system with the carrier Alamine 336, increasing the temperature of the receiving solution over the range of 30°C to 80°C resulted in increased citric acid flux. The extraction of citric and maleic acids from kiwi fruit juice was carried out by Schaefer and Hossain (1996). Alamine 336 and Aliquat 336 were compared for acid removal, and Almine 336 was recommended due to higher permeability at a pH of 3.3, which is the pH of untreated kiwi juice.

Liquid polyorganosiloxanes functionalized with amine, ether, ester, and alkyl organo-functional groups were used for the recovery of lactic acid and ethyl lactate (Yahaya and Brisdon, 2000). In this study, lactic acid transport was poor, but stable operation of 100 hours was observed for ethyl lactate with ether and amine functionalized polyorganosiloxanes. Using TOA as a carrier with a PVDF support, Juang et al. determined that the rate controlling steps for lactic acid transport were the interfacial chemical reaction and membrane diffusion (Juang et al., 1997). The ionic liquid tetradecyl(trihexyl)phosphonium bis(2,4,4-trimethylpentyl) phosphinate (IL-104) was dissolved in *n*-dodecane and used in an SLM that was in a spiral channel flat-sheet module (Martak et al., 2008). In this study, the transport of lactic acid due to the back transport of water from the receiving phase was reported for the first time. SLM systems for the enantiomer separation of L,D-lactic acid has also been reported (Hadik et al., 2005; Yang and Chung, 2007). Studies using the chiral selector *N*-3,5-dinitrobenzoyl-l-alanine octylester in HFSLMs reported that when the solvent toluene was removed from the SLM by a nitrogen purge, the chiral selector remained in the membrane pores and resulted in increased enantioselectivity but a decrease in flux (Hadik et al., 2005). The use of a modified carbonate form of Aliquat 336 was another approach to the separation of D,L-lactic acid mixtures, in which L-lactic acid was recovered with a separation factor of 2.3 (Yang and Chung, 2007). High purity aconitic acid was recovered from cane molasses solutions by an SLM process. The carrier TBP was able to selectively extract aconitic acid from a feed solution containing oxalic, maleic, and citric acids (McMurray and Griffin, 2002).

#### 16.2.3.3 Amino Acids

L-Phenylalanine is an important precursor to the artificial sweetener Aspartame (Thien et al., 1988) and its separation by SLMs has been of interest. The extraction of phenylalanine hydrochloride achieved with various crown ethers was studied (Bryjak et al., 1991). The extractant bis(2-ethylhexyl) phosphoric acid (D2EHPA) was studied for phenylalanine separation in an HFSLM (Choi et al., 2003) and a spiral channel SLM (Kertesz et al., 2004). Aliquat 336 has also been used as an extractant in various studies (Molinari et al., 1992; Campbell et al., 1994; Adarkar et al., 1997).

A variety of SLM systems have been applied to the enatioseparation of D,L-phenylalanine. An SLM process for the optical resolution of amino acids was proposed, which contained an encapsulated surfactant-protease complex in the membrane phase (Miyako et al., 2005). L-Phenylalanine was enantioselectively esterfied by the protease complex. After the L-phenylalanine ethyl ester diffused through the membrane phase, an α-chymotrypsin complex in the receiving phase catalyzed ester hydrolysis, re-forming the initial L-Phenylalanine. Resolution of a racemic mixture of phenylalanine was achieved with a separation factor of 1.8, with the chiral carrier *N*-decyl-(L)-hydroxyproline, which was complexed with copper ions (Clark et al., 2005). A two-stage continuous SLM process was tested for the enatioseparation of leucine and phenylalanine (Maximini et al., 2006). After five separation steps, leucine products consisting of 99% D-enantiomer and 99% L-enantiomer could be produced.

#### 16.2.3.4 Sugars

SLMs have been used to recover sugars from aqueous solutions. Of particular interest is the selective recovery of fructose over glucose during the production of high fructose syrup. An extensive review from 2004 summarizes the early work done in this field, and showed that boronic acid carriers have the potential to selectively extract fructose over glucose (Duggan, 2004). More recent work by Duggan et al. (2008) showed that the addition of Aliquat 336 with the carrier 2-(aminomethyl)-phenylboronic acid increased fructose flux and increased selectivity over glucose in an FSSLM system. The boronic acid derivative, 4-[8-(2-nitrophenoxy)octyloxycarbonyl] benzeneboronic acid, was used in both flat-sheet and hollow-fiber SLM configurations for the selective extraction of fructose from fermentation broths (Di Luccio et al., 2000). The HFSLM achieved longer stability and selectivity up to 20, compared with only 14 in the FSSLM system.

Carriers other than boronic acids have also been used in SLM systems for the facilitated transport of sugars. A resorcinarene carrier, a derivative of calixarene, was used in an FSSLM system and obtained higher selectivity for pentoses compared with fructoses, which is not typically seen with boronic acid carriers (Rhlalou et al., 2005). An SLM system composed of methyl cholate dissolved in cyclohexane and supported with a PVDF flat sheet was studied for the extraction of various sugars and was found to be stable for a period up to 23 days (Hassoune et al., 2008).

### 16.2.4 Textile Dye Removal from Wastewater

It has been recently demonstrated that SLM processes can be used to remove colored and toxic dyes from waste streams in the textile industry. The first work in this area was reported by Muthuraman and Palanivelu (2006a) and investigated the removal of the dye astacryl golden yellow from aqueous solutions in FSSLMs containing palm, sunflower, and coconut oils. The proposed transport mechanism was based on Type 1 facilitated transport and involved the neutrally charged dye diffusing into the oil phase forming an oil-dye complex, followed by diffusion of the oil-dye complex through the oil membrane phase, and concluded with a stripping step accomplished by $H_2SO_4$.

The carrier tributylphosphate has also been used in PTFE FSSLMs for the removal of dyes from textile waste streams. The removal of the anionic dyes C.I. Reactive Yellow 125, C.I. Reactive Red 158, and C.I. Reactive Red 159 was reported (Muthuraman and Palanivelu, 2006b). Optimal removal of the dyes occurred when a low feed pH of 1.0 was used, along with a stripping solution of 0.1 M NaOH. In another study, silk dyes, acid red 10B and acid pink BE, were recovered with efficiencies of 94.2% and 85.7%, respectively (Hajarabeevi et al., 2007).

### 16.2.5 Sample Treatment for Analytical Testing

Sample enrichment in analytical chemistry applications is possible with SLMs. Extensive reviews on this field are available, in which the majority of the literature has been focused on bioanalytical and environmental sample testing (Jonsson and Mathiasson, 2000; Lee et al., 2008, Pedersen-Bjergaard and Rasmussen, 2008). The use of SLMs for this application is commonly referred to as liquid-phase mircoextraction with three phases (Lee et al., 2008). Systems can be traditional SLMs, but more commonly only contain a single hollow fiber. A single hollow fiber is closed off at one end, and an organic membrane solution is embedded in the porous support, as shown in Figure 16.3. A small amount of acceptor solution is placed in the lumen of the hollow fiber. The loaded fiber is then placed in the sample, and the target analyte is concentrated in the acceptor solution.

### 16.2.6 Crystal Growth

SLMs have been used as biomimetic processes for the synthesis and design of crystals. An example of this application is the formation of strontium carbonate crystals (Sun et al., 2006). This SLM system consists of an organic membrane solution of the carrier *o*-phenanthroline dissolved in chloroform and is embedded in a porous support of sterified fibrin. The carrier combines with $Sr^{2+}$, which is present in the feed solution. After diffusing through the membrane, the $Sr^{2+}$ is captured by $CO_3^{2-}$, which is present in the stripping phase, forming crystals. The crystal morphology can be changed by adding different crystal modifiers to the stripping solution. Similar processes have been applied for the creation of $MnCO_3$ (Sun et al., 2008), $SrWO_4$ (Dong et al., 2008), and $CaCO_3$ crystals (Wu et al., 2004).

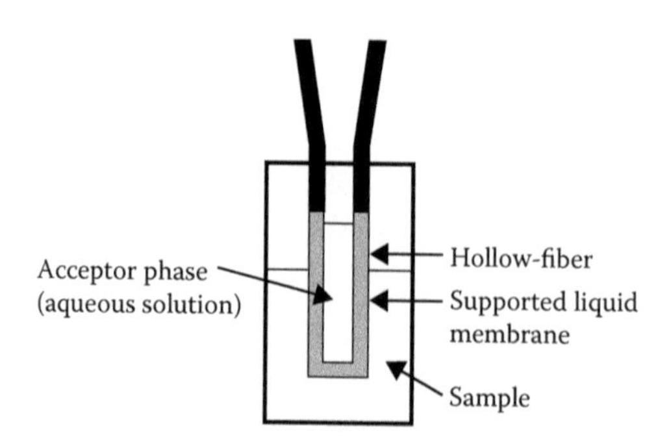


**FIGURE 16.3** Single hollow-fiber supported liquid membrane used for sample enrichment in analytical applications. (Reprinted from Pedersen-Bjergaard, S. and Rasmussen, K.E., *J. Chromatogr. A* 1184, 132–142. Copyright 2008, with permission from Elsevier.)

## 16.3 STABILITY

SLMs have seen limited commercial success due to the lack of long-term stability. The main mechanisms for SLM instability will be addressed, and recent approaches have been taken to solve this problem. Six mechanisms for SLM instability were indentified by Kemperman et al. (1996) and include:

- Pressure difference over the membrane
- Solubility of the liquid membrane phase in aqueous feed and strip solutions
- Wetting of pores in polymer support by aqueous phases
- Blocking of membrane pores by water blockage or carrier precipitation
- Osmotic pressure difference over the membrane
- Emulsification of the liquid membrane phase from lateral shear forces

Emulsification of the liquid membrane phase and the osmotic pressure difference over the membrane are considered to be the two major mechanisms of SLM instability (Dreher and Stevens, 1998). The other mechanisms can be avoided by properly designing and operating an SLM system.

There have been many attempts to improve SLM stability, including:

- Use of room temperature ionic liquids (RTILs) in SLMs (Fortunato et al., 2005)
- Reimpregnation of the liquid membrane phase in the support (Nakao et al., 1987; Tanigaki et al., 1988; Teramoto et al., 2000b)
- Use of composite supports that have a barrier or stabilization layer

For the last attempt, the barrier layer has been formed by applying hydrophilic coatings on hydrophobic supports (Wijerset al., 1998; He et al., 2004), interfacial polymerization (Kemperman et al., 1998; Wang et al., 1998), phase-inversion (Yang et al., 2007), and plasma polymerization (Yang et al., 2000).

## 16.4 SUPPORTED LIQUID MEMBRANES WITH STRIP DISPERSION

To overcome the problem of instability with traditional SLMs, Ho et al. developed SLMs with strip dispersion (Ho and Poddar, 2001; Ho, 2003). This technique maintains a constant supply of organic membrane solution to the SLM to ensure long-term stability. The SLM with a strip dispersion system developed by Ho is shown in Figure 16.4, and an enlarged view is shown in Figure 16.5. As

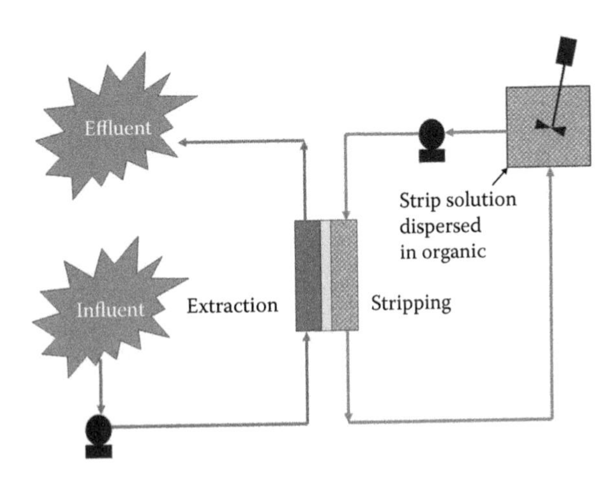


**FIGURE 16.4** Schematic diagram of the supported liquid membrane with strip dispersion process. (Reprinted with permisssion from Ho, W.S.W. and Wang, B.B., *Ind. Eng. Chem. Res.*, 41, 381–388. Copyright 2002 American Chemical Society.)

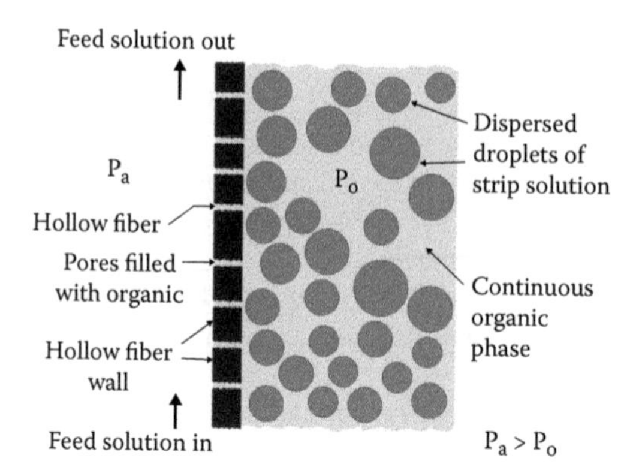


**FIGURE 16.5** Enlarged view of the supported liquid membrane with strip dispersion. (Reprinted with permission from Ho, W.S.W. and Wang, B.B., *Ind. Eng. Chem. Res.*, 41, 381–388. Copyright 2002 American Chemical Society.)

shown in these figures, an aqueous strip solution is dispersed in an organic membrane solution by a mixer, and passed onto one side of a membrane support. When a microporous hydrophobic support is used, the organic phase of the dispersion becomes imbedded in the pores of the support, typically hollow-fiber modules, forming a stable SLM. Stability is maintained by having a constant supply of organic membrane solution to the pores. An aqueous feed solution is passed on the other side of the membrane support, and the target species is extracted into the organic solution by a selective carrier. The target species is then stripped by the aqueous strip solution. For final recovery of the target species, the mixer is turned off or a settler is used for the loaded strip dispersion (not shown), and the dispersion quickly separates.

## 16.5 THEORY

### 16.5.1 Carrier-Facilitated Transport Models

Mass transfer in the SLMs with strip dispersion occurs through carrier-facilitated transport as previously described in Section 16.1.1. Flux through an SLM may be expressed as Equation 16.1 (Ho and Wang, 2002):

$$j = K_o\left[c_f - \left(\frac{m_s}{m_f}\right)c_s\right] \tag{16.1}$$

When $m_f$ is much greater than $m_s$, as when strong acids are used in the aqueous strip solution, the flux can be described by Equation 16.2:

$$j = K_o c_f \tag{16.2}$$

where the overall mass transfer resistance is described by Equation 16.3 and consists of the feed-side resistance, the interfacial resistance due to the complexation/extraction reaction, the membrane phase resistance, the interfacial resistance due to the decomplexation/stripping reaction, and the strip-side resistance (Ho and Wang, 2002).

$$\frac{1}{K_o} = \frac{1}{k_a} + \frac{1}{m_f k_e} + \frac{1}{m_f k_m} + \frac{1}{m_f k_s} + \frac{1}{(m_f/m_s)k_{as}} \tag{16.3}$$

### 16.5.2 Proton Transfer and Its Influence on Feed-Side pH

The removal of metals using acidic extractants results in the transfer of protons from the stripping solution to the aqueous feed solution. As discussed in Section 16.1.1, this can be classified as counter facilitated transport. Ho et al. (2001) reported the influence of this proton transfer on the feed solution pH, as a function of distance in a hollow-fiber module. When feed pH values are above 3, low module utilization efficiencies occurred due to significant reduction in the feed pH.

## 16.6 APPLICATIONS

Applications for SLMs with strip dispersion are similar to those found for traditional SLMs. Heavy metal removal and recovery has been the focus of current literature. Promising applications in biochemical processing have been demonstrated.

### 16.6.1 Metal Removal and Recovery

#### 16.6.1.1 Chromium Removal and Recovery

Ho and Poddar (2001) demonstrated the removal and recovery of Cr (VI), in the form of chromic acid, using commercial size hollow-fiber modules. In a two-step process, Cr (VI) was removed from 100 to 0.05 ppm in the acidic feed solution, and concentrated to 200,000 ppm in the aqueous strip solution. The organic membrane solution contained Amberlite LA-2, 1-dodeconol, and PLURONIC L31 in Isopar L. Stripping was achieved with NaOH. Sulfuric acid was found to compete with chromic acid for complexation with the amine carrier at Cr (VI) concentrations less than 100 ppm.

A mathematical model for the removal and recovery of Cr (VI) was optimized by Ortiz et al. (2003). The proposed configuration involves a two-step process similar to that described by Ho and Poddar (2001) and consisted of an organic membrane solution that was formed by Alamine 336, 1-dodeconol, and PLURONIC L31 in Isopar L. Optimal operating conditions were reported, and simulations were compared with experimental results.

#### 16.6.1.2 Copper Removal and Recovery

Laboratory and pilot plant studies on the removal and recovery of copper were reported by Ho et al. (2001, 2002). In pilot plant studies with a 4"-diameter hollow-fiber module using feed solutions from the Berkeley Pit (Butte, Montana), copper in the feed was reduced from 150 to < 0.1 ppm using the carrier LIX 973N. The aqueous strip solution was concentrated to 3,700 ppm, which is suitable for copper metal recovery via electrowinning, with no significant contamination from other metals present in the feed. Results from the pilot plant study are shown in Figure 16.6.

The recovery of copper catalyst from the wet peroxide oxidation process was reported by Ortiz et al. (Urtiaga et al. 2005, 2006). LIX 622N in kerosene was used as the membrane solution and was stripped with sulfuric acid. The final aqueous strip solution, containing copper sulfate, was found to be nonecotoxic and suitable for recycle to the oxidation reactor (Urtiaga et al., 2006). One hollow-fiber module using SLMs with strip dispersion was compared with nondispersive solvent extraction using two modules, one for extraction and one for stripping.

#### 16.6.1.3 Zinc Removal and Recovery

Ho et al. presented laboratory and scale-up results for the removal and recovery of zinc with the carrier Cyanex 301(Ho and Wang, 2001; Ho et al., 2002). The feed solution was from the Berkeley Pit (Butte, Montana) and was previously treated for copper removal. After a 4-hour run time in recycle mode, 121 L of feed solution was reduced from 526 to 0.65 ppm, while achieving a strip solution concentration of 20,000 ppm, which is suitable for reuse.

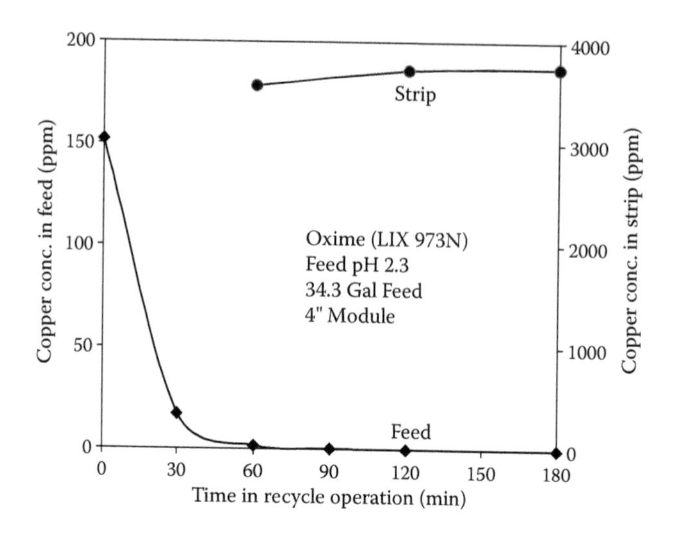


**FIGURE 16.6** Results from an SLM with strip dispersion pilot plant study for the recovery of copper. (Ho, W.S.W.: *N.Y. Acad. Sci.* 984. 97–122. 2003. Copyright Wiley-VCH Verlag GmBH and Co. kGaA. Reproduced with permission.)

The selective removal and recovery of zinc from copper containing feed solutions was studied by He et al. using a flat-sheet polypropylene support (He et al., 2006). An optimum pH of 2 to 2.5 was reported for the separation of zinc from copper using the carrier D2EHPA in kerosene.

#### 16.6.1.4 Cobalt Removal and Recovery

A system for the removal and recovery of cobalt was reported by Ho (Ho, 2003). An organic solution containing Cyanex 301, 1-dodecanol, and Isopar L was employed to effectively recover cobalt at high concentrations. A low feed pH of ~2 was used to avoid proton transfer over the length of the hollow-fiber module, as discussed in Section 16.5.2.

#### 16.6.1.5 Gold Removal and Recovery

The removal and recovery of gold from alkaline cyanide solutions using SLM with strip dispersion has been studied by Sonawane et al. using the organic extractant LIX-79 (*N,N*-bis(2-ethyl hexyl) guanidine) in *n*-heptane (Sonawane et al., 2008). Selectivity of metal cyanide salts was reported as followed: $Au(CN)_2^- > Zn(CN_4^{2-}) > Ag(CN)_2^- > Ni(CN)_4^{2-} > Fe(CN)_6^{2-} > Cu(CN)_4^{3-}$.

#### 16.6.1.6 Cadmium Removal and Recovery

The advantages of SLMs with strip dispersion over traditional SLMs are exhibited in the removal and recovery of cadium (Gu et al. 2006; He et al., 2007). In the selective removal of cadium from zinc, the loss of membrane solution was reported to be 0.6% to 1.4% for SLM with strip dispersion and 12% to 14% for SLM (Gu et al. 2006). The low loss of membrane solution suggests long stability. Increased recovery, higher selectivity, and increased flux were also reported. Similar advantages were reported for the simultaneous removal of cadium and cyanide from simulated electroplating rinse water (He et al., 2007). An organic solution containing the carriers trialkylphosphine oxide (TRPO) and D2EHPA in kerosene removed $CN^-$ and Cd (II) from initial concentrations of 30 and 20 ppm to 0.05 and 0.02 ppm, respectively. In another study, cadium was recovered from HCl solutions in an SLM with strip dispersion system with the carrier Cyanex 923 (Sastre et al., 2008). It was reported that Cd (II) was selectively transported from binary solutions containing Cr (VI) and Cu (II), but not from solutions containing Fe (III) and Zn (II).

### 16.6.2 Nuclear Waste Management

#### 16.6.2.1 Strontium Removal

New extractants were synthesized for the removal of radioactive strontium (Sr-90) (Ho and Wang, 2002). The extractants were branched alkyl phenylphosphonic acids and were more effective in strontium removal than the conventional extractant D2EHPA.

#### 16.6.2.2 Uranium Removal and Recovery

Uranium recovery was achieved in an SLM with strip dispersion system, consisting of the carrier TBP dissolved in *n*-paraffin hydrocarbon and a stripping solution of 0.01 M $HNO_3$ (Roy et al., 2008). When using an untreated oxalate supernatant as the feed solution, U (VI) was recovered with an efficiency of 84%. Plutonium, which was present in the feed, was extracted by TBP but was not recovered in the strip solution.

### 16.6.3 Antibiotic Removal and Recovery

Ho (2003) recovered penicillin G with an efficiency of 93% using the secondary amine Amberlite LA-2 in Isopar L. With a feed solution concentration of 8,840 ppm, the aqueous strip solution containing sodium carbonate was concentrated to 41,000 ppm. The recovery of the antibiotic cephalexin has also been demonstrated with the carrier Aliquat 336 (Vilt and Ho, 2009). Various experimental parameters such as strip dispersion mixing rate, Aliquat 336 concentration in the organic membrane phase, KCl and citrate buffer concentrations in the aqueous strip solution, volume of aqueous strip solution, and feed and strip dispersion flow rates were investigated. High extraction and recovery rates were achieved when maintaining a proper pH in the aqueous strip solution combined with an excess of KCl. Typical results using an Aliquat 336 concentration of 2.5 wt% (48 mM) have shown >99% extraction, recovery rates of 96% to 98%, and enrichment ratios of 1.6 to 3.3 from a feed solution containing 15 mM of cephalexin. Mass transfer resistances were analyzed, and showed the resistance from the extraction reaction to be dominant.

## 16.7 CONCLUDING REMARKS

SLMs have been demonstrated in a wide variety of applications. Industrial use of SLMs has been limited due to poor long-term stability. The SLM with strip dispersion addresses the problem of stability combined with simple operation. Applications for SLM with strip dispersion are similar to those for traditional SLMs, and have focused on heavy metal and nuclear waste removal and recovery. The removal and recovery of fermentation products and biochemicals is expected to be an area of focus in the future. The technique of SLM with strip dispersion suggests that long-term stability should be maintained, but long-term continuous tests have yet to be published.

## REFERENCES

Adarkar, J.A., S.B. Sawant, J.B. Joshi, and V.G. Pangarkar. 1997. Extraction of amino acids using an immobilized liquid membrane. *Biotechnol. Prog.* 13: 493–496.

Aguilar, J.C., M. Sanchez-Castellanos, E. Rodriguez de San Miguel, and J. de Gyves. 2001. Cd (II) and Pb (II) extraction and transport modeling in SLM and PIM systems using Kelex 100 as carrier. *J. Membr. Sci.* 190: 107–118.

Alguacil, F.J., A.G. Coedo, and M.T. Dorado. 2000. Transport of chromium (VI) through a Cyanex 923-xylene flat-sheet supported liquid membrane. *Hydrometallurgy* 57: 51–56.

Alguacil, F.J. 2000. Mechanistic study of active transport of copper(II) from ammoniacal/ammonium carbonate medium using LIX 973N as a carrier across a liquid membrane. *Hydrometallurgy* 61: 177–183.

Alguacil, F.J., and P. Navarro. 2001. Permeation of cadmium through a supported liquid membrane impregnated with CYANEX 923. *Hydrometallurgy* 61: 137–142.

Alguacil, F.J., M. Alonso, and A.M. Sastre. 2001. Modeling of mass transfer in facilitated supported liquid membrane transport of copper (II) using MOC-55 TD in Iberfluid. *J. Membr. Sci.* 184: 117–122.
Alguacil, F.J., and S. Martinez. 2001. Solvent extraction of Zn(II) by Cyanex 923 and its application to a solid-supported liquid membrane system. *J. Chem. Technol. Biotechnol.* 76: 298–302.
Alguacil, F.J. 2002. Facilitated transport and separation of manganese and cobalt by a supported liquid membrane using DP-8R as a mobile carrier. *Hydrometallurgy* 65: 9–14.
Alguacil, F.J., M. Alonso, and A.M. Sastre. 2002. Copper separation from nitrate/nitric acid media using Acorga M5640 extractant. Part II. Supported liquid membrane study. *Chem. Eng. J.* 85: 265–272.
Alguacil, F.J., and M. Alonso. 2003. Chromium (VI) removal through facilitated transport using CYANEX 923 as carrier and reducing stripping with hydrazine sulfate. *Environ. Sci. Technol.* 37: 1043–1047.
Alguacil, F.J., and M. Alonso. 2005. Separation of zinc(II) from cobalt (II) solutions using supported liquid membrane with DP-8R(di(2-ethylhexyl) phosphoric acid) as a carrier. *Sep. Purif. Technol.* 41: 179–184.
Alguacil, F.J., M. Alonso, and A.M. Sastre. 2005. Facilitated supported liquid-membrane transport of gold (I) and gold (III) using Cyanex 921. *J. Membr. Sci.* 252: 237–244.
Alguacil, F.J., M. Alonso, and A. Lopez-Delgado. 2006. Membrane-based extraction of nickel(II) using the mixture Acorga M5640 and DP-8R as carrier. *J. Braz. Chem. Soc.* 17: 839–843.
Alonso, M., A. Lopez-Delgado, A.M. Sastre, and F.J. Alguacil. 2006. Kinetic modeling of the facilitated transport of cadmium (II) using Cyanex 923 as ionophore. *Chem. Eng. J.* 118: 213–219.
Alpoguz, H.K., A. Kaya, and I. Sener. 2007. Carrier-mediated transport of Hg(II) through bulk and supported liquid membranes. *J. Macromol. Sci. A* 44: 1061–1068.
Ansari, S.A., P.K. Mohapatra, D.R. Prabhu, and V.K. Manchanda. 2006. Transport of americium(III) through a supported liquid membrane containing *N,N,N′,N′*-tetraoctyl-3-oxapentane diamide (TODGA) in *n*-dodecane as the carrier. *J. Membr. Sci.* 282: 133–141.
Ansari, S.A., P.K. Mohapatra, D.R., Raut, V.C. Adya, S.K. Thulasidas, and V.K. Manchanda. 2008. Separation of Am(III) and trivalent lanthanides from simulated high-level waste using a hollow fiber-supported liquid membrane. *Sep. Purif. Technol.* 63: 239–242.
Arous, O., A. Gherrou, and H. Kerdjoudj. 2004. Removal of Ag(I), Cu(II) and Zn(III) ions with a supported liquid membrane containing cryptands as carriers. *Desalination* 161: 295–303.
Ata, O.N., A.V. Bese, S. Colak, B. Donmez, and A. Cakici. 2004. Effect of parameters on the transport of zinc ion through supported liquid membrane. *Chem. Eng. Process.* 43: 895–903.
Ata, O.N. 2005. Modeling of copper ion transport through supported liquid membrane containing LIX 984. *Hydrometallurgy* 77: 269–277.
Ata, O.N., and S. Colak. 2005. Modeling of zinc transport through a supported liquid membrane. *Hydrometallurgy* 80: 155–162.
Bansal, B., X.D. Chen, and M.M. Hossain. 2005. Transport of lithium through a supported liquid membrane of LIX54 and TOPO in kerosene. *Chem. Eng. Process.* 44: 1327–1336.
Belkhouche, N.-E., M.A. Didi, R. Romero, J.A. Joensson, and D. Villemin. 2006. Study of new organophosphorus derivates carriers on the selective recovery of M(II) and M(III) metals, using supported liquid membrane extraction. *J. Membr. Sci.* 284: 398–405.
Berends, A.M., and G.J. Witkamp. 2001. Removal of aluminum from pickling bath liquids by tertiary and quaternary amine extractants. *Solvent Extr. Ion Exch.* 19: 473–490.
Bhandare, A.A., and A.P. Argekar. 2002a. Transport of mercury (II) ion through a supported liquid membrane containing a triisobutylphosphine sulfide(Cyanex 471X) as a mobile carrier. *J. Chem. Technol. Biotechnol.* 77: 811–816.
Bhandare, A.A., and A.P. Argekar. 2002b. Separation and recovery of platinum and rhodium by supported liquid membranes using bis(2-ethylhexyl)phosphoric acid (HDEHP) as a mobile carrier. *J. Membr. Sci.* 201: 233–237.
Bryjak, M., P. Wieczorek, P. Kafarski, and B. Lejczak. 1991. Transport of amino acids and their phosphonic acid analogs through supported liquid membranes containing macrocyclic carriers. Experimental parameters. *J. Membr. Sci.* 56: 167–180.
Bukhari, N., M. Chaudry, M. Ashraf, and M. Mazhar. 2004. Cobalt (II) transport through triethanolamine-cyclohexanone supported liquid membranes. *J. Membr. Sci.* 234: 157–165.
Buonomenna, M.G., R. Molinari, and E. Drioli. 2002. Selective mass transfer of iron (III) in supported liquid membrane using highly acidic extractants, 3-phenyl-4-acyl-5-isoxazolones. *Desalination* 148: 257–262.
Campbell, M.J., R.P. Walter, R. Singleton, and C.J. Knowles. 1994. Investigation of the stability and selectivity of phenylalanine transport across a supported liquid membrane. *J. Chem. Technol. Biotechnol.* 60: 263–273.

Canet, L., M. Ilpide, and P. Seta. 2002. Efficient facilitated transport of lead, cadmium, zinc, and silver across a flat-sheet-supported liquid membrane mediated by lasalocid A. *Sep. Sci. Technol.* 37: 1851–1860.

Chaudry, M.A., N. Bukhari, M. Mazhar, and W. Abbasi. 2007. Coupled transport of chromium(III) ions across triethanolamine/cyclohexanone based supported liquid membranes for tannery waste treatment. *Sep. Purif. Technol.* 55: 292–299.

Choi, J.W., K.S. Cho, B.K. Oh, I.J.Youn, J. Jeong, S. Park, and W.H. Lee. 2001. Numerical simulation of separation of cobalt and nickel using hollow fiber supported liquid membrane (HFSLM). *J. Ind. Eng. Chem.* 7: 230–240.

Choi, J.W., K.S. Cho, B.K. Oh, Y.K. Kim, I.J. Youn, and W.H. Lee. 2003. Separation and concentration of L-phenylalanine using hollow fiber supported liquid membrane. *J. Ind. Eng. Chem.* 9: 294–300.

Choi, D.W., and Y.H. Kim. 2005. Chromium(VI) removal in a semi-continuous process of hollow fiber membrane with organic extractants. *Korean J. Chem. Eng.* 22: 894–898.

Clark, J.D., B. Han, A.S. Bhown, and S.R. Wickramasinghe. 2005. Amino acid resolution using supported liquid membranes. *Sep. Purif. Technol.* 42: 201–211.

Cooper, C.A., Y.S. Lin, and M. Gonzalez. 2004. Separation properties of surface modified silica supported liquid membranes for divalent metal removal/recovery. *J. Membr. Sci.* 229: 11–25.

de Gyves, J., and E. Rodriguez de San Miguel. 1999. Metal ion separations by supported liquid membranes. *Ind. Eng. Chem. Res.* 38: 2182–2202.

Di Luccio, M., B.D. Smith, T. Kida, C.P. Borges, and T.L.M. Alves. 2000. Separation of fructose from a mixture of sugars using supported liquid membranes. *J. Membr. Sci.* 174: 217–224.

Dong, F.-Q., Q.-S. Wu, D.-M. Sun, and Y.-P. Ding. 2008. Morphology-tunable synthesis of $SrWO_4$ crystals via biomimetic supported liquid membrane (SLM) system. *J. Mater. Sci.* 43: 641–644.

Dreher, T.M., and G.W. Stevens. 1998. Instability mechanisms of supported liquid membranes. *Sep. Sci. Technol.* 33: 835–853.

Duggan, P.J. 2004. Fructose-permeable liquid membranes containing boronic acid carriers. *Aust. J. Chem.* 57: 291–299.

Duggan, P.J., T.A. Houston, M.J. Kiefel, S.M. Levonis, B.D. Smith, and M.L. Szydzik. 2008. Enhanced fructose, glucose and lactose transport promoted by a lipophilic 2-(aminomethyl)-phenylboronic acid. *Tetrahedron* 64: 7122–7126.

Fontas, C., C. Palet, V. Salvado, and M. Hidalgo. 2000. A hollow fiber supported liquid membrane based on Aliquat 336 as a carrier for rhodium(III) transport and preconcentration. *J. Membr. Sci.* 178: 131–139.

Fontas, C., V. Salvado, and M. Hidalgo. 2003. Selective enrichment of palladium from spent automotive catalysts by using a liquid membrane system. *J. Membr. Sci.* 223: 39–48.

Fontas, C., R. Tayeb, S. Tingry, M. Hidalgo, and P. Seta. 2005. Transport of platinum(IV) through supported liquid membrane (SLM) and polymeric plasticized membrane (PPM). *J. Membr. Sci.* 263: 96–102.

Fortunato, R., C.A.M. Afonso, J. Benavente, E. Rodriguez-Castellon, and J.G. Crespo. 2005. Stability of supported ionic liquid membranes as studied by X– ray photoelectron spectroscopy. *J. Membr. Sci.* 256: 216–223.

Friesen, D.T., W.C. Babcock, D.J. Brose, and A.R. Chambers. 1991. Recovery of citric acid from fermentation beer using supported-liquid membranes. *J. Membr. Sci.* 56: 127–141.

Fu, S.S., H. Mastuyama, and M. Teramoto. 2004a. Ce (III) recovery by supported liquid membrane using olyethylene hollow fiber prepared via thermally induced phase separation. *Sep. Purif. Technol.* 36: 17–22.

Fu, S.S., M. Teramoto, and H. Mastuyama. 2004b. Uphill transport of Ce (III) by supported liquid membranes containing octyl (phenyl)-*N,N*-diisobutylcarbamoylmethylphosphine oxide in 2-nitrophenyl octyl ether. *Sep. Sci. Technol.* 39: 517–538.

Gega, J., W. Walkowiak, and B. Gajda. 2001. Separation of Co(II) and Ni(II) ions by supported and hybrid liquid membranes. *Sep. Purif. Technol.* 22: 551–557.

Gherrou, A., H. Kerdjoudj, R. Molinari, and E. Drioli. 2001a. Modelization of the transport of silver and copper in acidic thiourea medium through a supported liquid membrane. *Desalination* 139: 317–325.

Gherrou, A., H. Kerdjoudj, R. Molinari, and E. Drioli. 2001b. Facilitated co-transport of Ag (I), Cu (II), and Zn (II) ions by DB18C6 and DA18C6 crown ethers as carriers: interface behavior on ion transport. *Sep. Sci. Technol.* 36: 2293–2308.

Gherrou, A., H. Kerdjoudj, R. Molinari, and E. Drioli. 2002a. Removal of silver and copper ions from acidic thiourea solutions with a supported liquid membrane containing D2EHPA as carrier. *Sep. Purif. Technol.* 28: 235–244.

Gherrou, A., H. Kerdjoudj, R. Molinari, and E. Drioli. 2002b. Facilitated co-transport of Ag(I), Cu(II), and Zn (II) ions by using a crown ether as carrier: influence of the SLM preparation method on ions flux. *Sep. Sci. Technol.* 37: 2317–2336.

Gherrou, A., and H. Kerdjoudj. 2002. Removal of gold as $Au(Tu)^{2+}$ complex with a supported liquid membrane containing macrocyclic polyethers ligands as carriers. *Desalination* 144: 231–236.

Ghosh, A.C., S. Borthakur, M.K. Roy, and N.N. Dutta. 1995. Extraction of cephalosporin C using supported liquid membrane. *Sep. Technol.* 5: 121–126.

Gill, J.S., H. Singh, and C.K. Gupta. 2000. Studies on supported liquid membrane for simultaneous separation of Fe(III), Cu(II) and Ni(II) from dilute feed. *Hydrometallurgy* 55: 113–116.

Gu, S., Y. Yu, D. He, and M. Ma. 2006. Comparison of transport and separation of Cd (II) between strip dispersion hybrid liquid membrane (SDHLM) and supported liquid membrane (SLM) using tri-*n*-octylamine as carrier. *Sep. Purif. Technol.* 51: 277–284.

Hadik, P., L.-P. Szabo, E. Nagy, and Z. Farkas. 2005. Enantioseparation of D,L-lactic acid by membrane techniques. *J. Membr. Sci.* 251: 223–232.

Hajarabeevi, N., I.M. Bilal, S. Amalraj, and K. Palanivelu. 2007. Textile anionic dyes recovery using tri-*n*-butyl phosphate as carrier through supported liquid membrane. *J. Environ. Sci. Eng.* 49: 33–40.

Hassoune, H., T. Rhlalou, and J.-F. Verchere. 2008. Studies on sugars extraction across a supported liquid membrane: complexation site of glucose and galactose with methyl cholate. *J. Membr. Sci.* 315: 180–186.

He, D., and M. Ma. 2000a. Effect of paraffin and surfactant on coupled transport of cadmium (II) ions through liquid membranes. *Hydrometallurgy* 56: 157–170.

He, D., and M. Ma. 2000b Kinetics of cadmium(II) transport through a liquid membrane containing tricapryl amine in xylene. *Sep. Sci. Technol.* 35: 1573–1585.

He, D., X. Luo, C. Yang, M. Ma, and Y. Wan. 2006. Study of transport and separation of Zn(II) by a combined supported liquid membrane/strip dispersion process containing D2EHPA in kerosene as the carrier. *Desalination* 194: 40–51.

He, D., S. Gu, and M. Ma. 2007. Simultaneous removal and recovery of cadmium (II) and $CN^-$ from simulated electroplating rinse wastewater by a strip dispersion hybrid liquid membrane (SDHLM) containing double carrier. *J. Membr. Sci.* 305: 36–47.

He, T., L.A.M. Versteeg, M.H.V. Mulder, and M. Wessling. 2004. Composite hollow fiber membranes for organic solvent-based liquid–liquid extraction. *J. Membr. Sci.* 234: 1–10.

Ho, W.S.W., T.K. Poddar, R. Pusic, and J. Roller. 1999. Unique Membrane Technology for Removal/Recovery of Metals from Wastewaters and Process Streams. *American Electroplaters and Surface Finishers Society/EPA Conference for Environmental Excellence*. Lake Buena Vista, FL.

Ho, W.S.W., B. Wang, T.E. Neumuller, and J. Roller. 2001. Supported liquid membranes for removal and recovery of metals from waste waters and process streams. *Environ. Prog.* 20: 117–121.

Ho, W.S.W., and T.K. Poddar. 2001. New membrane technology for removal and recovery of chromium from waste waters. *Environ. Prog.* 20: 44–52.

Ho, W.S.W., T.K. Poddar, and T.E. Neumuller. 2002. Removal and recovery of copper and zinc by supported liquid membranes with strip dispersion. *J. Chin. Inst. Chem. Engrs.* 33: 67–76.

Ho, W.S.W., and B.B. Wang. 2002. Strontium removal by new alkyl phenylphosphonic acids in supported liquid membranes with strip dispersion. *Ind. Eng. Chem.* Res. 41: 381–388.

Ho, W.S.W. 2003. Removal and recovery of metals and other materials by supported liquid membranes with strip disperion. *Ann. N. Y. Acad. Sci.* 984: 97–122.

Huang, D., K. Huang, S. Chen, S. Liu, and J. Yu. 2008. Rapid reaction-diffusion model for the enantioseparation of phenylalanine across hollow fiber supported liquid membrane. *Sep. Sci. Technol.* 43: 259–272.

Jeong, J., J.-C. Lee, and W. Kim. 2003. Modeling on the counteractive facilitated transport of Co in Co-Ni mixtures by hollow-fiber supported liquid membrane. *Sep. Sci. Technol.* 38: 499–517.

Jonsson, J.A., and L. Mathiasson. 2000. Membrane-based techniques for sample enrichment. *J. Chromatogr. A* 902: 205–225.

Juang, R.-S., and L.-J. Chen. 1997. Transport of citric acid across a supported liquid membrane containing various salts of a tertiary amine. *J. Membr. Sci.* 123: 81–87.

Juang, R.-S., S.-H. Lee, and R.-C. Shiau. 1997. Mass-transfer modeling of permeation of lactic acid across amine-mediated supported liquid membranes. *J. Membr. Sci.* 137: 231–239.

Juang, R., S. Lee, and R. Shiau.1998. Carrier-facilitated liquid membrane extraction of penicillin G from aqueous streams. *J. Membr. Sci.* 146: 95–104.

Juang, R.-S., H.-C. Kao, and W.-H. Wu. 2004a. Liquid membrane transport and separation of $Zn^{2+}$ and $Cd^{2+}$ from sulfate media using organophosphorus acids as mobile carriers. *J. Chem. Technol. Biotechnol.* 79: 140–147.

Juang, R.-S., H.-C. Kao, and W.-H. Wu. 2004b. Analysis of liquid membrane extraction of binary Zn(II) and Cd(II) from chloride media with Aliquat 336 based on thermodynamic equilibrium models. *J. Membr. Sci.* 228: 169–177.

Kedari, C.S., S.S. Pandit, and A. Ramanujam. 1999. Selective permeation of plutonium (IV) through supported liquid membrane containing 2-ethylhexyl 2-ethylhexyl phosphonic acid as ion carrier. *J. Membr. Sci.* 156: 187–196.

Kemperman, A.J.B., D. Bargeman, T. Van Den Boomgaard, and H. Strathmann. 1996. Stability of supported liquid membranes: state of the art. *Sep. Sci. Technol.* 31: 2733–2762.

Kemperman, A.J.B., H.H.M. Rolevink, T. Van Den Boomgaard, and H. Strathmann. 1998. Stabilization of supported liquid membranes by interfacial polymerization top layers. *J. Membr. Sci.* 138: 43.

Kertesz, R., S. Schlosser, and M. Simo. 2004. Mass-transfer characteristics of a spiral-channel SLM module in pertraction of phenylalanine. *Desalination* 163: 103–117.

Kim, J.K., J.S. Kim, Y.G. Shul, K.W. Lee, and W.Z. Oh. 2001. Selective extraction of cesium ion with calix[4] arene crown ether through thin sheet supported liquid membranes. *J. Membr. Sci.* 187: 3–11.

Kocherginsky, N.M., Q. Yang, and L. Seelam. 2007. Recent advances in supported liquid membrane technology. *Sep. Purif. Technol.* 53: 171–177.

Kozlowski, C.A., and W. Walkowiak. 2005. Applicability of liquid membranes in chromium(VI) transport with amines as ion carriers. *J. Membr. Sci.* 266: 143–150.

Kumar, A., and A.M. Sastre. 2000. Hollow fiber supported liquid membrane for the separation/concentration of gold (I) from aqueous cyanide media: modeling and mass transfer evaluation. *Ind. Eng. Chem. Res.* 39: 146–154.

Lakshmi, D.S., P.K. Mohapatra, D. Mohan, and V.K. Manchanda. 2004. Uranium transport using a PTFE flat-sheet membrane containing Alamine 336 in toluene as the carrier. *Desalination* 163: 13–18.

Lee, C.J., H.J. Yeh, W.J. Yang, and C.R. Kan. 1993. Extractive separation of penicillin G by facilitated transport via carrier supported liquid membranes. *Biotechnol. Bioeng.* 42: 527–534.

Lee, C.J., H.J. Yeh, W.J. Yang, and C.R. Kan. 1994. Separation of penicillin G from phenylacetic acid in a supported liquid membrane system. *Biotechnol. Bioeng.* 43: 309–313.

Lee, J.-C., J. Jeong, J.T. Park, I.J. Youn, and H.-S. Chung. 1999. Selective and simultaneous extractions of Zn and Cu ions by hollow fiber SLM modules containing HEH(EHP) and LIX84. *Sep. Sci. Technol.* 34: 1689–1701.

Lee, J.-C., J. Jeong, K.-S. Chung, and M. Kobayashi. 2004. Active facilitated transport and separation of Co in Co-Ni sulfate solution by hollow fiber supported liquid membrane containing HEH(EHP). *Sep. Sci. Technol.* 39: 1519–1533.

Lee, J., H.K. Lee, K.E. Rasmussen, and S. Pedersen-Bjergaard. 2008. Environmental and bioanalytical applications of hollow fiber membrane liquid-phase microextraction: a review. *Analytica Chimica Acta* 624: 253–268.

Lv, J., Q. Yang, J. Jiang, and T.S. Chung. 2007. Exploration of heavy metal ions transmembrane flux enhancement across a supported liquid membrane by appropriate carrier selection. *Chem. Eng. Sci.* 62: 6032–6039.

Madaeni, S.S., and H.R.K. Zand. 2005. Selective transport of bismuth ions through supported liquid membrane. *Chem. Eng. Technol.* 28: 892–898.

Marchese, J., J.L. Lopez, and J.A. Quinn. 1989. Facilitated transport of benzylpenicillin through immobilized liquid membrane. *J. Chem. Tech. Biotechnol.* 46: 149–159.

Marchese, J., and M. Campderros. 2004. Mass transfer of cadmium ions in a hollow-fiber module by pertraction. *Desalination* 164:141–149.

Marchese, J., F. Valenzuela, C. Basualto, and A. Acosta. 2004. Transport of molybdenum with Alamine 336 using supported liquid membrane. *Hydrometallurgy* 72: 309–317.

Martak, J., S. Schlosser, and S. Vlckova. 2008. Pertraction of lactic acid through supported liquid membranes containing phosphonium ionic liquid. *J. Membr. Sci.* 318: 298–310.

Matsumoto, M., T. Ohtani, and K. Kondo. 2007. Comparison of solvent extraction and supported liquid membrane permeation using an ionic liquid for concentrating penicillin G. *J. Membr. Sci.* 289: 92–96.

Maximini, A., H. Chmiel, H. Holdik, and N.W. Maier. 2006. Development of a supported liquid membrane process for separating enantiomers of *N*-protected amino acid derivatives. *J. Membr. Sci.* 276: 221–231.

McMurray, S.H., and G.J. Griffin. 2002. Extraction of aconitic acid from mixtures of organic acids and cane molasses solutions using supported liquid membranes. *J. Chem. Technol. Biotechnol.* 77: 1262–1268.

Miyako, E., T. Maruyama, F. Kubota, N. Kamiya, and M. Goto. 2005. Optical resolution of various amino acids using a supported liquid membrane encapsulating a surfactant-protease complex. *Langmuir* 21: 4674–4679.

Mohapatra, P.K., A.K. Pandey, and V.K. Manchanda. 1999. Facilitated transport of americium (III) from nitric acid media using 3-phenyl-4-benzoyl-5-isoxazolone and tri-*n*-octyl phosphine oxide in dodecane as the carrier. *Radiochimica Acta* 84: 147–152.

Mohapatra, P.K., and V.K. Manchanda. 2003. Liquid membrane based separations of actinides and fission products. *Indian J. Chem.* 42A: 2925–2938.

Mohapatra, P.K., D.S. Lakshmi, D. Mohan, and V.K. Manchanda. 2004. Selective transport of cesium using a supported liquid membrane containing di-*t*-butyl benzo 18 crown 6 as the carrier. *J. Membr. Sci.* 232: 133–139.

Mohapatra, P.K., D.S. Lakshmi, D. Mohan, and V.K. Manchanda. 2006a. Uranium pertraction across a PTFE flatsheet membrane containing Aliquat 336 as the carrier. *Sep. Purif. Technol.* 51: 24–30.

Mohapatra, P.K., D.S. Lakshmi, and V.K. Manchanda. 2006b. Diluent effect on Sr (II) extraction using di-*tert*-butyl cyclohexano 18 crown 6 as the extractant and its correlation with transport data obtained from supported liquid membrane studies. *Desalination* 198: 166–172.

Molinari, R., L. De Bartolo, and E. Drioli. 1992. Coupled transport of amino acids through a supported liquid membrane. I. Experimental optimization. *J. Membr. Sci.* 73: 203–215.

Molinari, R., T. Poerio, R. Cassano, N. Picci, and P. Argurio. 2004. Copper (II) removal from wastewaters by a new synthesized selective extractant and SLM viability. *Ind. Eng. Chem. Res.* 43: 623–628.

Molinari, R., P. Argurio, and F. Pirillo. 2005. Comparison between stagnant sandwich and supported liquid membranes in copper (II) removal from aqueous solutions: flux, stability and model elaboration. *J. Membr. Sci.* 256: 158–168.

Molinari, R., A. Caruso, P. Argurio, and T. Poerio. 2006. Diclofenac Transport through Stagnant Sandwich and Supported Liquid Membrane Systems. *Ind. Eng. Chem. Res.* 45: 9115–9121.

Muthuraman, G., and K. Palanivelu. 2006a. Transport of textile dye in vegetable oils based supported liquid membrane. *Dyes and Pigments* 70: 99–104.

Muthuraman, G., and K. Palanivelu. 2006b. Removal of CI reactive yellow 125, CI reactive red 158 and CI reactive red 159 dyes from aqueous solution with a supported liquid membrane containing tributylphosphate as carrier. *J. Text. Inst.* 97: 341–347.

Nakao, M., K. Takahashi, and H. Takeuchi. 1987. A method for continuous operation of supported liquid membranes. *J. Chem. Eng. Jpn.* 20: 326–328.

Nowier, H.G., N. El-Said, and H.F. Aly. 2000. Carrier-mediated transport of toxic elements through liquid membranes. Transport of Cd (II) from high salinity chloride medium through supported liquid membrane containing TBP/cyclohexane. *J. Membr. Sci.* 177: 41–47.

Ortiz, I., M. Fresnedo San Roman, S.M. Corvalan, and A.M. Eliceche. 2003. Modeling and optimization of an emulsion pertraction process for removal and concentration of Cr(VI). *Ind. Eng. Chem. Res.* 42: 5891–5899.

Panja, S., R. Ruhela, S.K. Misra, J.N. Sharma, S.C. Tripathi, and A. Dakshinamoorthy. 2008. Facilitated transport of Am(III) through a flat-sheet supported liquid membrane (FSSLM) containing tetra(2-ethyl hexyl) diglycolamide (TEHDGA) as carrier. *J. Membr. Sci.* 325: 158–165.

Parhi, P.K., and K. Sarangi. 2008. Separation of copper, zinc, cobalt and nickel ions by supported liquid membrane technique using LIX 84I, TOPS-99 and Cyanex 272. *Sep. Purif. Technol. 59*: 169–174.

Park, S.-W., H.-I. Jung, T.-Y. Kim, and J.-W. Lee. 2004. Effect of rheological properties on mass transfer of Cr(VI) through a supported liquid membrane with non-Newtonian liquid. *Sep. Sci. Technol. 39*: 781–797.

Park, S.-W., B.-S. Choi, S.-S. Kim, and J.-W. Lee. 2006. Facilitated transport of organic acid through a supported liquid membrane with a carrier. *Desalination* 193: 304–312.

Pedersen-Bjergaard, S., and K.E. Rasmussen. 2008. Liquid-phase microextraction with porous hollow fibers, a miniaturized and highly flexible format for liquid–liquid extraction. *J. Chromatogr. A* 1184: 132–142.

Perez, M.E.M., J.A. Reyes-Aguilera, T.I. Saucedo, M.P. Gonzalez, R. Navarro, and M. Avila-Rodriguez. 2007. Study of As(V) transfer through a supported liquid membrane impregnated with trioctylphosphine oxide (Cyanex 921). *J. Membr. Sci.* 302: 119–126.

Prakorn, R., S. Eakkapit, P. Weerawat, H. Milan, and P. Ura. 2006. Permeation study on the hollow-fiber supported liquid membrane for the extraction of cobalt (II). *Korean J. Chem. Eng.* 23: 117–123.

Prapasawat, T., P. Ramakul, C. Satayaprasert, U. Pancharoen, and A.W. Lothongkum. 2008. Separation of As(III) and As(V) by hollow fiber supported liquid membrane based on the mass transfer theory. *Korean J. Chem. Eng.* 25: 158–163.

Reyes-Aguilera, J.A., M.P. Gonzalez, R. Navarro, T.I. Saucedo, and M. Avila-Rodriguez. 2008. Supported liquid membranes (SLM) for recovery of bismuth from aqueous solutions. *J. Membr. Sci.* 310: 13–19.

Ramakul, P., T. Prapasawad, U. Pancharoen, and W. Pattaveekongka. 2007. Separation of radioactive metal ions by hollow fiber-supported liquid membrane and permeability analysis. *J. Chinese Inst. Chem. Eng.* 38: 489–494.

Rathore, N.S., J.V. Sonawane, A. Kumar, A.K. Venugopalan, R.K. Singh, D.D. Bajpai, and J.P. Shukla. 2001. Hollow fiber supported liquid membrane: a novel technique for separation and recovery of plutonium from aqueous acidic wastes. *J. Membr. Sci.* 189: 119–128.

Rathore, N.S., J.V. Sonawane, S.K. Gupta, A.K. Pabby, A.K. Venugopalan, R.D. Changrani, and P.K. Dey. 2004. Separation of uranium and plutonium from aqueous acidic wastes using a hollow fiber supported liquid membrane. *Sep. Sci. Technol*. 39: 1295–1319.

Raut, D.R., P.K. Mohapatra, S.A. Ansari, and V.K. Manchanda. 2008. Evaluation of a calix[4]-bis-crown-6 ionophore-based supported liquid membrane system for selective 137Cs transport from acidic solutions. *J. Membr. Sci.* 310: 229–236.

Rawat, N., P.K. Mohapatra, D.S. Lakshmi, A. Bhattacharyya, and V.K. Manchanda. 2006. Evaluation of a supported liquid membrane containing a macrocyclic ionophore for selective removal of strontium from nuclear waste solution. *J. Membr. Sci.* 275: 82–88.

Rhlalou, T., M. Ferhat, M.A. Frouji, D., Langevin, M. Métayer, and J.F. Verchère. 2000. Facilitated transport of sugars by a resorcinarene through a supported liquid membrane. *J. Membr. Sci.* 168: 63–73.

Rockman, J.T., E. Kehat, and R. Lavie. 1997. Thermally enhanced extraction of citric acid through supported liquid membrane. *AIChE J.* 43: 2376–2380.

Roy, S.C., J.V. Sonawane, N.S. Rathore, A.K., Pabby, P. Janardan, R.D. Changrani, P.K. Dey, and S.R. Bharadwaj. 2008. Pseudo-emulsion based hollow fiber strip dispersion technique (PEHFSD): optimization, modeling and application of PEHFSD for recovery of U (VI) from process effluent. *Sep. Sci. Technol.* 43: 3305–3332.

Sangtumrong, S., P. Ramakul, C. Satayaprasert, U. Pancharoen, and A.W. Lothongkum. 2007. Purely separation of mixture of mercury and arsenic via hollow fiber supported liquid membrane. *J. Ind. Eng. Chem.* 13: 751–756.

Sarangi, K., and R.P. Das. 2004. Separation of copper and zinc by supported liquid membrane using TOPS-99 as mobile carrier. *Hydrometallurgy* 71: 335–342.

Sastre, A.M., F.J. Alguacil, M. Alonso, F. Lopez, and A. Lopez-Delgado. 2008. On cadmium (II) membrane-based extraction using Cyanex 923 as carrier. *Solvent Extr. Ion Exc.* 26: 192–207.

Sawant, S.R., J.V. Sonawane, A.K. Venugopalan, P.K. Dey, A. Kumar, and J.N. Mathur. 2003. Membrane based U (VI) transport across supported liquid membrane from aqueous acidic media. *Ind. J. Chem. Tech.* 10: 531–538.

Sawant, S.R., J.V. Sonawane, A.K. Pabby, A.K. Venugopalan, P.K. Dey, and B. Venkataramani. 2004. Transport of Pu (IV) across supported liquid membrane from nitric acid medium using Cyanex-923 as the mobile receptor. *Ind. J. Chem. Tech.* 11: 548–554.

Schaefer, A., and M. M Hossain. 1996. Extraction of organic acids from kiwi fruit juice using a supported liquid membrane process. *Bioprocess Eng.* 16: 25–33.

Scovazzo, P., A.E. Visser, J.H. Davis, R.D. Rogers, C.A., Koval, D.L. DuBois, and R.D. Noble. 2002. Supported ionic liquid membranes and facilitated ionic liquid membranes. *ACS Symposium Series* 818: 69–87.

Shailesh, S., P.N. Pathak, P.K. Mohapatra, and V.K. Manchanda. 2008. Role of diluents in uranium transport across a supported liquid membrane using di(2-ethylhexyl)isobutyramide as the carrier. *Desalination* 232: 281–290.

Shamsipur, M., G. Azimi, and S.S. Madaeni. 2000. Selective transport of zinc as Zn (SCN)42-ion through a supported liquid membrane using $K^+$-dicyclohexyl-18-crown-6 as carrier. *J. Membr. Sci.* 165: 217–223.

Shamsipur, M., S.Y. Kazemi, G. Azimi, S.S. Madaeni, V. Lippolis, A. Garau, and F. Isaia. 2003. Selective transport of silver ion through a supported liquid membrane using some mixed aza-thioether crowns containing a 1,10-phenanthroline sub-unit as specific ion carriers. *J. Membr. Sci.* 215: 87–93.

Shamsipur, M., O.R. Hashemi, and V. Lippolis. 2006. A supported liquid membrane system for simultaneous separation of silver(I) and mercury(II) from dilute feed solutions. *J. Membr. Sci.* 282: 322–327.

Shukla, J.P., A. Kumar, R.K. Singh, and R.H. Iyer. 1996. Separation of radiotoxic actinides from reprocessing wastes with liquid membranes. *ACS Symposium Series* 642: 391–408.

Shukla, J.P., C.S. Kedari, and G.R. Dharmapurikar. 1998. Selective permeation of plutonium(IV) through a supported liquid membrane containing Tri-iso-amyl phosphate as an ionophore. *J. Nucl. Sci. Technol.* 35: 419–424.

Singh, S.K., S.K. Misra, M. Sudersanan, A. Dakshinamoorthy, S.K. Munshi, and P.K. Dey. 2007. Carrier-mediated transport of uranium from phosphoric acid medium across TOPO/n-dodecane-supported liquid membrane. *Hydrometallurgy* 87: 190–196.

Sirkar, K.K. 1996. Hollow fiber-contained liquid membranes for separations: an overview. *ACS Symposium Series* 642: 222–238.

Sirman, T., D.L. Pyle, and A.S. Grandison. 1990. Extraction of citric acid using a supported liquid membrane, in *Separation in Biotechnology 2*. London: Elsevier.

Sonawane, J.V., A.K. Pabby, and A.M. Sastre. 2008. Pseudo-emulsion based hollow fiber strip dispersion: a novel methodology for gold recovery. *AIChE J.* 54: 453–463.

Sriram, S., and V.K. Manchanda. 2002. Transport of metal ions across a supported liquid membrane (SLM) using dimethyldibutyltetradecyl-1,3-malonamide (DMDBTDMA) as the carrier. *Solvent Extr. Ion Exch.* 20: 97–114.

Sun, D.-M., Q.-S. Wua, and Y.-P. Ding. 2006. A novel method for crystal control: synthesis and design of strontium carbonate with different morphologies by supported liquid membrane. *J. Appl. Cryst.* 39: 544–549.

Sun, D.-M., D.-Z. Zhu, and Q.-S. Wu. 2008. Synthesis and design of $MnCO_3$ crystals with different morphologies by supported liquid membrane. *J. Chem. Crystallogr.* 38: 949–952.

Swain, B., K. Sarangi, and R. Prasad Das. 2004a. Separation of cadmium and zinc by supported liquid membrane using TOPS-99 as mobile carrier. *Sep. Sci. Technol.* 39: 2171–2188.

Swain, B., K. Sarangi, and R.P. Das. 2004b. Effect of different anions on separation of copper and zinc by supported liquid membrane using TOPS-99 as mobile carrier. *J. Membr. Sci.* 243: 189–194.

Swain, B., K. Sarangi, and R. Prasad Das. 2006. Effect of different anions on separation of cadmium and zinc by supported liquid membrane using TOPS-99 as mobile carrier. *J. Membr. Sci.* 277: 240–248.

Swain, B., J. Jeong, J.-C. Lee, G.-H. Lee. 2007a. Extraction of Co(II) by supported liquid membrane and solvent extraction using Cyanex 272 as an extractant: a comparison study. *J. Membr. Sci.* 288: 139–148.

Swain, B., J. Jeong, J.-C. Lee, and G.-H. Lee. 2007b. Separation of Co(II) and Li(I) by supported liquid membrane using Cyanex 272 as mobile carrier. *J. Membr. Sci.* 297: 253–261.

Tanigaki, M., M. Ueda, and W. Eguchi. 1988. Facilitated transport of zinc chloride through hollow fiber supported liquid membrane. Part 2. Membrane stability. *Sep. Sci. Technol.* 23: 1161–1169.

Tayeb, R., A. Zaghbani, S. Tingry, P. Seta, and M. Dhahbi. 2007. Carrier-mediated transport of Cr(III) across Lasalocid A-NPOE supported liquid membrane. *Desalination* 204: 234–240.

Teramoto, M., S.S. Fu, K. Takatani, N. Ohnishi, T. Maki, T. Fukui, and K. Arai. 2000a. Treatment of simulated low level radioactive wastewater by supported liquid membranes: uphill transport of Ce (III) using CMPO as carrier. *Sep. Purif. Technol.* 18: 57–69.

Teramoto, M., Y. Sakaida, S.S. Fu, N. Ohnishi, H. Matsuyama, T. Maki, T. Fukui, and K. Arai. 2000b. An attempt for the stabilization of supported liquid membrane. *Sep. Purif. Technol.* 21: 137–144.

Thien, M.P., T.A. Hatton, and D.I.C. Wang.1988. Separation and concentration of amino acids using liquid emulsion membranes. *Biotechnol. Bioeng.* 32: 604–615.

Tripathy, S.S., K. Sarangi, and R.P. Das. 2002. Extraction of cadmium(II) by supported liquid membrane using TOPS-99 as mobile carrier. *Sep. Sci. Technol.* 37: 2897–2911.

Tsikas, D., E. Kaltsidou-Schottelius, and G. Brunner.1992. Hollow fiber-supported liquid membranes for the extraction of penicillins and the synthesis of 6-aminopenicillanic acid. *Chemie Ingenieur Technik* 64: 545–548.

Uheida, A., Y. Zhang, and M. Muhammed. 2004. Transport of palladium (II) through hollow fiber supported liquid membrane facilitated by nonylthiourea. *J. Membr. Sci.* 241: 289–295.

Ura, P., R. Prakorn, P. Weerawat, and H. Milan. 2006. Feasibility study on the separation of uranium and thorium by a hollow fiber supported liquid membrane and mass transfer modeling. *J. Ind. Eng. Chem.* 12: 673–681.

Urtiaga, A.M., A. Alonso, I. Ortiz, J.A. Daoud, S.A. El-Reefy, S. Perez de Ortiz, and T. Gallego. 2000. Comparison of liquid membrane processes for the removal of cadmium from wet phosphoric acid. *J. Membr. Sci.* 164: 229–240.

Urtiaga, A., M.J. Abellan, J.A. Irabien, and I. Ortiz. 2006. Use of membrane contactors as an efficient alternative to reduce effluent ecotoxicity. *Desalination* 191: 79–85.

Valenzuela, F., M.A. Vega, M.F. Yanez, and C. Basualto. 2002. Application of a mathematical model for copper permeation from a Chilean mine water through a hollow fiber-type supported liquid membrane. *J. Membr. Sci.* 204: 385–400.

Van de Voorde, I., L. Pinoy, and R.F. De Ketelaere. 2004. Recovery of nickel ions by supported liquid membrane (SLM) extraction. *J. Membr. Sci.* 234: 11–21.

Venkateswaran, P., and K. Palanivelu. 2005. Studies on recovery of hexavalent chromium from plating wastewater by supported liquid membrane using tri-n-butyl phosphate as carrier. *Hydrometallurgy* 78: 107–115.

Venkateswaran, P., A.N. Gopalakrishnan, and K. Palanivelu. 2007. Di(2-ethylhexyl)phosphoric acid-coconut oil supported liquid membrane for the separation of copper ions from copper plating wastewater. *J. Environ. Sci.* 19: 1446–1453.

Vilt, M.E., and W.S.W. Ho. 2009. Supported liquid membranes with strip dispersion for the recovery of cephalexin. *J. Membr. Sci.* 342: 80–87.

Wang, Y.C., Y.S. Thio, and F.M. Doyle. 1998. Formation of semi-permeable polyamide skin layers on the surface of supported liquid membranes. *J. Membr. Sci.* 147: 109–116.

Wijers, M.C., M. Jin, M. Wessling, and H. Strathmann. 1998. Supported liquid membranes modification with sulfonated poly(ether ether ketone). Permeability, selectivity and stability. *J. Membr. Sci.* 147: 117.

Wu, Q.-S., D.-M. Sun, H.-J. Liu, and Y.-P. Ding. 2004. Abnormal polymorph conversion of calcium carbonate and nano-self-assembly of vaterite by a supported liquid membrane system. *Crystal Growth & Design* 4: 717–720.

Yahaya, G.O., B.J. Brisdon, and R. England. 2000. Facilitated transport of lactic acid and its ethyl ester by supported liquid membranes containing functionalized polyorganosiloxanes as carriers. *J. Membr. Sci.* 168: 187–201.

Yang, Q., and N.M. Kocherginsky. 2006. Copper recovery and spent ammoniacal etchant regeneration based on hollow fiber supported liquid membrane technology: from bench-scale to pilot-scale tests. *J. Membr. Sci.* 286: 301–309.

Yang, Q., J. Jiang, T.-S. Chung, and N.M. Kocherginsky. 2006. Experimental and computational studies of membrane extraction of Cu(II). *AIChE J.* 52: 3266–3277.

Yang, Q., T.-S. Chung, Y. Xiao, and K. Wang. 2007. The development of chemically modified P84 Co-polyimide membranes as supported liquid membrane matrix for Cu(II) removal with prolonged stability. *Chem. Eng. Sci.* 62: 1721–1729.

Yang, Q., and T.-S. Chung. 2007. Modification of the commercial carrier in supported liquid membrane system to enhance lactic acid flux and to separate L,D-lactic acid enantiomers. *J. Membr. Sci.* 294: 127–131.

Yang, X.J., A.G. Fane, J. Bi, and H.J. Griesser. 2000. Stabilization of supported liquid membranes by plasma polymerization surface coating. *J. Membr. Sci.* 168: 29–37.

Yang, X.J., A.G. Fane, and S. MacNaughton. 2001. Removal and recovery of heavy metals from wastewaters by supported liquid membranes. *Water Sci. Technol.* 43: 341–348.

Yassine, T. 2000. Separation of 99mTc from 99Mo by using TOPO-kerosene supported liquid membrane. *J. Radioanal. Nucl. Chem.* 246: 665–669.

Zaghbani, A., R. Tayeb, M. Dhahbi, M. Hidalgo, F. Vocanson, I. Bonnamour, P. Seta, and C. Fontas. 2007. Selective thiacalix[4]arene bearing three amide groups as ionophore of binary Pd(II) and Au (III) extraction by a supported liquid membrane system. *Sep. Purif. Technol.* 57: 374–379.

Zhang, B., and G. Gozzelino. 2003. Facilitated transport of Fe (III) and Cu (II) ions through supported liquid membranes. *Colloid Surf. A.* 215: 67–76.

# 17 Ionic Liquid-Based Supported Liquid Membranes

*Michiaki Matsumoto*

## CONTENTS



## 17.1 INTRODUCTION

Liquid membranes are thin liquid layers that separate two miscible phases. They became attractive because diffusivity in the liquid phase is generally three or four orders of magnitude higher than in solid membranes. Liquid membranes are separated into three types: bulk, emulsion, and supported. Bulk-liquid membranes are limited to laboratory use because such configuration is inadequate for practical use. Emulsion liquid membranes were originally proposed by Li (1968) and had an advantage of high permeation rates based on a high surface area per unit volume. Processes utilizing emulsion liquid membranes are more complex because emulsification and demulsification steps are needed before and after the permeation step using multiple emulsions. A supported liquid membrane (SLM), which is a porous solid structure impregnated with liquid, has also been studied. The SLM process has several advantages over other liquid membrane systems, such as simple configuration, low energy consumption, and a reduction in the amount of solvent used. The general characteristics and the current picture of SLM systems are reviewed in other chapters. This chapter focuses on ionic liquids as impregnating liquids. An SLM system using ionic liquids has some variations: ionic liquids diluted by solvents (Marták et al., 2008), the membranes of ionic liquids on porous alumina membranes (Huang et al., 2008), and composite membranes with cellulose triacetate (Jung et al., 2010) and polydimethylsiloxane (Izák et al., 2008). Since most previous research has been conducted using supported ionic liquid membranes (SILMs) absorbed into polymeric membranes, only this type will be described in this chapter. Here we briefly describe the characteristics of ionic liquids.

An ionic liquid is a liquid at room temperature that only consists of ions. Ionic liquids have bulky and asymmetric organic cations, such as 1-alkyl-3-methylimidazolium ([Rmim]), 1-alkylpyridinium, *N*-methyl-*N*-alkylpyrrolidinium, tetraalkylammonium, and tetraalkylphosphonium ions. The common anionic parts of ionic liquids are halides, nitrate, acetate, hexafluorophosphate ([$PF_6$]), tetrafluoroborate ([$BF_4$]), trifluorimethylsufonate ([OTf]), and bis(trifluoromethanesulfonyl)imide ([$Nf_2T$]). Figures 17.1 and 17.2 show the molecular structures of typical cations and the anion parts

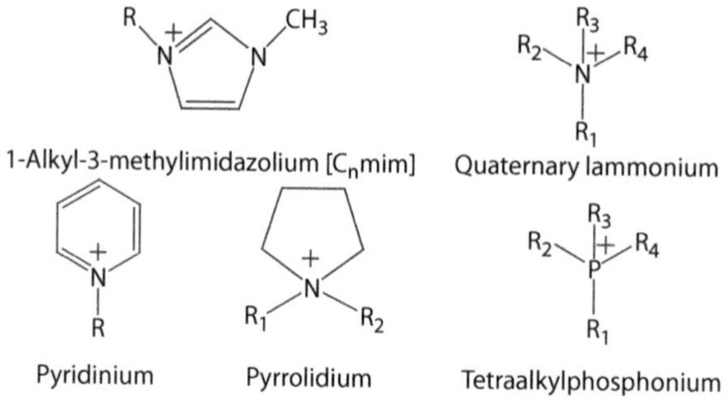

1-Alkyl-3-methylimidazolium [$C_n$mim]

Quaternary lammonium

Pyridinium

Pyrrolidium

Tetraalkylphosphonium

**FIGURE 17.1** Cation parts of ionic liquids.

of ionic liquids, respectively. They have extremely low vapor pressure and in many cases show high thermal stability. The miscibility of ionic liquids with water and organic solvent and such physicochemical properties as densities and viscosities can be controlled by changing the structures of cation and anion moieties. Table 17.1 shows a typical example of the effect of the anion parts in 1-methyl-3-butylimidazolium salt on the physicochemical properties (Olivier-Bourbigou and Magna, 2002). Currently, a number of practical applications are being considered. Plechkova and Seddon (2008) described how ionic liquids are leaving academic labs and entering a wide variety of industrial applications. Why do we use ionic liquids as membrane components in SLMs? In SLMs, membrane instability is recognized as a serious problem. Using an ionic liquid as an impregnating liquid is believed to be one efficient method for preventing membrane degradation. One reason the membrane can be stabilized is that the evaporation of impregnated solvents can be prevented due to the extremely low vapor pressure of the ionic liquids. Another possible reason is that the loss of membrane liquids under a cross-membrane pressure difference is depressed since ionic liquids are immobilized by a large van der Waals force and are not easily pushed out because of their high viscosities (Zhao et al., 2010).

Up to now, three classes of compounds have been transported by SILMs: (1) gases such as $CO_2$, (2) hydrophobic organic compounds such as aromatic hydrocarbon, and (3) hydrophilic biomolecules such as organic acids and amino acids. The latter two classes are concerned with liquid separation. In this chapter, the literature using SILMs is separately surveyed into gas and liquid separation systems, and the support material and permeation performance are reviewed as a whole. In addition, the integration of SILMs with an enzymatic reaction is briefly introduced as an example of an advanced SILM. However, analytical applications using ionic liquids are not included in this chapter. Please refer to review articles from this area: Pandey, 2007 and Han and Armstrong, 2007.

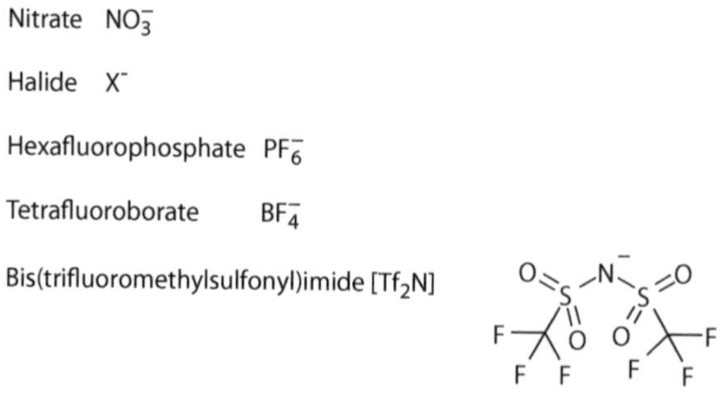

Nitrate $NO_3^-$

Halide $X^-$

Hexafluorophosphate $PF_6^-$

Tetrafluoroborate $BF_4^-$

Bis(trifluoromethylsulfonyl)imide [$Tf_2N$]

**FIGURE 17.2** Anion parts of ionic liquids.

**TABLE 17.1**
**Properties of 1-Butyl-3-Methylimidazolium Salt**

| Anion | Melting Point (°C) | Density (g cm$^{-3}$) | Viscosity (mPa s) | Conductivity (S m$^{-1}$) |
|---|---|---|---|---|
| $BF_4^-$ | −82 | 1.17 | 233 | 0.173 |
| $PF_6^-$ | −61 | 1.37 | 312 | 0.146 |
| $CF_3SO_3^-$ | 16 | 1.290 | 90 | 0.37 |
| $CF_3CO_2^-$ | −50 | 1.209 | 73 | 0.32 |
| $Tf_2N^-$ | −4 | 1.429 | 52 | 0.39 |

*Source:* Olivier-Bourbigou, H. and Magna, L., 2002, reproduced by permission of Elsevier.

## 17.2 SILMS FOR GAS/VAPOR SEPARATIONS

Research on using SILMs for gas separation dates to 1995 (Quinn et al.). A review was written on liquid membrane-based gas/vapor separation including SILMs by Krull et al. (2008). They claimed that ionic liquid or liquid molten salts improve membrane stability, although use of ionic liquids is still in the minority. As shown in Table 17.2, which summarizes recent SILM studies on gas and vapor separations after the above review, interest is increasing in using ionic liquids in gas separations, and only over the last two years use of ionic liquids has become in the majority. From Table 17.2, the present main task is the separation of $CO_2$ from the gas mixture due to high $CO_2$ solubility. $CO_2$ capture with ionic liquids was previously discussed in detail (Bara et al., 2009; Hasib-ur-Rahman et al., 2010). Other main tasks are dehumidification, the facilitated transport of olefin, and the vapor separation of aromatic hydrocarbons.

**TABLE 17.2**
**Gas/Vapor Separation with SILM**

| Gas/Vapor | Ionic Liquids | Support Materials | Reference |
|---|---|---|---|
| $CO_2/N_2$ | [bmim][$BF_4$] | Hydrophilic PES | Zhao, 2010 |
| $H_2$, $N_2$, $CO_2$, $CH_4$ | 12 ionic liquids including imidazolium, ammonium, phosphonium, and sulfonium | Hydrophobic PVDF | Cserjesi, 2010 |
| Benzene/cyclohexane | [$C_n$mim][$PF_6$], T, B | Hydrophilic PVDF | Matsumoto, 2009 |
| $CO_2$ ,$O_2$, $N_2$, $CH_4$ | [$C_n$mim][$Tf_2N$] | PES | Bara, 2009 |
| $CO_2$, $SO_2$ | [mim][Ac] [bim][Ac] | PVDF | Luis, 2009 |
| $CO_2$, $H_2$ | [$C_3NH_2$mim][$Tf_2N$] | Cross-linked nylon | Myers, 2008 |
| $CO_2$, $CH_4$ | [Bmim][$BF_4$] | PVDF | Park, 2009 |
| $CO_2$, $CH_4$, $N_2$ | [emim][$BF_4$],[emim][dca] [emim][$CF_3SO_3$],[emim][$Tf_2N$], [bmim][BETI] | Hydrophobic PTFE | Scovazzo, 2009 |
| $H_2$, $N_2$, $CO_2$ | VACEM type | Hydrophobic PVDF | Cserjesi, 2009 |
| $H_2$, $N_2$, $CO_2$ | [Cnmim][$PF_6$], [bmim][$BF_4$] | Hydrophobic PVDF | Neves, 2009 |
| $H_2O$, $CH_4$ | [emim][$Tf_2N$],[emim][dca], [emim][$BF_4$], [N(4)111][$Tf_2N$] | PES | Scovazzo, 2010 |
| Toluene/cyclohexane | [bmim][$PF_6$] | PVDF | Wang, 2008 |
| $H_2O$/propanol | | | |
| $H_2O$/ethanol | | | |
| $CO_2/N_2$ | [bmim][$N(CN)_2$], [bmim][$C(CN)_3$], [emim [$B(CN)_4$] | Hydrophilic PES | Mahurin, 2010 |

## 17.3 SILMS FOR LIQUID SEPARATIONS

As mentioned above, ionic liquids can be used as hydrophobic or hydrophilic membrane components by tuning the cation and anion parts. Therefore, both hydrophilic and hydrophobic compounds can be permeated through SILMs depending on the ionic liquids used. Table 17.3 summarizes the transport of liquid compounds with SILMs. In some papers, to enhance the permeation efficiency and selectivity, specific carriers were added to the membrane solution of ionic liquids. Although carriers are normally used for conventional organic solvents as the membrane solution, in some cases, the solubility of the extractants (tri-*n*-octylphosphine oxide and tri-*n*-octylamine) used in the metal extraction to the ionic liquid was reported to be poor (Matsumoto et al., 2004). On the other hand, ionic liquids can dissolve calixarenes that have little solubility in aliphatic solvents (Matsumoto et al., 2010b).

## 17.4 SILM PREPARATION

In the SLM process, the membrane solution plays the key role in permeation and selectivity. The role of membrane supports is also important in membrane permeation and its stability. The liquid membrane stability and the mechanical stability depend on such microstructures as pore shape, pore size, and tortuosity of the membrane support (Dżygıel and Wieczorek, 2009). Many types of polymeric and inorganic microporous membrane supports have been studied for the immobilization of membrane solution. However, as seen in Tables 17.2 and 17.3, few supports, including polyvinylidene fluoride (PVDF), polyethersulfonate (PES), and nylon have been frequently used as SILMs. PVDF membrane mainly acts as an inert solid support without establishing strong interactions with the substrates or with the ionic liquids because it displayed a low tendency to undergo van der Waals interactions (Branco et al., 2002). Hydrophilic polypropylene is also used as a flat-sheet membrane (Branco et al., 2002) and hollow-fiber membranes (Matsumoto et al., 2010a).

**TABLE 17.3**
**Liquid Separation with SILM**

| Substances | Ionic Liquids | Support Materials | Reference |
|---|---|---|---|
| Alkylamine, seven organic compounds | [bmim][$PF_6$] | Hydrophilic PVDF | Branco, 2002a,b |
| NaCl, thymol blue | [$C_n$mim][$PF_6$], [$C_n$mim][$BF_4$] | Hydrophilic PVDF | Fortunato, 2004 |
| Amino acid, amino acid ester | [$C_8$nmim][$PF_6$], | Hydrophilic PVDF | Fortunato, 2005 |
| Stability test | [$C_8$nmim][X], | Hydrophilic PVDF<br>Nylon | Fortunato, 2005; Hernández-Fernández, 2009b; de los Ríos, 2009 |
| Benzene, toluene, xylene | [$C_n$nmim][$PF_6$] | Hydrophilic PVDF | Matsumoto, 2005 |
| Benzene/cyclohexane | [$C_n$nmim][$PF_6$] | Hydrophilic PVDF | Matsumoto, 2006a,b, 2007b |
| Organic nitrogen and sulfer compounds | | | |
| Penicillin G | Aliquat 336 | Hydrophilic PVDF | Matsumoto, 2007a, 2010a |
| Lactic acid | | | |
| Saccharides | Aliquat 336, carrier calixarene | Hydrophilic PVDF | Matsumoto, 2010b |
| Toluene | [$C_8$mim][Cl], carrier Ag | Bulk | Chakraborty and Bart, 2007 |
| Rare earth metals | [$C_4$mim][$Tf_2N$], carrier DODGA | Hydrophobic PVDF | Kubota 2010 |
| Butanol, butyric ester | [$C_n$nmim][$PF_6$] | Nylon | Hernández-Fernández, 2007a, 2009c; de los Ríos, 2008a,b, 2009, 2010 |

The general immobilization method of ionic liquids to prepare SILMs is that a porous membrane is placed in the ionic liquids, allowing the membrane to soak up the ionic liquids. Recently, two methods for preparing SILMs using pressure- or vacuum-based techniques were examined (Hernández-Fernández et al., 2009b). The pressure-based technique immobilized the highest amount of ionic liquids within the membranes and offered the highest SILM stability. Immobilization under vacuum was concluded to be suitable for ionic liquids with low viscosity.

In the SLM process, the most popular modules are the hollow fiber and flat-sheet types. The main target in the design of membrane contactors is to maximize the mass transfer rate by producing as much interfacial area as possible and reducing the disadvantages of the two fluid phases to be contacted (Román et al., 2009). From this point, hollow fibers were applied in the separation of gas mixtures as well as liquid mixtures because it is difficult to scale up flat-sheet SLMs for industrial applications. Most of the research listed in Tables 17.2 and 17.3 was conducted using flat-sheet SILMs, perhaps caused by the difficulty of the immobilization of ionic liquids within the hollow-fiber membranes. Normally an organic membrane solution is pumped through the hollow fiber system, and the pores in the fibers are filled with the organic solution. Then the source and the receiving aqueous solutions are fed to the inner and outer sides of the hollow fibers to push the organic solution out; the membrane solution in the pores remains. This immobilization is considered unsuitable for ionic liquids due to their high viscosity. This is one subject to practically realize the SILM process.

## 17.5 EFFECT OF IONIC LIQUIDS ON SELECTIVITY AND PERMEABILITY

Permeation through SILMs follows the solution-diffusion mechanism (Bara et al., 2009b; Scovazzo, 2009), where permeability, $P$, is the product of its solubility, $S$, and diffusivity, $D$, in the membrane (Equation 17.1).

$$P = D \cdot S \tag{17.1}$$

Selectivity, $\alpha_{ij}$, between the two species concerned is defined as the ratio of permeabilities (Equation 17.2).

$$\alpha_{ij} = \frac{P_i}{P_j} = \frac{S_i}{S_j} \cdot \frac{D_i}{D_j} \tag{17.2}$$

For most pairs of gas ($CO_2/N_2$, $CO_2/CH_4$, etc.) and liquid organic compounds (benzene/cyclohexane, toluene/heptane), it can be assumed that $D_i \approx D_j$ (Bara et al., 2009b; Scovazzo, 2009). Therefore, solubility is the main factor in determining the permeation selectivity.

Some studies on choice of ionic liquid for separating $CO_2$ gas have been conducted (Bara et al., 2008; Scovazzo, 2009; Raeissi and Peters, 2009). According to the Camper and Kilaru viscosity models, the solubility is expressed by Equations 17.3 and 17.4 (Scovazzo, 2009).

$$S(\text{mol/L}) = \left[\left\{\exp(\alpha + \frac{\beta}{(V_{IL})^{4/3}}) - 1\right\} V_{IL}\right]^{-1} \tag{17.3}$$

where $\alpha$ and $\beta$ are the gas-specific parameters and $V_{IL}$ is the molar volume of the ionic liquid:

$$\ln S \approx B + C(\delta_{IL})^2 \tag{17.4}$$

and

$$\delta_{IL} = \left\{ \frac{K_v RT}{V_{IL}} \ln\left( \frac{1\times10^{-9}\mu V_{IL}}{hN_A} \right) \right\}^{1/2} \tag{17.5}$$

where B and C are gas-specific parameters, $\delta_{IL}$ and $\mu$ are solubility parameters and the kinetic viscosity of the ionic liquid, $h$ is the Planck constant, $N_A$ is Avogadro's constant, and $K_v$ is a proportional constant. From the above models, the molar volume and viscosity are considered.

Since viscosity has little effect on selectivity, the relationship between $CO_2/N_2$ selectivity and molar volume is shown in Figure 17.3, where the universal model represents the regression results expressed as:

$$\alpha_{CO_2,N_2} = \frac{2895}{(V_{IL})^{0.865}} \tag{17.6}$$

The experimental data in Figure 17.3 are well correlated with Equation 17.6 and the Camper model. Hence the selectivity of gas separation through SILMs was found to be determined by the molar volume of the ionic liquids.

Scovazzo (2009) also examined the effect of ionic liquids on permeability. He argued that the solute diffusivity in ionic liquid is less dependent on viscosity and more dependent on solute size than the conventional Stokes–Einstein model, and that correlating the diffusivities of all ionic liquids is difficult. However, since at least both diffusivity and solubility are functions of molar volume and viscosity, permeability is also a function of molar volume and viscosity. Figure 17.4 shows the relationship between the permeability and the viscosity of ionic liquids. This plot shows the dominance of viscosity in determining $CO_2$ permeability in SILMs.

Meindersma et al. (2005) studied the effect of ionic liquids on the separation of toluene from heptane by extraction. Their experimental results with toluene/heptane selectivity with 10 (v/v)% toluene in the feed at 40°C are shown in Figure 17.5 as a function of the distribution coefficient of toluene. In the figure, the vertical and horizontal lines represent the values using sulfolane as an extraction solvent, which is a conventional solvent for extracting aromatic compounds. From Figure 17.5, selectivity is found to be much larger than unity. It is well known that ionic liquids extract more

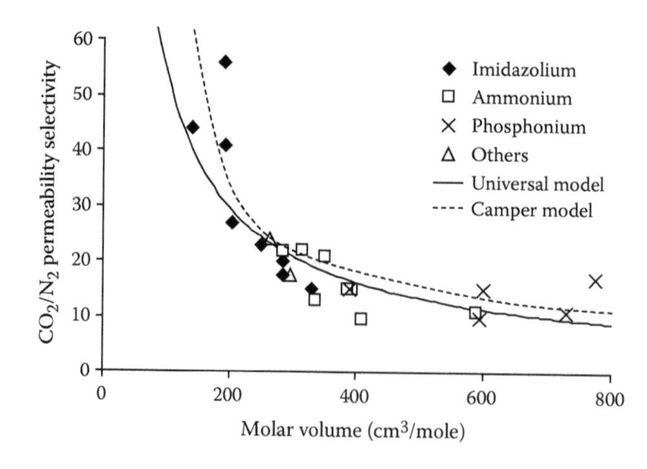


**FIGURE 17.3** Effect of molar volume of ionic liquids on $CO_2/N_2$ permeability selectivity. (Reprinted from *J. Membr. Sci.*, 343, Scovazzo, P., 199–211, Copyright 2009, with permission from Elsevier.)

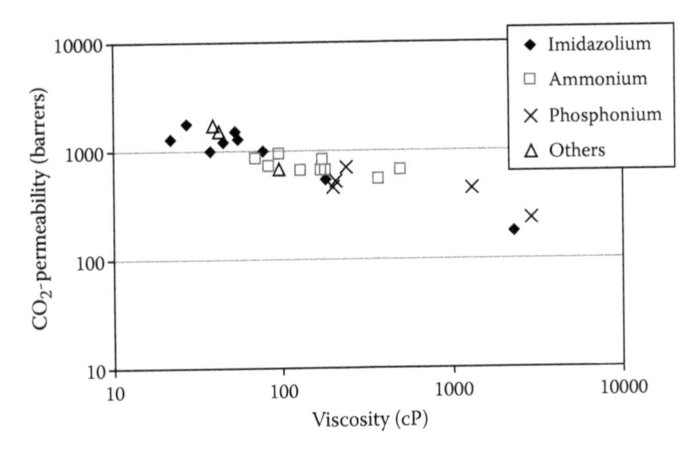


**FIGURE 17.4** Effect of viscosity of ionic liquids on $CO_2$ permeability. (Reprinted from *J. Membr. Sci.*, 343, Scovazzo, P., 199–211, Copyright 2009, with permission from Elsevier.)

aromatic compounds than aliphatic compounds because of the π-ionic interaction between ionic liquids and aromatic compounds (Meindersma et al., 2005). The following conclusions about the effect of ionic liquids were experimentally obtained (Meindersma et al., 2005; Anjan 2006): (1) a shorter alkyl group on imidazolium cations results in higher toluene/heptane selectivity because a longer alkyl chain trends to interact with aliphatic compounds; (2) the selectivity of ionic liquids with identical alkyl chain length and identical anions increases in the following order: pyridinium > imidazolium > phosphonium ~ quaternary ammonium > isoquinolinium; and (3) the effect of the anionic part on selectivity is not explained by the molar volumes of the anionic parts alone. In general, the effect of the anion parts on selectivity follows this trend: R-$SO_4$ > $BF_4$ > $PF_6$ > $Tf_2N$ > X. There is no systematic and theoretical research on the effect of ionic liquids on the permeability of liquid compounds.

## 17.6 EFFECT OF SOLUTES ON SELECTIVITY AND PERMEABILITY

As described above, based on the solution-diffusion model the permeability is expressed as Equation 17.1. The diffusivity in the ionic liquid phase is calculated by the empirical Wilke–Chang equation:

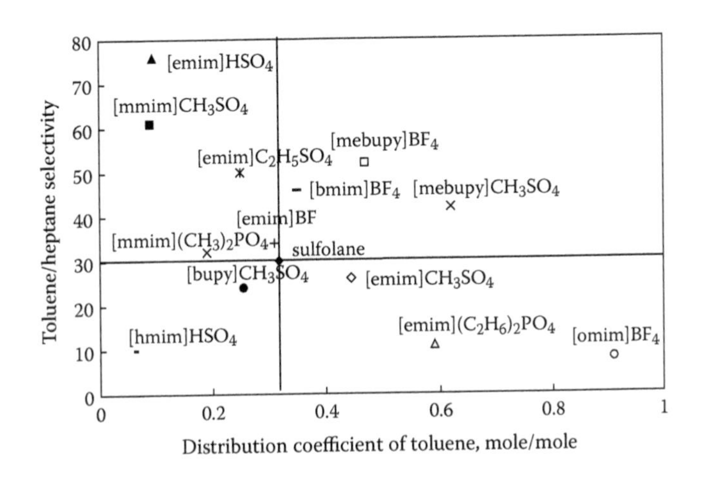


**FIGURE 17.5** Relationship between toluene/heptane selectivity and distribution coefficient of toluene. (Reprinted from *Fuel Process. Technol.*, 87, Meindersma, G.W., 59–70, Copyright 2005, with permission from Elsevier.)

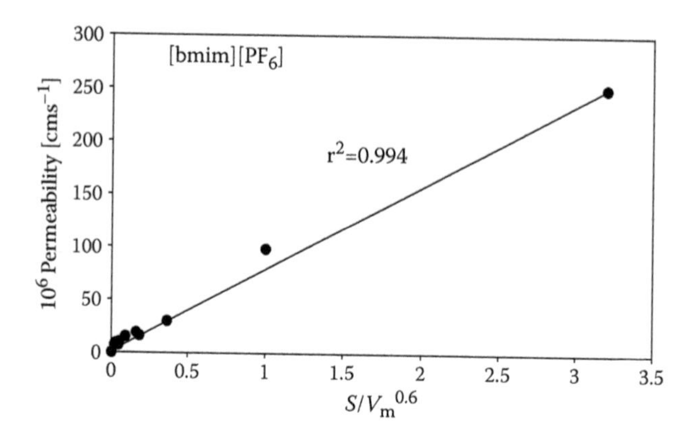


**FIGURE 17.6** Relationship between permeability of various organic compounds and $S/V_m^{0.6}$. (Reprinted from *J. Membr. Sci.*, 307, de los Rios, A.P., et al., 225–232, Copyright 2008a, with permission from Elsevier.)

$$D = 7.4 \times 10^{-8} \frac{\chi^{0.5} M^{0.5} T}{\eta V_m^{0.6}} \tag{17.7}$$

where $M$, $\chi$, and $\eta$ are the molecular weight, the association parameter, and the viscosity of the ionic liquid, respectively. $V_m$ is the molecular volume of the solute at boiling point and $T$ is the temperature.

Equation 17.8 is derived from Equations 17.1 and 17.7:

$$P = C \frac{S}{V_m^{0.6}} \tag{17.8}$$

where $C$ is a constant at a fixed temperature and a specific ionic liquid.

de los Ríos et al. (2008a) examined the validity of Equation 17.8; that is, the solution-diffusion model, for 16 different organic compounds using $[C_4mim][PF_6]$. Figure 17.6 shows a plot in relation to Equation 17.8. A straight line passes through the origin. Therefore selectivity is easily predicted by the following equation.

$$\alpha_{ij} = \frac{S_i / V_{m,i}^{0.6}}{S_j / V_{m,j}^{0.6}} \tag{17.9}$$

## 17.7 SILM STABILITY

Although SLMs have been widely studied for the separation and enrichment of various compounds and have many merits over other separation techniques, SLMs are not often used in large scale. The main reason is the membrane instability caused by the loss of the membrane solution from the membrane pore. Instability is usually attributed to the solubilization or evaporation of membrane liquid in the adjacent phase or its displacement from the pores under transmembrane pressure and/or osmotic pressure differences across the membrane (Wang et al., 2008). Such many factors as the support structure and the nature of the membrane liquid that influences SLM stability have been studied. Previous research clarified that organic solvents with high viscosity and high organic-water interfacial tension enhance stability (Dżygiel and Wieczorek, 2009). Ionic liquids fit these conditions. Furthermore, when the adjacent phases of the SILM are gas or vapor, ionic liquids with negligible vapor pressure may prevent the loss of membrane liquid from the membrane by evaporation and dissolution. Hanioka et al. (2008) reported a mixed gas separation of $CO_2/CH_4$ under humid conditions with a functionized ionic liquid membrane for 260 days with no observable change in permeance and selectivity. In cases of the separation of liquid compounds, although many papers characterize SILMs as more stable than SLMs, scant direct data exist for the membrane duration and the time courses of the flux and selectivity at present. Fortunato et al. (2004) reported the retention of $[C_4mim][PF_6]$ in membrane pores during 230 hours without leaking using water as a feed/receiving phase. de los Ríos et al. (2010) reported that selectivity remained constant during eight consecutive runs of 48-hour runs, each of which used $[C_4mim][BF_4]$ as an ionic liquid and hexane as a feed/receiving phase. Although the ionic liquids form two-phase with water or organic solvents, they are partially miscible in hydrocarbon, alcohol, and water. Hence, the ionic liquids physically immobilized in the membrane pores might eventually be lost, even if it takes much longer for the degradation of the membrane function than for immobilization using a conventional organic solvent. Therefore, for the practical separation of liquid compounds, we considered retaining ionic liquids in the pores of the membrane by the chemical interaction between ionic liquids and the support or using a polymer inclusion membrane (PIM) (Nghiem et al., 2006) to prevent direct contact between the ionic liquid and the adjacent liquid phase.

## 17.8 INTEGRATION OF SILM WITH ENZYMATIC REACTION

A novel membrane bioreactor system combining membrane transport using SILMs and enzymatic reactions was briefly introduced as an example of an advanced SILM. Such enzymes as lipase have been proved to be catalytically active in ionic liquids without much denaturation (Moniruzzaman et al., 2010). Miyako et al. (2003a) employed an SILM for a lipase-facilitated transport system of ibuprofen. *S*-ibuprofen is selectively converted to *S*-ibuprofen ester with lipase CRL in the feed phase. The resultant ester, having diffused across the SILM, is hydrolyzed to the original *S*-ibuprofen by another lipase PPL dissolved in the receiving phase (Figure 17.7). They also described SILM as stable up to 120 hours, while the isooctane membrane was broken after 75 hours (Miyako et al., 2003b). Recently, the kinetic resolution of 1-phenylethanol was reported based on a similar principle (Hernández-Fernández et al., 2007b, 2009a).

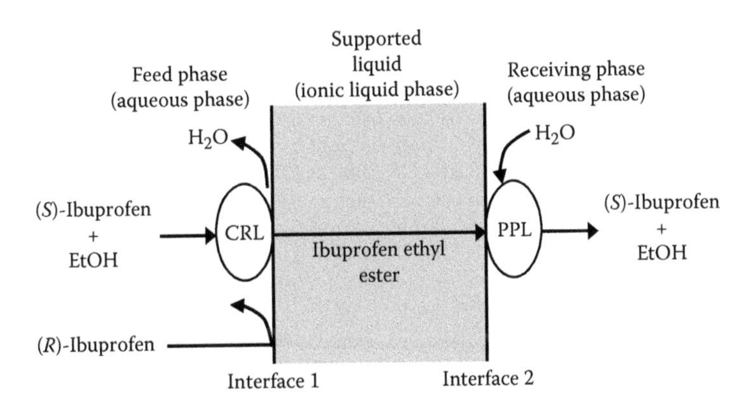


**FIGURE 17.7** Schematic diagram of enantioselective transport of *S*-ibuprofen through a lipase-facilitated SILM. CRL, lipase to promote esterification reaction; PPL, lipase to promote hydrolysis reaction. (From Miyako, E., et al., 2003a, *Chem. Commun.* 2926–2927; reproduced by permission of the Royal Society of Chemistry.)

## 17.9 FUTURE PERSPECTIVES

Research using SILMs is almost 10 years old, but exponential growth is just beginning. Over the past 2 years many papers have been published, especially concerning gas separation. Practical applications of SLM processes have been limited due to membrane instability. Many configurations using a membrane contactor have been developed to overcome problems related to membrane instability (Dżygıel and Wieczorek, 2009). Using ionic liquids as a membrane solution is a simple answer to such problems. Especially in gas/vapor separation, the nonvolatile nature of ionic liquids should allow for significant improvement of the current processes and the development of new approaches to gas/vapor separation, and a fair amount of effort will be needed for ionic liquids to become commercially viable (Scovazzo, 2009). In liquid separation, the problems of membrane instability remain despite the considerable improvement compared with SLMs using the conventional organic solvent. The physical dissolution of ionic liquids to adjacent liquid phases is in principle inevitable. When the dissolution of ionic liquids is not negligible, countermeasures are required. On this occasion, PIMs will become an excellent alternative. They are formed by casting a solution that contains an ionic liquid and a base polymer such as CTA and poly (vinyl chloride) to form a thin, flexible, stable film. Although diffusivity is expected to be lower in PIMs than in SILMs, this disadvantage can be easily offset by creating a much thinner membrane than its traditional SLM counterpart (Nghiem et al., 2006).

In addition to the above technological problems, other problems remain. The reputation of ionic liquids as an environmentally friendly solvent is based on their negligible vapor pressure. This opposite nature indicates the possibility of the accumulation and enrichment of ionic liquids in the environment. Therefore, accurate knowledge about the behavior of ionic liquids in terrestrial environments is required as well as the properties of ionic liquids as solvents (Pham et al., 2010).

## REFERENCES


Anjan, S.T., 2006. Ionic liquids for aromatic extraction: Are they ready? 2006. *Chem. Eng. Prog.* 2(12): 3–39.

Bara, J.E., C.J. Gabriel, T.K. Carlisle, et al. 2009a. Gas separations in fluoroalkyl-functionalized room-temperature ionic liquids using supported liquid membranes. *Chem. Eng. J.* 14: 43–50.

Bara, J.E., T.K. Carlisle, et al. 2009b. Guide to $CO_2$ separations in imidazolium-based room-temperature ionic liquids. *Ind. Eng. Chem. Res.* 48: 2739–2751.

Branco, L.C., J.G. Crespo, and C.A.M. Afonso. 2002a. Highly selective transport of organic compounds by using supported liquid membranes based on ionic liquids. *Angew. Chem. Int. Ed.* 41: 2771–2773.

Branco, L.C., J.G. Crespo, and C.A.M. Afonso. 2002b. Studies on the selective transport of organic compounds by using ionic liquids as novel supported liquid membranes. *Chem. Eur. J.* 8: 3865–3871.

Chakraborty, M., and H. Bart. 2007. Highly selective and efficient transport of toluene in bulk ionic liquid membranes containing $Ag^+$ as carrier. *Fuel Process. Technol.* 88: 43–49.

Cserjési, P., N. Nemestóthy, A. Vass, et al. 2009. Study on gas separation by supported liquid membranes applying novel ionic liquids. *Desalination* 245: 743–747.

Cserjési, P., N. Nemestóthy, and K. Bélafi-Bakó. 2010. Gas separation properties of supported liquid membranes prepared with unconventional ionic liquids. *J. Membr. Sci.* 349: 6–11.

Dżygıel P., and P.P. Wieczorek. 2009. Supported liquid membranes and their modifications: definition, classification, theory, stability, application and perspective, pp. 73–140, in *Liquid Membranes*. V.S. Kislik, ed., Oxford: Elsevier.

Fortunato, R., C.A.M. Afonso, M.A.M. Reis, et al. 2004. Supported liquid membranes using ionic liquids: Study of stability and transport mechanisms. *J. Membr. Sci.* 242: 197–209.

Fortunato, R., M.J. González-Muñoz, M. Kubasiewicz, et al. 2005. Liquid membranes using ionic liquids: The influence of water on solute transport. *J. Membr. Sci.* 249: 153–162.

Fortunato, R., C.A.M. Afonso, J. Benavente, et al. 2005. Stability of supported ionic liquid membranes as studied by X-ray photoelectron spectroscopy. *J. Membr. Sci.* 256: 216–223.

Han, X., and D.W. Armstrong. 2007. Ionic liquids in separations, *Acc. Chem. Res.* 40: 1079–1086.

Hanioka, S., T. Maruyama, T. Sotani, et al. 2008. $CO_2$ separation facilitated by task-specific ionic liquids using a supported liquid membrane. *J. Membr. Sci.* 314: 1–4.

Hasib-ur-Rahman, M., M. Siaj, and F. Larachi. 2010. Ionic liquids for $CO_2$ capture. *Chem. Eng. Process.* 49: 313–322.
Hernández-Fernández, F.J., A.P. de los Ríos, M. Ribio, et al. 2007a. A novel application of supported liquid membranes based on ionic liquids to the selective simultaneous separation of the substrates and products of a transesterification reaction. *J. Membr. Sci.* 293: 73–80.
Hernández-Fernández, F.J., A.P. de los Ríos, F. Thomás-Alonso, et al. 2007b. Integrated reaction/separation processes for the kinetic resolution of rac-1-phenylethanol using supported liquid membranes based on ionic liquids. *Chem. Eng. Process.* 46: 818–824.
Hernández-Fernández, F.J., A.P. de los Ríos, F. Thomás-Alonso, et al. 2009a. Kinetic resolution of 1-phenylethanol integrated with separation of substrates and products by a supported ionic liquid membrane. *J. Chem. Technol. Biotechnol.* 84: 337–342.
Hernández-Fernández, F.J., A.P. de los Ríos, F. Thomás-Alonso, et al. 2009b. Preparation of supported ionic liquid membranes: Influence of the ionic liquid immobilization method on their operational stability. *J. Membr. Sci.* 341: 172–177.
Hernández-Fernández, F.J., A.P. de los Ríos, F. Thomás-Alonso, et al. 2009c. Improvement in the separation efficiency of transesterification reaction compounds by the use of supported ionic liquid membranes based on the dicyanamide anion. *Desalination* 244: 122–129.
Huang, J., H. Luo, C. Liang, et al. 2008. Advanced liquid membranes based on novel ionic liquids for selective separation of olefin/paraffin via olefin-facilitated transport. *Ind. Eng. Chem. Res.* 47: 881–888.
Izák, P., K. Schwarz, W. Ruth, et al. 2008. Increased productivity of *Clostridium acetobutylicum* fermentation of acetone, butanol, and ethanol by pervaparation through supported ionic liquid membranes. *Appl. Microbiol. Biotechnol.* 78: 597–602.
Jung, S., J. Palgunadi, J.H. Kim, et al. 2010. Highly efficient metal-free membranes for the separation of acetylene/olefin mixtures: pyrrolidium-based ionic liquids as acetylene transport carrier. *J. Membr. Sci.* 354: 509–519.
Krull, F.F., C. Fritzmann, and T. Melin. 2008. Liquid membranes for gas/vapor separations. *J. Membr. Sci.* 325: 509–519.
Kubota, F., Y. Shimobori, Y. Koyanagi, et al. 2010. Uphill transport of rare-earth metals through a highly stable supported liquid membrane based on ionic liquid. *Anal. Sci.* 26: 289–290.
Li, N.N. 1968. Separating hydrocarbon with liquid membranes. US Patent 3,410,794.
Luis, P., L.A. Neves, C.A.M. Afonso, et al. 2009. Facilitated transport of $CO_2$ and $SO_2$ through supported ionic liquid membranes (SILMs). *Desalination* 245: 485–493.
Mahurin, S.M., J. Lee, G.A. Baker, et al. 2010. Performance of nitrile-containing anions in task-specific ionic liquids for improved $CO_2/N_2$ separation. *J. Membr. Sci.* 353: 177–183.
Marták, J., Š. Schlosser, and S. Vlčková. 2008. Pertraction of lactic acid through supported liquid membranes containing phosphonium ionic liquid. *J. Membr. Sci.* 318: 298–310.
Matsumoto, M., K. Mochiduki, K. Fukunishi, et al. 2004. Extraction of organic acids using imidazolium-based ionic liquids and their toxicity to *Lactobacillus rhamnosus*. *Separ. Purif. Technol.* 40: 97–101.
Matsumoto, M., Y. Inomoto, and K. Kondo. 2005. Selective separation of aromatic hydrocarbons through supported liquid membranes based on ionic liquids. *J. Membr. Sci.* 246: 77–81.
Matsumoto, M., K. Ueba, and K. Kondo. 2006a. Separation of benzene/cyclohexane mixture through supported liquid membranes with an ionic liquid. *Solv. Extr. Res. Develop., Japan* 13: 51–59.
Matsumoto, M., M. Mikami, and K. Kondo. 2006b. Separation of organic nitrogen compounds by supported liquid membranes based on ionic liquids. *J. Japan Pet. Inst.* 49: 256–261.
Matsumoto, M., T. Ohtani, and K. Kondo. 2007a. Comparison of solvent extraction and supported liquid membrane permeation using an ionic liquid for concentrating penicillin G. *J. Membr. Sci.* 289: 92–96.
Matsumoto, M., M. Mikami, and K. Kondo. 2007b. Selective permeation of organic sulfur and nitrogen compounds in model mixture of petroleum fraction through supported ionic liquid membranes. *J. Chem. Eng. Japan* 40: 1007–1010.
Matsumoto, M., K. Ueba, and K. Kondo. 2009. Vapor permeation of hydrocarbons through supported liquid membranes based on ionic liquids. *Desalination* 241: 365–371.
Matsumoto, M., W. Hasegawa, K. Kondo, et al. 2010a. Application of supported ionic liquid membranes using a flat sheet and hollow fibers to lactic acid recovery. *Desal. Water Treat.* 14: 37–46.
Matsumoto, M., T. Nobuyasu, and K. Kondo. 2010b. Effect of carriers on the transport of saccharides by supported ionic liquid membranes. *Solv. Extr. Res. Develop., Japan* 17: 249–253.
Meindersma, G.W., A. Podt, and A.B. de Haan. 2005. Selection of ionic liquids for the extraction of aromatic hydrocarbons from aromatic/aliphatic mixtures. *Fuel Process. Technol.* 87: 59–70.

Miyako, E., T. Maruyama, N. Kamiya, et al. 2003a. Enzyme-facilitated enantioselective transport of (S)-ibuprofen through a supported liquid membrane base on ionic liquids. *Chem. Commun.* 2926–2927.

Miyako, E., T. Maruyama, N. Kamiya, et al. 2003b. Use of ionic liquids in a lipase-facilitated supported liquid membrane. *Biotechnol. Lett.* 25: 805–808.

Moniruzzaman, M., K. Nakashima, N. Kamiya, et al. 2010. Recent advances of enzymatic reactions in ionic liquids. *Biochem. Eng. J.* 48: 295–314.

Myers, C., H. Pennline, D. Luebke, et al. 2008. High temperature separation of carbon dioxide/hydrogen mixtures using facilitated supported ionic liquid membranes. *J. Membr. Sci.* 322–331.

Neves, L.A., N. Nemestóthy, V.D. Alves, et al. 2009. Separation of biohydrogen by supported ionic liquid membranes. *Desalination* 240: 311–315.

Nghiem, L.D., P. Mornane, I.D. Potter, et al. 2006. Extraction and transport of metal ions and small organic compounds using polymer inclusion membranes (PIMs). *J. Membr. Sci.* 281: 7–41.

Olivier-Bourbigou, H., and L. Magna. 2002. Ionic liquids: perspectives for organic and catalytic reactions. *J. Mol. Catal. A.* 182/183: 419–437.

Pandey, S. 2007. Analytical application of room-temperature ionic liquids: A review of recent efforts. *Anal. Chim. Acta.* 556: 38–45.

Park, Y., B. Kim, Y. Byun, et al. 2009. Preparation of supported ionic liquid membranes (SILM) for the removal of acidic gases from crude natural gas. *Desalination* 236: 342–348.

Pham, T.P.T., C. Cho, and Y. Yun. 2010. Environmental fate and toxicity of ionic liquids: A review. *Water Res.* 44: 352–372.

Plechkova, N.V., and K.R. Seddon. 2008. Applications of ionic liquids in the chemical industry. *Chem. Soc. Rev.* 37: 123–150.

Quinn, R., J.B. Appleby, and G.P. Pez. 1995. New facilitated transport membranes for the separation of carbon dioxide from hydrogen and methane. *J. Membr. Sci.* 104: 139–146.

Raeissi, S., and C.J. Peters. 2009. A potential ionic liquid for $CO_2$ separating gas membranes: selection and gas solubility studies. *Green Chem.* 11: 185–192.

de los Ríos, A.P., F.J. Hernández-Fernández, M. Ribio, et al. 2008a. Prediction of the selectivity in the recovery of transesterification reaction products using supported liquid membranes based on ionic liquids. *J. Membr. Sci.* 307: 225–232.

de los Ríos, A.P., F.J. Hernández-Fernández, F. Thomás-Alonso, et al. 2008b. On the importance of the nature of the ionic liquids in the selective simultaneous of the substrates and products of a transesterification reaction through supported ionic liquid membranes. *J. Membr. Sci.* 307: 233–238.

de los Ríos, A.P., F.J. Hernández-Fernández, F. Thomás-Alonso, et al. 2009. Stability studied of supported liquid membranes based on ionic liquids: Effect of surrounding phase nature. *Desalination* 245: 776–782.

de los Ríos, A.P., F.J. Hernández-Fernández, H. Presa, et al. 2009. Tailoring supported ionic liquid membranes for the selective separation of transesterification reaction compounds. *J. Membr. Sci.* 328: 81–85.

de los Ríos, A.P., F.J. Hernández-Fernández, M. Rubio, et al. 2010. Highly selective transport of transesterification reaction compounds through supported liquid membranes containing ionic liquids based on the tetrafluoroborate anion. *Desalination* 250: 101–104.

Román M.F.S., E. Bringas, R. Ibañez, et al. 2009. Liquid membrane technology: Fundamentals and review of its application. *J. Chem. Technol. Biotechnol.* 85: 2–10.

Scovazzo, P., D. Havard, M. McShea, et al. 2009. Long-term continuous mixed-gas dry fed $CO_2/CH_4$ and $CO_2/N_2$ separation performance and selectivities for room temperature ionic liquid membranes. *J. Membr. Sci.* 327: 41–48.

Scovazzo, P. 2009. Determination of the upper limits, benchmarks, and critical properties for gas separations using stabilized room temperature ionic liquid membranes (SILMs) for the purpose of guiding future research. *J. Membr. Sci.* 343: 199–211.

Scovazzo, P. 2010. Testing and evaluation of room temperature ionic liquid (RTIL) membranes for gas dehumidification. *J. Membr. Sci.* 355: 7–17.

Wang, B., J. Lin, F. Wu, et al. 2008. Stability and selectivity of supported liquid membranes with ionic liquids for the separation of organic liquids by vapor permeation. *Ind. Eng. Chem. Res.* 47: 8355–8360.

Zhao, W., G. He, L. Zhang, et al. 2010. Effect of water in ionic liquid on the separation performance of supported ionic liquid membrane for $CO_2/N_2$. *J. Membr. Sci.* 350: 279–285.

# 18 Solving Challenging Industrial Separation Problems through Electrodialysis

*S. Sridhar, R. Surya Murali, and A. A. Khan*

## CONTENTS

## 18.1 INTRODUCTION

### 18.1.1 Background

Electrodialysis (ED) is a membrane process in which ionic species present in aqueous solutions can be separated or concentrated by application of electrical potential across charged polymeric membranes. Compounds that can be separated are inorganic and organic salts, mineral and organic acids, alkalis, and even hazardous chemicals such as sodium azide, which can ionize in aqueous medium. There are two different types of ion-exchange membranes: cation exchange membrane (CEM) where the polymer matrix carries negatively charged groups and anion exchange membrane (AEM) that contain positively charged groups. This process is widely used for desalination of brackish water, production of potable water, and table salt. In general, AEMs and CEMs are placed alternatively in the ED stack whereas in certain specific cases only one type of membrane is used. Bipolar membranes are advanced versions that have a CEM laminated on one side and an AEM on the other side of the supporting fabric. Bipolar membranes possess the unique water splitting ability that results in the production of acids and bases during the desalination process, enhancing overall economy.

The history of ED began in 1890 when Maigrot and Sabates tried to demineralize sugar syrup. But the term electrodialysis was proposed for the first time in the beginning of the 20th century when attempts were successfully made to purify sugar syrup using soluble zinc or iron anode with simultaneous ozonation of the solutions (Schollomeyer, 1902). The term electrodialysis refers to the combination of two processes—electrolysis, which refers to passage of an electric current through an ionic solution for separation of chemical species at electrodes, and dialysis, wherein a solute is driven across a semipermeable membrane due to concentration gradient. Research on modern membranes began way back in 1925 through studies on the properties of colloidal membranes (Molau, 1981). With the advent of synthetic ion exchange membranes based on phenol formaldehyde, polycondensation resins, crosslinked polystyrene, and perfluorinated ionomers, ED could compete with conventional separation techniques such as distillation, adsorption, and reverse osmosis. With growing improvement in thermal and chemical stability of the membranes, ED could later be introduced into food and pharmaceutical industries.

### 18.1.2 Theoretical Aspects

#### 18.1.2.1 Principle of Electrodialysis

##### *18.1.2.1.1 Conventional Configuration with Membranes of Opposite Charges*

The principle of conventional unipolar ED is described in Figure 18.1. Electrically charged CEM and AEM are placed alternatively between cathode and anode for the removal of ions from feed solution. A set of one anion and one cation membrane forms a cell pair. The space between two membranes forms a chamber or compartment. The compartments from where the ions migrate (or feed is introduced) are called the diluate chambers and compartments that collect these ions are the concentrate chambers. Diluate and concentrate chambers alternate throughout the stack just as cation and anion membranes. The feed solution is introduced in the diluate chambers and finally gets depleted off the ions. When direct current is applied, the positively charged ions migrate to the cathode and the negatively charged ions migrate toward the anode. CEM allows cations to pass through, which are then retained by the AEM. Similarly, anions pass through the AEM but get retained by the CEM. This results in enrichment and depletion of ions in the alternate chambers. Gases liberated at anode and cathode are continuously removed by an electrode rinse solution. Depending on

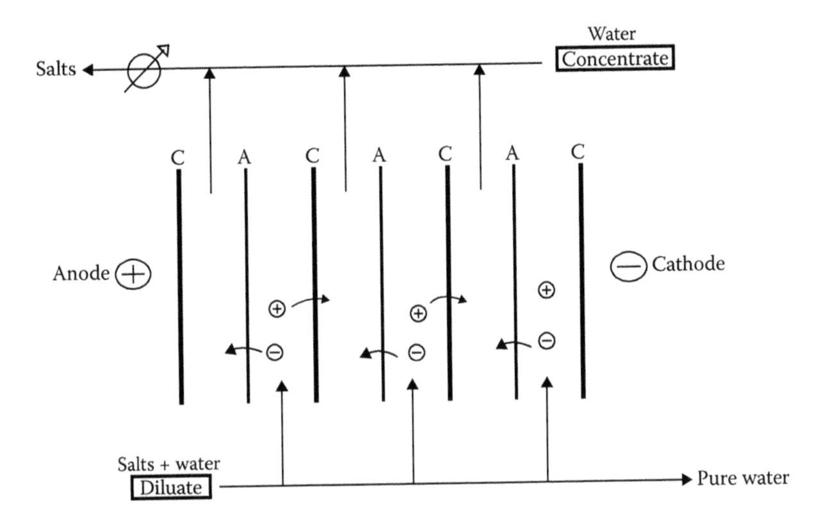


**FIGURE 18.1** Principle of conventional electrodialysis.

the value of the isolated compound, the diluate is generally the product and the concentrate is either a by-product or effluent.

#### *18.1.2.1.2 Configuration of Membranes with One Type of Charge*

Unipolar ED with membranes of one type of charge has its own applications such as deacidification of fruit juices (Rautenbach and Albrecht, 1989). The principle is explained in Figure 18.2. In this process either AEMs or only CEMs are stacked. Figure 18.2a shows only CEMs placed between the two electrodes for replacement of $Ca^{+2}$ ions present in hard water with $Na^{+}$ ions for water softening. Feed that contains the bivalent cations and the solution that provides the monovalent cations are pumped into alternate chambers. When electric potential is applied, the monovalent ions start moving toward the cathode through CEMs and replace the bivalent ones. Gases liberated at cathode are removed with rinsing solution. Similarly, when only AEMs are stacked, multivalent anions can be replaced. In Figure 18.2b, AEMs are placed between electrodes with grape juice and NaOH solutions distributed in alternate chambers. The acidic nature of grape juice is reduced by exchange of $C_6H_5O_7^-$ ions with $OH^-$ through AEMs.

#### *18.1.2.1.3 Configuration with Bipolar Membranes*

ED based on bipolar membranes (EDBMs) is an efficient process for generating acids and bases without undesirable by-products. Bipolar membrane consists of a CEM and an AEM laminated together and have the ability to split the water molecule into protons and hydroxyl ions. Figure 18.3 represents a three-compartment EDBM process. Salt solution (MX) is passed in the diluate chamber between individual CEMs and AEMs, which are also stacked along with the bipolar membranes. The $M^+$ ions migrate through the CEM but get retained at the AEM side of the bipolar membrane where they combine with the hydroxyl ions to form a basic MOH solution. Similarly, acid solution HX gets formed at the cation side of the bipolar membrane due to contact between $H^+$ ions and the migrating $X^-$ ions coming through AEM. The desalted water exits the diluate chambers and can be recycled or reused. Acid and alkali exit the other two compartments. Similarly, two-compartment EDBM can also be employed by combining bipolar membranes with either CEMs or AEMs.

### 18.1.2.2 Permselectivity

Ion permselectivity is the affinity of an ion toward a membrane and its exchange rate in comparison to other ions present in the solution. In general, a standard ion is taken as reference such as chloride

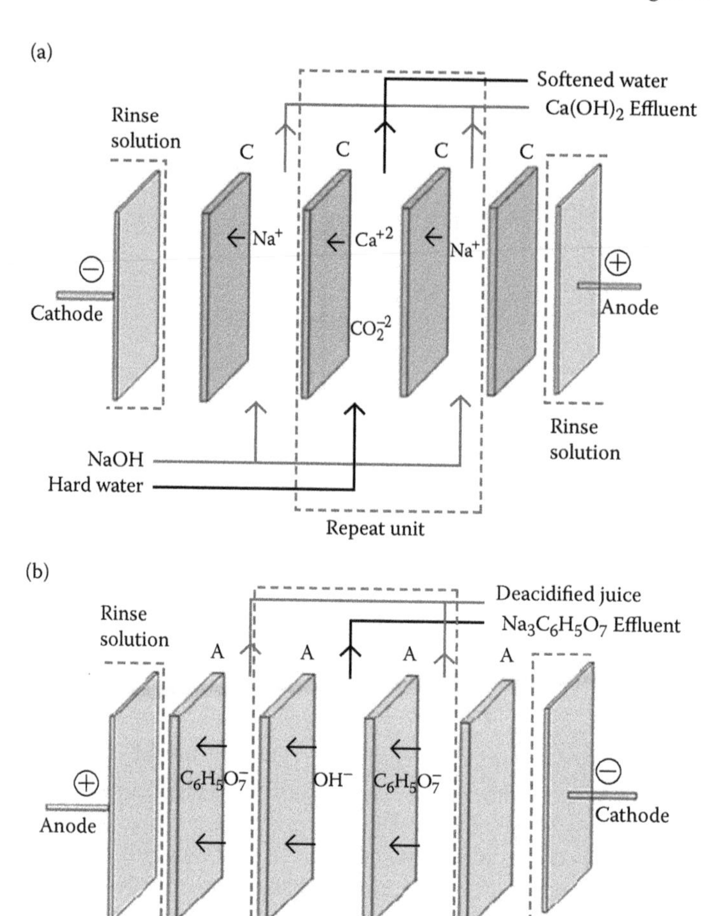


**FIGURE 18.2** Principle of (a) cation membrane electrodialysis and (b) anion membrane electrodialysis.

ion for anion exchange and sodium for cation exchange. The ratio of the exchange rate of any other specific ion with respect to the reference ion is determined. When an ion exchange membrane is placed in an aqueous electrolyte solution, it gets swollen due to water sorption and its functional group absorbs counter ions present in the solution. Application of potential difference generates a flux $J_i$ of counter ions across the membrane, which is given by Nernst–Planck equation, neglecting diffusion and convection terms (Tanaka, 2007).

$$J_i = -Z_i U_i \bar{C}_i \frac{dE}{dx} \tag{18.1}$$

where $\bar{C}_i$ is the ion concentration in the membrane phase during electrodialysis (equiv/cm$^3$), $U_i$ is the mobility of ions in the membrane (cm$^2$/(d$E$/d$x$)$s$), and (d$E$/d$x$) is the potential gradient.

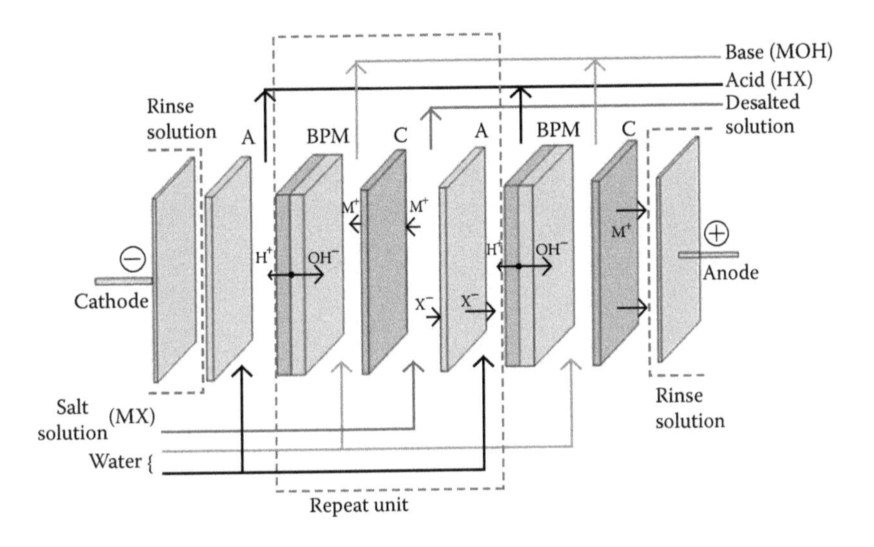


**FIGURE 18.3** Principle of bipolar electrodialysis.

Transport number $t_i$, of an ion $i$, is given by:

$$t_i = \frac{Z_i J_i}{\sum Z_i J_i} = \frac{Z_i^2 U_i \bar{C}_i}{\sum Z_i^2 U_i \bar{C}_i} \tag{18.2}$$

If ions $A$ and $B$ are exchanged across the membrane, then the transport number of $A$ relative to $B$ is given by permselectivity $P_B^A$

$$P_B^A = \frac{t_A / t_B}{C_A / C_B} = \frac{Z_A U_A \left(\bar{C}_A / \bar{C}_B\right)}{Z_B U_B (C_A / C_B)} \tag{18.3}$$

The ion exchange equilibrium constant $K_B^A$ is expressed as:

$$K_B^A = \frac{\left(\bar{C}_A / \bar{C}_B\right)}{(C_A / C_B)} \tag{18.4}$$

where $C_A$ and $C_B$ are the concentrations of ions $A$ and $B$, respectively, on the diluate side of the membrane in equiv/cm$^3$. Permselectivity can be expressed in terms of equilibrium constant as:

$$P_B^A = \frac{Z_A U_A}{Z_B Z_B} K_B^A \tag{18.5}$$

#### 18.1.2.3 Energy and Efficiency

The energy required in electrodialysis is the electric current that enables the exchange of ions across the membranes from one solution to the other. Electric power is also necessary for pumping the concentrate, diluate, and rinse solutions continuously into the stack from storage tanks. In practice, electrical energy utilized for a cell is more than the calculated theoretical energy due to ohmic resistances, boundary layer effect, and concentration polarization. Current utilization efficiency is the measure of how effective ions are exchanged across the membrane for a given applied current. The efficiency is primarily influenced by the stack design, flow rates, and concentration of ions in the feed. For a given stack and feed composition, the current utilization efficiency (η) is given by:

$$\eta = \frac{Z_i F q_f (C_{id} - C_{od})}{iN} \tag{18.6}$$

where $Z_i$ is the charge number, $F$ is Faraday's constant (96,485 Amp-s/mol), and $q_f$ is feed flow rate (L/s), $C_{id}$ is diluate inlet concentration (mol/L), $C_{od}$ is diluate outlet concentration (mol/L), $i$ is applied current in amps, and $N$ is the number of cell pairs. To minimize operating cost in a commercial stack, a current efficiency of >80% is desirable. Low current efficiencies indicate water splitting in the diluate or concentrate streams, inefficient or corroded electrodes, or back diffusion from the concentrate to the diluate.

### 18.1.3 Applications of ED

The early applications of ED were desalination for recovery of salt from seawater and production of ultrapure water. ED has played a major role in the area of wastewater minimization and especially in metal finishing and electroplating industries (Robert, 1995). Due to high conductivity of metal ions, ED is effectively used to separate metals like nickel, gold, silver, copper, cadmium, platinum, zinc, tin, and other metals from electroplating solutions. Concentration and purification of chromate by a two-stage ED process is well established. Removal of boron, fluoride, and nitrate from ground water besides separation of Pb and Mn (II) ions from wastewater are some popular applications. In more recent times, ED has been employed for concentration of dilute solutions, separation of electrolytes and nonelectrolytes, as well as production of alkalis and acids from their salts. This technique has also been applied for demineralization of sugar and milk products such as whey, treatment of amino acid and blood, removal of trace inorganic contaminants from aqueous process streams, concentration of mineral acids, preparation of isotonic solutions, and stabilization of wine (Sadrzadeh et al., 2007). Production of organic acids is one of the major applications of ED in both food and chemical industries (Chuanhui et al., 2007; Mathieu, 2002).

With the advent of bipolar membranes, ED has been used for the economical production of formic acid, acetic acid, propionic acid, lactic acid, p-toluenesulfonic acid, salicylic acid, itaconic acid, vitamin C, gluconic acid, and citric acid (Huang et al., 2007; Nikbakht et al., 2007), as well as dairy products such as bovine milk casein. ED has also been applied to concentrate organic salts like lysine, gluconate, propionate, pyruvate, and lactate and for the demineralization of organic acids.

ED has offered an improvised way to deacidify tropical fruit juices without affecting their aroma and flavor compared to conventional precipitation methods that involve addition of extra agents or ion-exchange resins that generate effluents during regeneration (Vera et al., 2009). Based on the specific type of ion and ionization potential, operating conditions can be altered for separation of monovalent, divalent, and multivalent ions from aqueous solutions at various stages of the process (Sadrzadeh et al., 2007). ED has been combined with conventional equilibrium driven processes to achieve desired product purities and integrated with other membrane processes such as microfiltration to facilitate water recovery from paper industry effluent (Natraj et al., 2007). Some of the commercial ED plants commissioned for various applications are listed in Table 18.1 along with details of the membranes used.

**TABLE 18.1**
**Examples of Industrial Applications of Electrodialysis**

| Membrane Suppliers | Membrane Trade Name | Application | Details of Installed Plant[a] |
|---|---|---|---|
| Eurodia,[b] France | Neosepta CM1 (CEM)<br>Neosepta AM1 (AEM) | Recovery of salts from concentrated brine | Pilot plant |
| Eurodia, France | CMX-Sb (CEM)<br>ACS (AEM) | Nitrate removal from drinking water | Austria<br>3320 $m^3$/day |
| Asahi Glass, Japan | Selemion | Desalination of brackish water | Commercial plant |
| Eurodia, France | CMX-Sb (CEM)<br>AXE-01 (AEM) | Removal of melassigenic ions ($Na^+$, $K^+$, $Ca^{2+}$) from beet sugar syrups | Pilot plant<br>24 $m^3$/day |
| Du Pont, United States, and Asahi Glass, Japan | Nafion 450 (CEM)<br>Selemion AMP (AEM) | Treatment of metal finishing water | Pilot plant |
| Du Pont, United States, and Solvay, United States | Nafion®117 (CEM)<br>ARA (AEM) | Recovery of sulfuric acid from effluents | Commercial plant |
| FuMA-Tech GmbH, Germany | Fumasep FKB (CEM)<br>Fumasep FAB (AEM) | Synthesis of [EMIM]-based ionic liquids | 1166.5 kg/day |
| Eurodia, France | Neosepta CM1 (CEM)<br>Neosepta AM1 (AEM) | Demineralization of concentrated whey | United States<br>6000 L/h |
| Eurodia, France | Neosepta CM1 (CEM)<br>Neosepta AM1 (AEM) | Production of boiler feed water by reverse ED | Netherlands<br>1472 $m^3$/day |

[a] Available information regarding scale of operation, plant location, and capacity.
[b] The main shareholder of Eurodia Industrie is Tokuyama Corporation, Tokyo, Japan.

## 18.2 MEMBRANE FORMULATION

### 18.2.1 CEM

CEMs are negatively charged membranes that contain functional groups such as $SO_3^-$, $COO^-$, $PO_3^-$, $HPO_3^-$ or $AsO_3^-$. Most of the commercially available CEMs consist of styrene (St) grafted onto a polymer substrate and subsequent modification by sulfonation (Tokuyama Soda and Asahi Glass, Japan) using reagents such as chlorosulfonic acid, acetyl sulfate, or sulfuric acid.

Styrene (St)-divinylbenzene (DVB) based CEMs are widely applied for electrodialysis and can prepared by the method described by Jin et al., 2002. A solution containing 88 wt% to 99 wt% of St, 0 wt% to 10 wt% of DVB, and 0.01 wt% to 0.04 wt % of benzoyl peroxide is prepared. A reinforcing material like polyvinyl chloride (PVC) is immersed in the monomer solution for 3 hours at room temperature to sufficiently absorb the monomer. The monomer filled films are placed between glass plates and sealed by aluminum tape to prevent evaporation followed by heating at 80°C for 0.5 to 10 hours to enable polymerization. Copolymers of St-DVB are sulfonated with concentrated sulfuric acid or chlorosulfonic acid resulting in sulfonic acid groups getting attached to the paraposition of the aromatic ring. Figure 18.4 shows the reaction scheme for synthesis of St-DVB-based CEM.

Radiation-induced graft polymerization is considered an important method for modification of chemical and physical properties of polymeric materials. In order to obtain various kinds of functional polymers, grafting of monomers containing reactive functional groups can be tried. Phosphonic and carboxylic groups are induced by radiation grafting method onto films such as polyethylene (PE), polypropylene (PP), and so forth.

Chemical stability of CEMs could be improved by perfluorination of the polymer backbone. Perfluorocarbon type CEM can be used in strong oxidizing mediums and high temperatures.

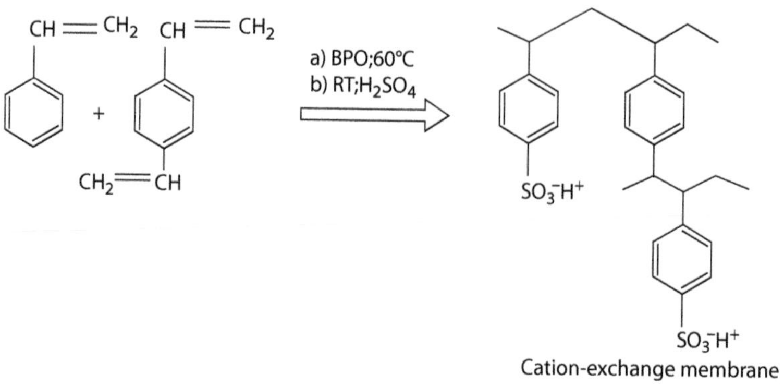

**FIGURE 18.4** Synthesis of styrene-DVB-based cation exchange membrane. (Adapted from Winston Ho, W.S., and K.K. Sarkar, eds., *Membrane Handbook,* New York: Van Nostrand Reinhold, 1992. With permission from Elsevier.)

Nafion® membrane developed by Du Pont is one of the most famous perfluorinated CEMs, which were introduced in 1966. Some of the perfluorinated CEMs such as Aciplex® (Asahi Chemicals), Flemion® (Asahi Glass), and Dow Xus® (Dow Chemical) were developed later. Figure 18.5 gives the general structure of perfluorinated CEM (Helen et al., 2006). The values of $m$, $n$, and $x$ in the structure differ among the various commercial membranes available in the market. For Nafion 117, $m > 1$, $n = 2$, $x = 5$–$13.5$, and $y = 1{,}000$, whereas in the case of Flemion, $m = 0$, 1 and $n =$ 1–5. For Aciplex, $m = 0$, 3, $n = 2$–5, and $x = 1.5$–14, and for Dow membrane, $m = 0$, $n = 2$, and $x =$ 3.6–10.

## 18.2.2 AEM

AEM contains positively charged groups such as primary, secondary, and tertiary amino groups, tertiary sulfonium groups, quaternary ammonium groups ($NH_4^+$), and quaternary phosphonium and other groups that provide a positive fixed charge in aqueous solutions. The AEMs are often prepared by amination of halomethylated polymer using various amines and diamines. Polymeric membranes having chloromethyl groups are generally prepared by copolymerization of chloromethyl styrene with other vinyl or divinyl monomers (Winston Ho and Sarkar, 1992). Figure 18.6 reveals the method for preparation of St-DVB-based AEM, the aminated forms of which have been commercialized by Tokuyama Soda (Neosepta) and Asahi Glass.

To avoid the use of carcinogenic chemicals such as chloromethylmethylether (CME) or bis-chloromethylether (BCME), Friedel–Crafts chloroacetylation can be employed (Laskin et al., 1971; Taylor and Laughlin, 1976). Membrane prepared through chloroacetylation of poly (2, 6-dimethyl-1, 4-phenylene oxide) (PPO) was reported by Liang et al. in 2006. Figure 18.7 gives the procedure for synthesis of PPO-based AEM. Once the base polymer has been chloromethylated

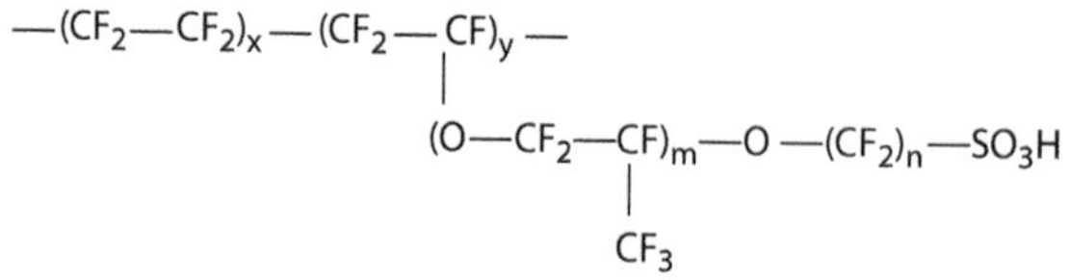

$$-(CF_2-CF_2)_x-(CF_2-\underset{\substack{|\\(O-CF_2-\underset{\substack{|\\CF_3}}{CF})_m-O-(CF_2)_n-SO_3H}}{CF})_y-$$

**FIGURE 18.5** General structure of perfluorocarbon type CEM. (Adapted from Helen, M., B. Viswanathan, L. Himakumar, et al. 2006. Strategies for the design of membranes for fuel cells. *Res. Signpost*).

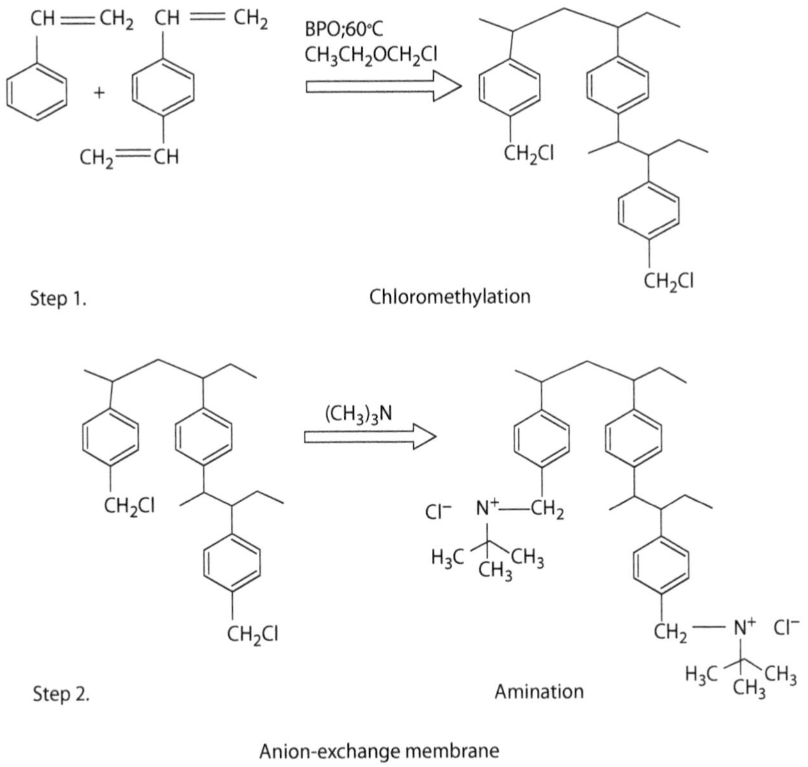

**FIGURE 18.6** Preparation of styrene-DVB-based anion exchange membrane. (Reprinted from *Membrane Handbook*, Winston Ho, W.S., and K.K. Sarkar, Copyright 1992, with permission from Elsevier.)

or chloroacetylated, films are prepared by the solution casting method and functionalized by quaternary amino group in a trimethyl amine aqueous solution. Radiation-induced grafting methodology is also used for preparing AEM. Vinylbenzyl chloride or vinyl pyridines are grafted into fluorine containing polymer films such as poly(vinylidene fluoride) (PVDF), fluorinated ethylene propylene (FEP), perfluoroalkoxyethylene (PFA), poly(tetrafluoroethylene) (PTFE), and so forth,

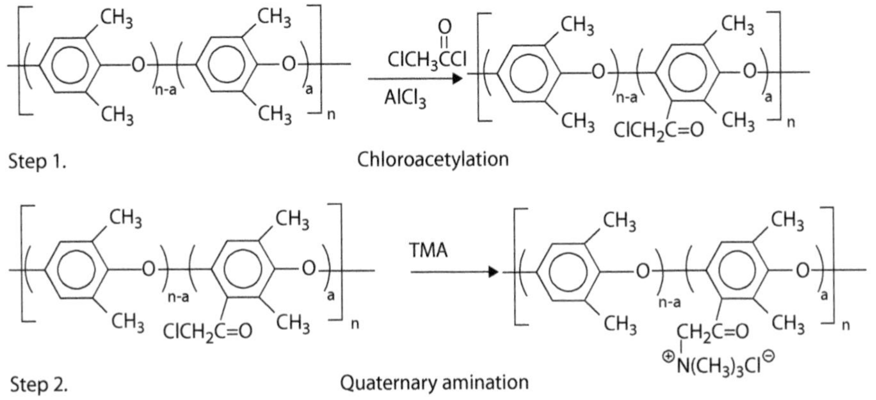

**FIGURE 18.7** Preparation of PPO-based anion exchange membrane. (With kind permission from Springer Science & Business Media: *Journal of Membrane Science,* 286, 2006, 185–192, Liang, W., et al.)

followed by subsequent reaction with tertiary amino groups to yield AEM. Chloromethylated polymers are generally used to produce strongly alkaline AEMs. Highly aminated products can be obtained by the reaction of poly(epichlorohydrin) or poly(epichlorohydrin-co-ethyleneoxide) with tertiary 1,4-diazobicycle-(2.2.2)-octane in dipolar aprotic solvent medium (Patrick, 1998). These AEMs are inert in acidic medium but susceptible in alkaline solutions since quaternary ammonium groups present in the membrane are decomposed by alkali as per Hofmann degradation reaction. Various trials have been made to improve the stability of AEMs by incorporation of anion exchange groups; namely, 1-benzyl-1-azonia-4-azabicyclo-(2, 2, 2)-octane hydroxide (Bernd et al., 1990), benzyl trimethyl ammonium (Toshikatsu et al., 1996), and N,N,$N^{1}$,$N^{11}$-tetramethyl hexane diammonium (TMHDA) (Sata et al., 2004). Preparation of perflourinated AEM is challenging due to the strong electron-donor properties of the fluorine atom. However, AEM with excellent durability was prepared by chemical modification of a fluorinated CEM such as Nafion by transformation of its sulfonic acid groups into quaternary ammonium groups (Kiyohide et al., 1986; Sata, 2004).

### 18.2.3 Bipolar Membranes

Bipolar membrane is a composite that consists of a layered ion exchange structure composed of a CEM (fixed negative charges) on one side and an AEM (fixed positive charges) on the other side. The three major components in a bipolar membrane are anion exchange layer, cation exchange layer, and a contacting region between two layers. Water dissociation into hydroxide ions and protons occurs at the contact region. Rate of water dissociation is enhanced for fixed electric potential difference across the contact region, including a catalyst since activation energy for the overall dissociation reaction is reduced (Salvador et al., 1998). The contact region also influences the long-term stability of a bipolar membrane. For some membranes the contact region is also a part of the ion selective layer.

**TABLE 18.2**
**Manufacturers of Commercial Bipolar Membranes**

| Membrane Trade Name | Anion Permeable Layer | Cation Permeable Layer | Manufacturer |
|---|---|---|---|
| AQ-BA-06 (PS) | Polystyrene (PS)/vinylbenzoyl chloride co-polymer with diamines | Sulfonated polystyrene and Kraton G | Aqualytics Inc., United States |
| AQ-BA-04 (PSf) | Polysulfone (PSf) with tertiary and quaternary amines | Sulfonated PSf | -do- |
| BP-1 | Aminated polysulfone | Poly(styrene sulfonate) crosslinked with divinylbenzene in a poly(vinyl chloride) matrix | Tokuyama, Tokyo, Japan |
| Selemion BP-1 | Styrene/divinylbenzene co-polymer with quaternary amines | Perfluorinated polymer with sulfonic acid groups | Asahi Glass, Japan |
| B-17 | Perfluorinated polymer with secondary and quaternary amines | Perfluorinated polymer with sulfonic acid groups | Tosoh, Tokyo, Japan |
| MB-1, MB-3 | Ion exchange resin with secondary amines (MB-1) or quaternary amines (MB-3) in polyethylene binder | Ion exchange resin with sulfonic acid (MB-1) or phosphoric acid (MB-3) groups | Russia |
| BMI-9000 | Gel polystyrene crosslinked by divinylbenzene and attached with quaternary ammonium groups | Gel polystyrene crosslinked by divinylbenzene and attached with sulphonic acid groups | Membranes International Inc., United States |

Preparation of a bipolar membrane requires an efficient contact of the cation exchange layer and anion exchange layer. Various methods are initiated such as lamination, pressing, gluing of the CEM and AEM together, solution casting of an ion exchange membrane over another of opposite charge (Kemperman, 2000), or functionalizing both sides of a membrane (Moussaoui and Hurwitz, 1998). Of all these methods, solution casting seems to be attractive because it is simple, inexpensive, and retains all the desired prosperities for commercial application such as good mechanical strength, ability to operate at high current density, high permselectivity, low resistance, and so forth (Hao et al., 2001). Some of the commercial bipolar membranes are listed in Table 18.2 (Friedrich, 2001).

## 18.3 EXPERIMENTAL PROCEDURES AND PROTOCOLS

### 18.3.1 Laboratory Setup

The experimental setup is shown in Figure 18.8a. The system consists of three storage tanks (one each for feed), concentrate, and electrode wash solutions. Each tank is connected to chemically resistant centrifugal pumps having magnetically coupled drive and propylene wetted parts. Control

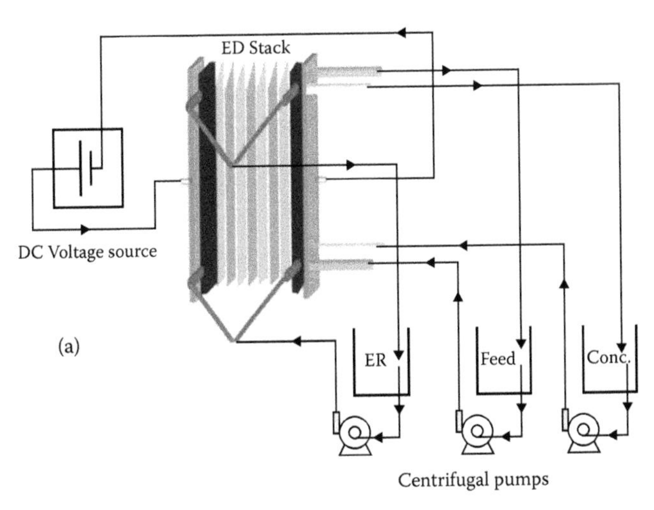

**FIGURE 18.8** (a) Schematic diagram of an ED experimental setup and (b) photograph of a typical laboratory ED system.

valves and bypass is provided to adjust the flow-rates independently in each line. Braided PVC tubing is used for circulation of liquids throughout the system. Perforations in the edges of the membranes and the gaskets match with each other to provide two pairs of internal hydraulic manifolds to carry fluids into and out of the compartments. One pair communicates with the depletion compartments (diluate) and the other with the enrichment compartments (concentrate). At both the ends of the membrane stack electrode plates made of stainless steel coated with titanium or metal alloys are placed. Energy is provided through a regulated DC power supply. Turbulence promoters made of flexible PVC wire mesh are used to reduce the concentration polarization. A digital conductivity meter is used to determine the concentration of ions in diluate and concentrate streams at different intervals of time. A pictorial representation of a typical laboratory ED system is provided in Figure 18.8b.

The tanks are washed thoroughly and system cleaned with distilled water for 30 minutes. Feed solution of known volume (say 10 L) and composition is taken in Tank 1. Tap water of volume equal to or less than that of the diluate is introduced into Tank 2 to facilitate conductivity and ion exchange. 2.5% w/v sodium bisulphate solution is taken in Tank 3 to rinse the electrodes. After filling the tanks with the respective solutions, they are pumped through the electrodialysis stack at controlled flow rates. Control valves are adjusted to maintain equal flow rates in both the diluate and concentrate compartments ensuring almost equal pressure drop. After stabilizing the flow, an electric potential is applied across the stack to attain a specific current density for a desired period of time. A voltage of around 0 to 100 V is sufficient to achieve currents in the range of 0 to 20 amps due to the high conductivity of the diluate solution. However, voltage corresponding to limiting current density (LCD) should not be exceeded to avoid wastage of electric power. Samples of outlet streams are collected at regular intervals to determine conductivity (salt concentration) of both diluate and concentrate solutions using the digital conductivity meter. Once the conductivity of the diluate reaches the desired level, the operation is stopped.

### 18.3.2 Analytical Methods

Feed properties such as temperature, viscosity, degree of dissociation of salts, conductivity, total dissolved salts (TDSs), and so forth should be determined. During the processing of an aqueous solution by ED, performance is determined by recording conductivity, total dissolved solids (TDSs) and pH.

#### 18.3.2.1 Conductivity

Electrical conductivity is the conductance of a unit cube of solution present between two electric probes or electrodes and is most commonly measured by a conductivity cell that consists of two electrodes made of platinum with an area of 1 cm$^2$ separated by a distance of 1 cm. Different configurations of cells are characterized by a cell constant K that is a function of distance between the electrodes, electrode area, and electric field. These electrodes are placed in the solution whose conductivity is to be measured and a transducer measures the conductance of the probe. The relationship between conductance and conductivity is given by:

$$\text{Conductivity} = \text{Conductance} \times \text{K factor} \tag{18.7}$$

The unit of conductivity is Siemens per centimeter (S/cm).

#### 18.3.2.2 Total Dissolved Salts

TDS is the amount of solids (salts) present in the solution in dissolved form. TDS estimation using a sand bath heater is one method of measuring dissolved salts in any solution. A known volume (say 10 mL) of the solution is taken in a crucible that is then place on a sand bath. Temperature of the sand bath is gradually increased to 105°C by a magnetic hot plate. The solution is allowed to vaporize slowly until the residue becomes dry. During heating, volatile components and water present in the solution evaporate whereas the salts remain. Weights of the empty crucible ($D_1$) and that containing the residue ($D_2$) are recorded in mg. The procedure is repeated until consistent weight readings are obtained. TDS of the solution is determined from:

$$\text{TDS (mg/L)} = [(D_2 - D_1)/10] \times 1000 \tag{18.8}$$

### 18.3.3 Effect of Operating Parameters

The performance of an electrodialysis plant is determined by a set of fixed and variable process parameters such as stack construction, feed properties, flow rates, voltage, and current.

#### 18.3.3.1 Voltage and Current

Voltage is the electrical potential that is applied between two electrodes and is the driving force in ED. For a given feed concentration, increasing the voltage across the cell increases current and decreases the operation time in achieving the desired diluate concentration. After a certain time the rate of reduction in diluate concentration slows down with simultaneous lowering in the current between the electrodes due to depletion in ion concentration. Figure 18.9 displays the relation between voltage and current density. Increase in voltage results in a proportional increase in current density until an intermediate point is reached where maximum current density is attained and further increase in voltage does not cause any significant rise in current. This point in the curve where a change in slope can be observed is called limiting current density (LCD) and application voltage above a value corresponding to the LCD results in the current getting wasted and subsequently a reduction in process efficiency. The LCD is a function of flow rate and feed composition. Once ion concentration at the membrane surface in the diluate channel tends to be zero; few charges are available to carry the electrical current that reaches a minimum value.

#### 18.3.3.2 Flow Rate

At constant concentration, temperature, and voltage, increasing the flow rate improves cell performance. This is on the basis of so-called Nernst idealization, assuming a complete mixed zone at the center of the solution and a completely static zone in the boundary layer adjacent to the membrane. Exchange of ions will take place in the bulk solution, boundary layer, and membrane. Increasing the solution velocity decreases the thickness of boundary layer, which results in more rapid exchange of ions and a decrease in ED operation time. However, at very high flow rates, the process may become retarded due to low residence time that prevents ions from moving across membranes.

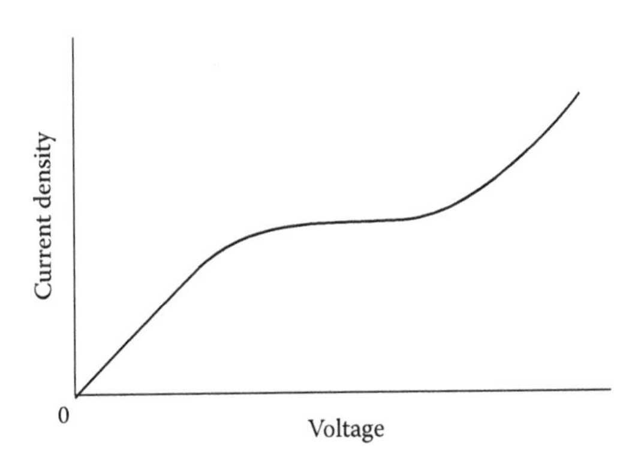


**FIGURE 18.9** Relation between voltage and current density for an ion exchange membrane.

#### 18.3.3.3 Feed Composition

General feed properties to be considered are salt concentration, temperature, and nature of ions. Separation performance increases with increasing concentration (ionic strength) of feed solution, which enhances driving force and improves ion mobility. On the other hand, at constant voltage the total operating time increases with rising feed concentration. If the feed solution contains ions of different valencies, ions possessing lower valency exchange faster than multivalent ones. The larger the ion size, the more is the hindrance to its transport through the membrane owing to low diffusivity. Increasing temperature has a positive impact on the cell performance as it enhances energy of the ions.

### 18.3.4 Characterization of Electrodialysis Membranes

#### 18.3.4.1 Ion Exchange Capacity

The ion exchange capacity (IEC) indicates the number of milliequivalents of ions in 1 g of the dry polymer (Smitha et al., 2003). The membranes were soaked in 50 ml of 0.01 N sodium hydroxide solution for 12 hours at ambient temperature. Then 10 ml of solution was titrated against 0.01 N sulfuric acid. The sample was regenerated with 1 M hydrochloric acid, washed with water, and dried to a constant weight. The IEC was calculated according to the equation

$$\mathrm{IEC} = \frac{\mathrm{B} - \mathrm{P} \times 0.01 \times 5}{\mathrm{m}} \tag{18.9}$$

where IEC is the ion exchange capacity, $B$ is the amount of 0.01 N sulfuric acid used to neutralize blank, $P$ is the amount of 0.01 N sulfuric acid used to neutralize the membranes, 0.01 is the factor used to account for normality of sulfuric acid/NaOH, 5 is the factor corresponding to the ratio of the amount of NaOH taken to dissolve the polymer to the amount used for titration, and m is the sample mass in grams.

#### 18.3.4.2 Water Sorption and Membrane Swelling

Swelling of ion exchange membrane in solvent occurs because of ion exchange groups in the polymer. The extent of swelling depends on the membrane pretreatment, nature of polymer matrix, IEC, species of reinforcing fabric, and degree of crosslink. In order to determine their interaction, weighed samples of circular pieces of polymer films were soaked in water and methanol. The films were taken out after carefully wiping out excess water to estimate the amount absorbed at the particular time $t$. The film was then quickly placed back in the solvent. The process was repeated until the films attained steady state as indicated by constant weight after a certain period of soaking time. The degree of swelling is calculated from:

$$\text{Degree of swelling} = \frac{M_s}{M_d} \tag{18.10}$$

where $M_s$ is the mass of the swollen polymer in grams and $M_d$ is the mass of the dry polymer in grams.

The percent sorption is obtained from:

$$\%\ \text{Sorption} = \frac{M_s - M_d}{M_d} \times 100 \tag{18.11}$$

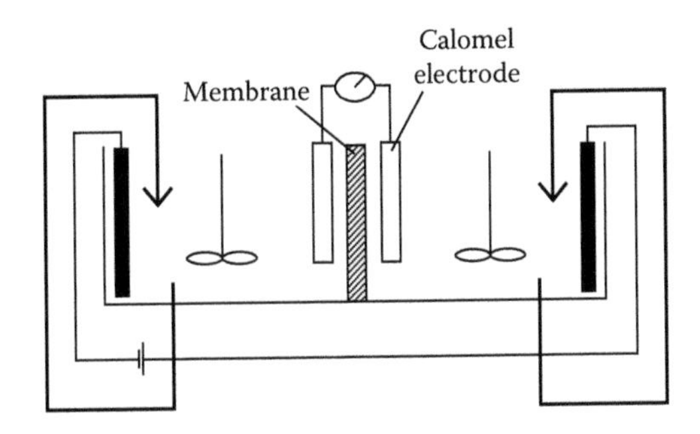


**FIGURE 18.10** Experimental setup for determining the electrical resistance of ion exchange membranes.

#### 18.3.4.3 Electrical Resistance

The electrical resistance of an ion exchange membrane is an important property that determines the energy required for the ED process. The membrane area resistance is given by $\Omega$ cm$^2$. The electrical resistance of an ion exchange membrane is measured with a two-compartment cell shown in Figure 18.10. It consists of platinum electrodes that act as anode and cathode, and calomel electrodes for measuring the potential difference between the solutions. Salt solution of given concentration is placed in two compartments of the cell. A current density (I) is applied and the potential difference $P_0$ between the electrodes without any membrane is measured for certain current densities at a constant temperature of 25°C. Before determining the potential difference, the membrane should be equilibrated with the salt solution and then placed between calomel electrodes and potential difference $P_1$ is measured to obtain the same current density. Electric resistance of the membrane for a given current density is measured by:

$$R = (P_1 - P_0)/I \tag{18.12}$$

where $R$ is electrical resistance that should be low in the absence of the membrane. To increase the accuracy of measurement, 0.5 N NaCl or KCl standard solution is used (Sata et al, 2004).

#### 18.3.4.4 Transport Number

In an aqueous solution, salts will ionize and certain current is carried by each of the ion. Transport number ($t_j$) of a specific ion is the fraction of the electric current carried by the ion ($i_j$).

$$t_j = i_j/I \tag{18.13}$$

where $I$ is the total current. The membrane to be analyzed is equilibrated in 0.3 M KCl and then placed between two compartments as shown in Figure 18.11, which contain 0.1 M KCl in one

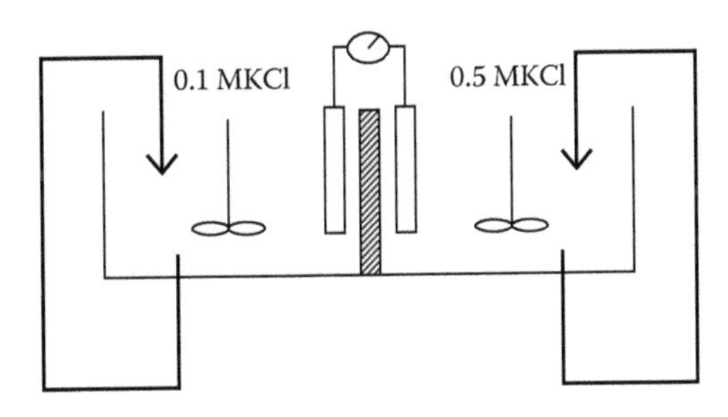


**FIGURE 18.11** Experimental setup for determining the transport number of ion exchange membranes.

compartment and 0.5 M KCl in the other. Ag-AgCl electrodes are placed in the two compartments to measure the potential difference. The transport number (Sata, 2004) can be obtained by:

$$E = 2\, t_{+(app)}\, RT/F\, (\ln(a^{i}/a^{ii})) \tag{18.14}$$

where E is the potential difference measured, R the gas constant, T the absolute temperature, F Faraday's constant, $a^{i}$ and $a^{ii}$ are activities of two solutions separated by the membrane, and $t_{+(app)}$ the apparent transport number as the water transport is not taken into consideration. When very diluate solutions are used, apparent transport number value will be close to ideal value.

The transport number of the anion, $t_{-}$, can be found from the relation:

$$t_{+} + t_{-} = 1 \tag{18.15}$$

#### 18.3.4.5 Proton Conductivity

The proton conductivity of water-equilibrated CEMs can be determined by a four-electrode electrochemical impedance spectroscopy (EIS) method using a PGSTAT20 frequency analyzer (by EcoChemie B.V. Spectra) and recorded between 1 MHz and 0.1 Hz with 10 points per decade at maximum perturbation amplitude of 10 mV. The impedance spectra are fitted on the basis of an equivalent circuit. Zview 2.1b software from Scribner Associates Inc. was used for curve fitting. To measure the temperature dependence of the conductivity, the cell is placed in a sealed, tempered, double-walled vessel and the temperature recorded in close proximity to the membrane with a K-type thermocouple. To avoid changes in the humidification levels (100% R.H) during the measurements, a Teflon bowl filled with water is placed at the bottom of the vessel. Measurements are carried out in a conductivity cell at temperatures ranging from 30°C to 150°C (Smitha et al., 2005).

#### 18.3.4.6 Mechanical Strength

The equipment used for carrying out the test is a Universal Testing Machine (e.g., Shimadzu make, model AGS-10kNG) with an operating head load of approximately 5 kN. Cross-sectional areas of the membrane samples of known width and thickness are calculated. The films are then placed between the grips of the testing machine. The grip length is generally 5 cm and the speed of testing is set at the rate of 12.5 mm/min. Tensile strength is then calculated using the equation:

$$\text{Tensile Strength} = \text{Maximum Load/Cross-sectional area } (\text{N/mm}^2) \tag{18.16}$$

## 18.4 DESIGN

The performance of electrodialysis process in any module is limited by concentration polarization and membrane fouling. Concentration polarization is the depletion of ion concentration near membrane surface compared to bulk solution. Moreover, the concentration of the retained non-ionizing molecules such as organic compounds increases near the membrane surface with process time. Concentration polarization causes increase in energy consumption by lowering the rate of ion removal from the solution. Fouling occurs due to scale formation through adsorption of solute particles onto the surface or plugging of the pores. Efficient module design may help in overcoming these problems to some extent.

### 18.4.1 Module Geometry

The building block of a membrane system is called the module. Electrodialysis modules are commonly implemented in the form of plate and frame stack and recently spiral wound module.

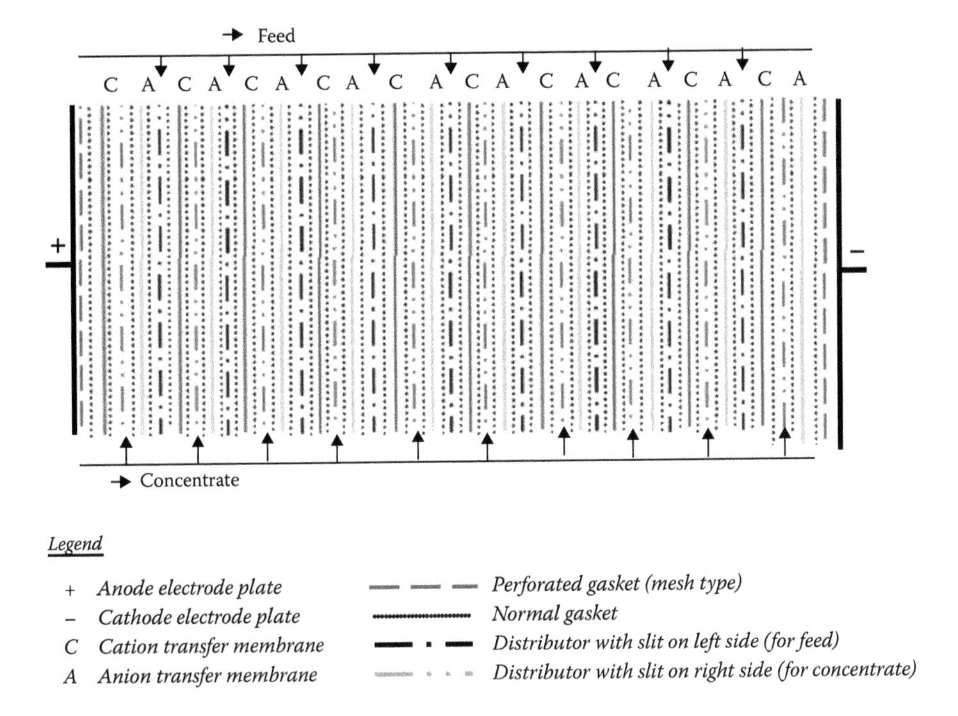


**FIGURE 18.12** Schematic diagram of plate and frame type module.

#### 18.4.1.1 Plate and Frame

A schematic diagram of plate and frame ED stack is shown in Figure 18.12. It consists of an AEM, a CEM, a gasket, and a spacer that separates the ion exchange membranes. A spacer is placed in between gaskets, which contains ducts at two opposite corners for directing the feed and concentrate to the corresponding chambers. A set of two membranes, two gaskets, and one spacer constitutes a cell pair. This repeated arrangement of cell pairs is held in between oppositely charged electrodes to make up an ED stack. Feed and concentrate are allowed to flow in alternate chambers. Gases that are liberated at electrodes are continuously removed with rinsing solution. Any leakages in the stack between feed, concentrate, and rinse solutions are prevented with proper sealing using high vacuum grease. For commercial ED stack design there are two types of spacers: tortuous-path and flat-sheet (Noble and Stern, 1995). Flow path shown in Figure 18.13a indicates a direction of flow that is more

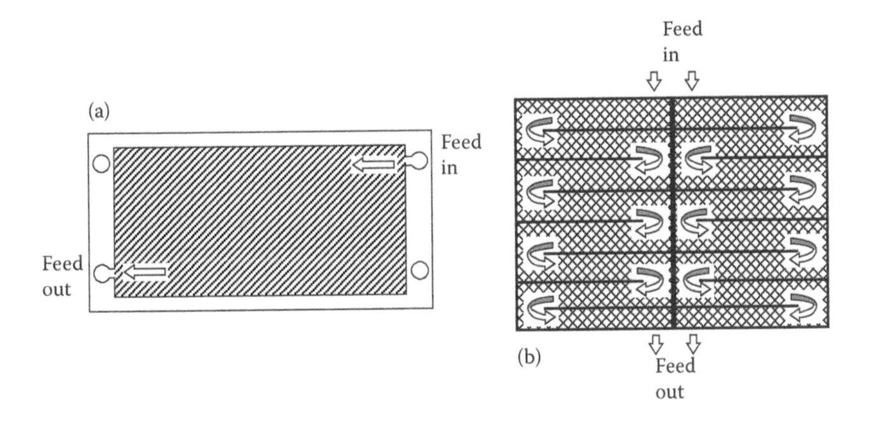


**FIGURE 18.13** Schematic diagrams of (a) a flat-sheet ED spacer and (b) a tortuous-path spacer.

or less straight from the entrance to the exit in a flat sheet spacer. The entrance ports and exit ports are equal in number. In the tortuous-path spacer shown in Figure 18.13b, the solution flows in a narrow channel by making a 180° bend at the entrance and exit of the compartment, consequently increasing the residence time of the solution within each compartment.

#### 18.4.1.2 Spiral Wound

A spirally wound electrodialysis module mainly consists of four parts: the core, spiral wound, membrane, and shell. Both core and shell act as electrodes with electrical connections. Lengthy flat sheets of anion and cation selective membranes are made like pouches with adhesive binding at the edges. This arrangement is spirally wound and fitted into an annular space between core and shell. Spacers are located between the membranes in the pouch and also adjacent to the electrodes. A felted mat or woven cloth of synthetic fiber is a good spacer (Karn, 1980). Feed and concentrate flow in the alternate channels. Two ends of the shell are sealed with removable lids that contain provisions for inlet and outlet of solutions. Rinse solution flows between electrodes and the membrane module to remove the liberated gases. Hydraulic arrangements for spiral wound module are parallel flow and cross flow (Wen et al., 1995). Figure 18.14 shows the schematic diagram of flow pattern in a spiral wound module.

### 18.4.2 Process Flow Diagram for Commercial ED

The simple process flow diagram (PFD) of a commercial scale ED plant is shown in Figure 18.15. The list of equipment includes a DC power supply, ED stack, feed tank, prefilter assembly, diluate tank, concentrate tank, rinse tank, heat exchangers, and centrifugal pumps for transport of the three aqueous streams from storage vessels to the stack. Feed solution from the feed tank is processed through the prefilter to remove any solid particles present. A brief description of the process was already given in Section 18.3.1. To maintain the temperature of the diluate, concentrate, and rinse solutions coming out of the ED stack, heat exchangers can be used.

### 18.4.3 Capital and Operating Costs

The capital cost of an ED system depends on several factors such as membrane area requirement, pump capacities, hardware (electrodes and stack) and thyristor or rheostat for DC power supply. Membrane area required for the process is a strong deciding factor for capital cost. Capital cost will be approximately 2 to 2.5 times the membrane area. The amount of feed processed per square meter area of the membrane stack per hour with the lab scale or pilot plant is noted and accordingly the scale-up ratio can be obtained by knowing the feed capacity of the commercial plant. For example, if the quantity of feed processed in the laboratory is 5 L/h with 1 $m^2$ membrane area, then the scale-up ratio for processing 1,000 L/h of effluent would give 200 $m^2$ as the area requirement, provided the same linear velocity is maintained.

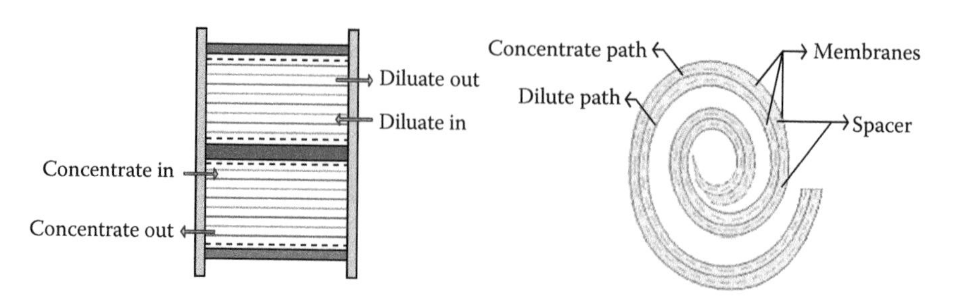


**FIGURE 18.14** Schematic diagram of flow pattern in a spiral wound module.

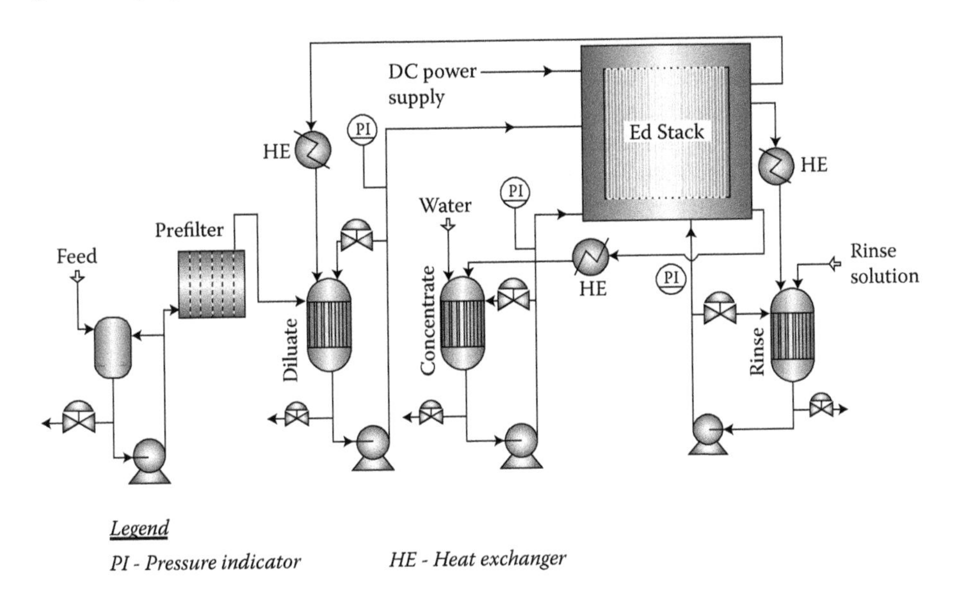


**FIGURE 18.15** Process flow diagram of a commercial ED plant.

The operating cost includes energy requirement, membrane replacement cost, and chemical consumption. Energy consumption is broadly categorized into pumping costs and DC power. During the process several chemicals such as HCl, NaOH, EDTA, and sodium metabisulfate are required for cleaning and storing the membranes apart from reagents for electrode rinsing. Energy consumption for pumps assuming 60% efficiency is given by the following equation:

$$E(KWh) = \frac{H \times Q}{360 \times 0.6} \quad (18.17)$$

where E is the energy (KWh), H is the head (m), and Q is the flowrate ($m^3/h$). Actual energy required for electrodialysis i.e., DC power consumed in kWh is given by:

$$\text{Power} = \frac{\sqrt{3} VI \cos\phi}{1000} \quad (18.18)$$

Robust state-of-the-art membranes cost approximately \$200 to \$250 per $m^2$ and a life of 3 to 5 years, subject to nature of the effluent, may be assumed before replacement is required. Other running cost components that have to be considered are utilities, labor, depreciation, interest on capital investment, prefilter cartridges replacement, and so forth. The energy consumption for seawater desalination is generally in the range of 5 to 10 $kWh/m^3$ of desalinated water having a salt concentration of 0.5 g/L at a current density of 300 to 600 $A/m^2$. The desalinated water cost is approximately \$1/$m^3$ assuming one kWh of power to cost \$0.05.

## 18.5 CASE STUDY AND FUTURE PROSPECTS

### 18.5.1 Case Study on Acetic Acid Recovery

Bulk drug industries produce effluents that contain a mixture of salts (1%–3%), water and a reasonable concentration of organic solvent (5%–20%) used as medium for production of important drugs.

Due to the presence of the salts and other hazardous ionizing compounds, the effluents cannot be processed by conventional methods to recover the solvent. Instead they are disposed after further treatment in the effluent treatment plant (ETP), resulting in huge losses of solvent besides extra expenditure for chemical oxygen demand (COD) reduction arising from presence of organic solvents in the effluent. ED offers a promising solution to this problem as it can be employed to remove the ionizable compounds after which the desalted effluent could be sent for evaporation/distillation to remove water and recover the solvent for recycle. Figure 18.16 represents a hybrid ED-distillation process scheme that enables solvent recovery in the bulk drug industry. The economics of the process depends upon the solvent and salt concentration in the feed. High solvent and low salt concentrations makes the process economically feasible.

In the case of effluents containing weakly ionizable organic acids such as $CH_3COOH$, the judicious control of ED operating conditions (DC potential, current, etc.) renders exchange of the faster permeating salts such $K_2SO_4$ and NaCl into the concentrate tank, leaving behind a diluate tank enriched in $CH_3COOH$ due to differences in the ionization potentials. The diluate can then be distilled to recover the organic acid. Hydrochloric acid (HCl) is a widely generated byproduct in reactions that produce fine chemicals and drug intermediates. The most feasible method to remove HCl for product isolation is ED. In the agrochemical industry, pesticide/herbicide production yields aqueous reaction liquor containing the product such as glyphosate (a herbicide) and a salt (KCl). ED using bipolar membranes not only isolates the herbicide but also yields KOH and HCl as valuable byproducts.

Acetic acid is a valuable industrial solvent used in the manufacture of acetic anhydride, vinyl acetate, synthetic fabrics, fungicides, and pharmaceutical products. A hybrid ED-distillation process can be employed to recover acetic acid present in industrial effluents from food, chemical, and pharmaceutical industries. Apart from recycling of the acid, ED helps in reducing the load of COD on the ETP. In the present work, acetic acid was recovered from an aqueous effluent supplied by a local industry that contained 12% of the acid, 5% NaCl, 5% $K_2SO_4$ 0.3% $H_2SO_4$, and water (balance). Experiments were carried out on a bench-scale system using an electrodialysis stack of 1 $m^2$ effective area that comprised of 10 cell pairs of AEM and CEM. The acetic acid present in diluate and concentrate was analyzed by titration with standardized NaOH solution. The flow rates of diluate, concentrate, and rinse solutions were kept equal (25 mL/s) in order to maintain constant pressure drop across the membranes. The concentrate was more enriched with $H_2SO_4$ and the two salts

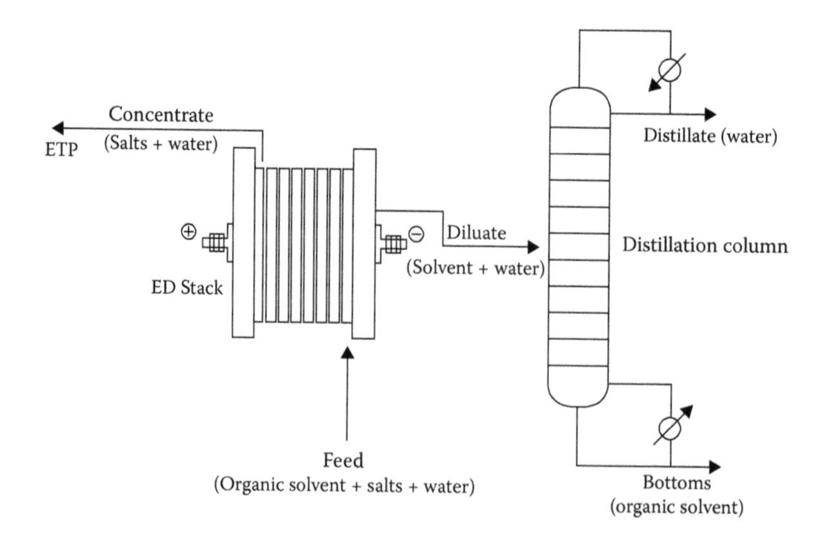


**FIGURE 18.16** Electrodialysis-distillation hybrid process for solvent recovery in chemical/bulk drug industries.

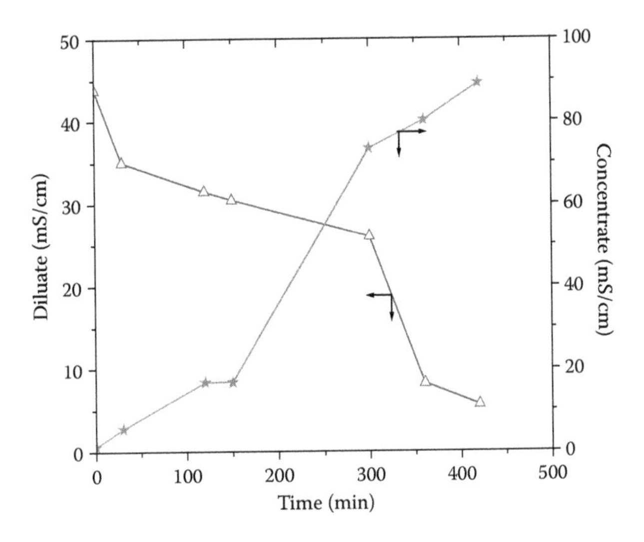


**FIGURE 18.17** Effect of ED operation time on conductivity of diluate and concentrate.

due to their greater ionization potential compared to weakly ionizing acetic acid. Figure 18.17 shows the gradual decrease of diluate conductivity with a simultaneous rise in concentrate conductivity as a function of time. ED could be stopped at a point of time where the $H_2SO_4$ and the salts' concentrations in diluate were low enough to enable corrosion-free and smooth distillation for water removal and recovery of acetic acid. Figure 18.18 shows that the voltage was high in the beginning due to low initial concentrate conductivity, which would increase resistance within the stack but gradually be reduced once all the three solutions became considerably conductive. After the majority of ions were separated from the diluate, the voltage increased again. In contrast, the current increased steadily to reach an optimum value and then came down as conductivity in alternate diluate chambers decreased with time. However, an average voltage of 20 V DC and an average current of 8 amps could be considered for lowering the conductivity of the effluent from 74.5 to 4.5 mS/cm to enable recovery of at least 70% of the acid.

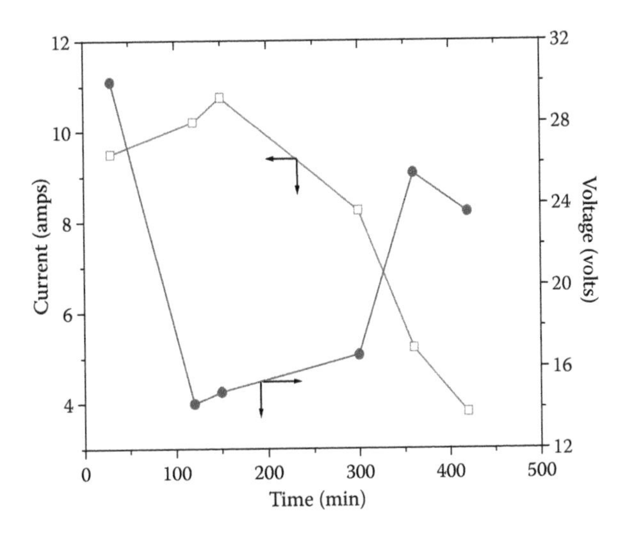


**FIGURE 18.18** Variation of current and voltage with time.

### 18.5.2 Future Prospects

In the past few decades ED's potential has mostly been realized for desalination of brackish water and demineralization of whey. However, its future lies in solving challenging separation problems in the chemical and bulk drug industry to recover solvents, pharmaceuticals, and agrochemicals. Hybrid processes involving ED for desalination or deacidification of industrial effluents followed by evaporation or distillation to isolate valuable products have exhibited tremendous potential. Combination of ED with reverse osmosis (RO) for further concentration of RO brines could be exploited for improved recovery of water and organics (Gurtler et al., 2008) with simultaneous reduction in evaporator load and energy savings (Pellegrino et al., 2007). The ED-RO combine could be employed to recover and concentrate ammonia from manure.

Reverse electrodialysis (RED) has attracted keen interest due to the possibility of generating renewable energy in the form of sustainable electricity by controlled mixing of a dilute salt solution such as surface water with a concentrated electrolyte solution such as sea water (Post et al., 2009). A power density of 460 mW/m$^2$ at a linear flow velocity of 0.54 cm/s has been reported by Turek and Bandura (2007), using the RED technique. Ion conductive spacers could be fabricated for increased power generation in RED.

ED by bipolar membranes has enhanced process economy due to the production of useful byproducts such as acids and bases. Mixed matrix ED membranes comprisedof resin-filled bipolar membranes are being used for production of tartaric acid (Zhang et al., 2009). State-of-the-art ED membranes are chemically and mechanically more robust than ever, but the cost of fabrication, especially in case of bipolar membranes, needs to be brought down for ED to become more acceptable in developing countries. Electrodialysis using ultrafiltration membranes (EDUF) is a recent development wherein some ion exchange membranes are replaced in the ED stack by ultrafiltration membranes to fractionate value-added products from complex feedstocks on the basis of differences in electrical charge and molecular weights. This technique can been applied for recovery of polyphenols and bioactive peptides (Poulin et al., 2007). The focus of researchers and industrialists in the future would be to make innovations in the design of ED stack components and development of membranes with improved ion transport and antifouling properties.

## REFERENCES

Bernd, B., H. Strathmann, and F. Effenberger. 1990. Anion-exchange membranes with improved alkaline stability. *Desalination* 79: 125–144.

Chuanhui, H., X. Tongwen, Z. Yaping, et al. 2007. Application of electrodialysis to the production of organic acids: State-of-the-art and recent developments. *J. Membrane Sci.* 288: 1–12.

Francois Poulin, J., J. Amiot, and L. Bazinet. 2007. Improved peptide fractionation by electrodialysis with ultrafiltration membrane: Influence of ultrafiltration membrane stacking and electrical field strength. *J. Membrane Sci.* 299: 83–90.

Friedrich, G.W. 2001. *Bipolar Membrane Electrodialysis.* Ph.D. thesis, University of Twente.

Gunther Molau, E. 1981. Heterogeneous ion-exchange membranes. *J. Membrane Sci.* 8: 309–330.

Gurtler, B.K., T.A. Vetter, E.M. Perdue, et al. 2008. Combining reverse osmosis and pulsed electrical current electrodialysis for improved recovery of dissolved organic matter from seawater. *J. Membrane Sci.* 323: 328–336.

Hao, J.H., L.X. Yu, C.X. Chen, et al. 2001. Preparation of bipolar membranes (II). *J. Appl. Polym. Sci.* 82: 1733–1738.

Huang, C., T. Xu, Y. Zhang, et al. 2007. Application of electrodialysis to the production of organic acids: State-of-the-art and recent developments. *J. Membrane Sci.* 288: 1–12.

Helen, M., B. Viswanathan, L. Himakumar, et al. 2006. Strategies for the design of membranes for fuel cells. *Res. Signpost* 1–42.

Jan Post, W., V.M. Hubertus Hamelers, and J.N. Cees Buisman. 2009. Influence of multivalent ions on power production from mixing salt and fresh water with a reverse electrodialysis system. *J. Membrane Sci.* 330: 65–72.

Karn, W.S. 1980. Spirally Wound Electrodialysis Cell. U.S. Patent 4225403.

Kemperman, A.J.B. 2000. *Handbook on Bipolar Membrane Technology*. The Netherlands: Twente University Press.

Kiyohide, M., T. Etsuko, K. Sugimoto, et al. 1986. Novel anion exchange membranes having fluorocarbon backbone: Preparation and stability. *J. Appl. Polym. Sci.* 32: 4137–4143.

Laskin, S., M. Kusschner, R.T. Drew, et al. 1971. Tumors of the respiratory tract induce by inhalation of bis(chloromethyl)ether. *Arch. Environ. Health* 23: 135–136.

Liang. W., X. Tongwen, and Y. Weihua. 2006. Fundamental studies of a new series of anion exchange membranes: Membranes prepared through chloroacetylation of poly(2,6-dimethyl-1,4-phenylene oxide) (PPO) followed by quaternary amination. *J. Membrane Sci.* 286: 185–192.

Mathieu, B. 2002. Production of organic acids by bipolar electrodialysis: realizations and perspectives. *Desalination* 144: 157–162.

El Moussaoui, R., and H. Hurwitz. 1998. Single-film membrane-process for obtaining it and use thereof. U.S. Patent 5,840,192.

Natraj, S.K., S. Sridhar, I.N. Shaikha, et al. 2007. Membrane based microfiltration/electrodialysis hybrid process for the treatment of paper wastewater. *Sepa. Purif. Technol.* 57: 185–192.

Nikbakht, R., M. Sadrzadeh, and T. Mohammadi. 2007. Effect of operating parameters on concentration of citric acid using electrodialysis. *J. Food Eng.* 83: 596–604.

Noble, R.D., and S.A. Stern. 1995. *Membrane Separation Technology: Principles and Applications*. The Netherlands: Elsevier Science B.V.

Patrick, A. 1998. *Strongly Alkaline Anion Exchange Membranes and Process for Producing the Same*. U.S. Patent 5746917.

Pellegrino, J., C. Gorman, and L. Richards. 2007. A speculative hybrid reverse osmosis/electrodialysis unit operation. *Desalination* 214:11–30.

Rautenbach, R., and R. Albrecht. 1989. *Membrane Processes*. New York: John Wiley & Sons.

Robert, B.L. 1995. *Separation Processes in Waste Minimization*. New York: Taylor & Francis.

Sadrzadeh, M., A. Razmi, and T. Mohammadi. 2007. Separation of monovalent, divalent and trivalent ions from wastewater at various operating conditions using electrodialysis. *Desalination* 205: 53–61.

Salvador, M., R. Patricio, and A. Antonio. 1998. Electric field-assisted proton exchange and water dissociation at the junction of a fixed-charge bipolar membrane. *Chem. Phys. Lett.* 294: 406–412.

Sata, T. 2004. *Ion Exchange Membranes: Preparation, Characterization Modification and Application*. Cambridge: The Royal Society of Chemistry.

Sata. T., M. Tsujimoto, T. Yamaguchi, and K. Matsusaki. 2004. Anion-exchange membranes containing diamines: Preparation and stability in alkaline solution. *J. Membrane Sci.* 244: 25–34.

Schollmeyer, G. 1902. Reinigung von Znckersaften durch Elektro-dialyse und mit Ozon. German Patent No. 136670.

Smitha, B., S. Sridhar, and A.A. Khan. 2003. Synthesis and characterization of proton conducting polymer membranes for fuel cells. *J. Membrane Sci.* 225: 63–76.

Smitha, B., S. Sridhar, and A.A. Khan. 2005. Proton conducting composite membranes from polysulfone and heteropolyacid for fuel cell applications. *J. Polym. Sci.: Part B: Polymer Physics* 43: 1538–1547.

Tanaka, Y. *Ion Exchange Membranes, Volume 12: Fundamentals and Applications*. 2007, Elsevier.

Taylor, L.D., and P. McLaughlin. 1976. Preparations and reactions of quaternary salts of bis(chloromethyl) ether. *J. Appl. Polym. Sci.* 20: 2225–2228.

Toshikatsu, S., M. Tsujimoto, T. Yamaguchi, et al. 1996. Change of anion exchange membranes in an aqueous sodium hydroxide solution at high temperature. *J. Membrane Sci.* 112: 161–170.

Turek, M., and B. Bandura. 2007. Renewable energy by reverse electrodialysis. *Desalination* 205: 67–74.

Vera, E., J. Sandeaux, F. Persin, et al. 2009. Modeling of clarified tropical fruit juice deacidification by electrodialysis. *J. Membrane Sci.* 326: 472–483.

Winston Ho, W.S., and K.K. Sarkar, eds. *Membrane Handbook*. New York: Van Nostrand Reinhold, 1992.

Wen, T., G.S. Solt, and Y.F. Sun. 1995. Spirally wound electrodialysis (SpED) modules. *Desalination* 101: 79–91.

Yong-Jin, C., K. Moon-Sung, and M. Seung-Hyeon. 2002. A new preparation method for cation-exchange membrane using monomer sorption into reinforcing materials. *Desalination* 146: 287–291.

Zhang, K., M. Wang, D. Wang, et al. 2009. The energy-saving production of tartaric acid using ion exchange resin-filling bipolar membrane electrodialysis. *J. Membrane Sci.* 341: 246–251.

# 19 Hemodialysis Membranes
## *History, Properties, and Future Development*

*William H. Van Geertruyden, Zhongping Huang, and William R. Clark*

## CONTENTS



## 19.1 FUNCTION AND EARLY HISTORY OF HEMODIALYSIS MEMBRANES

The human kidney regulates water volume in the blood, maintains hemostasis, and removes toxins and byproducts of human metabolic functions such as urea and creatinine. Diseases of the kidney are caused primarily by acute infection, hypertension, and diabetes. There are typically three treatment options for patients suffering from chronic kidney disease (CKD) or end-stage renal disease (ESRD): kidney transplantation, peritoneal dialysis (PD), and hemodialysis (HD). Hemodialysis refers to the extracorporeal filtering of blood through an artificial kidney for the purpose of maintaining or supplementing kidney function. It is the primary treatment modality for CKD and ESRD patients. In this process, a blood circuit is formed when blood is drawn from a patient and circulated continuously past a membrane to remove small and middle molecular weight (MW) solutes normally passed by the kidney. The artificial kidney is comprised of several components, none of which is more critical to the survival of the patient than the dialysis membrane. It is the limiting step in determining the performance of a dialysis session.

Georg Haas is generally regarded as the first to perform the first human dialysis treatment in 1924 by using celloidin (cellulose-trinitrate) tubes 1.2m in length immersed in a dialysate bath and placed in a glass cylinder (Paskale, 2001). Beginning in the 1940s with Willem Kolff and lasting through much of the 20th century, cellulose-based membranes have been used in hemodialysis.

Despite their performance in removing small middle molecular weight solutes, cellulose-based membranes (regenerated cellulose, cuprammonium cellulose, and cuprammonium rayon) were replaced by modified cellulose membranes in the 1960s due to their better biocompatibility and improved removal of larger molecular weight solutes.

## 19.2 STATE-OF-THE-ART SYNTHETIC POLYMER MEMBRANES FOR HEMODIALYSIS

In the 1970s and 1980s, the poor biocompatibility of cellulose membranes became more of a concern regarding the unexpected effects in dialysis patient. In addition, the high requirement of removal of middle and low molecule weight protein as well as the requirement of high filtration rate led to the development of many membranes that are made from synthetic materials (polysulfone, polyether-sulfone, polyamide, etc.). These membranes not only have acceptable biocompatibility, but also have higher convective and diffusive transport characteristics. Today, these synthetic hollow-fiber membranes are housed in a hemodialyzer cartridge as shown in Figure 19.1. Typically, a vascular access is made by a catheter or arterial/venous fistula via the forearm and the process can last up to 4 hours. Blood pumped from the patient is forced through the individual hollow-fiber membranes as a dialysate fluid is pumped countercurrent to the opposite side of the blood.

Compared to the cellulosic membrane, synthetic polymer membranes are thicker (>20 μm) and may be either symmetric or asymmetric in structure. Different manufacturing processes allow asymmetric membrane to have two or three layers with different pore sizes and distributions. Ideally, a membrane that has different sieving coefficient or selectivity on both the inner and outer surface of the membrane to prevent unexpected harmful solutes (e.g., endotoxin fragments) from back-transfer into the bloodstream is desired. To achieve this goal, a multiple-layer structure membrane was developed (Lysaght, 1998; Soltys et al., 2000).

The dual-skinned hollow-fiber membrane is one type of asymmetric membrane with thin skins on both the inner (lumen) and outer surface of annular macroporous matrix to form the so-called

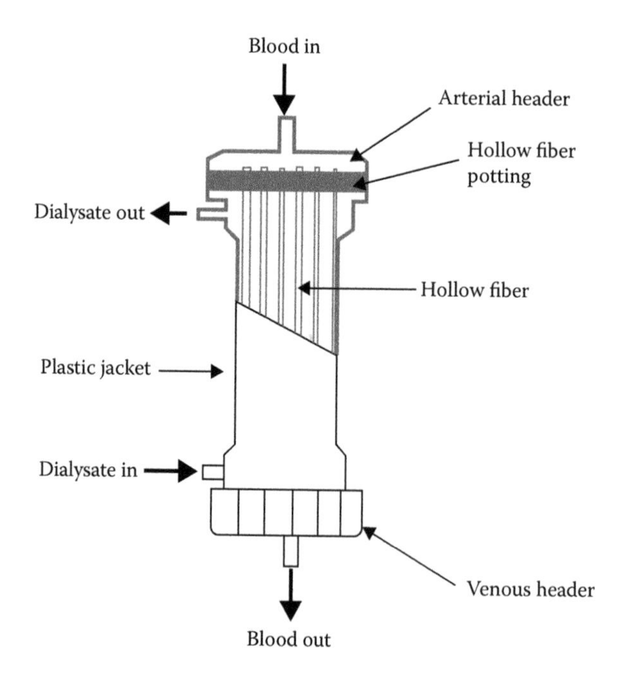


**FIGURE 19.1** Schematic diagram of a hemodialyzer cartridge.

three-layer membrane. The directional character of the convective flow across the membrane results in different sieving coefficient of dual skin membrane.

### 19.2.1 Solute Removal Mechanisms in Polymer Membranes

The primary purpose of dialysis for renal failure patients is to reestablish the body's water solute concentrations and plasma water balance that cannot be achieved by the natural kidney. In order to remove substances from the blood with a dialyzer, a solute must be eliminated from the blood side through the membrane to the dialysate side by means of diffusion, convection, and/or absorption (membrane binding).

#### 19.2.1.1 Solute Diffusion

A uremic toxin is characterized by different molecular weight, degree of protein binding, volume of distribution, and charge. The most expansively studied group of these toxins is the low molecular weight nitrogenous waste products (Clark et al., 2000). These toxins have high diffusivity of mass transfer characteristics due to their low molecular weight and lack of protein binding (Clark, 2001).

Molecules will move randomly in all directions within the solution by means of diffusion. Statistically, this movement will cause the solute transfer from the region of high concentration to the low concentration. This process is governed by Fick's law:

$$J = -DA\frac{dc}{dx} = -DA\frac{\Delta c}{\Delta x} \tag{19.1}$$

where $J$ is the diffusive mass transfer rate (mg/sec), $D$ is the diffusivity (cm$^2$/sec), $\Delta c$ is the concentration difference (mg/mL), and $\Delta x$ is the distance (mm).

This equation indicates that the mass flux ($J$) of a substance over a short distance ($\Delta x$) is proportional to its concentration difference ($\Delta c$) over this short distance and the area ($A$) of transportation front. $D$ is the diffusivity and its value depends on the type of substance and the operating temperature. The minus sign indicates that the mass transfer is from the higher concentration region to the lower region opposite to the concentration difference.

Equation 19.1 is the general form of diffusive mass transfer equation that is very similar to the heat transfer equation governed by Fourier's law. For a particular device such as a hemodialyzer, it is easy to express its solute transfer in a less general form so that the mechanism of solute mass transfer can be easy evaluated, and therefore the operating conditions can be adjusted to achieve high performance of the device. In general, $\Delta x$ is constant in the general design of a dialyzer; so the major variables are the concentration difference and the area. Equation 19.1 can be written as:

$$J = -K_o A \Delta c \tag{19.2}$$

where $K_o$ is the overall mass transfer coefficient with the unit of cm/sec and is defined as:

$$K_o = \frac{J/A}{-\Delta c} = \frac{unitflux}{drivingforce} \tag{19.3}$$

$K_oA$ is very useful parameter in comparison to hemodialysis efficiency. It is defined as the mass transfer-area coefficient with the unit of cm$^3$/sec. Equation 19.3 can also be written as:

$$unitflux = \frac{J}{A} = \frac{-\Delta c}{1/K_o} = \frac{-\Delta c}{R_o} \tag{19.4}$$

Equation 19.4 indicates that the unit mass transfer flux of a substance is proportion to the driving force and is inversely proportional to some resistance ($R_o$). This form of the equation is very similar to the heat flux equation, which is proportional to its driving force (temperature difference) and is inversely proportional to resistance (heat conductivity).

In hemodialysis, according to the above equations, the diffusive transportation of solutes can be improved by either increasing the driving force or decreasing the overall resistance across the membrane. The overall resistance is the index of the difficulty in getting from the center of the bloodstream to the center of the dialysate stream (Jacobs et al., 1996). It is the sum of resistances in the blood side, dialysate side, and the membrane itself

$$R_o = R_B + R_M + R_D \tag{19.5}$$

where $R_o$ is the total mass transfer resistance, $R_B$ is the blood-side resistance, $R_M$ is the membrane resistance, and $R_D$ is the dialysate resistance.

From Equation 19.5, we can easily find that the largest resistance is the control term, improving performance of dialysis. From fluid mechanics, the blood-side and dialysate-side resistances can be conceptualized to reside in a small layer of fluid just adjacent to the membrane surface in each compartment. This small layer is called the boundary layer and the main task to achieve high performance of dialysis is to reduce the width of both layers. In general, increasing the fluid flow rate in both compartments can reduce the resistance by increasing the shear rate, but there is an optimal combination, especially for dialysate flow rate, to achieve optimal dialysate perfusion of the fiber bundle.

The other resistance related to the membrane itself depends on membrane structure, such as membrane thickness, pore size, pore distribution, and diffusive permeability. To reduce this part of membrane resistance, several steps were made both in cellulosic membranes and in synthetic membranes. The thin wall thickness of cellulosic membranes and high porosity of synthetic membrane decrease membrane resistance. With this decrease in membrane mass transfer resistance, the other two resistances (blood-side and dialysate-side diffusive resistance) are relatively important and are the main considerations on which to focus.

#### 19.2.1.2 Solute Convection

During dialysis treatment, not only the uremic toxin is taken from blood to dialysate and useful substances (e.g., calcium, acetate) are back-transferred into blood, but excess water is also taken from the patient for volume control. This process occurs by means of ultrafiltration, which is the flow of solvent across a membrane in response to a pressure difference. Because of solvent drag, convective solute removal occurs with ultrafiltration of plasma water. Unlike diffusive transportation of substances, the driving force for this process is the hydrostatic pressure gradient, called transmembrane pressure (TMP). In order to measure this process in dialysis, it is desirable to define an operational coefficient (equivalent to $K_o$ in diffusive transport) with linear proportion to flux and pressure gradient. This coefficient is called the ultrafiltration coefficient $K_{uf}$, which measures the water permeability of dialysis membrane and is defined as follows:

$$K_{uf} = \frac{Q_{uf}}{P_B - P_D} = \frac{water\_flux}{transmembrane\_pressure} \tag{19.6}$$

$K_{uf}$ is the ultrafiltration coefficient with unit of mL/min/mmHg, $Q_{uf}$ is the water flux across the membrane or ultrafiltration rate, and $P_B$ and $P_D$ are the average pressure on the blood compartment and dialysate compartment.

Because of the solvent drag force, the solutes in the blood side will be transferred into the dialysate side, which is termed solute permeability. The solute permeability differs from water permeability

**TABLE 19.1**
**Diffusive and Convective Effect in Different Hemodialysis Treatment Methods**

| Treatment Type | Diffusive | Convective |
|---|---|---|
| Low-flux dialysis | High | Low (minimal) |
| High-flux dialysis | High | Medium |
| Hemodiafiltration | High | High |
| Hemofiltration | None | Maximum |

for different molecular weight solutes. For small solutes, the ratio of water and solute flux across the membrane is 1, but this is not the case for large molecular weigh solutes that are always lower than 1. Therefore, from this point, we can find that the convective solute removal depends on both membrane structures (pore size and pore distribution, etc.) and solutes.

Usually, a high flux hemodialyzer has greater convective mass transport than a low flux hemodialyzer because of their pore size difference. But the elimination of solutes is also contributed by the different treatments applied (Table 19.1). Since hemofiltration is pure convective transportation of uremic solute, it has maximum convective mass transport without diffusive mass transport.

One fact regarding the convective transportation across the membrane is the back-transfer of similar-sized molecules from dialysate to blood, which possibly is the cytokine-induced substance (CIS). So, the design of dialysis membrane structure mainly focuses on this issue to prevent CIS transfer but also to achieve maximum convection to total solute removal in order to better simulate the nature human kidney clearance.

Comparing diffusion to convection, solute molecular size, membrane structure, and even dialysis duration will alter dominant of solute removal in each other. Small MW solute removal is mainly mediated by diffusion, while large MW solute is primarily mediated by convection if dialysis treatment involves two removal mechanisms.

As has been discussed above, various artificial kidney treatment techniques have an effect on the solute removal mechanism. The low flux hemodialysis treatment is always related to low flux hemodialyzer, which is made of cellulose material. Because of its relative thinness and a uniform pore distribution, cellulose membrane has high diffusive transportation of urea and creatinine. On the other hand, the more open pores and thicker fiber wall of polymer membrane make it possible to deliver high water flux and high solute convective transportation. However, convection and diffusion always interact with each other and is very difficult to distinguish during modern hemodialysis. For example, in hemodiafiltration, the diffusive and convective transportation is coupled/interacted with each other. As treatment progresses, concentration gradient decreases, such that the instantaneous rate of diffusive mass transfer falls. This dissipation occurs more rapidly for small solutes. Therefore, as treatment time increases, small molecular removal becomes inefficient while middle/large solute removal is relatively preserved.

#### 19.2.1.3 Adsorption

The last important solute transport mechanism in artificial kidney is adsorption. Though the designs of membranes mainly focus on how to increase diffusive and convective removal characteristics, some membranes do have the additional benefit of removing solutes by adsorption. On the other hand, because of the adsorption characteristics, some large molecular weight protein (albumin, immunoglobulin) will deposit on the membrane surface to form the so-called second membrane or protein cake. These formations will change the membrane permeability and influence the mass transfer of uremic solutes. For small molecular weight solutes, this second membrane has little effect on its removal characteristic, but for large molecular solutes, the influence is serious. Some have found that the formation of this second membrane occurs within 20 minutes after dialysis therapy (Rockel et al., 1986). In fact, membrane characteristics (charge, hydrophobicity, porosity, etc.) will affect the formation and structure of this secondary membrane. As for the importance of dialyzer reuse process, the reprocessing chemical may also alter the membrane properties and consequently affect the structure and thickness of secondary membrane.

## 19.3 FUTURE MEMBRANES FOR HEMODIALYSIS

As was stated earlier, the performance of the dialysis membrane determines the quality and duration of the treatment. Improving the efficiency of maintenance dialysis can be accomplished in part by the development of new dialysis membranes, whereby the survival rate of the patient may be improved. The nonuniformity of pore distribution, irregularities in pore shapes and size, and the limited reusability of current dialysis membranes has led to extensive research in the area of membranes for hemodialysis.

### 19.3.1 Nanoporous Anodic Aluminum Oxide

Nanoporous anodic aluminum oxide (AAO) has received attention for use as a membrane in applications such as gas separation, drug delivery, and bone fixation (Gorokh et al., 2006; Briggs et al., 2004; Darder et al., 2005; Dawai et al., 2003). The nanoporous AAO membrane structure is inherently different from those of other ceramic based membranes produced by sol-gel, for example. Nanoporous AAO membranes have a self-ordered pore arrangement (Figure 19.2a and 2b) with high pore densities that are essential to maximize permeation and molecular flux across the membrane in a fluid separation application (Chen et al., 2005). AAO membranes with this highly uniform and self-organized nanoporous structure are an ideal choice over the contemporary cellulose-based and synthetic polymer membranes in the hemodialysis application as well. There are several advantages of AAO membranes over membranes currently used, including high porosity and uniform pore size from 5 up to 300 nm, high hydraulic conductivity (water permeability), uniform distribution of pores (less than 1% variation), excellent pore structure, and high resistance to chemical and temperature degradation. Nanoporous AAO is created as a thin-film oxide during the electrochemical process of anodization of an aluminum substrate. The aluminum substrate is placed in an acid electrolyte as the anode in an electrochemical cell setup. As the electrical potential is raised, a nanoporous array of aluminum oxide is grown at a specified rate. Membrane characteristics, such as pore size, interpore spacing, and membrane thickness are highly dependent on the anodization parameters, the details of which are shown elsewhere (Masuda and Fakuda, 1995; Masuda and Satoh, 1996; Masuda and Hasegwa, 1997). AAO membranes have been successfully manufactured in both sheet and tube form.

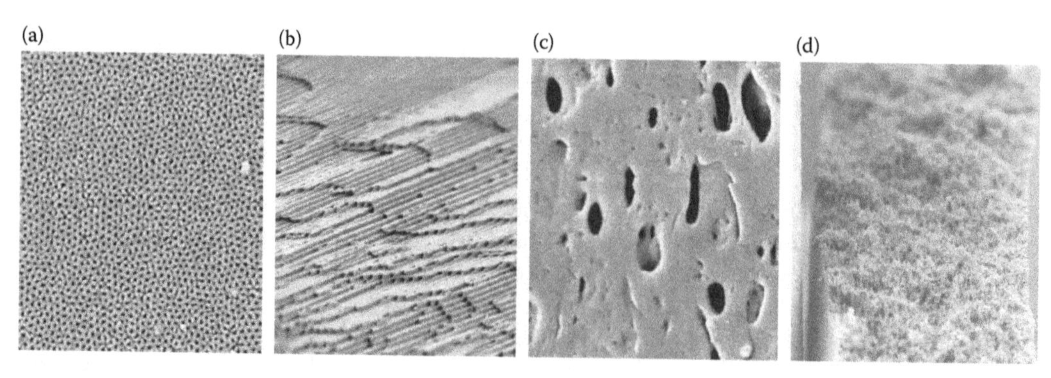


**FIGURE 19.2** Surface view (a,c) and cross-sectional view (b,d) of a state-of-the-art polyethersulfone (c,d) dialysis membrane and a nanoporous AAO membrane (a,b).

#### 19.3.1.1 High Porosity and Uniform Pore Size

Figure 19.2a and 2b shows high resolution scanning electron microscope (SEM) images of the morphology of an AAO membrane. In comparison, a polyethersulfone membrane currently used as a dialysis membrane is shown in Figure 19.2c and 2d. The pore sizes on the surface of this polyethersulfone dialysis membrane are not uniform, and the regularity of pore shapes is also unsatisfactory. Some pores appear oval in shape and some appear as slits. In contrast, the pore shape on the AAO membrane surface is regular in shape and is uniformly circular. Additionally, the cross section of the AAO membrane reveals straight nanochannels, whereas the polymer membrane exhibits a tortuous structure.

The determinants of convective solute removal are primarily the sieving properties of the membrane used and the ultrafiltration rate (Henderson, 1996). The mechanism by which convection occurs is termed solvent drag. If the molecular dimensions of a solute are such that transmembrane passage occurs, the solute is swept ("dragged") across the membrane in association with ultrafiltered plasma water. Thus, the rate of convective solute removal can be modified either by changes in the rate of solvent (plasma water) flow or by changes in the mean effective pore size of the membrane. If a straight cylindrical pore model is considered (such as the case in this ceramic membrane), the fluid flow along the length of a cylinder is governed by the Hagen–Poiseuille equation (Bird and Stewart, 1960):

$$\Delta P = \frac{Q\rho\mu L}{\pi r^4} \text{ or } Q = \frac{\pi r^4 \Delta P}{\rho\mu L} \tag{19.7}$$

where $\Delta P$ is the pressure gradient across the membrane (transmembrane pressure), $Q$ is the flow rate or ultrafiltration rate across the membrane, $\rho$ is the density of the fluid, $L$ is the length of pore channel, and $r$ is the radius of pore. So, the rate of ultrafiltration is directly related to the fourth power of the pore radius at a constant transmembrane pressure. In other words, the convective transfer of solute is determined by the fourth power of the pore radius.

The diffusive properties of a dialysis membrane are determined mainly by the porosity and pore size (Clark and Ronco, 2001). Based on a cylindrical pore model (Clark and Gao, 2002), membrane porosity is directly proportional to both the number of pores and the fourth power of the pore radius ($r^4$). Therefore, diffusive permeability is strongly dependent on pore size. Studies over the past 15 years suggest a direct relationship between delivered urea-based HD doses and patient outcome (Gotch and Sargent, 1985; Owen et al., 1993; Parker et al., 1994; Collins et al., 1994; Hakim et al., 1994). Since the elimination of low-molecular weight (MW) nitrogenous waste products is mainly obtained by diffusion through a dialysis membrane, higher porosity will achieve better elimination of these uremic toxins. In other words, AAO membranes with higher porosity can deliver a higher urea-based HD dose than polymer membranes given the same time.

#### 19.3.1.2 High Hydraulic Conductivity (Water Permeability)

A clear trend in hemodialysis today is the increasing use of convective therapies (i.e., hemofiltration (HF) and hemodiafiltration (HDF)). This is largely because, in comparison with high-flux HD, these convective therapies provide (a) significantly higher clearance of relatively large uremic toxins (e.g., β2-microglobulin), and (b) improved hemodynamic stability. β2-microglobulin is a low molecular weight (11,800 daltons) protein found on the membranes of virtually all body cells. Free β2-microglobulin is a product of cell breakdown. It is secreted by the renal glomeruli, then absorbed and catabolized by the renal tubular cells. Decreased glomerular filtration is associated with high serum levels of β2-microglobulin. Dialysis-related amyloid that can be found in patients with amyloid bone disease is largely a consequence of high β2-microglobulin levels in dialysis (Huang et al., 2004). Due to the potential for amyloid disease, the removal efficiency of β2-microglobulin in hemodialysis plays a very important role in patient outcome. The removal mechanism of these solutes is dominated by convective transport related to hydraulic permeability.

A previous study by Huang et al. showed that hydraulic conductivity (water permeability) of a sheet AAO membrane was approximately twice that of a polyethersulfone dialysis membrane (Huang et al., 2004). As stated previously, polyethersulfone dialysis membranes have an irregular pore structure (both inner and outer surfaces) and a wide pore size distribution, whereas AAO membranes have a highly monodisperse pore size distribution. The more uniform the pore size of the membrane, the higher the hydraulic conductivity of the membrane. Enhanced convective transfer of middle and large molecular weight solutes can therefore be achieved by using alumina membranes with these properties.

#### 19.3.1.3 Excellent Pore Structure and High Resistance to Chemical and Temperature Degradation

Cellulose and synthetic polymer-based membranes are often reused in dialysis equipment. The potential disadvantages of reused dialyzers include deposition of blood elements onto the dialyzer membrane, exposure of patients to reprocessing chemicals, qualitative or quantitative changes in dialyzer membrane surface area and permeability, and loss of structural integrity of the dialyzer (Stoncek et al., 1984). In addition, reuse of the dialyzer could potentially decrease the clearance of uremic toxins and may lead to a decrease in the delivered dose of dialysate (Held et al., 1987). These disadvantages are mainly associated with the structure and materials used in current dialysis membranes. When comparing the cross section of two membranes, nanoporous AAO and polyethersulfone, large differences in their structures are evident (Figure 19.2b and 2d). The channel or path from the interior to the exterior surface of the polyethersulfone dialysis membrane is not straight. It is unclear if there are true channels or paths from one side to the other. Instead, the cross section of the polymer membrane resembles a spongelike material. Accordingly, in practice there might be some blood or blood fragments left inside this structure even after a thorough cleaning with chemical reagents. In contrast, the channel or path connecting both surfaces of an AAO membrane is straight and smooth. Besides the above structure limitation, heat disinfecting methods cannot be used on cellulose or synthetic polymer dialyzers due to their inadequate temperature resistance. In contrast, the high temperature resistance of AAO membranes makes them more ideal for heat disinfecting processes, thereby significantly increasing the reusability.

#### 19.3.1.4 Uniform Pore Distribution

Hemodialysis membranes currently used in dialysis equipment do not have a cylindrical pore structure with monodispersed pore sizes; rather, they exhibit a wide pore size distribution and a tortuous structure. The different pore size distribution will have a significant influence on the membrane's sieving properties. Figure 19.3 shows pore size distributions (a) and solute sieving coefficient

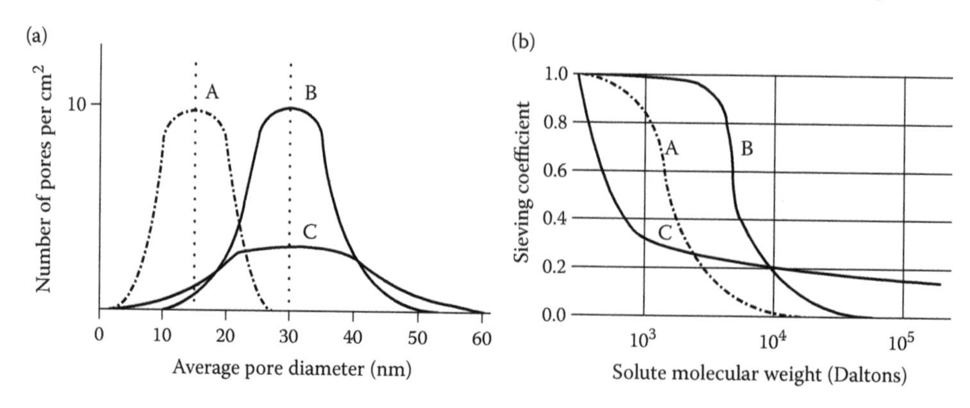


**FIGURE 19.3** Pore size distribution (a) and solute sieving coefficient (b) profiles for three hypothetical dialysis membranes. (Reprinted with permission from Clark, W. and Ronco, C., 2001, *Nephrol. Dial. Transplant.* 16(Suppl. 5): 56–60.)

**TABLE 19.2**
**Solute Diffusive Permeability, $P_s$ (cm/s)**

| Solute | Alumina Membrane | PES Membrane | % Increase |
|---|---|---|---|
| Urea | 2.48E-03 | 9.5E-4 | 161 |
| Creatinine | 1.57E-03 | 6.57E-4 | 139 |
| Vancomycin | 1.16E-03 | 4.4E-4 | 164 |

profiles (b) for three hypothetical dialysis membranes (Clark and Ronco, 2001) The hypothetical membrane represented by curve A has a large number of relatively small pores while the membrane represented by curve B has a large number of relatively large pores. Based on the narrow pore size distribution, the solute sieving coefficient versus molecular weight profile shown in Figure 19.3b for both membranes has the desirable sharp cutoff, similar to that of a human kidney.

The hypothetical membrane represented by curve C exhibits a pore size distribution that is unfavorable from both a diffusive transport and sieving perspective. The broad distribution of pores explains not only the "early" drop-off in sieving coefficient at relatively low MW but also the tail effect at high MW. This latter phenomenon will cause unacceptably high albumin loss during hemodialysis. The membrane represented in Figure 19.2c will likely show a pore size distribution very similar to the undesirable situation shown in curve C of Figure 19.3a.

Experimental studies have been conducted on minimodule hemodialyzers to measure the hydraulic conductivity, diffusive permeability, rejection coefficient/sieving coefficient, and clearance of solutes (urea, creatinine, vancomycin, inulin, and albumin) of nanoporous AAO tube membranes (Attaluri et al., 2009; Huang et al., 2008; Belwalkar et al., 2008). For the nanoporous alumina membrane, the measured hydraulic permeability was: $Lp = 8.7 \times 10^{-9}\ m \times s^{-1} \times Pa^{-1}$ ($K = 30.3 \times 10^{-15}\ m^2 \times s^{-1} \times Pa^{-1}$), while for a polyethersulfone (PES) membrane, it has been reported as: $Lp = 5.02 \times 10^{-10}\ m \times s^{-1} \times Pa^{-1}$ ($K = 15.06 \times 10^{-15}\ m^2 \times s^{-1} \times Pa^{-1}$) (Liao et al. 2005). The hydraulic conductivity of the AAO tube membrane is therefore approximately twice that of the PES membrane.

The solutes diffusive permeability results are shown in Table 19.2 and Figure 19.4. The diffusive permeability for three investigated solutes passing through the alumina membrane is approximately 140% higher than for the PES membrane (Liao et al., 2005). The difference is most likely due to the difference in pore structure that allows easier passage of solutes through the straight channels on the AAO membrane.

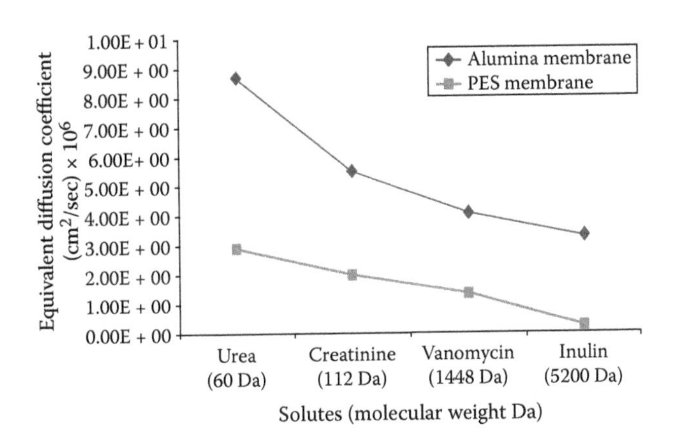


**FIGURE 19.4** Equivalent diffusion coefficient versus molecular weight.

**TABLE 19.3**
**Solute Sieving Coefficient**

| Solute | $R_{obs}$ | SC |
|---|---|---|
| Urea | 0.014 | 0.98 |
| Creatinine | 0.002 | 0.99 |
| Vancomycin | 0.044 | 0.95 |
| Inulin | 0.047 | 0.95 |
| Albumin | – | <0.003 |

Sieving coefficients for middle molecular solutes (vancomycin and inulin) have been shown to be very high for nanoporous AAO while the sieving coefficient for albumin was very low, as shown in Table 19.3. These results demonstrate that nanoporous AAO membranes can significantly increase the clearance of middle molecular weight solutes through high convective mass transfer capacity while preventing any loss of albumin.

Solutes clearance results are shown in Table 19.4. Because of the size of minimodule alumina membranes dialyzers, the in vitro experiments were conducted with a total test solution of 1 liter under a blood flow rate, dialysate flow rate, and ultrafiltration rate of 15 ml/min, 20 ml/min, and 2 ml/min, respectively. In order to make a valid comparison, the results are also expressed in terms of solute reduction ratio, especially for urea reduction ration (URR), which is frequently used in clinical evaluation. The URR calculation was based on a scaled-down in vitro condition from a regular dialysis system with volume 6 liters and blood flow rate of 400 ml/min. The URR for the alumina membrane dialyzer was 63% higher than the clinical accepted value (0.22) for average 1-hour hemodialysis.

### 19.3.2 "Smart" Membranes and Other Approaches

Thus far, commercially available extracorporeal artificial kidneys have supplemented only one major component of the kidney: removal of small molecular weight solutes from the blood. However, there is research being performed to enhance the performance of artificial kidneys in its metabolic, endocrine, and immune roles. The renal assist device (RAD) or bioartificial kidney has been developed by researchers at the Cleveland Clinic and University of Michigan with this goal in mind (Fissell, 2006). The RAD process utilizes human or animal renal cortical epithelial cells and seeded within hollow-fiber bioreactors. The RAD acts as a metabolically active replacement for the renal proximal tubule.

Another approach to developing novel membranes for hemodialysis has come from utilizing a MEMS-based approach. Fissell et al. have demonstrated a nanoporous polysilicon membrane using microfabrication techniques to produce ~9 nm wide slits ~45 μm in length with a 2-μm separation between slits (Fissell et al., 2002, 2009). While the porosity of these membranes may not match

**TABLE 19.4**
**Solute Clearance or Reduction Ratio**

| Solute | Clearance (ml/min) | Reduction Ratio/Hour |
|---|---|---|
| Urea | 9.03 ± 0.15 | 0.36 ± 0.01 |
| Creatinine | 8.96 ± 0.15 | 0.36 ± 0.01 |
| Vanomycin | 7.81 ± 0.18 | 0.31 ± 0.01 |
| Inulin | 6.88 ± 0.31 | 0.28 ± 0.01 |

AAO membranes and thus be used in intermittent dialysis treatment, their use as a membrane implantable device has gained interest.

## REFERENCES

Attaluri, A., Z. Huang, A. Belwalkar, et al. 2009. Evaluation of nano-porous alumina membranes for hemodialysis application. *ASAIO J.* 55(3): 217–223.

Belwalkar, A., E. Grasing, W. Van Geertruyden, et al. 2008. Effect of processing parameters on pore structure and thickness of anodic aluminum oxide (AAO) tubular membranes. *J. Membrane Sci.,* 319: 192–198.

Bird, R., W. Stewart, and E.N. Lightfoot in *Transport Phenomena*, 1st ed., R.B. Bird, W.E. Stewart, and E.N. Lightfoot, eds. New York: John Wiley and Sons. 1960, pp. 34–70.

Briggs, E., A. Walpole, P. Wilshaw, et al. 2004. Formation of highly adherent nano-porous alumina on Ti-based substrates: A novel bone implant coating, *J. Mater. Sci. Mater. Med.* 15: 1021–1029.

Chen, W., J. Yuan, and X. Xia. 2005. Characterization and manipulation of the electroosmotic flow in porous anodic alumina membranes. *Anal. Chem.* 77: 8102–8108.

Clark, W., and J.H. Shinaberger. 2000. Effect of dialysate-side mass transfer resistance on small solute removal in hemodialysis. *Blood Purif.* 18: 260–263.

Clark, W., and L.W. Henderson. 2001. Renal vs. continuous vs. intermittent therapies for removal of uremic toxins. *Kidney Int.* 59 (Suppl. 78): S298–S303.

Clark, W., and C. Ronco. 2001. *Nephrol. Dial. Transpl.* 16(Suppl. 5): 56–60.

Clark, W., D. Gao, 2002. *Contrib Nephrol* 137: 70–77.

Collins, A., J. Ma, A. Umen, et al. 1994. *Am. J. Kidney Dis.* 23: 272–282.

Darder, P. A., M. Hernandez-Velez, E. Manova, et al. 2005. Encapsulation of enzymes in alumina membrane of controlled pore size. *Thin Solid Films* 495: 321–326.

Dawai, G., V. Yadavalli, M. Paulose, et al. 2003. Controlled molecular release using nanoporous alumina capsules. *Ther. Micro. Nanotechnol., Biomed. Microdevic.* 5: 75–80.

Fissell, W., A. Dubnishevab, A. Eldridgeb, et al. 2009. High-performance silicon nanopore hemofiltration membranes. *J. Membrane Sci.* 326: 58–63.

Fissell, W. 2002. Initial characterization of a nanoengineered ultrafiltration membrane. *J. Am. Soc. Nephrol.* 13: 602A.

Fissell, W. 2006. Developments towards an artificial kidney. *Expert Rev. Med. Devic.* 3: 155–165.

Gorokh, G., A. Mozalev, D. Solovei, et al. 2006. Anodic formation of low-aspect-ratio porous alumina films for metal-oxide sensor application. *Electrochim. Acta* 52: 1771–1780.

Gotch, F., and J. Sargent. 1985. A mechanistic analysis of the National Cooperative Dialysis Study (NCDS). *Kidney Int.* 28: 526–534.

Hakim, R., J. Breyer, N. Ismail, et al. 1994. *Am. J. Kidney Dis.* 23: 661–669.

Held, P., M. Pauly, and L. Dinmond. 1987. *J. Amer. Med. Assoc.* 257: 645–650.

Henderson, L. 1996. Biophysics of ultrafiltration and hemofiltration. In *Replacement of Renal Function by Dialysis*, 4th ed., C. Jacobs, C. M. Kjellstrand, K.M. Koch, and J.F. Winchester, eds. Dortdrecht: Kluwer Academic Publishers, pp. 114–145.

Huang, Z., W. Zhang, and S. Tang. 2004. ASME IMECE 2004 Paper #59404.

Huang, Z., A. Attaluri, A. Belwalkar, et al. 2008. An Experimental Study of Transport Properties of Ceramic Membranes for Use in Hemodialysis, Proceedings of the ASME 2008 Summer Bioengineering Conference, Marco Island, Florida.

Humes, H., D. Buffington, S. MacKay, et al. 1999. Replacement of renal function in uremic animals with a tissue-engineered kidney. *Nat. Biotechnol.* 17: 451–455.

Jacobs, C., C. M. Kjellstrand, K. M. Koch, et al. 1996. *Replacement of Renal Function by Dialysis*, 4th ed. Kluwer Academic Publishers.

Liao, Z., E. Klein, C. Poh, et al. 2005. Measurement of hollow fiber membrane transport properties in hemodialyzers. *J. Membrane Sci.* 256: 176–183.

Lysaght, M. J. 1998. Hemodialysis membrane in transition. *Contrib Nephrol.* 61: 1–17.

Masuda, H., and K. Fukuda. 1995. "Ordered metal nanohole arrays made by a two-step replication of honeycomb structures of anodic alumina." *Science* 268: 1466.

Masuda, H., and M. Satoh. 1996. Fabrication of gold nanodot array using anodic porous alumina as an evaporation mask. *Jpn. J. Appl. Phys.* 35: L126.

Masuda, H., and F. Hasegwa.1997. Self-ordering of cell arrangement of anodic porous alumina formed in sulfuric acid solution. *J. Electrochem. Soc.* 144: L127–L130.

Owen, W., N. Lew, Y. Liu, et al. 1993. *N. Engl. J. Med.* 329: 1001–1006.
Parker, T., L. Husni, W. Huang, et al. 1994. *Am. J. Kidney Dis.* 23: 670–680.
Paskalev, D. 2001. Georg Haas (1886–1971): The Forgotten Hemodialysis Pioneer. *Dialysis & Transplantation.* 30(12): 828.
Rockel, A., J. Hertel, P. Fiegel, et al. 1986. Permeability and secondary membrane formation of a high flux polysulfone filter. *Kidney Int.* 30: 429–432.
Soltys, P.J., A. Zydney, J.K. Leypoldt, et al. 2000. Potential of dual-skinned, high-flux membranes to reduce backtransport in hemodialysis. *Kidney Int.* 58: 818–828.
Stoncek, D., P. Keshaviah, P. Craddock, et al. 1984. *J. Lab. Clin. Med.* 104: 304–311.

# 20 Separation of Homogeneous Liquid Mixtures by Pervaporation

*S. Sridhar and K. Sunitha*

## CONTENTS

## 20.1 OVERVIEW OF PERVAPORATION

### 20.1.1 Introduction

Membrane-based processes have now become an integral part of our day-to-day life. The rapid development of membrane technology is due to the large number of practical applications possible on an industrial scale. Membranes differ widely in their structure, function, and mode of operation depending on the process application. Criteria for membrane development include intrinsic characteristics of efficiency, simplicity, flexibility, environmental compatibility, and scale-up. Membranes can contribute significantly to process intensification, an interesting strategy offered for realizing a sustainable industrial growth that is compatible with desirable quality of life.

Conventional industrial separation processes are based on energy-intensive methods, which create an increasing market for membrane technologies and thus motivating researchers and industrialists in developing reliable design strategies that aim to optimize economic performance of membrane separation systems (Halwagi, 1992; Qi and Henson, 2000). Current membrane separation technology as a whole has grown in importance because of increasingly stringent requirements for product purity (Porter, 1990). Among the concentration-driven processes, the pervaporation (PV) technique has been studied extensively because of its potential for solving difficult separation problems (Fleming and Slater, 1992). PV is a rapidly developing membrane process with the potential for replacing or complementing distillation for purification of industrial solvents.

PV differs from other membrane processes because a phase change occurs across the barrier (Hickey et al., 1992). The name "pervaporation" is coined from the terms permeation and evaporation. A particular species in the liquid state is transported due to the driving force of chemical potential from one side of the nonporous membrane to the other side where it evolves into a vapor state (Figure 20.1). Separation is based on differences in sorption and diffusion properties of the permeating components and membrane permselectivity (Mulder and Smolders, 1984; Lee, 1975). PV is used mostly where distillation is difficult, hazardous, or uneconomical.

In PV, separation is influenced by the chemical nature and morphology of the dense nonporous membrane and depends upon the extent of interaction between the polymer and the permeating species. The volatile solute diffuses from the bulk feed to the membrane surface where preferential sorption takes place followed by diffusion through the matrix to the permeate side where desorption occurs. If the boundary layers at the feed side and the permeate side are considered, then a five-step mechanism for mass transfer may be proposed that involves (1) diffusion of solute from the bulk feed to the liquid-membrane interface, (2) preferential sorption into the membrane upstream layer, (3) diffusion through the membrane thickness to the permeate side, (4) desorption into vapor at the downstream side, and (5) diffusion from the membrane-vapor boundary layer to the bulk vapor phase.

### 20.1.2 Origin and Environmental Impact

The PV technique was first described by Kober in 1917 when he noticed liquid evaporating through a tightly closed collision bag suspended in air (Kober, 1917; Karlsson and Tragardh, 1994). Quantitative work on PV was done on ethanol/water mixture (Heisler et al., 1956) as well as separation of benzoic acid, hydroquinone, and citric acid from aqueous solutions. The process was found to be energy-efficient as long it involved removal of the minor feed constituent through the membrane. The environmental impact of the process is relatively nil and there is no addition or discharge of hazardous chemicals or dissipation of heat.

### 20.1.3 Pervaporation Theory

Nonporous (dense) membranes are used in PV since this process does not function by molecular sieving or convective flow. A solution-diffusion model was described for a homogeneous dense membrane for the first time by Binning (Binning et al., 1961). The driving force for mass transfer is established through a difference in vapor pressures of the permeants across the membrane. PV

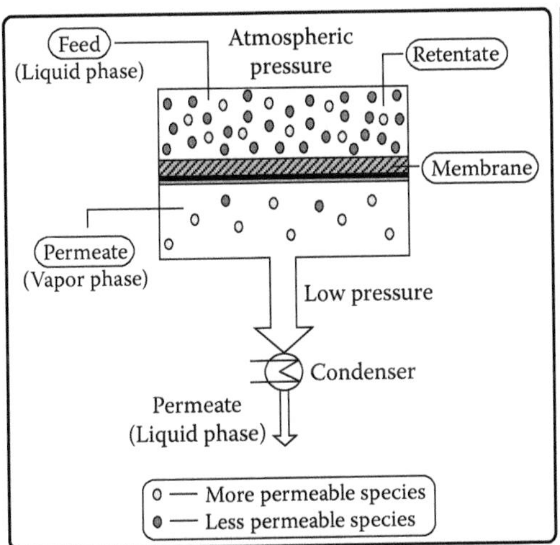


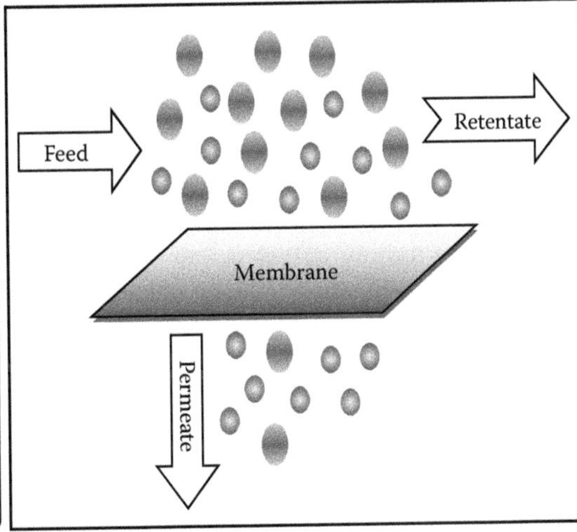


**FIGURE 20.1** Basic principle of pervaporation.

is governed by the polarity differences among feed components and separation is obtained since the membrane has the ability to transport one particular component more readily than the other(s) present in the feed solution due to preferential affinity. Rate of mass transfer depends on membrane thickness, temperature, and pressure conditions as well as properties of the permeating species such as vapor pressure, polarity, and molecular size and shape. Sorption is a thermodynamic parameter that occurs when the solution is brought in contact with the surface of the dry membrane. Swelling occurs due to affinity between the polymer and the liquid. Equilibrium is reached after a certain period. The permeation is described by Fick's law:

$$J_i = -D_i \frac{dc_i}{dt} \tag{20.1}$$

where $J_i$ is the flux, $D_i$ is the diffusivity, $c_i$ is the concentration of component i within the membrane, and t the transmembrane distance.

*Activation energy for PV transport:* The phase change of the permeating species is one of the most distinguishing features of PV. However, except for the obvious fact that sufficient energy needs to be supplied to prevent a temperature drop during PV, enthalpy for vaporization of permeate is supplied from the feed itself, which gradually experiences a drop in temperature. The temperature dependence of flux generally exhibits an Arrhenius-type relation given by:

$$J = J_o \exp\left(\frac{E}{RT}\right) \tag{20.2}$$

where $E$ is activation energy for permeation.

Based on solution-diffusion theory, the flux is given by:

$$J = \frac{J_o}{l}\left(P_o - P_l\right) \tag{20.3}$$

where $p_0$ and $p_l$ are the partial vapor pressure of the permeant in liquid feed and permeate vapor, respectively, and $l$ is the membrane thickness. $P$ is the permeability coefficient of the membrane and is related to the solubility coefficient ($S$) and diffusivity coefficient ($D$):

$$P = D * S \tag{20.4}$$

where $D$ and $S$ normally depend on temperature. The extent of interaction between the polymer and permeating species is an important factor to be considered. Two structural parameters that affect the permeability of the feed species are the nature of a polymer that is determined by its glass transition temperature and degree of crystallinity measured by X-ray diffraction. Diffusion is controlled by the probability of forming sufficient free volume within the membrane to enable mobility. The diffusion coefficient $D$ is a kinetic parameter that may be related to this probability in the simplest case by:

$$D = RTA_d \exp\left(\frac{B_d}{V_f}\right) \tag{20.5}$$

where $R$ is the universal gas constant, $T$ is the absolute temperature, $B_d$ is the amount of free volume required, $A_d$ is a parameter related to the size and shape of the permeant, and $V_f$ is the fractional free volume.

### 20.1.4 Vapor Permeation and Membrane Distillation

Vapor permeation (VP), also known as evapomeation, is a membrane-based process that can be used to separate mixtures of saturated vapors. In vapor permeation, the feed is vapor and no phase change occurs during the process, which makes it less complex. VP has proven to be especially suitable for the purification of top products of fractionation columns that can be charged directly as feed. Although VP is less sensitive to concentration polarization, mass transfer is sensitive to the degree of superheating and frictional losses at the feed side.

Membrane distillation (MD) is a separation method that uses a nonwetting, microporous membrane and is also known as membrane evaporation/thermopervaporation. The porous membrane separates the liquid and vapor phases and the pores are not wetted by the liquid. Separation by MD is based on the relative volatility of various components in the feed solution. The feed in liquid phase is exposed to one side of the membrane just as in PV process and components of higher volatility pass through the membrane pores by a convective or diffusive mechanism. There are no interactions between the membrane and the liquids present in the feed.

### 20.1.5 Solution Diffusion Model and Mass Transfer Coefficient

Solution-diffusion is a common mechanism for dense membranes (Figure 20.2) that follows the five steps described in Section 20.1.1. Diffusion of a particular component through the membrane depends on several physical and chemical factors such as size and shape of the liquid feed molecules as well as liquid-liquid (coupling effect) and polymer-liquid interactions. Vaporization of the permeated component is rapid when the partial pressure is kept at a low value at the permeate side. According to the solution-diffusion model, high selectivity is obtained in PV by selective sorption of one particular component followed by its diffusion through the barrier. In general, high solubility leads to high diffusion due to the following reasons:

1. A greater quantity of preferentially permeating component enters the polymer, swells it, and promotes the free rotation of the polymer segments about the chain axis that reduces the activation energy required for diffusion.
2. The free volume in the polymer increases, allowing diffusion of the noninteracting component through the membrane since the amorphous phase is swollen. As a result competitive diffusion takes place leading to poor selectivity. PV is more effective and economical than distillation when the separation is performed for small volatility differences (close boiling points) and azeotropic mixtures.

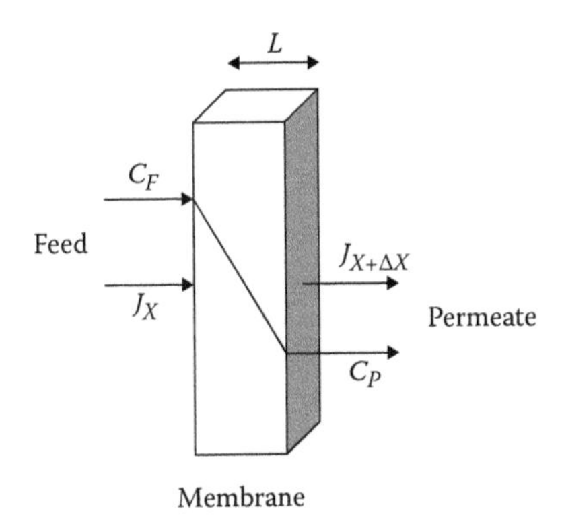


**FIGURE 20.2** Solution-diffusion model.

However, liquid and gas permeation both follow Fick's first law of diffusion, where the time for attaining steady state is inversely proportional to membrane thickness (Binning, 1961):

$$J_x = \frac{D(C_F - C_P)}{L} \tag{20.6}$$

where $J_x$ is the flux of the permeating component, $D$ is the diffusion coefficient, $(C_F - C_P)$ is the concentration difference across the membrane, and $L$ is the thickness of the membrane.

The resistance in the series model can be used to describe mass transfer. Figure 20.3 exhibits the concentration profile for the organic solute based on film theory. In this model, the permeate flux is related to the concentration gradient between the two bulk phases through an overall mass transfer coefficient:

$$J_i = k_l V_l (x_f - x^*) \tag{20.7}$$

where $J_i$ is the organic flux of component i, $k_l$ is the liquid boundary layer mass transfer coefficient, $V_l$ is the total molar volume of the feed, and $x_f$ and $x^*$ are the mole fractions of the component in the feed and at the membrane-fluid interface, respectively. Similarly, the flux through the permeate vapor phase boundary layer is expressed as:

$$J_i = K_v (p^* - p_b) \tag{20.8}$$

where $K_v$ is the vapor boundary layer mass transfer coefficient and $p$ is the partial vapor pressure. This leads to a simple flux expression where $J_i$ could explicitly be related to the operating conditions such as downstream pressure:

$$J_i = \frac{Q_i \left( p_i^{sat} \gamma_i \, x_b - p_b \right)}{L} \tag{20.9}$$

where $Q_i$ is the overall permeability, $L$ is the membrane thickness, $\gamma_i$ is the activity coefficient, and $p_i^{sat}$ is the saturated vapor pressure.

Similarly,

$$J_i = \frac{Q_i^m \left( p_i^{sat} \gamma_i \, x_i - p^* \right)}{L} \tag{20.10}$$

Here $Q_i^m$ is the membrane vapor permeability.

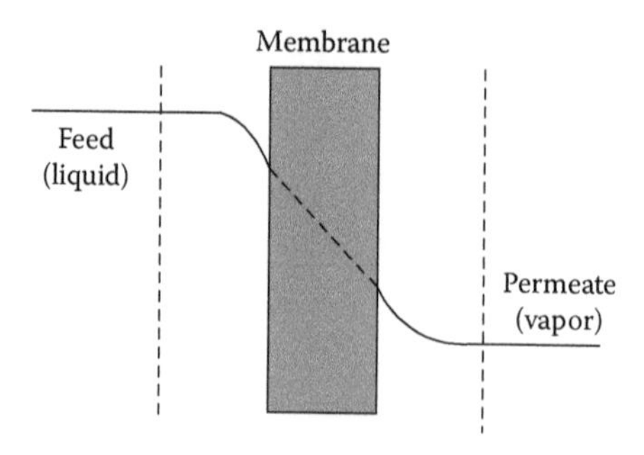


**FIGURE 20.3** Concentration profile across the membrane.

Combining the four equations (Equations 20.7 through 20.10) leads to:

$$\frac{L}{Q_i} = \frac{p_i^{sat}}{k_l \rho_l} + \frac{L}{Q_i^m} + \frac{1}{k_v} \tag{20.11}$$

By rearranging the terms in the above equation the following expression is obtained:

$$\frac{1}{Q_i \, H_i' / L} = \frac{1}{k_l} + \frac{L}{Q_i^m H_i'} + \frac{1}{k_v H_i'} \tag{20.12}$$

where $H_i = p_i^{sat} \gamma_i^\infty$ and $H_i' = p_i^{sat} \gamma_i^\infty / \rho_l$ are Henry's law constants, $\gamma_i^\infty$ is the infinite dilution activity coefficient, and $p_i^*$ is the partial vapor pressure at equilibrium. This equation is further simplified as:

$$\frac{1}{k_i} = \frac{1}{k_l} + \frac{L}{Q_i^m} + \frac{1}{k_v H_i'} \tag{20.13}$$

Here $k_i = Q_i H_i' / L$ gives the overall mass transfer coefficient (Raghunath and Hwang, 1992; Nijhuis, 1991). By assuming thermodynamic equilibrium at the membrane-fluid interfaces, the inverse of overall mass transfer coefficient gives the overall resistance.

### 20.1.6 Performance Parameters and Advantages

PV is a rate-controlled separation process that is addressed by three issues: membrane productivity (flux), selectivity (separation factor), and stability. Flux is the amount of a specific component permeating per unit area per unit time for a given membrane:

$$J = \frac{Q}{A * T} \tag{20.14}$$

where $J$ (kg/m$^2$ h) is the flux and $Q$ (kg) is the weight of the component that permeated through an effective area $A$ (m$^2$) in time $T$ (h). The selectivity can be determined from the feed and permeate compositions as follows:

$$\alpha = \frac{X_i * Y_j}{X_j * Y_i} \tag{20.15}$$

where $X$ refers to feed and $Y$ refers to permeate, and i and j are the mole fractions of the slower and faster permeating components, respectively. When the separation factor is unity, no separation occurs; when it approaches infinity, the membrane becomes perfectly "semipermeable". Permeability of the components across the membrane is the molar flux per unit pressure and is given by:

$$P_E = \frac{J}{Mw * P} \tag{20.16}$$

where $P_E$ is the permeability, $J$ is the molar flux, Mw the molecular weight, and $P$ the vapor pressure of the particular component. Industries have a widespread attraction toward PV as an energy-efficient technique for separating close boiling and azeotropic liquid mixtures.

PV offers many advantages when compared to conventional distillation:

- It reduces the demand for energy, because only the liquid fraction to be separated is vaporized.
- It can be operated without consuming any sorbents.
- It does not release any emissions into the environment.
- It can be applied for recovery of hazardous and heat sensitive compounds.
- It is not limited by azeotropic compositions.
- It is flexible and has compact modular design.
- It is safe for the operator.

The most attractive feature of PV is its ability in separating azeotropic, close boiling, and heat-sensitive mixtures. Unlike distillation, where separation is based on the boiling point differences of the components, PV does not require high temperatures for operation and in case of hazardous and explosive materials, ambient temperature can be maintained. PV can be operated continuously without consuming any extra agents for recovery of solvents. PV systems have additional advantages of flexibility and compactness which is ideal for varying feed compositions and desirable product purities especially in the pharmaceutical industry, where different solvents are used in drug synthesis and effluents change significantly in composition from one batch to another.

### 20.1.7 Types of Pervaporation Membranes and Their Suppliers

Membranes for PV can be classified on the basis of the nature of material from which they are made, such as polymeric, inorganic, and mixed matrix.

#### 20.1.7.1 Polymeric Membranes

Polymers constitute the bulk of PV membranes used worldwide. The most widely commercialized hydrophilic membrane is composite poly(vinyl alcohol) (PVA) cast on a polyacrylonitrile (PAN) ultraporous substrate for the dehydration of rectified spirit to absolute ethanol. The most common hydrophobic/organophilic membrane is poly(dimethyl siloxane) (PDMS), also known as silicone rubber, which finds application for extraction of volatile organic compounds (VOCs) from aqueous solutions and industrial effluents. Commercial suppliers include CM-Celfa, Switzerland, as well as Sulzer Chemtech GMBH and GKSS Research Center based in Germany, who fabricate modules for dehydration of alcohols, ketones, carboxylic acids, and ethers as well as extraction of volatile organic compounds (VOCs) from aqueous streams.

#### 20.1.7.2 Inorganic Membranes

Inorganic membranes are versatile and offer better resistance to chemical attack and high temperatures. Metal membranes are stable at temperatures ranging from 500°C to 800°C and many ceramics are usable even at 1000°C, which makes them attractive for further research. In general, the types of inorganic materials used are ceramics, glass, metals, and zeolites. Zeolites are useful materials for producing membranes with crystalline pores. The term zeolite was first proposed in 1756 by Swedish geologist A. F. Cronstedt, and it designates a variety of hydrated microporous alumino silicates with a general formula $M_{x/n}.[(AlO_2)_x (SiO_2)_y].zH_2O$ and a framework structure based on channels and cavities. More than 135 types of natural and synthetic zeolites have been reported. In general the physical and chemical properties of zeolites are determined by the Si/Al ratio. Transport occurs by adsorption in pore walls and molecular sieving. Commercial silica membranes coated on ceramic supports for dehydration applications at ambient and high temperatures (up to 250°C) are supplied by Pervatech BV, Netherlands, as well as Sulzer Chemtech GMBH. Commercial NaA type zeolite membranes, fabricated by Mitsui Engineering, Japan, have shown enhanced thermal and mechanical stability over polymeric ones. Other materials used for making inorganic membranes are titania and zirconia.

#### 20.1.7.3 Mixed Matrix Membranes

To improve the separation performance of PV membranes, the incorporation of inorganic materials such as zeolites, carbon nanotubes, or carbon molecular sieves into the polymeric membranes produces mixed matrix membranes (MMMs), which combine useful molecular sieving properties of zeolites with the desirable mechanical robustness and flexibility of polymers. Mass transport occurs through both solution-diffusion mechanism and molecular sieving. Incorporation of nonporous fillers in polymers is also possible, which creates a tortuous path for permeation (low flux) but improved separation. Sulzer Chemtech supplies a mixed matrix hydrophobic membrane (code 1070) for extraction of low-molecular-weight organic solvents from aqueous solutions.

### 20.1.8 Applications of Pervaporation

PV has several industrial applications that could be classified into three categories on the basis of feed composition and nature of membrane used:

1. Dehydration of aqueous-organic mixtures (hydrophilic PV)
2. Removal of trace organic compounds from aqueous mixtures (hydrophobic PV)
3. Separation of organic-organic mixtures (organophilic PV)

Figure 20.4 displays the wide and varied applications of PV with appropriate examples.

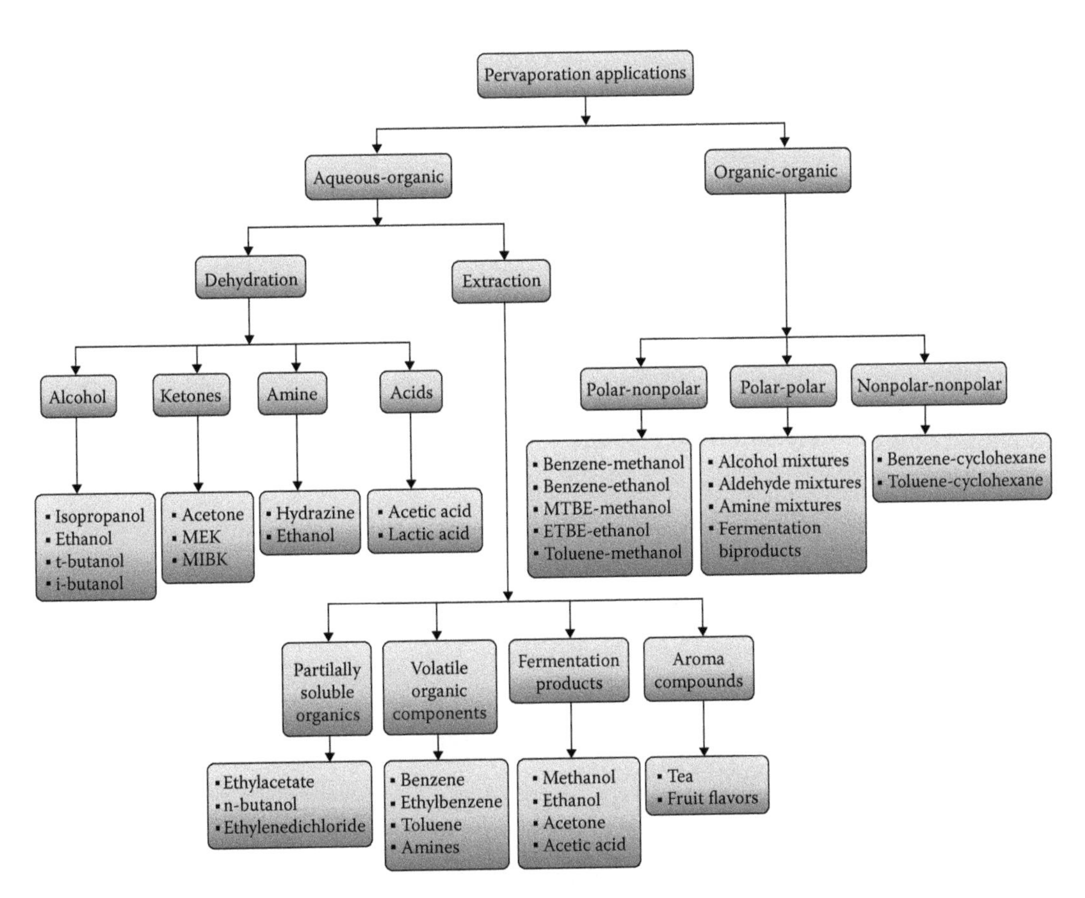


**FIGURE 20.4** Liquid mixture systems that could be separated by pervaporation.

#### 20.1.8.1 Hydrophilic Pervaporation

The target compound to be separated here is water, which is generally present as part of an azeotropic mixture or in small concentrations with a closely boiling organic compound or multicomponent organic mixture. Examples of membrane materials used for dehydration are PVA, poly(acrylic acid) (PAA), poly(vinyl pyrrolidone) (PVP), PAN, chitosan (CS), cellulose derivatives, and so forth. Examples of some applications are:

- Breaking of azeotropes: ethanol-water (95.6–4.4 wt%), isopropanol-water (87.5 wt%–12.5 wt%), dioxane-water (82 wt%–18 wt%)
- Dehydration of multicomponent mixtures (ethylaceteate + butylacetate + water)
- Solvent dehydration (dimethyl formamide, acetone, acetic acid, etc. containing $\leq$10 wt% water)
- Continuous removal of water from esterification reactions (ethyl ester)
- Dehydration of hazardous compounds (Hydrazine, Mono Methyl Hydrazine (MMH), Unsymmetrical Dimethyl Hydrazine (UDMH)

#### 20.1.8.2 Hydrophobic Pervaporation

The permeant here is a single compound or a mixture of volatile organics that are present in small percentages to trace concentrations (ppm level) in a bulk aqueous medium.

Some examples of membrane materials used here are PDMS, polybutadiene, poly(trimethylsilyl-1-propyne) (PTMSP), and poly(vinylidene fluoride) (PVDF). Examples of some applications are:

- Industrial effluent treatment
- Extraction of expensive aroma compounds and flavors from food and cosmetic industries
- Separation of volatile products and by-products such as alcohols, ketones, aldehydes, and so forth from fermentation broths
- Removal of pollutants such as trihalomethanes (THMs) and methyl-*tert*-butyl ether (MTBE) from ground water
- Recovery of solvents such as methanol, ethanol, ethyl acetate, and tetrahydrofuran (THF) from pharmaceutical industry process streams

#### 20.1.8.3 Organophilic Pervaporation

The most difficult separation by PV is to isolate one organic compound from another, which even scientists in the United States which researchers have defined as one of the most challenging research areas. These mixtures include combination of polar/nonpolar, aromatic/alicyclic, and isomeric compounds. Examples of membrane materials that can be used here are poly(ether imide) (PEI), PVA, PDMS, and poly(ether-*block*-amide) (PEBAX).

Some of the applications are:

- Separation of methanol from MTBE or toluene
- Separation of ethanol from Ethyl-*tert*-butyl ether (ETBE) or benzene
- Separation of benzene from cyclohexane
- Separation of a mixture of alcohols, aldehydes, or amines

## 20.2 SYNTHESIS AND CHARACTERIZATION OF PERVAPORATION MEMBRANES

### 20.2.1 Membrane Selection

The most crucial decision in PV is the choice of the right membrane material to treat a particular feed system. Selection can be done in two ways: either conducting trial and error with the available polymers for a certain liquid system, or following a systematic procedure of choosing a few promising polymers as described in Sections 20.2.1.1 to 20.2.1.5.

#### 20.2.1.1 Literature Survey

Literature searches can be done in two ways: identifying a particular polymer that has different PV applications, or based on a liquid mixture system of interest the types of membranes reported are found. This confined literature can be saved for further investigations to facilitate membrane modification. The component that is to be separated from the system should be present in lower concentrations. This particular component can be either water or an organic component. For the removal of water from an organic/water system a hydrophilic polymer should be selected, whereas in a case where the organic compound is to be separated, a hydrophobic/organophilic polymer will be more suitable.

#### 20.2.1.2 Feed Mixture Composition

Feed mixture composition involves studying the nature of the components present in the feed and the desired permeant that would help in identifying the type of membrane material to be tested.

#### 20.2.1.3 Solubility Parameter Theory

Solubility parameter theory is an approach to predict polymer-liquid interactions with the help of the polymer structure and polarity of the liquids that need to be separated. The solubility parameter (δH) defined by Hildebrand and Scott (1950) measured cohesive energy density, or the strength of the intermolecular forces (cohesive energy) holding molecules together in the liquid phase (Mandal and Pangarkar, 2002; Price and Shillcock, 2002). Membrane polarity contributes to good interaction if one of the components of the feed mixture is polar.

Cohesive energy is the net effect of all the interatomic/molecular interactions including van der Waals interactions, covalent bonds, ionic bonds, hydrogen bonds, electrostatic interactions, induced dipole, and permanent dipole interactions (Hancock et al., 1997). The total cohesive energy ($E_{\text{total}}$) that holds a liquid together is approximated by the sum of the energy required to overcome atomic dispersion forces ($E_{\text{d}}$), interactions between permanent dipoles of adjacent molecules (polar interaction) ($E_{\text{p}}$), and hydrogen bonds between molecules (proton donor/acceptor) ($E_{\text{h}}$) (Mandal and Pangarkar, 2002; Hansen,1969; Hansen, 2000):

$$E_{\text{total}} = E_{\text{d}} + E_{\text{p}} + E_{\text{h}} \tag{20.17}$$

These three components lie as vectors along the orthogonal axes in three-dimensional phase where the end point of the vector represents the solubility parameter. The distance Δ between the end points of the vectors representing the polymer and solvent is given by the equation:

$$\Delta = \left[\left(\delta_{\text{d}}^{\text{s}} - \delta_{\text{d}}^{\text{p}}\right)^2 + \left(\delta_{\text{p}}^{\text{s}} - \delta_{\text{p}}^{\text{p}}\right)^2 + \left(\delta_{\text{h}}^{\text{s}} - \delta_{\text{h}}^{\text{p}}\right)^2\right]^{1/2} \tag{20.18}$$

Here *s* and *p* refer to the solvent and the polymer, respectively. The smaller the value of Δ, the greater is the affinity between polymer and liquid (Villaluenga et al., 2003). This method avoids relying on graphic representations of solubility behavior and allows numerical values to be used solely.

#### 20.2.1.4 Chemical, Thermal, and Mechanical Stability

The polymer material should be chemically inert with respective to the liquid feed mixture. This test can be carried out by soaking a piece of the membrane in the feed medium for a considerable amount of time (3 to 4 days) to observe if there is partial dissolution or degradation. The polymer should also be capable of functioning at higher temperatures since some of the feed mixtures,

**TABLE 20.1**
**Important Polymers for Pervaporation and Their Chemical Structures**

| Polymer | Structure | Nature |
|---|---|---|
| Polyethylene | $-[CH_2-CH_2]_n-$ | Hydrophobic |
| Polypropylene | $-[CH_2-CH(CH_3)]_n-$ | Hydrophobic |
| Polybutadiene | $-[CH=CH-CH_2-CH_2]_n-$ | Hydrophobic |
| Sodium alginate | 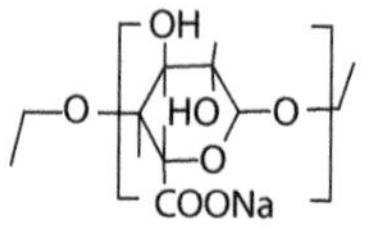 OH, HO, O, O, O, COONa | Hydrophilic |
| Cellulose triacetate | 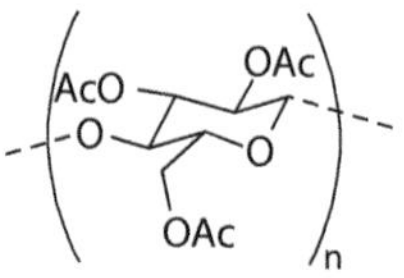 AcO, OAc, O, O, OAc, n | Hydrophilic |
| Polystyrene | $-[CH_2-CH_2(C_6H_5)]_n-$ | Hydrophobic |
| Poly(vinyl alcohol) | $-[CH(OH)-CH_2]_n-$ | Hydrophilic |
| Poly(acrylonitrile) | $-[CH(CN)-CH_2]_n-$ | Hydrophilic |
| Poly(vinylidene fluoride) | $-[CF_2-CH_2]_n-$ | Hydrophobic |
| Poly(vinyl acetate) | $-[CH(OOCCH_3)-CH_2]_n-$ | Hydrophilic |
| Poly(tetrafluoroethylene) | $-[CF_2-CF_2]_n-$ | Hydrophobic |
| Poly(methyl methacrylate) | $-[CH(CH_3)(COOCH_3)-CH_2]_n-$ | Hydrophilic |
| Nylon 66 | $-[NH(CH_2)_6NHCO(CH_2)_4CO]_n-$ | Hydrophilic |
| Poly(phenylene oxide) | 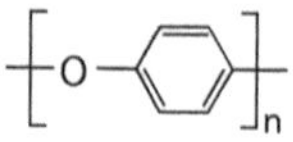 $-[O-C_6H_4]_n-$ | Hydrophilic/ Hydrophobic |

*(continued)*

**TABLE 20.1 Continued**
**Important Polymers for Pervaporation and Their Chemical Structures**

| Polymer | Structure | Nature |
|---|---|---|
| Poly(acrylic acid) | $-[CH_2-CH(COOH)]_n-$ | Hydrophilic |
| Poly(vinyl pyrrolidone) | $-[CH(N\text{-pyrrolidone})-CH_2]_n-$ | Hydrophilic |
| Poly(vinyl chloride) | $-[CH_2-CH_2(Cl)]_n-$ | Hydrophobic |
| Poly(ethylene terephthalate) | $-[OCH_2CH_2OOC-C_6H_4-CO]_n-$ | Hydrophilic |
| Chitosan | H, $CH_2OH$, O, H, OH, $NH_2$, H, O | Hydrophilic |
| Polyamide | $-(HN-C_6H_4-NH-C(=O)-C_6H_4-C(=O))_n-$ | Hydrophilic |
| Polyimide | $-[N(CO)_2C_6H_2(CO)_2N-C_6H_4-O-C_6H_4]_n-$ | Hydrophilic |
| Poly(dimethyl siloxane) | $-[Si(CH_3)_2-O]_m-$ | Hydrophobic |
| Polysulfone | $-[C_6H_4-SO_2-C_6H_4-O-C_6H_4-C(CH_3)_2-C_6H_4-O]_n-$ | Hydrophilic/ hydrophobic |

especially distillates, could be available at elevated temperatures. The selected polymer should have a film-forming property and the prepared membrane must be subjected to a bending test or tensile strength measurement to investigate mechanical stability. Selectivity would be a more vital factor since the flux can be enhanced by decreasing the thickness to prepare a thin-film composite membrane that is the commercial form consisting of a ≤2 μm thick dense skin layer of the selected polymer supported on a thicker but porous substrate.

### 20.2.2 Polymers for Pervaporation Application

Polymers shown in Table 20.1 are utilized for preparing PV membranes that are classified into hydrophilic and hydrophobic polymers. By tendency, glassy polymers are suitable for making water-selective membranes used for solvent dehydration and rubbery polymers are favorable to the selective removal of organic compounds from water. Ionic polymers contain groups that are neutralized by counter ions and may be viewed as crosslinked polyelectrolytes. Ionic membranes can be subdivided into cationic and anionic types that are normally water-selective. However, polymer materials for dehydration should maintain a proper balance between hydrophilicity and degree of swelling to maximize selectivity.

### 20.2.3 Membrane Preparation

#### 20.2.3.1 Substrate Preparation by Phase Inversion

The ultraporous hollow fiber or flat-sheet substrates for membrane casting are prepared by dissolution of polymers such as polysulfone or PVDF in nonvolatile solvents such as dimethyl formamide or n-methyl pyrrolidone (NMP), which is cast on a suitable fabric (nonwoven polyester) followed by phase inversion in a nonsolvent bath (usually water) for 5 minutes, resulting in the polymer being transformed from liquid to a solid state (Figure 20.5). This is a versatile technique wherein the pore diameter and pore size distribution can be controlled. Porous substrates are also prepared by sintering, stretching, leaching, and track etching techniques to prepare composites.

#### 20.2.3.2 Solution Casting and Solvent Evaporation

The apparatus shown in Figure 20.6 is used for membrane preparation through solution casting and solvent evaporation. Homogeneous bubble-free solutions of polymers in appropriate low-boiling solvents are cast to the desired thicknesses on a clean glass plate followed by complete evaporation of the solvent. Thickness of the membrane is fixed using a doctor's blade to vary the gap between the movable metallic bar and glass plate. The glass plate is made to slide through the gap to spread the solution uniformly. The volatile solvent evaporates, leaving behind the defect-free dense polymer membrane for further applications. Apart from solution casting, dip coating, plasma polymerization, interfacial polymerization, and in situ polymerization can also be used to prepare PV membranes.

#### 20.2.3.3 Dip Coating

Most of the composite membranes are prepared by the dip coating technique to obtain a dense top layer. An ultraporous substrate in the form of a hollow fiber or a flat sheet is immersed in a coating solution containing the desired polymer for a specific duration and then removed from the bath, which results in a thin layer of polymer solution adhering to the substrate. The fiber or sheet is dried at room temperature or in an oven. The Navier-Stokes equation is used to describe the thickness of the coating layer:

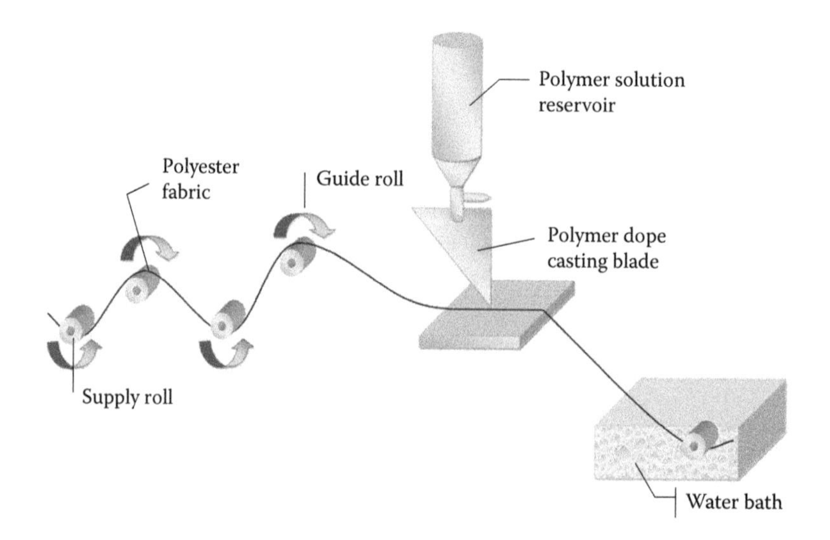


**FIGURE 20.5** Schematic of phase inversion process for preparation of ultraporous substrate.

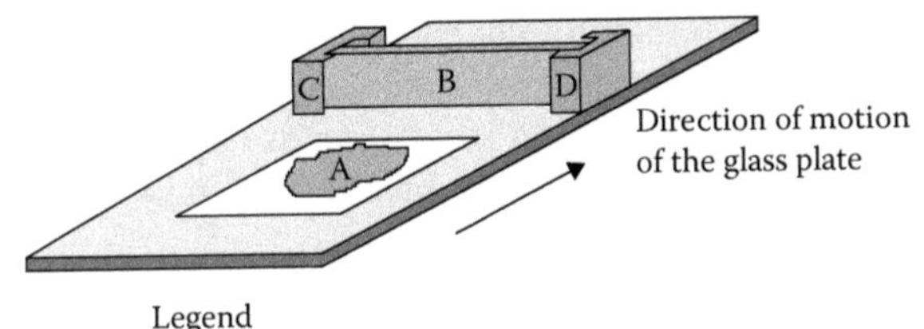


Legend
A - Polymer solution on a glass plate
B - Movable metallic bar
C & D - Metallic supports

**FIGURE 20.6** Schematic of a membrane casting apparatus.

$$T_{\text{equi}} = \frac{2}{3}\sqrt{\frac{\eta V}{\rho g}} \tag{20.19}$$

where $T_{\text{equi}}$ is the equilibrium thickness, $V$ is the coating velocity, ρ is the viscosity, and $g$ represents gravity.

#### 20.2.3.4 Interfacial Polymerization

Interfacial polymerization is another method used for creating a nonporous top layer on a porous substrate through a polymerization reaction that occurs between two reactive monomers at the interface of two immiscible solvents. The porous substrate is initially saturated in an aqueous solution containing a reactant such as a diamine, followed by immersion in an immiscible organic solvent such as hexane containing a reagent like an acid chloride. A dense top layer forms on the porous substrate at the interface between water and the organic solvent when the two monomers react with each other, as shown in Figure 20.7, after which the composite is heated to crosslink the selective layer.

### 20.2.4 Modification of Pervaporation Membranes

The techniques for controlling the balance between hydrophilicity and hydrophobicity have been discussed in the literature (Huang, 1991). Improvement in membrane performance is often achieved

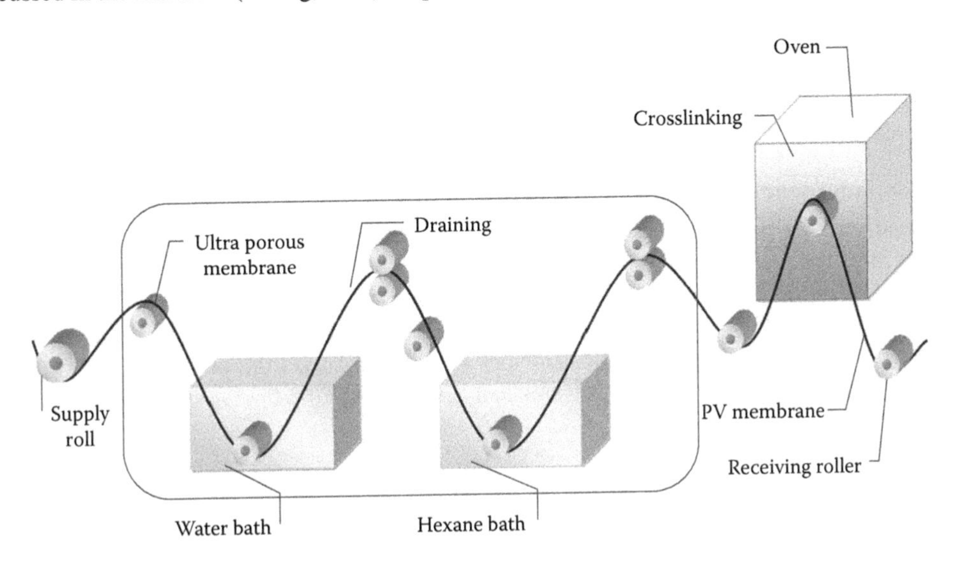


**FIGURE 20.7** Schematic illustration of interfacial polymerization to fabricate pervaporation membrane.

by polymer modification, for which several techniques are in vogue, such as crosslinking, grafting, blending, copolymerization, and incorporation of inorganic materials. Membrane modification is done to improve flux and selectivity. Since an increase in flux usually accompanies a decrease in selectivity, the modification procedure needs to be fine-tuned to achieve optimum performance.

#### 20.2.4.1 Crosslinking

Crosslinking is done to serve two purposes: to make the polymer insoluble in the feed mixture and decrease the degree of membrane swelling to enhance selectivity. Most of the PV membranes are crosslinked before being utilized for any separation application. Crosslinks can be ionic or covalent and are created by external reagents, heat, or irradiation. Some of the well-known covalent crosslinkers are aldehydes and isocyanates while ionic reagents are mineral acids such as $H_2SO_4$ and $H_3PO_4$. Polymers that possess reactive groups such as –OH, –COOH, or $-NH_2$ can be easily crosslinked. The degree of crosslinking in a polymer can be controlled by varying reaction time, temperature, and reagent concentration, and is estimated from the ion exchange capacity (IEC).

#### 20.2.4.2 Blending

A physical mixture of polymers that are not bonded covalently is known as a polymer blend. Blending is an ideal technique to accumulate the properties of different polymers into a single membrane. Optimum blending ratios can be determined by mixing various amounts of hydrophilic polymer with a hydrophobic polymer and measuring the permeability and selectivity (Park et al., 1994; George et al., 1999). Both homogeneous (polymers miscible in all compositions) and heterogeneous (polymers not miscible in all compositions) blends can be used in membrane preparation. In most cases, homogeneous blends are used as they possess sufficient chemical and mechanical stability. Acid-base combinations such as CS-PVA and SA- poly(ethyleneimine) are ionically bonded and display enhanced water permeability with increased rejection of organic solvent, known as the salting out effect, making them suitable for PV-based dehydrations.

#### 20.2.4.3 Grafting, Copolymerization, and Plasma Etching

Grafting by irradiation is a versatile technique for modifying polymer films. If a monomer contains a functional group that can react with a functional group in the polymer, chemical grafting can occur. On the other hand, copolymerization is the covalent bonding of two monomers to improve the properties of the resultant polymer in terms of chemical, mechanical, and thermal stabilities as well as permeation characteristics (Tanihara et al., 1995). Examples of some copolymers are polyimide and Pebax, which find wide application as membrane materials. Etching can be used to deposit plasma over different substrates and create selective layers made from a wide variety of materials. However, etching is still in the developmental stage and is yet to be used for making commercial PV membranes.

#### 20.2.4.4 Interpenetrating Polymer Network

An interpenetrating polymer network (IPN) is a material containing two polymers in a network-like structure. The conditions for preparing an IPN are that the two polymers are synthesized and/or crosslinked in the presence of the other, have similar kinetics, and the polymers are not phase-separated. Even if only one polymer is crosslinked the combination can still be considered a semi-IPN. IPNs are distinguishable from blends, block copolymers, and graft copolymers in two ways: (1) IPNs swell but do not dissolve in solvents and (2) the creep and flow is suppressed (Lee et al., 1991).

### 20.2.5 Membrane Characterization

Membranes need to be characterized for certain properties such as pore size and distribution, free volume, crystallinity, intermolecular interactions, and morphological nature.

#### 20.2.5.1 Thermal Gravimetric Analysis

Thermal gravimetric analysis (TGA) is a simple analytical technique that measures the weight loss (or weight gain) of a material as a function of temperature. As materials are heated, they either lose or gain weight by reacting with the atmosphere in the testing environment. Since weight loss and gain are disruptive processes to the sample material or batch, knowledge of the magnitude and temperature range of those reactions are necessary. This allows for the effective quantitative analysis of thermal reactions that are accompanied by mass changes. Examples include evaporation, decomposition, gas absorption, and dehydration to name a few. Changes can be measured over a temperature ranging from ambient to 900°C. This technique is useful in evaluating the thermal stability of a PV membrane that may be exposed to high temperature feeds.

#### 20.2.5.2 X-ray Diffraction

Highly crystalline polymers do not dissolve easily in many organic solvents due to a lack of flexible groups, which prevent swelling. The crystallites act as physical crosslinks, keeping the polymer tightly packed. Since dissolution generally occurs in the amorphous part of the polymer, the degree of crystallinity has a large influence on sorption of the feed in the membrane. Highly crystalline polymers have lower permeability than amorphous polymers. In polymers of the type $(—CH_2—CHR—)_n$ the size of the side group (R) plays an important role in predicting polymer crystallinity. X-ray diffraction (XRD) is an important key tool for investigating properties of polymers and composites in relation to their solid-state structures. Atomic arrangement and degree of crystallinity can be studied. The change in intersegmental distance, also known as d-spacing, can be determined before and after modification of the PV membrane.

#### 20.2.5.3 Transmission Electron Microscopy

A transmission electron microscope (TEM) is an immensely valuable and versatile instrument in material science for investigating the fine structure of materials. Here a beam of electrons is generated and allowed to transmit through the sample, which is then detected by an electron microscope to yield information about the local structure (dislocations), average structure, and chemical composition. This technique would be especially useful in studying the distribution of inorganic fillers such as carbon nanotubes and zeolites in a polymer matrix and the pore structure of purely inorganic PV membranes.

#### 20.2.5.4 Fourier Transform Infrared Radiation

Fourier transform infrared radiation (FTIR) is a qualitative and quantitative analysis that detects the chemical bonds, molecular structure, and functional groups in a sample by producing an infrared absorption spectrum. When a sample absorbs infrared light, the absorbed light energy is converted into atomic bond vibrations. Specific groupings of bonded atoms, located within a material's molecular structure, absorb infrared light within loosely defined frequency ranges. The defined absorption frequency range remains fairly constant and is not significantly dependant on the material type. FTIR of a PV membrane helps in studying intermolecular interactions and transport mechanisms. Modification in the PV membrane could be confirmed through shifts in peaks, absence of old peaks, emergence of new peaks, and reduction in transmission.

#### 20.2.5.5 Scanning Electron Microscopy

Scanning electron microscopy (SEM) is a method that uses electrons rather than light to form an image. SEM also produces images of high resolution, which means that closely spaced features can be examined at high magnification. Morphology of surface, cross-section, pore diameter, and pore size distribution can be analyzed.

#### 20.2.5.6 Atomic Force Microscopy

Atomic force microscopy (AFM) is a tool for determining surface topography of a sample at subnanometer resolution. This provides the specific interaction between the molecules and intrinsic properties.

#### 20.2.5.7 Differential Scanning Calorimetry

Differential scanning calorimetry (DSC) is a quantitative analytical tool for measuring the physical and chemical changes in materials, since polymer properties change with thermal conditions. DSC records the rate of energy absorption as a function of the sample's temperature. The area under the DSC peak represents the energy of the associated thermal transition. The DSC spectra of homopolymers and blend membranes are generally recorded over the temperature range of 25°C to 200°C at the heating rate of 5°C/min in hermetically sealed aluminum pans. Membrane samples are allowed to attain steady state with the solvents and the sample pan is conditioned in the instrument before running the experiment. One of the main objectives of this characterization method is the determination of the glass transition temperature ($T_g$) that specifies the critical temperature up to which a membrane can be exposed as the polymer passes from glassy to rubbery state.

#### 20.2.5.8 Tensile Strength

Tensile strength and % elongation at break are intensive properties that reveal the mechanical stability of a PV membrane. The tensile strength of a membrane is the maximum amount of stress that it can be subjected to before failure. The maximum load applied in breaking a tensile test piece divided by its original cross-sectional area is measured in Newtons/mm$^2$ or tons/ in$^2$:

$$R_m = \frac{P_{max}}{A_o} \quad (20.20)$$

where $P_{max}$ is the maximum load and $A_o$ is the cross-sectional area of the sample. PV membranes should possess adequate tensile strength and % elongation in order to have potential for industrial application.

#### 20.2.5.9 Ion Exchange Capacity

Ion exchange capacity (IEC) is used to determine the total number of groups available for interaction in a membrane sample through estimation of the number of milliequivalents of ions in 1g of the dry polymer. To determine IEC, specimens of similar weight are soaked in 50 mL of 0.01 N sodium hydroxide solution for 12 h at ambient temperature. Then, 10 mL of soaked solution is pipetted out and titrated with 0.01 N sulfuric acid. The membrane samples are regenerated with 1 M hydrochloric acid, washed free of acid with water, and dried to a constant weight. The IEC is calculated according to the equation:

$$\text{IEC} = \frac{(B-P)*0.01*5}{m} \quad (20.21)$$

where $B$ is amount of sulfuric acid used to neutralize blank sample in mL, $P$ is amount of sulfuric acid used to neutralize the membrane in mL, 0.01 is the normality of the sulfuric acid, 5 is the factor corresponding to the ratio of the amount of NaOH taken to dissolve the polymer to the amount used for titration, and m is sample mass in g.

#### 20.2.5.10 Sorption Studies

Sorption studies give a measure of the affinity between a membrane and liquid. Percentage sorption is calculated by the equation:

$$\%\ \text{Sorption} = \frac{M_s - M_d}{M_d} * 100 \quad (20.22)$$

where $M_s$ and $M_d$ are the mass of swollen and dry membranes, respectively.

The degree of swelling can be calculated using the equation:

$$\text{Degree of Swelling} = \frac{M_s}{M_d} \quad (20.23)$$

## 20.3 EXPERIMENTAL PROCEDURES AND PROTOCOLS

### 20.3.1 Installation and Operation of a Laboratory Pervaporation Unit

A schematic of the bench scale unit is provided in Figure 20.8. This PV cell consists of two bell-shaped column reducers made of glass that are held together with external flanges and bolts. The top part is used as the feed chamber and the bottom part as the permeate chamber. The membrane is fixed between the two chambers by means of poly(tetrafluoroethylene) (PTFE) gaskets and tie rods on a porous polished stainless steel (SS) plate which has an embedded sintered porous SS screen facing the feed side. The membrane is fixed in such a way that the active layer is exposed to the feed solution and is held tightly in its place by vacuum grease to ensure a leakproof arrangement. A desired volume (say 70–100 mL) of the feed solution is introduced into the feed chamber of the cell and stirred continuously using a glass stirrer attached with PTFE impeller and driven by an overhead stirring motor to minimize concentration polarization. The feed solution is maintained at atmospheric pressure and the downstream side is evacuated continuously by a vacuum pump. The level of vacuum is checked using a McLeod gauge (tilting type) connected to the manifold via a rubber hose. The active surface area of the membrane varies between 20 to 50 $cm^2$ depending upon the diameter of the bell-shaped reducers, and the pressure at the downstream side could be varied between 0.05–20 mmHg or more depending on the flux. Due to the pressure drop in the downstream side, the permeate is obtained in the form of vapor and collected in condenser traps cooled by liquid nitrogen or dry ice/acetone mixture. The membrane is generally soaked in the feed solution overnight to ensure equilibrium. The initial permeate that comes through immediately as liquid is discarded and then the PV experiment is started to collect permeate over a period of 6 to 8 h. The amount of permeate collected is weighed using an accurate balance to determine the flux. The feed and the permeate samples are then analyzed for their composition by gas chromatography (GC)/refractive index (RI) to estimate the selectivity.

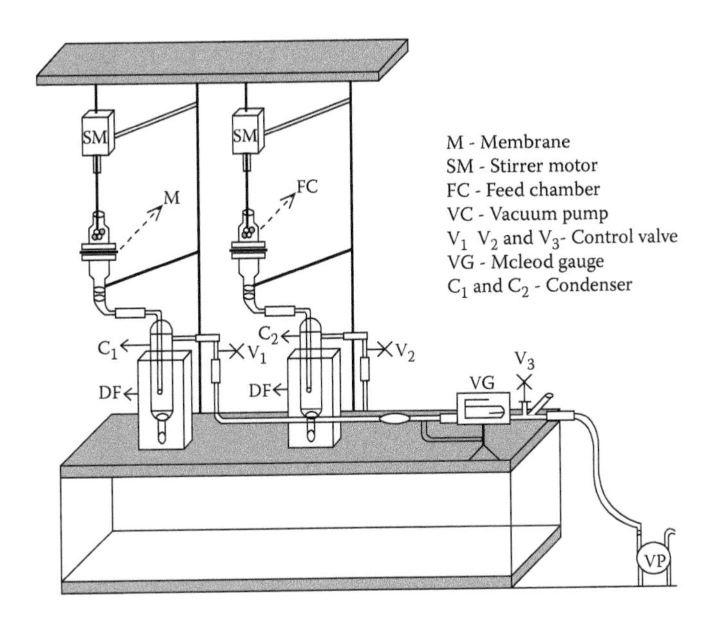


**FIGURE 20.8** Schematic of PV experimental manifold.

### 20.3.2 Analysis of Feed and Permeate

#### 20.3.2.1 Gas Chromatography

The (GC) consists of an injection port, a column, and a detector. A carrier gas such as helium flows through the column fixed inside an oven wherein the temperature is maintained at a desirable value and programmed over a specific range. A sample size of 0.1 to 0.2 μL is injected by syringe into this chamber through a port covered by a rubber septum. Once inside, the sample gets vaporized and is carried through the column by the carrier gas. As the mixture travels through this column, its components get adsorbed and desorbed at different rates depending upon the interaction between the nature of packing inside the column and volatility of the injected components. Thus the compounds exit the instrument through a detector after different time intervals. The detector sends an electronic message to the recorder that responds through a change in the baseline, and which normally represents carrier gas flow and is a straight line, resulting in appearance of peaks. Compounds can be identified on the basis of their retention times. The flow rate of the gas influences how rapidly a compound will travel through the column; the faster the flow rate, the lower the retention time. Generally, the flow rate is kept constant throughout a run.

#### 20.3.2.2 Refractive Index

Feed and permeate samples can also be analyzed by determining the RI of pure components as well as mixtures containing known concentrations of the faster permeating component. A graph is plotted between component concentration and RI (Figure 20.9). Then the RI of an unknown sample such as permeate is noted and the corresponding concentration is determined from the graph.

### 20.3.3 Effect of Operating Parameters in Pervaporation

After the selection of a membrane for a particular feed system, the effect of variable parameters mentioned below need to be studied to optimize membrane performance on a commercial level:

1. Feed composition
2. Feed temperature
3. Permeate pressure
4. Membrane thickness

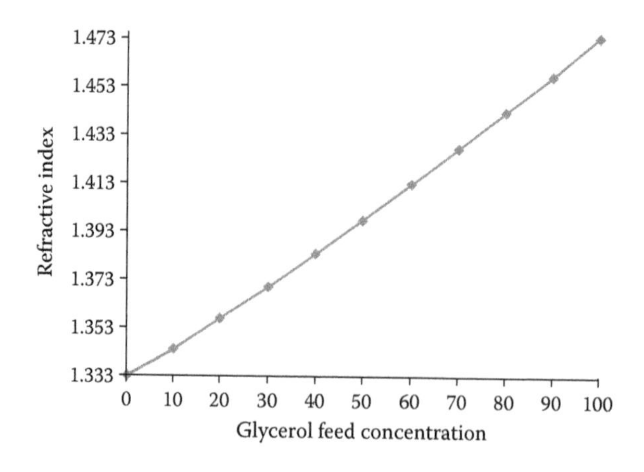


**FIGURE 20.9** Standard refractive index (RI) graph for glycerol/water system.

#### 20.3.3.1 Feed Composition

The change in feed concentration has a direct impact on the performance of the membrane. In the case of dehydration through a hydrophilic membrane, flux increases with feed water concentration. In the case of extraction of an organic component from water through a hydrophobic membrane, the membrane swells with increasing organic concentration, resulting in greater flux but lower selectivity. In general, the selectivity increases with decreasing concentration of the preferentially permeating component due to lower swelling.

#### 20.3.3.2 Feed Temperature

Flux in PV process is directly proportional to the operating temperature due to an increase in activity of the feed molecules and consequently the diffusion rate. The selectivity also strongly depends on the temperature since the former varies inversely with the flux. Temperature has an Arrhenius type effect on permeability of PV membrane, which is given by the equation:

$$J = J_0 \exp\left(\frac{E}{\mathrm{RT}}\right) \tag{20.24}$$

where $E$ is the activation energy for permeation, $R$ the universal gas constant, $T$ the absolute temperature, $J$ the flux of the permeating component, and $J_o$ is a constant.

#### 20.3.3.3 Permeate Pressure

A rise in permeate pressure (decrease in vacuum) favors the transport of the more volatile component in the feed and thus selectivity depends on vapor pressures of the feed constituents. Flux improves with decreasing permeate pressure (increasing vacuum).

#### 20.3.3.4 Membrane Thickness

Thickness of the membrane affects the flux but selectivity is expected to be independent of this parameter. Therefore on a commercial scale the thickness of the selective top layer of the composite membrane is kept as low as possible to enable maximum throughput.

## 20.4 DESIGN OF PERVAPORATION MODULES

The composition and morphology of the membranes is a key to effective use of membrane technology. The choice of the membrane strongly depends on the type of application. In order to apply membranes on an industrial scale, large areas are required. Membranes are never applied as one flat sheet, because this large surface often results in high investment costs. That is why modular configurations are fabricated to enable a large membrane surface to be housed in the smallest possible volume. Membranes are implemented in several types of modules. The choice for a certain kind of membrane system is determined by a great number of aspects, such as its cost, packing density, risk of fouling, and cleaning opportunities. There are now mainly four categories of modules that are used extensively for commercial applications:

1. Tubular
2. Plate and frame
3. Hollow fiber
4. Spiral wound

Capillary and pillow-shaped configurations are not discussed in this chapter.

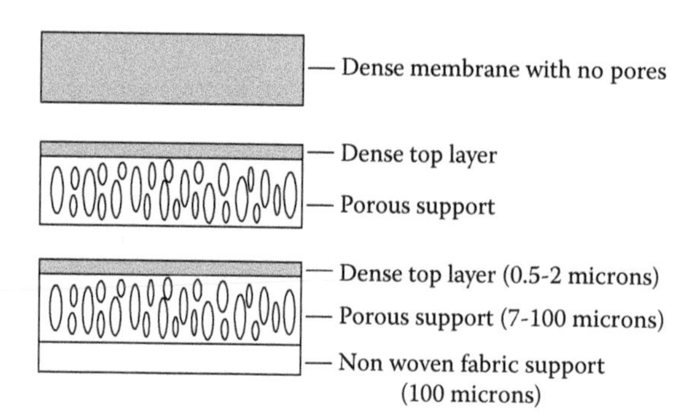


**FIGURE 20.10** Types of membrane morphologies for scale-up.

### 20.4.1 Membrane Morphology for Scale-Up

According to the structural characteristics that influence PV performance, several different membrane structures such as porous, dense, and asymmetric membranes are prepared. A sound knowledge of membrane morphology is required to select the right one. On a laboratory scale, dense unsupported membranes can be used, but to attain commercial viability, membranes must be either in composite or asymmetric form since these provide adequate mechanical strength to the membrane in addition to the thin selective top layer that enables high flux. Asymmetric membranes are prepared by phase inversion technique after partial evaporation of the volatile solvent used to make the polymer dope, followed by immersion in a nonsolvent bath after controlled evaporation in atmosphere for a specific but short duration (Section 20.2.3.1). A thin-film composite membrane (Figure 20.10) consists of three layers and is prepared by casting a polymer solution onto an ultraporous substrate (45–50 μm) made of PAN, PVDF, poly(ether sulfone)(PES), or polysulfone PSF, which is supported on a nonwoven textile fabric support (100 μm thick). The selective polymer layer is only 0.5–2 μm thick to allow maximum flux at minimum fabrication cost.

### 20.4.2 Tubular Module

The schematic of a tubular membrane module (Figure 20.11) shows that the membranes are located inside a tube, as they are not self-supporting. The tube is made up of a special kind of material such as ceramic to support the actual membrane. Feed flow in this type of module is usually inside the tubes since the membranes are coated on the inner circumference of the tube. Tubular membranes have a diameter of about 5 to 15 mm. This type has a drawback of low packing density, which results in capital costs. An example of a tubular membrane for PV is the silica membrane supplied by Pervatech BV, Netherlands.

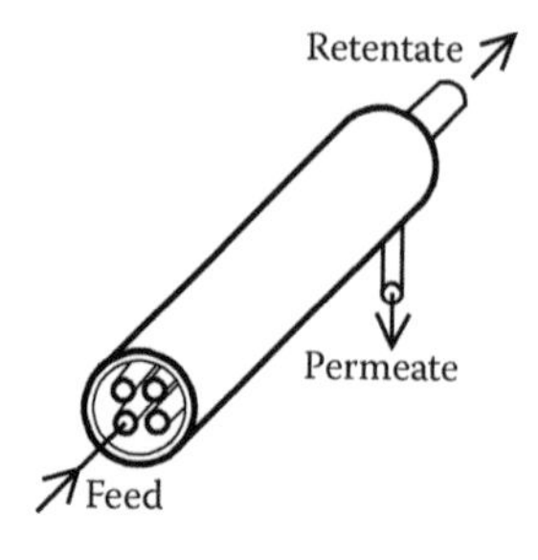


**FIGURE 20.11** Schematic of a tubular membrane module.

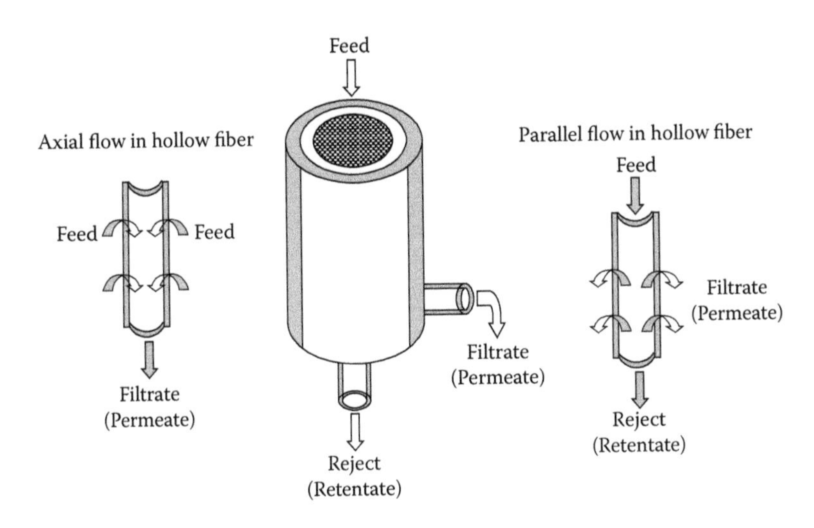


**FIGURE 20.12** Schematic of a hollow-fiber membrane module.

### 20.4.3 Hollow Fiber

This type of module (Figure 20.12) contains thousands of hollow fibers in a compact volume and sealed at the end and housed within a shell. The feed solution flows either in radial or parallel mode with respect to the fibers depending on the permeate direction. Thus, configuring membranes into hollow fibers enables the packaging of large areas of membrane in very compact volumes. Higher membrane packing density in a module leads to lower production costs. Hollow fibers for PV can be produced by coating ultraporous hollow fibers of PAN, PES, or PSF with solutions of PVA, chitosan, PVA, CS, sodium alginate (SA), and so forth. The surface areas obtained with HF membranes could go up to 10,000 $m^2/m^3$.

### 20.4.4 Plate and Frame

A plate and frame module is similar to that of a filter press (Figure 20.13). The flat membranes are placed in a sandwich-like fashion with their feed sides facing each other. A suitable distributor is placed in between each feed and permeate compartment. The number of membrane sets needed for a given surface area furnished with feed and vacuum chamber distributors, sealing gaskets, and two end plates constitute one plate and frame module, and several such modules can be arranged in series or parallel. The main advantage of this module is its flexibility toward different separation tasks; for instance, replaceable feed distributors for viscous media or in cases where high turbulence is needed. These modules are used for both organophilic as well as hydrophilic separations.

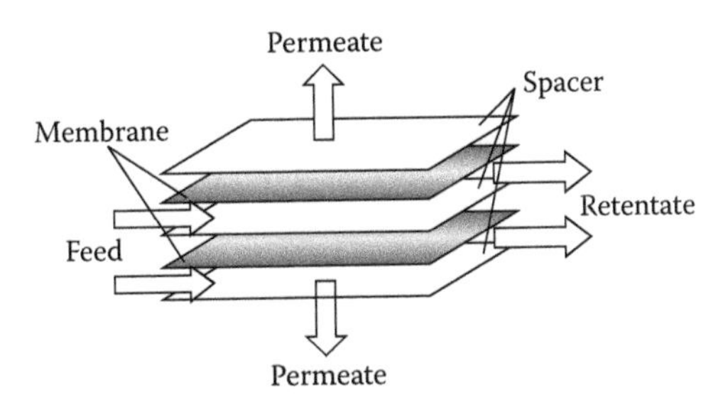


**FIGURE 20.13** Schematic of a plate and frame module.

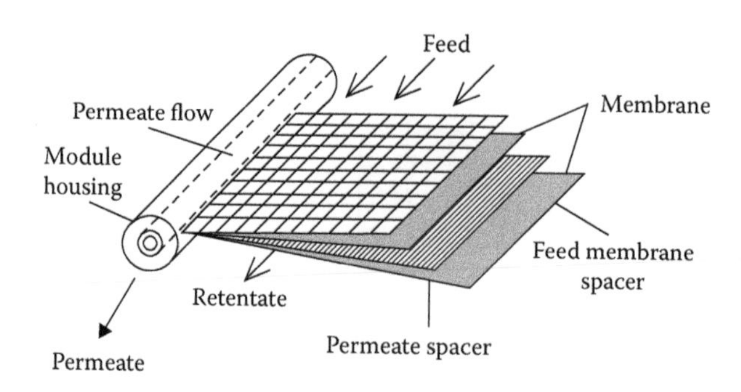


**FIGURE 20.14** Schematic of a spiral wound module.

However, these modules offer low surface area per unit volume. Sulzer Chemtech and CM-Celfa manufacture plate and frame PV modules.

### 20.4.5 Spiral Wound

The spiral wound module is a flat-sheet membrane sandwiched between feed and permeate spacers and wrapped around a central perforated tube that acts as the permeate channel (Figure 20.14). Membranes are manufactured as flat sheets and the membrane envelop inclusive of feed and permeate distributors is sealed on to the center tube and rolled around it to pack a large area into a small volume. Spiral modules are more compact and cheaper than the plate and frame modules, replacement is easy, and modularity allows scale-up merely by adding more units in parallel. They possess moderately high surface area per unit volume (900 $m^2/m^3$). Distributors between membrane layers promote turbulent flow to ensure low fouling and longer life. The problem that arises for PV application is the dissolution of the epoxy glue at the sealing in hot alcohol and lack of uniform vacuum in the permeate side. However, use of polyurethane adhesive and improved modular design could overcome these problems.

## 20.5 CASE STUDY AND CHALLENGES FOR THE FUTURE

### 20.5.1 Case Study on Isobutanol Extraction

A PV modular cell designed and fabricated in-house (Figure 20.8) and described in Section 20.3.1 was used for the extraction of the partially water-soluble isobutanol (IBOH) from aqueous solutions. A feed solution of about 100 ml containing 0–7% of IBOH in water was charged into the PV cell affixed with PDMS membrane coated on PES substrate. The membrane was prepared by casting 5% PDMS solution on the substrate and crosslinked by heating for 18 h in an electric oven at 110°C. The IBOH feed was kept under stirring and maintained at atmospheric pressure after soaking for about 6 h and then operated under vacuum. The permeate sample was collected after 4 to 6 h. Membrane performance was evaluated for varying feed organic concentration. The collected permeate contained two layers due to poor solubility of IBOH in water. Separate analysis of the top and bottom layers was done by GC and the results are displayed in Table 20.2.

The top layer was enriched in pure organic component and the bottom layer was predominantly water with some concentration of IBOH.

The top layer of permeate was found to be pure IBOH, indicating a selectivity of infinity in the organic-enriched phase that is easily separated from the aqueous bottom layer using a simple separating funnel. The bottom layer having small concentrations of IBOH can be recycled back to the PV unit to prevent any losses. Flux was observed to increase with feed IBOH concentration as seen

**TABLE 20.2**
**Effect of Feed Composition on PDMS Membrane Performance for Isobutanol Extraction**

| Feed Composition (wt%) | | Permeate Composition (wt%) | | | | Wt% of Top Layer (in Permeate) | Flux (kg/m² h) | | Selectivity to IBOH | |
|---|---|---|---|---|---|---|---|---|---|---|
| | | Top Layer | | Bottom Layer | | | Top layer | Bottom layer | Top layer | Bottom layer |
| IBOH | Water | IBOH | Water | IBOH | Water | | | | | |
| 6.91 | 93.09 | 100 | 0 | 5.42 | 94.58 | 7.7 | 0.03 | 0.36 | ∞ | 0.77 |
| 5.76 | 94.24 | 100 | 0 | 16.76 | 83.24 | 4.25 | 0.008 | 0.18 | ∞ | 3.29 |
| 1.69 | 98.31 | 100 | 0 | 3.43 | 96.57 | 4.7 | 0.007 | 0.14 | ∞ | 2.05 |
| 0.33 | 99.67 | 100 | 0 | 3.25 | 96.75 | 4.8 | 0.002 | 0.04 | ∞ | 10.17 |

in Table 20.2 and Figure 20.15 due to enhanced swelling of the organophilic PDMS membrane that preferentially sorbs IBOH. Selectivity in the aqueous bottom layer was observed to reduce (Figure 20.15b) since swelling causes plasticization of the membrane, allowing more water molecules to escape into the downstream side.

A process flow diagram (PFD) is presented in Figure 20.16 to depict a commercial PV system for extraction of organics such as IBOH that have poor solubility in water. The main accessories of the PV system are feed tank, circulating and recycling pumps, cartridge prefilter, preheater, plate and frame membrane module, condenser, phase separator, and vacuum pump. The feed velocity is controlled by a circulation pump. The permeate side of the module is provided with a vacuum gauge. The system is attached to a vacuum pump to reduce the pressure at the permeate side. The feed is prefiltered to remove any suspended particles and preheated to a temperature of 40°C to 50°C to account for any drop in temperature arising from enthalpy losses caused by continuous permeate vaporization. The plate and frame membrane modules are housed in a vacuum chamber where the permeate gets vaporized followed by its condensation. A phase separator is provided from which the organic-rich product (top layer) is collected. A permeate recycle option is provided for the clean solution to reenter the feed chamber. The reject from the membrane is collected as nearly pure water that could be reused in one of the process steps of the solvent recovery unit or main chemical process plant.

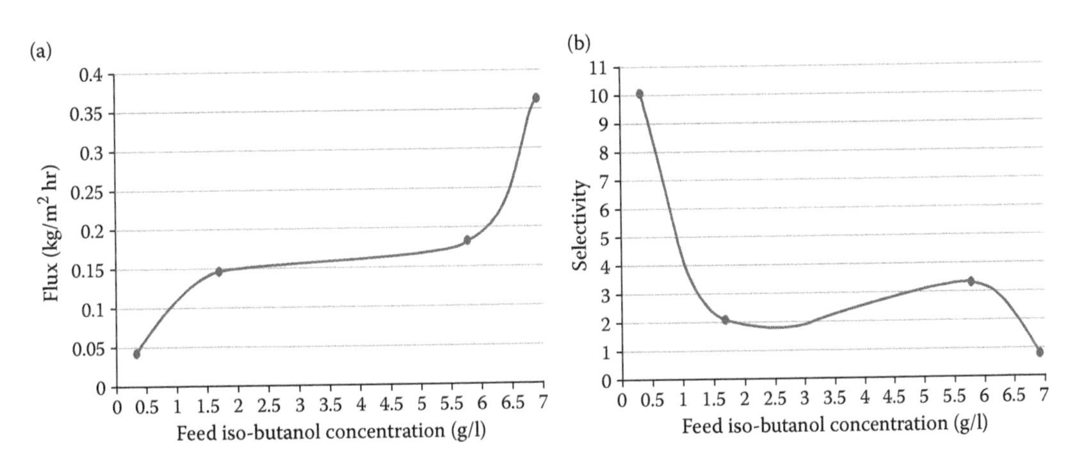


**FIGURE 20.15** Effect of feed composition on (a) flux and (b) selectivity of aqueous bottom layer during isobutanol extraction through PDMS membrane.

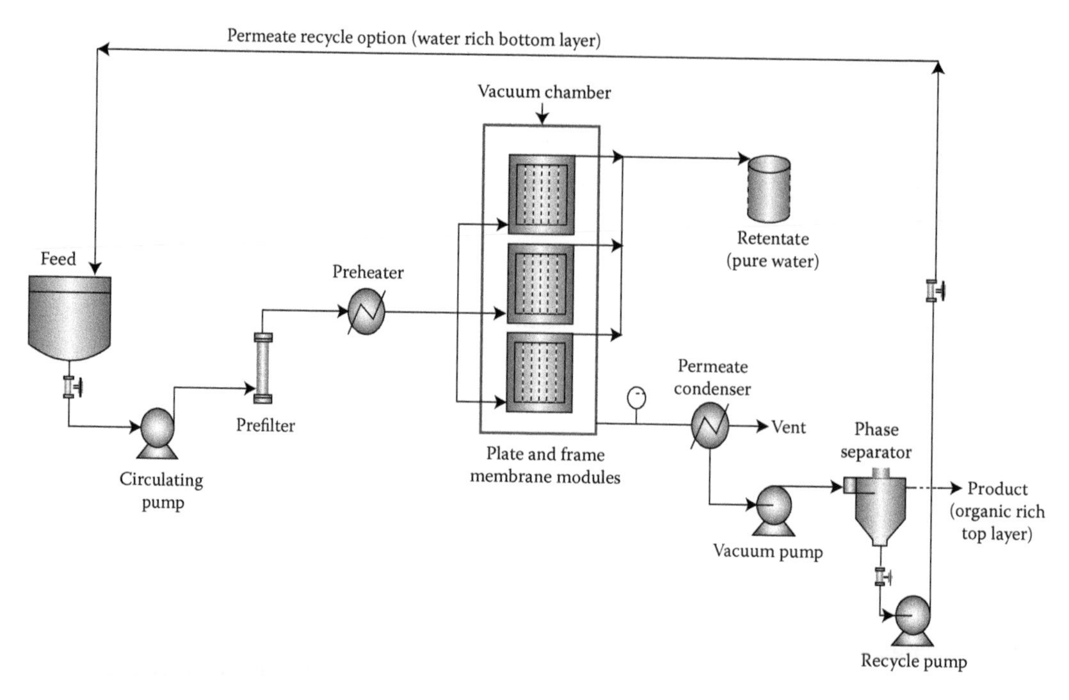


**FIGURE 20.16** Process flow diagram for extraction of organics having poor solubility in water.

### 20.5.2 Challenges Ahead for Pervaporation Technology

PV being an emerging technology has the potential to efficiently recover alcohols, organic solvents, and other biofuels. For PV to be economically and technically viable, especially in developing countries such as India, the following issues need to be addressed under future research programs:

1. Development of highly selective and permeable membranes made from new polymers, zeolite/ceramic composites and mixed matrix materials and their scale-up into modular configurations
2. Synergy of performing both alcohol recovery and dehydration using fractional condensation technology and heat integration
3. Reduction of capital investment for PV systems especially through improved design of membrane module and system
4. Optimized integration of PV with fermentation, esterification, and other reactions to enhance product yield or allow PV aided removal of inhibitors
5. Longer trials with actual feeds to assess membrane life and module stability

## REFERENCES

Binning, R.C., R.J. Lee, J.F. Jennings, et al. 1961. Separation of liquid mixtures by permeation. *Ind. Eng. Chem.* 53: 45–50.

El-Halwagi, M.M. 1992. Synthesis of reverse osmosis networks for waste reduction. *AIChE J.* 38: 1185–1198.

Fleming H.L., C.S. Slater, W.S.W. Ho, et al. 1992. *Pervaporation in Membrane Handbook.* New York: Van Nostrand Reinhold. Chapter 10, p. 105.

George, S.C., K. Prasad, J.P. Misra, et al. 1999. Separation of alkane-acetone mixtures using styrene-butadiene rubber/natural rubber blend membranes. *J. Appl. Polym. Sci.* 74: 3059–3068.

Hancock, B.C., P. York, and R.C. Rowe. 1997. The use of solubility parameters in pharmaceutical dosage form design. *Int. J. Pharm.* 148: 1–21.

Hansen, C.M. *Hansen Solubility Parameters. A User's Handbook.* 2000. Boca Raton, Florida: CRC Press LLC.
Hansen, C.M. The universality of the solubility parameter. 1969. *Ind. Eng. Chem. Product Res. Develop.* 8: 2–11.
Heisler, E.G., A.S. Hunter, J. Siciliano, et al. 1956. Solute and temperature effects in the pervaporation of aqueous alcoholic solutions. *Science* 124: 77–79.
Hickey, P.J., F.P. Juricic, and C.S. Slater. 1992. The effect of process parameters on the pervaporation of alcohols through organophilic membranes. *Separ. Sci. Technol.* 27: 843–861.
Huang, R.Y.M., ed. *Pervaporation Membrane Separation Processes.* 1991. Amsterdam: Elsevier.
Karlsson, H.O.E., and G. Tragardh. 1994. Pervaporation of aroma compounds: Models and experiments. *Food Eng.* Lund University, Sweden.
Kober, P.A. 1917. Pervaporation, perstillation and percrystallization, *J. Am. Chem. Soc.* 39: 944–948.
Lee, C.H. 1975. Theory of reverse osmosis and some other membrane permeation operations. *J. Appl. Polym. Sci.*19: 83–95.
Lee, Y.K., I.S. Sohn, E.J. Jeon, et al. 1991. IPN membranes for the pervaporation of ethanol-water mixture. *Polym. J.* 23: 427–433.
Mandal, S., and V.G. Pangarkar. 2002. Separation of methanol–benzene and methanol–toluene mixtures by pervaporation: Effects of thermodynamics and structural phenomenon. *J. Membrane Sci.* 201: 175–190.
Mulder, M.H.V., and C.A. Smolders. 1984. On the mechanism of separation of ethanol/water mixtures by pervaporation. Part I. Calculations of concentration profiles. *J. Membrane Sci.* 17: 289–307.
Nijhuis, H.H., M.H.V. Mulder, and C.A. Smolders. 1991. Removal of trace organics from aqueous solutions. Effect of membrane thickness. *J. Membrane Sci.* 61: 99–111.
Park, H.C., R.M. Meertens, M.H.V. Mulder, et al. 1994. Pervaporation of alcohol-toluene mixtures through polymer blend membranes of poly(acrylic acid) and poly (vinyl alcohol). *J. Membrane Sci.* 90: 265–274.
Porter, M.C. 1990. *Handbook of Industrial Membrane Technology.* Park Ridge, New Jersey: Noyes Publications, p. 61.
Price, G.J., and I.M. Shillcock. 2002. Inverse gas chromatographic measurement of solubility parameters in liquid crystalline systems. *J. Chromatogr. A* 964: 199–204.
Qi, R., and A. Henson. 2000. Membrane system design for multicomponent gas mixtures via mixed-integer non-linear programming. *Comput. Chem. Eng.* 24: 2719–2737.
Raghunath, B., and S.T. Hwang. 1992. Effect of boundary layer mass transfer resistance in the pervaporation of dilute organics. *J. Membrane Sci.* 65: 147–161.
Tanihara, N., N. Umeo, T. Kawabata, et al. 1995. Pervaporation of organic liquid mixtures through poly(ether imide) segmented copolymer membranes. *J. Membrane Sci.* 104: 181–192.
Villaluenga, J.P.G., M. Khayet, P. Godino, et al. 2003. Pervaporation of toluene/alcohol mixtures through a coextruded linear low-density polyethylene membrane. *Ind. Eng. Chem. Res.* 42: 386–391.

# 21 Carbon Dioxide–Selective Membranes

*Bishnupada Mandal and W. S. Winston Ho*

**CONTENTS**



## 21.1 INTRODUCTION

The separation of carbon dioxide from other gases such as $H_2$, $N_2$, CO, and $CH_4$ has great importance in many industrial areas such as purification of synthesis gas to obtain high-purity $H_2$ for fuel cells, the removal of $CO_2$ from natural gas for natural gas sweetening, and the separation of $CO_2$ from flue gas for the greenhouse gas emission reduction. There are different ways to separate $CO_2$, including absorption using aqueous solutions of conventional, nonsterically hindered amines (monoethanolamine, diethanolamine, triethanolamine, *N*-methyldiethanolamine and their blends) (Mandal et al., 2001, 2003, 2004; Mandal and Bandyopadhyay, 2005, 2006a, 2006b; Paul et al., 2009a, 2009b, 2009c), absorption using aqueous solutions of sterically hindered amines (Sartori et al., 1987, 1994; Ho et al., 1990a, 1990b; Shulik et al., 1996), and pressure swing adsorption (PSA) with different adsorbents (e.g., zeolites or activated carbon), and so forth. The main drawback of these processes is that they are energy-intensive.

Separations using synthetic membranes have been widely adopted for environmental and energy applications since the 1960s (Ho and Sirkar, 1992). The membrane-based process has the most advantages due to operational simplicity, light weight and space efficiency, and low energy consumption. During the last half-century, membrane separation technology has been used in a wide spectrum of applications in chemical industry and human life, ranging from gas separation, water desalination, artificial kidneys, and controlled drug delivery. The key commercial requirements for the membrane include the following criteria: (1) high permeation rate for the more permeable component, (2) high selectivity, (3) manufacturing reproducibility, (4) economical manufacturing process, (5) long-term stability under operating conditions, and (6) tolerance to contaminants and moderate temperature excursions. Of these, high permeation rate and high selectivity are clearly the most basic. The higher the selectivity, the lower the driving force required to achieving a given separation, and therefore the lower the operating cost of the membrane system. The higher the flux, the smaller the required membrane area, and therefore the lower the capital cost of the membrane system, assuming equivalent manufacturing costs per unit of membrane area (Koros and Pinnau, 1994).

In most current membrane processes, the separation mechanisms are based on solution, diffusion, and/or sieving. With these types of membranes, an increase in selectivity is often accompanied by a decrease in flux, and vice versa (Gottschlich et al., 1988; Ho and Sirkar, 1992). However, facilitated transport membranes offer an attractive method of achieving high selectivity while also maintaining high flux. This type of membrane is based on the facilitated transport mechanism that combines diffusion with the reversible reaction of a targeted component with reactive carriers inside the membrane (Ho and Sirkar, 1992).

For facilitated transport, reactive carriers are mainly two types: (1) the mobile carrier (move freely across the membrane) and (2) the fixed carrier (limited mobility around its equilibrium position). The membranes with these two types of carriers are shown in Figure 21.1. In a mobile carrier membrane, the mobile carriers react with the targeted component on the feed side of the membrane, move across the membrane, and release this component on the permeate side. The carrier-component complex diffuses in parallel with the molecular diffusion of the component. As a result, the transport of this component is augmented or facilitated. The other nonreactive components diffuse across the membrane down their concentration gradients via the solution-diffusion mechanism. Their transport is not affected by the facilitated transport. In a fixed carrier membrane, the targeted component reacts with one carrier site and then hops to the next unreacted carrier site along the direction of the concentration gradient via the "hopping" mechanism (Cussler et al., 1989).

Facilitated transport membranes have several advantages over the conventional membranes based on the solution-diffusion mechanism: (1) they are often highly selective, especially at low concentration driving forces, (2) high permeability can be achieved when the concentration driving force is low, and (3) both high permeability and high selectivity can be maintained at the same time.

Facilitated transport membranes have been studied for a long time. Today, the potential applications of facilitated transport membranes include acid gas removal, metal recovery, olefin/paraffin separation, and $O_2/N_2$ separation (Ho and Sirkar, 1992). This chapter reviews the recent development of carbon dioxide–selective membranes for environmental and energy applications.

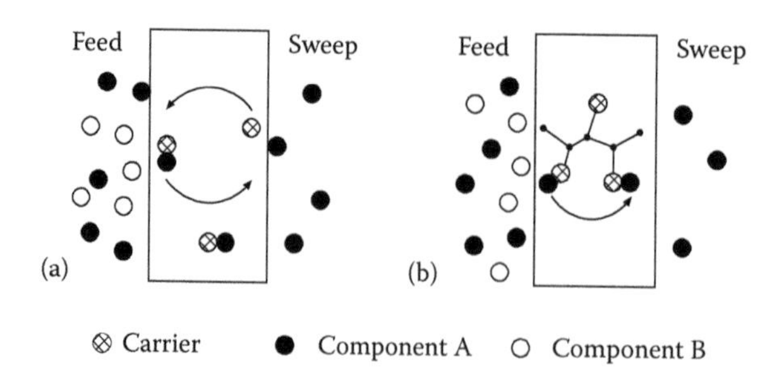


**FIGURE 21.1** Schematic of facilitated transport membranes: (a) membrane with mobile carriers; (b) membrane with fixed carriers.

## 21.2 CARBON DIOXIDE–SELECTIVE MEMBRANES

$CO_2$-selective membranes investigated so far are based on either the solution-diffusion mechanism or facilitated transport mechanism. Facilitated transport membranes are very attractive for the removal of $CO_2$ from synthesis gas, natural gas, and flue gas. For membranes based on the solution-diffusion mechanism, it is challenging to achieve a high $CO_2/H_2$ selectivity. This is due to the fact that $H_2$ exhibits higher diffusivity coefficient than $CO_2$. As a result, in most conventional polymeric membranes, the $CO_2/H_2$ selectivities are usually less than one (Lin and Freeman, 2005). Recently, crosslinked PEG membranes were reported to show a reasonable $CO_2/H_2$ selectivity (~30) but at an extremely low temperature of –20°C (Lin et al., 2006). On the other hand, facilitated transport membranes can achieve very high $CO_2/H_2$ selectivities. Tee et al. (2006) and Zou and Ho (2006) recently reported a $CO_2/H_2$ selectivity ranging from 100 to 450 using membranes containing both fixed and mobile carriers. High-purity hydrogen is essential for fuel cell application (Song, 2002). $CO_2$-selective membranes have the potential to obtain high-purity $H_2$ by separating $CO_2$ from the synthesis gas to enhance the water gas shift reaction (Huang et al., 2005a; Zou et al., 2007; Zou and Ho, 2007; Huang and Ho, 2008) or to be followed by methanation (Ledjeff-Hey et al., 2000), thus producing high-purity $H_2$ on the high-pressure feed side.

### 21.2.1 Facilitated Transport Carbon Dioxide–Selective Membranes

Facilitated transport $CO_2$-selective membranes reported in the literature include supported liquid membranes (SLMs), ion-exchange membranes, and membranes with reactive carriers. Ward and Robb (1967) immobilized an aqueous solution of bicarbonate-carbonate into a porous support and obtained a $CO_2/O_2$ separation factor of about 1,500. Meldon et al. (1977) investigated the facilitated transport of $CO_2$ through an immobilized alkaline liquid film. Their experimental results confirmed that weak acid buffers significantly increased the $CO_2$ transport. However, SLMs have two major problems: loss of solvent and loss or degradation of carriers. The loss of solvent is caused by its evaporation, especially at a high temperature and/or its permeation through support under a high transmembrane pressure. The loss of carrier occurs when the carrier solution is forced to permeate through support (washout), and the degradation of carriers is led by the irreversible reaction of the carrier with impurities in the feed gas stream (LeBlanc et al., 1980; Way et al., 1987).

To overcome the instability of the SLMs, LeBlanc et al. (1980) reported the ion exchange facilitated transport membrane. Ionic carriers were retained inside the in the ion-exchange membranes by electrostatic forces, therefore minimizing the washout of carriers. Way et al. (1987) and Yamaguchi et al. (1996) used perfluorosulfonic acid ionomer cation-exchange membranes containing amines as the carriers. The ion-exchange membrane used by Langevin et al. (1993) was sulfonated styrene-divinylbenzene in a fluorinated matrix, and the transport model based on the Nernst-Planck equation was developed. Matsuyama et al. (1994, 1996) grafted acrylic acid and methacrylic acid on different substrates and used various diamines, diethlylenetriamine and triethylenetetramine, as the carriers. They also blended poly(acrylic acid) with poly(vinyl alcohol) to prepare membranes and introduced monoprotonated ethylenediamine into the membranes by ion-exchange and used it as the carrier (Matsuyama et al., 2001). Other approaches were also proposed to solve the instability problems of SLMs. Quinn et al. (1995) developed membranes consisting of molten salt hydrates, which were nonvolatile and immobilized in microporous polypropylene supports. Teramoto et al. (2001, 2002, 2004) developed a bulk flow liquid membrane, in which a carrier solution was forced to permeate through the membrane and then was recycled continuously. In addition to amines, reactive carriers included ionic liquids (Ilconich et al., 2007; Bara et al., 2008, 2009; Hanioka et al., 2008) and poly(amidoamine) dendrimers (Kovvali and Sirkar, 2001; Wang et al., 2007; Duan et al., 2008).

Membranes with reactive carriers bonded in the membrane matrices were reported by several researchers and were believed to have better stability than SLMs and ion-exchange membranes. Yamaguchi et al. (1995) developed membranes with poly(allylamine) and compared them with ion-exchange membranes containing amines as the counter ions. Matsuyama et al. (1999) heat-treated poly(vinyl alcohol)-polyethylenimine membranes to improve their stability and to increase the amount of polyethylenimine retained inside the membranes, which increased the water content, therefore increasing the diffusivity of the carrier complex. Quinn and Laciak (1997) developed polyelectrolyte membranes based on poly (vinylbenzyltrimethylammonium fluoride) (PVBTAF) and achieved a $CO_2/H_2$ selectivity of 87 at 23°C. They also blended fluoride-containing organic and inorganic salts, such as CsF, into the PVBTAF membranes and obtained a $CO_2$ permeance more than four times of that of PVBTAF (Quinn et al., 1997). Other membranes include poly(*N*-vinyl-γ-sodium aminobutyrate) (Zhang et al., 2002), polyvinylamine (Kim et al., 2004), chitosan (El-Azzami and Grulke, 2009a), poly(ethylene glycol) (PEG)-grafted polyinoic materials (Hu et al., 2006), and interfacially polymerized films (Shil and Ho, 2006; Zhao et al., 2006).

Recently, aminoacids (Tee et al., 2006; Zou and Ho, 2006; Yegani et al., 2007; El-Azzami and Grulke, 2009b) and alkanolamines (Francisco et al., 2007) as mobile carriers were incorporated into hydrophilic polymers. Ho and his coworkers synthesized crosslinked poly(vinyl alcohol) membranes containing polyamines as fixed carriers and aminoacid salts as mobile carriers (Ho, 1997, 2000, 2008; Tee et al., 2006; Zou and Ho, 2006, Mandal and Ho, 2007; Zou et al., 2008; Bai and Ho, 2009). They reported membranes containing dimethylglycine salts and polyethylenimine, and found that both $CO_2$ permeability and $CO_2/H_2$ selectivity of the membranes increased as temperature increased in the temperature range of 50°C to 100°C (Tee et al., 2006). Recently, they reported on membranes containing amines up to about 180°C (Zou and Ho, 2006).

Models for facilitated transport on different systems have been developed and studied. Analytical and numerical methods were used to do the modeling and to compare with experimental results (Ward, 1970; Smith and Quinn, 1979; Donaldson and Lapinas, 1982; Gottschlich et al., 1988; Cussler et al., 1989; Noble, 1990, 1991, 1992; Ho and Dalrymple, 1994; Hong et al., 1996; Kang et al., 1996).

Selectivity and permeability are the two important parameters to characterize the separation performance of a membrane as mentioned before. They are defined as follows (Ho and Sirkar, 1992):

$$\text{Selectivity } (\alpha) \frac{y_i / y_j}{x_i / x_j} \tag{21.1}$$

where $y_i$ and $x_i$ are the mole fractions of component $i$ in the permeate and retentate side, and $y_j$ and $x_j$ are component $j$ in the permeate and retentate side, respectively.

$$\text{Permeability } (P_i) = \frac{N_i}{\Delta p_i / l} \tag{21.2}$$

The common unit of permeability, $P_i$, is Barrer, which is $10^{-10}$ cm$^3$ (STP)·cm/(cm$^2$·s·cm Hg). ($\Delta P_i/l$) is referred to as the permeance, and its common unit is the gas permeation unit (GPU = $10^{-6}$ cm$^3$ (STP)/(cm$^2$·s·cm Hg)).

### 21.2.2 Carbon Dioxide–Selective Polymeric Membranes

This section reviews our recent work on new $CO_2$-selective polymeric membranes for the facilitated transport of $CO_2$. Transport properties of the membranes synthesized, including $CO_2$ flux and permeability, $H_2$ flux, $CO_2/H_2$, $CO_2/N_2$, and $CO_2$/CO selectivities, were studied. The effects of feed pressure, water concentration, and temperature on transport properties were investigated.

#### 21.2.2.1 Synthesis of $CO_2$-Selective Membranes

$CO_2$-selective polymeric membranes with the thin-film-composite structure were prepared by casting an aqueous solution onto microporous BHA Teflon supports (thickness: 60 μm, average pore size: 0.2 μm, BHA Technologies, Kansas City, MO) or microporous polysulfone supports (thickness: about 60 μm excluding nonwoven fabric support, average pore size: 0.05 μm, NL Chemical Technology, Mount Prospect, IL, and GE Infrastructure Vista, CA). The aqueous solution was prepared from poly(vinyl alcohol), formaldehyde (crosslinking agent), potassium hydroxide, 2-aminoisobutyric acid (AIBA) potassium salt, and poly(allylamine) (Zou and Ho, 2006; Mandal and Ho, 2007). The active dense layer was about 20 to 80 μm thick. The thickness of a membrane to be mentioned hereafter refers to the thickness of the active layer. The membranes used in the present study contained 50 wt% PVA, 18.3 wt% KOH, 20.7 wt% AIBA-K, and 11.0 wt% poly(allylamine), unless otherwise indicated. Formaldehyde equivalent to a 60 mol% of crosslinking degree was added into the casting solution (Zou and Ho, 2006; Mandal and Ho, 2007).

#### 21.2.2.2 Transport Mechanism

The membranes synthesized contained both AIBA-K and $KHCO_3$-$K_2CO_3$ (converted from KOH) as the mobile carriers, and poly(allylamine) as the fixed carrier for $CO_2$ transport. 2-aminoisobutyric acid-potassium salt is a sterically hindered amine, and its reaction with $CO_2$ is depicted in Equation 21.3 (Sartori et al., 1987). Poly(allylamine) contains sterically unhindered, primary amino groups, and the reaction of these groups is shown in Equation 21.4 (Sartori et al., 1987). The reaction mechanism of the $CO_2$ with $KHCO_3$-$K_2CO_3$ was presumably similar to that of potassium carbonate promoted by sterically hindered amine described in Equation 21.5 (Sartori and Savage, 1983; Shulik et al., 1996):

$$R-NH_2+CO_2+H_2O \rightleftharpoons R-NH_3^+ + HCO_3^- \quad (21.3)$$

$$2R-NH_2+CO_2 \rightleftharpoons R-NH-COO^- + R-NH_3^+ \quad (21.4)$$

$$CO_3^{2-}+CO_2+H_2O \rightleftharpoons 2HCO_3^- \quad (21.5)$$

The transfer of $CO_2$ across the membrane is enhanced by the facilitated transport with the reactions mentioned above, and the total flux equation for the $CO_2$ transport in the membranes can be expressed as (Ho and Dalrymple, 1994):

$$N_A = D_A\left(\frac{C_{A|p1}-C_{A|p2}}{l}\right)+D_{AB}\left(\frac{C_{AB|p1m}-C_{AB|p2m}}{l}\right) \quad (21.6)$$

In Equation 21.6, the first term on the right-hand side is the flux due to the physical solution-diffusion mechanism, while the second term is the flux contributed by the chemical reaction. The nonreacting gases such as $H_2$, $N_2$, CO, and $CH_4$ do not have chemical association with the carriers and therefore can only be transported by the physical mechanism, which is limited by their low solubility on the highly polar sites in the membranes (Ho and Sirkar, 1992; Quinn et al., 1995; Kim et al., 2004). For these nonreacting gases, the flux equation in the membrane is the first term on the right-hand side of Equation 21.6 only.

#### 21.2.2.3 Transport Measurement

The gas permeation tests were conducted by using a permeation gas permeation apparatus (Zou and Ho, 2006). Two feed gases were used for the gas permeation tests: one consisting of 20% $CO_2$, 40% $H_2$, and 40% $N_2$, and the other consisting of 17% $CO_2$, 1.0% CO, 45% $H_2$, and 37% $N_2$ (all on dry basis). The second composition was used to simulate the composition of the synthesis gas from autothermal reforming of gasoline with air. Argon was used as the sweep gas for the ease of gas chromatography analysis.

Gas flow rates were controlled by Brooks flow-meters (Brooks Instrument, Hatfield, PA). The feed and the sweep gas rates were kept typically at 60 and 30 cc/min, respectively. A proper amount of water was pumped into two vessels using two Varian Prostar 210 pumps (Varian, Palo Alto, CA) to control the water contents of the feed gas and the sweep gas, respectively, before they entered the permeation cell. The pressure of the retentate was controlled by a back-pressure regulator, and the pressure on the permeate side was set close to atmospheric pressure via a near-ambient pressure regulator. The compositions for both the retentate and permeate gases were analyzed using an Agilent 6890N gas chromatograph with two thermal conductivity detectors (TCDs) (Agilent Technologies, Palo Alto, CA). The GC columns used were SUPELCO Carboxen® 1004 micropacked columns (Sigma-Aldrich, St. Louis, MO). Each of the membrane permeation measurements was taken after the membrane had been exposed to the feed and permeate streams under a specific condition (temperature, pressure, and water rate) for at least 6 hours, which allowed for steady-state permeation. From each permeation measurement, the transport properties including $CO_2$ flux and permeability and $CO_2/H_2$ selectivity as well as $H_2$ flux were obtained.

#### 21.2.2.4 Effects of Feed Pressure on Separation Performance

The effects of feed pressure on $CO_2$ flux and permeability, $H_2$ flux, and $CO_2/H_2$ selectivity were investigated using a membrane with a thickness of ~60 μm on the BHA microporous Teflon support (Zou and Ho, 2006). The feed gas composition was 20% $CO_2$, 40% $H_2$, and 40% $N_2$. Figure 21.2 illustrates the effects of feed pressure on $CO_2$ flux and permeability. As illustrated in this figure, $CO_2$ flux increased first linearly with the feed pressure and then approached a constant value. This can be explained with the carrier saturation phenomenon. As described by Ho and Dalrymple (1994), when the partial pressure of $CO_2$ is equal to or higher than a critical $CO_2$ partial pressure, $p_{1c}$, the carrier saturation occurs, in which the concentration of $CO_2$-carrier reaction product attains its maximum value, $C_{AB,max}$, and becomes constant. In other words, further increase in the partial pressure of $CO_2$ will not increase the concentration of $CO_2$-carrier reaction product. This can be expressed as follows:

$$C_{AB\,|\,p1} = H_{AB\,|\,p1}\, p_1 = C_{AB,\,max} = \text{constant when } p_1 \geq p_{1c} \tag{21.7}$$

The flux of $CO_2$ due to the solution-diffusion mechanism is negligible compared to the facilitated transport flux. Therefore, the total $CO_2$ flux becomes constant eventually as the feed pressure increases.

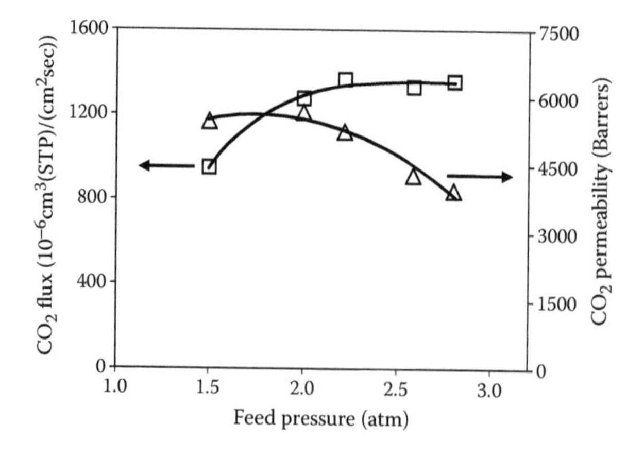


**FIGURE 21.2** Effects of feed pressure on $CO_2$ flux and permeability. (□) $CO_2$ flux; (Δ) $CO_2$ permeability; 110°C; water rates = 0.03/0.03 cc/min (feed/sweep); membrane thickness = 60 μm. (Reprinted from *J. Membrane Sci.*, 286, Zou, J., and W.S.W. Ho, 310–321, Copyright 2006, with permission from Elsevier.)

$CO_2$ permeability also decreased when the feed pressure increased, as depicted in Figure 21.2. This can also be explained by using the carrier saturation phenomenon. Increasing the $CO_2$ partial pressure, $p_1$, will not further increase the $CO_2$ flux since all the carriers have already reacted with $CO_2$ and attained the maximum capacities. Hence in order to maintain the equality in Equation 21.2, $CO_2$ permeability will decrease. However, unlike $CO_2$ flux, $H_2$ flux increased linearly with the feed pressure (Zou and Ho, 2006). This is because $H_2$ does not react with carriers. Therefore, the flux increased linearly with the feed pressure.

Figure 21.3 depicts the effect of feed pressure on $CO_2/H_2$ selectivity (Zou and Ho, 2006). As depicted in this figure, the $CO_2/H_2$ selectivity dropped as the pressure increased. Again, this can be explained using the carrier saturation phenomenon described earlier that $CO_2$ permeability reduces as the pressure increases. As a result, the selectivity decreases as the pressure increases since $H_2$ permeability usually does not change with pressure significantly.

#### 21.2.2.5 Effects of Water Content on Separation Performance

The effects of water content on the membrane separation performance at 120°C and 150°C were investigated (Zou and Ho, 2006). The feed gas composition was 20% $CO_2$, 40% $H_2$, and 40% $N_2$. Figure 21.4 depicts the $CO_2$ permeability as a function of the water concentration on the sweep side. As the water concentration on the sweep side increased, $CO_2$ permeability increased almost linearly. When the water content was increased from 58% to 93% (mol), the $CO_2$ permeability at 120°C increased from 3,700 to as high as 8,200 barrers, while the permeability at 150°C increasing from 920 to 2,700 barrers. These increases are presumably due to two reasons: (1) higher water content on the sweep side raised the water retention inside the membrane, thus resulting in increasing the mobility of both mobile and fixed carriers and the reaction rates of $CO_2$ with the carriers (Equations 21.3 through 21.5), and (2) higher water content on the sweep side diluted the permeated $CO_2$ concentration on the sweep side, thus resulting in increasing the driving force for the $CO_2$ transport. The increase of $CO_2$ permeability with increasing water content in the gas was also reported by Quinn and Laciak (1997) and Quinn et al. (1997).

Figure 21.5 shows the $CO_2/H_2$ selectivity as a function of the water concentration on the sweep side at 120°C and 150°C (Zou and Ho, 2006). The $CO_2/H_2$ selectivities at both temperatures increased as the water content on the sweep side increased. At the sweep water content of 93 mol%, the $CO_2/H_2$ selectivities at 120°C and 150°C reached 450 and 270, respectively. These increases could be explained by the rise of $CO_2$ permeation rate while the transport of $H_2$ was not significantly affected by the increase of the water content.

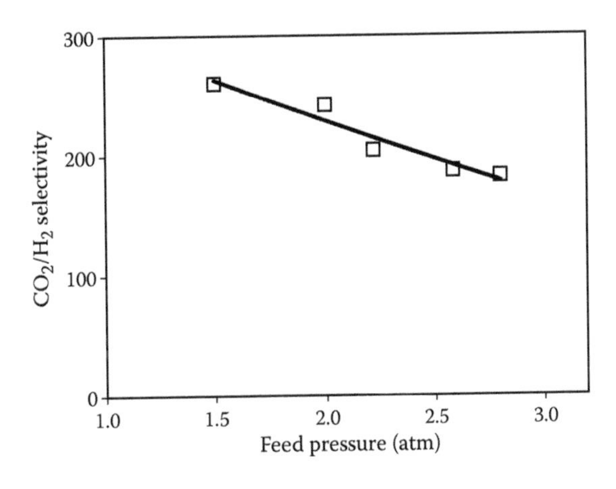


**FIGURE 21.3** Effect of feed pressure on $CO_2/H_2$ selectivity. At 110°C with water rates = 0.03/0.03 cc/min (feed/sweep), membrane thickness: = 60 μm. (Reprinted from *J. Membrane Sci.*, 286, Zou, J., and W.S.W. Ho, 310–321, Copyright 2006, with permission from Elsevier.)

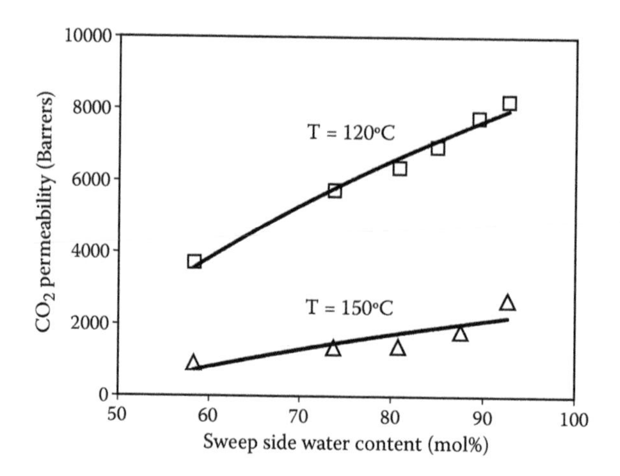


**FIGURE 21.4** $CO_2$ permeability versus water content on the sweep side. (□) T = 120°C; (Δ) T = 150°C; feed water content = 41 mol%; feed pressure = 2.0 atm; membrane thickness = 60 μm. (Reprinted from *J. Membrane Sci.*, 286, Zou, J., and W.S.W. Ho, 310–321, Copyright 2006, with permission from Elsevier.)

The water content on the feed side also had significant effects on $CO_2$ permeability and $CO_2/H_2$ selectivity (Zou and Ho, 2006). Both $CO_2$ permeability and $CO_2/H_2$ selectivity increased as the water content in the feed stream increased. This might be explained by the case that the higher water content on the feed side raised the water retention inside the membrane and therefore increased the mobility of both mobile and fixed carriers and the reaction rates of $CO_2$ with the carriers. As a result, the $CO_2$ transport was enhanced while the transport of $H_2$ was not significantly affected.

#### 21.2.2.6 Effects of Temperature on Separation Performance

The effects of temperature on $CO_2$ permeability, $CO_2/H_2$, $CO_2/N_2$, and $CO_2/CO$ selectivity were studied in the temperatures range of 100°C to 180°C (Zou and Ho, 2006). Figure 21.6 illustrates the effects of temperature on $CO_2$ permeability and $CO_2/H_2$ selectivity using a 26-μm membrane containing 40.0 wt% PVA, 20.0 wt% KOH, 20.0 wt% AIBA-K, and 20.0 wt% poly(allylamine) (Zou and Ho, 2006). As shown in this figure, the $CO_2$ permeability and $CO_2/H_2$ selectivity all decreased

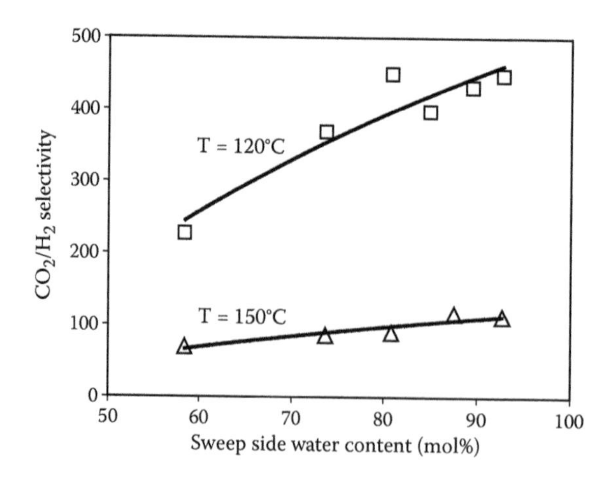


**FIGURE 21.5** $CO_2/H_2$ selectivity versus water content on the sweep side. (□) T = 120°C; (Δ) T = 150°C; feed water content = 41 mol%; feed pressure = 2.0 atm; membrane thickness = 60 μm. (Reprinted from *J. Membrane Sci.*, 286, Zou, J., and W.S.W. Ho, 310–321, Copyright 2006, with permission from Elsevier.)

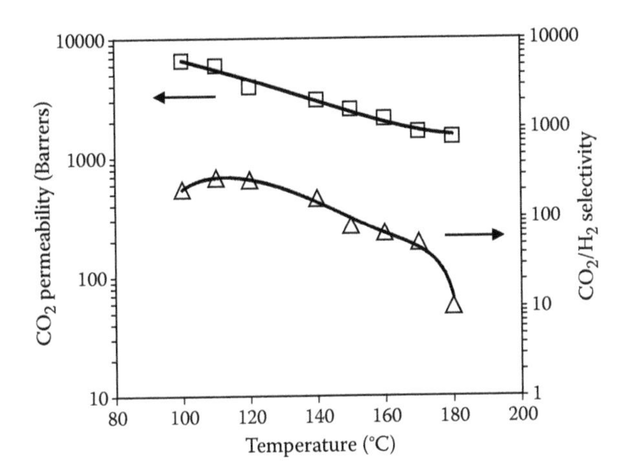


**FIGURE 21.6** Effects of temperature on $CO_2$ permeability and $CO_2/H_2$ selectivity. (□) $CO_2$ permeability; (Δ) $CO_2/H_2$ selectivity; feed pressure = 2.1 atm; with increasing water rates at elevated temperatures; membrane thickness = 26 μm. (Reprinted from *J. Membrane Sci.*, 286, Zou, J., and W.S.W. Ho, 310–321, Copyright 2006, with permission from Elsevier.)

as temperature increased. This was due to the fact that $CO_2$ permeability decreased as temperature increased, while the transport of both $H_2$ was not affected by temperature significantly. The $CO_2/H_2$ selectivity was reduced slightly as the temperature increased to 170°C. But the selectivity reduced significantly to about 10 at 180°C mainly due to the significant swelling of the membrane, thus resulting in a sharp increase of $H_2$ permeability at this high temperature.

### 21.2.3 $CO_2$-Selective Membrane Applications

This section reviews our recent work on new $CO_2$-selective polymeric membranes for two applications: (1) water gas shift membrane reactor in fuel processing for fuel cells and (2) $CO_2$ capture from flue gas.

#### 21.2.3.1 $CO_2$-Selective Water Gas Shift Membrane Reactor in Fuel Processing for Fuel Cells

Hydrogen is the preferred fuel for fuel cells for the emerging energy conversion approach in most cases, especially for proton-exchange membrane fuel cells (PEMFCs). However, how to obtain high-purity hydrogen is the key issue for PEMFC applications. Hydrogen is mainly produced by steam reforming of natural gas followed by the water gas shift (WGS) reaction (Equation 21.8) (Kroschwitz and Howe-Grant, 1995). The resulting synthesis gas mainly contains $H_2$, CO, and $CO_2$. The CO concentration in synthesis gas needs to be reduced to less than 10 ppm, since even a very small amount of CO deteriorates fuel cell performance by poisoning platinum, the electrocatalyst in PEMFCs (Ahmed and Krumpelt, 2001; Song, 2002). There are several options for the CO cleanup. One option is to develop a WGS membrane reactor, in which a membrane removes either $H_2$ or $CO_2$ from the reactor to shift the reversible WGS reaction forward so that the CO concentration can be further converted/reduced.

$$CO + H_2O \rightleftharpoons CO_2 + H_2 \ (\Delta H = -41.16 \text{ kJ/mol}) \tag{21.8}$$

Several researchers have achieved high CO conversions beyond the equilibrium conversions or even close to 100% by using palladium or other inorganic $H_2$-selective WGS membrane reactors (Uemiya et al., 1991; Damle et al., 1994; Basile et al., 1996; Xue et al., 1996; Criscuoli et al., 2000; Giessler et al., 2003; Tosti et al., 2003; Brunetti et al., 2006; Iyoha et al., 2007; Battersby et al., 2008). However, preparing thin, flawless, and durable $H_2$-selective membranes remains a challenge for the commercial application of this type of membrane reactor (Armor, 1998). With $CO_2$-selective membranes (Ho, 1997, 2000, 2008; Tee et al., 2006; Zou and Ho, 2006; Mandal and Ho, 2007; Bai and Ho, 2009) removing $CO_2$ continuously, a $CO_2$-selective WGS membrane reactor provides a promising approach of decreasing CO concentration and enhancing hydrogen production at temperatures (~150°C) in line with proton-exchange membrane fuel cells. In comparison with the $H_2$-selective membrane reactor, the $CO_2$-selective WGS membrane reactor is more advantageous because (1) an $H_2$-rich product is recovered at a high pressure (feed gas pressure) and (2) air and/or steam can be used to sweep the permeate, $CO_2$, out of the reactor to obtain a high driving force for the separation. The first advantage eliminates a need for an unwanted compressor for compressing the hydrogen product back to its original feed pressure for an $H_2$-selective WGS membrane reactor.

Huang et al. (2005a) developed a one-dimensional nonisothermal model to simulate the reaction and transport process for the $CO_2$-selective WGS membrane reactor. The modeling results have shown that $H_2$ enhancement and CO reduction to 10 ppm or lower are achievable for autothermal reforming synthesis gas via $CO_2$ removal. With this model, the authors elucidated the effects of system parameters on the membrane reactor performance. Using the membrane synthesized and the commercial WGS catalyst, a CO concentration of less than 10 ppm was obtained in the $H_2$ product in WGS membrane reactor experiments and verified the model developed (Huang et al., 2005b; Zou et al., 2007).

Ho and his students set up a rectangular flat-sheet membrane reactor using their $CO_2$-selective membrane and the commercial $Cu/ZnO/Al_2O_3$ catalyst to conduct water gas shift (WGS) membrane reactor experiments and to demonstrate the WGS membrane reactor concept (Huang et al., 2005b; Zou et al., 2007; Huang and Ho, 2008). This reactor had well-defined gas flow and velocity for both the feed and sweep sides, and it was suitable for modeling and scale-up work. It had a width of about 7 inches (17.8 cm) and a length of about 7.5 inches (19.1 cm), with an effective membrane area of about 340 $cm^2$. The gas mixture of 1% CO, 45% $H_2$, 17% $CO_2$, and 37% $N_2$ was used as the feed gas, and argon was used as the sweep gas for the ease of gas chromatography (GC) analysis. The rationales for this CO level are twofold: (1) it can be readily produced from commercial WGS reactors and (2) it requires $CO_2$ removal for its reduction via WGS reaction.

The performance of the membrane reactor was investigated by varying the feed flow rate. Figure 21.7 shows the results obtained from this membrane reactor for the syngas at 150°C and 2 atm (Zou et

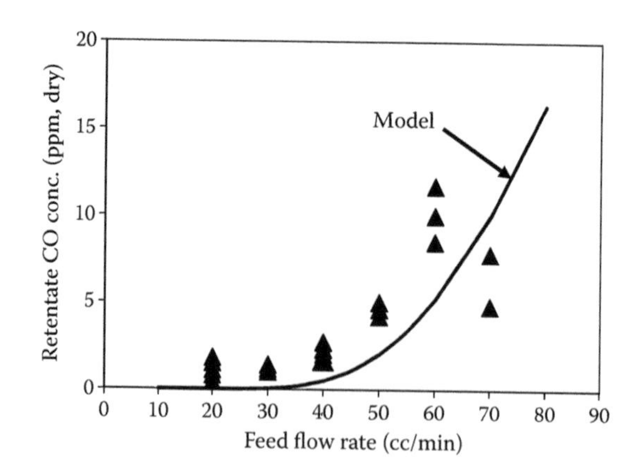


**FIGURE 21.7** Exit dry CO concentration versus feed flow rate in the rectangular WGS membrane reactor. Feed gas: 1% CO, 17% $CO_2$, 45% $H_2$, 37% $N_2$; T = 150°C, $p_f$ = 2.0 atm; $p_s$ = 1.0 atm; feed/sweep molar flow rates = 1/1 (dry basis). (Reprinted with permission from *Ind. Eng. Chem. Res.*, 46, Zou, J., et al., 2272–2279, Copyright 2007, American Chemical Society.)

al., 2007). As shown in this figure, the CO concentration in the exit stream (i.e., the $H_2$ product) was <10 ppm (on the dry basis) for the various feed rates of the syngas at 20, 30, 40, 50, 60, and 70 cc/min (with ~30% steam in the syngas). The nonisothermal model was modified for the flat-sheet membrane configuration of the reactor. In the model calculation, it was assumed that the membrane had a $CO_2$ permeability of 6500 barrers, $CO_2/H_2$ selectivity of 40, and negligible $N_2$ and CO permeation, which were reasonable based on previous experimental results (Zou and Ho, 2006; Zou et al., 2007; Huang et al., 2008). As shown in Figure 21.7, the data agreed reasonably well with the prediction by the model.

Zou and Ho (2007) also demonstrated $CO_2$ removal to drive WGS reaction and to achieve <10 ppm CO concentration in the treated hydrogen product. In their work, they developed a process combining carbon dioxide removal by using a polymeric membrane with subsequent water gas shift (WGS) reaction to purify hydrogen for fuel cells. The rectangular membrane permeation cell with well-defined countercurrent gas flows was used to study the $CO_2$ removal. A feed gas consisting of 1% CO, 17% $CO_2$, 45% $H_2$, and 37% $N_2$ was used to simulate the synthesis gas from autothermal reforming of gasoline with air. With this permeation cell running at 120°C, the $CO_2$ concentration in the gas mixture was reduced from 17% to as low as 10 ppm, resulting in more than 99.5% of $CO_2$ removed. Then, with another feed gas consisting of 1.19% CO, 0.10% $CO_2$, 53.87% $H_2$, and 44.84% $N_2$ used to simulate the synthesis gas after the $CO_2$-removal step, a reactor packed with a commercial low-temperature WGS catalyst was operated at 140°C to 150°C to convert CO to $H_2$. With such a low $CO_2$ concentration in the feed gas, the reversible WGS reaction was shifted forward so that the CO concentration was decreased from 1.19% to less than 10 ppm (on dry basis), which met the requirement of proton-exchange membrane fuel cells. The WGS reactor had a gas hourly space velocity of 7650 $h^{-1}$ at 150°C, and the $H_2$ concentration in the exit was more than 54 mol% (on dry basis).

#### 21.2.3.2 $CO_2$ Capture from Flue Gas

In recent years, the increasing public concern over global warming has concentrated on the man-made emissions of greenhouse gases. In the United States, 82.4% of the total greenhouse emissions consisted of $CO_2$ from the combustion of fossil fuels (U.S. Department of Energy, 2005). For the application of membranes in $CO_2$ capture from the flue gas, Hirayama et al. (1999) reported that a $CO_2/N_2$ selectivity of greater than 70 and a minimum $CO_2$ permeability of 100 barrers were required for an economic operation. As an alternative to conventional polymeric membranes, facilitated transport membranes have shown to be more promising in satisfying these goals.

Huang et al. (2008) studied $CO_2$ capture with a feed gas simulating the flue gas by using the novel polymeric facilitated transport membrane. The highest $CO_2/N_2$ selectivity of 493 and the highest $CO_2$ permeability of 6196 barrers were observed at 110°C. In the $CO_2$ capture experiments from a gas mixture with $N_2$, a permeate $CO_2$ dry concentration of greater than 98% was obtained by using steam as the sweep gas. The effects of feed flow rate and sweep-to-feed molar ratio on the membrane separation performance were investigated. A one-dimensional isothermal model was established to examine the performance of a hollow-fiber membrane module composed of the described $CO_2$-selective membrane. The modeling results show that a $CO_2$ recovery of greater than 95% and a permeate $CO_2$ dry concentration of above 98% are achievable from a 1,000 standard cubic feet per minute (SCFM) or 21.06 mol/s flue gas stream with a 2 ft (0.61 m) hollow-fiber module containing 980,000 fibers.

## 21.3 CONCLUSIONS

We have reviewed the recent developments of $CO_2$-selective polymeric membranes for environmental and energy applications. $CO_2$-selective membranes containing both mobile and fixed carriers in crosslinked poly(vinyl alcohol) have been discussed. The membranes have shown high $CO_2$ permeability as well as high $CO_2/H_2$, $CO_2/N_2$, and $CO_2$/CO selectivities up to 170°C. The $CO_2$ permeability and $CO_2/H_2$ selectivity reduced with increasing feed pressure, which could be explained with the carrier saturation phenomenon. Both the permeability and selectivity increased significantly with increasing water contents in both feed and sweep. Results from WGS membrane reactor experiments have shown carbon monoxide reduction to 10 ppm as well as significant hydrogen enhancement via $CO_2$ removal. The data has been in good agreement with model prediction. A process combining carbon dioxide removal by using a polymeric membrane with subsequent water gas shift (WGS) reaction to achieve <10 ppm CO for the purification of hydrogen for fuel cells was also developed. In addition, in the $CO_2$ capture experiments from a gas mixture with $N_2$, a permeate $CO_2$ dry concentration of greater than 98% was obtained by using steam as the sweep gas.

## NOMENCLATURE

| | |
|---|---|
| $C_A$ | $CO_2$ concentration (mol/cm$^3$) |
| $C_{AB}$ | $CO_2$-carrier reaction product concentration (mol/cm$^3$) |
| $D_A$ | diffusivity coefficient for $CO_2$ (cm$^2$/s) |
| $D_{AB}$ | diffusivity coefficient for $CO_2$-carrier reaction product (cm$^2$/s) |
| $H_{AB}$ | Henry's law constant for $CO_2$-carrier reaction product (mol/cm$^3$/atm) |
| $\Delta H$ | heat of reaction (J/mol) |
| $l$ | membrane thickness (cm) |
| $N_i$ | steady-state flux of component $i$ (mol/cm$^2$/s) |
| $P_i$ | permeability of component $i$ (barrer) |
| $p$ | pressure (atm) |
| $p_1$ | $CO_2$ partial pressure on the high pressure side of membrane (atm) |
| $p_2$ | $CO_2$ partial pressure on the low pressure side of membrane (atm) |
| $p_{1c}$ | critical $CO_2$ partial pressure at which carrier saturation occurs (atm) |
| $p_{1m}$ | $CO_2$ partial pressure in the membrane on the high pressure side of membrane (atm) |
| $p_{2m}$ | $CO_2$ partial pressure in the membrane on the low pressure side of membrane (atm) |
| $\Delta p_i$ | partial pressure difference of component $i$ |
| $x$ | component mole fraction in the retentate stream |
| $y$ | component mole fraction in the permeate stream |

## GREEK LETTER

| | |
|---|---|
| $\alpha$ | selectivity |

## SUBSCRIPTS

| | |
|---|---|
| $i$ | component $i$ |
| $j$ | component $j$ |

## ACKNOWLEDGMENTS

We would like to thank the National Science Foundation, the Office of Naval Research (Grant Nos. N000140810343 and N000140810547), the Ohio Department of Development (Wright Center of Innovation Grant No. 342-0561), and the Ohio State University for the financial support of this work. Part of this material is based on work supported by the National Science Foundation under Grant No. 0625758 and 1033131.

## REFERENCES

Ahmed, S., and M. Krumpelt. 2001. Hydrogen from hydrocarbon fuels for fuel cells. *Int. J. Hydrogen Energy* 26: 291.

Armor, J.N. 1998. Applications of catalytic inorganic membrane reactors to refinery products. *J. Membr. Sci.* 147: 217.

Bai, H., and W.S.W. Ho. 2009. New carbon dioxide-selective membranes based on sulfonated polybenzimidazole (SPBI) copolymer matrix for fuel cell applications. *Ind. Eng. Chem. Res.* 48 (5): 2344–2354.

Bara, J.E., C.J. Gabriel, T.K. Carlisle, et al. 2009. Gas separations in fluoroalkyl-functionalized room-temperature ionic liquids using supported liquid membranes. *Chem. Eng. J.* 147 (1): 43–50.

Bara, J.E., C.J. Gabriel, E.S. Hatakeyama, et al. 2008. Improving $CO_2$ selectivity in polymerized room-temperature ionic liquid gas separation membranes through incorporation of polar substituents. *J. Membr. Sci.* 321 (1): 3–7.

Basile, A., A. Criscuoli, F. Santella, and E. Drioli. 1996. Membrane reactor for water gas shift reaction. *Gas Sep. Purif.* 10: 243.

Battersby, S., M.C. Duke, S. Liu, et al. 2008. Metal doped silica membrane reactor: Operational effects of reaction and permeation for the water gas shift reaction. *J. Membr. Sci.* 316 (1+2): 46–52.

Brunetti, A., G. Barbieri, E. Drioli, et al. 2006. Porous stainless steel supported silica membrane for WGS reaction. *Desalination* 200 (1–3): 681–683.

Criscuoli, A., A. Basile, and E. Drioli. 2000. An analysis of the performance of membrane reactors for the water-gas shift reaction using gas feed mixtures. *Catal. Today* 56: 53.

Cussler, E.L., R. Aris, and A. Bhown. 1989. On the limit of facilitated diffusion. *J. Membr. Sci.* 43: 149.

Damle, A.S., S.K. Gangwal, and V.K. Venkataraman. 1994. Simple model for a water gas shift membrane reactor. *Gas Sep. Purif.* 8 (2): 101–106.

Donaldson, T.L., and A.T. Lapinas. 1982. Secondary flux enhancement in two-carrier facilitated transport. *Chem. Eng. Sci.* 37: 715.

Duan, S., F.A. Chowdhury, T. Kai, et al. 2008. PAMAM dendrimer composite membrane for $CO_2$ separation: Addition of hyaluronic acid in gutter layer and application of novel hydroxyl PAMAM dendrimer. *Desalination* 234 (1–3): 278–285.

El-Azzami, L.A., and E.A. Grulke. 2009a. Parametric study of $CO_2$ fixed carrier facilitated transport through swollen chitosan membranes. *Ind. Eng. Chem. Res.* 48 (2): 894–902.

El-Azzami, L.A., and E.A. Grulke. 2009b. Carbon dioxide separation from hydrogen and nitrogen: Facilitated transport in arginine salt-chitosan membranes. *J. Membr. Sci.* 328 (1–2): 15–22.

Francisco, G.J., A. Chakma, and X. Feng. 2007. Membranes comprising of alkanolamines incorporated into poly(vinyl alcohol) matrix for $CO_2/N_2$ separation. *J. Membr. Sci.* 303 (1+2): 54–63.

Giessler, S., L. Jordan, J.C. Diniz da Costa, et al. 2003. Performance of hydrophobic and hydrophilic silica membrane reactors for the water gas shift reaction. *Sep. Purif. Technol.* 32: 255.

Gottschlich, D.E., D.L. Roberts, and J.D. Way. 1988. A theoretical comparison of facilitated transport and solution-diffusion membrane modules for gas separation. *Gas Sep. Purif.* 2: 65.

Hanioka, S., T. Maruyama, T. Sotani, et al. 2008. $CO_2$ separation facilitated by task-specific ionic liquids using a supported liquid membrane. *J. Membr. Sci.* 314 (1–2): 1–4.

Hirayama, Y., Y. Kase, N. Tanihara, et al. 1999. Permeation properties to $CO_2$ and $N_2$ of poly(ethylene oxide)-containing and crosslinked polymer films. *J. Membr. Sci.* 160: 87.

Ho, W.S.W. 1997. Membranes comprising salts of aminoacids in hydrophilic polymers. U.S. Patent 5,611,843.

Ho, W.S.W. 2000. Membranes comprising aminoacid salts in polyamine polymers and blends. U.S. Patent 6,099,621.

Ho, W.S.W. 2008. Membranes, methods of making membranes, and methods of separating gases using membranes. U.S. Patent Application Public. No. US-2008-0168900.

Ho, W.S.W., and D.C. Dalrymple. 1994. Facilitated transport of olefins in $Ag^+$-containing polymer membranes. *J. Membr. Sci.* 91: 13–25.

Ho, W.S.W., and K.K. Sirkar (eds). 1992. *Membrane Handbook*. New York: Chapman & Hall. Boston: Kluwer Academic Publishers, reprint edition, 2001.

Ho, W.S.W., G. Sartori, and E.L. Stogryn. 1990a. Absorbent composition containing a severely-hindered amine mixture with amine salts and/or aminoacid additives for the absorption of $H_2S$. U.S. Patent 4,961,873.

Ho, W.S.W., E.L. Stogryn, and G. Sartori. 1990b. Absorbent composition containing a severely-hindered amine mixture for the absorption of $H_2S$. U.S. Patent 4,894,178.

Hong, J., Y.S. Kang, J. Jang, et al. 1996. Analysis of facilitated transport in polymeric membrane with fixed site carrier. 2. Series RC circuit model. *J. Membr. Sci.* 109: 159.

Hu, X., J. Tang, A. Blasig, et al. 2006. $CO_2$ permeability, diffusivity and solubility in polyethylene glycol-grafted polyionic membranes and their $CO_2$ selectivity relative to methane and nitrogen. *J. Membr. Sci.* 281 (1+2): 130–138.

Huang, J., L. El-Azzami, and W.S.W. Ho. 2005a. Modeling of $CO_2$-selective water-gas-shift membrane reactor for fuel cell. *J. Membr. Sci.* 261 (1–2): 67–75.

Huang, J., and W.S.W. Ho. 2008. Effects of system parameters on the performance of $CO_2$-selective WGS membrane reactor for fuel cells. *J. Chin. Ins. Chem. Engr.* 39 (2): 129–136.

Huang, J., J. Zou, and W.S.W. Ho. 2005b. A modeling and experimental study of $CO_2$-selective water-gas-shift membrane reactor for fuel cells. In *Proceedings of Topical Conference on Fuel Cells Technology*, AIChE Annual Meeting, Cincinnati, OH, Oct. 30–Nov. 4, 2005, Paper 501b.

Huang, J., J. Zou, and W.S.W. Ho. 2008. Carbon dioxide capture using a $CO_2$-selective facilitated transport membrane. *Ind. Eng. Chem. Res.* 47: 1261–1267.

Ilconich, J., C. Myers, H. Pennline, et al. 2007. Experimental investigation of the permeability and selectivity of supported ionic liquid membranes for $CO_2$/He separation at temperatures up to 125°C. *J. Membr. Sci.* 298 (1+2): 41–47.

Iyoha, O., R. Enick, R. Killmeyer, et al. 2007. $H_2$ production from simulated coal syngas containing $H_2S$ in multi-tubular Pd and 80 wt% Pd-20 wt% Cu membrane reactors at 1173K. *J. Membr. Sci.* 306 (1–2): 103–115.

Kang, Y.S., J. Hong, J. Jang, et al. 1996. Analysis of facilitated transport in solid membranes with fixed site carriers 1. Single RC circuit model. *J. Membr. Sci.* 109: 149.

Kim, T., B. Li, and M. Hagg. 2004. Novel fixed-site-carrier polyvinylamine membrane for carbon dioxide capture. *J. Polym. Sci. Part B: Polym. Phys.* 42: 4326.

Koros, W.J., and R. Mahajan. 2000. Pushing the limits on possibilities for large scale gas separation: which strategies. *J. Membr. Sci.* 175: 181.

Koros, W.J., and I. Pinnau. 1994. Membrane formation for gas separation process. In *Polymeric Gas Separation Membranes*, D.R. Paul and Y.P. Yampol'skii (eds.). Boca Raton, FL: CRC Press.

Kovvali, A.S., and K.K. Sirkar. 2001. Dendrimer liquid membranes: $CO_2$ separation from gas mixtures. *Ind. Eng. Chem. Res.* 40 (11): 2502–2511.

Kroschwitz, J.I., and M. Howe-Grant (eds.). 1995. *Encyclopedia of Chemical Technology*, 4th ed. Vol. 13, p. 927. New York: John Wiley & Sons.

Langevin, D., M. Pinoche, E. Selegny, et al. 1993. $CO_2$ facilitated transport through functionalized cation-exchange membranes. *J. Membr. Sci.* 82: 51.

LeBlanc, O.H., W.J. Ward, S.L. Matson, et al. 1980. Facilitated transport in ion-exchange membranes. *J. Membr. Sci.* 6: 339.

Ledjeff-Hey, K., J. Roes, and R. Wolters. 2000. $CO_2$-scrubbing and methanation as purification system for PEFC. *J. Power Sources* 86: 556.

Lin, H., and B.D. Freeman. 2005. Materials selection guidelines for membranes that remove $CO_2$ from gas mixture. *J. Mol. Struct.* 739: 57.

Lin, H., E.V. Wagner, B.D. Freeman, et al. 2006. Plasticization-enhanced hydrogen purification using polymeric membranes. *Science* 311: 639.

Mandal, B., and S.S. Bandyopadhyay. 2005. Simultaneous absorption of carbon dioxide and hydrogen sulfide into aqueous blends of 2-amino-2-methyl-1-propanol and diethanolamine. *Chem. Eng. Sci.* 60: 6438–6451.

Mandal, B., and S.S. Bandyapadhyay. 2006a. Simultaneous absorption of $CO_2$ and $H_2S$ into aqueous blends of N-methyldiethanolamine and diethanolamine. *Env. Sci. Technol.* 40: 6076–6084.

Mandal, B., and S.S. Bandyapadhyay. 2006b. Absorption of carbon dioxide into aqueous blends of 2-amino-2-methyl-1-propanol and monoethanolamine. *Chem. Eng. Sci.* 61: 5440–5447.

Mandal, B., A.K. Biswas, and S.S. Bandyapadhyay. 2003. Absorption of carbon dioxide into aqueous blends of 2-amino-2-methyl-1-propanol and diethanolamine. *Chem. Eng. Sci.* 58: 4137–4144.

Mandal, B., A.K. Biswas, and S.S. Bandyapadhyay. 2004. Selective absorption of $H_2S$ from gas streams containing $H_2S$ and $CO_2$ in aqueous solutions of N-methyldiethanolamine and 2-amino-2-methyl-1-propanol. *Sep. Purifi. Technol.* 35: 191–202.

Mandal, B.P., M. Guha, A.K. Biswas, et al. 2001. Removal of $CO_2$ by absorption in mixed amines: Modeling of absorption in aqueous MDEA/MEA and AMP/MEA solutions. *Chem. Eng. Sci.* 56: 6217–6224.

Mandal, B., and W.S.W. Ho. 2007. Synthesis gas purification by polymeric membranes containing fixed and mobile carriers. *Int. J. Chem. Sci.* 5: 1938–1946.

Matsuyama, H., A. Terada, T. Nakagawara, et al. 1999. Facilitated transport of $CO_2$ through polyethylenimine/poly(vinyl alcohol) blend membrane. *J. Membr. Sci.* 163: 221.

Matsuyama, H., M. Teramoto, and K. Iwai. 1994. Development of a new functional cation-exchange membrane and its application to facilitated transport of $CO_2$. *J. Membr. Sci.* 93: 237.

Matsuyama, H., M. Teramoto, K. Matsui, et al. 2001. Preparation of poly(acrylic acid)/poly(vinyl alcohol) membrane for the facilitated transport of $CO_2$. *J. Appl. Polym. Sci.* 81: 936.

Matsuyama, H., M. Teramoto, H. Sakakura, et al. 1996. Facilitated transport of $CO_2$ through various ion exchange membranes prepared by plasma graft polymerization. *J. Membr. Sci.* 117: 251.

Matuleviclus, E.S., and N.N. Li. 1975. Facilitated transport through liquid membranes. *Sep. Purif. Methods* 4: 73.

Meldon, J.H., K.A. Smith, and C.K. Colton. 1977. The effect of weak acids upon the transport of carbon dioxide in alkaline solutions. *Chem. Eng. Sci.* 32: 939.
Noble, R.D. 1990. Analysis of facilitated transport with fixed site carrier membranes. *J. Membr. Sci.* 50: 207.
Noble, R.D. 1991. Facilitated transport mechanism in fixed site carrier membranes. *J. Membr. Sci.* 60: 297.
Noble, R.D. 1992. Generalized microscopic mechanism of facilitated transport in fixed site carrier membranes. *J. Membr. Sci.* 75: 121.
Paul, S., A.K. Ghoshal, and B. Mandal. 2009a. Kinetics of absorption of carbon dioxide into aqueous solution of 2-(1-piperazinyl)-ethylamine. *Chem. Eng. Sci.* 64: 313–321.
Paul, S., A.K. Ghoshal, and B. Mandal. 2009b. Kinetics of absorption of carbon dioxide into aqueous blends of 2-(1-piperazinyl)-ethylamine and N-methyldiethanolamine. *Chem. Eng. Sci.* 64: 1618–1622.
Paul, S., A.K. Ghoshal, and B. Mandal. 2009c. Absorption of carbon dioxide into aqueous solutions of 2-piperidineethanol: Kinetics analysis. *Ind. Eng. Chem. Res.* 48(3):1414–1419.
Quinn, R., J.B. Appleby, and G.P. Pez. 1995. New facilitated transport membranes for the separation of carbon dioxide from hydrogen and methane. *J. Membr. Sci.* 104: 139.
Quinn, R., and D.V. Laciak. 1997. Polyelectrolyte membranes for acid gas separations. *J. Membr. Sci.* 131: 49.
Quinn, R., D.V. Laciak, and G.P. Pez. 1997. Polyelectrolyte-salt blend membranes for acid gas separations. *J. Membr. Sci.* 131: 61.
Sartori, G., and D.W. Savage. 1983. Sterically hindered amines for $CO_2$ removal from gases. *Ind. Eng. Chem. Fundam.* 22: 239.
Sartori, G., W.S.W. Ho, D.W. Savage, et al. 1987. Sterically-hindered amines for acid-gas absorption. *Sep. Purif. Methods* 16: 171.
Sartori, G., W.S.W. Ho, W.A. Thaler, et al. 1994. Sterically-hindered amines for acid gas absorption. *Spec. Publ.-R. Soc. Chem.* 153: 205.
Shil, G., and W.S.W. Ho. 2006. Synthesis and characterization of interfacially polymerized membranes for $CO_2$ separation. *J. Environ. Eng. Mgmt.* 16 (4): 233–241 (2006).
Shulik, L.J., G. Sartori, W.S.W. Ho, et al. 1996. A novel, $V^{+5}$-stable $K_2CO_3$ promoter for $CO_2$ absorption. *Sep. Sci. Technol.* 31: 1663.
Smith, D.R., and J.A. Quinn. 1979. The prediction of facilitation factors for reaction augmented membrane transport. *AIChE J.* 25: 197.
Song, C. 2002. Fuel processing for low-temperature and high-temperature fuel cells: Challenges, and opportunities for sustainable development in the 21st century. *Catal. Today* 77: 17.
Tee, Y.H., J. Zou, and W.S.W. Ho. 2006. $CO_2$-selective membranes containing dimethylglycine mobile carriers and polyethylenimine fixed carrier. *J. Chin. Inst. Chem. Engrs.* 37: 37.
Teramoto, M., S. Kitada, N. Ohnishi, et al. 2004. Separation and concentration of $CO_2$ by capillary-type facilitated transport membrane module with permeation of carrier solution. *J. Membr. Sci.* 234: 83.
Teramoto, M., N. Takeuchi, T. Maki, et al. 2001. Gas separation by liquid membrane accompanied by permeation of membrane liquid through membrane physical transport. *Sep. Purif. Technol.* 24: 101.
Teramoto, M., N. Takeuchi, T. Maki, et al. 2002. Facilitated transport of $CO_2$ through liquid membrane accompanied by permeation of carrier solution. *Sep. Purif. Technol.* 27: 25.
Tosti, S., A. Basile, G. Chiappetta, et al. 2003. Pd-Ag membrane reactors for water gas shift reaction. *Chem. Eng. J.* 93: 23.
Uemiya, S., N. Sato, H. Ando, et al. 1991. The water gas shift reaction assisted by a palladium membrane reactor. *Ind. Eng. Chem. Res.* 30: 585.
U.S. Department of Energy. 2005. Emissions of greenhouse gases in the United States 2004. ftp://ftp.eia.doe.gov/pub/oiaf/1605/cdrom/pdf/ggrpt/057304.pdf.
Wang, Z., M. Li, Y. Cai, et al. 2007. Novel $CO_2$ selectively permeating membranes containing PETEDA dendrimer. *J. Membr. Sci.* 290 (1+2): 250–258.
Ward, W.J. 1970. Analytical and experimental studies of facilitated transport. *AIChE J.* 16: 405.
Ward, W.J., and W.L. Robb. 1967. Carbon dioxide-oxygen separation: facilitated transport of carbon dioxide across a liquid film. *Science* 156: 1481.
Way, J.D., R.D. Noble, D.L. Reed, et al. 1987. Facilitated transport of $CO_2$ in ion exchange membranes. *AIChE J.* 33: 480.
Xue, E., M. O'Keeffe, and J.R.H. Ross. 1996. Water-gas shift conversion using a feed with a low steam to carbon monoxide ratio and containing sulphur. *Catal. Today* 30: 107.
Yamaguchi, T., L.M. Boetje, C.A. Koval, et al. 1995. Transport properties of carbon dioxide through amine functionalized carrier membranes. *Ind. Eng. Chem. Res.* 34: 4071.
Yamaguchi, T., C.A. Koval, R.D. Noble, et al. 1996. Transport mechanism of carbon dioxide through perfluorosulfonate ionomer membranes containing an amine carrier. *Chem. Eng. Sci.* 51: 4781.

Yegani, R., H. Hirozawa, M. Teramoto, et al. 2007. Selective separation of $CO_2$ by using novel facilitated transport membrane at elevated temperatures and pressures. *J. Membr. Sci.* 291 (1+2): 157–164.

Zhang, Y., Z. Wang, and S.C. Wang. 2002. Selective permeation of $CO_2$ through new facilitated transport membranes. *Desalination* 145: 385.

Zhao, J., Z. Wang, J. Wang, et al. 2006. Influence of heat-treatment on $CO_2$ separation performance of novel fixed carrier composite membranes prepared by interfacial polymerization. *J. Membr. Sci.* 283 (1+2): 346–356.

Zou, J., and W.S.W. Ho. 2006. $CO_2$-selective polymeric membranes containing amines in crosslinked poly(vinyl alcohol). *J. Membr. Sci.* 286: 310–321.

Zou, J., and W.S.W. Ho. 2007. Hydrogen purification for fuel cells by carbon dioxide removal membrane followed by water gas shift reaction. *J. Chem. Eng. Japan* 40 (11): 1011–1020.

Zou, J., J. Huang, and W.S.W. Ho. 2007. $CO_2$-selective water gas shift membrane reactor for fuel cell hydrogen processing. *Ind. Eng. Chem. Res.* 46: 2272–2279.

Zou, J., J. Huang, and W.S.W. Ho. 2008. Facilitated transport membranes for environmental, energy and biochemical applications. In *Advanced Membrane Technology and Applications*, N.N. Li, A.G. Fane, W.S.W. Ho, and T. Matsuura (eds.). New York: John Wiley & Sons, pp. 721–754.

# 22 Gas Absorption of $CO_2$ and $H_2S$ Using Membrane Contactors

*Meng-Hui Li, Alvin R. Caparanga, and Allan N. Soriano*

**CONTENTS**



## 22.1 INTRODUCTION

The removal of carbon dioxide ($CO_2$) and hydrogen sulfide ($H_2S$) from industrial and domestic gas mixtures is an important industrial process. The simultaneous separation of $CO_2$ and $H_2S$ from crude natural gas, refinery gas, and coal gas produces a product gas with improved heating value, reduced volume, and decreased tendency to corrode pipelines and containers during transport, distribution, and storage. The removal of $H_2S$ from feed gas stream in ammonia synthesis prevents poisoning of the catalyst by the sulfide. $H_2S$ is toxic, and when present in the fuel gas, on combustion forms $SO_2$, which is a major contributor to acid rain formation.

$CO_2$ in the atmosphere has been associated with global climate change. The reported increase in $CO_2$ level in the atmosphere is mainly due to the flue gas released from power plants that burn fossil fuels, particularly coal. Technologies being tested and developed for the separation and capture of $CO_2$ from flue gas and for the separation of $H_2S$ from fuel gas or off-gas have been projected to offer economic and environmental benefits. $CO_2$ separated from flue gas has been found to have increasing potential for further use. It can be used as a raw material in the beverage industry when purified. A high-purity $CO_2$ recovered from flue gas has been utilized to promote plant growth in greenhouses and increase crop yield (Feron and Jansen, 1995; Klaasen et al., 2005). Captured $CO_2$ may be injected into geologic formations, which can arrest $CO_2$ for thousands of years (Klara et al., 2003). Since $CO_2$ is a major contributor to the global warming effect of greenhouse gases (GHGs) emitted to the atmosphere from anthropogenic activities (Carapellucci and Milazzo, 2003; Yamasaki, 2003; Figueroa et al., 2008), the separation and capture of $CO_2$ helps mitigate global climate change concerns.

To enhance $CO_2$ sequestration to reduce total emission of GHGs into the atmosphere (Yang et al., 2008) is an option that seems to complement the requirements of the by-product end-use of $CO_2$. Despite the difficulty of developing a separation scheme for postcombustion capture of $CO_2$ from coal-fired power plants, due to the low partial pressure of $CO_2$ in the flue gas at ambient conditions (<16 kPa), the postcombustion $CO_2$ capture still has the highest potential for reducing $CO_2$ concentration in GHG emissions (Figueroa et al., 2008). Some current and emerging technologies for postcombustion capture of $CO_2$ may be sorted into three general categories: absorption, adsorption, and membrane separation. Classification of **absorption** processes depends on the liquid absorbent system used: (a) *absorption in amine-based systems* (Abadanes et al., 2004; Al-Juaied and Rochelle, 2006; Idem et al., 2006; Jassim and Rochelle, 2006; Rao and Rubin, 2006; Figueroa et al., 2008), (b) *capture in carbonate-based systems* (Rochelle et al., 2006; Oexmann et al., 2008; Knuutila et al., doi:10.1016/j.ijggc.2008.06.006), (c) *wet-scrubbing in ammonia-based systems* (Yeh et al., 2005; Resnik et al., 2006), (d) *absorption in dual-alkali systems* (Huang et al., 2001), (e) *capture in enzyme-based systems* (Trachtenberg et al., 1999; Boa and Trachtenberg, 2006; Figueroa et al., 2008), and (f) *absorption in ionic liquids* (Kanel, 2003; Zhang et al., 2006b; Huttenhuis et al., 2007; Figueroa et al., 2008).

Adsorption processes include (g) *capture in solid sorbents* (Nelson et al., 2006; Wang et al., 2008), (h) *adsorption on metal organic frameworks* (Figueroa et al., 2008), (i) *adsorption using mesoporous molecular sieves* (Huang et al., 2003; Gray et al., 2005; Xu et al., 2005), (j) *adsorption by activated carbon* (Maroto-Valer et al., 2005; Pevida et al., 2008), and (k) *adsorption on lithium compounds* (Fauth et al., 2005; Kato et al., 2005).

**Membrane separation** is primarily based on selective gas permeation through polymeric (Xu et al., 2002; Shekhawat et al., 2003; Powell and Qiao, 2006; Favre, 2007; Yang et al., 2008), inorganic carbon, alumina, silica, zeolite, mixed-matrix and hybrid, and facilitated transport membranes (Shekhawat et al., 2003; Yang et al., 2008). Some of these processes, particularly absorption, may also be applied to selective separation of $H_2S$ or simultaneous separation of $CO_2$ and $H_2S$.

Research on developing technologies for separation of $CO_2$ from flue gas and removal of $H_2S$ and $CO_2$ from crude natural gas have been inspired by the stringent demands of product quality, environmental legislation, energy efficiency, and cost reduction requirements so that the technology can compete with those already proven and adopted, such as some of those enumerated. Hybrid processes, such as the combination of membrane separation and gas absorption as a compact system called membrane contactor, have been a popular option. The membrane contactor exploits the advantages of membrane separation and gas absorption while reducing the corresponding limitations as separate processes.

Membrane separation alone (e.g., gas permeation) has been considered inappropriate for $CO_2$ removal from flue gas because of incomplete or unclear arguments on the technology (Davidson and Metz, 2005). There are reports of high specific energy requirements (for feed mixtures with <10% $CO_2$) (Herzog, 2001; Feron and Jansen, 2002; Corti et al., 2004), which become comparable with those of conventional gas absorption as the $CO_2$ composition of the feed reaches 20%. The resistance of the membrane material to high temperature is also limited. Membrane units with good permeability, however, have compact sizes that are approximately 4 to 6 times smaller than conventional gas absorption columns due to overall $CO_2$ transfer rate.

**TABLE 22.1**
**Advantages and Disadvantages of Membrane Gas Absorption**

| Advantages | Consequences |
|---|---|
| Gas and liquid flows on opposite sides of the HF can be manipulated independently. | Flooding, foaming, channeling, and entrainment are avoided. |
| The membrane device is compact with no moving parts. | The membrane is more economical; it consumes less energy and space. |
| The membrane contactor is modular. | Linear scale-up is possible. By using fewer or more modules, it can be operated over a wide range of capacities. |
| Interfacial area is constant and known a priori. | Prediction of membrane contact performance is easier. |
| Liquid holdup is low. | This is an attractive feature when using expensive liquid absorbents. |
| **Disadvantages** | **Consequences** |
| The membrane fibers have small diameters and small channels around the fibers. | The gas and liquid flows are normally laminar, which result in low mass transfer rate. Turbulent flow is possible at the expense of high power consumption. |
| The presence of membrane provides additional mass-transfer resistance. | Additional resistance results in lower mass transfer rate. |
| Membranes are subject to fouling. | A minor problem, this can shorten the life of the membrane so that the cost of periodic membrane replacement can be an issue. |

Conventional gas absorption uses a packed tower, spray tower, venturi scrubber, or bubble column, where gas mixtures are usually contacted with and usually dispersed in the liquid, resulting in low mass-transfer rate. On the positive side, the absorption process can offer very high selectivity and high driving force for mass transport even at very low concentrations. The membrane offers a flexible modular efficient device that can be custom designed, modified, and integrated in existing installations. Microporous hollow-fiber membrane contactor (HFMC) modules, for example, offer surface area per unit volume that is an order of magnitude greater than what packed towers and other conventional columns can offer (i.e., three- to tenfold reduction in size (Matson et al., 1983; Falk-Pedersen and Dannstrom, 1997; Feron and Jansen, 1997). This large area is sustained at very low flow rates, where packed towers would not normally operate, and at very high flow rates, where packed towers would flood. The large area offered by membranes promotes good mass transfer. On the contrary, the membrane itself contributes additional mass-transfer resistance due to its mere presence and the possible wetting of the pore surfaces that may be caused by formation of stagnant liquid film in the pores of the membrane. In this hybrid process, this disadvantage is usually eclipsed by the more pronounced positive effect of the membrane's large area on the overall mass transfer. The other operational advantages and disadvantages of membrane gas absorption (Gabelman and Hwang, 1999; Li and Chen, 2005) are summarized in Table 22.1.

## 22.2 PRINCIPLES OF MEMBRANE GAS ABSORPTION

HFMC configuration resembles that of a shell-and-tube heat exchanger except that the hollow fibers are arranged without well-defined pitch common in the heat exchanger type. The gas phase (gas stream to be treated) may flow through the fiber (lumen side) while the liquid phase (liquid absorbent) is on the shell side or vice versa. Either way, the components of the gas phase with affinity to the absorption liquid will diffuse through the (ideally) gas-filled pores of the membrane to the other side where the liquid will absorb them physically or via chemical reactions. Figures 22.1 and 22.2 illustrate the process in a single hollow fiber.

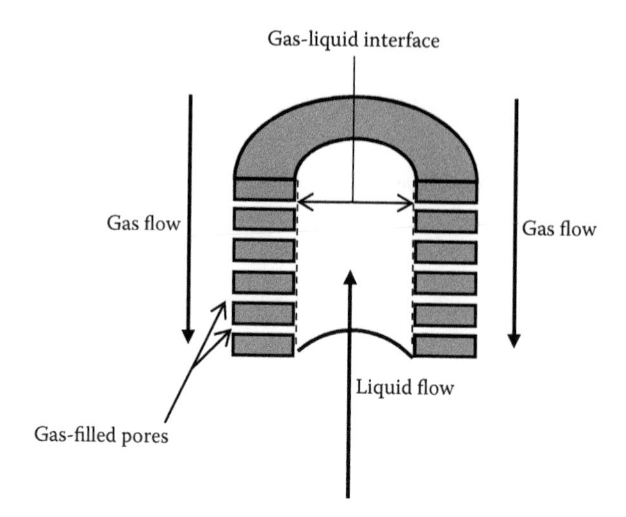


FIGURE 22.1 Gas flow (lumen side) and liquid flow (shell side) in a single hollow fiber.

Membrane gas absorption (MGA) (i.e., gas absorption using membrane contactor), is basically a gas-liquid contacting separation process (Qi and Cussler, 1985). Except for the different sources of additional resistance, the mass transfer involved in MGA is governed by the same principles as those of conventional gas-liquid contacting operations. The driving force in MGA, which is based on the concentration gradient across the membrane wall, is the same as that of conventional gas-liquid separations in tray or packed columns. The absorption, likewise, is limited by physical or chemical equilibrium. While the presence of membrane affects the overall mass-transfer rate and absorption efficiency, it does not affect equilibrium. Circulating the liquid through the contactor after removing the product of chemical reaction from the liquids will shift the equilibrium to the liquid-phase side, which favors good mass transfer. This concept is not new; however, there have been recent developments in the design of the contactors that have resulted in better efficiency and greater capacity.

### 22.2.1 Role of Membrane

The membrane provides support and a large surface for the gas phase and the liquid phase to come in direct contact so that the desired component (the solute) of the gas phase is allowed to transfer

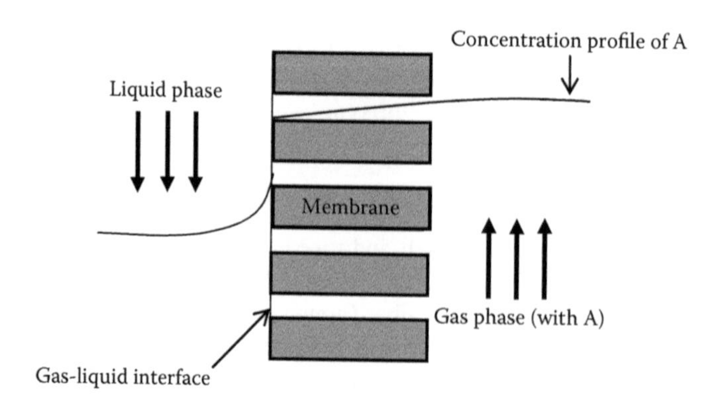


FIGURE 22.2 Concentration profile of solute (A) in a single hollow fiber.

into the liquid phase without dispersing one phase into the other. The membrane also keeps the two phases separated to prevent contamination and avoid additional mass-transfer resistance due to liquid leakage into the pores (i.e., wetting).

It is essential that the membrane wall is hydrophobic enough not to be wetted by high-surface tension liquids such as water or aqueous absorbent solutions. The membrane's hydrophobicity and small pores prevent the liquid absorbent from passing through the membrane wall so that the gas phase and liquid phase remain separated from each other. The wetting phenomenon is influenced by several factors such as pore size, pressure difference across the membrane, and the liquid–membrane interaction. The minimum pressure required to force the liquid into the pore of a hydrophobic membrane is calculated using the Young-Laplace equation (Kim and Yang, 2000; Kumar et al., 2002; Klaasen et al., 2005):

$$\Delta P = \frac{2\gamma}{r_{max}} \cos\theta \tag{22.1}$$

where $\Delta P$ is the breakthrough pressure, $\gamma$ the surface tension of absorption liquid, $\theta$ the contact angle between the liquid phase and the liquid material, and $r_{max}$ the maximum radius of pore in the membrane. A membrane surface with contact angle greater than 90° is hydrophobic. A good membrane contactor is one with high breakthrough pressure, and to prevent wetting, the system should be operated at a pressure lower than the breakthrough pressure.

## 22.3 LIQUID ABSORBENT SYSTEMS FOR $CO_2$ AND $H_2S$ SEPARATION

The liquid absorbent, not the membrane, determines the process selectivity in MGA. A summary of some liquid absorbent systems for gas absorption of $CO_2$ and $H_2S$ in HFMCs tested in recent studies or used in current practice is given in Table 22.2. It is believed that selection of such varied absorption liquid systems has been based on some general criteria for selecting liquid absorbent system for gas absorption in membrane contactors as follows:

1. The absorbent should have **high solubility or reactivity** with the target solute component of the gas phase. The physical or chemical affinity of the solute to the liquid absorbent suppresses the mass-transfer resistance of the liquid phase, leading to higher mass flux.
2. For a hydrophobic membrane material, the liquid absorbent chosen should have high enough **surface tension** to reduce the liquid leakage into the membrane pores. It has been observed that stagnant liquid layer in the pores, even at marginal wetting, has higher resistance than that of the liquid (Rangwala, 1996; Evren, 2000).
3. The choice of **suitable combination of absorption liquid and membrane material** is very important since their chemical compatibility determines the long-term stability of the membrane module. The combination should be such that the pores of the membrane remain gas-filled at an excess pressure of preferably greater than 1 bar on the liquid side. The wetting of membrane can be caused by the reaction of the liquid absorbent with the membrane, resulting in changes in surface or pore morphology and reduction in breakthrough pressure (Barbe et al., 2000; Kumar et al., 2002; Wang et al., 2004c).
4. Liquid absorbents should have **low vapor pressure**; otherwise, its evaporation at high temperatures, where chemical absorption for irreversible reaction is better, will produce vapor that may be forced through the pores and penetrate through the membrane into the gas phase (Kim and Yang, 2000). The liquid should, at the same time, be **thermally and chemically stable** over a wide range of temperatures to avoid decomposition.
5. The liquid absorbent should be **easily regenerated** for a reduced power requirement.

**TABLE 22.2**
**Some Liquid Absorption Systems Used in Recent Studies on $CO_2$ and $H_2S$ Absorption in HF Membrane Contactors**

| Gas Phase System | Liquid Absorbent System (aq.)/ Membrane Material | Absorption Effectiveness/Absorption Rate | Reference |
|---|---|---|---|
| $CO_2$ (1%–15%): $N_2$ | AMP + PZ (1 M AMP; 0.1–0.4 M PZ)/PVDF | Increased at increased liquid and gas flow rates and PZ concentration | (Lin et al., 2008) |
| $CO_2$: $N_2$ | MDEA, MDEA+AMP, MDEA+PZ/NR | Enhanced in the presence of PZ or AMP; highest with PZ; gas and liquid flows had limited effect | (Lu et al., 2007) |
| $CO_2$ (15%): $N_2$ | MEA (3 M)/PP | Increased with decreased gas flow rate | (Bottino et al., 2008) |
| $CO_2$ (20%): $N_2$ | DEA (2 M)/PVDF, PP | Decreased due to wetting | (Zhang et al., 2008) |
| $CO_2$ (10%): $CH_4$ | DEA, NaOH/PP | Absorption rate increased with increased absorbent concentration and liquid velocity and decreased gas flow rate; increased temperature had negative effect | (Al-Marzouqui et al., 2008b) |
| $CO_2$ (10%): $CH_4$ | DEA/PP | Absorption rate increased with increased absorbent concentration and liquid velocity and decreased gas flow rate; increased temperature had negative effect | (Al-Marzouqui et al., 2008a) |
| $CO_2$: $N_2$ | Amino acid salt/PTFE | Increased at increased liquid flow rate and amine concentration | (Kumar et al., 2002) |
| $CO_2$: $N_2$: $O_2$ | PG, MEA, MDEA/PP | >90% in PG, better than in MEA or MDEA; increased at increased liquid and gas flow rates, PG concentration and $CO_2$ fraction in gas feed | (Yan et al., 2007) |
| $CO_2$: $CH_4$ | NaOH(+NaCl), MEA/ PVDF | Better in NaOH; increased at high concentration | (Atchariyawut et al., 2007) |
| $CO_2$: $N_2$ | MEA+MDEA/PP | Absorption rate and efficiency increased at increased MEA content and increased liquid flow rate | (Gong et al., 2006) |
| $CO_2$ (20%): $N_2$ | DEA (2 M)/PP | Rate increased as $CO_2$ volume fraction in feed increased | (Wang et al., 2005) |
| $CO_2$ (20%): $N_2$ | DEA (2 M)/PP | $CO_2$ flux increased at increased gas velocity; inlet gas velocity has no significant effect on absorption rate | (Zhang et al., 2006a) |
| $CO_2$ (40%): $N_2$ | MEA, MDEA, AMP/ PTFE | Absorption rate increased but separation efficiency decreased at high temperature; higher efficiency but moderate absorption rate with AMP | (Kim and Yang, 2000) |
| $H_2S$, $CO_2$ | $K_2CO_3$/PP | $CO_2$ absorption controlled by liquid velocity; $H_2S$ absorption controlled by combined gas and membrane resistance | (Dindore et al., 2005) |
| $H_2S$ (1000 ppm): $CO_2$ (5%–23 vol%) | $Na_2CO_3$ (2 M)/PVDF | 100 % $H_2S$ removal with <5 vol % $CO_2$ in the permeate | (Wang et al., 2004a) |
| $H_2S$ (17.9–1159 ppm): $N_2$ | $Na_2CO_3$ (2 M)/PVDF | 100% removal at short residence time (<0.1 s) and low L/G ratio | (Wang et al., 2002) |
| $H_2S$ (17.2 ppm): $N_2$ | NaOH (10 M)/PVDF | Improved as gas velocity is decreased | (Li et al., 2000; Kong and Li, 2001) |
| $H_2S$ (16–24 ppm): $N_2$ | NaOH (10 M)/ polysulfone | Not reported; radial $H_2S$ gradient negligible unless membrane resistance approaches zero | (Li et al., 1998) |
| $H_2S$ (100 ppm): air | $H_2O$/PP | Removal efficiency increased by up to 89% at increased gas flow rate | (Boucif et al., 2007) |

*Note:* NR, not reported.

### 22.3.1 Alkanolamines

Based on the above criteria, alkanolamines have been selected for removal of $CO_2$ and $H_2S$ in most applications and investigations. The high reactivity of alkanolamines with $CO_2$ and $H_2S$, which results in effective absorption, is the major reason for selection. Widely used and studied alkanolamines are monoethanolamine (MEA) for systems with low $H_2S$ and $CO_2$ concentrations, diethanolamine (DEA) for systems with COS and $CS_2$ in addition to $H_2S$ and $CS_2$, diglycolamine (DGA) for systems with relatively high concentrations of $H_2S$ and $CO_2$, *n*-methyldiethanolamine (MDEA) for selective absorption of $CO_2$ in the presence of $H_2S$, and di-2-propanolamine (DIPA) for selective absorption of $H_2S$ and 2-amino-2-methyl-1-propanol (AMP) (Kohl and Nielsen, 1997). The sterically hindered primary amine AMP is preferred to other amines since its bulkiness (i.e., amine attached to tertiary carbon atom) inhibits the formation of carbamates, resulting in faster reaction rate and higher $CO_2$-loading capacity. The effectiveness of aqueous alkanolamines to absorb $CO_2$ and $H_2S$ has been evaluated individually in aqueous solutions and as mixed absorbent systems such as mixed amines and glycol-amine. Mixed amines have been favored over monoamines in absorption (Xiao et al., 2000; Liao and Li, 2002; Gong et al., 2006) and have been shown to improve absorption rate and reduce energy requirement for regeneration of the absorbent (Seo and Hong, 1996). Absorption activators or accelerators such as piperazine (PZ) have also been mixed with the amine systems to improve rate of absorption (Seo and Hong, 2000; Lin et al., 2008).

The chemical reactions that occur following absorption of $CO_2$ in aqueous solution of primary or secondary amine are usually described in terms of the two-step zwitterion mechanism as follows (Caplow, 1968; Dankwerts, 1979):

$$CO_2 + RNH_2 \underset{k_{ar}}{\overset{k_{af}}{\rightleftharpoons}} RNH_2^+COO^- \tag{22.2}$$

$$RNH_2^+COO^- + B \xrightarrow{k_b} RNHCOO^- + BH^+ \tag{22.3}$$

The zwitterion formed in the first step is subsequently deprotonated by a base B in the second step. In aqueous alkanolamine solutions, the base can be the water, the hydroxyl ion, or the amine itself (Blauwhoff et al., 1984). For example, if the base is an amine ($RNH_2$), then Equation 22.3 becomes

$$RNH_2^+COO^- + RNH_2 \xrightarrow{k_{amine}} RNHCOO^- + RNH_3^+.$$

This mechanism has been used in some recent studies to model absorption of $CO_2$ or $H_2S$ in HFMC (Wang et al., 2004b; Zhang et al., 2006a; Lin et al., 2008). This mechanism and the kinetic data available for corresponding amines are important in the design of a good MGA system.

### 22.3.2 Amino Acid Salts

Alkanolamines wet polyolefins; this makes them generally not suitable for polypropylene membranes. Aqueous solutions of amino acid salts, traditionally used as a promoter for conventional gas-treatment solvents, have been identified as a possible alternative to aqueous alkanolamines in polyolefin membrane contactors (Feron and Jansen, 1995). $CO_2$ Removal Absorption Liquid (CORAL), developed by TNO Environment Energy and Process Innovation of The Netherlands, is based on amino acid salts (Kumar et al., 2002). It has similar absorption characteristics as aqueous alkanolamines and does not wet polypropylene membranes. In a recent study, potassium glycinate (PG), whose aqueous solutions have higher surface tension than water, was used in the absorption of $CO_2$ from $CO_2$-$N_2$-$O_2$ mixture in a polypropylene HFMC and was found to perform better as absorption liquid than MEA and MDEA (Yan et al., 2007).

### 22.3.3 Aqueous Alkali Solutions

In conventional gas absorption technologies developed for the acid gases, $CO_2$ and $H_2S$, the liquid absorbents are generally based on hot $K_2CO_3$ solutions (usually with a sterically hindered-amine promoter), ambient-temperature $Na_2CO_3$ or $K_2CO_3$ solutions, and ambient-temperature free caustic solutions (Kohl and Nielsen, 1997). These absorbents have been used, respectively, for high-pressure removal of $CO_2$ or $H_2S$, removal of $H_2S$ from coke-oven gas with vacuum regeneration and removal of trace quantities of $CO_2$ or $H_2S$ from gases. There are some recent studies on the application of alkali solutions containing $Na_2CO_3$, $K_2CO_3$ or NaOH for the absorption-removal of $H_2S$ in HFMC (Li et al., 1998; Li et al., 2000; Kong and Li, 2001; Wang et al., 2002, 2004a; Dindore et al., 2005).

The following chemical equations describe the chemical reactions that take place following physical absorption of $CO_2$ as shown in Equations 22.4 through 22.8 (Suchdeo and Schultz, 1974; Hikita et al., 1976) or of $H_2S$ as shown in Equations 22.4 through 22.6 and Equations 22.9 through 22.12 (Dindore et al., 2005) in aqueous $K_2CO_3$:

$$CO_2 + OH^- \underset{k_{1r}}{\overset{k_{1f}}{\rightleftarrows}} HCO_3^- \tag{22.4}$$

$$HCO_3^- + OH^- \underset{k_{2r}}{\overset{k_{2f}}{\rightleftarrows}} CO_3^{2-} + H_2O \tag{22.5}$$

$$H_2O + H_2O \underset{k_{3r}}{\overset{k_{3f}}{\rightleftarrows}} H_3O^+ + OH^- \tag{22.6}$$

$$CO_2 + 2H_2O \underset{k_{4r}}{\overset{k_{4f}}{\rightleftarrows}} HCO_3^- + H_3O^+ \tag{22.7}$$

$$\text{Overall reaction: } CO_2 + CO_3^{2-} + H_2O \rightarrow 2HCO_3^- \tag{22.8}$$

$$H_2S + OH^- \underset{k_{6r}}{\overset{k_{6f}}{\rightleftarrows}} HS^- + H_2O \tag{22.9}$$

$$HS^- + OH^- \underset{k_{7r}}{\overset{k_{7f}}{\rightleftarrows}} S^{2-} + H_2O \tag{22.10}$$

$$H_2S + HCO_3^- \underset{k_{8r}}{\overset{k_{8f}}{\rightleftarrows}} HS^- + H_2O + CO_2 \tag{22.11}$$

$$\text{Overall reaction: } H_2S + CO_3^{2-} \rightarrow HS^- + HCO_3^- \tag{22.12}$$

Literature values of kinetic parameters for these systems are summarized in Dindore et al. (Dindore et al., 2005). These equations and kinetic parameters are good baseline data for the design of an effective membrane contactor for gas absorption of $CO_2$ or $H_2S$ in aqueous $K_2CO_3$ with an appropriate membrane material.

### 22.3.4 Ionic Liquids

Ionic liquids (ILs) are highly stable organic salts that are liquid at low temperatures. ILs are composed entirely of organic cations and either organic or inorganic anions. Their low vapor pressure and tunable polarity makes them a novel solvent. Recently, ILs have attracted considerable attention due to their notably exceptional affinity with $CO_2$ (e.g., remarkable solubility of $CO_2$ in imidazolium-based ILs (Soriano et al., 2008; Cadena et al., 2004; Kumelan et al., 2004; Chen et al., 2006). Likewise, poly(ionic liquid) materials have been synthesized to produce a system that can absorb $CO_2$ at a rate 6 to 7.6 times faster than room-temperature ILs (Tang et al., 2005). However, the application of ILs in HFMC has been stalled by the solvents' expensive cost. At present, they are instead used as task-specific ILs in which one (possible) application is $CO_2$ separation in supported liquid membrane (SLM) facilitated by a task-specific IL (Shekhawat et al., 2003; Hanioka et al., 2008).

## 22.4 MEMBRANE MATERIALS

On the basis of the roles of the membrane in HFMC discussed in Section 22.2.1 and criteria (2) and (3) in Section 22.3, the different membrane materials suited for application in $CO_2$ and $H_2S$ removal using MGA are polypropylene (PP), polyethylene (PE), polytetrafluoroethane (PTFE), and polyvinylidinefluoride (PVDF). PP and PTFE are mostly used in $CO_2$ absorption employing aqueous alkanolamines or amino acid salt as liquid absorbent (Kim and Yang, 2000; Kumar et al., 2002; Wang et al., 2005; Gong et al., 2006; Zhang et al., 2006a; Yan et al., 2007; Al-Marzouqui et al., 2008a; Bottino et al., 2008; Zhang et al., 2008). One study, however, demonstrated that PTFE is superior to PP over time when used with alkanolamines (MEA and AMP) to absorb $CO_2$ (deMontigny et al., 2006b). PVDF is mostly applied in $H_2S$ absorption or selective absorption of $H_2S$ or $CO_2$ from systems containing both gases using aqueous alkali solutions as liquid absorbent (Li et al., 2000; Kong and Li, 2001; Wang et al., 2002, 2004a).

The long-term stability of the membrane depends on the ability of the membrane material to withstand certain conditions. Some factors that can be considered to affect the membrane's long-term stability are fouling and chemical and thermal stability of the membrane material. Fouling may not be much a concern in the laboratory scale operations. In an actual industrial setting, it cannot be ignored because suspended particulate matters in the gas feed (e.g., unfiltered flue gas) may deposit on the pore surface and eventually clog the pores.

The chemical stability of the membrane as emphasized in item (3) in Section 22.3 considerably affects the membrane's useful life. A chemically stable membrane material is resistant to surface erosion by $CO_2$-loaded liquid absorbent, which can affect morphological changes on the surface of the microporous membrane and, consequently, make the material vulnerable to chemical attack (Barbe et al., 2000).

In some applications, the gas feed to the HFMC (i.e., flue gas) may be at some elevated temperatures. It is for this reason that the membrane material must be of a certain degree of thermal stability. The glass transition temperature (*Tg*) determines the thermal stability of amorphous membrane materials. Common membrane materials for HFMC for $CO_2$ absorption are semicrystalline polymers with the following *Tg* values: PTFE (126°C~130°C), PP (–15°C ~ –10°C), and PE (–125°C ~ –120°C). The low *Tg* values could have caused instability and wetting problems reported for PE and PP. The chemical structure of PTFE (i.e., four fluoride side groups on a flexible polyvinyl chain) explains the polymer's high *Tg* and enhanced chemical and thermal stability (Li and Chen, 2005). For high-temperature applications, a polymer membrane material with high *Tg* is usually recommended. PVDF (*Tg*: –45°C ~ –35°C) is mostly applied to treat gas streams with $H_2S$ employing strong alkali in HFMC. The *Tg* value may be too low but it is chemical compatibility between the alkali and the membrane material (PVDF) that is more important in this case. In a recent study, the highly permeable but nonporous polyvinyltrimethylsilane (PVTMS) membrane (*Tg* = 155°C) performed well in the separation of $CO_2$ from biohydrogen in a gas absorption membrane contactor (not hollow fiber) with aqueous $K_2CO_3$ (Modigell et al., 2008).

## 22.5 PREVENTION OF WETTING

As already discussed, wetting of the membrane pores may be contributing largely to the effective membrane resistance to mass transfer. One study reported relatively low percentage of pore wetting

(~13%) contributed to 98% of the membrane resistance and 20% to 50% of total resistance to mass transfer (Mavroudi et al., 2006). Another study reported a 20% reduction in overall mass transfer coefficient even for marginally wetted pores (i.e., 5% wetted) (Wang et al., 2005). The effect of wetting on reduction in mass transfer coefficient is discussed in the next section.

Solving the wetting problem by decreasing the liquid absorbent concentration, as suggested by the Young-Laplace equation (Equation 22.1), compromises the absorption efficiency. A better option would be to increase the breakthrough pressure by changing the membrane properties such as using membrane with smaller pore size and by widening the polarity difference between the liquid absorbent and membrane material. Experimental breakthrough pressures for a number of aqueous alkanolamine solutions with PTFE membrane have been reported (Kumar et al., 2002). It has also been pointed out that enlargement of pore entrances of PP membranes due to nonwetting intrusion of water meniscus into some pores resulted in reduced breakthrough pressure (Evren, 2000; Kumar et al., 2002), a result that could be predicted using Equation 22.1.

The following summary lists measures to avoid wetting in membrane contactors (Li and Chen, 2005):

(a) **Using hydrophobic membranes** for aqueous liquid absorbents, which leads to large contact angles
(b) **Surface modification of membrane** to improve its nonwettability
(c) **Using composite membrane** with top stabilization layer on the liquid-side that is highly permeable to the solute
(d) **Selection of denser hollow-fiber membrane** that can offer greater flexibilities in feed-gas pressures
(e) **Selection of liquids with suitable** surface tension to reduce (or eliminate) the tendency of liquid leakage into the pores of hydrophobic membrane
(f) **Optimizing operating condition** such as liquid pressure to prevent gas dispersion into the liquid without causing (partial) wetting of the membrane, for improved absorption performance.

## 22.6 MASS TRANSFER

### 22.6.1 Factors Affecting Mass Transfer

For $CO_2$ or $H_2S$ absorption in HFMC involving a chemical reaction, the mass transfer will be influenced by the following:

(a) *Nature and concentration level of the absorption liquid.* The chemical affinity of the liquid absorbent with $CO_2$ or $H_2S$ determines the reaction kinetics and mechanism. The overall kinetics depends on the concentration level of the active component of the liquid absorbent.
(b) *Solubility of $CO_2$ or $H_2S$ in the absorption liquid.* If the solubility of $CO_2$ or $H_2S$ in the absorbent liquid is high then the mass transfer resistance in the liquid phase is low, resulting in favorable mass transfer.
(c) *Operating temperature.* At higher temperature, mass transfer is faster in the liquid phase because high temperatures usually favor chemical kinetics; however, the elevated temperature would also decrease the solubility of either $CO_2$ or $H_2S$ (i.e., thermodynamically unfavorable). The temperature at which trade-off between kinetics and thermodynamics is optimized is usually desired.
(d) *Fluid dynamics.* Flow conditions such as gas and liquid flow rates, flow configurations (i.e., parallel, countercurrent, or cross-flow), and choice of fluid to flow in the lumen side or shell side may influence mass transfer.

(e) *Membrane type and characteristics.* As discussed in Sections 22.4 and 22.5, the membrane has to be hydrophobic, sufficiently porous, and should not interact with the liquid to lower the mass transfer resistance due to its mere presence in the setup.

(f) *$CO_2$- or $H_2S$-loading in the liquid absorbent.* $CO_2$ or $H_2S$ may be present in poorly regenerated absorbents. This will limit mass transfer.

(g) *Partial pressure of $CO_2$ or $H_2S$ in the feed gas.* Local saturation of active components in the liquid may occur; this will limit mass transfer due to decreased effective driving force.

These factors have been considered in some recent studies on absorption of $CO_2$ and/or $H_2S$ in HFMC, such as those presented in Table 22.2.

## 22.6.2 The Film Model

In the nonwetted mode operation of HFMC the gas–liquid interface is at the pore mouth of the membrane adjacent to the liquid side. Careful control of the pressure across the membrane should preserve this immobile interface. Mass transfer in MGA consists of the following consecutive steps: (1) diffusion from bulk gas phase to the membrane pore entrance, (2) diffusion through membrane pore from the pore entrance to the gas-liquid interface at the pore mouth, and (3) absorption into the liquid phase via diffusion with or without chemical reaction. Normally used to describe the trans-membrane and interphase mass transfer in HFMC, the film specifies three principal resistances to mass transfer such as encountered in the gas film, the membrane, and the liquid film (see Figure 22.3). In nonwetted mode MGA with chemical reaction, the gas–film resistance is as important as the liquid–film resistance, particularly when solubility is not high. Based on the concentration and/or partial pressure driving force across the membrane, the molar flux of component A per unit fiber length, $J_A$, is:

$$J_A = K_L\left(\frac{P_{A,G}}{m} - \frac{P_A^*}{m}\right) = k_G(P_{A,G} - P_{A,M}) = k_M(P_{A,M} - P_{A,i}) = k_L E\left(\frac{P_{A,i}}{m} - \frac{P_A^*}{m}\right) \tag{22.13}$$

where $k_L$, $k_M$, and $k_G$ are the mass transfer coefficients of the liquid film (m/s), membrane (mole·s$^{-1}$·m$^{-2}$·Pa) and gas film (mole·s$^{-1}$·m$^{-2}$·Pa), respectively; $K_L$ is the overall liquid mass transfer

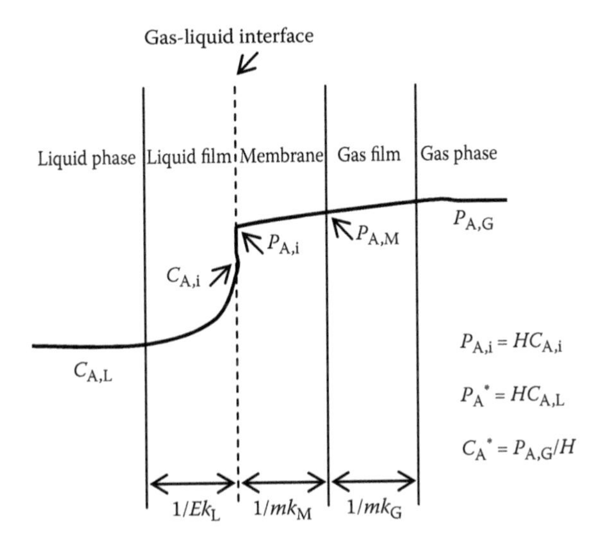


**FIGURE 22.3** Concentration profile and resistances in a single hollow fiber (film model).

coefficient (m/s); $d_i$, $d_o$, and $d_m$ are the respective inside, outside, and log-mean diameters of the fiber (m); $P_{A,G}$, $P_{A,M}$ and $P_{A,L}$ are the respective partial pressures of A (Pa) in the gas phase, gas-membrane interface, and gas-liquid interface; $C_{A,G}$, $C_{A,i}$, and $C_{A,L}$ are the respective molar concentrations of A (mol·m$^{-3}$) in the gas phase, gas–liquid interface, and liquid phase; $m$ is the distribution coefficient of A between the gas phase and the liquid phase (Pa·m$^3$·mol$^{-1}$), which is equal to the Henry's law constant $H$ for purely physical absorption; and $E$ is the enhancement factor due to chemical reaction ($E = 1$ for purely physical absorption). Equation 22.13 is based on the following assumptions: steady state condition, discontinuous concentration profile at the gas–liquid interface, equilibrium at the gas–liquid interface, film diffusion across gas–liquid interface (i.e., no velocity gradient), negligible effect of pore morphology on mass transfer, uniform pore size distribution and membrane thickness, well-mixed bulk liquid, and the same driving forces for physical and chemical absorption. The equilibrium at the interface is described by the Henry law (i.e., $P_{A,i} = HC_{A,i}$) so that $C^*_{A,L} = \frac{P_{A,L}}{H} = 0$.

When liquid flows in the lumen side, the overall resistance to mass transfer across the membrane based on the liquid film is evaluated using Equation 22.14 (Kreulen et al., 1993). When gas flows in the lumen side, Equation 22.15 gives the equation for overall mass transfer coefficient. Both equations are derived from Equation 22.13.

$$\frac{1}{K_L} = \frac{1}{mk_G}\left(\frac{d_i}{d_o}\right) + \frac{1}{mk_M}\left(\frac{d_i}{d_m}\right) + \frac{1}{k_L E} \tag{22.14}$$

$$\frac{1}{K_L} = \frac{1}{mk_G}\left(\frac{d_o}{d_i}\right) + \frac{1}{mk_M}\left(\frac{d_o}{d_m}\right) + \frac{1}{k_L E} \tag{22.15}$$

It should be noted that the limitations of Equations 22.14 or 22.15 are based on fact that the bulk liquid phase is not guaranteed inside a small-diameter fiber, that the driving forces for physical and chemical absorption are not always the same, and that loading increases in the axial direction.

### 22.6.3 Differential Equation-Based Model

In most studies, the liquid phase has been designed to flow in the lumen side. Contrary to the well-mixed bulk liquid assumed in the film model, the liquid flow through the hollow fiber in the differential equation-based model is laminar with a parabolic velocity profile. The mass balance of component A and fully developed laminar-flow velocity profile of liquid in the lumen in cylindrical coordinates are given as Equations 22.16 and 22.17, respectively.

$$v_A \frac{\partial C_A}{\partial z} = D_A\left[\frac{1}{r}\frac{\partial}{\partial r}\left(r\frac{\partial C_A}{\partial r}\right)\right] - R_A \tag{22.16}$$

$$v_A(r) = 2v_L\left[1 - \left(\frac{r}{r_i}\right)^2\right] \tag{22.17}$$

where $R_A$ is the reaction rate of A, $v_A$ is the velocity profile, $D_A$ is the diffusivity coefficient, $v_L$ is the average liquid velocity, and $r_i$ is the fiber inside radius. Equations 22.16 and 22.17 are based on the following assumptions: single component absorption from a gas mixture (i.e., $CO_2$ or $H_2S$), constant physical properties, equilibrium described by Henry's law, steady-state isothermal flow, no axial diffusion, fully developed laminar flow in the tube side, and perfectly cylindrical symmetry. In some recent studies, mathematical models derived from on Equations 22.16 and 22.17, with boundary conditions that are appropriate to the corresponding scenarios, have been developed and subsequently simulated and the results checked against experimental data (Li et al., 1998; Kim and Yang, 2000; Mavroudi et al., 2003; Wang et al., 2004b; Dindore et al., 2005; Wang et al., 2005; Gong et al., 2006; Mavroudi et al., 2006; Zhang et al., 2006a; Boucif et al., 2007; Lu et al., 2007; Al-Marzouqui et al., 2008b; Al-Marzouqui et al., 2008a; Keshavarz et al., 2008a; Keshavarz et al., 2008c, b; Zhang et al., 2008). The initial and boundary conditions used to solve Equations 22.16) and 22.17 are as follows:

$$C_A = C_{A,in} \quad \text{at } z = 0 \text{ and } 0 \le r \le r_i \tag{22.18}$$

$$\frac{\partial C_A}{\partial r} = 0 \quad \text{at } r = 0, \text{ and } 0 \le z \le L \tag{22.19}$$

$$D_A\left(\frac{\partial C_A}{\partial r}\right) = \frac{K_{ext}}{m}\left(C_A^* - C_{A,i}\right) \quad \text{at } r = R \text{ and } 0 \le z \le L \tag{22.20}$$

where $K_{ext}$, calculated by Equation 22.21, is the external mass transfer coefficient (mole·s$^{-1}$·m$^{-2}$·Pa) that takes into account membrane wall and shell-side resistance.

$$\frac{1}{K_{ext}} = \frac{1}{k_G}\frac{d_i}{d_o} + \frac{1}{k_M}\left(\frac{d_i}{d_m}\right) \tag{22.21}$$

### 22.6.4 Overall Mass Transfer Coefficient

Equation 22.22 has been used to experimentally estimate the overall mass transfer coefficient ($K_L$) in HFMC. In the equation, $V_L$ is the liquid volumetric flow rate, $S$ is the gas-liquid interfacial area, $C_{A,in}$ and $C_{A,out}$ are the respective inlet and outlet concentrations of the solute (A) in the absorbent liquid, and $\Delta C_m$ is the log-mean concentration driving force, which is evaluated using Equation 22.23, where $P_{A,in}$ and $P_{A,out}$ are the inlet and outlet partial pressures of A in the gas phase and $H$ is Henry's law constant.

$$K_L = \frac{V_L}{S}\frac{C_{A,out} - C_{A,in}}{\Delta C_m} \tag{22.22}$$

$$\Delta C_m = \frac{\left(\frac{P_{A,in}}{H} - C_{A,out}\right) - \left(\frac{P_{A,out}}{H} - C_{A,in}\right)}{\ln\dfrac{\frac{P_{A,in}}{H} - C_{A,out}}{\frac{P_{A,out}}{H} - C_{A,in}}} \tag{22.23}$$

Equation 22.22 could be modified to estimate the overall mass transfer coefficient based on pressure drop in the gas phase and the assumption of ideal plug-flow at both liquid side (lumen) and gas side (shell) (Yeon et al., 2003). The result is Equation 22.24.

$$K_L \cdot A = \frac{mR}{\Delta P_m} \tag{22.24}$$

where $R$ is the $CO_2$ absorption rate per unit contactor volume and $\Delta P_m$ is the log-mean pressure driving force. Table 22.3 shows typical values of overall mass transfer coefficients reported by recent studies. It should be noted that the overall mass transfer coefficient for $H_2S$ absorption in aqueous alkali is up to twofold greater than absorption of $CO_2$ in aqueous alkanolamines in membrane gas contactor regardless of fiber material used.

The overall (volumetric) mass transfer coefficient ($K_Ga$), which is commonly used in absorption in packed columns (Aroonwilas and Veawab, 2004), may also be used to measure membrane gas absorption performance as it was successfully used as a basis for comparing the performance of $CO_2$ absorption in membrane gas absorption and packed column (deMontigny et al., 2006a; Yan et al., 2008). Such coefficient is calculated as follows:

$$K_G a = \frac{G_I}{P\left(y_{A,G} - y_{A,G}^*\right)} \frac{dY_{A,G}}{dz} \tag{22.25}$$

where $G$ is the inert gas flow rate, $Y_{A,G}$ is the mole ratio of $CO_2$ to inert gas component, $dY_{A,G}/dz$ is the solute concentration gradient, and $P$ is the total pressure.

### 22.6.5 Liquid-Side Mass Transfer Coefficient

For purely physical absorption in HFMC with the liquid phase in the lumen side, the two most commonly used models to estimate liquid side mass transfer coefficient ($k_L$) are as follows:

$$Sh = \frac{k_L d}{D_A} = \sqrt[3]{\left(3.37^3 + 1.62^3\ Gz\right)} \qquad \text{for } 10 < Gz < 20 \tag{22.26}$$

$$Sh = \frac{k_L d}{D_A} = 1.62\sqrt[3]{Gz} \qquad \text{for } Gz > 20 \tag{22.27}$$

where $Sh$ is the Sherwood number, $Gz$ is the Graetz number $\left(Gz = \frac{v_L d_i^2}{D_A L}\right)$, $d_i$ is the inside diameter of hollow fiber, $D_A$ is the molecular diffusivity of $CO_2$ (or $H_2S$) and $v_L$ is the liquid-phase velocity. Equation 22.26 (Kreulen et al., 1993) was derived for physical absorption under constant gas–liquid interface conditions while Equation 22.27 (Nii and Takeuchi, 1994; Rangwala, 1996) was based

**TABLE 22.3**
**Typical Values of Overall Mass Transfer Coefficient for $CO_2$ and $H_2S$ Absorption in Membrane Gas Contactor**

| Membrane Material (Average Pore Size) | Gas/Liquid System | Overall Mass Transfer Coefficient, m·s$^{-1}$ | Reference |
|---|---|---|---|
| PP (0.04 μm) | $CO_2$-$N_2$/DEA | 1.25–2.7 × $10^{-4}$ | Wang et al., 2005 |
| PTFE (0.80 μm) | $CO_2$-$N_2$/AMP | 1.5–2.5 × $10^{-4}$ | Kim and Yang, 2000 |
| PP (0.20 μm) | $CO_2$/CORAL | 0.8–3.9 × $10^{-3}$ | Feron and Jansen, 1997 |
| PVDF (0.04 μm) | $CO_2$-$H_2S$-$N_2$/$Na_2CO_3$ | $CO_2$: 1.5–2.5 × $10^{-4}$<br>$H_2S$: 0.01–0.0144 | Wang et al., 2004a |
| PVDF (0.04 μm) | $H_2S$-$N_2$/$Na_2CO_3$ | 0.0134–0.0164 | Wang et al., 2002 |
| PP (0.06 μm) | $H_2S$-$N_2$/$Na_2CO_3$ | 1.0 1.5 × $10^{-4}$ | Boucif et al., 2007 |

on the assumption of fully developed laminar liquid flow through the fibers. The limitations of the applicability of the two equations are discussed in the review paper of Li and Chen (Li and Chen, 2005).

The correlations used for evaluating liquid-side mass transfer coefficient when the liquid phase flows on the shell side are generally expressed in the form

$$Sh = f(\mathrm{Re}, Sc) = a\mathrm{Re}^m\, Sc^n \tag{22.28}$$

where $Re$ is the Reynolds number, $Sc$ is the Schmidt number, and $a$, $m$, and $n$ are constants. Yang and Cussler presented several correlations in the form of Equation 22.28 for $CO_2$ absorption in water as absorbent inside, outside in parallel flow, and outside in cross-flow (Yang and Cussler, 1986). Another correlation, originally developed for a different separation process, which incorporated the hydraulic diameter ($d_h$), effective length of the membrane ($L$), and packing fraction of the module ($\phi$) in the equation (Prasad and Sirkar, 1988), was recently considered to estimate mass transfer coefficient for absorption of $CO_2$ using alkanolamine solutions in hollow-fiber membrane contactor (Lu et al., 2007):

$$Sh = 5.85(1-\phi)\frac{d_h}{L}\mathrm{Re}^{0.60}\, Sc^{0.33} \tag{22.29}$$

### 22.6.6 Gas-Side Mass Transfer Coefficient

The correlations for predicting gas-side mass transfer coefficients in membrane contactors may also be of the form of Equation 22.28. Some of the correlations of this form presented and discussed in detail in a literature (Gabelman and Hwang, 1999) may be useful for specific systems involving gas absorption of $H_2S$ and/or $CO_2$ in MGA. It should be emphasized that the contribution of the gas-phase resistance is significant only at low gas velocities.

### 22.6.7 Membrane Mass Transfer Coefficient

In studies on gas absorption of $CO_2$ in membrane contactors, the local individual mass transfer coefficient based on the (nonwetted) membrane, $k_M$, is predicted using either of the following equations:

$$k_M = \frac{D_{eff}\varepsilon}{\tau\Delta r} \quad \text{(Mavroudi et al., 2006)} \tag{22.30}$$

$$k_M = \frac{2D_{eff}\varepsilon}{\tau d_i \ln\left(\dfrac{d_o}{d_i}\right)} \quad \text{(Lin et al., 2008)} \tag{22.31}$$

where $\varepsilon$, $\tau$, $\Delta r$, $d_i$, and $d_o$ are porosity, tortuosity, thickness, and inside diameter and outside diameter, respectively, of the membrane. The effective diffusivity in gas-filled pores, $D_e$, is estimated from the Knudsen diffusivity ($D_{Knud}$) and binary diffusivity ($D_G$) of $CO_2$ in gas phase, when pore diameter is between $1 \times 10^{-7}$ to $1 \times 10^{-5}$ m, as follows:

$$\frac{1}{D_{eff}} = \frac{1}{D_{Knud}} + \frac{1}{D_G} \tag{22.32}$$

When the membrane pores are wetted, even partially, the effective resistance of the membrane is magnified. The mass transfer coefficient is usually evaluated in terms of fraction of pores wetted ($x$) (Mavroudi et al., 2003) or fraction of depth of liquid penetration (β) (Lin et al., 2008):

$$\frac{1}{k_M} = (1-x)\frac{\tau\Delta r}{D_G} + x\frac{\tau\Delta r}{D_L} \tag{22.33}$$

$$\frac{1}{k_M} = \frac{\beta}{k_{M,liquid\ filled}} + \frac{1-\beta}{k_{M,gas\ filled}} \tag{22.34}$$

Typical values of membrane mass transfer coefficients reported by recent studies ranges from $4.77 \times 10^{-4}$ to 7.5 $4.77 \times 10^{-4-}$ m·s$^{-1}$ for a system consisting of $CO_2$-$N_2$ absorbed aqueous AMP+PZ and MDEA+PZ systems in PVDF HFMC with average pore size of 0.2 μm (Lin et al., 2008). For a system consisting of $H_2S$-$N_2$ absorbed in aqueous NaOH in polysulfone HFMC, the value reported was 0.0125 to 0.025 m·s$^{-1}$; however, for the same system in polyethersulfone HFMC, the value was approximately $5 \times 10^{-4-}$ m·s$^{-1}$ (Li et al., 1998).

### 22.6.8 Enhancement Factor

Enhancement factor ($E$) is defined by the following equation:

$$E = \frac{J_{A,chem}}{J_{A,phys}} \tag{22.35}$$

where the numerator refers to the absorption flux or rate in the liquid in the presence of a chemical reaction and the denominator is the flux or rate of purely physical absorption. Traditional models, (e.g., the film model), have limited applications to approximate $E$ in membrane contactors because there may be no well-mixed bulk due to small fiber diameter, low-Re flow, and zero velocity-gradient in the mass transfer zone. Differential equations are used instead to describe the absorption in a membrane contactor as already discussed in Section 22.6.3. However, a well-mixed bulk phase may be approximated when the gas-liquid contact time is short, as in traditional mass transfer models. In this light, one study used the following models for evaluating E in a $CO_2$ absorption system in membrane contactor with alkanolamine absorbents (Lu et al., 2007):

$$E = \frac{Ha^2}{2\left(E_\infty^* - 1\right)} + \sqrt{\frac{Ha^2}{4\left(E_\infty^* - 1\right)^2} + \frac{E_\infty^* Ha^2}{E_\infty^* - 1} + 1} \tag{22.36}$$

where Hatta number ($Ha$) and asymptotic infinite enhancement factor ($E_\infty^*$) are defined, respectively, as:

$$Ha = \frac{1}{k_L^o} + \sqrt{k_{ov,rate} D_A} \tag{22.37}$$

$$E_\infty^* = \left(1 + \frac{C_{B,in} D_B}{m C_{A,i} D_A}\right)\left(\frac{D_A}{D_B}\right)^{0.50} \tag{22.38}$$

where $k_L^o$ is the physical mass transfer coefficient, $k_{ov,rate}$ is the overall reaction rate constant based on reaction mechanism and inlet conditions, $C_{B,in}$ is the concentration of alkanolamine in the inlet absorbent stream, $D_A$ and $D_B$ are the respective $CO_2$ and alkanolamine diffusion coefficients in aqueous solution, and $C_{A,i}$ is the concentration of $CO_2$ in gas–liquid interface.

## 22.7 RECOMMENDATION FOR FUTURE RESEARCH

To this end, it can be concluded that membrane gas absorption is an option for recovery and reuse of $CO_2$ from flue gas, biogas, and off-gas, removal of acid gases from fuel gas and natural gas mixtures, and treatment of gases as end-of-pipe technology or to recover products as integrated process solution that can be applied for removal of $H_2S$, $CO_2$ and other potential pollutants from flue gas or off-gas to meet emission standards. However, the industrial-scale adoption of the proposed technology is still uncertain in the near future. While there have been claims of successful separation of the target components ($CO_2$ and/or $H_2S$) in the laboratory scale, some operating MGA facilities based on membrane gas absorption, and increasing interest in using MGA for separation of $CO_2$, $H_2S$ or other gases, there are still important things usually left out in the studies. It should be emphasized that these things are vital in the large-scale design of an operational MGA facilities. The $SO_x$ and $NO_x$ compounds that may be present in the flue gas and the $CH_4$, CO or $CO_2$ present in fuel gas may compete with compete with $CO_2$ and $H_2S$, respectively, during reaction with the active component of the liquid absorbent. Further, these compounds may interact with the membrane material. High temperatures were rarely considered in studies on $CO_2$ absorption. It should also be emphasized that flue gases are emitted at high temperatures. It is unfortunate that although high temperatures favor high absorption rates via chemical reaction, liquid vaporization that may promote pore wetting can also take place. The thermal stability of the membrane should also be considered. The search for novel membrane materials for this purpose may be another avenue for research. Continued research, development, demonstration, and more studies on the economic feasibility of the process when applied in different industries will surely help the development of the technology for large-scale adoption in the future.

## REFERENCES


Abadanes, J.C., E.S. Rubin, and E.J. Anthony. 2004. Sorbent cost and performance in $CO_2$ capture systems. *Ind. Eng. Chem. Res.* 43: 3462–3466.

Al-Juaied, M., and G.T. Rochelle. 2006. Absorption of $CO_2$ in aqueous diglycolamine. *Ind. Eng. Chem. Res.* 45: 2473–2482.

Al-Marzouqui, M., M.H. El-Naas, S.A.M. Marzouk, et al. 2008a. Modeling of $CO_2$ absorption in membrane contactors. *Sep. Purif. Technol.* 59: 286–293.

Al-Marzouqui, M., M. El-Naas, S. Marzouk, et al. 2008b. Modeling of chemical absorption of $CO_2$ in membrane contactors. *Sep. Purif. Technol.* 62: 499–506.

Aroonwilas, A., and A. Veawab. 2004. Characterization and comparison of the $CO_2$ absorption performance into single and blended alkanolamines in a packed column. *Ind. Eng. Chem. Res.* 43: 2228–2237.

Atchariyawut, S., R. Jiraratananon, and R. Wang. 2007. Separation of $CO_2$ from $CH_4$ by using gas-liquid membrane contacting process. *J. Membrane Sci.* 304: 163–172.

Barbe, A.M., P.A. Hogan, and R.A. Johnson. 2000. Surface morphology changes during initial usage of hydrophobic, microporous polypropylene membranes. *J. Membrane Sci.* 172: 149–156.

Blauwhoff, P.M.M., G.F. Versteeg, and W.P.M. van Swaaij. 1984. A study on the reaction between $CO_2$ and alkanolamines in aqueous solutions. *Chem. Eng. Sci.* 39: 207–235.

Boa, L., and M.C. Trachtenberg. 2006. Facilitated transport of $CO_2$ across a membrane: comparing enzyme, amine, and alkaline. *J. Membrane Sci.* 280: 330–334.

Bottino, A., G. Capanelli, A. Comite, et al. 2008. $CO_2$ removal from a gas stream by membrane contactor. *Sep. Purif. Technol.* 59: 85–90.

Boucif, N., E. Favre, D. Roizard, et al. 2007. Hollow fiber membrane contactor for hydrogen sulfide odor control. *AIChE J.* 54: 122–131.

Cadena, C., J.L. Anthony, J.K. Shah, et al. 2004. Why is $CO_2$ so soluble in imidazolium-based ionic liquids? *J. Am. Chem. Soc.* 126: 5300–5308.

Caplow, M. 1968. Kinetics of carbamate formation and breakdown. *J. Am. Chem. Soc.* 90: 6795–6803.

Carapellucci, R., and A. Milazzo. 2003. Membrane systems for $CO_2$ capture and their integration with gas turbine plants. *P. I. Mech. Eng. A–J. Pow.* 217: 505–517.

Chen, Y., S. Zhang, X. Yuan, et al. 2006. Solubility of $CO_2$ in imidazolium-based tetrafluoroborate ionic liquids. *Thermochim. Acta* 441: 42–44.

Corti, A., D. Fiaschi, and L. Lombardi. 2004. Carbon dioxide removal in power generation using membrane technology. *Energy* 29: 2025–2043.

Dankwerts, P.V. 1979. The reaction of $CO_2$ with ethanolamines. *Chem. Eng. Sci.* 34: 443–446.

Davidson, O., and B. Metz. 2005. Special report on carbon dioxide capture and storage. In *International Panel on Climate Change*. Geneva, Switzerland.

deMontigny, D., D. Aboudheir, P. Tontiwachwuthikul, et al. 2006a. Using a packed-bed column model to simulate the performance of membrane absorber. *Ind. Eng. Chem. Res.* 45: 2580–2585.

deMontigny, D., P. Tontiwachwuthikul, and A. Chakma. 2006b. Using polypropylene and polytetrafluorethylene membranes in a membrane contactor for $CO_2$ absorption. *J. Membrane Sci.* 277: 99–107.

Dindore, V.Y., D.W.F. Brilman, and G.F. Versteeg. 2005. Modeling of cross–flow membrane contactors: mass transfer with chemical reactions. *J. Membrane Sci.* 255: 275–289.

Evren, V. 2000. A numerical approach to the determination of mass transfer performances through partially wetted microporous membranes: transfer of oxygen to water. *J. Membrane Sci.* 175: 97–110.

Falk-Pedersen, O., and H. Dannstrom. 1997. Separation of carbon dioxide from offshore gas turbine exchaust. *Energ. Convers. Manage.* 38: S81–S86.

Fauth, D.J., E.A. Frommel, J.S. Hoffman, et al. 2005. Eutectic salt promoted lithium zirconate: Novel high temperature sorbent for $CO_2$ capture. *Fuel Process. Technol.* 86: 1503–1521.

Favre, E. 2007. Carbon dioxide recovery from post-combustion processes: Can gas permeation membranes compete with absorption? *J. Membrane Sci.* 294: 50–59.

Feron, P.H.M., and A.E. Jansen. 1995. The production of carbon dioxide from flue gas by membrane gas absorption. *Energ. Convers. Manage.* 38: S93–S98.

Feron, P.H.M., and A.E. Jansen. 1997. The production of carbon dioxide from flue gas by membrane gas absorption. *Energ. Convers. Manage.* 38: S93–S98.

Feron, P.H.M., and A.E. Jansen. 2002. $CO_2$ separation with polyolefin membrane contactors and dedicated absorption liquids: Performance and prospects. *Sep. Purif. Technol.* 27: 231–242.

Figueroa, J.D., T. Fout, S. Plasynski, et al. 2008. Advances in $CO_2$ capture technology—The U.S. Department of Energy's Carbon Sequestration Program. *Int. J. Greenh. Gas Con.* 2: 9–20.

Gabelman, A., and S.T. Hwang. 1999. Hollow fiber membrane contactors. *J. Membrane Sci.* 159: 61–106.

Gong, Y., Z. Wang, and S. Wang. 2006. Experiments and simulation of $CO_2$ removal by mixed amines in a hollow fiber membrane module. *Chem. Eng. Process.* 45: 652–660.

Gray, M.L., Y. Soong, K.J. Champagne, et al. 2005. Improved immobilized carbon dioxide capture sorbents. *Fuel Process. Technol.* 86: 1449–1455.

Hanioka, S., T. Maruyama, T. Sotaniet al. 2008. $CO_2$ separation facilitated by task-specific ionic liquids using a supported liquid membrane. *J. Membrane Sci.* 314: 1–4.

Herzog, H. 2001. What future for carbon capture and sequestration? *Environ. Sci. Technol.* 35: 148–153.

Hikita, H., S. Asai, and T. Takatsuka. 1976. Absorption of carbon dioxide into aqueous sodium hydroxide and sodium carbonate and bicarbonate solutions. *Chem. Eng. J.* 11: 131–141.

Huang, H.P., Y. Shi, W. Li, et al. 2001. Dual alkali approaches for the capture and separation of $CO_2$. *Energ. Fuels* 15: 263–268.

Huang, H.Y., R.T. Yang, D. Chinn, et al. 2003. Amine-grafted MCM-48 and silica xerogel as superior sorbents for acidic gas removal from natural gas. *Ind. Eng. Chem. Res.* 42: 2427–2233.

Huttenhuis, P.J.G., N.J. Agrawal, J.A. Hogendoorn, et al. 2007. Gas solubility of $H_2S$ and $CO_2$ in aqueous solutions of N-methyldiethanolamine. *J. Petrol. Sci. Eng.* 55: 122–134.

Idem, R., M. Wilson, P. Tontiwachwuthikul, et al. 2006. Pilot plant studies of the $CO_2$ capture performance of aqueous MEA and mixed MEA/MDEA solvents at the University of Regina $CO_2$ Capture Technology Development Plant and the Boundary Dam $CO_2$ Capture Demonstration Plant. *Ind. Eng. Chem. Res.* 45: 2414–2420.

Jassim, M.S., and G.T. Rochelle. 2006. Innovative absorber/stripper configurations for $CO_2$ capture by aqueous monoethanolamine. *Ind. Eng. Chem. Res.* 45: 2465–2472.

Kanel, J.S. Overview: industrial application of ionic liquids for liquid extraction. In *Chemical Industry Vision 2020 Technology Partnership Workshop*. New York, 2003.

Kato, M., K. Nakagawa, K. Essaki, et al. 2005. Novel $CO_2$ absorbents using lithium-containing oxide. *Int. J. Appl. Ceram. Tech.* 2: 467–475.

Keshavarz, P., J. Ayatollahi, and J. Fathikalajahi. 2008a. Mathematical modeling of gas-liquid membrane contactors using random distribution of fibers. *J. Membrane Sci.* 325: 98–108.

Keshavarz, P., J. Fathikalajahi, and J. Ayatollahi. 2008b. Analysis of $CO_2$ separation and simulation of a partially wetted hollow fiber membrane contactor. *J. Hazard. Mater.* 152: 1237–1247.

Keshavarz, P., J. Fathikalajahi, and J. Ayatollahi. 2008c. Mathematical modeling of the simultaneous absorption of carbon dioxide and hydrogen sulfide in a hollow fiber membrane contactor. *Sep. Purif. Technol.* 63: 145–155.

Kim, Y.S., and S.M. Yang. 2000. Absorption of carbon dioxide through hollow fiber membranes using various aqueous absorbents. *Sep. Purif. Technol.* 21: 101–109.

Klaasen, R., P.H.M. Feron, and A.E. Jansen. 2005. Membrane contactors in industrial applications. *Chem. Eng. Res. Des.* 83: 234–246.

Klara, S.M., R.D. Srivastava, and H.G. McIlvred. 2003. Integrated collaborative technology development for $CO_2$ sequestration in geologic formations—United States Department of Energy R&D. *Energ. Converg. Manage.* 44: 2699–2712.

Knuutila, H., H.F. Svendsen, and M. Anttila. 2009. $CO_2$ capture from coal-fired power plants based on sodium carbonate slurry; a systems feasibility and sensitivity study. *Int. J. Greenh. Gas Con.* 3:143–151.

Kohl, A.L., and R.B. Nielsen. 1997. *Gas Purification.* Fifth Edition. Houston, TX: Gulf Publishing Company.

Kong, J., and K. Li. 2001. An improved gas permeation method for characterizing and predicting the performance of microporous asymmetric hollow fibre membranes used in gas absorption. *J. Membrane Sci.* 182: 271–281.

Kreulen, H., C.A. Smolders, G.F. versteeg, et al. 1993. Microporous hollow fiber membrane modules as gas-liquid contactors. Part 1. Physical mass transfer processes. A specific application: Mass transfer in highly viscous liquids. *J. Membrane Sci.* 78: 197–216.

Kumar, P.S., J.A. Hogendoorn, P.H.M. Feron, et al. 2002. New absorption liquids for the removal of $CO_2$ from dilute gas streams using membrane contactors. *Sep. Purif. Technol.* 57: 1639–1651.

Kumelan, J., A.P. Kamps, D. Tuma, et al. 2004. Solubility of $CO_2$ in the ionic liquid [bmim][$CH_3SO_4$] and [bmim][$PF_6$]. *J. Chem. Eng. Data* 51: 1802–1807.

Li, J.L., and B.H. Chen. 2005. Review of $CO_2$ absorption using chemical solvents in hollow fiber membrane contactors. *Sep. Purif. Technol.* 41: 109–122.

Li, K., J. Kong, and X. Tan. 2000. Design of hollow fibre membrane modules for soluble gas removal. *Chem. Eng. Sci.* 55: 5579–5588.

Li, K., D. Wang, C.C. Koe, et al. 1998. Use of asymmetric hollow fibre modules for elimination of $H_2S$ from gas streams via a membrane absorption method. *Chem. Eng. Sci.* 53: 1111–1119.

Liao, C.H., and M.H. Li. 2002. Kinetics of absorption of carbon dioxide into aqueous solutions of monomethanolamine + n-methyldiethanolamine. *Chem. Eng. Sci.* 57: 4569–4582.

Lin, S.H., P.C. Chiang, C.F. Hsieh, et al. 2008. Absorption of carbon dioxide by the absorbent composed of piperazine and 2-amino-2-methyl-1-propanol in PVDF membrane reactor. *J. Chinese Inst. Chem. Eng.* 39: 13–21.

Lu, J.G., Y.F. Zheng, M.D. Cheng, et al. 2007. Effects of activators on mass transfer enhancement in a hollow fiber contactor using activated alkanolamine solutions. *J. Membrane Sci.* 289: 138–149.

Maroto-Valer, M.M., Z. Tang, and Y. Zhang. 2005. $CO_2$ capture by activated and impregnated anthracites. *Fuel Process. Technol.* 86: 1487–1502.

Matson, S.L., J. Lopez, and W.J. Ward. 1983. Separation of gases with synthetic membranes. *Chem. Eng. Sci.* 38: 503–512.

Mavroudi, M., S.O. Kaldis, and G.O. Sakellaropoulos. 2003. Reduction of $CO_2$ emissions by a membrane contacting process. *Fuel* 82: 2153–2159.

Mavroudi, M., S.O. Kaldis, and G.O. Sakellaropoulos. 2006. A study of mass transfer resistance in membrane gas-liquid contacting processes. *J. Membrane Sci.* 272: 103–115.

Modigell, M., M. Schumacher, V. Teplyakov, et al. 2008. A membrane contactor for efficient $CO_2$ removal in biohydrogen production. *Desalination* 224: 186–190.

Nelson, T.O., D.A. Green, P.D. Box, et al. Production of concentrated $CO_2$ from flue gas using dry regenerable carbonate solvents in thermal-swing process. Fifth Annual Conference on Carbon Sequestration, Alexandria, VA, 2006.

Nii, S., and H. Takeuchi. 1994. Removal of $CO_2$ and/or $SO_2$ from gas streams by membrane absorption method. *Gas Sep. Purif.* 8: 107–114.

Oexmann, J., C. Hensel, and A. Kather. 2008. Post-combustion $CO_2$-capture from coal-fired power plants: Preliminary evaluation of an integrated chemical absorption process with piperazine-promoted potassium carbonate. *Int. J. Greenh. Gas Con.* 2: 539–552.

Pevida, C., M.G. Plaza, B. Arias, et al. 2008. Surface modification of activated carbons for $CO_2$ capture. *Appl. Surf. Sci.* 254: 7165–7172.

Powell, C.E., and G.G. Qiao. 2006. Polymeric $CO_2/N_2$ gas separation membranes for the capture of carbon dioxide from power plant flue gases. *J. Membrane Sci.* 279: 1–49.

Prasad, R., and K.K. Sirkar. 1988. Dispersion-free solvent extraction with microporous hollow-fiber modules. *AIChE J.* 34: 177–188.

Qi, Z., and E.L. Cussler. 1985. Microporous hollow fibers for gas absorption: I. Mass transfer in the liquid. *J. Membrane Sci.* 23: 321–332.

Rangwala, H.A. 1996. Absorption of carbon dioxide into aqueous solutions using hollow fiber contactors. *J. Membrane Sci.* 112: 229–240.

Rao, A.B., and E.S. Rubin. 2006. Identifying cost-effective $CO_2$ control levels for amine-based $CO_2$ capture systems. *Ind. Eng. Chem. Res.* 45: 2421–2429.

Resnik, K.P., W. Garber, D.C. Hreha, et al. A parametric scan for regenerative ammonia-based scrubbing for the capture of $CO_2$. 23rd Annual International Pittsburgh Coal Conference, Pittsburgh, PA, 2006.

Rochelle, G., E. Chen, R. Dugas, et al. Solvent and process enhancements for $CO_2$ absorption/stripping. 2005 Annual Conference on Capture and Sequestration, Alexandria, VA, 2006.

Seo, D.J., and W.H. Hong. 1996. Solubilities of carbon dioxide in aqueous mixtures of diethanolamine and 2-amino-2-methyl-1-propanol. *J. Chem. Eng. Data* 41: 258–260.

Seo, D.J., and W.H. Hong. 2000. Effect of piperazine on the kinetics of carbon dioxide with aqueous solutions of 2-amino-2-methyl-1-propanol. *Ind. Eng. Chem. Res.* 39: 2062–2067.

Shekhawat, D., D.R. Luebke, and H.W. Pennline. A review of carbon dioxide selective membranes—A topical report. National Energy Technology Laboratory, U.S. Department of Energy, 2003.

Soriano, A.N., B.T. Doma Jr., and M.H. Li. 2008. Solubility of carbon dioxide in 1-ethyl-3-methylimidazolium 2-(2-methoxyethoxy) ethylsulfate. *J. Chem. Thermodyn.* 40 (12):1654–1660.

Suchdeo, S.R., and J.S. Schultz. 1974. The permeability of gases through reacting solutions: The carbon dioxide–bicarbonate membrane system. *Chem. Eng. Sci.* 29: 13–23.

Tang, J., H. Tang, W. Sun, et al. 2005. Poly(ionic liquid)s: A new material with enhanced and fast $CO_2$ absorption. *Chem. Commun.* 3325–3327.

Trachtenberg, M.C., C.K. Tu, R.A. Landers, et al. 1999. Carbon dioxide transport by proteic and facilitated transport membranes. *Life Support Biosphere Sci.* 6: 293–302.

Wang, D., W.K. Teo, and K. Li. 2002. Removal of $H_2S$ to ultra-low concentrations using an asymmetric hollow fibre membrane module. *Sep. Purif. Technol.* 27: 33–40.

Wang, D., W.K. Teo, and K. Li. 2004a. Selective removal of trace $H_2S$ from gas streams containing $CO_2$ using hollow fibre membrane modules/contractors. *Sep. Purif. Technol.* 35: 125–131.

Wang, M., C.G. Lee, and C.K. Ryu. 2008. $CO_2$ sorption and desorption efficiency of $Ca_2SiO_4$. *Int. J. Hydrog. Energ.* 33: 6368–6372.

Wang, R., D.F. Li, and D.T. Liang. 2004b. Modeling of carbon dioxide capture by three typical amine solutions in hollow fiber membrane contactors. *Chem. Eng. Prog.* 43: 849–856.

Wang, R., D.F. Li, C. Zhou, et al. 2004c. Impact of DEA solutions with and without $CO_2$ loading on porous polypropylene membranes intended for use as contactors. *J. Membrane Sci.* 229: 147–157.

Wang, R., H.Y. Zhang, P.H.M. Feron, et al. 2005. Influence of membrane wetting on $CO_2$ capture in microporous hollow fiber membrane contactors. *Sep. Purif. Technol.* 46: 33–40.

Xiao, J., C.W. Li, and M.H. Li. 2000. Kinetics of absorption of carbon dioxide into aqueous solutions 2-amino-2-methyl-1-propanol + monoethanolamine. *Chem. Eng. Sci.* 55: 161–175.

Xu, X., C. Song, B.G. Miller, et al. 2005. Adsorption separation of carbon dioxide from flue gas of natural gas-fired boiler by novel nanoporous "molecular basket" adsorbent. *Fuel Process. Technol.* 86: 1457–1472.

Xu, Z.K., C. Dannenberg, J. Springer, et al. 2002. Novel poly(arylene ether) as membranes for gas separation. *J. Membrane Sci.* 205: 23–31.

Yamasaki, A. 2003. An overview of $CO_2$ mitigation options for global warming—Emphasizing $CO_2$ sequestration options. *J. Chem. Eng. Japan* 36: 361–375.

Yan, S., M. Fang, W. Zhang, et al. 2008. Comparative analysis of $CO_2$ separation from flue gas by membrane gas absorption technology and chemical absorption technology in China. *Energ. Convers. Manage.* 49: 3188–3197.

Yan, S.P., M.X. Fang, W.F. Zhang, et al. 2007. Experimental study on the separation of $CO_2$ from flue gas using hollow fiber membrane contactors without wetting. *Fuel Process. Technol.* 88: 501–511.

Yang, H., Z. Xu, M. Fan, et al. 2008. Progress in carbon dioxide separation capture: A review. *J. Environ. Sci.* 20: 14–27.

Yang, M.C., and E.L. Cussler. 1986. Designing hollow-fiber contactors. *AIChE J.* 32: 1910–1916.

Yeh, J.T., K.P. Resnik, K. Rygle, et al. 2005. Semi-batch absorption and regeneration studies for $CO_2$ capture by aqueous ammonia. *Fuel Process. Technol.* 86: 1533–1546.

Yeon, S.H., B. Sea, Y.I. Park, et al. 2003. Determination of mass transfer rates in PVDF and PTFE hollow fiber membranes for $CO_2$ absorptions. *Sep. Purif. Technol.* 38: 271–293.

Zhang, H.Y., R. Wang, D.T. Liang, et al. 2006a. Modeling and experimental study of $CO_2$ absorption in a hollow fiber membrane contactor. *J. Membrane Sci.* 279: 301–310.

Zhang, H.Y., R. Wang, D.T. Liang, et al. 2008. Theoretical and experimental studies of membrane wetting in the membrane gas-liquid contacting process for $CO_2$ absorption. *J. Membrane Sci.* 308: 162–170.

Zhang, S., Y. Chen, F. Li, et al. 2006b. Fixation and conversion of $CO_2$ using ionic liquids. *Catal. Today* 115: 61–69.

# 23 Membrane Reactor
## *Concept, Applications, and Prospects*

*Kaushik Nath and Mathur S. Bhakhar*

**CONTENTS**



### 23.1 INTRODUCTION

Membrane technologies have been established as very effective and commercially attractive options for separation and purification processes. Not only have many different membrane separation processes been developed during the past half-century but the existing processes are also being constantly improved in order to enhance their economic competitiveness. Significant improvements are currently being made in the development of new membrane materials with higher selectivity and/or permeability, in the fabrication methods for high-flux asymmetric or composite membranes, in membrane module construction, and in process design. Membrane reactors (MRs) find their applications, both in commercial as well as laboratory scale, in a variety of chemical, biochemical, petrochemical, and environmental systems. Unlike traditional membrane separation processes, where the separator follows the reactor as a separate unit, the membrane reactor offers a synergic integration of reaction and separation into a single process unit. In such an integrated process, the membrane is used as an active participant in a chemical transformation for enhancing the reaction rate, selectivity, and yield. Thus, not only does the membrane plays the role as a separator but also as part of the reactor. The obvious advantages of this approach include a more compact design, better conversion for equilibrium limited reaction, substantial energy savings, and less capital investment.

The two most common applications in the combination of a membrane and a chemical reactor are product removal and reactant purification. In a membrane reactor, one of the products can be eliminated in situ continuously. Removal of a product increases the residence time for a given volume of reactor and drives equilibrium-limited reactions toward completion. Even a small percentage of equilibrium conversion can result in a huge economic gain for industrial processes employing chemical or biochemical reactions. Moreover, for gas phase decomposition reactions, membrane reactors can fundamentally change the pressure dependence of conversion so that the reactions are preferentially performed at high pressures rather than low. Higher pressures allow much smaller reactors and more efficient purification. Membrane reactors can also be advantageous for sequential endothermic and exothermic reactions, by using the product extraction to promote heat transfer.In case of reactant purification the membrane is selective to one reactant, and thus allows separating it in the same device used to make the chemical reaction. Common examples are based on ceramic membranes selective to oxygen, which can be used for catalytic partial oxidation of natural gas. With the advent of new inorganic materials and high-temperature membrane systems, there have been plenty of research and development activities in the field of membrane reactor technology over the last few decades. However, the application of membrane reactors has been largely demonstrated at the laboratory scale, mostly for hydrogenation, dehydrogenation, decomposition, and oxidation reactions including partial oxidation and oxidative coupling of methane.

Although some small industrial installations already exist, the concept has yet to find widespread industrial applications. One of the important factors hindering further commercial development of MRs are the membrane themselves, their support and the surrounding modules (performance, stability, and cost), which still need optimization and new developments. The membrane can also be used to prevent catalyst deactivation or byproduct formation by removing intermediates and products from the part of the reactor where reaction takes place. In the biotechnology field, membrane reactors are utilized for the removal of metabolites from the reactor in production of various biochemicals or for the immobilization of biological components such as proteins or cells. This review presents the principles of membrane reactors, functional perspectives of membrane, type of membranes in membrane reactors, and various important reactions carried out in membrane reactor from the unique perspective of investigators who have made significant contributions to their fields.

## 23.2 MAJOR FUNCTIONS OF MEMBRANES IN MEMBRANE REACTOR

The concept of combining membranes and reactors is manifested in various configurations, whereby membrane is capable of performing a wide variety of useful functions. In the simplest form of a membrane reactor the membrane is used as a contactor that separates the catalyst from the reaction medium. It is often used in cell culture and fermentation processes such as the enzymatic degradation of pectin in fruit juice. In the second type of membrane reactor the membrane shows the selective mass-transport properties, and is used to shift the equilibrium of a chemical reaction by selectivity removing the reaction products, for example, in dehydrogenation or oxygenation reactions such as the dehydrogenation of *n*-butane. The third type of membrane reactor combines the membrane contactor and separation function, such as in enzyme catalyzed deesterification reactions. These three types of membrane reactors are schematically represented in Figure 23.1. Sirkar et al. (1999) reviewed extensively the functional perspective of membranes in membrane reactors. Some of the major functions of membrane in a membrane reactor are as follows:

(a) Membrane as a contactor

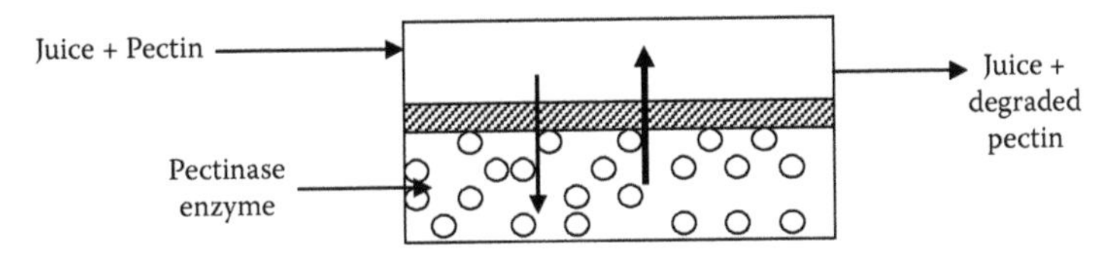


Pectin → galacturonic acid

(b) Membrane as a separating barrier

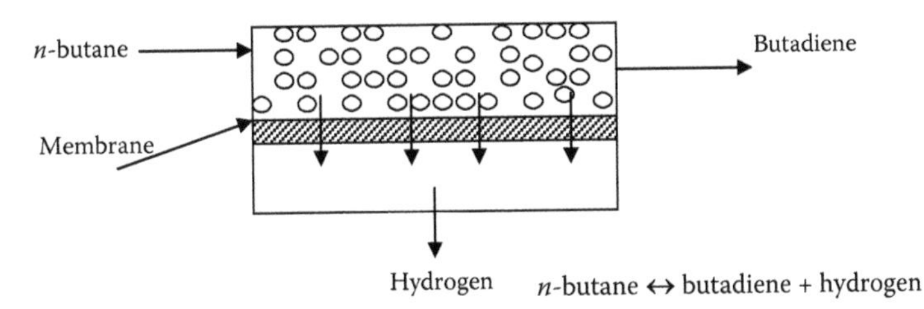


(c) Membrane as contacting and separating barrier

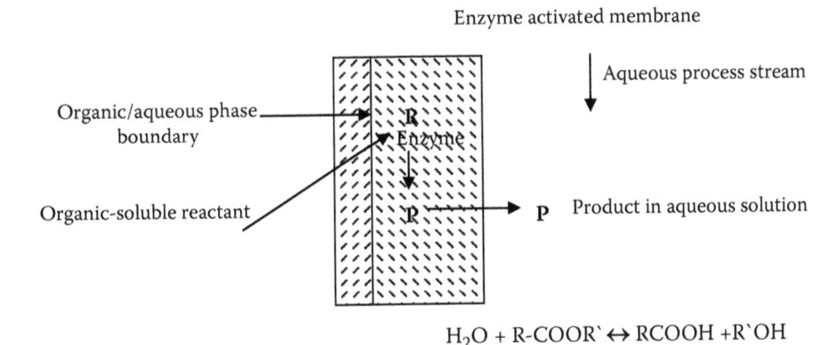


**FIGURE 23.1** (a–c) Three general types of membrane reactor. (Reproduced with permission from Nath, K., *Membrane Separation Processes*, Prentice Hall of India, New Delhi, 2008.)

1. As a separator, to remove the product(s) to increase the reaction conversion by shifting the reaction equilibrium
2. As a separator of reactants from a mixture of feed introduced into a reactor
3. As a distributor where the controlled addition of reactant(s) limits side reactions
4. As an active contactor, where the controlled diffusion of reactants to the catalyst can lead to an engineered catalytic reaction zone
5. As a catalyst itself
6. As a reactor itself

In the first two cases, the membrane is usually catalytically inert and is coupled with a conventional fixed bed of catalyst placed on one of the membrane sides. In the following section various functions of membrane in a membrane reactor are briefly discussed.

### 23.2.1 Separation of Products

The separation of product(s) from a reaction mixture is usually carried out to purify, enrich, or concentrate the products. Product removal also results in increased conversion of a number of equilibrium limited reactions, such as alkane dehydrogenation and selective extraction of the hydrogen produced.

Other $H_2$ producing reactions such as the water gas shift, the steam reforming of methane, and the decomposition of $H_2S$ and HI have been also successfully investigated with the MR extractor mode. To explain the separation of products, let us consider the elementary reversible reaction (Figure 23.2)

$$A + B \Leftrightarrow C + D$$

Here, the product D is required to be removed as a permeate through the membrane. If the compound C also permeates to some extent through the membrane, the permeate stream is enriched in species D vis-à-vis the product species C: permeation of D leads to partial purification of the product C in the reactor outlet stream. The removal of D facilitates shifting of the reversible reaction equilibrium toward forward (right) direction, leading to higher equilibrium conversion of A and B to C and D. Additionally, if there is an undesirable side reaction of the following type takes place

$$B + D \Leftrightarrow E$$

The separation of product D from the reaction mixture minimizes the loss of reactant B to the undesirable side reaction. This increases the selectivity of conversion to product C (or D). Dehydrogenation of ethanol in a Pd membrane reactor is an ideal example of above types of reaction whereby the product hydrogen is withdrawn through the Pd membrane to shift the reaction to the right.

$$\text{Desirable reaction: } C_2H_5OH \Leftrightarrow CH_3CHO + H_2$$

$$\text{Side reaction: } C_2H_5OH + CH_3CHO \Leftrightarrow CH_3COOC_2H_5 + H_2$$

The $H_2$ permselectivity of the membrane and its permeability are two important factors controlling the efficiency of the process. In the partial oxidation of methane intermediate products such as methanol or formaldehyde can be removed via a membrane. This can prevent further oxidation of these products into CO and $CO_2$. Sometimes in fermentation reactions accumulation of one of the products could be inhibitory to the growth of microorganism. Hence removal of product from the fermentation broth via a membrane can considerably reduce product inhibition and increase volumetric productivity of the fermentor. Pressure-driven membrane separation processes such as reverse osmosis, nanofiltration, ultrafiltartion, and pervaporation have also been employed in different membrane reactor systems. Vasudevan et al. (1987) described a membrane sandwich reactor in which the *Saccharomyces cerevisae* (ATCC 4126) cells were effectively placed between an ultrafiltration (UF) membrane and a reverse osmosis (RO) membrane to increase the product ethanol concentration. Here the reactor was fed with a solution of glucose at a high pressure from the UF membrane side and the product solution was forced out through the RO membrane. In esterification reaction, due to the formation of water along with ester, it is required to add a large excess of alcohol to drive the forward equilibrium. However, adding a polyvinyl alcohol- (PVA)

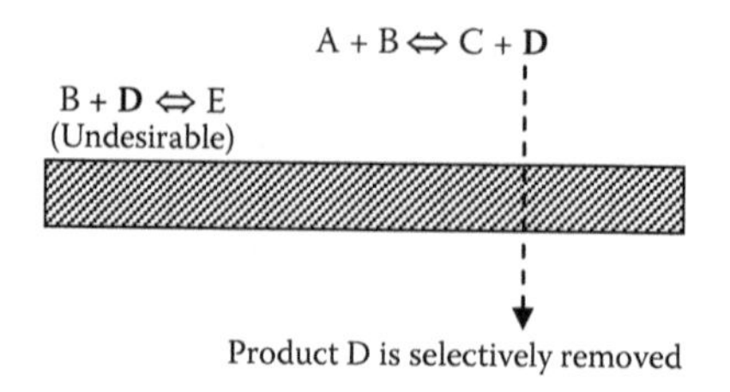


**FIGURE 23.2** Membrane for removal of products.

based water selective pervaporation membrane to esterification reactor allows one to shift the equilibrium to the right, thereby reducing the need for excess alcohol beyond that needed for solubilization of acid.

### 23.2.2 Separation of Reactants

In this case the membrane is selective to one reactant, and thus allows separating it in the same device used to make the chemical reaction. The main examples are based on ceramic membranes selective to oxygen, which can be used for catalytic partial oxidation of natural gas. Membranes allow $O_2$ transport from air to a lower $O_2$ partial pressure side without allowing $N_2$ to be transported through the membrane. Under such conditions partial oxidation of $CH_4$ to syngas can be carried out at a high temperature (800°C) without contaminating the reaction gas mixture with $N_2$. Another example of separation of reactants by membrane is the treatment of organic pollutants from wastewater prior to biodegradation. Industrial wastewater often contains a variety of priority pollutants that are not very susceptible to biodegradation. These industrial wastewaters are frequently released from organic synthesis and contain high salts, extreme pHs, and residual catalysts, all of which adversely affect the growth of microbial cultures used for biodegradation. Brookes and Livingston (1994) have employed a silicone capillary membrane-based device to extract organic priority pollutants from these demanding wastewaters. The biological reaction medium is circulated between a bioreactor and one side of this membrane device. On the other side of the silicone capillary membranes flows the wastewater. Priority organic pollutants (aniline, 4-chloroaniline, 3,4-dichloroaniline, etc.) from the wastewater (having high pH 9-11) are partitioned through the silicone membrane into the biological reaction medium. Successful pilot studies have been conducted using this technique.

### 23.2.3 Controlled Addition of Reactants

The membrane reactor can be used to control the reaction pathway in various gas phases, gas-liquid, and liquid phase reactions. This can be accomplished by localized introduction of reactant, controlled addition of nutrients in biochemical reactions, or simply by adjustment of reactant ratios. The distributor mode is generally used for consecutive parallel reaction systems such as partial oxidation or oxidative dehydrogenation of hydrocarbons or oxidative coupling of methane. For these applications the membrane separation of alkane from $O_2$ is generally used to control the supply of $O_2$ in a fixed bed of catalyst in order to bypass the flammability area, to optimize the $O_2$ profile concentration along the reactor and to maximize the selectivity in the desired oxygenate product. It also helps mitigate temperature rise as a result of exothermic reactions. In such reactors, the $O_2$ permselectivity of the membrane is an important economic factor because air can be used instead of pure $O_2$. In the oxidative dehydrogenation of ethane to ethylene, controlling the ratio of ethane to oxygen is important to maintain selectivity and conversion (with respect to $CO_2$, CO, etc.). Tonkovich et al. (1996) studied the oxidative dehydrogenation of ethane to ethylene at 600°C using $\alpha$-$Al_2O_3$ membrane packed with a magnesium oxide catalyst. The study reveals that at low-to-moderate ethane to oxygen feed ratios (<1.5), the ethylene yield in the membrane reactor exceeds that in a cofeed plug flow reactor (PFR) operated under the same conditions by a factor of 3. Lee et al. (1993) carried out simultaneous biodegradation of toluene and p-xylene in a completely mixed aerated bioreactor using the microorganism *Pseudomonas putida*. Under aerobic conditions, the microorganisms can utilize toluene and xylene as carbon sources. Toluene and p-xylene were introduced into the reactor in a controlled manner through silicone capillary membranes. The removal efficiency of these species increased at the beginning with an increase in the transfer rate of the pollutant mixture and with increased rate of agitation. But the efficiency decreased after certain point since the limiting substrate shifted from carbon to $O_2$. For given impeller speeds, the membrane can be designed to control the rate of introduction of organic pollutants to the biomass-containing medium.

### 23.2.4 The Membrane as an Active Contactor

*In the active contactor mode*, the membrane acts as a diffusion barrier and does not need to be permselective but catalytically active. The concept can be used with a forced-flow mode or with an opposing reactant mode. The forced-flow contactor mode, largely investigated for enzyme-catalyzed reactions, can also be applied to the total oxidation of volatile organic compounds. The opposing reactant contactor mode applies to both equilibrium and irreversible reactions, if the reaction is sufficiently fast compared to transport resistance (diffusion rate of reactants in the membrane). In such a case a small reaction zone forms in the membrane (if sufficiently thick and symmetric) in which reactants are in stoichiometric ratio. Triphasic (gas/liquid/solid) reactions, which are limited by the diffusion of the volatile reactant (e.g., olefin hydrogenation), can also be improved by using this concept. Indeed the volatile reactant does not have to diffuse through a liquid film, as far as a gas/liquid interface is created inside the pores, in direct contact with the catalyst.

### 23.2.5 The Membrane as a Catalyst Itself

In most catalytic membrane reactors, where membrane is not inert and is made up of an inherently catalyst material, membrane can play a dual role of a catalyst and a separator at the same time. In this type of application the membrane material serves as both separator and catalyst, and controls the two most important functions of the reactor. As in the previous case such porous catalytic membranes are used as active contactors to improve the access of the reactants to the catalyst. A number of meso- and microporous inorganic membrane materials are found in literature for their intrinsic catalytic properties such as alumina, titania, zeolites with acid sites, rhenium oxide, LaOCl, $RuO_2$-$TiO_2$ and $RuO_2$-$SiO_2$, VMgO, or La-based perovskites. In most contactors, the catalytic membrane does not need to be permselective but needs to be highly active for the considered reaction, to contain a sufficient quantity of active sites, to have a sufficiently low overall permeability, and to operate in the diffusion controlled regime. In most cases, new synthesis methods have to be developed for preparing these catalytically active membranes, namely when the optimum catalyst composition is complex.

Palladium, a known catalyst for hydrogenation and dehydrogenation reaction, can also occlude hydrogen selectively. Thus, for dehydrogenation reactions, a palladium membrane simultaneously acts as a catalyst and allows the product hydrogen to be removed through the membrane and obtained in purer form. In the case of an esterification reaction involving a carboxylic acid such as oleic acid and an alcohol such as methanol in the presence of a cation exchange membrane, the membrane can act as an acid catalyst by providing $H^+$ ions according to the following reaction:

$$R_1COOH + R_2OH = R_1COOR_2 + H_2O$$

Here, protons, the counter ions in the cation exchange membrane are introduced from one side of the membrane. On the other side of the membrane surface, which is exposed to the reactants acid and alcohol, a layer of protons appears. These protons catalyze the esterification reaction. No separate acid catalyst is required, eliminating the need for catalyst separation after reaction. The cation exchange membrane can also separate water that is produced in the esterification, thereby facilitating the forward equilibrium. The acid form of the Nafion membrane is also capable of catalyzing the esterification of methanol or *n*-butanol to methyl acetate or butyl acetate, respectively.

### 23.2.6 The Membrane as a Reactor Itself

Catalysts are used frequently in membrane reactors. There may be two different modes of operations; first, the membrane may physically segregate the catalyst in the reactor or have the catalyst immobilized in the porous/microporous structure or on the membrane surface. The membrane having the catalyst immobilized in/on it functions almost in the same way as a catalyst particle in a reactor. In all such configurations the bulk flow of the reaction mixture along the reactor length occurs while diffusion of the reactants/products occurs generally in a direction normal to the porous/microporous catalyst. When the bulk flow of a reaction mixture takes place through the membrane from one membrane surface to the other, the membrane is the reactor. Generally, the membrane in such a case is porous/microporous to reduce the pressure drop for practical flow rates. The convective motion of the reaction mixture through the membrane created by an applied pressure drop across the membrane thickness may involve Knudsen diffusion, Poiseuille flow, or a transitional regime for gaseous reaction mixtures (Sirkar et al., 1999). Such a reactor can be identified as a pore flow-through reactor (PFTR).

## 23.3 TYPES OF MEMBRANES USED IN MEMBRANE REACTORS

Both polymeric and inorganic membranes find their uses in various membrane reactor appliances. Polymeric membranes used in the earlier membrane reactors had a high degree of solute permeability and diffusivity. The concept of a membrane reactor was first successfully proven in bioreactor applications with polymeric membranes and appropriate enzymes as catalyst. But the use of polymeric membranes is limited to low-temperature operations. They also offer less resistance to hostile environments such as organic solvents and large pH fluctuations. In contrast, inorganic membranes have the considerable thermal and chemical stability and high mechanical strength compared to their polymeric counterparts. However, the progress of inorganic membranes had been relatively dormant until the development of thin composite metal membranes and ion-conducting ceramic membranes. These membranes, unlike polymeric membranes, offer virtually complete separation. Inorganic membranes also allow regeneration through the oxidative removal of carbonaceous species at 400°C to 500°C and therefore have longer life expectancies than the polymeric membranes. Therefore, inorganic membranes are considered as superior candidates to polymeric membranes in various industrial membrane reactor applications.

Most of the possible applications of membrane reactors in the petroleum and petrochemical industries deal with gaseous catalytic reactions. Thus, gas permeation membranes are the primary component of membrane reactors. Gas permeation membranes have different types of physical structure: dense, microporous, or asymmetric, which is a combination of the two. Separating properties of dense membranes are a function of the solubility and diffusivity of each gaseous component in the membrane material. Microporous membranes generally follow four mechanisms: Knudsen diffusion, surface diffusion, capillary condensation, or molecular sieving.

Different membrane shapes can be used such as flat discs, plates, corrugated systems, tubes (dead-end or not), capillaries, hollow fibers, or monolithic multichannel elements for ceramic membranes, but also foils, spirals, or helix for metallic membranes. The shape of the separative element induces a specific surface/volume ratio for the reactor, which needs to be maximized, typically above 500 $m^2/m^3$, for industrial applications. Apart from the evident need for low cost, resistant, and efficient membranes for the process, highly permeable membranes are required for all applications. This parameter is directly related to the membrane structure, which can be dense or porous and defines the transport mechanisms. The following are the different types of inorganic membranes available in the literature:

- Precipitated oxides
- Zeolite-based
- Metallic membranes
- Ion-conducting ceramic membranes
- Carbon

### 23.3.1 PRECIPITATED OXIDES

Precipitated oxides membranes can be produced by precipitating fine oxide particles onto a mesoporous support. By producing small particles and packing them tightly together on top of a mesoporous support, a pore structure with <20 A° micropores is possible to obtain. However, such densely packed oxide particles are often subject to surface chemistry of the oxide, delamination, and occasional flaws in the surface. These could be avoided by making the layer thicker although at the expense of membrane permeability. These oxide compositions are grown at intermediate temperatures where the oxides are still in a metastable form and subject to additional phase changes or reactivity of outer surface hydroxyl groups with water or other feed components. Heating such materials above and below their preparation temperature can often cause collapse or changes in pore size distribution. These hurdles have limited the approach of precipitation (as a means of controlling pore size) to laboratory operations. In addition, such membranes are not easily scaled to large membrane devices.

### 23.3.2 Zeolite-Based

Zeolites are microporous aluminosilicate materials having well-defined, uniform pore system of molecular dimensions due to their porous crystalline structure. Zeolites are relatively stable at high temperatures, can be acidic or basic in nature, and can exhibit hydrophilic or organophilic properties. They can be size- and shape-selective, more easily allowing straight-chain than branched hydrocarbons to pass through, for example. When used for catalytic reactions, they can be intermediate- or product shape selective as well. However, separations using zeolite membranes are not always based simply on the size or shape of the diffusing species. When interactions between the surface and the diffusing molecules are important, adsorption or surface diffusion can dominate the transport. In these cases, separations where the larger molecules preferentially pass through the membrane can occur.

Zeolite membranes can be synthesized by two major processes, namely one-step hydrothermal synthesisi and secondary (seeded) growth. Hydrothermal synthesis is an in situ process where the support is placed in direct contact with the alkaline precursor. In secondary growth synthesis an existing zeolite phase (or embedding microcrystals of zeolite) is attached to the support followed by hydrothermal growth of the applied seeds. Both these techniques, mentioned above, involve nucleation and crystal growth steps, although nucleation sites are created and applied separately and with more control to the surface of the support during secondary synthesis. Zeolite membranes prepared by these methods usually have a very good separation factor, but the permeance is too low for practical applications. Thus the most challenging work in the field of zeolite membranes is to prepare membranes with high permeance while keeping the separation factor high.

The most widely studied zeolite structure for membrane synthesis is the MFI type. They are relatively inexpensive to synthesize in the laboratory, and because of their approximately 0.55 nm pore size, MFI-type zeolites have been studied extensively for use as separation devices. Silicalite-1 is an MFI-type structure whose framework consists of only silicon and oxygen atoms. ZSM-5 zeolites have exactly the same structure, but some of the silicon atoms are replaced by aluminum atoms. These aluminum atoms in the framework lead to a net negative charge, and therefore must have cations associated with them. Na-ZSM-5 (with sodium as the cation) and H-ZSM-5 (with hydrogen as the associated cation) are the most common forms of this structure.

MFI zeolite membranes (silicalite-1, ZSM-5), on flat or tubular supports, have been investigated extensively for application in gas separation, catalytic reactors, and pervaporation. Silicalite-1 has for instance been applied in the separation of *n*-hexane from its branched isomers in a hydroisomerization membrane reactor (McLeary et al., 2006). The linear alkanes are preferentially adsorbed by the zeolite and supplied to a packed bed of catalyst. Various zeolite membranes, their synthesis process, and salient features are summarized in Table 23.1.

**TABLE 23.1**
**Zeolite Membranes and Their Salient Features**

| Membrane Type | Synthesis Methods | Salient Features | Specific Use | Reference |
|---|---|---|---|---|
| Zeolite NaA | Secondary growth on the external surface of α-alumina tubular supports using a semicontinuous system. | Displays a separation factor of 3,600 at a water permeation flux of 3.8 kg/h $m^2$. | Pervaporation of ethanol/water mixtures | Pina et al., 2004 |
| ZSM-5 | Liquid phase hydrothermal synthesis on tubular commercial supports and subsequently ion-exchanged with Pt. | N/A | Combustion of *n*-hexane present at low concentrations in air | Aguado et al., 2005 |
| Zeolite NaA | Seeded hydrothermal synthesis onto the inner side of porous α-alumina tubular supports using a semi-continuous system. | Selectivities up to 16,000 at fluxes of 0.50 kg $m^{-2}$ $h^{-1}$ at 323 K for the pervaporation of 90:10 (wt.%) ethanol/water mixtures. | Pervaporation of ethanol/water mixtures | Pera-Titus et al., 2006 |
| Silicalite-1 (MFI) membrane | Synthesized on a porous α-$Al_2O_3$ tubular support by in situ nucleation and secondary growth. | The membrane combines high permeances (1.6–8.0 μmol $m^{-2}$ $s^{-1}$ $Pa^{-1}$) and also good separation properties (the $H_2/SF_6$ permselectivity value is equal to 24 at room temperature). | Pervaporation of an ethanol/water mixture | Algieri et al., 2003 |
| NaX and NaY zeolite (Faujasite) membrane | Grown hydrothermally on the surface of a porous cylindrical support. | The Si/Al ratio of these NaX and NaY zeolites determined by atomic absorption spectrophotometry was 1.3 and 2.1 respectively. Thickness of the membrane was about 20–30 μm. | Benzene/cyclohexane and benzene/n-hexane separation | Kita et al., 2001 |
| $Al_2O_3$ supported silicalite-1 membranes | In situ hydrothermal synthesis of zeolite layer within the macropores of alumina ceramic supports (tubes/discs). | More stable zeolite membranes are produced, avoiding outer separating layers which are prone to mechanical damage. These membranes have also been shown to have good thermal stability up to 500°C, with no evidence of crack formation. | Gas separation | Romanos et al., 2001 |

(*continued*)

**TABLE 23.1 Continued**
**Zeolite Membranes and Their Salient Features**

| Membrane Type | Synthesis Methods | Salient Features | Specific Use | Reference |
|---|---|---|---|---|
| Zeolite-A | Synthesized by secondary growth on the surface of macroporous α-alumina and stainless steel tubular supports, previously seeded with zeolite A crystals. | Higher separation factors were achieved (e.g., ideal selectivity $H_2/C_3H_8$ of 17.35 at 323 K, and $H_2/C_3H_8$ separation factor of 7.28 at the same temperature). | Separation of $H_2/N_2$ and $H_2/C_3H_8$ mixtures | Moron et al., 2002 |
| X-type zeolite | Prepared by a template-free method on porous tubular supports. | The total flux was 2.7 kg/m$^2$ h, and the 1,3-propanediol/glycerol selectivity was 41. | Separation of 1,3-propanediol from glycerol in aqueous mixtures by pervaporation; at 308 K | Li et al., 2002 |
| Silicalite-1 with α- $Al_2O_3$ support | Seeded synthesis, by two successive secondary for 24 h, from clear solution at 175°C. | Primarily c-oriented crystals. Membrane thickness: 25–40 μm. | Separation of aromatic isomer vapors | Xomeritakis and Tsapatsis, 1999 |
| Silicalite-1 with silicon wafer support | Seeded secondary growth, hydrothermal. | Primarily a- and b-oriented crystals. Membrane thickness: 240–1,200 nm. | Separation of mixture of C4, C6, and xylene isomer | Hedlund et al., 2002 |

### 23.3.3 Metallic Membranes

Dense metal membranes are commercially available in a variety of compositions and can be made into large devices. However, there are only two major types of metallic membranes available in the literature: one is palladium-based and other is silver-based. The palladium-based membranes are the most frequently used metal membranes owing to their very high hydrogen selectivity (Nagamete and Inoue, 1985, 1986). They have been used for hydrogenation, dehydrogenation, dehydrogenation-oxidation, and steam reforming reactions. The exclusive transport mechanism toward hydrogen is generally believed to be due to the interaction of hydrogen atoms and palladium metal. Molecular hydrogen is dissociated and chemisorbed on one side of the membrane and dissolved in the palladium matrix. The atomic hydrogen then diffuses to the other side of the membrane. Here, the hydrogen atoms recombine and desorb as molecular hydrogen.

Palladium membranes can be prepared by various methods such as electroless plating, sputtering, spray pyrolysis, chemical vapor deposition (CVD), and electrochemical vapor deposition (EVD). In electroless plating, thin palladium-based composite membranes can be produced by dipping porous support materials such as ceramic, metal, or glass in a bath of palladium salt complex solution and a reducing agent. For uniform coating, the support material must be clean, sensitized, and activated before dipping; Palladium deposition takes place on the support surface by autocatalyzed reduction of metastable metallic complexes. In the case of palladium deposition, amine complexes such as $[Pd(NH_3)_4]Cl_2$ are generally used in the presence of a reducing agent such as hydrazine or sodium hypophosphate (Zaman and Chakma, 1994). In the sputtering method, atoms are removed from a suitable target and deposited on the support surface in the form of a thin film by means of rapid ion bombardment from a high-energy plasma (Konno et al., 1988). Thin films of binary and ternary alloys of palladium with manganese, cobalt, ruthenium, tin, and lead were deposited on polymeric, porous stainless steel and oxide supports by Gryaznov et al. (1993). Sputter deposition on polymer membranes usually produces very fragile and defective membranes that cannot function at high temperatures. In spray pyrolysis, a solution of metal salts is sprayed into a heated gas stream and pyrolyzed. Spray pyrolysis is applied for the production of fine metals or metal oxide particles. Li et al. (1993) produced palladium-silver alloy membrane by using spray pyrolysis of a palladium nitrate and silver nitrate solution in a hydrogen-oxygen flame. The thickness of the alloy membrane was 1.5 to 2 μm and the separation factor of hydrogen to nitrogen was approximately 24 at 773 K. The CVD method involves deposition of a desired component in the vapor state by means of a chemical reaction. This technique can be used to deposit films of $SiO_2$, $TiO_2$, $Al_2O_3$, and $B_2O_3$ on different porous support materials such as vycor tubes by reacting the respective chloride precursors with water at 100 °C to 800°C. EVD is a variation of the CVD technique. Here the porous substrate separates the mixture of chloride vapors ($ZCl_3$, $YCl_3$, etc.) and an oxygen source (water vapor or oxygen). Initially, the reactants from both sides of the substrate interdiffuse into the pores and form the solid oxides, as in the CVD process. When the pores are closed, oxygen ions are conducted across the solid oxide and the oxide film grows on the chloride side. Membranes with films of solid oxides deposited by the EVD techniques have been used in various catalytic chemical reactions.

There are several limitations with palladium membranes (Armor, 1998). Alloys of Pd can undergo surface inhomogeneity during long term operation and they are also badly affected by hydrogen embrittlement. This is a phenomenon in which dissolved hydrogen tend to cause different elongation of metallic film on glass support, leading it to fracture after repeated pressure and thermal cycling. Moreover, palladium has sensitivity toward traces of iron, which causes pinholes. In order to avoid the embrittlement, alloying of Pd with group IB metals such as silver is generally done. Another advantage of alloying is that the mechanical strength can be higher than for pure palladium.

The only other metal membrane that has been studied in membrane reactors is silver alloy, which permits oxygen permeation (Anshits et al., 1989; Gryaznov et al., 1986). Based on its poor performance, the silver alloy membranes have not been actively pursued for use in membrane reactors or separation devices. Development of these membranes has been limited by fabrication capabilities and availability of a wide composition of alloys. Pd membranes can undergo phase changes that can cause catastrophic failure of the membrane due to expansion of the lattice resulting in microcracks in the bulk metal. These phase changes are very pressure- and temperature-dependent. In the 1960s commercially manufactured Pd diffusers were used to extract $H_2$ from waste process gas streams.

However, they present the problem of hydrogen embrittlement, a phenomenon in which dissolved hydrogen tend to cause different elongation of metallic film on glass support, leading it to fracture after repeated pressure and thermal cycling. In order to avoid the embrittlement, alloying of Pd with group IB metals such a silver is generally done. Another advantage of alloying is that the mechanical strength can be higher than for pure palladium. In general, a Pd membrane becomes brittle after certain cycles of $\alpha$-$\beta$ hydride transformations due to the accompanied lattice expansion. In palladium-silver alloys, for example, the lattice has already been expanded by the silver atoms, and the PdAg lattice is less influenced by hydrogen and thus less brittle than the pure Pd lattice. An optimal value of the hydrogen permeation rate is reached for a silver content around 25 wt. % Ag (Uemiya et al., 1991).

### 23.3.4 Ion-Conducting Ceramic Membranes

Various gases, for example, oxygen, can be separated from air or other feed streams by systems of ion-conducting ceramic membranes. Such ceramic membranes exhibit ion conductivity at temperatures well above 500°C, generally in a range of about 600°C and 1100°C. An important consideration of such membranes is that the transport rate of the gas through the membrane must take place at a sufficient rate to make the separation economically attractive. Ceramic membrane materials useful in separating oxygen, in general are mixed conductors that possess both oxygen ion conduction and electronic conduction in either single-phase or dual-phase states. The driving

**TABLE 23.2**
**Carbon Membranes Prepared from Various Precursors**

| Precursor | Process | Salient Features | Reference |
|---|---|---|---|
| Phenolic resin | By pyrolysis (700°C, under vacuum) of a phenolic resin film supported on the inner face of a porous alumina tube | • A layer (thickness: 2 μm).<br>• The effective micropore size was estimated to be around 4.4 Å. | Centeno and Fuertes, 2001 |
| Poly(vinylidene chloride-co-vinyl chloride) | Pyrolysis of a polymeric film supported over a macroporous carbon substrate | • Composite carbon membrane, thickness 0.8 μm.<br>• High selectivities for the separation of permanent gases like $O_2/N_2$ system (Selectivity≈14 at 25°C). | Centeno and Fuertes, 2000. |
| Polymer latex | Carbonization | The estimated pore size of the membrane is between 5.0 and 5.5 Å (diameter). | Rao and Sircar, 1993 |
| Polyimide, P84 | Carbonization 550°C–800°C under vacuum environment | • The highest selectivity was attained by carbon membranes pyrolyzed at 800°C.<br>• P84, among four other polyimides carbon membranes exhibited the highest separation efficiency for $CO_2/CH_4$ separation. | Tin et al., 2004 |
| Cellulose from wood pulp | Cellulose hydrolysis with trifluoroacetic acid (TFA), at 30°C, 2 bar pressure | Better separation performance for selected gas pairs. | Lee and Hag, 2006 |
| Coal tar pitch | Carbonization of a coal tar pitch precursor, which is heat-polymerized and preoxidized under different conditions | The ideal separation factor of $H_2/CO_2$ on the carbon–carbon composite membrane prepared can reach 16.2. | Liang et al., 1999 |
| Furfuryl alcohol | Vapor deposition polymerization and slow carbonization at 600°C and by controlling polymerization-carbonization cycle | Supported carbon membrane. The permeance was 0.6–2.5 for $H_2$, 0.27–0.58 for $CO_2$, and 0.08 for $O_2$, all in MPU (1 MPU=$10^{-8}$ mol/m$^2$ Pa s). | Wang et al., 2000 |
| Cellulose and hemicellulose | Carbonization in the temperature range of 400°C–700°C | • Increased hydrogen/methane permselectivity at the expense of permeability.<br>• Can tolerate light hydrocarbons and separated hydrogen with a permeability of about 480 barrer and hydrogen/methane permselectivity >1000. | Grainger and Hagg, 2007 |
| Asymmetric polyimide hollow fiber membrane | Pyrolysis in the temperature range of 600°C to 1000°C | High performance of gas permeability and selectivity.<br>Membranes prepared by pyrolyzing at over 700°C had permeation rates of hydrogen ranging from $10^{-4}$ to $10^{-3}$ cm$^3$ (STP)/(cm$^2$ s cmHg). | Kusuki et al., 1997 |

force of the overall oxygen transport rate for the membrane is the different oxygen partial pressure applied across the membrane. Since the membrane is dense and gas-tight, the direct passage of oxygen molecules is blocked. Oxygen ions, however, can migrate selectively through the membrane. Dissociation and ionization of oxygen occurs at the membrane cathode surface where electrons are picked up from near surface electronic states. The flux of oxygen ions is charge-compensated by a simultaneous flux of electronic charge carriers. When the oxygen ions arrive at the opposite anode surface of the membrane, the individual ions release their electrons and recombine again to form oxygen molecules, which are released in the permeate stream.

The permeation rate through a nonporous ceramic membrane is controlled by two major factors: (1) the solid-state ionic transport rate within the membrane and (2) the ion surface exchange rate on either side of the membrane. The flux of the gas to be separated usually can be increased by reducing the thickness of the membrane until its thickness reaches a critical value. At above the critical value, the flux is controlled by both the ion surface exchange kinetics and solid-state ionic transport rate.

### 23.3.5 Carbon

Carbon membrane, with properties superior to those typically found in polymeric materials, is a new and high-performance carbonaceous material for gas separation. Carbon membrane can be produced by carbonizing various carbon-containing materials, such as thermosetting resin, pitch, coal, graphite, and plants, under inert atmosphere or vacuum, thus producing a dense layer having a microporous sieving network. Granular carbon materials have been used commercially for years for molecular size separation by adsorption. Using vapor deposition techniques, it is possible to grow extended layers of carbon sieving surface. Asymmetric molecular sieve carbon membrane is prepared by conventional pyrolysis of thermosetting polymeric film, followed by unequal oxidation. Morphology, pore size distribution, and gas separation characteristics of the membrane are important considerations. Rao and Sircar (1993) have demonstrated the use of carbon coated supports for recovery of $H_2$ from targeted refinery streams. In these selective surface-flow membranes, $H_2$ is rejected from the membrane and it allows one to recover high-pressure $H_2$ in an enriched retentate stream (the portion of the feed that is rejected by the membrane and not allowed to permeate). Tubular carbon molecular sieve membranes have recently been reported for the separation of gaseous mixtures. It is still difficult to achieve these as continuous, strong tubes free of microcracks. These carbon membranes are in very limited supply and often very small, which limits their evaluation and application. Table 23.2 gives a brief account of different types of carbon membrane prepared from various precursors.

## 23.4 TYPES OF MEMBRANE REACTORS

Membrane reactors are most commonly used when a reaction involves some form of catalyst, and there are two main types of these membrane reactors: the inert membrane reactor (IMR) and the catalytic membrane reactor (CMR). IMR allows catalyst pellets to flow with the reactants on the feed side (usually the inside of the membrane), and is known as an inert membrane reactor with catalyst on the feed side (IMRCF). In this kind of membrane reactor, the membrane does not participate in the reaction directly; it simply acts as a barrier to the reactants and some products. The role of the membrane is to remove products or to add reactants. The catalyst is located apart from the membrane structure. It may have two types of configurations (Coronas and Santamaria, 1999). In one configuration, IMR allows the permeation of products and in the other both permeation of products and reaction coupling take place. Removal of hydrogen in dehydrogenation reaction, decomposition of $H_2S$, and preferential permeation of hydrogen are examples of the first type. Sometimes equilibrium displacement can be enhanced through reaction coupling. In this case, on both sides of the membrane complementary processes are run that use either the permeated species (chemical coupling, e.g., dehydrogenation/hydrogenation or dehydrogenation/combustion reactions), or the heat generated in the reaction (thermal coupling, exothermic/endothermic processes). The reactions often use different catalysts, which would be packed on opposite sides of the membrane tube.

A CMR has a membrane that has either been coated with or is made of a material that contains catalyst, which means that the membrane itself participates in the reaction. Some of the reaction products (those that are small enough) pass through the membrane and exit the reactor on the permeate side. Porous ceramic membranes can be made, in whole or in part, of alumina, silica, titania, zirconia, or zeolites, materials that are catalytically active under suitable operating conditions. In this case, intrinsically active membranes are obtained, where the membrane itself is the catalyst. When the membrane is not catalytically active under reaction conditions, an active phase can be deposited on it using the membrane material as a support. Impregnation, ion-exchange, CVD, and sol-gel techniques are commonly employed for this purpose. Again, most of the differences with conventional catalyst preparation procedures using the same techniques stem from the existence of a continuous porous membrane structure.

Additionally some other configurations such as the inert membrane-packed bed reactor (IMPBR), packed bed catalytic membrane reactor (PBCMR), catalytic nonpermselective membrane reactor (CNMR), and inert semipermeable membrane reactor (ISMR) are also found in literature. In most of the cases of dehydrogenation/decomposition systems, the use of membrane reactors has been necessitated from the increased conversion and/or selectivity. The reactor has been operated in IMPBR, CMR, or PBCMR mode. ISMR operation with solid oxide membranes is used for reactions such as the decomposition of water or carbon dioxide while CMR or PBCMR mode would be suitable for the partial oxidation of methane. CNMR operation seems particularly attractive in environmentally important reactions, because this has the potential to eliminate completely the slip of the undesirable reactant. CNMR also provides a novel contacting method for two-phase reactions (Zaman and Chakma, 1994).

There are several other examples in the current literature that can be considered as inorganic membrane reactors with attractive properties. These include catalytic particle traps for car exhaust treatment, trifunctional membrane reactor for water treatment by ozonation, zeolite encapsulated catalyst, and solid oxide fuel cells and electrolysis cells. When separation and catalytic functions have not been performed by the same membrane material, the catalytic material may only cover the grains of the support, which may or may not be covered with a permselective top-layer membrane. Starting from formulated sols and macroporous supports with adapted tortuosity, porosity, and pore size, this method yields catalytic contactors with low-pressure drop and high reactivity. This concept is typically used in catalytic particle traps and in four-way catalysts. Such ceramic contactors can be designed to perform only the oxidation of soot, CO, and $C_xH_y$ whereas $NO_x$ are treated separately. In the four-way catalyst, the ceramic contactor should continuously perform the oxidation of fly-ash within its pores, the removal or reduction of noxious gases, and the oxidation of CO and hydrocarbons.

When a selective extraction is also required, the deposited catalytic material can be coupled with a permselective top-layer membrane. This concept, leading to compact efficient multifunctional systems, has been developed recently in several groups for both liquid and gas phase applications. A typical example of composite catalytic reactor for the steam reforming of methane is reported by Tsuru et al. (2006). Zeolite membrane-encapsulated catalyst (ZMEC) is an original concept of a microdesigned membrane reactor. In such a reactor the permselective membrane is coated on the catalyst grains. The permselective membrane controls the traffic of both reactants and products to and from the catalyst. The concept revealed an original way to use zeolite membranes in catalytic reactors while limiting the influence of defects (nonzeolite pores) on large-scale membranes (Nishiyama et al., 2004).

## 23.5 IMPORTANT REACTIONS CARRIED OUT IN MEMBRANE REACTORS

Membrane reactors hold much promise for enhanced conversion in thermodynamically limited reactions (dehydrogenation, esterification, etc.) where removal of one or more products through the -membrane could favor the forward equilibrium toward product generation. In the past 20 years or so, many of these problems have been addressed and the field of membrane reactors has expanded to include a wide variety of applications and reaction type. In the following section we will briefly touch upon some of the major reactions reported in recent literature where membrane reactors have been successfully applied.

### 23.5.1 DEHYDROGENATION

Almost all dehydrogenation reactions are thermodynamically limited. As dehydrogenation reactions are endothermic in nature, they are favored at high temperatures. This might lead to significant occurrence of side reactions and to catalyst deactivation by coking, thus reducing the selectivity toward the desired reaction products. In membrane reactors, equal conversions can be obtained at lower temperatures compared to fixed-bed reactors. In the case of equilibrium- limited dehydrogenation reaction, the close coupling of the reaction with selective permeation of hydrogen through the membrane significantly increased the conversion of the reaction at constant temperature. Reduction of the operating temperature while keeping the same conversion of the dehydrogenation reaction is also possible in a tile palladium membrane reactor due to the shift of equilibrium toward the hydrogen-producing side. In the meantime, high-purity hydrogen could be obtained from the permeate side of tile membrane reactor. Moreover, dehydrogenetaion causes an increase of the total number of gas molecules and therefore can be driven toward higher conversion by reduction of pressure. But this would lead to higher reactor volumes. Membrane reactors with more compact module configuration compared to fixed-bed reactors (equal conversion implied) could be an answer to this problem. A large number of reactions in membrane catalysis deal with hydrogen producing reactions such as dehydrogenation of alkanes. In refinery operations there are many more types of dehydrogenations that are considered with the membrane reactor system. These include dehydrogenation of cyclohexane to benzene, dehydrogenation of propane to propylene, reforming via dehydrogenation of heptane, dehydrogenation of ethyl benzene to styrene, oxidative dehydrogenation of butane to butadiene, and so forth. Table 23.3 summarizes some of the salient features of various dehydrogenation in membrane reactors.

### 23.5.2 OXIDATION

Many catalytic oxidation processes are carried out in membrane reactors comprised of dense metal oxides or ceramic membranes. Mixed oxides of the perovskite type display pronounced catalytic activity in oxidation reactions. Syn-gas, a mixture of CO and $H_2$, could be produced by selective partial oxidation of ethane or methane. However, the reactions are difficult to control because of the over oxidation of alkane to carbon dioxide at the initial high temperature. The use of membrane reactor facilitates the separation of methane/ethane from the oxygen, thus providing an extra margin of safety in avoiding explosive mixtures of these two reactants (Armor, 1998). For many oxidation reactions, the selectivity toward desired products can be enhanced considerably by using membrane reactor. Another reaction that can be carried out in membrane reactors is the oxidative coupling of methane. The catalytic oxidative coupling of methane is one of the most extensively studied subjects in catalysis in an effort to utilize the abundant resources of natural gas. One of the objectives of oxidative coupling of methane is to achieve high C2 selectivity at higher methane conversion compared to the conventional packed-bed reactors. Various others oxidation reactions that are studied using membrane reactors include oxidation of benzene to phenol, gas phase partial oxidation of toluene,

**TABLE 23.3**
**Membrane Reactor Applications in Dehydrogenation Reactions**

| Type of Dehydrogenation Reaction | Operating Conditions | Reactor Configuration | Membrane Type | Salient Features | Reference |
|---|---|---|---|---|---|
| Catalytic oxidative dehydrogenation of ethane | γ-alumina supported vanadium oxide catalyst | Fluidized bed membrane reactor (FLBMR) | Microporous silica based | Maximum ethylene yield observed in the FLBMR was 37% compared to 35% for the PBMR. Beside a high productivity, for the FLBMR a broader favorable operation range with respect to the oxygen–hydrocarbon ratio was observed. | Ahchieva et al., 2005 |
| Catalytic dehydrogenation of propane to propylene | Cat: Platinum loaded alumino silicate molecular sieve; temperature: 773 K–848 K | PBMR | Microporous silica based and Pd thin film supported by a porous ceramic substrate | Propylene yield of 39.6%, at liquid hourly space velocities (LHSV) of 3 and temp 823 K. | Collins et al., 1996 |
| Dehydrogenation of propane to propylene | Pt/K/Sn/$Al_2O_3$ packed catalyst; temperature of 350°C–550°C; transmembrane pressure difference: 1 psi | Lysothermal high-temperature shell-and tube; CMR | Pd-coated γ-$Al_2O_3$ membrane | Leads to conversion twice as high as equilibrium conversion and six times as high as conversion with a conventional reaction system at 500°C. | Chang et al., 2002 |
| Dehydrogenation of propane to propylene | Packed with Pt/γ-$Al_2O_3$ | PBMR | Composite membrane of thin Pd film and porous ceramic substrate | Enhanced conversion of propane. | Anwu et al., 1996 |
| Oxidative dehydrogenation of propane to propene | Catalyst (V—Mg—O or Ni based) | CMR | Commercial porous alumina membranes | Increase in propene yields at low $C_3H_8/O_2$ ratio. | Pantazidis et al., 1995 |

| | | | | | |
|---|---|---|---|---|---|
| Oxidative dehydrogenation of butane to butadiene | V/MgO catalyst; 550°C | CMR | Alumina impregnated with MgO | Feed configurations that lead to a low partial pressure of oxygen also give rise to increased selectivity. | Pedernera et al., 2002 |
| Oxidative dehydrogenation of butane to butadiene and butane | A fixed-bed of V–Mg–O catalyst | Uniform and nonuniform inert membrane reactor | N/A | Selectivity of up to 61.1% and a yield of 42.2% were obtained. | Ge et al., 2001 |
| Dehydrocyclization of heptane | Commercial Pt/$Al_2O_3$ catalyst; 400°C 17 atm | PFR | Interstage Pd/Ag membrane | 65% higher toluene yield than a nonmembrane unit. | Ali and Baiker, 1996 |
| Dehydrogenation of cyclohexane to benzene | 0.5% Pt/$Al_2O_3$ catalyst, 200°C and 1 atm pressure using an argon stream saturated with cyclohexane vapor | Tubular reactor | Pd/Ag membrane | N/A | Itoh, 1987 |
| Dehydrogenation of ethyl benzene to styrene | Commercial K promoted iron oxide catalyst | Two-stage packed bed reactor followed by a membrane reactor | Tubular 40A° alumina mesoporous membrane | 4% yield enhancement compared with nonmembrane process. | Gallahar et al., 1993 |
| Oxidative dehydrogenation of methanol to formaldehyde | Fe-Mo oxide catalyst; 200-250°C | PBMR | Nonperm selective 316L stainless steel | Increased selectivity and yield due to distributed/controlled reactant addition. | Diakov et al., 2002 |
| Oxidative dehydrogenation of methanol to formaldehyde | Cu/P catalyst | CMR | Palladium/ceramic membrane; porous alumina membrane | Yield increased by 15% with the palladium/ceramic membrane, and 5% with the porous alumina membrane. | Deng and Wu, 1994 |
| Oxidative dehydrogenation of methanol | Pt catalyst | CMR with a composite multilayered ceramic tube impregnated with catalyst | Ceramic membrane | Hydrogen is a main product of reaction for mixtures rich in methanol and lean in oxygen for temperatures up to 300°C and 1 bar pressure. | Brinkmann et al., 2001 |

**TABLE 23.4**
**Membrane Reactor Applications in Oxidation Reactions**

| Type of Oxidation Reaction | Operating Conditions | Reactor Configuration | Membrane Type | Salient Features | Reference |
|---|---|---|---|---|---|
| Selective oxidation of methane to syn-gas ($CO/H_2$ mixture) | Rh-based reforming catalyst; temperature: 850°C | N/A | Hollow tube membrane of SrFeCo oxides | >98% $CH_4$ conversion with 90% CO selectivity was achieved. | Balachandran et al., 1997 |
| Partial oxidation of methane to syn-gas | 875°C LiLaNiO/γ-Al2O3 catalyst | PBMR | Dense ceramic membrane | 94% methane conversion and higher than 95% CO selectivity. | Wang et al., 2003 |
| Partial oxidation of ethane to syn-gas | 800°C–900°C; LiLaNiO/γ-$Al_2O_3$ | CMR | Dense ceramic membrane | Ethane conversion of 100%, the selectivity of CO was higher than 91%. | Wang et al., 2002 |
| Selective oxidation of ethane to ethylene | 825°C–875°C | CMR | Dense ceramic membrane | Ethylene yield of 56% with ethylene selectivity of 80%. | Akin and Lin, 2002 |
| Selective oxidation of carbon monoxide | Pt supported catalyst, 200°C; 4 bar pressure | CMR | Pt-Y zeolite membrane, tubular | Almost complete CO removal was achieved. | Bernardo et al., 2006 |
| Selective oxidation of carbon monoxide | 200°C–250°C Pt Y | CMR | Y-type eolita/porous γ-$Al_2O_3$ | Selective oxidation of CO—the CO conc. On the permeate side decreased below the detection limit. | Hasegawa et al., 2001 |
| Oxidation of benzene to phenol | <250°C | CMR | Pd on $Al_2O_3$ | Phenol selectivities of 80–97% at benzene conversions of 2–16% below 250°C; WHSV of 1.5 at 150°C. | Niwa et al., 2002 |
| Oxidative coupling of methane to get C2 hydrocarbons | Pervovskite catalyst BaCe0.8Gd0.2$O_3$ | Shell and tube membrane reactor | N/A | 26% conversion and and 62% selectivity were obtained with yield more than 16%. | Lu et al., 2000 |
| Gas phase partial oxidation of toluene | Vanadium catalyst with $Al_2O_3$-$TiO_2$ support | MR | Inorganic membrane deposited with vanadium | Selectivity to intermediate oxygenated products was higher for vanadium supported by $Al_2O_3$–$TiO_2$ and when the hydrocarbon to oxygen ratio was higher, the formation of carbon oxides was lowered. | Bottino et al., 2004 |
| Partial oxidation of 1 butene to maleic anhydride | 350°C surface velocity was 500 cm/h | CMR | Vycor® tube membrane | The maximum selectivity of 95% was obtained at 350°C and at this condition, oxygen permeability was almost four times higher than the reaction had not occurred. | Moon et al., 1997 |
| Oxidation of cyclohexane and *n*-dodecane | Catalyst: zeolite-encapsulated iron-phthalocyanine | CMR | Poly(dimethylsiloxane) polymer membrane | In the *n*-dodecane oxidation in a liquid phase catalytic membrane reactor, the alcohol and ketone products are exclusively recovered in the organic *n*-dodecane phase, demonstrating the possibility of the integrated reaction and separation in one single process unit. | Langhendries et al., 2000 |

**TABLE 23.5**
**Membrane Reactor Applications in Hydrogenation Reactions**

| Type of Oxidation Reaction | Operating Conditions | Reactor Configuration | Membrane Type | Salient Features | Reference |
|---|---|---|---|---|---|
| Continuous hydrogenation of sunflower seed oil | Pd catalyst | Three phase CMR | N/A | Selectivity increases with increasing temperature and decreasing hydrogen pressure. | Veldsink, 2001 |
| Catalytic hydrogenation of $CO_2$ into methanol | | CMR | Zeolite membrane | It is possible to obtain higher conversion of $CO_2$ and both higher methanol selectivity and yield with respect to conventional reactor. | Gallucci et al., 2004 |
| $CO_2$ hydrogenation to methanol | 210°C–230°C, Pd | CMR | MOR/ZSM-5/chabazite | Higher $CO_2$ conversion and $CH_3OH$ selectivity. | Barbieri et al., 2002 |
| Selected hydrogenation of conjugated dienes | Temp: 40°C; pressure: 0.1 MPa | CMR | PVP-Pd/CA and PVP-Pd/PAN hollow fibers | The selectivity for the formation of monoene and the conversion of diene can simultaneously be up to more than 91% in the membrane reactors. | Gao et al., 1995 |
| Selective hydrogenation of cyclopentadiene | Polymer anchored palladium catalysts, temperature: 40°C | CMR | Cellulose acetate hollow fiber | N/A | Gao et al., 1994 |
| Hydrogenation of 1,3 butadiene | 100°C, 1–41kPa | MR (ss) | Palladium membrane | The permeation rate was strongly dependent on butadiene pressure. At >40kpa pressure hydrogenation rate was over 10 times larger than that for at pressure 0.2 kPa. | Nagamoto and Inoue, 1986 |
| Hydrogenation of propene and selective hydrogenation of propyne | 30°C | MR | Polyamideimide membranes modified by titanium dioxide and treated further with palladium acetate | 100% conversion of propene at a maximum yield of 98% of propane. In the selective hydrogenation of 5% propyne in propene, a selectivity of 99% for propene at 100% conversion of propyne was achieved. | Ziegler et al., 2001 |
| Selective hydrogenation of 2-hexyne and hexadiene | Pt/$TiO_2$ catalyst | CMR | Microporous thin film membranes | Exhibits a hydrogenation activity significantly higher than those of comparable batch catalysts. | Lange et al., 1998 |

partial oxidation of 1-butene to maleic anhydride, selective oxidation of ethane to ethylene, and so forth. These are briefly summarized in Table 23.4.

### 23.5.3 Hydrogenation

Catalytic hydrogenation of olefins is studied extensively using membrane reactors (Table 23.5). Palladium among the metallic catalysts employed for hydrogenation, has a unique property of dissolving a large amount of hydrogen at room temperature and of making the hydrogen diffuse through it. Conversion of $CO_2$ into methanol by catalytic hydrogenation has been recognized as one of the most promising processes for stabilizing the atmospheric $CO_2$ level, and furthermore the methanol produced could be used as fuel or a basic chemical for satisfying the large demand worldwide. The use of membrane reactor helps increase $CO_2$ conversion into methanol with respect to a traditional catalytic system. Other hydrogenation processes using membrane reactors include hydrogenation of conjugated dienes into monoenes, cyclopentadiene, 1-3 butadiene to butane, selective hydrogenation of propyne to propane, and hydrogenation of nitrobenzene to aniline.

Hydrogenation is exothermic reaction, while dehydrogenation is an endothermic one. Therefore, in some membrane reactor applications the coupling of an endothermic reaction with an exothermic one can supply the required heat. Itoh and Wu (1997) considered the endothermic dehydrogenation of cyclohexane coupled with the oxidation of permeated hydrogen. Hydrogenation of vegetable oils to improve oxidative stability and change melting point has been an industrial process since the beginning of the 20th century.

Gobina and Hughes studied the dehydrogenation of *n*-butane on one side of the membrane using nitrogen, $CO/N_2$, and $O_2/N_2$ as sweep gases on the other side of the membrane (Gobina and Hughes, 1996). They obtained conversions up to 5 to 8 times higher than the calculated equilibrium value referring to the inlet concentration (5%). The largest values were obtained with $O_2$ and CO due to the coupling reactions with the permeated $H_2$.

## 23.6 MEMBRANE BIOREACTOR

Conventional bioreactors are commonly used in the production of organic chemicals, food products, pharmaceuticals, hormones, vitamins, and other biological products. There are two types: batch and continuous. In the former, the enzymes and microbial cells are used in their soluble or "free" form, whereas in the latter, immobilized biocatalysts are used after entrapping in a fixed bed. The batch reactors suffer from the disadvantages of less efficiency, low productivity, substrate depletion, and product inhibition. Although immobilization of enzymes or microbial cells is one way of overcoming the disadvantages of using the soluble or free biocatalyst, it too has certain limitations. Depending on the immobilization methods and the particular enzyme or microorganism, losses in activity of 10% to 90% are reported in the literature. Steric hindrance, enzyme-substrate orientation, and diffusional restriction problems may occur, which could affect activity and specificity, especially with macromolecular or colloidal substrates. An alternate approach to conventional immobilization methods and development of high-rate conversion processes is the membrane bioreactor, which uses synthetic semipermeable membranes of the appropriate chemical nature and physical configuration to localize the biocatalyst within the reaction vessel or separate it from the reaction mixture.

The membrane reactor can increase the efficiency (selectivity and yield) of cellular metabolic reactions by (1) influencing the concentration of primary substrate(s) in the immediate vicinity of the biomass, (2) by controlling the concentration of favorable and unfavorable coreactant species, and (3) by removing inhibitory metabolites. Additionally, the membrane can act as a controlled interfacial support that promotes more uniform dispersion of catalyst and reactants. There are obviously many other possible permutations on the general theme.

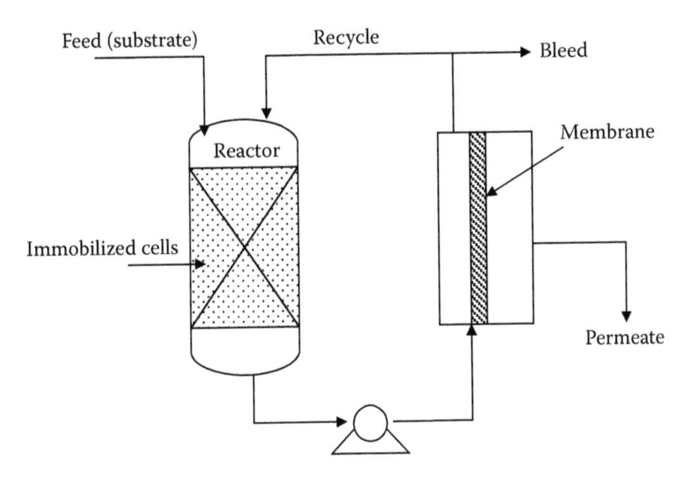


**FIGURE 23.3** Membrane recycle bioreactor. (Reproduced with permission from Nath, K., *Membrane Separation Processes*, Prentice Hall of India, New Delhi, 2008.)

### 23.6.1 Membrane Recycle Bioreactor

The basic principle of recycle membrane bioreactor is schematically shown in Figure 23.3. Typically, a reaction vessel, operated as a stirred tank reactor, is coupled in a semiclosed loop configuration via a suitable pump to a membrane module containing the appropriate semipermeable membrane. The enzyme or microbial cells are suspended in the main reaction vessel. Feed is pumped into the main reaction vessel at the same rate as the product (permeate) flow rate. The entire reaction mixture is pumped continuously through the membrane module, where the products of the reaction and any other compounds small enough to permeate through the membrane are removed. The membrane should be chosen to retain the biocatalyst while minimizing the retention of the product molecules. Since most enzymes are of the order of 10,000 to 100,000 MW, UF membranes with these MWCO Molecular weight cut off can be used. The recycle bioreactor is usually operated under completely mixed conditions, depending on the relative volume of the reaction vessel and the membrane module and the ratio of the recycling flow rate to flux. These reactors are more suited for substrate-inhibited reactions than product-inhibited reactions, when the conversion is high.

### 23.6.2 Plug-Flow Bioreactor

In a plug-flow bioreactor, the membrane separation unit is used as the bioreactor itself. The biocatalyst is entrapped on one side of the membrane, usually the permeate side, while the substrate is pumped in from the retentate side. The substrate diffuses across the membrane to react while the product and unconverted substrate diffuse back into the flowing stream on the tube side and be removed from the reactor. Hollow-fiber module is most widely used for this kind of bioreactor. Hollow fibers have high surface-area-to-volume ratios, important in any catalytic reaction. Several theoretical models for enzyme catalysis in hollow-fiber enzyme reactors (HFERs) have appeared in the literature (Cheryan and Mehaia, 1985). With the enzyme in the tube side and the feed pumped in through the shell side, the residence time distribution approximates a completely mixed model rather than the plug-flow model. With the biocatalyst in the shell side and feed pumped through the tube side, the residence time distribution approximates plug flow.

Novel membrane bioreactor configurations are continually being developed and discussed in the literature. Michaels et al. (1991) describes a hollow fiber module where the microbial culture is on the shell side. There is an inflow of nutrients and coreactants to the shell side (through a filtration device), and a shell-side outflow (also through a filtration device) bearing metabolites. The pollutant-containing stream is the feed to the lumen of the hollow fibers and the clean stream is the retentate. Stanley and Quinn (1987) provided an interesting analysis of a membrane reactor comprised of a reaction zone sandwiched between permselective membranes with different component selectivities. The object was to get greater conversion in a multireactant biochemical synthesis reaction under thermodynamically unfavorable conditions. Their analysis compares the membrane reactor with a conventional CSTR (continuous stirred tank reactor). The simple models contained therein can also be adapted to bioremediation systems. In general, most membrane applications of interest to bioremediation consist of a microbial mass in intimate contact with a membrane surface. The target solutes (probably organic) will be transported to the biomass where a decomposition reaction will occur, rendering the target solute innocuous.

## 23.7 ECONOMIC CONSIDERATIONS

Inorganic membranes have been applied to quite a large number of reactions. However, the work reported so far in the literature in many cases seems exploratory and tentative in nature and does not address economic feasibility of membrane reactors to the extent it is required to explore the commercial acceptance. The introduction of a membrane reactor is largely constrained by membrane costs and its availability. Unless production cost to the microporous membranes can be brought down considerably, opportunities will seem to be limited. For membrane reactor technology to be cost-effective, it must be shown that it can offer a combination of reduced energy and capital costs that will allow a payback of around 3 years when compared with competing technologies. If this includes the "sunken costs" of research and development programs, it can be seen that only very large processes, with substantial energy and capital benefits, will ever show a return on the initial investment.

A preliminary study of the limited number of cost assessments of zeolite membranes and zeolite membrane reactors, as reported in the literature, presents a not-so-impressive future of membrane reactor in terms of economic viability. A cost of about $3,000/m$^2$ for zeolite modules, of which 10% to 15% is contributed to the membrane itself, has been reported by Caro et al (2000). According to a feasibility study by Meindersma and de Haan (2002), for the separation of aromatics from naphtha feedstocks, the investments would not be economical unless the membrane module costs were reduced by a factor of 10 and the flux were increased by a factor of 25. Due to selectivity limitations of zeolite membranes, membrane processes are uneconomical and a considerable reduction in membrane costs is required to make such processes feasible (Tennison, 2001). Only very few processes would be able to tolerate costs of more than €1,000/m$^2$ and operation would also occur under far more demanding conditions than were generally used in modeling studies (Tennison, 2001). Apart from variable and fixed costs in processes, increasing consideration is required to be paid in future to the costs for minimizing the environmental damages of manufacturing. The disposal of toxic or dangerous waste, such as spent catalysts and chemicals, easily amounts to $250 to $750 per ton (Lange, 2001). It is therefore critical to minimize waste production and emission, both from a conservation responsibility as well as a cost perspective.

Membrane costs generally reflect a combination of the high processing costs associated with the production of inorganic membranes (for example, high temperatures and long cycle times) and the small production levels. However, current production technologies are unlikely to lead to any substantial economies of scale, because they are essentially batch processes. Although some R&D efforts are initiated to this direction, in terms of the development of hollow ceramic fiber systems for gas separation (Boudreau et al., 1999), there seems little prospect of these being applied to catalytic membrane reactors. The prospects for significant cost reductions appear bleak unless radically new approaches to both membrane and module design become available.

## 23.8 SUMMARY

The ability of membrane reactor to perform a wide variety of useful functions results in better material utilization, higher selectivity, less waste and pollution, and increased efficiency in separation and purification processes. They offer unique advantages in several specific applications yet are broadly adaptable to a wide range of uses. Significant progress has been made in the research and development of high-performance novel membrane materials, methods of preparation and characterization, reactor configurations, and understanding of the separation mechanisms to different reaction systems. But the future of catalytic membrane reactors depends on the availability of membranes that display not only excellent permeant purity, but also high flux at moderate cost and suitable stability in all phases of a typical load cycle. The issues of cocurrent and countercurrent flow, feed location and recycle, nature of sweep gas and its implications in terms of downstream separation, and effects on heat load require resolution. Moreover, considerable research and development on membrane lifetime, available space time, module design, membrane fouling, membrane poisoning, and membrane cost are essential. Based on these findings, an optimal design and an optimal choice of operating conditions should be sought. An economic analysis of a conceptual membrane reactor-based plant may then be made. This would point to the inadequacies in membrane properties and suggest areas for further improvement of the membrane materials. In the future, membrane cost and availability will be a primary factor that would determine the extent of their use in large-scale production.

## REFERENCES


Ahchieva, D., M. Peglow, S. Heinrich, et al. 2005. Oxidative dehydrogenation of ethane in a fluidized bed membrane reactor. *Appl. Catal. A-Gen.* 296(2): 176–185.

Armor, J.N. 1998. Applications of catalytic inorganic membrane reactors in refinery products. *J. Membrane Sci.* 147: 217–233.

Anshits, A.G., A.N. Shigapo', S.N. Vereshchagin, et al. 1989. Oxidative conversion of methane into C-2 hydrocarbons on silver membrane catalysts. *Kinet. Catal.*, 30(5): 1103.

Aguado S., J. Coronas, and J. Santamaría. 2005. Use of zeolite membrane reactors for the combustion of VOCs present in air at low concentrations. *Chem. Eng. Res. Des.* 83(3): 295–301.

Akin, F.T., and Y.S. Lin. 2002. Selective oxidation of ethane to ethylene in a dense tubular membrane reactor. *J. Membrane Sci.* 209(2): 457–467.

Algieri, C., P. Bernardo, G. Golemme, et al. 2003. Permeation properties of a thin silicalite–1 (MFI) membrane. *J. Membrane Sci.* 222: 181–190.

Anwu, L., Z. Hongbin, X. Guoxing, et al. 1996. Improvement of propane dehydrogenation with thin palladium-ceramic composite membrane. *Chinese J. Cat.* 17(3): 181–184.

Balachandran, U., J. Dusek, P. Maiya, et al. 1997. Ceramic membrane reactor for converting methane to syngas. *Catl. Today* 36: 265.

Barbieri, G., G. Marigliano, G. Golemme, et al. 2002. Simulation of $CO_2$ hydrogenation with $CH_3OH$ removal in a zeolite membrane reactor. *Chem. Eng. J.* 85(1): 53–59.

Bernardo, P., C. Algieri, G. Barbieri, et al. 2006. Catalytic zeolite membrane reactors for the selective CO oxidation. *Desalination* 200: 702–704.

Bottino, A., G. Capannelli, F. Cerutti, et al. 2004. Inorganic membrane reactors for the gas phase partial oxidation of toluene. *Chem. Eng. Res. Des.* 82(2): 229–235.

Boudreau, L.C., J.A. Kuck, and M. Tsapatsis. 1999. Deposition of oriented zeolite A films: In situ and secondary growth. *J. Mem. Sci.* 152(1): 41–59.

Brinkmann, T., S.P. Perera, and W.J. Thomas. 2001. An experimental and theoretical investigation of a catalytic membrane reactor for the oxidative dehydrogenation of methanol. *Chem. Eng. Sci.* 56(6): 2047–2061.

Brookes, P.R., and A.G. Livingston. 1994. Biotreatment of a point-source industrial wastewater arising in 3,4-dichloroaniline manufacture using an extractive membrane bioreactor. *Biotechnol. Prog.* 10: 65.

Caro J., M. Noack, P. Kolsch, et al. 2000. Zeolite membranes—State of their development and perspective. *Micropor. Mesopor. Mater.* 38(1): 3–24.

Centeno, T.A., and A.B. Fuertes. 2001. Carbon molecular sieve membranes derived from a phenolic resin supported on porous ceramic tubes. *Sep. Purific Technol*, 25(1–3): 379–384.

Centeno, T.A., and A.B. Fuertes. 2000. Carbon molecular sieve gas separation membranes based on poly(vinylidene chloride-co-vinyl chloride). *Carbon* 38(7): 1067–1073.
Chang, J.-S., H.-S. Roh, M.-S. Park, et al. 2002. Propane dehydrogenation over a hydrogen permselective membrane reactor. *Bull. Korean Chem. Soc.* 23(5): 674–678.
Collins, J.P., R.W. Schwartz, R. Sehgal, et al. 1996. Catalytic dehydrogenation of propane in hydrogen permselective membrane reactors. *Ind. Eng. Chem. Res.* 35: 4398.
Coronas, J., and J. Santamaria. 1999. Catalytic reactors based on porous ceramic membranes. *Catal. Today* 51: 377–389.
Deng, J., and J. Wu. 1994. Formaldehyde production by catalytic dehydrogenation of methanol in inorganic membrane reactors. *Appl. Catal. A-Gen.* 109(1): 63–76.
Diakov, V., B. Blackwell, and A. Varma. 2002. Methanol oxidative dehydrogenation in a catalytic packed-bed membrane reactor: experiments and model. *Chem. Eng. Sci.* 57(9): 1563–1569.
Gallucci, F., L. Paturzo, and A. Basile. 2004. An experimental study of $CO_2$ hydrogenation into methanol involving a zeolite membrane reactor. *Chem. Eng. Process.* 43(8): 1029–1036.
Gao, H., Y. Xu, S. Liao, et al. 1995. Catalytic polymeric hollow-fiber reactors for the selective hydrogenation of conjugated dienes. *J. Membrane Sci.* 106(3): 213–219.
Ge, S.H, C.H. Liu, and L.J. Wang. 2001. Oxidative dehydrogenation of butane using inert membrane reactor with a non-uniform permeation pattern. *Chem. Eng. J.* 84(3): 497–502.
Gobina, E., and R. Hughes. 1996. Reaction coupling in catalytic membrane reactors. *Chem. Eng. Sci.* 51: 3045–3050.
Grainger, D., and M.-B. Hägg. 2007. Evaluation of cellulose-derived carbon molecular sieve membranes for hydrogen separation from light hydrocarbons, *J. Membrane Sci.* 306: 1–2, 307–317.
Gryaznov, V.M., O.S. Serebryannikova, and Y.M. Serov. 1993. Preparation and catalysis over palladium composite membranes. *Appl. Catal. A.* 96: 15.
Gryamlov, V.M., V.I. Vedemikov, and S.G. Guliyanova. 1986. Participation of oxygen, having diffused through a silver membrane catalyst, in heterogeneous oxidation processes. *Kinet. Catal.* 27(1): 129.
Hasegawa, Y., K. Kusakabe, and S. Morooka. 2001. Selective oxidation of carbon monoxide in hydrogen-rich mixtures by permeation through a platinum-loaded Y-type zeolite membrane. *J. Membrane Sci.* 190(1): 1–8.
Hedlund, J., J. Sterte, M. Anthonis, et al. 2002. *Micropor. Mesopor. Mater.* 52(3): 179–189.
Itoh, N., and T.H. Wu. 1997. An adiabatic type of membrane reactor for coupling endothermic and exothermic reactions. *J. Membrane Sci.* 124: 213–222.
Konno, M., M. Shindo, S. Sugawara, et al. 1988. A composite palladium and porous aluminum oxide membrane for hydrogen gas separation. *J. Membrane Sci.* 37: 193.
Kusuki, Y., H. Shimazaki, N. Tanihara, et al. 1997. Gas permeation properties and characterization of asymmetric carbon membranes prepared by pyrolyzing asymmetric polyimide hollow fiber membrane. *J. Membrane Sci.* 134( 2): 245–253.
Kita, H., K. Fuchida, T. Horita, et al. 2001. Preparation of Faujasite membranes and their permeation properties. *Sep. Purif. Technol.* 25(1–3): 261–268.
Lange, C., S. Storck, B. Tesche, et al. 1998. Selective hydrogenation reactions with a microporous membrane catalyst, prepared by sol–gel dip coating. *J. Catal.* 175(2): 280–293.
Lange, J.P. 2001. Fuels and chemicals manufacturing; guidelines for understanding and minimizing the production costs. *CATTECH* 5(2): 82–95.
Lee, J.-Y., Y.-B. Choi, and H.-S. Kim. 1993. Simultaneous biodegradation of toluene and p-xylene in a novel bioreactor: Experimental results and mathematical analysis. *Biotechnol. Prog.* 9: 46.
Langhendries, G., G.-V. Baron, I.F.J. Vankelecom, et al. 2000. Selective hydrocarbon oxidation using a liquid-phase catalytic membrane reactor. *Catal. Today* 56(1–3): 131–135.
Liang, C., G. Sha, and S. Guo. 1999. Carbon membrane for gas separation derived from coal tar pitch. *Carbon* 37(9): 1391–1397.
Lie, J.A., and M.-B. Hägg. 2006. Carbon membranes from cellulose: Synthesis, performance and regeneration. *J. Membrane Sci.* 284(1–2): 79–86.
Li, S., V.-A. Tuan J.L. Falconer, et al. 2002. X-type zeolite membranes: Preparation, characterization, and pervaporation performance. *Micropor. Mesopor. Mater.* 53(1–3): 59–70.
Li, Z.Y., H. Maeda, K. Kusakabe, et al. 1993. Preparation of palladium-silver alloy membranes for hydrogen separation by the spray pyrolysis method. *J. Membrane Sci.* 78: 247.
Lu, Y., A.G. Dixon, W.R. Moser, et al. 2000. Oxygen-permeable dense membrane reactor for the oxidative coupling of methane. *J. Membrane Sci.* 170: 27–34.

McLeary, E.E., J.C. Jansen, and F. Kapteijn. 2006. Zeolite based films, membranes and membrane reactors: Progress and Prospects. *Micropor. Mesopor. Mater.* 90: 198–220.
Meindersma, G.W., and A.B. deHaan. 2002. Economical feasibility of zeolite membranes for industrial scale separations of aromatic hydrocarbons. *Desalination* 149(1–3): 29–34.
Moon, S., T. Kim, S. Park, et al. 1997. Vanadium pentoxide catalytic membrane reactor for partial oxidation of 1-butene. *Stud. Surf. Sci. Catal.* 110: 1231–1238.
Morón, F., M.P. Pina, E. Urriolabeitia, et al. 2002. Preparation and characterization of Pd-zeolite composite membranes for hydrogen separation. *Desalination* 147(1–3): 425–431.
Nagamoto, H., and H. Inoue. 1985. A reactor with catalytic membrane permeated by hydrogen. *Chem. Eng. Commun.* 34: 315.
Nagamoto H., and H. Inoue. 1986. The hydrogenation of 1,3 butadiene over a palladium membrane. *Bull. Chem Soc. Jpn.* 59: 3935–3939.
Nishiyama, N., K. Ichioka, D.-H. Park, et al. 2004. *Ind. Eng. Chem. Res.* 43: 1211.
Niwa, S.-I., M. Eswaramoorthy, J. Nair, et al. 2002. A one step conversion of benzene to phenol with a palladium membrane. *Science* 295: 105–107.
Pedernera, M., M.J. Alfonso, M. Menendez, et al. 2002. Simulation of a catalytic membrane reactor for the oxidative dehydrogenation of butane. *Chem. Eng. Sci.* 57(13): 2531–2544.
Pantazidis, A., J.A. Dalmon, and C. Mirodatos. 1995. Oxidative dehydrogenation of propane on catalytic membrane reactors. *Catal. Today* 25(3–4): 28, 403–408.
Pera-Titus, M., R. Mallada, J. Llorens, et al. 2006. Preparation of inner-side tubular zeolite NaA membranes in a semi-continuous synthesis system. *J. Membrane Sci.* 278(1–2): 401–409.
Pina, M.P., M. Arruebo, M. Felipe, et al. 2004. A semi-continuous method for the synthesis of NaA zeolite membranes on tubular supports. *J. Membrane Sci.* 244(1–2): 141–150.
Rao, M.B., and S. Sircar. 1993. Nanoporous carbon membrane for gas separation. *Gas Sep. Purif.* 7(4): 279–284.
Romanos, G.E., T.-A. Steriotis E.-S. Kikkinides, et al. 2001. Innovative methods for preparation and testing of $Al_2O_3$ supported silicalite-1 membranes. *J. Euro. Cer. Soc.* 21(2): 119–126.
Sirkar, K.K., P.S. Shanbhag, and A.S. Kovvali. 1999. Membrane in a reactor: A functional perspective. *Ind. Eng. Chem. Res.* 38: 3715–3737.
Tin, P.-S., T.-S. Chung, Y. Liu, et al. 2004. Separation of $CO_2/CH_4$ through carbon molecular sieve membranes derived from P84 polyimide. *Carbon.* 42(15): 3123–3131.
Tennison, S., 2000. Current hurdles in the commercial development of inorganic membrane reactors. *Membr. Technol.* 128: 4–9.
Tennison, S.R. 2001. Proceedings of International Workshop on Zeolitic and Microporous Membranes, July1–4, Purmerend. pp. 55–59.
Tonkovich, A.L.Y., R.B. Secker, E.L. Reed, et al. 1995. Membrane reactor/separator: A design for bimolecular reactant addition. *Sep. Sci. Technol.* 30 (7–9): 1609.
Tsuru, T., H. Shintani, T. Yoshioka, et al. 2006. A bimodal catalytic membrane having a hydrogen-permselective silica layer on a bimodal catalytic support: Preparation and application to the steam reforming of methane. *Appl. Catal. A-Gen.* 302: 78–85.
Uemiya, S., T. Matsuda and, E. Kikuchi. 1991. Hydrogen permeable palladium-silver alloy membrane supported on porous ceramics. *J. Membrane Sci.* 56: 325.
Vasudevan, M., T. Matsuura, G.K. Chotani, et al. 1987. Membrane transport and biocatalytic reaction in an immobilized yeast membrane reactor. *Ann. N.Y. Acad. Sci.* 506: 345.
Veldsink, J.W. 2001. Selective hydrogenation of sunflower seed oil in a three phase catalytic membrane reactor. *J. Am. Oil Chem. Soc.* 78(5): 443–448.
Wang, H., L. Zhang, and G.R. Gavalas. 2000. Preparation of supported carbon membranes from furfuryl alcohol by vapor deposition polymerization. *J. Membrane Sci.* 177(1–2): 25–31.
Wang, H., Y. Cong, and W. Yang. 2002. Partial oxidation of ethane to syngas in an oxygen-permeable membrane reactor *J. Membrane Sci.* 209(1): 143–152.
Wang, H., Y. Cong, and W. Yang. 2003. Investigation on the partial oxidation of methane to syngas in a tubular Ba0.5Sr0.5Co0.8Fe0.2O3–δ membrane reactor. *Catal. Today* 82(1–4): 157–166.
Xomeritakis, G., and M. Tsapatsis. 1999. Permeation of aromatic isomer vapors through oriented MFI-I type membranes made by secondary growth. *Chem. Mater.* 11(4): 875–878.
Zaman, J., and A. Chakma. 1994. Inorganic membrane reactors. *J. Membrane Sci.* 92: 1–28.
Ziegler, S., J. Theis, and D. Fritsch. 2001. Palladium modified porous polymeric membranes and their performance in selective hydrogenation of propyne, *J. Membrane Sci.* 187(1–2): 71–84.

# 24 Enzymatic Membrane Reactors in Applications of Membrane Separations Technology

## *Recent Advances*

*Gilbert M. Rios, M. P. Belleville, and D. Paolucci*

**CONTENTS**



## 24.1 INTRODUCTION

Membranes and enzymes are fundamental for all biological cells from the unicellular organisms to organisms a with higher level of evolution. Cell membranes are not only a physical boundary that separates the intracellular components from the extracellular environment, they also act as a selective barrier able to regulate what enters and exits the cell, thus facilitating the transport of materials needed for survival. The enzymes are found in all biological cells and are involved in almost all metabolic pathways. Their basic function is to increase the rate of reactions that would occur about a million times slower in their absence. Furthermore, they are selective for their substrates and speed up only a few reactions from among many possibilities.

Due to their intrinsic performance, both biological membranes and biological reactions have thus served as models for the development of industrial processes (i.e., membrane separation process and enzymatic reactors) since the middle of the 20th century. Currently the association of an enzymatic reactor with a membrane separation unit appears as a promising opportunity for human health and well-being, new environmentally friendly industrial processes, including good production in food, fine chemistry, or pharmaceutical industries, as well as new processes for environmental protection, original systems for energy production, and better-performing captors and sensors. In the following contribution, we endeavor to make a synthesis of the state of the art in this field.

## 24.2 ENZYMATIC MEMBRANE REACTOR CONCEPTS

Most of the time enzymatic reactions are carried out in classical batch reactors. At the end of the reaction the enzymes are inactivated before recovering the final products. The use of such reactors is relatively simple at any scale, with only the need to control the temperature and the pH. Nevertheless, this type of bioreactor presents a certain number of disadvantages, in particular when they must treat large quantities of raw material as is often the case in industrial practice: low productivity, high operating costs, loss of catalytic activity due to inactivation, and great variability of the quality of the products (Rios et al., 2004).

To overcome those limitations, some attempts to operate in a continuous way and reuse enzymes came out. At the beginning of the 20th century, the idea of immobilizing enzymes appeared as an interesting alternative. The objective was to ensure the localization of biological catalyst in a definite space area, the preservation of its global activity and reuse. Immobilization of enzymes in fixed or fluidized beds was proposed to conduct the reaction in a continuous mode. Fluidized beds are preferred since plugging and excessive pressure drop problems are reduced (Hamdane et al., 1988). However, a main disadvantage of the fluidized bed is the occurrence of mass transfer limitations due to diffusion at the surface or inside particles, which decreases conversion (Prazeres and Cabral, 1994). Moreover, if enzyme stability generally increases as a consequence of immobilization, specific activity more often drops due to structural modifications or hiding of active sites (Houng et al., 1992). Finally, the cost of immobilization must be accounted for.

By introducing membrane technologies, some of these disadvantages could be circumvented and the concept of enzyme immobilization could be considerably enlarged. The enzyme membrane reactor (EMR) is a specific mode for running continuous processes in which enzymes are separated from end products with the help of a selective membrane. Whatever the EMR configuration, the main objective is to ensure the complete rejection of the enzyme in order to maintain the full activity inside the reacting volume. Depending on the case, enzyme molecules may be freely circulating on the retentate side or immobilized onto the membrane surface or inside its porous structure. In the first case, the function of the membrane is to reject free enzymes in order to keep them inside the system (Figure 24.1a) and the reaction occurs on the retentate side. In the second process, the membrane becomes the catalyst media (Figure 24.1b) and reaction occurs during membrane crossing.

### 24.2.1 Free Enzyme Membrane Reactors

The free enzyme membrane reactor (FEMR), whose principle is described in Figure 24.1a, is particularly attractive to lead enzymatic reactions in which substrate and enzyme both present a size larger than those of the products. The apparatus is initially filled with enzyme and substrate solutions and then it is continuously fed with substrate while products are recovered in the permeate.

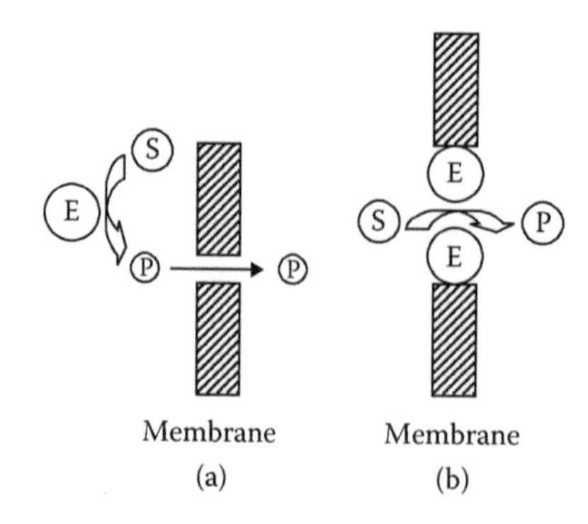


**FIGURE 24.1** Different membrane functions in an enzymatic membrane reactor: (a) rejection of the enzyme and (b) catalytic media. E, enzyme; S, substrate; P, product.

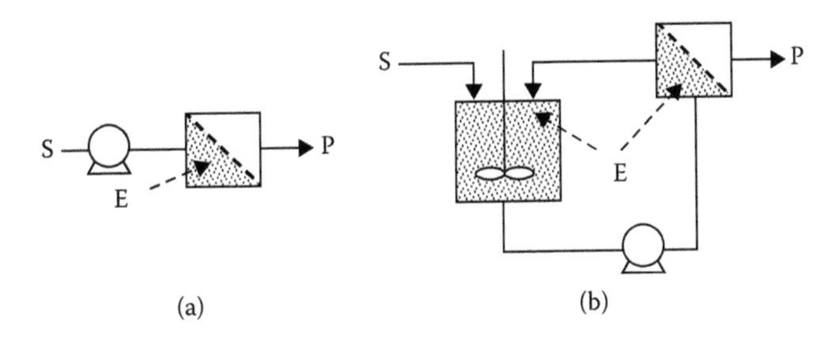


**FIGURE 24.2** Two types of immobilized enzyme membrane reactor (IEMR): (a) monophasic reactor and (b) biphasic reactor. E, enzyme; S, substrate; P, product.

As well, catalyst and nontransformed substrate are rejected by the membrane. Two types of reactors are possible. The first one is a dead-end cell reactor (DECR) (Figure 24.2a) where the whole solution is pumped toward the membrane. The second one, also called a continuous recycled membrane reactor (CRMR), works in a cross-flow filtration mode where the retentate is recycled in the reaction tank (Figure 24.2b). CRMR is generally preferred since it dramatically decreases the tendency toward concentration polarization and fouling.

The choice of the membrane is essential with regard to performance; in particular, enzyme, substrate, and product sizes must be accounted for. Because a lot of enzymes have a molecular weight between 10 and 80 kDa, ultrafiltration membranes with a molecular cutoff between 1 and 100 kDa are the most frequently used. In addition, it has also been reported by various authors that electrostatic or hydrophobic interactions between the biological molecules and the membrane could also influence the process performance (Rios et al., 2004). Mineral membranes are often preferred to organic barriers because they better endure high temperatures and pressures, thus allowing improved working and regeneration of the system (Paolucci-Jeanjean et al., 1999).

The most important advantages and drawbacks of CRMR with respect to the classical reactor configurations such as batch, fixed, or fluidized bed reactors with enzymes immobilized on various solid supports were listed by Rios et al. (2004) and reported in Table 24.1.

First, CRMR offers the advantage of a continuous mode and of the reuse of free enzymes, which ensures the economic viability of the process. In addition, inactivation and separation steps are deleted, which strongly reduces investment and processing costs. Regarding flexibility, it is worth noting that reaction and separation zones are placed in series and thus can be dealt with quite independently in order to optimize the whole process performance. Production may be modified by acting separately on temperature, pH, substrate and enzyme concentrations, fluid velocity, pressure, reactor volume, or membrane surface. A high catalyst concentration may be chosen that enables

**TABLE 24.1**
**Advantages and Drawbacks of FEMR**

| Advantages | Drawbacks |
|---|---|
| Continuous mode (substrate feeding) | Decrease of enzyme activity versus time (loss of catalyst, effect of shear stress, etc.) |
| Free enzyme | Heterogeneity of reaction conditions between the core of solution and the membrane surface |
| Retention and reuse of catalyst | Polarization layer and induced limitation |
| Reduction in substrate/product inhibition | Membrane fouling |
| Free enzyme end-product | |
| Control of product properties by enzyme and/or membrane choice | |
| Integrated process (single-step reaction/separation) | |

sharply increasing the reaction rate. In fact the main inconvenience of such a CRMR lies in flux limitations due to concentration polarization, membrane fouling, and in the possible decrease of enzyme activity over time.

It is worth noting that the use of CRMR is particularly recommended when the enzyme and substrate are larger than the products. If these conditions are not satisfied, enzymatic membranes (i.e., membranes onto which enzymes are immobilized) can represent more adequate tools for reaction processes. Such membranes, also referred to as active membranes, are at the heart of the so-called immobilized enzyme membrane reactors described below.

### 24.2.2 Immobilized Enzyme Membrane Reactors

Immobilized enzyme membrane reactors (IEMRs), whose principle is presented in Figure 24.1b, can be divided in two different systems. The first system is presented in Figure 24.3.

Feed is pumped toward the membrane module in which the substrate solution is flushed through the microporous medium by applying a transmembrane pressure. This guarantees particularly good contact conditions between substrate molecules and catalyst due to low pore size. From a structural point of view, the membrane can be considered as a specific macrosystem resulting from the assembly of swarms of microsystems: each pore could be looked upon as a particular microreactor. Thus, very efficient contacting conditions—a very high probability of capture of substrate molecules by catalyst as well as an improved contact time—can be obtained. With such a system, precise control of the reaction with minimized substrate and catalyst losses, faster reactions and higher yields, and cleaner products and lower operating costs can be expected (Paolucci-Jeanjean et al., 2005). Compared to fixed or fluidized bed reactors, IEMRs present three main advantages. First, as the product is continuously extracted from the system, IEMRs are more suitable when the product inhibits the reaction. Second, as the membrane thickness is very small compared to the bed length, the pressure drop is dramatically reduced. Third, the mass transfer is considerably increased, because, in the case of IEMRs, it is controlled by a convective mode instead of a diffusional one in the bead pores for the fixed and fluidized beds. This last difference is illustrated in Figure 24.4.

Well beyond its simple use as reactor, the IEMR can also act as a membrane contactor. In the last 10 years, interesting applications have been described in the literature. This is the case when one of the products shows low solubility into aqueous or organic solutions or when an undesirable by-product has to be extracted from the reaction device (Rios et al., 2004). This biphasic IEMR is presented in Figure 24.3b. The system involves two different vessels containing two nonmiscible solutions separated by an enzymatic membrane. The substrate diffuses from the feed solution, reacts with the biocatalyst, and the product diffuses toward the extracting solution with which it presents a very high affinity. The performance of such reactors is thus limited by mass transfer resistance. However the main advantages of biphasic IEMRs are that they do not require emulsion formation and the product can be obtained in a single phase.

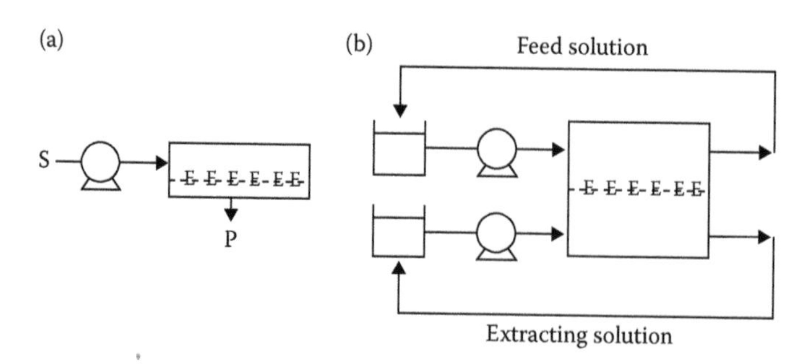


**FIGURE 24.3** Two types of immobilized enzyme membrane reactor (IEMR): monophasic reactor (a) and biphasic reactor (b) (E: enzyme, S, substrate, P: product).

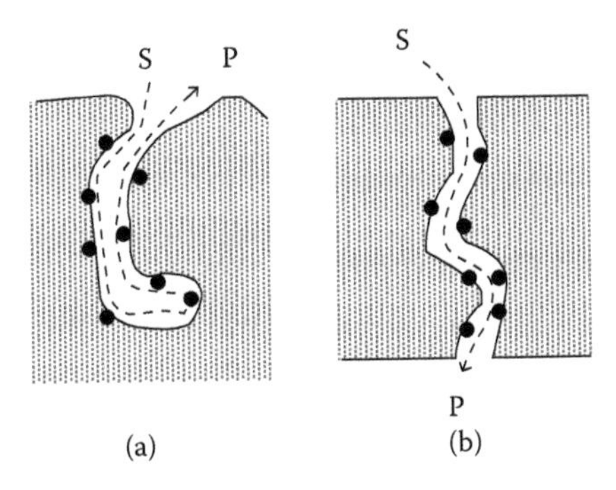


**FIGURE 24.4** Transfer mode in a bead pore (diffusion) – fixed or fluidized bed reactor (a) and in a membrane pore (convection) - IEMR (b) (•: enzyme, S, substrate, P: product).

### 24.2.3 Enzymatic Membrane Preparation

The idea of using a membrane as support for enzyme immobilization dates back more than 40 years. Since then, numerous papers have dealt with active membrane preparation. Basically, as shown in Figure 24.5, three different ways for preparing active membrane may be distinguished. The biocatalyst can be (1) entrapped within the membrane structure, (2) immobilized by gelification on the membrane surface, or (3) attached through covalent or noncovalent binding at the membrane surface.

Entrapment within the polymeric structure can be achieved by mixing the enzyme solution with the polymeric solution just before the membrane formation. The biocatalyst can be only physically entrapped or can be covalently linked to the polymer matrice to avoid enzyme leakage (Tan et al., 2002; Kanwar and Goswami, 2002). More recently, Torras et al. (2008) prepared active polysulfone membrane by adding active carbon previously loaded with dextranase before the polymeric film was cast. Alternatively, enzyme entrapment can be achieved by filtrating the enzyme solution from the support to the separating layer; the catalyst is thus retained in the porosity of the membrane support (Sousa et al., 2001; Xu et al., 2006; Wang et al., 2008). This procedure is especially attractive due to its simplicity and high enzyme loading; however, leaching risks are high. That is why such active membranes are generally used in biphasic configuration or when the feed solution permeates

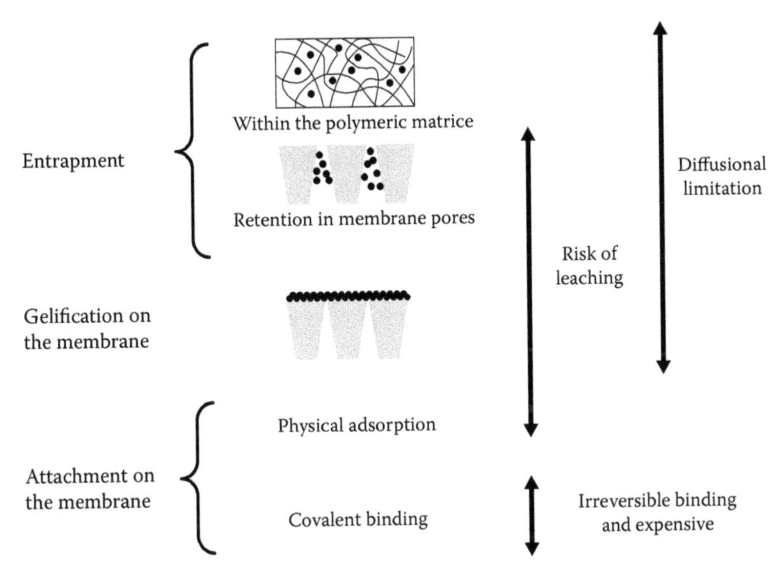


**FIGURE 24.5** Different types of active membrane preparation and their main drawbacks.

from the outside to the inside the membrane. To limit the risk of enzyme leakage, Hilal et al. (2004) suggest embedding cross-linked enzyme aggregates (CLEAs) within membrane porosity. The membrane pores are previously filled by filtration with enzymes and then the aggregates are formed by precipitation using a glutaraldehyde solution prepared in organic solvent. Whatever the procedures to entrap the enzyme within the membrane porosity, such active membranes are more convenient in the case of a reaction-limited regime rather than for a diffusion-limited regime. Immobilization by gelification on the membrane surface is simply obtained by filtration. Actually, if the pore diameter is properly chosen and the protein concentration is high enough, the rejected enzymes form a gel layer on the membrane surface (Sakaki et al., 2001; Trusek-Holownia and Noworyta, 2008). In order to improve the stability of this dynamic layer, Wang et al. (2008) cross-linked the enzyme molecules with glutaraldehyde. This immobilization method is interesting because it is very simple and leads to high enzyme loading. However, as with the entrapment method, most of the biocatalyst cannot be active due to diffusion limitation or steric obstruction that hinders access of the substrate to the catalytic site of enzyme. Wang et al. (2008) compared both immobilization methods with the same membrane reactor and observed that the performances are enhanced when the enzymes are on the surface.

The last method is enzyme attachment, which can take place by noncovalent binding (adsorption through hydrophobic or ionic interactions) or covalent binding on membrane surface. Enzyme adsorption is certainly the simplest and cheapest immobilization method because it can be achieved in one step (by immersion in enzyme solution or filtration) without the use of any activator. The method can be applied to various membrane materials from hydrophilic to hydrophobic ones; but it is generally reported and demonstrated (Shamel et al., 2007) that protein adsorption is higher on hydrophobic membranes. The major drawback of this method is linked to the risk of desorption due to the weakness of the binding force. Nevertheless, according to Tischer and Kasche (1999), immobilization via adsorption method is particularly appropriate when the membrane is used in nonaqueous solvents in which desorption phenomena may be overcome due to the low solubility of enzymes in organic solvents. Moreover, covalent attachment appears as an attractive alternative to prevent enzyme leakage since it provides virtually irreversible binding between the amino or carboxyl groups of the enzyme and functional groups of membrane through reactive groups such as carbodiimide, cyanogen bromide, and diazonium salts. It is also possible to use epichlorhydrin or a bifunctional agent such as glutaraldehyde. Most of the time, previous steps such as irradiation or chemical treatments are needed to activate the membrane surface, especially in the case of inorganic membranes. Thus covalent attachment appears as an expensive immobilization method. Moreover, the irreversibility of the linkage may be a serious drawback when the biocatalyst becomes inactive. However, a new method for enzyme immobilization onto ceramic membranes has been recently developed at the European Membrane Institute (IEM) (Belleville et al., 2001; Magnan et al., 2004). After coating the ceramic support by filtration with a polymer, the enzymes are covalently linked to the polymeric layer thanks to glutaraldehyde. Such active membranes present several advantages linked to their inorganic nature. They can be used under pressure and with a wide range of solvents, in particular in supercritical media (Lozano et al., 2004; Gumi et al., 2007) and the ceramic support can be easily regenerated when the enzymes are deactivated. Furthermore, the hydrophilic nature of the polymer coating offers a good environment to preserve enzyme toward deactivation by anhydrous conditions.

To conclude, there is no ideal route to prepare enzymatic membranes. Each method presents advantages and drawbacks. The choice has to be made on a case-by-case basis depending on the type of biocatalyst, membrane material, and reactor configuration.

## 24.3 EMR FOR PRODUCTION

Compared to the chemical route, biocatalysis processes are particularly attractive, especially in terms of pollution prevention, energy consumption, and high specificity of the reaction, which leads to high-quality products. As seen above, the association of an enzymatic reactor with a filtration unit could enhance the advantages of such processes. That is why many articles devoted to enzymatic membrane reactor applications in food, pharmaceutical, and fine chemistry areas are found in the literature.

### 24.3.1 The Food Industry

With regard to food applications, enzymes are mainly used in the beverage industry to improve fruit juice clarification and in the starch industry for sugar production, but there are also applications in the dairy and oil industries that deserve to be mentioned.

In the beverage industries, it has been demonstrated (Álvarez et al., 2000) that ultrafiltration and microfiltration represent attractive alternatives to conventional processes for juice clarification since there is no need for fining agents that generally represent a source of pollution. However, a pretreatment of juices and pulps with pectinases is usually recommended in order to reduce viscosity, improve permeate flux, and reduce membrane fouling. Most of the time this treatment is realised in a tank before juice filtration, but as reported in recent publications, EMRs, and more particularly FEMRs, could be used for the hydrolysis of pectins. According to Bélafi-Bakó et al. (2007), when this reaction is carried out in a FEMR, the product inhibition is avoided and the reactor productivity is enhanced compared to a batch system. Rodriguez-Nogales et al. (2008) studied a similar reaction during a long-term experiment (15 days) and observed a significant viscosity reduction (88% below the initial value), and suggested that both operations (pectin hydrolysis and juice clarification) can be achieved in a one-step operation with FEMR.

FERMs could also be useful in sweetener industry that produces different type of glucose syrups from starch hydrolysis. The use of α-amylases currently replaces acidic hydrolysis to convert the chains of amylose and amylopectin to smaller assimilable sugars. Paolucci-Jeanjean et al. (2000) showed that this process can be achieved in only one step in a FEMR using Termamyl™, a thermostable α-amylase supplied by NOVO Industry. The major drawbacks observed are the membrane fouling due to accumulation of high molecular weight products on one hand and enzyme inactivation on the other. However, in a more recent critical review, Sarbatly and England (2004) present different potential opportunities to prevent membrane fouling such as the use of particle as turbulence promoters or the use of an active membrane. In that latter case, the high molecular weight products that accumulate at the membrane surface are hydrolyzed and the degraded products can flow through the pore.

In the dairy industry the main applications of enzymes concern hydrolysis of whey proteins in order to produce functional and bioactive peptides on one hand and the lactose hydrolysis to improve milk digestibility on the other.

Enzymatic hydrolysis of whey proteins has demonstrated to be an excellent method to reduce their allerginicity and the peptides obtained are widely used as food ingredients in energy-providing drinks, hypoallergenic formulas, and enteral diets for children and sick adults. Although hydrolysis of protein is generally carried out in a batch reactor, this reaction can be achieved with CRMR. Recently, Cheyson et al. (2006a, b) studied the hydrolysis of whey with Protease N "Amano" (Amano Enzymes, Nagoya, Japan) in FEMR equipped with 10kDa polyethersulfone membranes (Biomax™ membrane, Milipore). They underlined the positive role of retentate temperature, which influences the solubility of protein and the viscosity of whey and thus leads to higher permeate fluxes. According to them, the reactor performances depend on the flux values and enzyme concentration. At low permeate flux as well as low enzyme concentration, enzyme activity is limited due to substrate inhibition whereas high permeate flux leads to enzyme leakage. To sum up, hydrodynamic properties of the membrane are crucial for reactor robustness.

Regarding lactose elimination, classical FEMR are not well adapted since substrate (lactose) and reaction products (glucose and galactose) have the same size. In order to obtain a high level of lactose hydrolysis, Foda and Lopez-Leiva (2000) showed that the residence time of lactose in the reactor has to be around 4 h. That is why they suggested filling the FEMR with a 4-h previously hydrolyzed solution that contained enzymes, oligosaccharide, and residual lactose, and then fed the reactor with fresh substrate. It is worth noting that the idea of feeding continuous closed-loop membrane reactors, filled at the start with a pretreated solution, offers many advantages to improve the control of system performance, as already shown in previous studies by Paolucci-Jeanjean et al. (2000). Alternatively, Grano et al. (2004) proposed to use a nonisothermal membrane reactor. As in biphasic reactors, the active membrane with immobilized β-galactosidase separates two substrate solutions. Thanks to the temperature difference applied across the membrane, a thermodialysis flux appears from the warm to cold side of the bioreactor and the lactose transformation occurs during the crossing. The reactor performance discussed in terms of percentage reduction of the production times is promising and comparable to those of batch and fluidized bed reactors previously reported in the literature (Roy and Gupta, 2003). However extrapolating from the lab-scale reactor to an industrial scale process will be probably hazardous. Another type of reactor is also proposed by Novalin et al. (2005). In that case the biocatalyst (i.e., β-galactosidase Maxilact™ or Gist-Brocades) is retained in the shell side of hollow-fiber module. Pasteurized skim milk is pumped through the hollow-fiber module (lumen side) whereas enzymatic solution is circulated inside the shell side. Due to diffusion gradient, lactose crosses the membrane to the shell side where the reaction takes place. The two major limitations (enzyme leakage and microbial growth in shell side) were solved by precoating the outside of the fiber surface with milk proteins and UV irradiation treatment of the enzymatic solution, respectively. This diffusion reactor that showed promising performances (a conversion rate around 80% within lab-scale operating conditions) needs to be investigated at the industrial scale.

Finally, the biotranformation of oils and fats in EMR is widely investigated in the literature. These bioreactors generally involve immobilized lipases or esterase that can catalyze a wide range of reactions such as hydrolysis, alcoholysis, transesterifications, aminolysis, and enantiomer resolution (Jaeger and Eggert 2002; Hasan et al., 2006). In the food area, the major application of EMRs is ester synthesis for the production of emulsifiers and aroma compounds. Due to hydrophobicity of reactant, reactions are carried out in organic media or involve the use of biphasic reactor. Examples of such reactors are discussed in the next part.

### 24.3.2 Fine Chemistry and Pharmaceuticals

Membrane systems and specially EMRs (for the production of chiral or even nonchiral molecules) are of great importance in the fine chemistry industry (Wöltinger et al., 2001).

Classical chemical processes generally lead to racemic mixtures that need further purification since some enantiomers exhibit different pharmacological activities and toxicity profiles. Thus the demand for enantiomerically pure or enriched organic compounds has increased rapidly since the last decade. Ong et al. (2008) reported an annual growth rate of 9.1% through 2008. Therefore, enantioselective biocatalysis is a key issue in the production of fine chemicals (amino acid, nonsteroid antiinflammatory drugs, chiral alcohols, vitamins, and antibiotics) for use in pharmaceuticals, agrochemicals, or cosmetics. Enzymes show impressive levels of stereospecificity, regioselectivity, and chemoselectivity and thus only catalyze reactions that lead to chiral products.

FEMRs have been used for the production of amino acid in laboratory and large-scale quantities. Degussa introduced a continuous acylase process employing EMR in 1981 and transferred this technology using soluble acylase to a production scale in 1982 (Wöltinger et al., 2005). Since then, a variety of proteinogenic (alanine, methionine, valine, tryptophan) and nonproteinogenic (O-benzylserine, norleucine, norvaline) amino acids have been prepared at Degussa in bulk quantities by resolution of the respective N-acetyl amino acids. From a racemic solution of N-acetyl-D,L-amino acid, the acylase catalyses by hydrolysis the production of L-amino acid whereas the N-acetyl-D-amino acid does not react. The L-amino acid is then isolated from the reaction mixture by ion chromatography and purified by crystallization.

In FEMRs the continuous product extraction displaces reaction equilibrium and thus enhances conversion rate. As an example, Xu et al. (2000) reported the use of lipases (Lipozyme IM, Novo Nordisk) for the production of structured lipids from medium chain triacylglycerols (MCT) and n-3 polyinsaturated fatty acids. They showed that the incorporation of polyunsaturated fatty acids into MCT was increased by about 15% over 80 h by using FEMR instead of a controlled batch reactor. They concluded that the elimination of the released medium chain fatty acids improved the incorporation of acyl donors into oils by forcing the reaction equilibrium to an upper level.

The negative effect of reaction by-product (i.e., water) is also observed during esterification reactions. In that case, ester synthesis is generally limited by water accumulation. To enhance the effectiveness of the process, Del Amor Villa and Wichmann (2005) suggested the coupling of an enzymatic reactor with a pervaporation unit. In this work, the reaction (synthesis of sugar fatty acid esters SFAEs) took place in an enzymatic membrane reactor where the catalyst remained retained by means of ultrafiltration. The permeate that contained products and unreact substrate was dehydrated by pervaporation before being recycled into the FEMR. It is also possible as suggested by Won et al. (2006) to replace the ultrafiltration membrane with a pervaporation one. In this way, the continuous extraction of water displaces the reaction equilibrium in favor of synthesis, increasing the conversion rate.

Since synthesis reactions generally involve low molecular weight substrates, and/or lead to products showing poor solubility in reaction medium, the use of biphasic IEMRs thus appears attractive. Such reactors have been successfully used to carry out the hydrolysis of babassu oil (Mercon et al., 2000) and palm oil (Knezevic et al., 2004) in order to produce fatty acids. In both studies, the enzymatic membrane was placed between two immiscible phases, the oil and an aqueous buffer where the fatty acids were released.

Besides oil hydrolysis applications, biphasic IEMRs were investigated for the resolution of racemic solutions. As an example, the resolution of (R,S) ibuprofen esters (1-heptyl-ibuprofen ester, 2-ethoxyethyl-ibuprofen esters) in biphasic IEMRs have been reported in different studies (Long et al., 2005; Wang et al., 2007). In practice, the racemic solution is recirculated along the active membrane where a lipase was immobilized by filtration. Due to its selectivity, the enzyme reacts preferentially with only one chiral compound, here the (S)-ibuprofen, which is 160 times medically more active than the (R)-form (which often causes side effects or toxicity). The same strategy was investigated for the resolution of naproxen esters by Giorno et al. (2007). They prepared two kinds of enzymatic membranes and tested them with different naproxen esters using triglycerides as reagents. Both membranes were simply obtained by filtrating the lipase solution through polyamide capillary membranes (Berghof, Germany), but in one case the aqueous lipase solution was replaced by an oil/water lipase emulsion. In that latter case the enantiomeric excess of the (S) naproxen acid was enhanced from 74% (in absence of emulsion) to 97% with methyl ester, and from 96% to 100% with butyl ester. Giorno et al. (2007) suggested that it was probably due to the presence of oil droplets that favored the mass transport of the hydrophobic substrate through the membrane.

However, due to high mass transfer resistance, the productivity of biphasic IEMRs is generally very low. They are thus limited to systems working in a reaction-limited regime. In other cases, the use of monophasic IEMRs will be more attractive.

Different synthesis reactions catalysed by α-chymotrypsin (Lozano et al., 2000) or lipase (Lozano et al., 2002; Magnan et al., 2004) have been successfully performed in organic solvent using a monophasic IEMR with a hybrid organic-inorganic catalytic membrane developed at IEM and briefly presented above. The enzyme stability was found particularly attractive since the half-life time of respectively 37 days and 202 days were obtained for α-chymotrypsin in the case of kyotorphin synthesis from N-benzoyl-L-tyrosine ethyl ester (BTEE) and L-argininamide (Arg-$NH_2$) in DMSO/ 20 mM phosphate buffer pH 7.0 (40:60 v/v) (Lozano et al., 2000), and for *Candida antarctica* lipase B (CALB) in the case of butyl butyrate synthesis from vinyl butyrate and 1-butanol in pure hexane solvent (Lozano et al., 2002).

These results confirm that the hydrophilic behavior of the inert protein insures a proper environment for the enzymes that are preserved from inactivation in nonaqueous media. Compared to a packed-bed filled with celite particles on which chymotripsyn had been previously adsorbed, the IEMR showed better performances in terms of conversion, selectivity, and enzyme stability for kyotorphin synthesis (Lozano et al., 2000).

Due to the drawbacks of the organic solvent and above all their toxicity, some authors tried to replace them by environmentally friendly solvents such as supercritical carbon dioxide (SC $CO_2$).

In a recent review, Knez (2009) presents the interest and limitations of using supercritical fluids as alternative solvents for biocatalytic processes. Despite the advantages of SC $CO_2$, which are the tunability of solvent properties and simple downstream processing features, only a few examples of EMRs using SC $CO_2$ as a reaction medium have been reported.

The first example concerns the hydrolysis of sunflower oil in a high-pressure FEMR (Knez and Habulin, 2002). In that case, enzymes were retained in the reactor thanks to a polysulfon membrane placed between two sintered plates and fitted in the reactor: a conversion close to 20% was obtained in the permeate. Unfortunately no comparison with a low-pressure reactor has been done. The second example is the butyl butyrate synthesis catalyzed by CALB in IEMR using SC $CO_2$ as solvent (Lozano et al., 2004). Several conditions of temperature and pressure were tested and results were compared to experiments performed in organic solvents (acetonitrile, acetone, and hexane). For both liquid and supercritical media, fluid density seems to be a key parameter, increasing the synthetic activity of the enzymatic membrane. All supercritical conditions gave a selectivity parameter (99%) higher than in organic solvents. As to enzymatic activity, it was also enhanced in supercritical medium since the lowest activity obtained in SC $CO_2$ was 33-fold higher than the value obtained with hexane, which was the more suitable organic solvent in this case. However it must be noted that the IEMR described here was used in a batch mode without membrane permeation, and therefore the interest of the system is thus limited. Indeed when the reaction is carried out in continuous dead-end filtration mode, the production of ester is enhanced compared to experiments carried out in batch and semibatch modes as shown by Gumi et al. (2007) for butyl laurate synthesis in SC $CO_2$. This is due to the convective transfer of substrate through the membrane, which is much more efficient than the diffusive one.

Finally, in the last example, Pomier et al. (2005) studied another type of high-pressure IEMR where SC $CO_2$ did not act as solvent media, but it was added to a very viscous vegetable oil (castor oil) in order to increase strongly fluidity and thus to allow membrane filtration. When crossing the enzyme grafted membrane, an enzymatic modification of oil (i.e., interesterification) occurred between castor oil triglycerides and methyl oleate giving notably methyl ricinoleate as product. The stability of the immobilized enzymes was checked during a 25-h period and a conversion around 30% was observed. It is worth noting that methyl ricinoleate concentration was higher in the permeate than in the retentate, showing thus that reaction happened mainly in the pores of the membrane where the contact between enzymes and substrates was much more favorable than at the internal surface.

In conclusion, the use of high-pressure EMRs is a very promising technology (SC $CO_2$ offers important advantages over organic solvents, such as ecological friendliness and ease of product recovery). However further investigations are needed to allow industrial development.

## 24.4 OTHER EMR APPLICATIONS

### 24.4.1 The Environment

The use of enzymatic systems for the treatment of water and sewage is still rather uncommon. What has already been said on mixed processes involving at the same time membranes and enzymes suggests however good prospects for the use of enzymatic technology in this field. As an introduction to the use of enzymes for such environmentally friendly processes, it is worth mentioning the recent publication of Ruggaber and Talley (2006). These authors explore the current state of use of extracellular enzymes in enhancing the bioremediation of recalcitrant substances and wastewater.

Extracellular enzymes such as proteases, lipases, and cellulases as well as peroxidases and laccases, either secreted from fungi or produced during a fermentation process, possess the ability to break down bonds within organic compounds and/or catalyze their transformation into less toxic and more biodegradable forms (degradation of TNT, phenols, PCBs, PAHs, dyes, etc.). Unlike many microorganisms, they remain effective in a wide range of pH and temperature and are not affected by toxic compounds. They have been shown to be effective to reduce pathogen counts, solid content, and to increase deflocculation in sludge.

As an example, a recent article of Lopez et al. (2004) presents an enzymatic membrane reactor (EMR) for the oxidation of xenobiotic compounds. The azo dye Orange II was selected as a model substrate of manganese peroxidase (MnP). The reaction was carried out in a stirred tank reactor coupled with a UF membrane. The membrane allowed the recovery of the enzymatic activity while both the parent dye and the degradation products could pass through. Different operational strategies (batch, fed-batch, and continuous) and parameters such as enzyme activity, $H_2O_2$ feeding rate, hydraulic retention time (in continuous operation), and dye loading rate were studied. In the best conditions, a continuous dye decolourization higher than 85% and minimal enzymatic deactivation were observed for 18 days, with an efficiency of 42.5 mg Orange II oxidized/MnP unit consumed.

In another article, Georgieva et al. (2009) studied the behavior and performance of a laccase from *Rhus vernicifera* that was immobilized on a polypropylene membrane chemically modified with chromic acid. Ethylenediamine and glutaraldehyde were used as spacer and bifunctional coupling agent, respectively. Phenol was used as substrate. At first the authors tried to discover how the immobilization procedures affected the enzyme reaction rate. Then the laccase catalytic behavior was studied in a non-isothermal bioreactor as a function of substrate concentration and temperature gradient.

Currently, even if laboratory and field studies have already shown the efficiency of enzymatic treatments, high production costs limit the widespread use of extracellular enzymes for remediation.

### 24.4.2 Microfluidic-Based Contactors/Sensors

The development of new contactors-sensors based on the concept of microfluidics is a fast-growing topic. Many papers dealing with membrane-based enzymatic sensors have been published during the last few years. Kapoli et al. (2008) used sensitive glutathione transferase (GST) for the detection of xenobiotics. More particularly a biosensor was assembled for the detection of malathion. The sensing scheme is based on the pH change occurring in a low buffer system, because of the ability of the xenobiotic to promote inhibition of the GST-catalyzing reaction. The enzyme was immobilized by crosslinking with glutaraldehyde and trapped behind a semipermeable membrane in front of the pH electrode. The results demonstrated that the immobilized enzyme behaved similar to a free enzyme.

In another study (Duong and Rhee, 2007) sensing membranes for the detection of glucose, lactate, and tyramine were successfully prepared by immobilizing enzymes and fluorophore on sol–gels. Glucose oxidase (GOD), lactate oxidase (LOD) and tyramine oxidase (TOD) were immobilized on individual or a mixture of different sol–gels (3-glycidoxypropyl-trimethoxysilane (GPTMS), methyl-triethoxysilane (MTES), and aminopropyl-trimethoxysilane (APTMS)). The oxidation of the substrates specifically catalyzed by the enzymes resulted in the reduction of the oxygen concentration, which changed the fluorescence intensity (FI) of the oxygen-sensitive ruthenium complex acting as the transducer. It is worth noting that, on the one hand, response of the sensing membranes depends on the composition of the sol–gels that were used as the supporting matrix for the immobilization, and on the other hand, these sensing membranes showed high stability due to covalent binding between the epoxy and amine groups of the sol–gels and enzymes, respectively.

On their side, Minakshi and Pundir (2008) developed an amperometric enzyme sensor for the determination of triglycerides (TGs) by mounting cellulose acetate (CA) membrane-bound commercial lipase (E.C.3.1.1.3), glycerol kinase (GK) (E.C.2.7.1.30), and glycerol-3-phosphate oxidase (GPO) (E.C.1.1.3.21) on a platinum working electrode connected to a potentiostat along with an Ag/AgCl reference electrode and Ag auxillary electrode. The biosensor measures the electrons generated from $H_2O_2$ under a potential

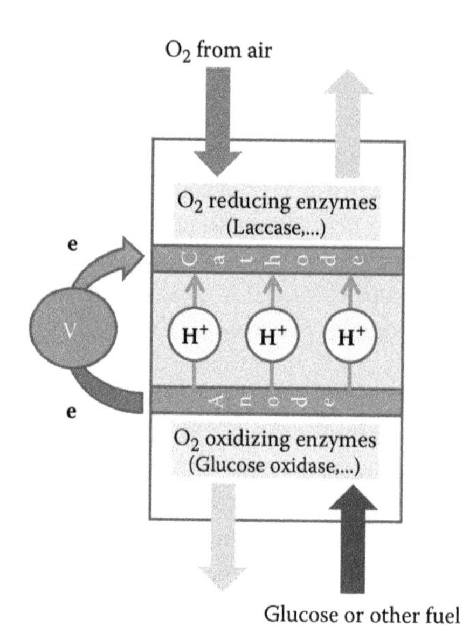


**FIGURE 24.6** Scheme of the EMPC: fuel conversion at anode, intermediate proton conducting membrane, reduction of oxygen to water at cathode.

of 0.4 V that in turn formed from triolein/triglycerides by coimmobilized lipase, GK and GPO. The concentration of triolein/TG was directly proportional to the current measured.

### 24.4.3 Energy-Related Systems

Not very far from these questions are those pertaining to miniaturized devices for energy production such as enzymatic biofuel cells. In recent years, different ideas and work within this area have emerged. Progress in miniature chip design has raised demands for implantable power sources in health care applications such as continuous glucose monitoring of diabetic patients. Pioneered by Adam Heller, miniaturized enzymatic biofuel cells convert blood sugars into electrical energy by employing for example glucose oxidase (GOx) on the anode and bilirubin oxidase on the cathode. Between the two electrodes there is a proton-conducting membrane. The principle of the enzymatic membrane fuel cell (EMFC) is shown in Figure 24.6 (Atanassov et al., 2007).

In order to match application demands it is crucial to increase lifetime and power output of the EMFCs as well as their miniaturization. As an example, in view of implant applications, Habrioux et al. (2008) developed a concentric glucose/O2 biofuel cell. Their device consisted of two carbon tubular electrodes where the enzymes (GOD and bilirubin oxidase (BOD)) were previously entrapped by an electrogenerated polypyrrole film. The electrodes, used as original conducting support for the compartment of the bioelectrodes and transport of dissolved oxygen via diffusive flow through the porosity, set up one in the other. This original concentric configuration offers the possibility to supply dissolved oxygen separately from the electrolyte in order to avoid secondary reactions. The results reported by Habrioux et al. (2008) thus demonstrated that the concentric design of this EMFC is a promising architecture for further development of microelectronic devices.

## 24.5 CONCLUSIONS

As clearly shown in the this contribution, many concepts now placed in the forefront by leading scientists and professionals as the most promising for our future are naturally tackled through the different applications that have been described, specifically microsystems and microfluidics, process and system intensification, clean energy production, and recycling and waste minimization. For these reasons we can assert that, with cells as an ultimate model, the breakthrough biomimetic approaches based on the use of enzymes and membranes undoubtedly represents a privileged direction to reach sustainable development.

## REFERENCES

Álvarez S., F.A. Riera, R. Álvarez, et al. 2000. A new integrated membrane process for producing clarified apple juice and apple juice aroma concentrate. *J.Food Eng.* 46: 109–125.

Atanassov, P., C. Apblett, S. Banta, et al. 2007. Construction of an amperometric enzymic sensor for triglyceride determination.*Electrochem. Soc. Interface* 16(2): 2–31.

Bélafi-Bakó, K., M. Eszterle, K. Kiss, et al. 2007. Hydrolysis of pectin by Aspergillus niger polygalacturonase in a membrane bioreactor. *J. Food Eng.* 78: 438–442.

Belleville, M.P., P. Lozano, J.L. Iborra, et al. 2001. Preparation of hybrid membranes for enzymatic reaction. *Sep. Purif. Technol.* 25: 229–233.

Cheison, S.C., Z. Wang, and S.-Y. Xua. 2006a. Hydrolysis of whey protein isolate in a tangential flow filter membrane reactor I. Characterisation of permeate flux and product recovery by multivariate data analysis. *J. Membrane Sci.* 283: 45–56.

Cheison, S.C., Z. Wang, and S.-Y. Xua. 2006b. Hydrolysis of whey protein isolate in a tangential flow filter membrane reactor II. Characterisation for the fate of the enzyme by multivariate data analysis. *J. Membrane Sci.* 286: 322–332.

Del Amor Villa, E.M., and R. Wichmann. 2005. Membranes in enzymatic synthesis of biotensides from renewable sources. *Catal. Today* 104: 318–322.

Duong, H.D., and J.I. Rhe. 2007. Preparation and characterization of sensing membranes for the detection of glucose, lactate and tyramine in microtiter plates. *Talanta* 72(4): 1275–1282.

Foda, M.I., and M. Lopez-Leiva. 2000. Continuous production of oligosaccharides from whey using a membrane reactor. *Process Biochem.* 35: 581–587.

Georgieva, S., T. Godjevargova, M. Portaccio, et al. 2008. Advantages in using non-isothermal bioreactors in bioremediation of water polluted by phenol by means of immobilized lactase from Rhus vernicifera. *J. Mol. Catal. B-Enzym.* 55(3–4): 177–184.

Grano, V., N. Diano, S. Rossi, et al. 2004. Production of low-lactose milk by means of nonisothermal bioreactors. *Biotechnol. Prog.* 20: 1393–1401.

Giorno, L., E. D'Amore, R. Mazzei, et al. 2007. An innovative approach to improve the performance of a two separate phase enzyme membrane reactor by immobilizing lipase in presence of emulsion. *J. Membrane Sci.* 295: 95–101.

Gumí, T., D. Paolucci-Jeanjean, M.P. Belleville, et al. 2007. Enzymatic membrane reactor involving a hybrid membrane in supercritical carbon dioxide. *J. Membrane Sci.* 297: 98–103.

Habrioux, A., G. Merle, K. Servat, et al. 2008. Concentric glucose/O2 biofuel cell. *Mat. Sci. Eng. C* 28: 932–938.

Hamdane, M., A.M. Wilhem, and J.P. Riba. 1988. Modelling of a fluidized bed immobilized enzyme reactor. Application to the hydrolysis of maltodextrins. *Chem. Eng. J.* 39: B25–B30.

Hasan, F., A.A. Shah, and A. Hameed. 2006. Industrial applications of microbial lipases. *Enzyme Microb. Tech.* 39: 235–251.

Hilal, N., R. Nigmatullin, and A. Alpatova. 2004. Immobilization of cross-linked lipase aggregates within microporous polymeric membranes. *J. Membrane Sci.* 238: 131–141.

Houng, J.Y., J.Y. Chiou, and K.C. Chen. 1992. Production of high maltose syrup using an ultrafiltration reactor. *Bioprocess Eng.* 8: 85–90.

Jaeger, K.-E., and T. Eggert. 2002. Lipases for biotechnology. *Curr. Opin. Biotech.* 13: 390–397.

Kanwar, L., and P. Goswami. 2002. Isolation of a Pseudomonas lipase produced in pure hydrocarbon substrate and its application in the synthesis of isoamyl acetate using membrane-immobilised lipase. *Enzyme Microb. Tech.* 31: 727–735.

Kapoli, P., I.A. Axarli, D. Platis, et al. 2008. Engineering sensitive glutathione transferase for the detection of xenobiotics. *Biosens. Bioelectron.* 24(3): 498–503.

Knez, Z. 2009. Enzymatic reactions in dense gases. *J. Supercrit. Fluid.* 47: 357–372.

Knez, Z., and M. Habulin. 2002. Compressed gases as alternative enzymatic-reaction solvents: A short review. *J Supercrit. Fluid.* 23(1): 29–42.

Knezevic, Z., G. Kukic, M. Vukozic, et al. 2004. Operating regime of a biphasic oil/aqueous hollow fibre reactor with immobilized lipase for oil hydrolysis. *Process Biochem.* 39: 1377–1385.

Long, W.S., P.C. Kow, A.H. Kamaruddin, et al. 2005. Comparison of kinetic resolution between two racemic ibuorofen esters in an enzymatic membrane reactor. *Process Biochem.* 40: 2417–2425.

López, C., M.T. Moreira, G. Feijoo, et al. 2004. Dye decolorization by manganese peroxidase in an enzymatic membrane bioreactor. *Biotechnol. Progr.* 20(1): 74–81.

Lozano, P., T. De Diego, M.P. Belleville, et al. 2000. A dynamic membrane reactor with immobilized α-chymotrypsin for continuous kyotorphin synthesis in organic media. *Biotech. Lett.* 22: 771–775.

Lozano, P., A.B. Pérez-Marin, T. De Diego, et al. 2002. Active membranes coated with immobilized Candida antarctica lipase B: Preparation and application for continuous butyl butyrate synthesis in organic media. *J. Membrane Sci.* 201: 55–64.

Lozano, P., G. Villora, D. Gomez, et al. 2004. Membrane reactor with immobilized Candida antarctica lipase B for ester synthesis in supercritical carbon dioxide. *J. Supercrit. Fluid.* 29: 121–128.

Magnan, E., I. Catarino, D. Paolucci-Jeanjean, et al. 2004 Immobilisation of lipase on a ceramic membrane: Activity and stability. *J. Membrane Sci.* 241: 161–166.

Mercon F., G.L. Sant Anna, and R. Nobrega. 2000. Enzyme hydrolysis of babassu oil in a membrane reactor. *J. Am. Oil Chem. Soc.* 77(10): 1043–1048.

Minakshi, K., and C.S. Pundir. 2008. Construction of an amperometric enzymic sensor for triglyceride determination. *Sens. Actuators B* 133(1): 251–255.

Novalin, S., W. Neuhaus, and K.D. Kulbe. 2005. A new innovative process to produce lactose-reduced skim milk. *J. Biotechnol.* 119: 212–218.

Ong, A.L., A.H. Haru Kamaruddin, S. Bhatia, et al. 2008. Enantioseparation of (R,S)-ketoprofen using Candida antarctica lipase B in an enzymatic membrane reactor. *J. Sep. Sci.* 31: 2476–2485.

Paolucci-Jeanjean, D., M.P. Belleville, G.M. Rios, et al. 1999. Why on earth can people need continuous recycle membrane reactors for starch hydrolysis? *Starch/Stärke* 51(1): 25–32.

Paolucci-Jeanjean, D., M.P. Belleville, and G.M. Rios. 2000. The effect of enzyme concentration and space time on the performance of a continuous recycle membrane reactor for one-step starch hydrolysis. *Biochem. Eng. J.* 5: 17–22.

Paolucci-Jeanjean, D., M.P. Belleville, and G.M. Rios. 2005. Biomolecule applications for membrane-based phase contacting systems: distribution, separation and reaction. A first state of the art. *Chem. Eng. Res. Des.* 83(A3): 302–308.

Pomier, E., J. Galy, D. Paolucci-Jeanjean, et al. 2005. A new reactor design combining enzyme, membrane and SC $CO_2$. Application to castor oil modification. *J. Membrane Sci.* 249: 127–132.

Prazeres, D.M.F., and J.M.S. Cabral. 1994. Review: Enzymatic membrane bioreactors and their applications. *Enzyme Microb. Tech.* 16: 738–750.

Rios, G.M., M.P. Belleville, D. Paolucci, et al. 2004. Progress in enzymatic membrane reactors—A review. *J. Membrane Sci.* 242: 189–196.

Rodriguez-Nogales, J.M., N. Ortega, M. Perez-Mateos, et al. 2008. Pectin hydrolysis in a free enzyme membrane reactor: An approach to the wine and juice clarification. *Food Chem.* 107: 112–119.

Roy, I., and M.N. Gupta. 2003. Lactose hydrolysis by Lactozym™ immobilized on cellulose beads in batch and fluidized bed modes. *Process Biochem.* 39: 325–332.

Ruggaber, T.P., and J.W. Talley. 2006. Enhancing mioremediation with enzymatic processes: A review. *Pract. Periodical Haz. Toxic Radioactive Waste Mgmt.* 10(2): 73–85.

Sakaki, K., L. Giorno, and E. Drioli. 2001. Lipase-catalyzed optical resolution of racemic naproxen in biphasic enzyme membrane reactors. *J. Membrane Sci.* 184: 27–38.

Sarbatly, R., and R. England 2004. Critical review of membrane bioreactor system used for continuous production of hydrolyzed starch. *Chem. Biochem. Eng. Q.* 18: 155–165.

Shamel, M.M., K.B. Ramachandran, M. Hasan, et al. 2007. Hydrolysis of palm and olive oils by immobilised lipase using hollow fibre reactor. *Biochem. Eng. J.* 34: 228–235.

Sousa, H.A., C. Rodrigues, E. Klein, et al. 2001. Immobilisation of pig liver esterase in hollow fibre membranes. *Enzyme Microb. Tech.* 29: 625–634.

Tan, T., F. Wang, and H. Zhang. 2002. Preparation of PVA/chitosan lipase membrane reactor and its application in synthesis of monoglyceride. *J. Mol. Catal. B-Enzym.* 18: 325–331.

Tischer, W., and V. Kasche. 1999. Immobilized enzymes: Crystals or carriers? *Tibtech.*17: 326–335.

Torras, C., D. Nabarlatz, G. Vallot, et al. 2008. Composite polymeric membranes for process intensification: Enzymatic hydrolysis of oligodextrans. *Chem. Eng. J.* 144: 259–266.

Trusek-Holownia, A., and A. Noworyta. 2008. Peptides removing in enzymatic membrane bioreactor. *Desalination* 221: 543–551.

Wang, Y., Y. Hu, J. Xu, et al. 2007. Immobilization of lipase with a special microstructure in composite hydrophilic CA/hydrophobic PTFE membrane for the chiral separation of racemic ibuporofen. *J. Membrane Sci.* 293: 133–141.

Wang, Y., J. Xu, G. Luo, et al. 2008. Immobilization of lipase by ultrafiltration and cross-linking onto the polysulfone membrane surface. *Bioresource Technol.* 99: 2299–2303.

Wöltinger J., Drauz K., and A.S. Bommarius. 2001. The membrane reactor in the fine chemicals industry. *Appl. Catal. A-Gen.* 221: 171–185.

Wöltinger, J., A. Karaus, W. Leuchtenberger, et al. 2005. Membrane reactors at Degussa. *Adv. Biochem. Eng. Biot.* 92: 289–316.

Won, K., J.K. Hong, K.-J. Kim, et al. 2006. Lipase-catalysed enantoselective esterification of racemic ibuprofen coupled with pervaporation. *Process Biochem.* 41: 267–269.

Xu, X., A. Skands, G. Jonsson, et al. 2000. *Biotechnol. Lett.* 22: 1667–1671.

Xu, J., Y. Wang, Y. Hu, et al. 2006. Candida rugosa lipase immobilized by a specially designed microstructure in the PVA/PTFE composite membrane *J. Membrane Sci.* 281: 410–416.

# 25 Membranes for Fuel Cell Application

## *Hybrid Organic and Inorganic Membranes*

*Sumedha Sharma and Kaustubha Mohanty*

**CONTENTS**



## 25.1 INTRODUCTION

Energy management and environmental protection are two parallel challenges of the present day that are chiefly attributed to the ever escalating economic and industrial growth that the world witnesses. On one hand we have the increasing energy requirement against the plummeting fossil fuel reserves while on the other hand is the key question of sustainability with burgeoning environmental pollution and issues such as global warming. While world energy estimates project that the current reserves may be sufficient for a few hundred years, the imbalanced demand-supply equation will be further unfavorably skewed due to our growing population and increasing energy requirements, leaving us with a few decades of energy supply only. However, even if our fossil reserves were sufficient, the pollution concerns arising out of their use make the search for alternative energy sources and utilization techniques even more imperative. Moreover, the growing industry for portable electronic devices and the increasing need for longer durability have led to widespread industrial and academic research to look beyond the aging lithium battery technology and start exploring fuel cells.

Fuel cells are electrochemical conversion devices that generate energy from fuel through a noncombustive process and have attracted recent research initiatives due to their benign emissions and higher electrical efficiency compared to heat engines in addition to their portability of use. They are based on a reverse electrolysis principle where the fuel is fed at the anode side and oxidant is fed at the cathode side. Half reactions take place at the two electrodes separated by an electrolyte. The mobile ions produced due to the half reaction pass through the electrolyte to the other electrode to participate in the half reaction of that electrode. Fuel cells are different from batteries due to their mode of continual reactant consumption requiring replenishment, while batteries store electrical energy chemically in a closed system. They do not require recharging, provide much longer operating life, are lighter or equal in weight per unit of power output, and provide much higher power-density compared to batteries. Fuel cells are of different types and may be classified depending on the operation temperature: low, medium, or high temperature; mobile ion; electrolyte: aqueous, molten, or solid; and sometimes the type of catalyst used: chemical or biological. Research trials over the world present a distinctive advantage and ease of portability when using a solid electrolyte fuel cell. The solid electrolyte used in these fuel cells is a membrane and the mobile ion is proton ($H^+$). The fuel dissociates at the anode and the protons are conducted through the membranes to the cathode. The electrons travel in an external circuit due to the electrically insulating

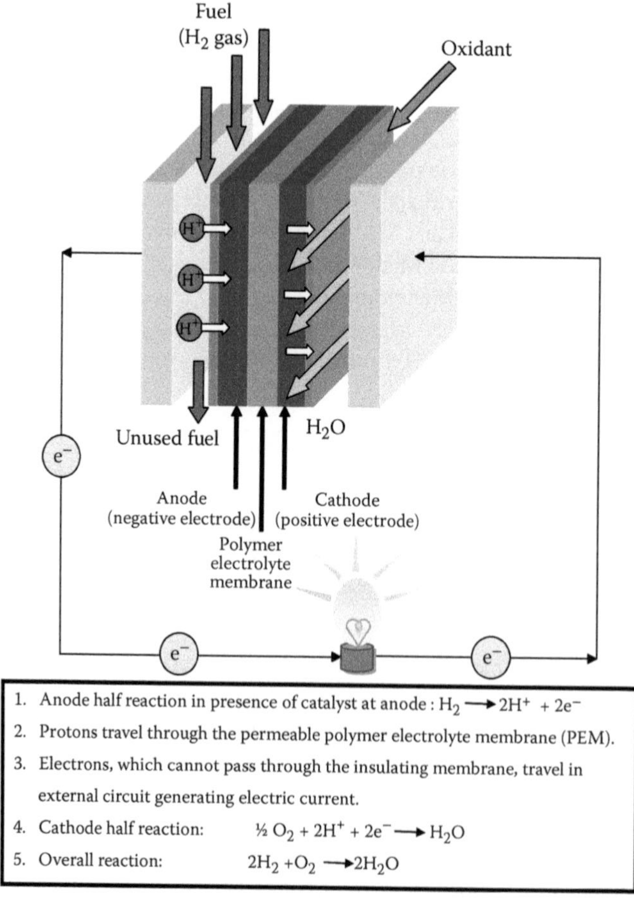


**FIGURE 25.1** Electrolyte membrane in a schematic hydrogen fuel cell.

nature of the membrane. At the cathode, the oxygen molecules from air react with the electrons that have traveled through the external circuit and the protons to form water as waste when pure hydrogen is used as fuel. Figure 25.1 shows the working of a typical proton exchange membrane fuel cell. Carbon dioxide and water are produced as waste at the cathode when other hydrocarbons such as methanol are used as fuel instead of hydrogen as in direct and indirect methanol fuel cells.

*Polymer electrolyte membrane fuel cells* (PEMFCs), commonly *proton exchange membrane fuel cells* and liquid methanol fed PEMFCs called *direct methanol fuel cells* (DMFCs) are the types that presently assume competitive importance from the application, particularly as portable, durable, and micropower sources, point of view. Hydrogen PEMFCs have found a place in vehicles and mobile application and in lower power combined heat and power systems. DMFCs on the other hand find the most suitable application in low-power portable electronic systems working for long time periods.

## 25.2 MEMBRANES AS FUEL CELL ELECTROLYTES

Membranes, in general, find their application in innumerable practical aspects in the world today. From purification for obtaining potable water to effluent treatment, industrial separation processes to specialized applications, membranes have penetrated various spheres of human and industrial application. The application of membranes as an electrolyte for fuel cells is one of the extraspecialized ones.

Membrane electrolyte in the fuel cells is usually sandwiched between the anode and the cathode and allows the transfer of mobile ions through it. Figure 25.1 illustrates the working of a polymer electrolyte membrane fuel cell.

### 25.2.1 Membrane Characteristics for Fuel Cell Applications

Membranes for proton exchange membrane fuel cells should possess the following characteristics:

- High ion conductivity while being electrically insulating
- Chemical, thermal, and mechanical stability
- Low permeability to gas and reactants
- Low cost and ready availability

Although varying with applications, common requirements for a polymer electrolyte membrane in direct methanol fuel cells include:

- Operation at high temperature
- Low methanol crossover (MCO) ($<10^{-6}$ mol min$^{-1}$ cm$^{-1}$) or low methanol diffusion coefficient in the membrane ($<5.6 \times 10^{-6}$ cm$^2$ s$^{-1}$ at 25°C) (Aricó et al., 2001)
- High ionic conductivity ($>80$ mS cm$^{-1}$) (Aricó et al., 2001)
- High chemical and mechanical durability, especially at a temperature of >80°C (for better CO tolerance
- Low ruthenium crossover (if the anode catalyst contains ruthenium)
- Low cost

Continuing research is focused toward optimizing the membrane performance in keeping with these requirements and lowering the cost in order to facilitate large-scale application.

### 25.2.2 Working of a Polymer Electrolyte Membrane

Scientists have long struggled to find and explain how this central component of the fuel cell, the proton exchange membranes, actually works. Knowledge of the working of membranes is important in order to develop new materials that may be less expensive and/or have improved properties. The structural conundrum was satisfyingly solved in 2008 when a team of researchers at the U.S. Department of Energy's Ames Laboratory proposed the working model of a fuel cell membrane.

This model, proposed by Ames Laboratory scientists Klaus Schmidt-Rohr and Qiang Chen, concentrates specifically on Nafion®, the most popular and widely used perfluorinated polymer film with the distinguishing characteristic of high selective permeability to water and protons. Schmidt-Rohr suggests that Nafion has a closely packed network of nanoscale cylindrical water channels running in parallel through the material. It is evidenced from nuclear magnetic resonance (NMR) studies that Nafion molecules consist of a rigid backbone structure with hairlike "defects" along the polymer chain. However, the arrangement of these molecules was not known. Some theories suggest a weblike network of channels while others propose spheroidal water clusters. To solve the structural mystery, Schmidt-Rohr focused on mathematical modeling of small-angle X-ray and neutron scattering, or SAXS/SANS. X-ray or neutron radiation is scattered by the sample and the resulting scattering pattern is analyzed to provide information about the size, shape, and orientation of the components of the sample on the nanometer scale. The theory proposed by Schmidt-Rohr and Qiang Chen is that these hydrophobic backbone structures cluster together to form long rigid cylinders about 2.5 nm in diameter with the hydrophilic hairlike structures toward the inside of water-filled tubes. Schmidt-Rohr has used a multidimensional Fourier transformation algorithm and was able to show that his model of long, densely packed channels closely matches the known scattering data of Nafion. The results from mathematical modeling of other proposed structures, in which the water clusters have other shapes, did not match the measured scattering curves. On the basis of these studies, although the cylinders in different parts of the material may not align perfectly, they do interconnect to create water channels passing through the membrane material, which can be tens of microns thick. It is this structure of relatively wide diameter channels, densely packed and running mostly parallel through the material that helps explain how water and protons can diffuse so easily through Nafion. Their model is also successful in explaining conductivity through membranes below the freezing point of water, as water continues to diffuse in the smaller diameter pores while it would freeze in larger channels.

### 25.2.3 Development of Membrane Material for a Fuel Cell

Polymer or hydrocarbon-based membranes were first tested as electrolytes in PEMFCs for Gemini space missions. Sulfonated phenol-formaldehyde resins, sulfonated poly-(styrene-divinylbenzene) copolymers, grafted polystyrene sulfonic acid membranes (*g*-PSSA), and so forth, were used as membrane materials initially but were found to be chemically weak. PEMFCs using these membranes showed poor performance and low lifetimes and durability. Further research for better materials resulted in a presently popular material, Nafion. Nafion is a perfluorinated sulfonic acid (PFSA) membrane that was developed in the mid-1960s by DuPont. It is based on an aliphatic perfluorocarbon sulfonic acid, and became popular due to its excellent physical properties, chemical stability, good conductivity, mechanical strength, and oxidative stability in both wet and dry states. A Nafion-based PEMFC was used for the NASA 30-day Biosatellite space mission (LaConti et al., 2003). Many commercially available PFSA membranes are synthesized by copolymerization of tetrafluoroethylene and perfluorinated vinyl ether sulfonyl fluoride, followed by hydrolysis of the sulfonyl fluoride groups (Banerjee and Curtin, 2004). The membranes are prepared from the as-synthesized material through extrusion or casting and hydrolysis techniques. Initial studies indicated a high durability up to 60,000 h of continuous operation at 43°C to 82°C for PEMFC stacks using Nafion 120, having a thickness of 250 μm and equivalent weight 1,200 (Steck, 1995). The perfluorinated backbone of Nafion imparts high chemical and thermal stability and the pendant sulfonic acid ($-SO_3H$) groups impart strong acidic characteristics that facilitate proton transport. The acid capacity of the membrane strongly influences the water uptake and the proton conductivity of the membrane is largely dependent on the water content.

However, Nafion membranes were eventually found to have several disadvantages. One major disadvantage associated with the use of Nafion is the drastic decrease of conductivity above 80°C due to evaporation of water, low mechanical strength at higher temperature, and moderate glass transition temperature. Parallel to this, there is an increasing demand for maximizing performance efficiency, improving water management and proton conductivity of PEMFCs by using thinner and lower equivalent weight PEMs (Borup et al., 2007), which guides the search for advanced membrane material. Also, Nafion membranes have several other disadvantages in DMFC applications such as high methanol crossover (MCO), high Ru crossover (for Pt-Ru anodes), and high cost in addition to the high humidification requirements. After crossover, the ruthenium redeposits on the cathode, decreasing the fuel cell performance. Methanol crossover, which reduces efficiency of oxygen reduction reaction by mixed potential effect, is the major detrimental factor for successful application of Nafion in DMFCs.

These chief disadvantages of the known ionomeric membranes in terms of low-proton conductivity at low-relative humidity, high reactant crossover, CO catalyst poisoning, water management, and poor mechanical properties above 130°C, prove to be the main obstacles to greater commercialization of polymer electrolyte fuel cells and have therefore attracted considerable research efforts. Operating temperature is a key limitation that seriously affects performance of PEMs. In hydrogen fuel cells, even trace amounts of carbon monoxide in the fuel may result in catalyst poisoning at low operating temperatures. To avoid this contamination, the hydrogen fuel must pass through a rigorous and expensive purification process that makes it an expensive alternative. Besides, operation at higher temperatures is preferred to derive the benefits of faster electrochemical reactions, reduction of CO catalyst poisoning, improved thermal management, simplified water management system, and flexibility of working with dried membranes (Costamagna et al., 2002; Curtin et al., 2004; Hogarth et al., 2005; Li et al., 2003; Mallant, 2003; Stamatin et al., 2006; Steele and Heinzel, 2001; Yang et al., 2001a).

These disadvantages led to increased research for developing new membrane materials that will improve performance and decrease cost. Four different methods of achieving this have been studied and proposed:

1. Developing membranes from new ionomers with desired properties
2. Developing membranes with suitable oxygenated acids (e.g., phosphoric or sulfuric acid) solubilized in polymer matrix bearing basic groups (e.g., polybenzimidazole)
3. Improving the characteristics of the known ionomeric membranes by dispersing low solubility materials (e.g., heteropolyacids) or insoluble particles such as metal oxides, phosphonates, or zirconium phosphates, in the polymer matrix
4. Developing membranes from nonproton-conducting polymeric matrix filled with ionomers or inorganic materials having high proton conductivity

The first two approaches have guided experimenting with materials besides the well-known Nafion while the latter two developments have led to what we call hybrid or composite membranes.

To overcome the material cost of Nafion, other polymers (Savadogo, 1998) and later their sulfonated derivatives were considered for PEM applications. Similarly, the disadvantage associated with Nafion in terms of temperature and water management led to the use of alternatives such as phosphoric acid doped polybenzimidazole (PBI), which can tolerate temperatures over 150°C. It also led to the development of new proton-conducting membranes such as poly(oxy-1,4-phenylene-oxy-1,4-phenylene carbonyl-1,4-phenylene) (PEEK) (Wang and Roovers, 1993; Kobayashi, 1994), poly(phenylquinoxalines) (PPQ) (Xing, 2004), poly(1,4-phenylene) (Kobayashi, 1998), poly(4-phenoxybenzoyl-1,4-phenylene) (PPBP), poly(phenylene sulfide) (Qi et al., 1998), and so forth. The chemical structures of some PFSA and hydrocarbon based copolymers used for fuel cell membranes are shown in Figure 25.2 (Borup et al., 2007). Even though many of the non-Nafion polymers are lower in cost and some also reduce methanol crossover, they do not show any significantly high ionic conductivity (Gil, 2004).

Nafion® ($m^3 \geq 1, n = 2, x = 5–13.5$)
Flemion® ($m = 0, 1, n = 1–5$)
Aciplex® ($m = 0, 3, n = 2–5, x = 1.5–14$)

PFSA membranes

$R_1, R_2, R_3, R_4$ = alkyls, halogens, OR, $CF = CF_2$, CN, $NO_2$, OH

BAM copolymers

ETFE grafted polystyrene acid

Sulfonated poly(ether ether ketone)

Sulfonated poly(arylene ether sulfone)

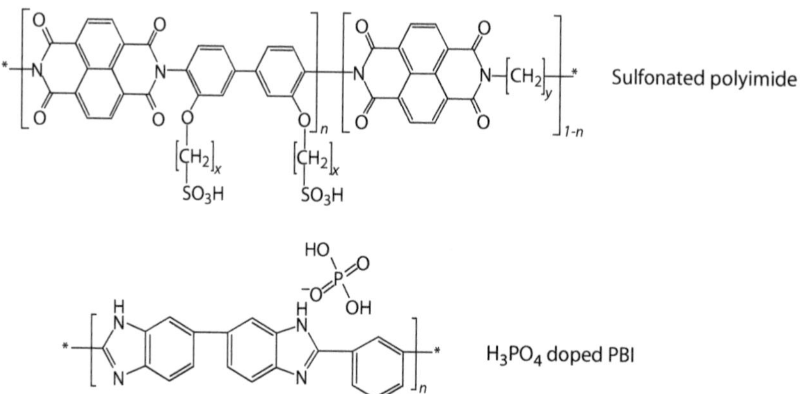


**FIGURE 25.2** Chemical structures of polymer electrolyte membranes. (Reprinted with permission from Borup, R., et al. *Chem. Rev.* 107: 3904–3951. Copyright 2007. American Chemical Society.)

Further search for membranes with improved properties has resulted in the development of hybrid materials or composite materials for membrane applications (Alberti and Casciola, 2003). Membranes containing metallic blocking layers have been proposed (Wang et al., 2006) and, organic–inorganic composite membranes containing Zr phosphonates (Watanabe, 2003), Sn-doped mordenites (Yasumoto, 1997), zeolites (Samant, 2004), and silica (Tricoli and Nanetti, 2003), have also been investigated. This research has guided PEMs technology toward using new proton exchangers based on solid acids or superacids supported on nanometric zirconia (LaConti et al., 2003) as alternatives to sulphonated or phosphorilated polyelectrolytes.

Details of the ongoing development in improving specific membrane properties are discussed below.

#### 25.2.3.1 Conductivity

Proton conductivity in membranes is the most important element for fuel cell performance. It is usually determined by "proton hopping" (Grottus) and migration of hydrated protons [$H^+(H_2O)_n$ species] (Arimura, 1999) and can be improved by the use of additives. Several different types of additives have been utilized for this purpose:

- Catalysts facilitating proton transport: Pd (Kim et al., 2003), Pt-Pd and Pt-Ag (Ma et al., 2003)
- Electronic conductors: electronically conducting polyaniline (Rodriguez et al., 2004)
- Water retention aid: zirconium phosphate (Vaivars et al., 2004; Yang et al., 2001b)
- Structure modifier: hydroxyl apatite (Park and Yamazaki, 2005) and zeolite (Tricoli and Nanetti, 2003)
- Acid sites: phosphotungstenic acid (Antonucci et al., 1999), molybdophosphoric and silicotungstinic acid (Dimitrova et al., 2002; Tazi and Savadogo, 2000)

#### 25.2.3.2 Methanol Crossover

In DMFCs, methanol crossover can be a major problem seriously affecting the fuel cell performance. Methanol crossover is the diffusion of the fuel methanol through the membrane without reaction. This adversely affects the fuel cell performance by poisoning of the cathode electrode, overall efficiency reduction due to poor fuel efficiency, and reduction in electrode potential due to methanol oxidation at the cathode. In order to reduce the diffusion of methanol, some physical or passive approaches can be used. These include increasing the membrane thickness, equivalent weight, cathode pressure and current density, and decreasing methanol concentration and operating temperature. However, most of these approaches adversely affect the DMFC performance and hence lack much potential in application.

Membranes based on sulphonated polymers (Rikukawa and Sanui, 2000) such as polyether sulphone (PES) (Ueda et al., 1993; Wang et al., 2002), polyimides with a six-member imide ring (Genies et al., 2001), poly(oxy-2,6-dimethyl-*p*-phenylene) (Kruczek and Matsuura, 1998), poly(ether ketone) (PEK) (Xiao et al., 2002), and PBI (Gieselman and Reynolds, 1992) lower the methanol crossover due to their inherent low compatibility with methanol and rigid backbone structure. A more effective approach is the use of inorganic additives to modify the structure of fluorinated and nonfluorinated (PBI, sPEEK, etc.) membranes. The use of additives can significantly reduce methanol crossover without decreasing proton conductivity. For example, impregnation of silica supported heteropolyacid into Nafion membrane has been found to show about 50% to 80% decrease in methanol crossover as compared to Nafion (Kim and Chang, 2007).

#### 25.2.3.3 Thermal Stability

Fluorinated membranes have a limited thermal flexibility. However, a wide range of operating temperatures is usually desired in order to make fuel cell systems more widely applicable. Moreover, a high operating temperature is preferred to take advantage of the following benefits:

- Improved kinetics of electrochemical reaction (methanol oxidation)
- Improved tolerance of the electrode catalyst to carbon monoxide poisoning
- Improved performance with enhanced proton conductivity
- Improved thermal management by possible use of cogenerated heat
- Simplification of design of cooling system
- Improved water management and flexibility of working with dry membranes

Proton-conducting hydrocarbon and heterocyclic-based polymer membranes are more stable at higher temperature. Examples of such membranes are those developed on the basis of acidic and basic polymers such as poly(ether ketones) (PEK), sulphonated PEEK (sPEEK), poly(arylene thio-ethilene sulfone) (Rodrigues et al., 2005), sulphonated polysulfones (sPSU) (Coplan and Goetz, 1983; Lufrano et al., 2000), polystyrene sulphonate (PSS) (Chen et al., 2004), PPQ and PBI (Samms et al., 1996; Staiti et al., 2001; Wainright et al., 1995; Wainright et al., 1997; Wang et al., 1996), and styrene grafted and sulfonated poly(vinylidene fluoride) PVDF-g-PSSA (Hietala et al., 1998; Lehtinen et al., 1998). However, another factor limiting the operating temperature is the water-assisted nature for most proton-conducting polymer membranes. Therefore, suitable additives are required that can improve water retention of the polymer membranes at a higher temperature. In situ impregnation of inorganic additives and inorganic nanocomposites of Nafion have been extensively investigated and have shown improved water retention as compared to Nafion (Jalani et al., 2005).

#### 25.2.3.4 Durability

Commercial applications of fuel cells demand long operation time and strength and durability of the membrane electrolyte, which become more disadvantages with Nafion membranes. Other hydrocarbon membranes have been developed that have an improved lifetime of 5,000 h (Polyfuel Web site, http://www.polyfuel.com) compared to Nafion. Modification of Nafion or other hydrocarbon membrane using inorganic additives also increases their durability. For example, modification of sPEEK membranes using silica fixes the heteropolyacid within the membrane and increases stability (Ponce et al., 2004) while crystalline additives such as $TiO_2$ and $WO_3$ have been found to impart additional strength to Nafion membranes (Shao et al., 2006).

## 25.3 HYBRID MEMBRANES

The chief purpose of using additives with membranes or introducing modifications in structure and morphology of the membranes is to achieve low material cost and improve membrane characteristics through lower reactant crossover, enhanced proton conductivity, good mechanical stability, and improved higher temperature performance. The proton conductivity and reactant crossover are both dependent on the water uptake of the membranes based on the clustering property and channel size and hence depend on the ionomer microstructure. The structure and functionality of membranes are usually improved by the addition of inorganic–organic and/or acidic–basic constituents to produce composite membranes (Neburchilov et al., 2007).

Often the composite is formed by dispersing the particles of suitable fillers in the ionomer matrix (Alberti and Casciola, 2003. Nano- and microcomposite membranes are prepared either by dispersing micro- or nano-sized filler particles in an ionomer solution followed by casting or by growing the filler particles within a preformed membrane or in an ionomer solution. The membranes are then usually obtained by film casting and solvent elimination (Alberti and Casciola, 2003). In organic–inorganic hybrids, the inorganic minerals precipitate in situ in a regular manner in the organic polymer matrix, and a strong interaction between the organic polymer and inorganic mineral is expected to result in a hybrid with markedly improved mechanical properties (Son et al., 2007).

The hybrid membranes, containing inorganic nanoparticles of solid acids in polymer matrix, are identified as a highly efficient family of proton conducting solid polymer electrolytes. Polymer composite membranes with inorganic additives of nanometric size were extensively studied because they are able to operate at higher temperatures than the pure polymers. (Baglio et al., 2005; Kim et al., 2005; Nunes et al., 2002) and have improved proton conductivity, water retention ability, and reactivity related to high surface area and mechanical support Stamatin et al., 2006).

### 25.3.1 How Do the Additives Help?

In a typical PEM structure there are two active groups. For example, in the case of Nafion, which is a sulfonated tetrafluoroethylene copolymer (Figure 25.3), the tetrafluoroethylene backbone of Nafion provides thermal and chemical stability to it and is hydrophobic in nature, while the sulfonic

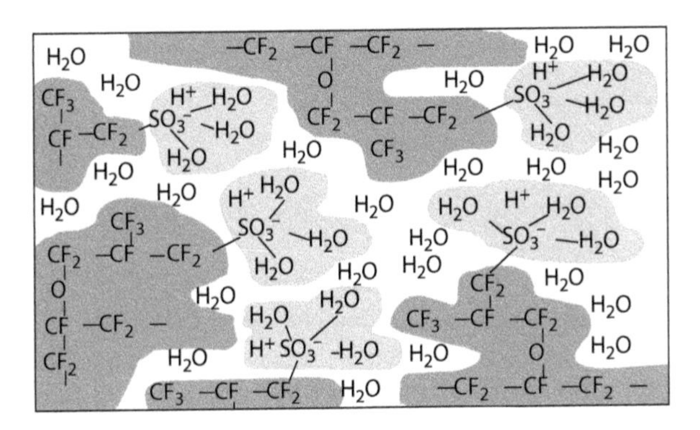


FIGURE 25.3 Schematic of the chemical makeup of Nafion membrane.

group, which is the side chain, is hydrophilic. Hence it helps to retain water near the membrane, which helps in increasing proton conductivity. In contact with water, clustering of the hydrophilic groups (Figure 25.4) occurs, which facilitates migration of protons by creating proton-conducting pathways. Thus water retention becomes an important property for membrane performance. This requirement proves to be a disadvantage for use of pure Nafion at higher operating temperatures. Nafion membranes also have some other disadvantages in DMFC applications such as high cost, low temperature limit, high humidification requirements, and high methanol and ruthenium (which is a catalyst at anode) crossover, which reduces the efficiency of the oxygen reduction reaction by the known mixed potential effect. This is the major barrier preventing successful use of Nafion membranes in DMFC (Neburchilov et al., 2007). Nearly 40% of methanol feed crosses the Nafion membrane to reach at cathode (Jung et al., 2003), which is not only a waste of fuel but also affects the fuel cell voltage in several ways (Panganin et al., 2005). Hence there arises the need for modifiers or additives.

The inorganic additives function by improving water retention at higher temperatures, improving high proton conductivity and reducing reactant crossover. Though the mechanism and chemistry involved in effecting this improvement varies with the type of additive used, the general principle remains the same (Larminie and Dicks, 2003). For example, zirconium is used as an additive for Nafion membranes with encouraging results. This is attributed to the high water retention capacity imparted by the high water affinity of hygroscopic zirconia. This facilitates proton conductivity through the hydrophilic part of the membrane. Zirconium also imparts higher crystallinity to the membrane. Since an increase in crystallinty results in a decrease in permeability, this additive gives a resultant low permeability to the reactant, thereby reducing crossover.

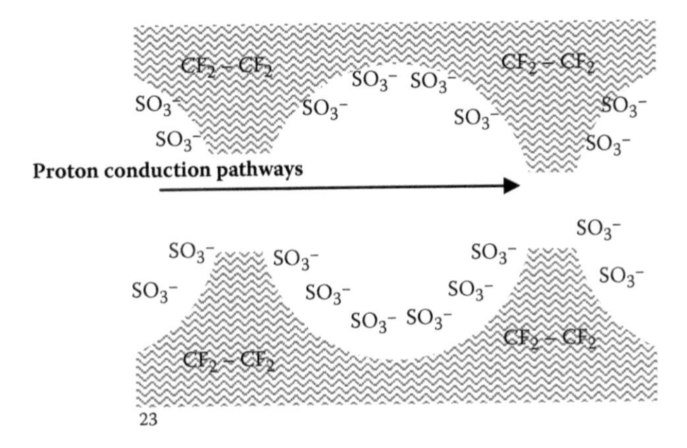


FIGURE 25.4 Cluster network model.

### 25.3.2 Organic-Inorganic Hybrid and Composite Membranes

Commercially, four types of membranes are being used:

- Nafion membranes
- Non-Nafion fluorinated membranes: phosphazenes, polybenzinimidazole, and PEEK (poly ether ether ketone)
- Composite fluorinated membranes, organic–inorganic composite, acidic–basic composite
- Composite nonfluorinated membranes, organic–inorganic composite, acidic–basic composite

This section discusses the various hybrid membranes investigated by several researchers.

#### 25.3.2.1 Nano-, Micro- and Macrocomposite Membranes

The properties of composite membranes depend on the nature of ionomer and particles used and on the size, amount, and homogeneity of dispersion and orientation of particles in the polymer matrix. In case of ionomers having distinct hydrophilic and hydrophobic domains such as in Nafion, the distribution of particles in the respective domains also influences the property of the composite. The filler influence on the properties of the original polymer if directly proportional to the interface interaction between the polymer and dispersed particles. Thus nanoparticles produce greater modification in characteristics than microparticles and when the properties of the original matrix are significantly influenced, the nanocomposites are also termed hybrid membranes.

For composite membrane preparation using high-proton conducting particles as fillers for non-proton conductive polymeric matrix, the particle congregation must be along continuous pathways to ensure good proton conductivity. These membranes are called macrocomposite as the organic and inorganic domains are distinctly separated. Macrocomposite membranes can be prepared by incorporating proton conducting particles in an insulating polymer matrix. However, they can be obtained only by using very high concentrations of inorganic materials that adversely influences the mechanical characteristics of the membrane. Therefore, usually the proton conduction pathways are obtained by filling a porous polymer structure (such as Teflon) with proton-conducting particles. The proton conductivity in these cases is additionally dependent on the porosity and tortuousness of the preformed pores of the polymer.

#### 25.3.2.2 Modified Nafion Membranes

Hybrid Nafion membranes are prepared by adding various types of additives such as $SiO_2$/PWA, $SiO_2$, $WO_3$, $TiO_2$, Pt/Zeolite, and *m*-montmorillonite to Nafion. The different additives have different character and hence showing different performance. Shao et al. have studied the effect of $SiO_2$/PWA (polytoungstonic acid), $SiO_2$, $WO_3$, and $TiO_2$ as additives with Nafion (Shao et al., 2006). By doing XRD (X-ray diffraction) analysis they had shown that $WO_3$ and $TiO_2$ are crystalline in nature where as $SiO_2$/PWAand $SiO_2$ are amorphous in nature. Hence $WO_3$ and $TiO_2$ give additional strength to the Nafion membrane. According to their experiment, $SiO_2$/PWA and $SiO_2$ are more hydrophilic in nature hence they have more water uptake. Also their area resistance is low for proton conductivity. Hence they have more proton conductivity as compared to $WO_3$ and $TiO_2$. Nanocomposite Nafion membranes containing 3 wt% of $SiO_2$ or $TiO_2$ prepared by bulk mixing of polymer solution with oxide nanocrystallites followed by casting have also been investigated. These membranes, 50 µm thick membranes and loaded with highly dispersed Platinum prepared through ion exchange technique were found to be very efficient at 80°C and effectively suppress $H_2$ and $O_2$ crossover without gas humidification (Watanabe et al., 1996; Watanabe et al., 1998). Metal-oxide loaded Nafion is more hydrophilic than unmodified Nafion. However, Pt-loaded membranes are too expensive for application and in the absence of Pt, the silica-loaded membranes show poor fuel performance when using anhydrous gases (Watanabe et al., 1996).

A significant enhancement of conductivity for Nafion 117 or recast Nafion is achieved in Nafion bifunctional silica composite membranes where sol–gel derived silica particles have both -SiOH and -$SO_3H$ functional groups (Wang et al., 2002). Due to the presence of sulphonated silica, the ion exchange capacity of these membranes increases linearly with silica loading. Concurrently the water content per gram of dry membrane also increases both at room temperature and at 80°C. However, the degree of hydration is barely affected and remains comparable with that of Nafion 117. These results also indicate that the driving force for membranes hydration could be the solvation of the ionization products of the highly acidic sulfonic function ($H^+$, $-SO_3^-$) and not the water interaction with the hydrophilic –SiOH groups.

Lin et al. have used 5 wt. % of montmorillonite with polyoxyproplene-diamine of various molecular weights as additives with Nafion membrane. They had observed that addition of this *m*-MMT to Nafion decreases methanol crossover but also decrease the proton conductivity due to putting more resistance to flow (Lin et al., 2007). Through experiments, Uchida et al. developed a new polymer electrolyte membrane for DMFC to minimize methanol crossover. Platinum nanocrystals were highly dispersed in a Nafion 117 film (Pt-PEM) to catalyze the oxidation of the crossover methanol. An appreciable increase in the cathode potential was observed, resulting from the reduced amount of methanol reaching the cathode (Uchida et al., 2002). Saccà et al. used $TiO_2$ as an inorganic filler with Nafion membrane and characterize at 80°C for polymer electrolyte fuel cells. According to their work, Nafion recast membrane shows ionic conductivity ranging from 120 to 140 mS/cm, while Nafion containing 3 wt% of $TiO_2$ composite membrane showed the values in the range 150 to 180 mS/cm. Nafion containing 3 wt% of $TiO_2$ has water uptake 29% compared to 20% for recast Nafion at 80°C (Saccà et al., 2005). From their work, it is clear that modified Nafion with 3 wt% $TiO_2$ is a better electrolyte as compared to pure Nafion. Jalani et al. synthesized Nafion nanocomposite membranes with zirconium oxide, silicon oxide and titanium oxide by an in situ method (Jalani et al., 2005). They characterized the membranes at 90°C and 120°C and different relative humidity. On the basis of water uptake, ion exchange capacity, and proton conductivity they found that zirconium oxide and silicon oxide as an additive to Nafion performs better as compared to titanium oxide. But the Nafion-titanium oxide membranes show better performance as compared to pure Nafion membrane.

Zirconium hydrogen phosphate modified Nafion has an insoluble ZrP entrapped in the pores (Stramatin et al., 2006). Nafion zirconium membrane is stable at $T = 150$°C with dry oxidant. The ZrP additive enhances water retention characteristics, improves the maximum working temperature, decreases the membrane resistance, and increases the dry weight and thickness of the membrane by 23% and 30%, respectively. Crystallinity and surface morphology play an important role in determining the conductivity of ZrP modified membranes. The surface area of the Nafion membrane increases by two orders of magnitude when modified by ZrP. Nafion zirconium membranes also have comparable proton conductivity to that of Nafion ($10^{-2}$ $Scm^{-1}$) at room temperature and 100% relative humidity.

Silica and molybdophosphoric acid modified Nafion membranes are prepared according to several methods by casting mixtures such as silica powder, dithenylsilicate (DPS) sol–gel reaction with Tetraethyl orthosilicate (TEOS), followed by solution casting of the Nafion solution phosphotungsticacid (PWA)-dopes composite silica/Nafion PWA and silica oxide. These membranes show good performance at T>100°C due to low levels of dehydration. Nafion membranes with 10 to 20 wt% DPS have a nanolayered microstructure that results in low methanol crossover (MCO). The proton conductivity increases with increasing DPS content. The modification of Nafion membranes through the addition of Molybdophosphoric acid (MoPh-a) has been shown to increase the proton conductivity 2 to 2.5 times, but with slightly increased MCO (Dimitrova et al., 2002).

Nafion-polyfurfuryl alcohol (PFA) nanocomposite membranes can be synthesized by in situ polymerisation of furfuryl alcohol within commercial Nafion membranes. Furfuryl alcohol is miscible with mixtures of water and alcohols (it penetrates into the hydrophilic channels of Nafion) and becomes hydrophobic following polymerization via acid catalysis. The chemically stable PFA is responsible for the low MCO through homogenous Nafion-PFA nanocomposite membranes at suitable PFA concentrations.

Heteropolyacids are electrocatalytically active and highly acidic and possess high proton conductivity in their hydrated form (Nakamura et al., 1981; Savadogo, 1992a; Savadogo, 1992b). Additionally the molecules are 1 nm in diameter and can be regarded as nanoparticles, and therefore are suitable to be used as fillers for membranes. Composite Nafion membranes are prepared by impregnation of preformed membranes with a heteropolyacid solution (Malhotra and Datta, 1997 and by mixing Nafion solution with a requisite amount of heteropolyacid followed by casting (Tazi and Savadogo, 2000; Tazi and Savadogo, 2001). Composite Nafion membranes with silicotungstic acid (STA), phosphotungstic acid (PTA), and phosphopolymolybdic acid (PMA) show improved proton conductivity and higher water uptake compared to Nafion 117 but lower tensile strength. Table 25.1 shows a comparison of properties of Nafion and heteropolyacid modified Nafion membranes.

Nafion-polypyrrole base membranes are a type of acid-base composite membranes studied. These can be prepared either by polypyrrole impregnation by in situ polymerization or polymerization in hydrogen peroxide with Fe(III) as the oxidizing agent. Membranes prepared using the latter technique has low MCO but high resistance and poor performance compared to Nafion due to poor electrode bonding.

#### 25.3.2.3 Other Hybrid Membranes

Modified polyvinylidene fluoride (PVDF) membranes (24% PVDF–16% $SiO_2$ with 60%–3 M $H_2SO_4$ (vol. %)) are nanoporous proton-conducting membranes (NP-PCM). These membranes utilize PVDF as the polymer binder to implant $SiO_2$ powder and acid into the polymer matrix to provide proton conductivity and can have very high surface area and thus two to four times higher ionic conductivity than Nafion®. Meanwhile, the MCO of PVDF membranes are two to four times lower than Nafion due to smaller pore sizes (1.5–3 nm compared to 3 nm for Nafion®) (Peled et al., 2000).

Impregnation of the NP-PCM pores with $Na_2SiO_3$ solution and hydrolyzing the silicate in sulfuric acid (forming silica gel) decreases the MCO further to almost an order of magnitude lower than Nafion®. These membranes operate in a wider temperature range than Nafion(from 0°C > 90°C). They are less sensitive to iron impurities (Fe tolerance >500 ppm) than Nafion®, which allows for the use of Pt–Fe catalysts or stainless steel fuel cell hardware. Modified PVDF membranes have good mechanical properties and stable dimensions that do not change with absorption of water.

Sulfonated poly (ether ether ketone) (sPEEK) membranes are often modified with silica and silanes. The silane modified SPEEK membrane has a lower methanol crossover but also lower conductivity than that modified by silica. The inorganic phase decreases the water and methanol crossover, in addition to fixing of the heteropolyacid to the membrane.

The modification of sPEEK by ZrP (10%) and PBI (5.6%) produces a membrane with improved chemical stability, unlike the unmodified one (Neburchilov et al., 2007).

Nanocomposite membranes prepared from sPEEK and 60 wt% heteropolyacid mixed with the polymer solution in dimethylacetamide (Zaidi et al., 2000) have been compared with unmodified

**TABLE 25.1**
**Comparison of Properties: Heteropolyacid Modified Nafion and Nafion 117**

| Property | Nafion 117 | Modified Nafion (Heteropolyacid Used) |
|---|---|---|
| Water uptake | 27% | 95% maximum (PMA) |
| Fuel performance at 80°C | 640 mA/cm$^2$ | Higher |
| Current density at 0.600V | | 940 mA/cm$^2$ (PMA) |
| Performance at higher temperature | | Stronger (PTA) |

sPEEK. The composite membranes have a higher glass transition temperature and much greater hydration at room temperature. Also the composite membranes are more conductive than the pure polymer, but the conductivity enhancement decreases with increasing degree of sulphonation.

sPEEK with Zr- sulfophenylene-phosphonate (SPP) composite membrane is prepared by mixing a 27 wt% SPEEK-WC solution in DMF with a suitable amount of the Zr (SPP) gel in DMF in order to obtain 20 wt% loading of Zr (SPP) in the membrane. This loading allows obtaining membranes with good performances in terms of proton conductivity and mechanical properties. After required pretreatment the membrane is prepared by casting on a glass plate, drying under controlled conditions, and finally treated with acid followed by neutralization and drying (Regina et al., 2006).

PBI is a basic polymer and acid doping results of PBI results in a single-phase polymer good oxidative and thermal stability and mechanical durability at higher temperatures between 100°C to 200°C. This property is utilized on preparing PBI based membranes cast from high molecular weight solution of PBI in dimethylacetamide and immersed in phosphoric acid solution. The chief advantages obtained over Nafion are higher proton conductivity, lower electroosmotic drag (Wang et al., 1996), high temperature operation and low gas humidification, low MCO, and lower cost (Reeve, 2002). However, the disadvantage of acid leaching in hot methanol solutions becomes crucial from an application point of view. Therefore, high molecular weight phosphotungstinic acid is mostly used to replace the low molecular weight acid.

Zirconium phosphate sulfophenylene-phosphonate [$Zr(HPO_4)O_3PC_6H_4SO_3H$] composite membranes absorb less water and methanol than the polymeric ones. Besides, water and methanol absorption does not increase significantly with temperature also. The composite membranes are reported to be less elastic and more brittle with respect to the polymeric ones. Methanol permeability through composite membranes is higher (about 30%) than that of the corresponding polymeric membranes (Regina et al., 2006).

Zirconium oxide is also used as a membrane additive that is incorporated via the hydrolysis of the zirconium tetrapropylate. This enables the preparation of hybrid membranes with decreased water swelling, proton conductivity, and water and methanol permeation. In contrast, the selectivity toward water/methanol increases with the inorganic content.

Hybrid membrane by using polyphosphomolybdic acid and polybenzimidazole (ABPBI–$PMo_{12}$) are prepared by dissolving ABPBI poly (2,5-benzimidazole) in methanesulfonic acid (MSA) and later adding the desired amount of $PMo_{12}$. The conductivity of the hybrid membrane is higher than that of the ABPBI membrane treated under the same conditions. As for the sulfonated derivative, both a larger uptake of acid and the contribution to conductivity of the inorganic acid clusters could be effective in increasing this performance. $PMo_{12}$ anionic clusters are additives that increase the phosphoric acid uptake. This is important not only to increase the conductivity, but also in order to avoid the leaching of the acid when the membranes operate as fuel cell electrolyte in PEMFC for long times. These membranes are stable up to 200°C and feature high conductivities at these temperatures, which makes them promising candidates for higher-temperature PEM fuel cells.

Hybrid membranes based on nanometer YSZ and polyacrylonitrile matrix is a new type of hybrid material reported to be obtained from polyacrylonitrile (PAN) by a thermo-oxidative process in centrifugal field. Nanometric oxide selected here (e.g., zirconium(IV) oxide-yttria stabilized, YSZ) is embedded in the polymer matrix. The polymer matrix made of pyridine ladder polymer has extremely low electronic conduction and a good affinity for proton transport through its pendant groups such as -OH or oxygen; therefore, the conduction mechanism is mainly governed by YSZ. These composites PAN-YSZ are reported to give a flexible and maneuverable membrane for PEFC where electrical properties are tunable by nano-oxide contribution while diffusion, water uptake, and other properties are determined by the polymer matrix. The structural properties, water uptake, reduced methanol permeability, and electrical conductivity compared with other membranes used in PEFCs are improved at higher temperatures (Stamatin et al., 2006).

Nowofol GmbH manufactures an irradiated and sulfinated polu(ethylene-alt-tetrafluoroethylene) (ETFE) polymer known commercially as ETFE-SA. The ETFE-SA membranes are prepared by irradiation followed by sulfonation without any grafts form other polymers. The semicrystalline nature of these membranes allows for cross-linking effects of the crystallites and results in low swelling of the membranes (Saarinen et al., 2005). These membranes are cheaper than Nafion115 and have the advantage of 10% lower MCO with 70% lower water and methanol uptake and a high lifetime of over 2000 h in a DMFC without performance loss. However, the proton conductivity is lower than Nafion®.

Hoku Scientific Inc. developed a butadiene/styrene rubber membrane that contains an inorganic cation exchange material, aluminosolicates-kaolin, and a silica-based binder tetraethylorthosilicate (TEOS). This membrane when tested in a PEM fuel cell showed 13% to 30% higher cell voltage than Nafion® membranes. The aluminosilicates here stabilize the proton conductivity while reducing the dimensional fluctuations caused by variations in electrochemical cell temperature and variations in water content of the composite membrane. The silica-based binder increases the cation exchange capacity of kaolinite and also increases the proton conductivity. The polymer-based binder provides higher mechanical strength and improved proton conductivity during operation (Taft and Kurano, 2003).

A distinct family of inorgano-organic polymers have also been developed by reacting polyethers such as poly(ethylene oxide) (PEO), poly(phenylene oxide) (PPO), and poly(tetramethylene oxide) (PTMO) with alkylosilanes (Homma et al., 1999; Homma et al., 2001). The reaction product consists of nano-sized silica interconnected by polymer chains. Membranes of these hybrids become proton-conducting in the presence of water vapor and when doped with PTA. PTMO-based membranes also show highest thermal stability based on similar molecular weights of the organic polymer.

#### 25.3.2.4 Modified Surface Membranes

Besides additives incorporated during membrane preparation, modification of the surface of the electrolyte is also found to be useful in improving the performance of fuel cells. Surface-modified Nafion membranes using plasma etching, palladium-sputtering, and Pd-layered Nafion membranes are very useful in reducing methanol crossover (Choi et al., 2001; Yoon et al., 2002). However, this technology is still in an improvement stage where problems of film cracking and instability are being studied. Kim et al. modified the Nafion surface by impregnating Pd nanophases. Such a modification could allow the selective transport of smaller water molecules or hydrogen ions (protons), while the passage of larger methanol molecules would be restricted (Kim et al., 2003). The Pd nanophases were used in combination with a Nafion117 membrane to restrict methanol crossover. The anode side of Nafion was modified by brushing a thin film of oleylamine-stabilized palladium, which can be easily dispersed in organic solvents. Only modifying the surface of Nafion can keep the bulk of the membrane intact, thus providing the best chance of maintaining the inherent conductivity of Nafion (Tian et al., 2007). Pu et al. modified the surface of nafion membrane by using palladium foil sandwiched between two Nafion polymer membranes, which resulted in selective transport of protons while eliminating methanol crossover to the cathode (Pu et al., 1995).

## 25.4 HYBRID MEMBRANE: THE RESULT OF HYBRIDIZATION ON IMPORTANT MEMBRANE PROPERTIES

The addition of inorganic components has a profound effect on the membrane characteristics. Hence, the hybrid membranes show an improvement in the desired membranes characteristics for fuel cell application. The relative advantages of hybrid membranes over the unmodified ones are chiefly attributed to the fact that the hybrid membranes show controllable physical properties, such as optical, electrical, and mechanical behaviors, by combining the properties of both organic polymers and inorganic compounds.

Some of these effects are discussed below.

### 25.4.1 Conductivity

The water uptake of the membranes typically determines the proton conductivity and methanol permeation based on the ionomer microstructure, cluster, and channel size (Neburchilov et al., 2007). The proton-conducting characteristics of membranes depend on the degree of porosity and on size and distribution (or tortuosity) of the preformed pores, as well as on the nature of the filler and the degree of pore filling. In the case of fibrous or lamellar particles, the proton conductivity also depends on their orientation inside the pores (Borup et al., 2007). Zirconium phospate (Gummaraju et al., 1996; Mauritz, 1998) is used as an additive that functions as a water retention amplifier while hydroxyapatite (Baradie et al., 2000) and zeolite (Apichatachutapan et al., 1996) act as structure modifier to improve the proton conductivity of membranes (Neburchilov et al., 2007).

For ionomers of low proton conductivity, the most positive results in proton conductivity have been reported for sulphonated polysulfone filled with $H_3Sb_3P_2O_{14}$ and for sPEEK membranes with zirconium phosphonate additives. Membranes modified by the addition of inorganic compounds such as $SiO_2$, MoPh, and ZrP have higher conductivity than Nafion®. Some of the reasons suggested for such high conductivity in these membranes (i.e., Nafionwith ZrP) are the higher proton mobility on the surface of ZrP particles and good water retention capabilities (Zoppi and Nunes, 1998). For high proton conducting ionomers such as perfluorinated polymers bearing sulfonic groups, additives such as ZrP, silica or heteropolyacid nanoparticles improve fuel cell performance at temperatures higher than 90°C to 100°C, even for relative humidity considerably lower than 100%. This improvement may not be directly attributed to an increase in number of proton carriers or particular proton acceptor properties of the inorganic particles. This is because while silica particles have negligible proton conductivity, –Si-O-Si- and –Si-OH groups are better proton acceptors than $-O_3POH$ groups. One explanation could be that the hydrophilic compounds act as water reservoirs. However, in a fuel cell, a proton diffuses toward the cathode with one or more water molecules (Zawodzinski et al., 1995) so the water reserve is constantly depleted if replacement occurs through either supply via humidified gas or by back diffusion of water produced at the cathode. It is also opined that the hydrophilic particles replace the loosely bonded water within the Nafion domain. Another opinion is that water interacting with the surface of the filler may be more likely to be retained than loosely bonded water in unmodified Nafion.

It is also suggested that mechanical reinforcement of the membrane due to inclusion of additives and hence a reduction in swelling property could significantly affect proton conductivity, particularly in the temperature range 90°C to 110°C, where high humidification conditions can be obtained at acceptable cell pressures (Alberti and Casciola, 2003).

### 25.4.2 Reactant Crossover

Methanol crossover amount to as much as 40% methanol losses have been reported (Tricoli et al., 2000) and hence this problem assumes a major concern in DMFC operation. MCO in Nafion-based membranes occurs via ion-cluster pores and ion channels within the hydrophobic polymer backbone (Neburchilov et al., 2007). Thus the change in membrane structure also changes the MCO. Use of inorganic additives in fluorinated and unfluorinated membranes is found to reduce the MCO without lowering the proton conductivity (Neburchilov et al., 2007). Water uptake of the membranes significantly governs the methanol permeation through the ionomer microstructures and clusters (Jiang et al., 2006; Miyake et al., 2001). For Nafion-silica membranes, for example, at low hydrophobic silica content the silica surrounds the hydrophilic ion clusters and ion channels. The newly formed tortuous Nafion structure alters the methanol transport. High silica content increases the contribution of the backbones to methanol permeability and increases proton conductivity and water uptake (Jiang et al., 2006). Since different membranes have been tested under different conditions, it is difficult to compare the MCO results obtained by different researchers. However, when a ratio of MCO for Nafion membrane to that for other membrane is used for comparison, the lowest MCO ratios have

**TABLE 25.2**
**Durability of Selected Membranes under Fuel Cell Operation**

| PEM | Application Type | Thickness (μm) | Temperature (°C)/ $P_{anode}/P_{cathode}$ | Humidification (% RH) | Life Test Conditions | Lifetime (h) | Reference |
|---|---|---|---|---|---|---|---|
| Nafion | $H_2$/air | 180 | 80 | | 1 A/cm$^2$ (startup), cont | 2,300–20,000 | Stucki et al., 1998 |
| | $H_2$/air | 25, 50 | 65/1 bar/1 bar | >100% | 0.8 A/cm$^2$, cont | >2,500 | Curtin et al., 2004 |
| Flemion | $H_2$/air | 50 | 80/0.1 MPa | 100% | 1 A/cm$^2$ cont | >4,000 | Endoh et al., 2002 |
| Gore PRIMEA 56[a] | $H_2$/air | 30 | 70/ambient | 100% | 0.8 A/cm$^2$ cont | >20,000 | Cleghorn et al., 2006 |
| BAM 3G[a] | $H_2$/air | | 80/2 atm | | | 14,000 | Steck and Stone, 1997 |
| BAM 3G | $H_2$/air | | 70/24 psig/24 psig | | 0.5 A/cm$^2$ | 4,061 | Wei et al., 1995 |
| Styrene sulfonic acid/Nafion composites | $H_2/O_2$ | >160 | 80/0.2 MPa/0.2 MPa | 100% | 0.3 A/cm$^2$ | >835 | Yu et al., 2003 |
| ETFE-based radiation grafted membrane | $H_2/O_2$ | 78 | 60/ambient | >100% | | <600 | Buchi et al., 1995 |
| | $H_2/O_2$ | 25 | 80/1 bar/1 bar | anode: 100%cathode: 0% | 0.5 A/cm$^2$ | >770 | Gubler et al., 2005 |
| | $H_2/O_2$ | 35 | 50/2 bar/3 bar | >100% | 0.2 A/cm$^2$ | ~1000 | Wang and Capuano, 1998 |
| Sulfonated poly(ether ether ketone) | $H_2/O_2$ | 40 | 50/ambient | | 0.05 V, cont | 4300 | Soczka-guth et al., 2002 |
| | $H_2/O_2$ | | 90 | | 0.5 mA/cm$^2$ | >1,000 | Rozière and Jones, 2003 |
| | DMFC | 60 | 120/1 bar/3 bar | | discont | 1,440 | |
| Sulfonated poly(arylene ether sulfone) | DMFC | 50 | 80/1 bar/1 bar | >100% | 0.5 V, cont | >3000 | Kim et al., 2006 |
| | $H_2/O_2$ | 40–50 | 80/ambient 90% | | 0.2 A/cm$^2$, cont | 5,000 | Aoki et al., 2006 |
| Sulfonated polyimide | $H_2/O_2$ | | 60/2 atm | | 0.25 A/cm$^2$ | >3,000 | Faure et al., 1997 |
| | $H_2/O_2$ | 50–60 | 80/1 bar/1 bar | 90% | 0.2 A/cm$^2$, cont | 5,000 | Asano et al., 2006 |
| PEMEAs Celtec-P (PBI-based)[a] | $H_2$/air | | 160/ambient | 0% | 0.2 A/cm$^2$ cont | >18,000 | Frank, 2003 |
| Acid-doped PBI | $H_2/O_2$ | | 150 | 0% | 0.5 V cont | 5,000 | Banerjee and Curtin, 2004 |

*Source:* Modified from Borup, R., et al. 2007. *Chem. Rev.* 107: 3904–3951.

been reported for poly(1-methyl pyrrole) modified Nafion, sPPZ, and for membranes developed by the companies Pall and 3P Energy. The corresponding MCO ratios are lower than Nafion by factors of 215, 123, 2.5, and 20 (Neburchilov et al., 2007).

### 25.4.3 Thermal Stability and Durability

Thermal stability of membranes is seen to generally improve with addition of inorganic additives to form hybrids. For example, Nafion-silica membranes show good performance at $T > 100°C$ due to low levels of dehydration. Nafion-zirconium membrane is stable at $T = 150°C$ with dry oxidant. Other hybrid membranes also indicate similar levels of performance. For example, hybrid membrane by using polyphosphomolybdic acid and polybenzimidazole (ABPBI–$PMo_{12}$) is stable up to 200°C and feature high conductivities at these temperatures. Asymmetric-based acrylic (AMPS) (Pei et al., 2006) and PBI (Wang et al., 1996) have been reported to have the best thermal stability and are able to operate up to 270°C and 160°C to 200°C, respectively. AMPS-based membranes also have lower MCO (by a factor of 14) and conductivity (by a factor of 2) than Nafion membranes.

A life test under normal operating conditions provides the most relevant PEM lifetime data. Table 25.2 summarizes the durability data of select membranes from fuel cell life tests. In passive DMFC, Polyfuel membranes have a lifetime of up to 5,000 h. As well, 35 μm thick ETFE-SA membranes show no decrease in performance for up to 2,000 h of testing. However, the overall efficiency of DMFC with ETFE membranes is about 40% to 65% lower than fuel cells with Nafion membranes (Saarinen et al., 2005). Stucki et al. reported that the life span of a fuel cell stack using Nafion 117 reached 15,000 h under 80°C continuous operations (Stucki et al., 1998) but the life times of similar PFSAs is significantly reduced to when thinner membranes and back-pressurized conditions are used (Borup, 2004; Curtin et al., 2004). The first, second, and third generation membranes (BAM1G, 2G, and 3G) from Ballard Advanced Materials Co. having chemical structure of sulfonated poly(phenylquinozaline)s, sulfonated poly(2,6-diphenyl-1,4-phenylene oxide)s, or poly(trifluorostyrene sulfonic acid)s, respectively (Steck and Stone, 1997) have also been investigated. The degradation of BAM1G membranes with lifetimes of some hundred hours at 70°C was considered to be due to oxidative decomposition leading to membrane embrittlement. BAM2G with electron-withdrawing functionalities (fluoride, bromide, and cyanide) introduced to improve oxidative stability also failed to achieve improved longevity. Later, sulfonated polytrifluorostyrene based BAM3G membranes were claimed to exhibit increased stability to main chain scission and achieved substantial durability in stack configurations as well as single cells (Steck and Stone, 1997; Wei et al., 1995).

Research on phosphoric acid doped PBIs, tested at elevated temperature (>150°C) demonstrated long lifetimes 5,000 h at 150°C (Li et al., 2004) and 6,000 h at 160°C (Schmidt, 2005) with low decay rates in cell voltage. The possibility of condensation in such systems associated with shutdown or idling makes it a difficult for them to operate under the transient conditions likely in automotive application. However, higher temperatures might have advantages for stationary applications with relatively modest load cycling requirements.

## 25.5 CONCLUSION: SELECTING THE BEST MEMBRANE

The entire basis for hybrid membranes research is the incentive of resolving the present-day bottlenecks in PEM fuel cell development and application by finding an improved low-cost alternative. This accounts for the huge volume of ongoing research in the area and the great number of papers and patents recently published. In several research results, significant improvements in fuel cell performance is found in terms of reduced reactant crossover, enhanced proton conductivity, and improved temperature stability and operational performance. Several other associated improvements with respect to these parameters are also achieved. For example, the problems of CO catalyst poisoning, water management, and mechanical stability are also dealt with. Today, fuel cells are being looked upon as highly prospective energy alternatives that will soon assume importance from the practical application perspectives. The market for mobile fuel cells with DMFC technologies is projected to reach $2.6 billion by 2012. The U.S. Department of Energy (DOE) has forecasted commercialization of portable fuel cell in 2010, and has also outlined targets for 2010 that will be important to achieve this forecast. The DOE targets are (Lightner, 2005):

- Power density: 100 W $l^{-1}$
- Current density: 1000 W h $l^{-1}$
- Lifetime: 5000 h
- Cost: \$3 $W^{-1}$

In 2009, the DOE reported that the manufacturing cost of an 80-kW automotive polymer electrolyte membrane (PEM) fuel cell system operating on direct hydrogen and projected to a manufacturing volume of 500,000 units per year was $61/kW. The DOE goal set in 2002 for cost reduction projected the cost for same system to be $30/kW (reported in 2002 dollars) by 2015 (Spendelow and Marcinkoski, 2009). The cost of Nafion typically ranges from \$600 to 1,200 $m^{-2}$ depending on the thickness, while other membranes such as sPPEK are cheaper at \$375 $m^{-2}$ (Agro et al., 2005). The cost of a membrane is a small fragment of the total cost of a fuel cell system and often smaller in comparison to the cost of catalyst and fuel, water, air, and thermal management. Based on a 50-kW PEMFC for automotive applications, Nafion membranes account for 20% of the cost of Nafion-based membrane electrode assemblies (MEAs) (Reeve, 2002). Nevertheless, economic considerations still drive the search for low-cost membrane material for commercialization of this technique. It is clear that popular membrane materials are insufficient to achieve this target. It also leads to the inference that for ensuring good practical applicability on a large scale, hybrid membranes become an imperative choice and an indispensable alternative.

However, practical results do not always replicate those envisaged during preparation and characterization of composite membranes. Also the practical application conditions are also not identical to the lab-scale research environment in most cases. A wide variety of ionomers and inorganic additives have been used but systematic and comparative research data on the effects of structure, size, distribution, orientation, and so forth of the various additives are lacking. Moreover, membrane properties have often been tested under varying operating conditions, making comparison of membrane performance difficult. Also, many researchers have used insoluble inorganic particles dispersed in the ionomeric matrices but not all have considered the structure, shape, dimensions, degree of crystallinity, and distribution within various regions of the hydrated ionomer. This has also led to significant discrepancies and even conflicting results while using the same type of additive. For example, α-ZrP, a widely investigated additive can be included in the polymer matrix in several ways, such as by in situ formation of ionomer solution or in situ precipitation in preformed membrane and as a preformed exfoliate lamellae or powdered preformed particles. In addition to the type of ionomer used, particle size, content, orientation of exfoliated particles, exfoliation degree, and so forth are some other variables that result in a different type of composite ZrP membrane with different properties. Hence a comparison between them is a challenging task. Therefore, although extensive research results are available, a definitive conclusion regarding the best materials for hybrid membrane does not immediately result. Such difficulties leave us a further scope for research in this area with emphasis on developing membranes suited to specific applications. The present-day challenge is therefore to investigate a suitable membrane type based on specific applications.

A comparison of various commercially available membranes for DMFC application gives a few potential choices for achieving the aforementioned DOE targets. These are (Neburchilov et al., 2007):

- 45 and 62 μm thick hydrocarbon membranes from Polyfuel exhibiting a 5,000 h lifetime at 40°C, 0.28 mV, in a passive DMFC at power densities of 80 and 60 mWcm$^{-2}$, respectively.

- 63 µm thick IonClad® R-4010 membranes (tetra-fluoroethylene poly(styrene sulfonic acid)) from Pall exhibits a factor of 3.6 lower methanol crossover than Nafion® and similar conductivity and stability for 1,000 h.

Several other composite fluorinated and nonfluorinated membranes with low cost, MCO and Ru-crossover (for Ru-Pt anodes), wider temperature range (80°C–180°C), and higher ionic conductivity compared to Nafion have been reported as suitable for DMFC application. In hydrogen PEMFCs, thinner membrane materials are preferred as they provide reduced ionic resistance and increased MEA performance. On the other hand, thin membranes result in increased MCO in DMFCs. The disadvantage of high MCO offsetting the advantage of lower ionic resistance, thicker membranes are preferred for DMFCs. Modified polymer membranes with suitable additives such as $SiO_2$, Zr, MoPh-a, and so forth, and acidic-basic composite membranes can be prepared to address specific operational requirements of each type of fuel cell. For example, if a composite ionomer–ZrP membrane is to be used for DMFC, the exfoliation degree, size, homogenous distribution, and particle orientation become important to achieve low MCO of pure ionomer membrane. It is suggested that this can be achieved by using a suitable preparation method of incorporating exfoliated Zr-P particles in ionomer solution followed by casting (Alberti and Casciola, 2003). If the same additive is to be used for a hydrogen or indirect methanol fuel cell, the suitable preparation method is in situ formation of Zr-P in organic solutions of Nafion or preformed membrane (Alberti and Casciola, 2003). Similarly, to prepare economic alternatives to Nafion for hydrogen fuel cells operating under 100°C, low-cost ionomers with high percentage of highly conducting additives such as sPPEK filled with superacid zirconium phosphonates and porous polymeric membranes filled with ZrP derivatives having proton conductivity higher than $10^{-2}$ S $cm^{-1}$ are suggested to be used (Alberti and Casciola, 2003). Moreover, there are other specific requirements in terms of membrane characteristics to be met from the application point of view. For example, one criterion PEM fuel cells are required to meet for automotive application is the ability to survive and start up at subfreezing temperatures. DOE's PEM fuel cell stack technical targets for the year 2010 included survivability at –40°C and startup to 50% rated power from –20°C in times as low as 30 s while consuming <5 MJ of energy inclusive of the shutdown conditioning and subsequent startup. Few publications report work in this area. The effect of freezing water on the property of fuel cell component materials including the membrane itself is expected to have significant effect on the performance and durability of the membranes during freeze/thaw cycling of fuel cells.

Hybrid inorganic–organic and composite membranes are therefore the most promising alternatives for both hydrogen PEMFCs and DMFCs. However, they must evolve further into cost-effective options for commercial application.

## REFERENCES


Agro, S., T. DeCarmine, S. DeFelice, et al. 2005. *Annual Progress Report of the DOE Hydrogen Program 2005*, p. 790. U.S. Department of Energy (DOE) website: http://www.hydrogen.energy.gov.

Alberti, G., and M. Casciola. 2003. Composite membranes for medium-temperature PEM fuel cells. *Annu. Rev. Mater. Res.* 33: 129–154.

Antonucci, P.L., A.S. Aricó, P. Cretí, et al. 1999. Investigation of a direct methanol fuel cell based on a composite Nafion®-silica electrolyte for high temperature operation. *Solid State Ionics* 125: 431–437.

Aoki, M., Y. Chikashige, K. Miyatake, et al. 2006. Durability of novel sulfonated poly(arylene ether) membrane in PEFC operation. *Electrochem. Commun.* 8: 1412–1416.

Apichatachutapan, W., R.B. Moore, and K.A. Mauritz. 1996. *J. Appl. Polym. Sci.* 62: 417–426.

Aricó, A.S., S. Srinivasan, and V. Antonucci. 2001. DMFCs: From fundamental aspects to technology development. *Fuel Cells* 1: 133–161.

Arimura, T., T. Ostrovskii, T. Okada, et al. 1999. The Effect of additives on the ionic conductivity performances of perfluoroalkyl sulfonated ionomer membranes. *Solid State Ionics* 118: 1–10.

Asano, N., M. Aoki, S. Suzuki, et al. 2006. Aliphatic/aromatic polyimide ionomers as proton conductive membrane for fuel cell applications. *J. Am. Chem. Soc.* 128: 1762–1769.

Baglio, V., A. Arico, A. Di Blasi, et. al. 2005. Nafion–$TiO_2$ composite DMFC membranes: Physico-chemical properties of the filler versus electrochemical performance. *Electrochim. Acta* 50: 1241–1246.

Banerjee, S., and D.E. Curtin. 2004. Nafion perfluorinated membranes in fuel cells. *J. Fluorine Chem.* 125: 1211–1216.

Baradie, B., J.P. Dodelet, and D. Guay. 2000. Hybrid Nafion®-inorganic membrane with potential applications for polymer electrolyte fuel cells. *J.Electroanal. Chem.* 489: 101–105.

Borup, R. 2004. *DOE Hydrogen Program Review*, 2004.

Borup, R., J. Meyers, B. Pivovar, et al. 2007. Scientific aspects of polymer electrolyte fuel cell durability and degradation. *Chem. Rev.* 107: 3904–3951.

Buchi, F.N., B. Gupta, O. Haas, et al. 1995. Study of radiation-grafted FEP-G-polystyrene membranes as polymer electrolytes in fuel cells. *Electrochim. Acta* 40: 345–353.

Chen, S.L., L. Krishnan, S. Srinivasan, et al. 2004. Ion exchange resin/poly styrene sulfonate composite membranes for PEM fuel cells. *J. Membrane Sci.* 243: 327–233.

Choi, W.C., J.D. Kim, and S.I. Woo. 2001. Modification of proton conducting membrane for reducing methanol crossover in a direct-methanol fuel cell. *J. Power Sources* 96: 411–414.

Cleghorn, S.J.C., D.K. Mayfield, et al. 2006. A polymer electrolyte fuel cell life test: 3 years of continuous operation. *J. Power Sources* 158: 446–454.

Costamagna, P., C. Yang, A.B. Bocarsly, et al. 2002. Nafion® 115/zirconium phosphate composite membranes for operation of PEMFCs above 100°C. *Electrochim. Acta* 47: 1023–1033.

Coplan, M.J., and G. Goetz. November 1, 1983. U.S. Patent 4,413,106.

Curtin, D.E., R.D. Lousenberg, T.J. Henry, et al. 2004. Advanced materials for improved PEMFC performance and life. *J. Power Sources* 131: 41–48.

Dimitrova, P., K.A. Friedrich, U. Stimming, et al. 2002. Modified Nafion-based membranes for use in direct methanol fuel cells. *Solid State Ionics* 150: 115–122.

Endoh, E., S. Terazono, and H. Widjaja. In *Electrochem. Soc. Abstracts*. Volume 2002-2. Salt Lake City, UT, October 20–24, 2002.

Faure, S., N. Cornet, G. Gebel, et al. 1997. 2nd International Symposium on New Materials for Fuel Cell and Modern Battery Systems, p. 818.

Frank, G. 2003. In *Proc. 2nd European PEFC Forum, Lucerne*, p. 1.

Genies, C., R. Mercier, B. Sillion, et al. 2001. Soluble sulfonated naphthalenic polyimides as materials for proton exchange membranes. *Polymer* 42: 359–373.

Gieselman, M.B., and J.R. Reynolds. 1992. water soluble polybenzimidazole based polyelectrolytes. *Macromolecules* 25: 4832–4834.

Gil, M., X.L. Ji, X.F. Li, et al. 2004. Direct synthesis of sulfonated aromatic poly(ether ether ketone) proton exchange membranes for fuel cell applications. *J. Membrane. Sci.* 234: 75–81.

Gubler, L., N. Prost, S.A. Gursel, et al. 2005. Proton exchange membranes prepared by radiation grafting of styrene/divinylbenzene onto poly(ethylene-*alt*-tetrafluoroethylene) for low temperature fuel cells. *Solid State Ionics* 176: 2849–2860.

Gummaraju, R.V., R.B. Moore, and K.A. Mauritz. 1996. *J. Polym. Sci.* 34: 2383–2392.

Hietala, S., M. Koel, E. Skoue, et al. 1998. Thermal stability of styrene grafted and sulfonated proton conducting membranes based on poly(vinylidene fluoride). *J. Mater. Chem.* 8: 1127–1132.

Hogarth, W.H.J., J.C. Diniz da Costa, and G.Q. (Max) Lu. 2005. Solid acid membranes for high temperature (>140°C) proton exchange membrane fuel cells. *J. Power Sources* 142: 223–237.

Honma, I., Y. Takeda, and J.M. Bae. 1999. Protonic conducting properties of sol-get derived organic/inorganic nanocomposite membranes doped with acidic functional molecules. *Solid State Ionics* 120: 255–264.

Honma, I., S. Nomura, and H. Nakajima. 2001. Protonic conducting organic/inorganic nanocomposites for polymer electrolyte membrane. *J. Membrane Sci.* 185: 83–94.

Jalani N.H., K. Dunn, and R. Datta. 2005. Synthesis and characterization of Nafion®-$MO_2$ (M = Zr, Si, Ti) nanocomposite membranes for higher temperature PEM fuel cells. *Electrochim. Acta* 51: 553–560.

Jiang, R., H.R. Kunz, and J.M. Fenton. 2006. Composite silica/Nafion® membranes prepared by tetraethylorthosilicate sol–gel reaction and solution casting for direct methanol fuel cells. *J. Membrane Sci.* 272: 116–124.

Jung, D.H., S.Y. Cho, D.H. Peck, et al. 2003. Preparation and performance of a Nafion®/montmorillonite nanocomposite membrane for direct methanol fuel cell. *J. Power Sources* 118: 205–211.

Kim, D.S., H.B. Park, J.W. Rhim, et al. 2005. Proton conductivity and methanol transport behavior of cross-linked PVA/PAA/silica hybrid membranes. *Solid State Ionics* 176: 117–126.

Kim, H., and H. Chang. 2007. Organic/iorganic hybrid membranes for direct methanol fuel cells. *J. Membrane Sci.* 288: 188–194.

Kim, Y.M., K.W. Park, J.H. Choi, et al. 2003. A Pd impregnated nanocomposite Nafion membrane for use in high-concentration methanol fuel in DMFC. *Electrochem. Commun.* 5: 571–574.

Kim, Y.S., J. Chlistunoff, and B. Pivovar. 2006. DOE Hydrogen Program Review, Arlington, VA.

Kobayashi, H., H. Tomita, and H. Moriyama. 1994. New metallic C60 compound: NaxC60(THF)y. *J. Am. Chem. Soc.* 116: 3153–3154.

Kobayashi, T., M. Rikukawa, K. Sanui, et al. 1998. Proton-conducting polymers derived from poly(ether-etherketone) and poly(4-phenoxybenzoyl-1,4-phenylene). *Solid State Ionics* 106: 219–225.

Kruczek, B., and T. Matsuura. 1998. Development and characterisation of homogeneous membranes made from high molecular weight sulfonated polyphenylene oxide. *J. Membrane Sci.* 146: 263–275.

LaConti, A.B., M. Hamdan, and R.C. McDonald. 2003. In *Handbook of Fuel Cells—Fundamentals, Technology and Application*, W. Vielstich, H.A. Gasteiger, and A. Lamm, (eds.). Vol. 3. New York: John Wiley & Sons.

Larminie, J., and A. Dicks. 2003. In *Fuel Cell Systems Explained*, Second Edition. Chichester, United Kingdom: John Wiley & Sons.

Lehtinen, T., G. Sundholm, S. Holmberg, et al. 1998. Electrochemical characterization of PVDF-based proton conducting membranes for fuel cells. *Electrochim. Acta* 43: 1881–1890.

Li, Q., R. He, J.O. Jensen, et al. 2003. Approaches and recent development of polymer electrolyte membranes for fuel cells operating above 100°C. *Chem. Mater.* 15: 4896–4915.

Li, Q., R. He, J.O. Jensen, et al. 2004. PBI-based polymer membranes for high temperature fuel cells—Preparation, characterization and fuel cell demonstration. *Fuel Cells* 4: 147–159.

Lightner, V. 2005. *7th Annual Small Fuel Cell Conference*. Washington, DC, April 27–29.

Lin, Y.F., C.Y. Yen, C.C.M. Ma, et al. 2007. Preparation and properties of high performance nanocomposite proton exchange membrane for fuel cell. *J. Power Sources* 165: 692–700.

Lufrano, F., G. Squadrito, A. Patti, et al. 2000. Sulfonated polysulfone as promising membranes for polymer electrolyte fuel cells. *J. Appl. Polym. Sci.* 77: 1250–1256.

Ma, Z.Q., P. Cheng, and T.S. Zhao. 2003. A palladium-alloy deposited Nafion membrane for direct methanol fuel cells. *J. Membrane Sci.* 215: 327–335.

Malhotra, S., and R. Datta. 1997. Membrane-supported nonvolatile acidic electrolytes allow higher temperature operation of proton-exchange membrane fuel cells. *J. Electrochem. Soc.* 144: L23–L26.

Mallant, R.K.A.M. 2003. PEMFC systems: The need for high temperature polymers as a consequence of PEMFC water and heat management. *J. Power Sources* 118: 424–429.

Mauritz, K.A. 1998. Organic-inorganic hybrid materials: Perfluorinated ionomers as sol-gel polymerization templates for inorganic alkoxides. *Mater. Sci. Eng. C* 6: 121–133.

Miyake, N., J.S. Wainright, and R.F. Savinell. 2001. Evaluation of a sol-gel derived Nafion/silica hybrid membrane for polymer electrolyte membrane fuel cell applications—II. Methanol uptake and methanol permeability. *J. Electrochem. Soc.* 148: A905–A909.

Nakamura, O., I. Ogino, and T. Kodama. 1981. Temperature and humidity ranges of some hydrates of high-proton-conductive dodecamolybdophosphoric acid and dodecatungstophosphoric acid crystals under an atmosphere of hydrogen or either oxygen or air. *Solid State Ionics* 3–4: 347–351.

Neburchilov, V., J. Martin, H. Wang, et al. 2007. A review of polymer electrolyte membranes for direct methanol fuel cells. *J. Power Sources* 169: 221–238.

Nolte, R., K. Ledjeff, M. Bauer, et al. 1993. Partially sulfonated poly(arlene ether sulfone)—A versatile proton conducting membrane material for modern energy conversion technologies. *J. Membrane Sci.* 83: 211–220.

Nunes, S.P., B. Ruffmann, E. Rikowski, et al. 2002. Inorganic modification of proton conductive polymer membranes for direct methanol fuel cells. *J. Membrane Sci.* 203: 215–225.

Panganin, V.A., E. Sitta, T. Iwasita, et al. 2005. Methanol crossover effect on the cathode potential of a direct PEM fuel cell. *J. Appl. Electrochem.* 35:1239–1243.

Park, Y.S., and Y. Yamazaki. 2005. Novel Nafion/hydroxyapatite composite membrane with high crystallinity and low methanol crossover for DMFCs. *Polym. Bull.* 53: 181–192.

Pei, H., L. Hong, and J.Y. Lee. 2006. Embedded polymerization driven asymmetric PEM for direct methanol fuel cells. *J. Membrane Sci.* 270: 169–178.

Peled, E., T. Duvdevani, A. Aharon, et al. 2000. A direct methanol fuel cell based on a novel low-cost nanoporous proton-conducting membrane. *Electrochem. Solid-State Lett.* 3: 525–528.

Polyfuel Web site: http://www.polyfuel.com.

Ponce, M.L., L.A.S. de A. Prado V. Silva, et al. 2004. Membranes for direct methanol fuel cell based on modified heteropolyacids. *Desalination* 162: 383–391.

Pu, C., W. Huang, K.L. Ley, et al. 1995. A methanol impermeable conducting composite electrolyte system. *J. Electrochem. Soc.* 142: L119–L120.

Qi, Z., M.C. Lefebvre, and P.G. Pickup. 1998. Electron and proton transport in gas diffusion electrodes containing electronically conductive proton-exchange polymers *J. Electroanal. Chem.* 459: 9–14.

Reeve, R.W. 2002. Update on status of Direct Methanol Fuel Cells, ETSU F/03/00232/REP, DTI/Pub URN 02/592.

Regina, A., E. Fontananova, E. Drioli, et al. 2006. Preparation and characterization of sulfonated PEEK-WC membranes for fuel cell applications A comparison between polymeric and composite membranes. *J. Power Sources* 160: 139–147.

Rikukawa, M., and K. Sanui. 2000. Polymer conducting polymer electrolyte membranes based on hydrocarbon polymers. *Prog. Polym. Sci.* 25: 1463–1502.

Rodriguez, G., J.I. Ladewig, B. Dicks, et al. 2004. Proceedings of the ARCCFN Annual Conference. Coffs Harbour, Australia, December 2–3, 2004, pp. 10–13.

Rodrigues, S., T. Reitz, T.D. Dang, et al. 2005. Proceedings of the Third International Energy Conversion Engineering Conference and Exhibit, San Francisco, California, August 15–18, 2005.

Rozière, J., and D.J. Jones. 2003. Non-fluorinated polymer materials for proton exchange membrane fuel cells. *Annu. Rev. Mater. Res.* 33: 503.

Saarinen, V., T. Kallio, M. Paronen, et al. 2005. New ETFE-based membrane for direct methanol fuel cell. *Electrochim. Acta* 50: 3453–3460.

Saccà, A., A. Carbone, E. Passalacqua, et al. 2005. Nafion–$TiO_2$ hybrid membranes for medium temperature polymer electrolyte fuel cells (PEFCs). *J. Power Sources* 152: 16–21.

Samant, P.V., and J.B. Fernandes. 2004. Enhanced activity of Pt(HY) and Pt–Ru(HY) zeolite catalysts for electrooxidation of methanol in fuel cells. *J. Power Sources* 125: 172–177.

Samms, S.R., S. Wasmus, and R.F. Savinell. 1996. *J. Electrochem. Soc.* 143: 1225–1232.

Savadogo, O. 1992a. The hydrogen evolution reaction in alkaline medium on nickel modified with $W0_4^{2-}$ or $Mo0_4^{2-}$. *Electrochim. Acta* 37: 1457–1459.

Savadogo, O. 1992b. *J. Electrochem. Soc.* 139: 1082–1087.

Savadogo, O. 1998. *J. New Mater. Electrochem. Syst.* 1: 47.

Schmidt, T.J. 2005. 208th Meeting Electrochem. Soc., Los Angeles, p 2197.

Shao, Z.G., H. Xu, M. Li, et al. 2006. Hybrid Nafion-inorganic oxides membrane doped with hereropolyacids for high temperature operation of proton exchange membrane fuel cell. *Solid State Ionics* 177: 779–785.

Soczka-guth, T., J. Baurmeister, G. Frank, et al. 2002. *Membranes for use in polymer electrolyte fuel cells or electrolyzers comprise a sulfonated aromatic polyether ether ketone of the formula (I).* U.S. Patent No. 6,355,149.

Son, D.-H., R.K. Sharma, Y.-G. Shul, et al. 2007. Preparation of Pt/zeolite–Nafion composite membranes for self-humidifying polymer electrolyte fuel cells. *J. Power Sources* 165: 733–738.

Spendelow, J., and J. Marcinkoski. 2009. Fuel cell system cost-2009. DOE Hydrogen Program Record. October 7, 2009.

Staiti, P., F. Lufrano, A.S. Aricó, et al. 2001. Sulfonated polybenzimidazole membranes—Preparation and physico-chemical characterization. *J. Membrane Sci.* 188: 71–78.

Stamatin, I., A. Morozana, K. Scott, et al. 2006. Hybrid membranes for fuel cells based on nanometer YSZ and polyacrylonitrile matrix. *J. Membrane Sci,* 277: 1–6.

Steck, A. 1995. In *Proc. 1st Int. Sym. New Materials for Fuel Cell Systems*, O. Savagodo, P.R. Roberge, and T.N. Veziroglu (eds.), p. 74.

Steck, A.E., and C. Stone. 1997. Proceedings of the Second International Symposium on New Materials for Fuel Cell and Modern Battery Systems, Montreal, July 6–10, 1997, p. 792.

Steele, B.C.H., and A. Heinzel. 2001. Materials for fuel cell technologies. *Nature* 414: 345–352.

Stucki, S., G.G. Scherer, S. Schlagowski, et al. 1998. *Appl. Electrochem.* 28: 1041.

Taft, K.M. III, and M.R. Kurano. Composite electrolyte for fuel cells. October 7, 2003. U.S. Patent No. 6,630,265.

Tazi, B., and O. Savadogo. 2000. Parameters of PEM fuel-cells based on new membranes fabricated from Nafion, silicotungstic acid and thiophene. *Electrochim. Acta* 45: 4329–4339.

Tazi B., and O. Savadogo. 2001. *J. New Mater. Electrochem. Systems* 4: 187–196.

Tian, A.H., J.Y. Kima, J.Y. Shi, et al. 2007. Surface modified Nafion membrane by oleylamine-stabilized Pd nanoparticles for DMFC applications. *J. Power Sources* 167: 302–308.

Tricoli, V., N. Carretta, and M. Bartolozzi. 2000. A comparative investigation of proton and methanol transport in fluorinated ionomeric membranes. *J. Electrochem. Soc.* 147: 1286–1290.

Tricoli, V., and F. Nanetti. 2003. Zeolite-Nafion composites as ion conducting membrane materials. *Electrochim. Acta* 48: 2625–2633.

Uchida, H., Y. Mizuno, and M. Watanabe. 2002. Suppression of methanol crossover and distribution of ohmic resistance in Pt-dispersed PEMs under DMFC Operation. *J. Electrochem. Soc.* 149: A682–A687.

Ueda, M., H. Toyota, T. Ochi, et al. 1993. Synthesis and characterization of aromatic poly(ether sulfone)s containing pendant sodium sulfonate groups. *J. Polym. Sci. Polym. Chem. Ed.* 31: 853–858.

Vaivars, G., N.W. Maxakato, T. Mokrani, et al. Zirconium phosphate based inorganic direct methanol fuel cell 2004. *Mater. Sci.* 10: 162–165.

Wang, F., and J. Roovers. 1993. Functionalization of poly(aryl ether ether ketone) (PEEK): synthesis and properties of aldehyde and carboxylic acid substituted PEEK. *Macromolecules* 26: 5295–5302.

Wang, F., M. Hickner, Y.S. Kim, et al. 2002. Direct polymerization of sulfonated poly(arylene ether sulfone) random (statistical) copolymers: Candidates for new proton exchange membranes. *J. Membrane Sci.* 197: 231–242.

Wang, H., and G.A. Capuano. Behavior of raipore radiation-grafted polymer membranes in h2/o2 fuel cells. 1998. *J. Electrochem. Soc.* 145: 780–784.

Wang, H., B.A. Holmberg, L. Huang, et al. 2002. Nafion-bifunctional silica composite proton conductive membranes. *J. Mater. Chem.* 12: 834–837.

Wang, J.T., R.F. Savinell, J. Wainright, et al. 1996. A $H_2/O_2$ fuel cell using acid doped polybenzimidazole as polymer electrolyte. *Electrochim. Acta* 41: 193–197.

Wang, L., D.M. Xing, Y.H. Liu, et al. 2006. $Pt/SiO_2$ catalyst as an addition to nafion/PTFE self-humidifying composite membrane. *J. Power Sources* 161: 61–67.

Wainright, J.S., J.T. Wang, D. Weng, et al. 1995. Acid-doped polybenzimidazoles: A new polymer electrolyte. *J. Electrochem. Soc.* 142: L121–L123.

Wainright, J.S., J. Wang, and M.H. Litt. 1997. In Proceedings of the Second International Symposium On New Materials for Fuel Cell and Modern Battery Systems. O. Savadogo, P.R. Roberge (eds.). Montreal, July 6–10, 1997.

Watanabe, M., H. Uchida, Y. Seki, et al. 1996. *J. Electrochem. Soc.* 143: 3847–3852.

Watanabe, M., H. Uchida, and M. Emori. 1998. *J. Electrochem. Soc.* 145: 1137–1141.

Watanabe, M., H. Uchida, K. Ohkubo, et al. 2003. Hydrogen purification for fuel cells: selective oxidation of carbon monoxide on Pt–Fe/zeolite catalysts *Appl. Catal. B* 46: 595–600.

Wei, J., C. Stone, and A.E. Steck. 1995. Trifluorostyrene and Substituted Trifluorostyrene Copolymeric Compositions and Ion-Exchange Membranes Formed Therefrom. U.S. Patent No. 5,422,411.

Xiao, G., G. Sun, D. Yan, et al. 2002. Synthesis of sulfonated poly(phthalazinone ether sulfone)s by direct polymerization. *Polymer* 43: 5335–5339.

Xing, P., G.P. Robertson, M.D. Guiver, et al. 2004. Sulfonated poly (aryl ether ketone)s containing naphthalene moieties obtained by direct copolymeriation as novel polymers for proton exchange membranes. *J. Polym. Sci. Polym. Chem.* 42: 2866–2876.

Yang, C., P. Costamagna, S. Srinivasan, et al. 2001a. Approaches and technical challenges to high temperature operation of proton exchange membrane fuel cells. *J. Power Sources* 103: 1–9.

Yang, C., S. Srinivasan, A.S. Aricó, et al. 2001b. Composite nafion/zirconium phosphate membranes for direct methanol fuel cell operation at high temperature. *Electrochem. Solid-State Lett.* 4:A31.

Yasumoto, E., K. Hatoh, and T. Gamou. December 30, 1997. Fuel Cell Device Equipped with Catalyst Material for Removing Carbon Monoxide and Method for Removing Carbon Monoxide. U.S. Patent No. 5,702,838.

Yoon, S.R., G.H. Hwang, W.I. Cho, et al. 2002. Modification of polymer electrolyte membranes for DMFCs using Pd film formed by sputtering. *J. Power Sources* 106: 215–223.

Yu, J., B. Yi, D. Xing, et al. 2003. Degradation mechanism of polystyrene sulfonic acid membrane and application of its composite membranes in fuel cells. *Phys. Chem. Chem. Phys.* 5: 611–615.

Zaidi, M.J., S.D. Mikhailenko, G.P. Robertson, et al. 2000. Proton conducting composite membranes from polyether ether ketone and heteropolyacids for fuel cell applications. *J. Membrane Sci.* 173: 17–34.

Zawodzinski, T.A., J. Davey, J. Valerio, and S. Gottesfeld. 1995. The water content dependence of electro-osmotic drag in proton-conducting polymer electrolytes. *Electrochim. Acta* 40: 297–302.

Zoppi, R.A., and S.P. Nunes. 1998. Electrochemical impedance studies of hybrids of perfluorosulfonic acid ionomer and silicon oxide by sol-gel reaction from solution. *J. Electroanal. Chem.* 445: 39–45.

# Index

Page numbers followed by *f* and *t* indicate figures and tables, respectively.

## D

## G



## H

## K



## L



## M

## N

## Q



## R



## S

## T

## U



## V

## W



## X



## Y



## Z

Milton Keynes UK
Ingram Content Group UK Ltd.
UKHW051905071024
449327UK00025B/2101